SOUTH WESTERN
Algebra 2
AN INTEGRATED APPROACH

GERVER, SGROI, CARTER, HANSEN
MOLINA & WESTEGAARD

JOIN US ON THE INTERNET
WWW: http://www.thomson.com
EMAIL: findit@kiosk.thomson.com   A service of I T P

SOUTH WESTERN
EDUCATIONAL
PUBLISHING

I T P An International Thomson Publishing Company

Cincinnati • Albany • Bonn • Boston • Detroit • London • Madrid • Melbourne • Mexico City • New York
Philadelphia • Pacific Grove • Paris • San Francisco • Singapore • Tokyo • Toronto • Washington

Editor-in-Chief	Peter McBride
Managing Editor	Eve Lewis
Project Manager	Enid Nagel
Editors	Tamara S. Jones, Timothy Bailey
Consulting Editor	Carl Kalota
Assistant Editor	Darrell E. Frye
Marketing Manager	Colleen J. Thomas
Marketing Coordinator	Dawn A. Zimmer
Mathematics Consultant	Carol Ann Dana
Production Coordinator	Patricia M. Boies
Manufacturing Coordinator	Jennifer D. Carles
Art Director	John Robb
Design Consultant	Elaine St. John-Lagenaur
Photographic Consultant	Devore M. Nixon
Photo Editing	Kathryn A. Russell/Pix Inc, Sam A. Marshall, Alix Roughen
Editorial Assistant	Mary Schwarz
Marketing Assistant	Laurie L. Brown
Editorial Development and Production	Gramercy Book Services, Inc.
Composition	New England Typographic Service, Inc.
Scans/Prepress/Imaging	Better Graphics, Inc.
Cover Design	Photonics Graphics

About the Cover The cover design is a collage of real world images from the chapter themes. How many can you find?

ISBN: 0-538-64421-4

1 2 3 4 5 6 7 8 VH 03 02 01 00 99 98 97 96

Printed in the United States of America

I(T)P®

International Thomson Publishing
South-Western Educational Publishing is an ITP Company. The ITP logo is a registered trademark used herein under License by South-Western Educational Publishing.

ABOUT THE Authors

Robert Gerver, Ph.D., has been a mathematics instructor at North Shore High School in Glen Head, New York for more than 15 years. During that time, he has also taught at several other institutions, grade 4 through graduate level. He has served on the New York Regents Competency Test Committee. He received his B.A. and M.S. degrees from Queens College of the City University of New York where he was elected to Phi Beta Kappa. He received his Ph.D. from New York University. In 1988, Dr. Gerver received the Presidential Award for Excellence in Mathematics Teaching for New York State.

Claudia Carter, mathematics teacher at the Mississippi School for Mathematics and Science, Columbus, Mississippi, has taught all levels of secondary mathematics and is an active advisor for Mu Alpha Theta. Ms. Carter is a Woodrow Wilson National Foundation Master Teacher and 1995 Tandy Scholar. She is past president of Mississippi Council of Teachers of Mathematics. She was the recipient of the Presidential Award for Excellence in Mathematics Teaching in 1989 and received Mississippi's Top Twenty Star Teacher Award in 1990, 1991, and 1992. Ms. Carter received her B.S. and M.A. degrees from the University of New Orleans.

David Molina, Ph.D., is the Glenadine Gibb Teaching Fellow in Mathematics Education and assistant professor in the Department of Curriculum and Instruction at the University of Texas in Austin. He has taught secondary mathematics and at present is researching curricular and instructional changes through technology. Dr. Molina is a consultant to school districts which are implementing systemic change. Dr. Molina received his B.S. degree from the University of Notre Dame and his M.A. and Ph.D. from the University of Texas, Austin.

Richard Sgroi, Ph.D., is mathematics director of Newburgh Enlarged City School District, Newburgh, New York. He has taught mathematics at all levels for more than 20 years. While teaching at North Shore High School, Glen Head, New York, he and his colleague Robert Gerver developed both *Dollars and Sense: Problem Solving Strategies in Consumer Mathematics* and *Sound Foundations: A Mathematics Simulation*. He and Dr. Gerver received recognition from the U.S. Department of Education's Program Effectiveness Panel for *Sound Foundations* to be included in Educational Programs That Work. Dr. Sgroi received his B.A. and M.S. degrees from Queens College of the City University of New York, and his Ph.D. from New York University.

Mary Hansen, a mathematics teacher at Independence High School, Independence, Kansas, teaches algebra and geometry. While teaching in Texas and North Carolina, Ms. Hansen was active in mathematics teachers staff development. She received her B.A. and M.S. degrees from Trinity University, San Antonio, Texas, where she was elected to Phi Beta Kappa.

Susanne Westegaard is a mathematics department chair of Montgomery-Lonsdale Public School, Montgomery, Minnesota where she has taught mathematics for more than 20 years. Ms. Westegaard has been on the Woodrow Wilson National Fellowship Traveling Team for several summers. She received the Distinguished Educator Award from the Minnesota Mutual Foundation and MAEF in May 1994 and the Presidential Award for Excellence in Teaching Mathematics in 1991. Ms. Westegaard received her B.S. degree from Yankton College, Yankton, South Dakota and her M.Ed. from the University of Minnesota.

Reviewers

Dianne Aydlett
Mathematics Teacher
Northeastern High School
Elizabeth City, North Carolina

Lee Bailey
Mathematics Resource Teacher
Brevard County Schools
Melbourne, Florida

Sue Ellen Baker
Chairperson, Mathematics Department
Hoover High School
Hoover, Alabama

Jacqueline Brannon Giles
Multicultural Consultant
Central College
Houston Community College System
Houston, Texas

George Bratton
Associate Professor of Mathematics
University of Central Arkansas
Conway, Arkansas

Peter Cassioppi
Mathematics Teacher
East High School
Rockford, Illinois

Nancy S. Cross
Mathematics Educator
Titusville High School
Titusville, Florida

Marilyn Dewoody
Chairperson, Mathematics Department
Fort Gibson High School
Fort Gibson, Oklahoma

Kevin Fitzpatrick
Department Chair, Mathematics
Greenwich High School
Greenwich, Connecticut

Bettye Forte
Executive Director Northeast Quadrant and
Mathematics Director
Fort Worth Independent School District
Fort Worth, Texas

Bruce Grip
Mathematics Teacher
Etiwanda High School
Rancho Cucamonga, California

Dale Johnson
Professor of Research
University of Tulsa
Tulsa, Oklahoma

Elaine Lewis
Mathematics Teacher
Evan Worthing High School
Houston, Texas

Allen Miedema
Mathematics Teacher
Woodinville High School
Woodinville, Washington

Martha Montgomery
Chairperson, Mathematics Department
Fremont Ross High School
Fremont, Ohio

Heikki Petaisto
Mathematics Department Head
Chino Valley High School
Chino Valley, Arizona

Marie Ritten
Mathematics Instructor
Stevens High School
Rapid City, South Dakota

Christopher R. White
Curriculum Director
Port Huron Area School District
Port Huron, Michigan

Connecting the World to Algebra

To the Student

You are about to embark on a most exciting algebra program!

South-Western Algebra bridges your skills and knowledge of today to what you will be doing in the future. In *South-Western Algebra*, you will connect algebra to topics such as geometry, earth science, biology, travel, ecology, entertainment, and personal finance.

Chapter themes help you to see how mathematics applies to daily life.

The ***Chapter Project*** helps you develop mathematical understanding and connect algebra to the real world. You can use the Internet to research your projects by using the address given in the Internet Connection.

The ***Data Activity*** and ***Explore Statistics*** lessons help you use data in various ways, including reading and interpreting charts and graphs.

AlgebraWorks showcases a specific job or career which requires the algebra you are learning. You will see the importance algebra has in everyday life!

The ***Problem Solving File*** prepares you with problem solving strategies that will help you on standardized tests. It will also sharpen your critical thinking and problem solving skills!

Algebra Workshops and ***Explore Activities*** help you understand ideas so you can become comfortable with algebra.

You will use ***calculators*** and ***computers*** as tools for learning and doing mathematics. Graphing calculators and computer spreadsheet applications show you how graphs communicate concepts about data.

South-Western Algebra will make you a confident problem solver who can clearly communicate in mathematics. You will feel empowered as you realize the endless opportunities that algebra brings to you.

Modeling and Predicting

THEME: Cars and Business

Applications and Connections

Technology and Other Tools

2 Real Numbers, Equations, and Inequalities

THEME: Engineering

Applications and Connections

Technology and Other Tools

3 Functions and Graphs

THEME: Personal Finance

Applications and Connections

Technology and Other Tools

4 Systems of Linear Equations

THEME: Manufacturing

5 Polynomials and Factoring

THEME: Profit and Loss

6 Quadratic Functions and Equations

THEME: Transportation

7 Inequalities and Linear Programming

THEME: Government

Technology and Other Tools

8 Exponents and Radicals

THEME: Aerospace

9 Exponential and Logarithmic Functions

THEME: Populations

10 Polynomial Functions

THEME: Amusement Parks

11 Rational Expressions and Equations

THEME: Photography

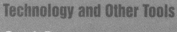

12 Conic Sections

THEME: Communication

Applications and Connections

Technology and Other Tools

13 Sequences and Series

THEME: Design

14 Probability and Statistics

THEME: Insuring America

Technology and Other Tools

15 Trigonometric Functions and Graphs

THEME: Navigation

16 Trigonometric Identities and Equations

THEME: Sound

Also...

1 Modeling and Predicting

Take a Look
AHEAD

Make notes about things that look new.
- Locate the Monthly Payment Formula. How many variables does this formula use? Which variable is used as an exponent?
- What types of matrix applications will you learn about in Lesson 1.3?
- What are the steps you use when solving a problem? Why is each step important?

Make notes about things that look familiar.
- Each lesson title in this chapter names a different mathematical tool. Make a list of these names. Put a check mark next to the ones you have used before.
- Skim the chapter to see how technology is used. In which lessons will a graphing calculator be used? Where might you use computer software?

DATA Activity

Sticker Shock

If you are looking forward to the day when you will own your own car, keep in mind that a car is one of the most expensive items to purchase and operate. You may suffer from "sticker shock" when you check out the prices of new models. You must also be prepared for the fixed and variable costs associated with car ownership. The table on the next page shows what some of these costs are and how they have changed over the last few years.

SKILL FOCUS

▶ Read and interpret data from a table.

▶ Add, subtract, multiply, and divide real numbers.

▶ Determine percent and percent change.

▶ Solve multistep problems.

▶ Write and graph equations.

In this chapter, you will see how:

• **CREDIT MANAGERS** investigate a consumer's bill-paying history and use formulas to determine payment schedules on loans.
(Lesson 1.1, page 11)

• **AUTOMOBILE MECHANICS** use computers to help them analyze engine problems.
(Lesson 1.2, page 18)

• **ACTUARIES** apply concepts of probability to determine the risk when insuring different groups of people and to set policy premiums accordingly.
(Lesson 1.4, page 31)

CARS & BUSINESS

Automobile Costs						
Item	1988	1989	1990	1991	1992	1993
Variable Costs (cents/mile)						
Gas and oil	5.20	5.20	5.40	6.70	6.00	6.00
Maintenance	1.60	1.90	2.10	2.20	2.20	2.40
Tires	0.80	0.80	0.90	0.90	0.90	0.90
Fixed Costs (dollars)						
Insurance	573	663	675	726	747	724
License and registration	139	139	151	165	169	179
Depreciation	1784	2094	2357	2543	2780	2883
Finance charge	565	626	680	779	832	696

1. What was the total variable cost for a car owner who drove 10,000 miles in 1991?

2. What was the percent change in maintenance cost for 1993 as compared to 1988?

3. What was the average finance charge for 1988–1993? Round to the nearest dollar.

4. In 1993, what percent of total fixed costs did depreciation represent? Round to the nearest tenth of a percent.

5. **WORKING TOGETHER** Find out the actual selling prices of two or three cars you like. Then assume each car will lose 15% of its original value with each passing year. Write an equation for the value of each car as a function of years. Graph each function and identify the type of graph. Compare and contrast the graphs for different cars.

Model Cars

Designing a new automobile may take years, during which the model or *prototype* goes through many changes. The process begins with drawings, now done on computers, to show the designer's ideas. At a later stage, the ideas take the form of a full-sized clay model, complete with bumpers and trim. Engineers measure the model with an electric scanner and use the information to create specifications for metal and plastic body parts.

PROJECT GOAL

To design and build model racing cars and analyze data about the models.

Getting Started

Work in groups of four or five students.

1. As a class, discuss the rules for the project. Decide if the cars will be constructed only of recycled (free) materials or set a limit on cost. Agree on a range of values for size and weight. Set a competition date.

2. With a group, collect some references on model-car building before you begin drawing plans. Discuss ideas and decide how project responsibilities will be shared.

3. Plan the forms you will use to keep accurate records about materials used, costs, and physical data about the models.

4. Name your car.

Internet Connection

www.swpco.com/
swpco/algebra2.html

PROJECT *Connections*

Lesson 1.2, page 17:
Submit drawings and plans for model car and use spreadsheets to organize data about work time, materials, and/or costs.

Lesson 1.4, page 30:
Use probability to describe different aspects of the model-car project.

Lesson 1.5, page 37:
Collect and analyze physical data about car models using measures of central tendency and boxplots.

Chapter Assessment, page 53:
Plan a class car show and rally. Use scatterplots and lines of best fit to investigate the relationship between car performance and a physical variable.

1.1 Use an Equation or Formula

Explore

- Gregory spent $274.74 on a set of four steel-belted radial tires for his truck. The total included $14.94 in sales tax.

    1. Let t represent the price of each tire. Explain why the equation $4t + 14.94 = 274.74$ models this purchase.

    2. What was the total price that Gregory paid for the four tires, not including sales tax? Explain how you found your answer.

    3. Find the price of each tire. Explain how you found your answer.

    4. Could you have used a different set of steps to find the price of each tire? Explain. Do you think one approach is more convenient than the other?

Build Understanding

- As you saw in Explore, you solve an equation by finding the values of the variables which make the equation true. The following properties of equality can be used to produce equivalent equations.

> **PROPERTIES OF EQUALITY**
>
> For all real numbers a, b, and c,
>
> If $a = b$, then $a + c = b + c$.   **Addition Property**
>
> If $a = b$, then $a - c + b - c$.   **Subtraction Property**
>
> If $a = b$, then $ac = bc$.   **Multiplication Property**
>
> If $a = b$, and $c \neq 0$,
>
> then $\dfrac{a}{c} = \dfrac{b}{c}$.   **Division Property**

A **literal equation** is an equation with more than one letter or variable. A **formula** is a type of literal equation. For example, $A = lw$ is the formula used to find the area of any rectangle. The methods for solving literal equations are the same as for equations with one variable.

EXAMPLE 1

CAR PRICING The *sticker price* that a car dealer charges for an item is given by the formula $s = c + m$, where c is the dealer's cost and m is the dealer's markup. A car carries a sticker price of $670 for air conditioning. If the dealer's markup is $134, what is the dealer's cost?

Solution

$$s = c + m$$ Write the formula.

$$670 = c + 134$$ Substitute known values.

$$670 - 134 = c + 134 - 134$$ Subtract 134 from both sides.

$$536 = c$$ Solve.

Check

$$670 \stackrel{?}{=} 536 + 134$$ Substitute the solution in the

$$670 = 670 \checkmark$$ original equation.

The dealer's cost is $536.   ◄

You may need to simplify an equation before you can solve it.

EXAMPLE 2

CAR LOANS The *Monthly Payment Formula* gives the amount a borrower must pay each month in order to repay a loan:

$$M = \frac{Pr(1 + r)^n}{(1 + r)^n - 1}$$

where M is the monthly payment, P is the amount borrowed, r is the monthly interest rate, and n is the number of payments to be made.

Megan borrowed $3000 for 3 years at a 9% annual interest rate to buy a used car. Determine

 a. Megan's monthly payment.
 b. The amount she must repay.

Solution

 a. The monthly interest rate is the annual interest rate divided by 12.

$$r = 0.09 \div 12 = 0.0075$$

The number of payments is 12 times the length of the loan in years.

$$n = 12(3) = 36$$

$$M = \frac{Pr(1 + r)^n}{(1 + r)^n - 1}$$ Write the formula.

$$= \frac{3000(0.0075)(1.0075)^{36}}{(1.0075)^{36} - 1}$$ Substitute known values.

$$= 95.40$$ Simplify using a calculator. Round to the nearest cent.

Megan's monthly payment is $95.40.

 b. $36(95.40) = 3434.4$

The amount she must repay is $3434.40. This amount is called the **deferred payment price**. To repay the loan, Megan must make 36 monthly payments of $95.40.   ◄

In Example 2, notice that the deferred payment price of $3434.40 is $434.40 more than the $3000 that Megan borrowed. The extra $434.40 is the **finance charge**, the amount the bank charges Megan for the service of providing her with the money she needs. The finance charge is the *cost* of the loan. The following example shows how finance charges are affected by the **amortization period**, the length of time a borrower takes to repay a loan.

EXAMPLE 3

COLLEGE LOAN Marcus borrows $5000 to finance his first year in college. His bank will loan him the money at an 8% annual rate and offers him the option of repaying the money over a four year or five year period.

a. Find the monthly payment for each amortization period.
b. Find the deferred payment price of each loan. Round answers to the nearest cent.
c. How much will Marcus save by paying off the loan in four years?

Solution

a. $M = \dfrac{Pr(1 + r)^n}{(1 + r)^n - 1}$, $P = 5000$, $r = \dfrac{0.08}{12} \approx 0.00667$

4-year repayment ($n = 48$): $M = \dfrac{5000(0.00667)(1.00667)^{48}}{(1.00667)^{48} - 1}$

$= 122.07$

The monthly payment on a 4-year loan is $122.07.

5-year repayment ($n = 60$): $M = \dfrac{5000(0.00667)(1.00667)^{60}}{(1.00667)^{60} - 1}$

$= 101.39$

The monthly payment on a 5-year loan is $101.39.

b. 4-year deferred payment price: $48(122.07) = 5859.36$
5-year deferred payment price: $60(101.39) = 6083.40$

The deferred payment price of a $5000 loan at an 8% annual rate is $5859.36 over four years and $6083.40 over five years.

c. The advantage of choosing a longer amortization period is that the monthly payments are smaller. The disadvantage is that the borrower must pay back a greater amount when the repayment period is longer. It will be to his advantage in the long run to repay the loan in four years rather than five.

COMMUNICATING ABOUT ALGEBRA

List advantages and disadvantages of a long amortization period.

Finance charge on 4-year loan: $5859.36 - $5000.00 = $859.36
Finance charge on 5-year loan: $6083.40 - $5000.00 = $1083.40
Additional cost of 5-year loan: $1083.40 - $859.36 = $224.04

By paying off the loan in four years, Marcus will save $224.04 in finance charges.

The Monthly Payment Formula can be used to write a formula for the amount that can be borrowed, given a specific monthly payment.

$$M = \frac{Pr(1 + r)^n}{(1 + r)^n - 1}$$ Monthly Payment Formula

$$M[(1 + r)^n - 1] = Pr(1 + r)^n$$ Multiply both sides by $[(1 + r)^n - 1]$.

$$\frac{M[(1 + r)^n - 1]}{r(1 + r)^n} = P$$ Divide both sides by $r(1 + r)^n$.

This formula is called the *Amount Formula*.

EXAMPLE 4

CAR LOAN Rita anticipates that she will be able to afford monthly payments of $150 on a car loan over the next three years.

  a. How much can she borrow at 10%?

  b. What if Rita could arrange to pay $160 per month for four years and she could find a 9% loan. How much more could she borrow?

Solution

  a. $M = 150, r = 0.1 \div 12 \approx 0.00833, n = 3(12) = 36$

$$P = \frac{M[(1 + r)^n - 1]}{r(1 + r)^n}$$ Write the formula.

$$= \frac{150[(1.00833)^{36} - 1]}{0.00833(1.00833)^{36}} \approx 4648.96$$

Rita could borrow about $4650.

  b. $M = 160, r = 0.09 \div 12 = 0.0075, n = 4(12) = 48$

$$P = \frac{160[(1.0075)^{48} - 1]}{0.0075(1.0075)^{48}} \approx 6429.57$$

$6429.57 - $4648.96 = 1780.61

By increasing her monthly payment to $160 and the amortization period to four years, and decreasing the interest rate on the loan to 9%, Rita could increase the loan amount by about $1780. ◄

TRY THESE

Name the properties of equality you would use, in the order that you would use them, to solve each equation. Do not solve the equation.

  1. $12x = 48$ **2.** $26 = n - 30$ **3.** $\frac{k}{5} + 14 = -18$ **4.** $7 = -5 + 8e$

Solve and check each equation.

  5. $m + 19 = 30$ **6.** $39 = 3p$ **7.** $-7 = -3 + y$ **8.** $15 = \frac{t}{3}$

  9. $2x - 7 = 11$ **10.** $12 = 6 + 3h$ **11.** $5c = 18 + 8 - 6$ **12.** $\frac{m}{4} + 7 = 26$

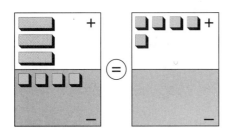

13. MODELING What equation is represented? Sketch the steps you would use to solve the equation.

14. WRITING MATHEMATICS Use the terms *equal* and *equivalent* in different sentences about equations. Explain how your sentences illustrate the differences between the terms.

15. CAR PRICING A car carries a sticker price of $890 for automatic transmission. If the dealer's cost is $712, what is the markup?

16. CAR LOAN Arturo's monthly bank payments on a 10%, two-year loan for a new car are $346.09. Find each amount.

    a. the amount borrowed
    b. the deferred payment price
    c. the finance charge

17. VEHICLE LOAN To finance the cost of a new recreational vehicle, Mr. and Mrs. Haney obtained a $30,000, 10-year loan from their bank at 8.5% annual interest. Find each amount.

    a. the monthly payment
    b. the deferred payment price
    c. the finance charge

PRACTICE

Solve and check each equation.

1. $n - 8 = 15$
2. $-3 = 5 + d$
3. $8x = -20$
4. $4x + 2 = 26$

5. $\frac{y}{5} = 8.4$
6. $5 = 9 - 2e$
7. $20k = 45 + 15$
8. $\frac{y}{10} - 13 = 3$

9. $4(2 + x) = 28$
10. $-3(9 - n) = -63$
11. $15 + k = 3(14 - 2)$
12. $2a + 9a + 13 = 35$

Solve for the given variable.

13. $P = R - C$, for R
14. $F = ma$, for a
15. $A = \frac{1}{2}bh$, for b

16. $F = \frac{9}{5}C + 32$, for C
17. $P = 2l + 2w$, for w
18. $T = g - (w + p)$, for w

19. PRICING The sticker price for a sound system on a new car is $408. The dealer's cost is $342.

    a. Write an equation relating sticker price, dealer cost, and markup.
    b. Solve the equation to find the markup.

20. CAR LOAN The monthly payments on a 6%, three-year loan for a used car are $97.35. Find each amount.

    a. the amount borrowed
    b. the deferred payment price
    c. the finance charge

21. CREDIT Kwan took out a five-year, $8000 education loan at 7.8% annual interest. Find each amount.

    a. the monthly payment
    b. the deferred payment price
    c. the finance charge

EXTEND

22. **WRITING MATHEMATICS** Write an equation with at least three terms that has as its solution $x = 12$. Explain how you found the equation.

23. Which of the following equations are equivalent to $5z - 3 = 17$?

    a. $5z - 5 = 15$ **b.** $5z = 14$ **c.** $5z + 7 = 27$ **d.** $z = 4$

Solve and check each equation.

24. $8x + 5 = 6x + 11$ 25. $-5 + 13h = 12h - 15$ 26. $4m + 9 = 7m + 24$

27. $6(x - 4) = 3(x + 2)$ 28. $8(5y + 3) = -7(3y + 14)$ 29. $2c - (3 - c) = 5c - 4$

30. $\dfrac{x}{2} + 7 = \dfrac{x}{3} - 2$ 31. $\dfrac{x + 5}{6} = \dfrac{x - 4}{3}$ 32. $8 + \dfrac{2x}{3} = \dfrac{3x}{4} + 7$

33. **CAR OPTIONS** A roof rack for which a car dealer paid $150 carries a sticker price of $177. Find the percent markup on the item.

34. **BUSINESS LOAN** To finance the cost of new sinks and hair dryers for her hair salon, Meg borrowed $8000 at 8% for three years. Find the percent of the loan amount that the bank earned in finance charges.

35. **CREDIT COST** The deferred payment price of a four-year, 8.4% loan is $4959.36. Find the amount of the loan. Round your answer to the nearest $100.

36. **HOME LOAN** You have decided to borrow $12,000 for a home-improvement loan. How much can you save on a 12.5% loan by repaying the money in four years rather than five?

THINK CRITICALLY

37. Leon said "If $x = 4x$ then I can divide both sides of the equation by x to get the equivalent equation $1 = 4$." What's wrong with Leon's reasoning?

38. Assume $a \neq 0$. Solve the equation $ax + b = 0$ for x. Explain how your result relates to solving one-variable equations in general.

MIXED REVIEW

Evaluate each expression for $x = 2$ and $y = -3$.

39. $2x - y$ 40. $\dfrac{3}{4}x + y - 5.5$ 41. $5x + 3y$ 42. $x^2 + (x + y)^3$

43. Find 27% of 150.   44. What percent of 25 is 60?

45. 87 is 75% of what number?   46. Find 0.2% of 500.

Solve and check each equation.

47. $3n = -21$ 48. $5e + 2 = 27$ 49. $2(m - 4) = 20$ 50. $6y - 5 = 9y + 13$

From 1908 to 1927, the Ford Motor Company produced the Model T Ford, the most successful automobile ever manufactured.

By the early 1920s, half of the cars sold in the world were Model Ts. The car was successful in part because it was durable and functional, as well as simple and inexpensive to operate. With the Model T, Americans learned how to use credit. Seventy-five percent of the Model Ts sold were purchased with borrowed money.

Today, borrowing is a way of life for many people. Americans owe more than $250 *billion* on cars alone. The typical person spends an average of about 6% of his or her income on monthly car payments. For this reason, the credit manager is one of the key figures in a successful car dealership. The credit manager investigates the bill-paying history of people who want to purchase cars, decides whether these potential borrowers are good risks, arranges credit-payment schedules, and maintains accurate records of payments.

Decision Making

1. Suppose you are a "typical" person making $180 car payments each month. What is your monthly income? annual income?

2. The 1990 census put the population of the United States at about 250 million. How much money do Americans owe per capita on cars?

The credit manager for Northside Auto approved the following loans at 8.875%.

Buyer	Amount Borrowed	Time
Rico Martelli	$17,800	5 years
Chang Liu	22,100	5 years
Katrina Langsley	15,200	4 years
Ahmad Baghaii	12,500	4 years
Thomas Jones	14,395	5 years

3. Compute the monthly payments for each buyer.

4. Find the deferred payment price of each car.

5. Find the total amount owed to Northside Auto by the five borrowers.

6. How much more will Northside Auto earn on these five cars by selling them on credit rather than for cash?

1.2 Use a Spreadsheet

Explore

SPOTLIGHT ON LEARNING

WHAT? In this lesson you will learn
- to organize and analyze data using spreadsheets.
- to use spreadsheets to compare the total costs of similar loans.

WHY? Computer spreadsheet programs help you organize data and compare costs of financing the purchase of a car and taking out business loans.

Marci has decided to buy a new car that sells for $12,450. She will make a **down payment**—a portion of the price paid at the time of purchase in cash—and finance the remainder of the cost with a four year, 10.8% loan. Before deciding whether to pay 10%, 20%, or 30% of the retail price of the car as a down payment, she wants to figure the total cost of the car including interest at each down payment.

1. Find the amount of each of the three possible down payments.

2. For each down payment, find the following.

    a. the amount that remains to be financed

    b. the monthly payment

    c. the deferred payment price of the loan

    d. the total cost of the car, including down payment

3. How much can she save by putting 20% down rather than 10%? by putting 30% down rather than 10%?

Build Understanding

CHECK UNDERSTANDING

Why does a shorter amortization period save money for the borrower?

Lenders generally require that you pay for a portion of your purchase in cash—the *down payment*—and borrow the remainder. As you saw in Explore, the total cost of the item depends on the amount of the down payment, the interest rate on the loan, and the length of the amortization period. A cost-conscious borrower can achieve substantial savings by increasing the down payment, or by finding a lower interest rate or shorter amortization period.

For a given purchase, calculating the many possible combinations of down payments, interest rates, and repayment periods in order to find which ones make the

most sense can be a time-consuming process. One way to simplify such calculations is to use a computer **spreadsheet**.

Financing an $8500 Purchase at 10%, 20%, and 30% Down

	A	B	C	D
1	Retail	Down	Amount of	Amount to be
2	Price	Payment,%	Down Payment	Financed
3	8500	10	=A3*(B3/100)	=A3−C3
4	8500	20	=A4*(B4/100)	=A4−C4
5	8500	30	=A5*(B5/100)	=A5−C5

Row 3

Column A

Cell B5

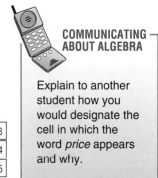
Computer spreadsheets are designed to operate using formulas and data. A spreadsheet contains **cells** designated by column and row. Columns are identified by letters in alphabetical order from left to right, and rows are identified by numbers in order from top to bottom.

In any cell, you can enter data or a formula. The data can be a **value** (numerical data) or a **label** (text). Values, but not labels, can be used in calculations. Labels are often used as titles for, or explanations of, the data in nearby cells. Entering a formula in a cell tells the spreadsheet program to perform a calculation, usually by referring to other cells that contain the required values. The symbol ∗ represents multiplication and the symbol / represents division.

EXAMPLE 1

Find the values for columns C and D of the down payment spreadsheet shown above.

Solution

Cell C3: A3 ∗ (B3 / 100) = $8500 \left(\dfrac{10}{100}\right)$ = 850

Cell C4: A4 ∗ (B4 / 100) = $8500 \left(\dfrac{20}{100}\right)$ = 1700

Cell C5: A5 ∗ (B5 / 100) = $8500 \left(\dfrac{30}{100}\right)$ = 2550

Cell D3: A3 − C3 = 8500 − 850 = 7650
Cell D4: A4 − C4 = 8500 − 1700 = 6800
Cell D5: A5 − C5 = 8500 − 2550 = 5950

The amounts of the down payments (Column C) are $850, $1700, and $2550. The amounts to be financed (Column D) are $7650, $6800, and $5950.

When you create a spreadsheet, you must enter formulas using appropriate symbols and parentheses. Most spreadsheet software programs require you to use the equal symbol = at the beginning of each formula entered.

EXAMPLE 2

Use computer symbols to write each algebraic expression.

a. $\dfrac{1}{a + b}$ **b.** $k - \dfrac{3(a + b)}{n}$ **c.** $\dfrac{Pr(1 + r)^n}{(1 + r)^n - 1}$

Solution

a. $\dfrac{1}{a + b} = 1 / (a + b)$

b. $k - \dfrac{3(a + b)}{n} = k - (3 * (a + b) / n)$

c. $\dfrac{Pr(1 + r)^n}{(1 + r)^n - 1} = (P * r * (1 + r) \wedge n) / ((1 + r) \wedge n - 1)$ ◄

With a spreadsheet, a buyer can easily compare the total financed prices of a planned purchase at a variety of down payments, interest rates, and/or amortization periods.

EXAMPLE 3

CAR LOAN Shauna has decided to purchase a new car priced at $14,270. She will put 10% or 20% down and finance the car with a 9.75% loan, to be repaid in four or five years. Design a spreadsheet showing the monthly payments and total financed price of the car under each of the possible options.

Solution

When designing the spreadsheet, be sure to include columns for the data that you know and for the information that you need to calculate. The following formulas were used to calculate the data in row 4:

    Column C: A4 * (B4 / 100)

    Column D: A4 − C4

    Column F: D4 * 0.008125 * 1.008125 $\wedge$ (12 * E4) / (1.008125 $\wedge$ (12 * E4) −1)

    Column G: 12 * E4 * F4 + C4

	A	B	C	D	E	F	G
1	Retail	Down	Down	Amount	Repayment	Monthly	Total
2	Price	Payment,	Payment	of Loan	Period,	Payments	Financed
3		%			yrs		Price
4	14270	10	1427	12843	4	324.19	16988.12
5	14270	10	1427	12843	5	271.30	17705.00
6	14270	20	2854	11416	4	288.17	16686.16
7	14270	20	2854	11416	5	241.15	17323.00

Data in the other rows are found using the same formulas but substitute the appropriate row number in the place row 4 appears in the above formulas.

Column G shows that the total price can be minimized by making the larger down payment and paying off the loan over four years. Shauna will pay $16,686.16 − $14,270.00 = $2,416.16 more than the retail price by the time she repays her loan. ◀

CHECK UNDERSTANDING

Why does 1.008125 appear in the formula for column F? Why does 12 ∗ E4 appear?

TRY THESE

Use computer symbols to write each algebraic expression.

1. $3n - 5$

2. k^4

3. $\dfrac{(a + h)}{2}$

4. $m - b^2$

Determine the values a computer will print in columns C and D.

		A	B	C	D
5.	1	8	3	=2∗(A1−B1)	=2∗A1−B1
6.	2	15	9	=0.6∗A2	=B2^2
7.	3	6.8	5.1	=B3/A3	=24−3.1∗B3
8.	4	2	12	=(B4−A4)/A4	=B4+3∗A4^3

For each empty cell, give the formula and the value the computer will print.

Area and Perimeter of a Rectangle

		A	B	C	D
	1	Length	Width	Area	Perimeter
9.	2	9	6		
10.	3	12	5		
11.	4	7.2	3.9		

Total Cost of Purchases

		A	B	C
	1	Cost of	Sales Tax,	Total Cost
	2	Items, $	%	
12.	3	3.20	4	
13.	4	5.75	6	
14.	5	8	5.25	

15. CAR PRICES In a car dealer's showroom, three cars showed sticker prices of $276 for power door locks. The dealer's cost for the locks was $240 for the first car, $225 for the second car, and $207 for the third car. Design a spreadsheet showing sticker price, dealer's cost, and markup for the locks. For the markup column, write the formulas and the values the computer will print.

16. WRITING MATHEMATICS In Lesson 1.1 you used a scientific calculator to find the monthly payments and deferred payments on specific loans. Explain the advantage of using a spreadsheet to calculate such payments. Give specific examples.

PRACTICE

Use computer symbols to write each algebraic expression.

1. $\dfrac{p}{x - y}$ **2.** $d^3 + v^2$ **3.** $4a - \dfrac{3}{c}$ **4.** $\dfrac{2hk^{5p}}{3(2 - b)^{5p}}$

Find the values a computer will print in columns C and D.

	A	B	C	D
5. 1	4	9	=A1+2*B1	=4*B1/2*A1
6. 2	20	6	=A2^(B2/3)	=B2/(A2/20)
7. 3	12.5	2.4	=2*A3–5*B3	=(A3–1.7)(B3+3)
8. 4	0.75	1.03	=256*A4^2	=(256*A4)^2

For each empty cell, give the formula and the value the computer will print.

Deferred Payment Price of a Loan

		A	B	C
	1	Repayment	Down	Deferred Payment
	2	Period, yr	Payment, %	Price
9.	3	2	189.64	
10.	4	3	152.40	
11.	5	4	115.16	

Amount to be Financed

		A	B	C	D
	1	Retail	Down	Amount of	Amount to be
	2	Price	Payment, %	Down Payment	Financed
12.	3	12340	10		
13.	4	12340	15		
14.	5	12340	20		

Monthly Payment of Five-Year Loan

		A	B	C	D
	1	Amount of	Annual	Monthly	Monthly
	2	Loan	Rate, %	Rate, %	Payment
15.	3	6825	6		
16.	4	6825	9		
17.	5	6825	12		

18. Recreation Stewart is buying a hot-air balloon priced at $27,500. He intends to make a down payment of $2,500, $5,000, or $7,500. Design a spreadsheet showing retail price, down payment, and amount to be financed. For the amount-to-be-financed column, write the formulas and the values the computer will print.

19. Geometry The area of a trapezoid with height h and bases b_1 and b_2 is $\dfrac{1}{2} h (b_1 + b_2)$. Design a spreadsheet showing the areas of trapezoids with bases 4 and 6 and heights 10, 15, and 20. Give the formula you use to find the area.

EXTEND

20. WRITING MATHEMATICS Look through newspapers to find an example of a retailer who offers different credit plans. Explain advantages and disadvantages of each plan. Select one item with an advertised price and compare the total financed price with each plan.

CREDIT Marsha is purchasing a grand piano that costs $19,600. She will make either a 15% or 30% down payment and repay the 8.4% loan in three or four years.

21. Predict the option that will be least expensive for Marsha.

22. Design and complete a spreadsheet that will allow her to compare the total financed price of the piano under all of the payment options.

23. Display the payment options in a bar graph using a y-scale from 20,000 to 25,000. If your spreadsheet program has graphing capabilities, use the program to create this graph.

THINK CRITICALLY

24. BUSINESS LOAN Kiran is buying a fast food franchise for $39,760. He wants to compare the total price of the franchise for down payments of 20% or 30%, loan rates of 9.6% or 10.8% annually, and amortization periods of 8 years or 10 years. Design and complete a spreadsheet showing all of these options.

MIXED REVIEW

Multiply.

25. $3(x - 5y)$ **26.** $4a(6a)$ **27.** $-2n(e - 3n)$ **28.** $(a + b)(a + b)$

Find the values a computer will print in columns C and D.

	A	B	C	D
29. 21	2	–3	=A21	=2–(2∗B21)
30. 22	3	12	=(B22–10)^A22	=B22/(A22+9)

31. STANDARDIZED TESTS Which of the following is a literal equation?

    A. $3x + 2 = x - 5$ **B.** $s = c(1 + r)$ **C.** $41 - 4 = 25 + 12$ **D.** $x^2 - 9 = 0$

PROJECT *Connection*   In this activity, your group will show the results of its design and planning efforts thus far.

1. Show drawings of the race car you want to build. Explain if the drawings are full-size or scaled up or down. Point out special features of your car. Accompany your drawings with documentation showing equations and calculations used to create the plans.

2. Assemble the materials you need and design a spreadsheet to organize information about costs, work time, and quantities. Use the formula features of the spreadsheet to determine total costs, time, and amount of materials needed to mass-produce cars.

3. Have your teacher verify that your plans meet project guidelines.

Career
Automobile Mechanic

The engines of the first automobiles were simple machines compared to the vehicles of today. With today's complex cars, more and more people turn to professional auto mechanics to repair their vehicles. At first, the job of the mechanic was primarily mechanical. The mechanic needed to understand the physical effect that each belt, valve, piston, pump, and hose had on every other engine part.

The electrical and computerized systems used today require a mechanic to know the principles of electronics and the mathematics of computers. Resistors are common elements of the circuit boards located within a car's electronic system. Resistance, measured in ohms, is the opposition of a material to the flow of electricity. When resistors are connected in parallel, there are separate conducting paths for the current. If two resistors in a parallel circuit have unequal resistances, more current flows through the path with lower resistance.

Decision Making

1. A computer uses the formula $a * b / (a + b)$ to calculate the total resistance of two resistances a and b connected in parallel.

   a. Find the total resistance of two resistors connected in parallel rated at 240 ohms and 300 ohms.

   b. Does the formula $a * b / (a + b)$ give the same result as the formula above? Explain.

2. Use computer symbols to write each algebraic expression.

   a. $\frac{5F}{4d^2}$, the pressure exerted on a piston of diameter d by a force F

   b. $\frac{RL}{d} - R$, the resistance of a resistor in a bridge circuit with a second resistor R and at a distance d from a ground in a cable of length L

   c. $2C + 1.57(D + d) + \frac{(D + d)}{4C}$ the length of a belt joining pulleys of diameter D and d with centers C units apart

3. An apprentice mechanic at Premier Auto Repair earns 50% of a master mechanic's wage for the first 1000 hours of work, 60% for the second 1000 hours, 70% for the third 1000 hours, and so on for 5000 hours. Design a spreadsheet showing total apprentice earnings for each 1000 hours of work and for 5000 hours of work, assuming master mechanic wages of $13, $14, $15, and $16 per hour.

Explore

	A	B	C
1	Down	Annual	Monthly
2	Payment	Interest Rate, %	Payment
3	1949	6	533.63
4	1949	9	557.80
5	3898	6	474.34
6	3898	9	495.82

The spreadsheet gives the monthly payments on four 3-year loan options for purchasing a $19,490 car. The options cover down payments of 10% ($1949) and 20% ($3898) and loan rates of 6% and 9%.

1. Copy the grid at the right. Write each monthly payment in the appropriate space in the grid.

	6% interest	9% interest
10% down		
20% down		

2. Make a new 2 × 2 grid showing the totals of monthly payments after three years. Explain how you found the totals.

3. Make a 2 × 2 grid showing the total financed price, including down payment, under each option. Explain how you found the totals.

Build Understanding

If you remove the labels from a table or spreadsheet, you have a rectangular arrangement of numbers, called a **matrix**. Brackets around the numbers are used to indicate a matrix like the one to the right. The plural of *matrix* is *matrices*.

$$A = \begin{bmatrix} 1949 & 6 & 533.63 \\ 1949 & 9 & 557.80 \\ 3898 & 6 & 474.34 \\ 3898 & 9 & 495.82 \end{bmatrix}$$

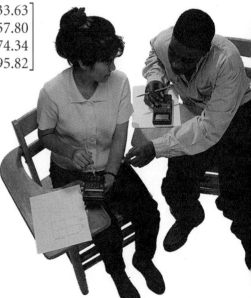

A matrix is usually named using a capital letter such as A or B. Each number in a matrix is called an **element** or an *entry*. The number of rows (horizontal) and the number of columns (vertical) in a matrix specify its **dimensions**. The dimensions of the above matrix are 4 × 3, (read "4 by 3.") and may be written $A_{4 \times 3}$. In general, a matrix with m rows and n columns is an $m \times n$ matrix and may be written

$$A_{m \times n}$$
rows   columns

A matrix can have only one row or only one column. A matrix with the same number of rows and columns is a **square matrix**.

Any matrix in which all the elements are zero is a **zero matrix**. Examples of a zero matrix are shown at the right.

$$\begin{bmatrix} 0 & 0 & 0 \\ 0 & 0 & 0 \end{bmatrix} \quad \begin{bmatrix} 0 & 0 & 0 \\ 0 & 0 & 0 \\ 0 & 0 & 0 \end{bmatrix}$$

If two matrices have the same dimensions, you can add or subtract the *corresponding elements* to find the sum or difference. **Corresponding elements** are the elements in the same position of each matrix. Matrices in which all the elements in corresponding positions are equal are **equal matrices**.

EXAMPLE 1

$$A = \begin{bmatrix} 9 & 11 & -6 \\ 5 & -20 & 0 \end{bmatrix} \quad B = \begin{bmatrix} -2 & 4 & -7 \\ 13 & 25 & 4 \end{bmatrix}$$

Find $A + B$ and $A - B$.

Solution

$$A + B = \begin{bmatrix} 9 + (-2) & 11 + 4 & -6 + (-7) \\ 5 + 13 & -20 + 25 & 0 + 4 \end{bmatrix} = \begin{bmatrix} 7 & 15 & -13 \\ 18 & 5 & 4 \end{bmatrix}$$

$$A - B = \begin{bmatrix} 9 - (-2) & 11 - 4 & -6 - (-7) \\ 5 - 13 & -20 - 25 & 0 - 4 \end{bmatrix} = \begin{bmatrix} 11 & 7 & 1 \\ -8 & -45 & -4 \end{bmatrix}$$

You can use a graphing calculator or computer software to add or subtract matrices.

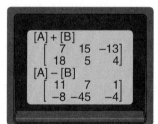

To multiply a matrix by a real number, multiply each element in the matrix by the number. Since real numbers are sometimes called *scalars*, multiplication of a matrix by a real number is called **scalar multiplication**.

EXAMPLE 2

$$C = \begin{bmatrix} 4 & 7.2 \\ 9 & -14.6 \end{bmatrix}$$

Find $5C$.

Solution

$$5C = \begin{bmatrix} 5(4) & 5(7.2) \\ 5(9) & 5(-14.6) \end{bmatrix} = \begin{bmatrix} 20 & 36 \\ 45 & -73 \end{bmatrix}$$

This problem can also be solved with a graphing calculator.

The commutative, associative, and distributive properties associated with real numbers also hold true for matrix addition and scalar multiplication.

One of the principal uses of matrices in business is to keep track of inventories.

EXAMPLE 3

CAR INVENTORIES Big Basin Motors has three dealerships. The matrices below represent each dealership's inventory as of March 1 of four models (two-door, four-door, luxury, and van) of three makes of cars.

	King Motors				**Basin Auto Sales**				**Nevada Vehicles**			
	2-dr	4-dr	Lux	Van	2-dr	4-dr	Lux	Van	2-dr	4-dr	Lux	Van
Tarpon	32	41	15	16	57	98	33	24	19	24	8	12
Glide	49	77	28	13	51	60	35	30	31	41	30	29
Nexel	8	40	20	11	2	21	22	25	5	8	15	17

a. Write a matrix showing the combined inventory of all three dealers.

b. The first matrix shows numbers of cars received at the dealerships from the manufacturer during March. The second matrix summarizes March sales. Write a matrix showing the combined end-of-the-month inventories of Tarpons, Glides, and Nexels.

March Shipments				**March Sales**			
2-dr	4-dr	Lux	Van	2-dr	4-dr	Lux	Van
23	35	10	12	20	9	5	7
34	3	11	11	21	30	10	9
4	16	17	8	23	13	19	10

c. The matrix gives the cost to the dealer of each model. Write a matrix giving the sticker price of each car, assuming the markup is 15%. Round to the nearest dollar.

	Dealer Cost			
	2-dr	4-dr	Lux	Van
	9445	11380	19500	18750
	8750	10435	12680	14100
	15250	18675	24490	21130

Solution

a.
$$\begin{bmatrix} 32 + 57 + 19 & 41 + 98 + 24 & 15 + 33 + 83 & 16 + 24 + 12 \\ 49 + 51 + 31 & 77 + 60 + 41 & 28 + 35 + 30 & 13 + 30 + 29 \\ 81 + 21 + 51 & 40 + 21 + 81 & 20 + 22 + 15 & 11 + 25 + 17 \end{bmatrix} = \begin{bmatrix} 108 & 163 & 56 & 52 \\ 131 & 178 & 93 & 72 \\ 15 & 69 & 57 & 53 \end{bmatrix}$$

b.
$$\begin{bmatrix} 108 & 163 & 56 & 52 \\ 131 & 178 & 93 & 72 \\ 15 & 69 & 57 & 53 \end{bmatrix} + \begin{bmatrix} 23 & 35 & 10 & 12 \\ 34 & 3 & 11 & 11 \\ 4 & 16 & 17 & 8 \end{bmatrix} - \begin{bmatrix} 20 & 9 & 5 & 7 \\ 21 & 30 & 10 & 9 \\ 23 & 13 & 19 & 10 \end{bmatrix} = \begin{bmatrix} 111 & 189 & 61 & 57 \\ 144 & 151 & 94 & 74 \\ -4 & 72 & 55 & 51 \end{bmatrix}$$

Notice stock of 2-door Nexels was short by 4, and these cars had to be back-ordered.

c. Multiply the cost matrix by 1.15.

$$1.15 \begin{bmatrix} 9445 & 11380 & 19500 & 18750 \\ 8750 & 10435 & 12680 & 14100 \\ 15250 & 18675 & 24490 & 21130 \end{bmatrix} = \begin{bmatrix} 10862 & 13087 & 22425 & 21563 \\ 10063 & 12000 & 14582 & 16215 \\ 17538 & 21476 & 28164 & 24300 \end{bmatrix}$$

◀

TRY THESE

Use matrices *A*, *B*, and *C* below for Exercises 1–6.

$$A = \begin{bmatrix} 5 & -5 \\ 4 & 7 \\ 11 & 0 \end{bmatrix} \qquad B = \begin{bmatrix} -6 & -10.5 \\ 4 & 8.5 \\ 20 & -3.5 \end{bmatrix} \qquad C = \begin{bmatrix} 12.2 & 15.2 \\ 0.2 & 2.2 \\ 1.2 & -15.6 \end{bmatrix}$$

1. Give the dimensions of *B*.   **2.** List the elements of *A*.

Find each matrix.

3. $A + B$ **4.** $C - B$ **5.** $6C$ **6.** $2(A - C)$

7. Show that $M + N = N + M$ for $M = \begin{bmatrix} 3 & 6 \\ 1 & 8 \end{bmatrix}$ and $N = \begin{bmatrix} 2 & 9 \\ 4 & 7 \end{bmatrix}$.

8. BOOK SALES Matrix *A* lists December 1 inventories of six best sellers at Page Books. Matrix *B* lists December sales of the six books. Matrix *C* lists new copies of the six books received from the publishers during December. Write a matrix expression showing the January 1 inventory of the books. Evaluate the expression.

$$A = \begin{bmatrix} 210 & 75 \\ 198 & 103 \\ 61 & 155 \end{bmatrix} \qquad B = \begin{bmatrix} 134 & 71 \\ 212 & 186 \\ 67 & 191 \end{bmatrix} \qquad C = \begin{bmatrix} 36 & 50 \\ 24 & 100 \\ 25 & 75 \end{bmatrix}$$

9. WRITING MATHEMATICS Write and solve a problem in your everyday life using matrices.

PRACTICE

Use matrices *A*, *B*, *C*, and *D* below for Exercises 1–9.

$$A = \begin{bmatrix} 7 & 7 & 4 \\ 13 & -2 & -8 \end{bmatrix} \quad B = \begin{bmatrix} 15 & 0 & -14 \\ 7 & 12 & -3 \end{bmatrix} \quad C = \begin{bmatrix} 0 & -5 & -6 \\ 20 & 5 & 2 \end{bmatrix} \quad D = \begin{bmatrix} 28 & 42 & \frac{4}{5} & -16 & -8 \end{bmatrix}$$

1. Give the elements of *C*.   **2.** Give the dimensions of *D*.

3. $A + C$ **4.** $8B$ **5.** $B - A$

6. $0.75D$ **7.** $-C + B$ **8.** $2B + 3A$

9. *A* plus the 2 × 3 zero matrix

10. Give an example of a column matrix. Give the dimensions of the matrix.

11. Let $J = \begin{bmatrix} 5 & 4 \\ -6 & 2 \end{bmatrix}$, $K = \begin{bmatrix} 2 & -3 \\ 5 & 4 \end{bmatrix}$, and $s = 3$. Show that $s(J + K) = sJ + sK$.

12. **AUTO LOANS** The matrix $\begin{bmatrix} 332.15 & 195.40 & 215.87 & 280.60 \\ 95.98 & 244.55 & 356.81 & 303.76 \end{bmatrix}$ gives the monthly payments of eight borrowers with three-year car loans from Sterling Auto Sales. Write a matrix showing the deferred payment prices of the cars, not including down payment.

COMPUTER SALES Matrices F, G, and H list the inventories of six models of computer printers at the Fairview, Gateway, and High Point branches of Office Central.

$$F = \begin{bmatrix} 21 & 19 & 15 \\ 30 & 8 & 6 \end{bmatrix} \qquad G = \begin{bmatrix} 15 & 16 & 11 \\ 21 & 4 & 6 \end{bmatrix} \qquad H = \begin{bmatrix} 35 & 36 & 27 \\ 25 & 13 & 18 \end{bmatrix}$$

13. Write a matrix giving the combined inventory of the printer models at all three locations.

14. Matrix S lists total sales at the three locations during the Fall Sale-a-Thon. Matrix N lists new stock of the models received from the manufacturers. Write a matrix showing the resulting combined inventory after the sale and after delivery of new stock.

$$S = \begin{bmatrix} 20 & 24 & 13 \\ 33 & 21 & 19 \end{bmatrix} \qquad N = \begin{bmatrix} 16 & 15 & 22 \\ 19 & 24 & 8 \end{bmatrix}$$

15. Office Central has 468 branches nationwide. Assume that every three branches of the company sold the same number of each printer model during the Sale-a-Thon as the Fairview, Gateway, and High Point branches sold. Write a matrix showing the total number of each model sold nationwide.

16. **WRITING MATHEMATICS** How can you tell whether two matrices can be added? How can you find the sum? Explain.

EXTEND

Determine whether the matrices in each pair are equal.

17. $\begin{bmatrix} 5 & 3 \\ -8 & 0 \end{bmatrix}$, $\begin{bmatrix} \dfrac{20}{4} & 6-3 \\ 2(-4) & 0+1 \end{bmatrix}$

18. $\begin{bmatrix} -1 & 2 \\ 3 & -4 \end{bmatrix}$, $\begin{bmatrix} 2-3 & \dfrac{12}{6} \\ 7-4 & -\dfrac{8}{2} \end{bmatrix}$

Solve each equation for A.

19. $5A = \begin{bmatrix} 14 & -8 & 20 \\ 55 & -28 & -60 \\ 2 & 0 & 100 \end{bmatrix}$

20. $A + \begin{bmatrix} 15 & 17 & -5 & 9 \\ 4 & 0 & 3 & -16 \end{bmatrix} = \begin{bmatrix} 9 & 2 & 7 & -1 \\ 4 & -11 & -12 & 5 \end{bmatrix}$

21. $4A + \begin{bmatrix} 5 & 6 & 1 \\ -3 & 2 & -8 \end{bmatrix} = \begin{bmatrix} 5 & 22 & -3 \\ 9 & -2 & -36 \end{bmatrix}$

22. $9A + \begin{bmatrix} 12 & 3 \\ -10 & -8 \end{bmatrix} = 2A + \begin{bmatrix} 19 & -18 \\ 4 & -43 \end{bmatrix}$

23. Is matrix subtraction commutative? Explain your answer, giving an example to illustrate your reasoning.

MATRIX PROPERTIES Use the notation illustrated at the right, where the element in ith row, jth column of matrix A is denoted a_{ij} and the element in the ith row and jth column of matrix B is denoted b_{ij}.

$$A = \begin{bmatrix} a_{11} & a_{12} & \cdots & a_{1n} \\ a_{21} & a_{22} & \cdots & a_{2n} \\ \cdot & \cdot & & \\ \cdot & \cdot & & \\ \cdot & \cdot & & \\ a_{m1} & a_{m2} & \cdots & a_{mn} \end{bmatrix} \qquad B = \begin{bmatrix} b_{11} & b_{12} & \cdots & b_{1n} \\ b_{21} & b_{22} & \cdots & b_{2n} \\ \cdot & \cdot & & \\ \cdot & \cdot & & \\ \cdot & \cdot & & \\ b_{m1} & b_{m2} & \cdots & b_{mn} \end{bmatrix}$$

24. What element is directly above a_{34}?

25. What element is directly to the left of b_{76}?

26. What are the dimensions of matrix A? of matrix B?

27. WRITING MATHEMATICS Explain why the dimensions of $A + B$ are $m \times n$.

28. Use matrices with the same dimensions to show the commutative property of addition.

29. Use matrices with the same dimensions to show the associative property of addition.

30. Use matrices with the same dimenstions to show the distributive property of scalar multiplication over addition.

THINK CRITICALLY

GEOMETRY Draw x- and y-coordinate axes on graph paper.

31. Plot three points (a, b), (c, d), and (e, f). Connect the points to form a triangle.

32. Choose two integers m and n. Find $\begin{bmatrix} a & c & e \\ b & d & f \end{bmatrix} + \begin{bmatrix} m & m & m \\ n & n & n \end{bmatrix}$. Using values in the top row as x-coordinates of points and values in the bottom row as y-coordinates, sketch the resulting triangle. Describe its relationship to the original triangle.

33. Find $2\begin{bmatrix} a & c & e \\ b & d & f \end{bmatrix}$ and sketch the resulting triangle. Describe its relationship to the original triangle.

34. Repeat Exercises 31–33 using different values. How is a triangle transformed through matrix addition and scalar multiplication?

MIXED REVIEW

Solve and check each equation.

35. $3x - 4 = -x + 8$

36. $4(c + 2) - 3(c - 3) = 11$

37. $4m - \dfrac{1}{2} = 2m + 1$

38. $\dfrac{5x}{3} - 8 = 4x - 15$

Evaluate if $A = [4 \quad 2 \quad -5 \quad 8 \quad -3 \quad 1]$ and $B = [11 \quad -4 \quad -6 \quad 0 \quad 5 \quad 2]$.

39. $A + B$

40. $-3B$

41. $B - A$

42. $2A + 3B$

43. STANDARDIZED TESTS Evaluate $6x^2 - 29x - 5$ for $x = 5$.

  A. 300 **B.** -120 **C.** 0 **D.** 150

1.4 Use Probability

Explore

The data in the table were compiled by an auto insurance company. Of 1000 drivers surveyed, 139 had been involved in accidents during the previous two years.

Ages and Genders of Drivers in 139 Accidents			
	under 25	**25–60**	**over 60**
Men	52	32	12
Women	27	9	7

1. Are men or women in the under-25 age category more likely to be involved in an accident? Explain.

2. Are men or women in the 25-and-over category more likely to be involved in an accident? Explain.

3. In which age group do most accidents seem to occur? Compared to a driver in another group, how much more likely is it that a driver in this category will be involved in an accident? Explain.

4. Suppose further research confirmed the data in this survey represent the driving population. If you were responsible for setting automobile insurance rates, how much would you charge a 22-year-old male for a policy for which a 63-year-old male driving the same car pays $500 a year? Explain.

Build Understanding

The **probability** of an event is a number from 0 to 1 that expresses the likelihood that the event will occur. An event that cannot occur has a probability of 0. An event that is certain to occur has a probability of 1. An event that is equally likely to occur or not occur has a probability of 0.5.

A probability can be determined theoretically or experimentally. Suppose you attempt to roll an even number with a number cube. There are six equally likely outcomes

1, 2, 3, 4, 5, 6

These represent the **sample space**, the set of all possible outcomes. Within the sample space, three outcomes are even numbers, the **event E** you are looking for

2, 4, 6

CHECK UNDERSTANDING

Give examples of events that have a probability of 0. A probability of 1. A probability of 0.5.

> **THEORETICAL PROBABILITY**
>
> The theoretical probability $P(E)$ that an event will occur is given by
>
> $$P(E) = \frac{\text{Number of outcomes in an event}}{\text{Number of outcomes in sample space}}$$

Therefore, P (rolling an even number) $= \frac{3}{6} = \frac{1}{2}$, 0.5, or 50%.

EXAMPLE 1

Find the theoretical probability that in three tosses of a penny, you will throw exactly two heads.

Solution

First, find the sample space. Make a list of all the possible outcomes when you throw a penny three times. One way to do this is to draw a **tree diagram**.

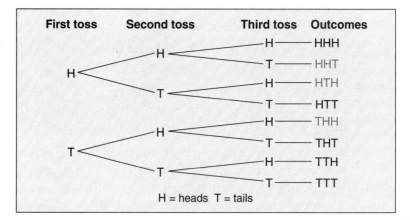

The sample space consists of eight outcomes. The event of throwing exactly two heads occurs three times, HHT, HTH, and THH. Therefore,

$$P \text{ (exactly two heads in three tosses)} = \frac{3}{8}, 0.375, \text{ or } 37.5\%. \quad \blacktriangleleft$$

Geometric probability involves using areas or lengths to determine the probability that an event will occur. Geometric probability has many real-world applications.

Suppose that a region A contains a smaller region B. The probability P that a randomly chosen point in A is in B is given by

$$P = \frac{\text{area of } B}{\text{area of } A}$$

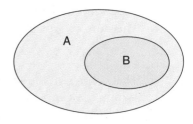

PROBLEM SOLVING TIP

You can use the strategy of making an organized list to solve Example 1. First list the possible outcomes of the first throw, H or T. For each of those outcomes, list the possible outcomes of the second throw. This gives HH, HT, TH, or TT. Finally, for each of those outcomes, list the possible outcomes of the third throw.

EXAMPLE 2

PARACHUTING The Crawford County fairgrounds is a rectangle 600-by-800 ft. The Exhibition Arena is a circle with radius of 120 ft. For the opening festivities, a skydiver will parachute into the fairgrounds. If the skydiver lands at a random point on the fairgrounds, what is the probability that the skydiver will be in the arena?

600 ft

800 ft

120 ft

Solution

$$P = \frac{\pi r^2}{lw} = \frac{\pi(120^2)}{800(600)} \approx 0.0942 \qquad \frac{\text{area of arena}}{\text{area of fairgrounds}}$$

The probability that the diver will land in the arena is about 9.4%. ◄

Theoretical probabilities are ideal probabilities. If you toss a quarter 1000 times, theoretically it should come up heads 50% of the time and tails 50% of the time. In reality, however, it would be unusual if you obtained *exactly* 500 heads and 500 tails. **Experimental probability** is a measure of what actually happens when an experiment is performed, a survey is conducted, or historical records are tabulated.

> **EXPERIMENTAL PROBABILITY**
>
> The experimental probability $P(E)$ that an event will occur is given by
>
> $$P(E) = \frac{\text{Number of times event occurs}}{\text{Number of trials}}$$

CHECK UNDERSTANDING

For a certain experiment a trial consists of tossing a penny three times. In 100 trials you obtain two heads 43 times. What is the experimental probability of throwing two heads in three tosses? How does this compare with the theoretical probability?

EXAMPLE 3

HIGHWAY STATISTICS A survey of drivers on a section of highway with a speed limit of 55 mi/h produced the results in the table. If the results apply in general to drivers on that section, what is the experimental probability that a randomly chosen driver will be driving more than 60 mi/h?

Speed Limit Survey (in mi/h)			
	under 50	50–60	over 60
Men	5	41	53
Women	11	32	21

Solution

Number of drivers exceeding 60 mi/h $= 53 + 21 = 74$
Number of drivers surveyed $= 5 + 11 + 41 + 32 + 53 + 21$
$\qquad\qquad\qquad = 163$

$P\text{ (exceeding 60 mi/h)} = \frac{74}{163} \approx 45.4\%$

The experimental probability is about 45.4%.

◄

Assume that the speed limit survey in Example 3 was representative of the entire population. Write a few sentences explaining how to find how many of 20,000 drivers on an interstate highway posted at 55 mph can be expected to exceed 60 mph.

The likelihood that an event will occur can be expressed as **odds**.

ODDS

Odds is the ratio of the number of ways an event can occur to the number of ways it could *fail* to occur.

$$\text{Odds} = \frac{\text{Number of outcomes in event}}{\text{Number of outcomes } not \text{ in event}}$$

EXAMPLE 4

What are the odds of rolling a number cube and getting a 2?

Solution

There is only one way the event can occur, by rolling a 2. There are 5 ways the event could fail to occur, by rolling 1, 3, 4, 5, or 6.

$$\text{Odds} = \frac{\text{Number of outcomes in event}}{\text{Number of outcomes } not \text{ in event}} = \frac{1}{5}$$

The odds of rolling a 2 are $\frac{1}{5}$, often stated as "1 to 5", or 1 : 5. ◄

TRY THESE

Use the spinner at the right for Exercises 1–10.

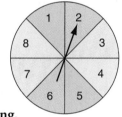

1. You want to spin a 5 or 6. List the outcomes in the event.

2. List the outcomes in the sample space.

Determine the theoretical probability that you will spin each of the following.

3. 5 or 6   4. an even number   5. 7

6. a number less than 4   7. a number   8. 19

9. What are the odds that you will spin a 5 or a 6?

10. Suppose you spin 200 times and obtain a 5 or a 6 fifty-eight times. What would be the experimental probability of spinning a 5 or a 6?

You roll a number cube twice.

11. Make a tree diagram showing the sample space.

12. Use the tree diagram to find the theoretical probability of rolling two odd numbers.

13. GEOMETRY A rectangular photo measuring 4 in. by 5 in. is pinned to a rectangular bulletin board measuring 18 in. by 30 in. What is the probability that a fly landing on the bulletin board will land on the photo?

14. WRITING MATHEMATICS Write a few sentences explaining the difference between experimental and theoretical probability. Give examples to illustrate your answer.

BIRTHDAYS Of 256 people surveyed, 32 were born in May.

15. What is the theoretical probability that a person chosen at random was born in May? Explain your reasoning.

16. Based on the survey, what is the experimental probability that a person chosen at random was born in May?

PRACTICE

Describe an event with the given probability.

1. 0   **2.** about 50%   **3.** 1

You throw two darts at the board. Assume both darts land in a section.

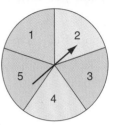

4. Make a tree diagram showing the sample space.

5. Use the tree diagram to find the theoretical probability of obtaining a sum of 7 or 8.

For Exercises 6–15 assume that 20 cards numbered from 1 to 20 are shuffled.

6. You want to draw a number that is divisible by 6. List the outcomes in the event.

7. List the outcomes in the sample space.

Give the theoretical probability that you will draw each of the following.

8. a number divisible by 6   **9.** a number divisible by 4   **10.** a number greater than 13

11. 30   **12.** 9 or 13   **13.** a number from 1 to 20

14. What are the odds that you will draw a number divisible by 6?

15. You draw a card 50 times, replacing and shuffling each time, and obtain a multiple of 6 eight times. What is the probability of drawing a multiple of 6?

16. GAMES If you throw a dart at random and it hits the board, what is the probability that you hit the center circle?

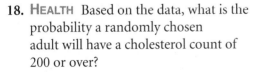

17. STATISTICS In a survey of 276 people, 23 reported that they were left-handed. What is the probability that a randomly chosen person is left-handed?

18. HEALTH Based on the data, what is the probability a randomly chosen adult will have a cholesterol count of 200 or over?

19. WRITING MATHEMATICS A TV weatherperson states that there is a 40% probability of rain tomorrow. Explain what the statement means.

Cholesterol Survey			
	under 200	**200–240**	**over 240**
Men	28	35	16
Women	19	24	20

EXTEND

20. The probability that an event will occur is $\frac{3}{5}$.

    a. What is the probability that the event will not occur?
    b. What are the odds that the event will occur?

21. The odds are 6 to 5 that a randomly chosen student in a certain group intends to go to college. What is the probability that a student in the group intends to go to college?

22. If you hit the board in Exercise 16 with a dart, what are the odds that you will hit the middle (shaded) ring?

THINK CRITICALLY

23. A penny is tossed 12 times. How many outcomes in the sample space consist of 11 heads and 1 tail? Explain.

24. Find the sum of the probability that a certain outcome will occur and the probability that it will not occur. Explain your reasoning.

MIXED REVIEW

Solve and check each equation.

25. $-3 = 5 - x$ **26.** $2 + 6x = -1$ **27.** $-6 + n = 3 - 2n$ **28.** $\dfrac{2b - 3}{5} = 7$

You throw a number cube. Determine the probability of each result.

29. 5   **30.** 1, 2, or 3   **31.** 2 or 5   **32.** 9

33. STANDARDIZED TESTS Which of the following matrices is equal to $3\begin{bmatrix} 2 & 6 \\ -1 & 4 \end{bmatrix}$?

A. $\begin{bmatrix} 6 & 6 \\ -3 & 12 \end{bmatrix}$ **B.** $\begin{bmatrix} \frac{2}{3} & 2 \\ -\frac{1}{3} & \frac{4}{3} \end{bmatrix}$ **C.** $\begin{bmatrix} 2 & 6 \\ -1 & 4 \end{bmatrix}$ **D.** $\begin{bmatrix} 6 & \frac{36}{2} \\ -\frac{9}{3} & 12 \end{bmatrix}$

PROJECT *Connection*   In this activity, you will investigate some ways to use probability to describe different aspects of the project.

1. Assume each group has the same chance of winning a model-car race. What is the probability that your group will win?

2. Have each group report on the progress of its model. What effect, if any, do the reports have on the probability you determined in Question 1? Explain.

3. Have each group tell what color their model will be. Make a frequency table and use the information to determine the experimental probability for each color. Use the probabilities to predict how many models of each color would be built by 300 student groups.

ALGEBRA WORKS

HELP WANTED

The insurance business is based on risk. An insurance company plans that the money it collects in premiums from its policy holders will be enough to settle all of the claims of the policy holders, pay the overhead on the business, and make a profit.

An insurance company cannot know in advance whether it will make any money when it charges you $2258 a year to insure your new sports car. By analyzing thousands of past claims, it can be confident that in the long run, the total in premiums that it charges its customers will be sufficient to produce a profit. To analyze past claims and determine how much to charge for policies, insurance companies employ *actuaries*, the mathematicians of the insurance business.

Decision Making

Statistics show that men are twice as likely to be involved in fatal accidents as women and that alcohol is involved in 48% of fatal accidents.

1. Of 43,500 automobile fatalities in 1992, about how many were men?

2. About how many of the deaths occurred in accidents involving alcohol?

There is a 30% probability that the driver in an accident is under 25.

3. What are the odds that a driver in an accident is under 25?

4. What is the probability that the driver in an accident is at least 25?

5. National Reliable Insurance Company charges you $2258 a year to insure your new Wildcat. The company charges another of its policyholders $862 annually for the same protection. List as many factors as you can that might differentiate you from the other policyholder and that may have resulted in a lower annual premium.

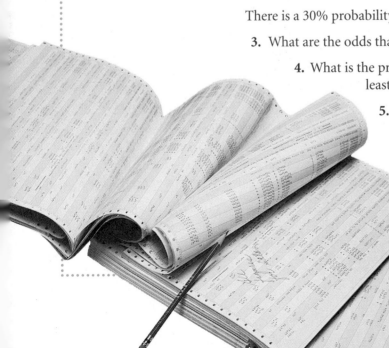

Explore

The matrix lists the sticker prices of six new car models being sold at Bargain Motors.

$$\begin{bmatrix} 11000 & 11100 \\ 10900 & 5000 \\ 10900 & 10500 \end{bmatrix}$$

The agency's newspaper ad reads: "Always a bargain at Bargain. In an age of expensive cars, the average price of our new models remains *under* $10,000."

1. Is Bargain's claim about the prices of its new models accurate? Explain.

2. Is the claim misleading? Explain.

3. Why is the average of the data not always a good representative of the complete set? Why is the average not a good representative of the data here?

4. Write an ad that gives a more truthful picture of the prices of new models at Bargain Motors.

Build Understanding

It is not always practical to display an entire set of data. Therefore, it is useful to have a few values to represent the set. Values that represent the overall "center" or middle are called **measures of central tendency**.

Among the most commonly used measures of central tendency are:

PROBLEM SOLVING TIP

You may find these memory devices useful:
• The median strip of a highway is in the *middle* of the road.
• The MOde is the element that occurs MOst Often.

• The **mean** is the average of a set of data. It is calculated by dividing the sum of the data by the number of items. The mean is an appropriate measure of central tendency when all of the data in a set are approximately equal. The mean may be an unrealistic view when the data include some values that are extremely high or low.

• The **median** of a set of data is the middle value when the data are arranged in numerical order. When there are two middle values, the median is the average of the two values. The median may be an appropriate measure of central tendency when the mean is not.

• The **mode** of a set of data is the element that occurs most often in the set. A set may have no mode, one mode, or several modes. The mode is an appropriate measure of central tendency when several items of data are the same.

EXAMPLE 1

SPORTS Following a performance on the parallel bars, a gymnast received the following scores from the judges: 8.8, 8.9, 8.8, 10.0, 8.8, 9.9, 8.8, 8.9, 9.0, 8.8.

Find each measure of central tendency.

a. mean

b. median

c. mode

d. Tell which measure best represents the data and explain why.

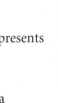

Solution

a. $\text{mean} = \dfrac{\text{sum of data}}{\text{number of items of data}}$

$= \dfrac{8.8 + 8.9 + 8.8 + 10.0 + 8.8 + 9.9 + 8.8 + 8.9 + 9.0 + 8.8}{10} = 9.07$

The mean is 9.07.

b. To find the median you must arrange the data in numerical order.

middle
↓   ↓

8.8, 8.8, 8.8, 8.8, $\boxed{8.8}$, $\boxed{8.9}$, 8.9, 9.0, 9.9, 10.0

Since there are two middle values, 8.8 and 8.9, the median is the average of the two middle values, $\dfrac{8.8 + 8.9}{2} = 8.85$.

c. The value 8.8 occurs five times, so the mode is 8.8.

d. The scores of 9.9 and 10.0 resulted in a mean of 9.07, a value that is greater than eight of the scores. Therefore, the mean is too great to be taken as a measure of central tendency.

The median of 8.85 is a better measure but is not actually included among the data. Since five of the ten judges awarded scores of 8.8, the mode might be the best measure of central tendency. ◄

The median tells you which value is in the middle of a set of data but it does not show you how the values in the set are distributed. Consider the following three sets of data.

$\{99, 100, 101\}$ $\{99, 100, 1 \text{ million}\}$ $\{-1 \text{ billion}, 100, 1000\}$

The number 100 is the median of each set. However, each set contains a different dispersion of numbers.

To better understand the *dispersion* of a set of data—the way the values are clumped together or spread out—you can use the following **measures of dispersion**.

- The **range** is the difference between the greatest and least values in the set.
- The **first** (or **lower**) **quartile** Q_1 is the median of the lower half of the data.
- The **second quartile** Q_2 is the median of the whole set of data.
- The **third** (or **upper**) **quartile** Q_3 is the median of the upper half of the data.

The quartiles divide an ordered data set into four quarters. You can incorporate these measures into a **boxplot**, sometimes called a *box-and-whisker plot,* to show how a set of data is dispersed. An asterisk shows any **outliers**, values far from the data set. **Whiskers** are lines that show the range of the data.

EXAMPLE 2

CONSUMERISM In a survey of eleven electronics stores, a consumer group turned up the following prices for Driveabout car phones.

124, 131, 169, 128, 129, 99, 149, 125, 175, 127, 219

Draw and interpret a boxplot of the data.

Solution

Arrange the data in numerical order to find the median and the quartiles.

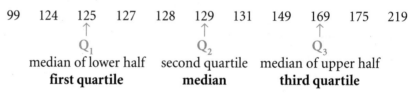

Draw a number line that includes the lowest and highest values. Then draw a box with ends that line up with the first and third quartiles. Draw a line through the box to show the median. Draw whiskers from the box to show the range of the data. The range is $219 - 99 = 20$.

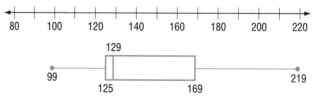

The box is tightly clustered in the interval from 125 to 129. Values from 129 to 169, the third quarter, are more widely dispersed. The values in the fourth quarter are more widely dispersed than those in the first quarter. Notice the length of the box represents the middle half of the data. A completed boxplot does not need to show the number line. ◄

CHECK UNDERSTANDING

Of the mean, median, mode, and range, which can you determine from a boxplot of a set of data?

PROBLEM SOLVING TIP

You may be able to use your graphing utility to draw boxplots.

Find the mean, median, mode, and range of each set of values.

1. 46, 28, 49, 53, 29

2. 8.5, 11.2, 9.3, 14.6, 15.0, 11.2, 9.3, 9.4

3. 488, 466, 517, 581, 404, 496

4. $-3.2, 5.5, -7, -2, 5.7, 0, 5.5, 10, -1$

5. **METEOROLOGY** The average monthly precipitation, January through December, in Galveston, Texas, in inches is: 3.0, 2.3, 2.1, 2.6, 3.3, 3.5, 3.8, 4.4, 5.8, 2.6, 3.2, 3.6.

   a. Find the mean, median, mode, and range of the data.

   b. Which measure of central tendency best represents the data? Explain.

Use the boxplot showing science test scores for Exercises 6–12.

6. What percent of the data is between 40 and 70?

7. What is the third quartile?   8. What is the least value?

9. What is the median?   10. What is the range?

11. What percent of the data is between 70 and 90?

12. What is the smallest interval that contains 25% of the data?

13. **CAR MILEAGE** The numbers of miles driven in one day by each of nine rental cars was 201, 108, 112, 203, 26, 181, 211, 209, 190. Draw and interpret a boxplot of the data.

14. **WRITING MATHEMATICS** Write a paragraph explaining why the mean can be misleading. Give an example of a set whose mean represents the set well and one whose mean does not represent the set well, and tell which measure would describe the data best.

PRACTICE

Find the mean, median, mode, and range of each set of values.

1. 20, 28, 9, 14, 21, 8, 33   2. 1.2, 1.1, 1.8, 0.9, 0.5, 1.5, 1.7, 1.5

3. 361, 314, 325, 320, 314, 336, 332, 361

4. $4, -7, -1, 5, -11, -2, 1, -1$

5. 6.68, 6.83, 6.65, 6.82, 6.77, 6.81

6. 9865, 9009, 9801, 9871, 9909, 9686, 9178

PHYSICAL FITNESS The resting pulse rates, in beats per minute, of the members of a soccer team are 53, 57, 68, 51, 58, 61, 79, 49, 53, 62, 71, 55, 51, 46, 60, 48.

7. Find the mean, median, mode, and range of the data.

8. Which measure of central tendency best represents the data? Give reasons for your answer.

9. Construct a boxplot to display the data.

10. ENTERTAINMENT The ages of the chess players invited to participate in a tournament are

19, 25, 27, 22, 13, 24, 29, 21, 22, 55, 18, 46, 23, 17, 23

Draw and interpret a boxplot of the data.

STATISTICS The boxplots summarize the results of the scores achieved by two algebra classes on the same test.

Class 1
42   98
75 84 85

Class 2
50   95
78 87 94

11. Which class had the highest score? the lowest? What were the scores?

12. Which class had the higher median? What was it?

13. In which class was the dispersion of grades in the middle 50% more even?

14. Which class achieved the better results on the test? Explain.

15. WRITING MATHEMATICS Conduct a survey in your class. Be sure to question at least eight people. Find the mean, median, and mode of the data, explaining how you found each. Then make a boxplot of your data.

EXTEND

The difference between the third and first quartiles is called the **interquartile range** (IQR). The IQR is used to identify extreme data items, called outliers. An **outlier** is any data item that is more than $1.5 \cdot$ IQR units below the first quartile or above the third quartile.

16. Determine the interquartile range for the set of chess players' ages in Exercise 10.

17. Indicate any ages in the set that would be classified as outliers. Show the outliers with an asterisk on your boxplot.

COMMERCE The table gives the average number of minutes of commercials on a city's radio stations.

18. Find the mean, median, and mode of the set of data.

19. Which measure of central tendency best represents the data? Give reasons for your answer.

Commercial Minutes per Hour	
Minutes	**Number of Stations**
39	1
41	3
42	4
45	4

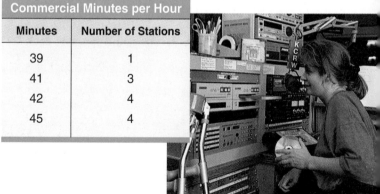

20. PHYSICAL FITNESS Janice has run a mean distance of 8 miles in each of her last nine workouts. How far must she go on her next run to raise her mean distance for ten workouts to 9 miles?

21. The median of the set of numbers 5, 12, 21, and n is 13. What is the mean of the numbers?

THINK CRITICALLY

HEIGHT SURVEY The heights in centimeters of a group of students are
160, 161, 164, 160, n, 161. Determine n such that the set of heights has

22. no mode   **23.** one mode   **24.** two modes

25. Give a set of ten values whose boxplot has no whiskers.

26. The mode of 5 one-digit numbers is 2. The median is 6. The mean is 5.2. What are the
numbers?

27. Jorge drove from Riceville to East Wilton at a mean speed of 50 mi/h and returned at a
mean speed of 60 mi/h. The distance was 300 miles in both directions. What was his
mean speed for the journey? Disregard any time he may have spent in either town.
Explain how you found your answer.

MIXED REVIEW

Let $A = \begin{bmatrix} 2 & 4 & -1 \\ -5 & 3 & 7 \end{bmatrix}$ and $B = \begin{bmatrix} -5 & 1 & 8 \\ -2 & 4 & 3 \end{bmatrix}$. Evaluate each expression.

28. $B - A$ **29.** $B + (\, 2 \times 3 \text{ zero matrix}\,)$ **30.** $-3A$ **31.** $A - 2B$

Find the mean, median, mode, first quartile, and third quartile of each set.

32. {6, 15, 13, 5, 8, 6, 9, 1, 11, 14]   **33.** {22, 26, 31, 25, 34, 22, 31, 36, 31, 29, 31, 30}

34. STANDARDIZED TESTS The range of a set of three numbers is 75. The lowest number is
57. Which of the following statements could be true?

     I. The mean is 81.
   II. The mode is 57.
 III. The median is 81.

     A. I only   **B.** II only   **C.** III only
     D. I and II   **E.** II and III   **F.** I, II, and III

PROJECT *Connection*   In this activity, you will collect and analyze data about each
model race car.

1. As a class, decide on the physical data that should be provided for each model. Some
possible variables include length, width, height, weight, and wheel diameter.

2. Decide which measure of central tendency would be best to use to report each statistic.
Give reasons for your choices.

3. Make a boxplot to summarize the data for each variable you used in Question 1. Discuss
the conclusions you can draw from examining the boxplots individually and together.

4. Discuss how you might analyze the data to predict which car(s) will perform best.

Explore

1. Draw x- and y-coordinate axes on a sheet of graph paper.

2. Find three ordered pairs (x, y) that satisfy the equation $y = \frac{2}{3}x + 1$. Graph each ordered pair. Draw a line connecting the points.

3. Choose two points P_1 and P_2 on the line you graphed in Question 2. Find the distances h and k that you must move right and up to go from P_1 to P_2. Calculate the ratio $\frac{k}{h}$.

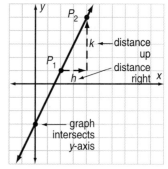

4. Give the y-coordinate of the point where the line you graphed intersects the y-axis.

5. Repeat Questions 2–4 for a different linear equation. In what form will you write the equation?

6. For each line that you graphed, compare your answers to Questions 3 and 4 with the equation of the line. Make a conjecture about the relationship between the numbers in a linear equation and the ratio and y-coordinate you found in Questions 3 and 4.

Build Understanding

A **linear equation** is an equation in which no variable has an exponent greater than one. The graph of a linear equation is a line. One form in which you can write a linear equation is $Ax + By = C$. This is called the **standard form**.

In earlier mathematics courses you learned to graph equations by making tables of values and then plotting and connecting the corresponding points as in Explore. That method will continue to be useful, particularly when you are graphing extremely complex equations.

Another way to graph a linear equation is to find the points at which the graph intersects the x- and y-axes.

- The **x-intercept** of a line is the value of x when $y = 0$.

- The **y-intercept** of a line is the value of y when $x = 0$.

EXAMPLE 1

Graph the equation $2x + 3y = 6$ using the x- and y-intercepts.

Solution

To find the x-intercept, let $y = 0$.

$$2x + 3(0) = 6 \qquad \text{Solve for } x.$$
$$2x = 6$$
$$x = 3$$

To find the y-intercept, let $x = 0$.

$$2(0) + 3y = 6 \qquad \text{Solve for } y.$$
$$3y = 6$$
$$y = 2$$

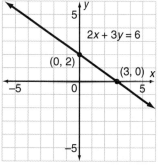

Plot the intercepts $(3, 0)$ and $(0, 2)$. Then connect the points. ◀

Another form in which a linear equation can be written is $y = mx + b$. This is called the **slope-intercept form**, where m is the slope of the line and b is the y-intercept.

For a given distance along a line, the **slope** of the line is the ratio of the number of units the line rises or falls vertically to the number of units the line moves from left to right horizontally. For a line containing two points (x_1, y_1) and (x_2, y_2),

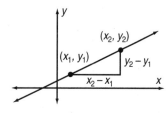

$$\text{slope} = \frac{\text{rise}}{\text{run}} = \frac{\text{change in } y}{\text{change in } x} = \frac{y_2 - y_1}{x_2 - x_1}$$

EXAMPLE 2

Graph the equation $2x + 3y = 6$ using the slope and y-intercept.

Solution

$$2x + 3y = 6 \qquad \text{Write in slope-intercept form.}$$
$$3y = -2x + 6$$
$$y = -\frac{2}{3}x + 2$$

The slope is $-\frac{2}{3}$ and the y-intercept is $(0, 2)$.

Plot the y-intercept. From the y-intercept, move 2 units up and 3 units left since the slope is negative. Plot a second point and connect the two points. ◀

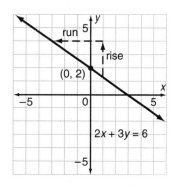

COMMUNICATING ABOUT ALGEBRA

In Example 2 could the second point be plotted by moving down 2 units and right 3 units? Explain.

THINK BACK

Slope can be expressed as $\frac{\text{rise}}{\text{run}}$.

A line with positive slope slants upward to the right. A line with negative slope slants upward to the left.

Linear equations can be used to model real life situations.

EXAMPLE 3

AUTOMOTIVE COSTS Jenna paid $30 for gasoline and oil for her car. Gasoline is $1.20 per gallon and oil is $2.00 per quart. What possible purchases could Jenna have made?

Solution

Let x represent the number of gallons of gasoline. Let y represent the number of quarts of oil. Then the cost is $1.2x + 2y = 30$.

Use a graphing utility to graph the equation. Any point on the line represents a combination of gasoline and oil with a total cost of $30. Use the TRACE feature to identify coordinates of different points. The three points that are identified represent 0 gallons of gas and 15 quarts of oil, 15 gallons of gas and 6 quarts of oil, and 25 gallons of gas and 0 quarts of oil.

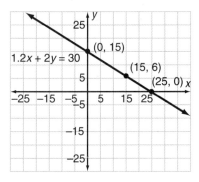

COMMUNICATING ABOUT ALGEBRA

In the solution to Example 3, do all the points of the line represent realistic solutions to the problem? Explain.

Graphing paired data as a **scatter plot** may show that the relationship between two real world quantities is approximately linear. In such situations, you can draw a **line of best fit** to model the linear relationship.

EXAMPLE 4

CONSUMERISM The table gives the price, highway mileage (mi/gal), number of cylinders and number of passengers of twelve new vehicles tested by a consumer group. Make a scatter plot of the price and highway mileage data, draw a line of best fit, and determine the equation of the line.

PROBLEM SOLVING TIP

In order to graph an equation with a graphing utility, it usually must be in slope-intercept form, $y = mx + b$.

Price	Mileage (mi/gal)	Number of Cylinders	Number of Passengers
7150	41.8	4	4
44,900	13.3	12	4
12,350	23.8	6	10
16,200	20.3	6	4
27,900	25.0	8	2
8150	35.4	4	6
22,000	24.4	8	5
32,750	21.8	8	6
37,000	12.3	8	12
15,900	28.0	6	8
8800	25.9	4	7
7400	31.0	4	2

Graph the data for price and highway mileage on a scatter plot. Then draw a line that as closely as possible models the linear pattern displayed by the points. To make the graph more compact, divide the prices by 1000. Graph the data.

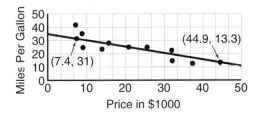

To find the slope of the line, choose two points on the line. For example,

$$P_1 = (7.4, 31) \quad \text{and} \quad P_2 = (44.9, 13.3)$$

$$\text{slope} = \frac{\text{change in } x}{\text{change in } y} = \frac{31 - 13.3}{7.4 - 44.9} = -0.472$$

Read the y-intercept from the graph: y-intercept ≈ 34.

Use the slope and y-intercept to write an equation of the line.

$$y = -0.472x + 34$$

Other lines and equations are possible depending on how the line is drawn to fit the data. ◄

The term **correlation** refers to the relationship between two sets of data. The scatter plot in Example 4 shows that, in general, as the price of a car goes up, the mileage goes down. A line of best fit for the relationship has a negative slope.

When one variable increases as the other decreases, the correlation between the variables is **negative**. When both variables increase or decrease together, the correlation is **positive**. When there is no apparent relationship between the variables, there is **no correlation.**

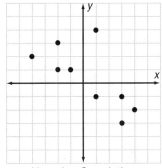

Negative Correlation

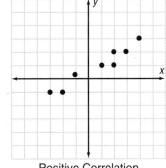

Positive Correlation

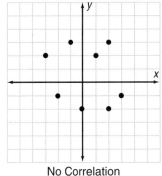

No Correlation

EXAMPLE 5

Tell whether you would expect the variables to show a positive correlation, a negative correlation, or no correlation.

  a. amortization period of a car loan and the deferred payment price
  b. amortization period of a car loan and the monthly payment amount
  c. amortization period of a car loan and the price of the car

Solution

  a. As the amortization period increases, the deferred payment price increases. The correlation is positive.

  b. As the amortization period increases, the monthly payments decrease. The correlation is negative.

  c. A buyer can choose any amortization period, regardless of the price of the car. There is no correlation between amortization period and price. ◄

A relationship can be positive or negative and strong or weak. The **coefficient of correlation** is a statistical measure of how closely data fit a line. Most graphing utilities can calculate the coefficient of correlation *r* for you. The coefficient of correlation ranges from -1 to 1. The closer the correlation coefficient is to -1 or 1, the stronger the relationship is between *x* and *y*. The closer the correlation coefficient is to 0, the weaker the relationship is between *x* and *y*.

EXAMPLE 6

Use a graphing utility to find the equation of the line of best fit and the coefficient of correlation for the data on car price and mileage from Example 4.

Solution

Enter the data in your graphing utility. Use the statistics feature to find the linear regression.

With each variable rounded to the nearest hundredth, the equation for the line of best fit for the data is

$$y = -0.53x + 35.83$$

The coefficient of correlation of -0.812 tells you that the data points lie reasonably close to the graph of the equation.

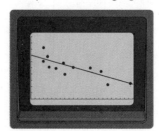 

◄

Find the x-intercept, y-intercept, and slope of the graph of each equation.

1. $2x + 5y = 10$ **2.** $y = -3x + 12$ **3.** $0.3x - 6y - 15 = 0$

4. Graph the equation $-2x + 7y = 14$ using intercepts.

5. Graph the equation $x - 2y = 6$ using the slope and y-intercept.

MOVIE TICKETS Adult tickets at Star Cinema cost $6. Children's tickets cost $4.
6. Write and graph an equation showing the numbers of adults and children who could have attended a film which brought in $480 in ticket sales.

7. Ninety children's tickets were sold. Use the graph to find the number of adult tickets sold. Explain how you found the answer.

SPORTS The table below gives the winning Olympic pole vault heights, in feet, from 1904 to 1988. Note that Olympics were not held in 1916, 1940, and 1944.

8. Make a scatter plot of the data and draw a line of best fit. (*Hint*: Let x represent the Olympics number starting with 1904 = 4.)

9. Find the equation of the line.

10. Describe the correlation between the variables.

11. Use your equation to predict the winning height for the 1992 Olympics.

12. The actual winning height in 1992 was 19.0 ft. How does your prediction compare?

13. Make a prediction for 1996 and find the winning height. How do the prediction and the actual height compare?

Year	Height	Year	Height	Year	Height	Year	Height	Year	Height
1904	11.5	1924	13.0	1948	14.1	1964	16.7	1980	19.0
1908	12.2	1928	13.8	1952	14.9	1968	17.7	1984	18.9
1912	13.0	1932	14.1	1956	15.0	1972	18.0	1988	19.8
1920	13.4	1936	14.3	1960	15.4	1976	18.0	1992	?

Tell whether you expect the following variables to show a positive correlation, a negative correlation, or no correlation.

14. the price of a car and the number of doors the car has

15. the speed that a car travels and the time it takes the car to travel 200 miles

16. the price of gasoline and the cost of a tank of gas

17. **WRITING MATHEMATICS** Two quick methods for sketching the graph of a line were described in the lesson. Describe how you would decide which method is better for different forms of equations.

PRACTICE

Find the x-intercept, y-intercept, and slope of the graph of each equation.

1. $y = \frac{1}{2}x - 2$

2. $4x + 5y = 10$

3. $2x - 4y + 3 = 0$

Graph the equation using intercepts.

4. $-3x + 2y = 12$

5. $x + y = 9$

6. $y = -2x + 4$

Graph the equation using the slope and y-intercept.

7. $y = 2x - 3$

8. $y = -\frac{3}{4}x + 1$

9. $2x - 5y + 15 = 0$

CIVIC AFFAIRS At the Carnegie Library book sale, hardcovers sold for $1 and paperbacks for $0.50. On the first day of the sale, receipts totaled $1500.

10. Write and graph an equation showing the numbers of hardcovers and paperbacks that might have been sold.

11. Seven hundred fifty hardcovers were sold. Use the graph to find the number of paperbacks sold. Explain how you found the answer.

SPORTS The table gives the winning Olympic men's 400-meter run times, in seconds, from 1904 to 1988. (Note that Olympics were not held in 1916, 1940, and 1944.)

12. Make a scatter plot of the data and draw a line of best fit.

13. Find the equation of the line.

14. Describe the correlation between the variables.

Year	Seconds	Year	Seconds	Year	Seconds	Year	Seconds	Year	Seconds
1904	49.2	1924	47.6	1948	46.2	1964	45.1	1980	44.6
1908	50.0	1928	47.8	1952	45.9	1968	43.6	1984	44.3
1912	48.2	1932	46.2	1956	46.7	1972	44.7	1988	43.9
1920	49.6	1936	46.5	1960	44.9	1976	44.3	1992	?

15. Use your equation to predict the winning time for the 1992 Olympics.

16. The actual winning time in 1992 was 43.5 seconds. How does your prediction compare with actual time?

17. Predict the winning time for the 1996 Olympics and compare with the actual time.

18. What does your equation predict for the winning time in the year 2500? Is this prediction realistic? Explain.

Do the following show positive correlation, negative correlation, or no correlation?

19. the diameter of a circle and the circumference of the circle

20. the width of a rectangle with an area of 10 in.^{2} and the length of the rectangle

21. your altitude above the earth's surface and the air pressure

22. a person's age and zip code

EXTEND

23. **WRITING MATHEMATICS** Give examples from your everyday life of a pair of variables that shows a positive correlation, a pair that shows a negative correlation, and a pair that shows no correlation.

24. A line intersects the y-axis at $(0, -7)$ and has a slope of 5. Find the equation of the line.

25. Lines that are parallel have the same slope. Find the equation of the line that contains the point $(-8, 1)$ and is parallel to the line $3x + 4y = 6$.

26. Lines that are perpendicular have slopes whose product is -1. Find the equation of the line that contains the point $(-3, 1)$ and is perpendicular to the line $y = \frac{1}{2}x - 9$.

27. **a.** Use the equation determined in Example 4 to approximate the mileage you could expect to get from cars that sell for $22,000 and $37,000.

    b. In each case, determine the difference between the predicted mileage and the actual value given in the data table. Express the difference as a percent of the actual value.

    c. Do you think the equation is a good prediction? Explain.

THINK CRITICALLY

28. Find the equation of the line that contains the points $(3, 7)$ and $(-3, -5)$.

29. What happens when you try to find the slope of the line that contains $(2, 4)$ and $(2, 6)$? Explain your answer.

PROPERTY DEPRECIATION A $600,000 building is depreciated 5% of its original value per year by its owner.

30. What is the amount of depreciation per year? per month?

31. Write a linear equation that models the value y of the building after x months.

32. Graph the equation you wrote. What restrictions must be placed on x and y for the graph to make sense in the real world?

33. What is the value of the building after 5 years?

34. How long before the building is fully depreciated (its value is zero)?

MIXED REVIEW

35. **STANDARDIZED TESTS** The odds that an event will occur are 9:5. What is the probability that the event will occur?

    A. $\frac{9}{14}$ **B.** $\frac{5}{9}$ **C.** $\frac{5}{14}$ **D.** $\frac{4}{5}$

Find the mean, median, mode, first quartile, and third quartile of each set.

36. $\{55, 57, 42, 44, 60, 58, 57, 51\}$

37. $\{3, 4, 1, 3, 5, 4, 0, 2, 6, 3, 4, 1, 1, 5\}$

Find the x-intercept, y-intercept, and slope of the graph of each equation.

38. $5x - 6y = 30$

39. $y = 5x - 10$

40. $2x + 2y + 8 = 0$

Problem Solving File

Solving Problems with Scientific Notation

Like pitching a baseball, solving problems is a skill that can be mastered only through long and patient practice. A skilled pitcher has a plan for dealing with each batter and an array of strategies for carrying out the plan. As you have acquired problem-solving experience in your previous mathematics courses, you have learned to use strategies such as drawing a diagram, finding a pattern, and guessing and checking. You have probably learned to employ the following four-step method when you attempt to solve a challenging problem.

Step 1 **UNDERSTAND** the problem. Analyze the information.

Step 2 **PLAN** an approach. Decide on the strategies, computations, and technology you want to use.

Step 3 **SOLVE** the problem. Apply the strategy. Carry out the work.

Step 4 **EXAMINE** the result. Check your answer. Look back at your work and ask yourself if the answer is reasonable, whether you could have solved the problem another way, and if you notice any general patterns that might apply to other problems.

The problem below involves some extremely large numbers. You can solve the problem using the four-step scheme and the strategy of rewriting the numbers in a more convenient form.

Problem

The following statistics describe motor vehicle use in the United States during 1991.

- Total number of cars and trucks: 185,000,000
- Total cost of gasoline used: $152,000,000,000
- Mean number of miles driven per gallon of gasoline: 16.9
- Mean number of gallons of gasoline used per vehicle: 683

Find the total distance traveled by all the motor vehicles during the year.

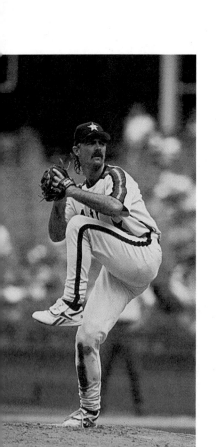

Explore the Problem

UNDERSTAND the problem.

1. What do you want to know? Of the data given, which can you use to answer the question? Which information is irrelevant?

PLAN an approach.

2. How can you operate on the relevant data to find the answer? What technology might be useful to carry out the operations?

SOLVE the problem.

Since a very large number is involved, you are less likely to make errors in computation if you first write all of the numbers in **scientific notation.**

> **SCIENTIFIC NOTATION**
>
> A number is written in scientific notation when it is expressed in the form $a \times 10^n$, where $1 \leq a < 10$, and n is an integer.

3. Use scientific notation to write each of the numbers you will be using in the computation. Explain the method you used.

4. Find the total distance traveled by all the motor vehicles during the year. Give the answer in scientific notation.

5. Rewrite the number using *standard* notation. Explain your method.

EXAMINE the result.

6. Is the answer reasonable? Explain how you can decide whether you might have made a computation error.

Investigate Further

7. After carrying out a computation involving numbers written in scientific notation, Antonio obtained the answer 462×10^{13}. How can he rewrite the answer in scientific notation? What is the answer?

8. Use the data on the previous page to write a fraction that expresses the portion of the total number of cars and trucks that is represented by one vehicle.

9. Evaluate the fraction, expressing the answer in scientific notation.

10. Write the answer using standard notation.

Of all the gasoline sold for cars and trucks in 1991, the portion represented by a single gallon was 0.00000000000787. The total cost of gasoline purchased was $152,000,000,000.

11. Write each number using scientific notation.

12. Find the mean price of a gallon of gas in 1991.

13. WRITING MATHEMATICS Use the data on cars given in this lesson to write and solve a problem using scientific notation.

ALGEBRA: WHO, WHERE, WHEN

The four-step problem-solving method was first described by the Hungarian-born mathematician George Pólya (1887–1985) in his classic book *How to Solve It,* published in 1945. Pólya emigrated to the United States in 1940 and spent most of his career teaching at Stanford University.

> **PROBLEM SOLVING PLAN**
>
> • Understand
> • Plan
> • Solve
> • Examine

PROBLEM SOLVING TIP

Use the ☐EE☐ or ☐EXP☐ key on your calculator to enter numbers in scientific notation. For example, $5.7 \times 10^9 \rightarrow 5.7$ EE 9

Apply the Strategy

14. INTERNATIONAL TRADE Between 1950 and 1990, Japan's motor vehicle production rose from 3.2×10^4 to 1.3×10^7 annually. How many times greater was Japan's annual production in 1990 than it was in 1950? Give the answer in standard notation.

PHYSICS Light travels at a velocity of 186,282 miles per second.

15. Find the length of 1 light-year, the distance that light travels in a 365-day year.

16. How does the total distance traveled by all the motor vehicles in the United States in 1991 compare with a light-year?

17. The farthest object visible to the naked eye is the Great Galaxy in the constellation of Andromeda, at a distance of about 13,600,000,000,000,000,000 miles. How long does it take light from the Great Galaxy to reach earth? Give the answer in standard notation.

18. How many feet does light travel in one *nanosecond* (1 billionth of a second)?

SOCIAL STUDIES Projections put the world population at about 6.2 billion in the year 2000. The land area of the earth is approximately 57.9 million square miles.

19. Find the projected population density in the year 2000 in persons per square mile.

20. Suppose that in the year 2500 each person is given a 10 ft $\times$ 10 ft plot of land. Find the world population in the year 2500.

REVIEW PROBLEM SOLVING STRATEGIES

TUNNELING THROUGH

1. A train 1 mile long travels through a tunnel 1 mile long at a rate of 1 mile per hour. How long will it take the train to pass completely through the tunnel? (Hint: Draw a picture.)

ON THE ROAD

2. Ramon Ortega is starting a sales trip during which he will drive 18,000 kilometers. According to his owner's manual, the new tires on his car are good for 12,000 kilometers. Ramon does not want to be bothered purchasing tires while on his trip.

   a. What is the minimum number of new spares Ramon should take along on the trip?

   b. What procedure should Ramon follow regarding his tires?

A JUICY PROBLEM

3. Nick poured himself a glass of cranberry juice and drank one sixth of it. He thought it was too tart, so he poured as much apple juice into the glass as he had drunk. He drank one-third of the mixture and then added apple juice to fill the glass again. Still trying to get the perfect taste, he drank half the mixture in the glass and added apple juice to fill the glass again. "Just right", he thought, as he drank the whole glass of this last mixture. How much of each type of juice did Nick drink?

   a. What are some strategies you can use to solve this problem?

   b. To help you focus on the relationships in this problem, at each step think about what part of the mixture was cranberry juice and what part was apple juice. Determine how much of each of those parts Nick drank.

CHAPTER REVIEW

VOCABULARY

Match the letter of the word in the right column with the description at the left.

1. a rectangular arrangement of numbers enclosed in brackets   **a.** cell

2. the most commonly occurring item in a set of data   **b.** mode

3. heads and tails when you toss a coin   **c.** matrix

4. the middle item in a set of data   **d.** sample space

5. a position in a spreadsheet where an item of data is stored   **e.** median

Lesson 1.1   USE AN EQUATION OR FORMULA   pages 5–11

- Use properties of equality to solve an equation. If the equation involves several operations, use addition or subtraction properties first. Then use multiplication or division properties.

- Use $M = \dfrac{Pr(1 + r)^n}{(1 + r)^n - 1}$ to find M, the monthly payment amount on a loan, where P is the amount borrowed, r is the monthly interest rate, and n is the number of payments made.

Solve and check.

6. $n - 3 = 5 + 5n$

7. $15 = \dfrac{b}{5} + 19$

8. $4x + 7 = 29$

9. $3(y - 6) = -4(-23 + 2y)$

10. Determine the monthly payment on a 4-year $10,000 car loan at 9.5% annual interest.

11. Determine the total amount to be repaid on the above loan.

Lesson 1.2   USE A SPREADSHEET   pages 12–18

- A cell of a spreadsheet is designated by the row and column in which it appears.

- Spreadsheet formulas use + for addition, − for subtraction, * for multiplication, / for division, ^ to indicate raising to a power, and parentheses to indicate that a quantity is to be evaluated separately.

12. For each empty cell, give the formula and the value the computer will print.

	A	B	C	D
1	Number of	Cost per	Cost before	Cost plus
2	items bought	item	tax	5% tax
3	7	5.29		
4	16	2.98		

Lesson 1.3 USE A MATRIX

- To add or subtract matrices with the same dimensions, add or subtract corresponding elements.

- To multiply a matrix by a real number, multiply each element by the number.

Given $A = \begin{bmatrix} 2 & -7 & 5 \\ 13 & 21 & -11 \end{bmatrix}$ and $B = \begin{bmatrix} -6 & -8 & 1 \\ 0 & 15 & 9 \end{bmatrix}$; find each value.

13. $A + B$ **14.** $A - B$ **15.** $5B$ **16.** $B - 2A$

Lesson 1.4 USE PROBABILITY
pages 25–31

- The theoretical probability P(E) that an event will occur is $\dfrac{\text{number of outcomes in event}}{\text{number of outcomes in sample space}}$.

You roll a number cube. Give the value(s).

17. sample space   **18.** P (2 or 3)   **19.** P (even number)

Lesson 1.5 USE STATISTICS
pages 32–37

- In a set of data, the mean or average is the sum of the data divided by the number of items, the median is the middle value (or average of two middle values) when the items are arranged in order, the mode is the item that occurs most often, and the range is the difference between the greatest and least values.

Determine the mean, median, mode, and range of each set of data.

20. 4, 7, 11, 3, 2, 13, 2   **21.** 177, 211, 156, 151   **22.** 3.5, 3.7, 3.2, 4.0, 3.7, 3.2

Lesson 1.6 USE A GRAPH
pages 38–45

- When a linear equation is written in the form $y = mx + b$, m is the slope and b is the y-intercept.

- When two sets of data increase or decrease together, the sets have a positive correlation. When one set increases as the other decreases, the sets have a negative correlation. When two sets have neither a positive nor a negative correlation, they have no correlation.

Determine the x-intercept, y-intercept, and slope of the graph of each equation.

23. $7x - 5 = y$ **24.** $4x + 2y = 12$ **25.** $3(x - 2y) + 5 = 4y + 3$

Tell whether the correlation is positive, negative, or does not exist.

26. height and area of a triangle   **27.** time needed to travel 100 miles and velocity

Lesson 1.7 SOLVING PROBLEMS WITH SCIENTIFIC NOTATION
pages 46–49

- A number is in scientific notation when it is expressed as $a \times 10^n$, where $1 \le a < 10$, and n is an integer.

Express using scientific notation.

28. 45,600,000   **29.** 0.000307   **30.** 9.1   **31.** 0.77777

Chapter Review **51**

CHAPTER ASSESSMENT

CHAPTER TEST

Solve each equation.

1. $6n + 11 = 3n + 23$ 2. $k + 20 = 5k + 44$

3. $\dfrac{3x}{5} = 15$

4. $\dfrac{7c}{8} - (c - 2) = 12$

5. $4(2x - 5) = 3(2x + 8)$

6. $3(a - 7) - 3(2a - 4) = 4(a + 3)$

7. $\dfrac{2x + 4}{3} - \dfrac{3x - 5}{5} = \dfrac{1}{2}$

Carla borrowed $8200 at 9% annual interest for 3 years to finance a new piano.

8. Determine the amount of her monthly payments.

9. Determine the total amount she repaid.

10. Determine the cost of the loan.

11. Write the expression $4m + \dfrac{p^2}{4}$ using computer symbols.

12. For each empty cell, give the formula and the value the computer will print.

	A	B	C	D
1	Rectangle	Rectangle	Area	Perimeter
2	Length	Width		
3	13	9		
4	8.5	6		

Find each matrix.

$$A = \begin{bmatrix} 3 & 4 \\ 2 & -6 \\ -5 & 3 \end{bmatrix} \qquad B = \begin{bmatrix} -5 & -8 \\ 9 & 9 \\ -4 & 8 \end{bmatrix}$$

13. $A + B$ 14. $B - A$ 15. $2A + 3B$

16. WRITING MATHEMATICS Provide a set of data for which the mean is not a good measure of central tendency and explain why it is not.

A spinner has equal sections numbered from 1 to 8. Determine each theoretical probability.

17. $P(3)$ 18. P (even number)

19. P (a number from 1 to 8)

20. What are the odds of spinning a number greater than 5?

21. In 50 spins of the spinner, 1 came up 9 times. What is the probability of spinning 1?

22. In successive games, a pitcher struck out 6, 2, 8, 8, 3, 10, 9, 4, and 4 batters. Determine the mean, median, mode, and range of the data.

23. A circle with radius 2 cm is inside a circle of radius 5 cm. What is the geometrical probability that a dart will land inside the small circle if it lands inside the large one?

24. Determine the x-intercept, y-intercept, and slope of the graph of $3x - 5y + 10 = 0$.

25. STANDARDIZED TESTS Which variable pairs have a negative correlation?

   I. velocity of travel and distance traveled
   II. days since last Thanksgiving and days till next Thanksgiving
   III. number of shirts you can buy for $50 and the cost per shirt

   A. I, II, and III   **B.** I and II
   C. II and III   **D.** I and III

Express using scientific notation.

26. 234,000,000,000   27. 0.000012

Express using standard notation.

28. 4.55×10^4 29. 8.02×10^{-7}

PERFORMANCE ASSESSMENT

EXPERIMENTAL VS. THEORETICAL Work with a partner. Use a number cube, spinner, or some other device for generating numbers randomly. Calculate the experimental probability of getting a number or set of numbers by conducting 20 trials with the device. Repeat the experiment. Compare the results of the two experiments with each other and with the theoretical probability of the event. Explain any differences you observe.

STATISTICS IN THE NEWS Work in groups of three. Each of you should gather data from newspaper stories relating to business, sports, and entertainment. Each set of data should lend itself to analysis using measures of central tendency. Share sets of data with the other members of your group and choose one set to represent each of the categories listed above. For each set, calculate and display statistics that represent each group appropriately. Write a short summary explaining why your group believes the statistics it chose to represent each set are appropriate.

DOTS PER BOOK Computer printers are classified by "dpi," the number of ink dots they print per inch both horizontally and vertically. Suppose your math textbook were printed on a printer rated at 600 dpi.

a. Find the dimensions of the printed portion of a page. Then determine the maximum number of dots the printer is capable of printing on a page.
b. Determine the maximum number of dots that could be printed in the entire book.
c. Assume that each of the 13 million high school students in the U.S. has six books the same size as your math book. Determine the maximum number of dots that could be printed on all of the pages combined.

SUBJECT PREFERENCES Do students who enjoy mathematics and science dislike social science and language? Do music and art students dislike biology and chemistry? Design a poll that allows classmates to express numerically preferences for two categories of courses, math/science and non-math/science. Make a scatter plot of your results and write a paragraph detailing your conclusions.

PROJECT ASSESSMENT

 Have a class model car show and rally.

1. Each group should prepare an information card to display with its model. Include the car's name, key measurements and features, number of work hours involved, and the names of the group members who were part of the project.

2. Plan a measurable performance test in which all completed models can participate. Since your models do not have an energy source, you will have to devise a way to set them in motion. Remember that whatever method you choose, you must be able to control it so that every model operates under the same conditions.

3. Record the results for each model. Investigate the relationship of each car's performance to another variable such as length or weight. Make a scatterplot and determine the line of best fit. Discuss the value of this line as a predictor of a car's performance.

CUMULATIVE REVIEW

Solve and check each equation.

1. $2x + 3 = 11$ 2. $\dfrac{z}{3} - 5 = 1$

3. **WRITING MATHEMATICS** To model the problem "Find three consecutive integers such that the sum of the first and third is 26." Joe wrote the equation $(x - 1) + (x + 1) = 26$ and Gil wrote $x + (x + 2) = 26$. They both got the same correct answer. Explain how each boy used the variable, and tell the integers.

4. For their home, the Leones are considering installing a swimming pool priced at $35,900. They intend to make a down payment of $5,000, 7,500, or $10,000. Design a spreadsheet showing retail price, down payment, and the amount to be financed. For the amount-to-be-financed column, write the formulas and the values the computer will print.

5. **STANDARDIZED TESTS** If matrix M is 3×4, what must be the dimensions of matrix N in order to determine $M + N$?

   A. no restrictions on N **B.** 3×4
   C. 3×4 or 4×3 **D.** 4×3

For Questions 6–8, use matrices A and B below.

$$A = \begin{bmatrix} 3 & 2 & -1 \\ -4 & 0 & 6 \end{bmatrix} \quad B = \begin{bmatrix} -2 & -5 & 0 \\ 1 & -3 & -4 \end{bmatrix}$$

6. Find $A + B$.   7. Find $-4B$.

8. Find $2A - B$.

Use the spinner to find the theoretical probability that in one spin you get:

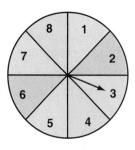

9. an even number

10. a number greater than 3

11. an odd number greater than 3

12. What are the odds of rolling a number cube and getting an odd number?

13. Find the mean, median, mode, and range of 26, 48, 29, 73, 49.

A survey asked people how many times during one month they watched TV news. Use the results shown in the boxplot to answer Questions 14–17.

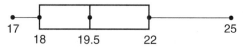

14. What is the range of the data?

15. What is the median?

16. What is the third quartile?

17. What percent of the data is between 17 and 18?

Find the x-intercept, y-intercept, and slope of the graph of each equation.

18. $y = 4x - 3$ 19. $2x + 3y = 7$

Graph the equation using intercepts.

20. $2x + 3y = 6$ 21. $3y = 4x - 12$

Graph the equation using the slope and y-intercept.

22. $y = 3x - 2$ 23. $3x - 2y + 8 = 0$

Do the following show a positive correlation, a negative correlation, or no correlation?

24. the base of a triangle with an area of 20 in.² and the height of the triangle

25. the measure of the side of an equilateral triangle and the perimeter of the triangle.

26. The smallest designated unit of time is called a *yoctosecond*, which is equivalent to 0.000 000 000 000 000 000 000 001 s. Write this measure in scientific notation.

STANDARDIZED TEST

STANDARD FIVE-CHOICE Select the best choice for each question.

1. In which pair of equations are the two equations *not* equivalent?

   A. $\frac{x}{5} = 3$

   $5 \cdot \frac{x}{5} = 3 \cdot 5$

   B. $\frac{z}{3} = \frac{1}{3}$

   $3 \cdot \frac{z}{3} = \frac{1}{3} \cdot \frac{1}{3}$

   C. $k + 7 = 9$

   $k + 7 - 7 = 9 - 7$

   D. $2m - 3 = 5$

   $2m - 3 = 5 + 3$

   E. $2a + 5 = 9$

   $\frac{2a + 5 - 5}{2} = \frac{9 - 5}{2}$

2. At Mario's Motors, Martha sold 20 cars in January and 35 cars in February. The percent of increase in her sales was

   A. 15%   **B.** 42.9%   **C.** 57.1%
   D. 75%   **E.** 80%

3. This spreadsheet shows areas of trapezoids with bases 4 and 6 and heights 10, 15, and 20.

	A	B	C	D
1	Base 1	Base 2	Height	Area
2	4	6	10	50
3	4	6	15	75
4	4	6	20	100

   The formula used to calculate the area 75 is

   A. 0.5 * C2 * (A3 + B3)
   B. 0.5 * C2 * (A3 * B3)
   C. 0.5 * C2 + (A3 + B3)
   D. 0.5 ^ C2 * (A3 + B3)
   E. 0.5 ^ C2 + (A3 + B3)

4. A *nanosecond* is one-billionth of a second of time. In scientific notation, this measure is

   A. 1.0×10^{9} **B.** 1.0×10^{-9}
   C. 1.0×10^{8} **D.** 1.0×10^{-8}
   E. 1.0×10^{-10}

5. If $A = \begin{bmatrix} -2 & 1 & 4 \\ 3 & 0 & -3 \end{bmatrix}$ and

   $B = \begin{bmatrix} 3 & -2 & -1 \\ -1 & 1 & -3 \end{bmatrix}$ then $B - 2A$ equals

   A. $\begin{bmatrix} -8 & 5 & 6 \\ 5 & -2 & 3 \end{bmatrix}$ **B.** $\begin{bmatrix} 7 & -3 & -5 \\ -4 & 6 & 5 \end{bmatrix}$

   C. $\begin{bmatrix} -7 & 1 & 3 \\ 7 & -4 & -9 \end{bmatrix}$ **D.** $\begin{bmatrix} 7 & -4 & -9 \\ -7 & 1 & 3 \end{bmatrix}$

   E. $\begin{bmatrix} -4 & 6 & 5 \\ 7 & -3 & -5 \end{bmatrix}$

For Questions 6–7, use {the digits 0–9}.

6. You want to draw a number that is divisible by 2 or by 3 but not by both. A list of the outcomes in the event is

   A. 2, 3, 4, 6, 8, 9   **B.** 2, 3, 4, 8, 9
   C. 0, 2, 3, 4, 5, 8, 9   **D.** 0, 2, 3, 4, 6, 8, 9
   E. 0, 2, 3, 4, 8, 9

7. The probability that the event will occur is

   A. $\frac{2}{3}$ **B.** $\frac{2}{5}$ **C.** $\frac{3}{5}$ **D.** $\frac{1}{2}$ **E.** $\frac{5}{9}$

8. Which of the following statements are true about the data set {7, 1, 6, 4, 2}?

   I. The mean is 4.
   II. The median is 4.
   III. The mode is 4.

   A. I and II   **B.** I and III   **C.** I only
   D. III only   **E.** I, II, and III

9. The slope and y-intercept of the line $2y = 3x - 5$ are

   A. 3 and -5 **B.** -3 and 5
   C. $\frac{3}{2}$ and 5   **D.** $\frac{3}{2}$ and -5

   E. $\frac{3}{2}$ and $-\frac{5}{2}$

10. The slope of the line $3y = 5x - 4$ is

   A. $-\frac{4}{3}$ **B.** $\frac{4}{3}$ **C.** $\frac{5}{3}$ **D.** $-\frac{5}{3}$ **E.** 5

2 Real Numbers, Equations, and Inequalities

Take a Look
AHEAD

Make a list of things that look new.
- Copy the names of properties that you do not think you have learned about before.
- In everyday usage, what is meant by a "compound sentence"? What different types of compound sentences are there?

Make a list of things that look familiar.
- What symbols and operations can you identify?
- What is meant by order of operations? Why are these rules inportant?
- Describe different types of graphs that can be used to display data.

DATA Activity

SKILL FOCUS
- Read and interpret tables.
- Add, subtract, multiply, and divide real numbers.
- Convert units of measurement.
- Write numbers in scientific notation.
- Use linear equations.

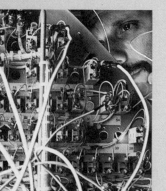

Electrical Resistance

Up to the late 1940s, electrical engineering dealt mostly with lighting and transmission of electrical power, but its scope grew rapidly with the need for electronic devices.

Electrical engineers must understand basic principles of electric currents and the properties of different materials. Metals are good conductors of electricity, although every conduction offers some opposition or **resistance** (measured in *ohms*, Ω) to the flow of electricity. The table on the next page shows standard gauge numbers and properties of copper wire.

Engineering

In this chapter, you will see how:

- **ENVIRONMENTAL ENGINEERS** use formulas to determine the effect of pollutants.
  (Lesson 2.3, page 73)

- **CIVIL ENGINEERS** use formulas to determine where to place expansion joints in highways.
  (Lesson 2.4, page 81)

- **SOUND ENGINEERS** use absolute value inequalities in their study of how humans and other animals respond to sound.
  (Lesson 2.5, page 88)

Gauge Number	Diameter, mm	Resistance, Ω/km
0	8.252	0.322
1	7.348	0.406
2	6.544	0.502
3	5.827	0.646
4	5.189	0.815
5	4.621	1.028
6	4.115	1.296
7	3.665	1.634
8	3.264	2.061
9	2.906	2.599
10	2.588	3.277
11	2.305	4.132
12	2.053	5.211

Use the table to answer the following questions.

1. Find the gauge number of wire that has approximately half the diameter of number 0 wire.

2. Determine the cross-sectional areas of number 0 wire and the wire you answered in Question 1 Round to two decimal places. How do the areas compare?

3. Which number wire is approximately four times as resistant as number 2 wire?

4. Determine the resistance of number 11 wire in terms of ohms per meter. Express your answer using scientific notation.

5. Suppose you wanted to wind a coil of number 7 wire that had a resistance of 1 ohm. How many meters of wire would you use? Round to one decimal place.

6. **WORKING TOGETHER** Gauge numbers run to 40. Plan a method using the given data to predict the approximate diameters of higher number wires. Use your method to predict a diameter for number 30 wire. Compare your method and prediction with other groups.

Engineering Your Career

Engineering is sometimes described as the application of scientific and mathematical knowledge to solve practical problems such as the design and construction of machinery, bridges, roads, waterways, or chemical processes. Engineering is divided into branches such as automotive, biomedical, chemical, civil, electrical, and mechanical engineering.

PROJECT GOAL

To explore physical and mathematical relationships that apply to engineering.

PROJECT Connections

Lesson 2.2, page 67:
Build a meter-stick balance.

Lesson 2.5, page 87:
Explore how to use absolute value equations to describe a balanced system of weights.

Lesson 2.6, page 93:
Experiment to discover an important law of balance.

Chapter Assessment, page 101:
Plan a class Engineering Day to report on career opportunities.

Getting Started

Work in groups.

1. Research different fields of engineering. Find information in the school or local library or your school guidance office. Create a list with as many different types of engineering careers as you can.

2. Each group member should select two or three of the engineering careers that he or she finds most interesting and research these. How many years of schooling are needed and what types of courses are taken? What are the responsibilities of each type of engineer and what projects might each work on?

3. For the Project Connections, assemble the following materials: meter stick, large paper clamp, 15 cm length of wire hanger, two 1 qt milk cartons, rocks or other material to weight down the containers, two unsharpened pencils, two small pieces of clay, two paper clips, assorted coins, clear tape, plastic sandwich bags, and scissors.

✉ Internet Connection

www.swpco.com/
swpco/algebra2.html

Think Back/Working Together

● Work with a partner. You each need two sheets of graph paper. Recall the **Pythagorean theorem** states that for a right triangle with legs *a* and *b* and hypotenuse *c*, $a^2 + b^2 = c^2$.

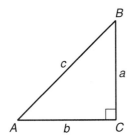

1. On a sheet of graph paper, draw a right triangle for which the legs are whole units. Label your triangle as shown above. Do not let your partner see your triangle.

2. Compute the length of the hypotenuse to the nearest hundredth using the Pythagorean theorem. Do not tell your partner the length.

3. On another sheet of graph paper, draw a number line. The length of each unit should be the same as the units on your triangle. Graph a point to indicate the length of your triangle's hypotenuse.

4. Exchange number lines with your partner. Draw a right triangle with a hypotenuse that length.

5. Compare the triangles you and your partner drew. Did you draw congruent triangles? If not, discuss reasons why.

Explore

In this activity, you locate square roots of numbers on the number line using a ruler and a compass.

6. Draw a number line. Leave about 4 cm between each number. Label zero as *A*.

COMMUNICATING ABOUT ALGEBRA

The sum of the lengths of the two legs of a right triangle must be greater than the length of the hypotenuse. If you know the length of leg *a* and hypotenuse *c*, how can you express the minimum length of leg *b*? How can you express the maximum length of leg *b*?

THINK BACK

THINK BACK

Recall that congruent triangles are triangles whose angles and sides are congruent to the corresponding angles and sides of each other.

THINK BACK

A *circle* is defined as the set of all points in a plane equidistant from a given point.

7. Start at A. Construct right triangle ABC with legs each 1 unit long. What is the length of the hypotenuse $\overline{AC}$? Explain.

8. Place the point of your compass at A. Place the compass pencil point at C. Draw an arc that intersects the number line. Label the point of intersection D. What is the length of $\overline{AD}$? What real number will you use to identify D?

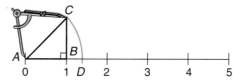

9. Construct right triangle ADE with the length of $\overline{AD}$ equal to $\sqrt{2}$ and $\overline{DE}$ equal to 1, as shown below. What is the length of hypotenuse $\overline{AE}$? Explain.

10. Place the point of your compass at A. Place the compass pencil point at E. Draw an arc that intersects the number line. Label the point of intersection F. What is the length of $\overline{AF}$? What real number will you use to identify F?

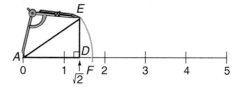

Make Connections

- Extend the process you used in Explore. Locate the following square roots on the number line. Then use a calculator to find the decimal approximation of each square root to the nearest hundredth.

  11. $\sqrt{4}$ **12.** $\sqrt{5}$ **13.** $\sqrt{6}$ **14.** $\sqrt{7}$ **15.** $\sqrt{8}$ **16.** $\sqrt{9}$ **17.** $\sqrt{10}$

  Use the points you located on the number line to locate each of the following. Explain your methods. Then use a calculator to find the decimal approximation of each to the nearest hundredth.

  18. $2\sqrt{2}$ **19.** $\sqrt{3} + \sqrt{2}$

  20. $\sqrt{10} - 3$ **21.** $3 + \sqrt{8}$

Summarize

22. **WRITING MATHEMATICS** You used a compass to construct a line segment equal in length to another line segment that was the hypotenuse of a triangle. What assures you that the process you used produces line segments of equal length?

23. **MODELING** Suppose you constructed a spiral made up of successive right triangles as shown. The first triangle has legs of length 1 unit and a hypotenuse of length $\sqrt{2}$ units. The second triangle has legs of length 1 unit and $\sqrt{2}$ units and a hypotenuse of length $\sqrt{3}$ units. The third triangle has legs of length 1 unit and $\sqrt{3}$ units and a hypotenuse of length $\sqrt{4}$ units.

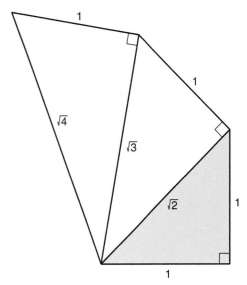

a. Continue the pattern. Write the lengths of the legs and the hypotenuse of the fourth, fifth, and sixth triangles.

b. What is the length of the hypotenuse of the eighth triangle in the spiral?

c. If the last triangle you construct has a hypotenuse with length $\sqrt{13}$ units, how many triangles will you have constructed?

24. **THINKING CRITICALLY** How many square roots of real numbers have decimal approximations from 2 to 3 on the number line? Explain your reasoning. First consider square roots of integers.

25. **THINKING CRITICALLY** Find a whole number n such that $4 < \sqrt{n} < 5$. Explain your reasoning.

26. **GOING FURTHER** Examine the decimal approximations for the square roots you found in Questions 11–17. What do you notice about the decimals produced? Do you think you would draw the same conclusion if the decimals were extended to more places?

Explore/Working Together

The following are subsets of the real numbers.

Natural numbers	$\{1, 2, 3, \ldots\}$
Whole numbers	$\{0, 1, 2, \ldots\}$
Integers	$\{\ldots, -3, -2, -1, 0, 1, 2, 3, \ldots\}$
Rational numbers	Numbers, such as $\frac{5}{6}$, $-\frac{4}{19}$, 0.2 and -7, that can be written as the ratio of two integers
Irrational numbers	Numbers, such as π, $\sqrt{3}$, and $-\sqrt{17}$, that cannot be written as the ratio of two integers

1. Work with a partner. Make a drawing to show the relationships between the sets of numbers above.

2. Find numbers that belong to more than one of the sets of numbers above. Name the sets to which each number belongs.

Build Understanding

A **rational number** is a number that can be written in the form $\frac{a}{b}$ where a and b are integers and b is not equal to 0.

EXAMPLE 1

Show that each number is rational.

a. 0.375   b. $0.\overline{35}$

Solution

a. $0.375 = \dfrac{375}{1000} = \dfrac{3}{8}$

Since $\frac{3}{8}$ is in the form $\frac{a}{b}$ and $b \neq 0$, 0.375 is a rational number.

b. Let $r = 0.353535\ldots$

$$100r = 35.353535\ldots \quad \text{Multiply both sides by 100.}$$
$$99r = 35 \quad \text{Subtract the first equation from the second.}$$
$$r = \frac{35}{99} \quad \text{Divide both sides of the equation by 99.}$$

Since $\frac{35}{99}$ is in the form $\frac{a}{b}$ and $b \neq 0$, $0.\overline{35}$ is a rational number. ◀

Numbers that cannot be expressed as a ratio of integers $\frac{a}{b}$ are called **irrational numbers**. Examples of irrational numbers are π, decimals that are both nonterminating and nonrepeating, and the square root of any number that is not a perfect square.

$$\pi \qquad 0.01001000100001\ldots \qquad \sqrt{6} \qquad \sqrt{119.53}$$

Recall that a positive real number has two square roots. The positive or **principal square root** is $\sqrt{n}$ and the negative square root is $-\sqrt{n}$. The symbol $\sqrt{}$ is called a **radical symbol** and n is called the **radicand**.

THINK BACK

Recall that numbers such as 1,4,9,16 and 25 are perfect squares. Their square roots are integers.

EXAMPLE 2

Show whether each number is rational or irrational.

a. $\sqrt{\dfrac{1}{64}}$ **b.** $-\sqrt{19}$

Solution

a. $\sqrt{\dfrac{1}{64}} = \sqrt{\left(\dfrac{1}{8}\right)^2} = \dfrac{1}{8}$ Both 1 and 64 are perfect squares.

Since $\dfrac{1}{8}$ is in the form $\dfrac{a}{b}$ and $b \neq 0$, $\sqrt{\dfrac{1}{64}}$ is a rational number.

b. $-\sqrt{19}$

Since 19 is not a perfect square, $-\sqrt{19}$ is an irrational number. ◄

CHECK UNDERSTANDING

What is the meaning of the prefix *tri*? Why do you think this property is called the trichotomy property?

The rational numbers and the irrational numbers together form the **real numbers**. The *trichotomy* or *comparison property* describes the possible relationships between any two real numbers a and b.

TRICHOTOMY PROPERTY

For all real numbers a and b, exactly one of the following is true:

$$a < b \qquad a = b \qquad a > b$$

EXAMPLE 3

Use $<$, $=$, or $>$ to compare $\sqrt{19}$ and $4.\overline{35}$.

Solution

Compare decimal forms using a calculator.

$$\sqrt{19} \qquad\qquad 4.\overline{35}$$
$$4.35889\ldots \qquad 4.3535\ldots \quad \text{Compare the first digits that are different.}$$

Since $8 > 3$, $\sqrt{19} > 4.\overline{35}$. ◄

PROBLEM SOLVING TIP

Use a calculator to find decimal forms of real numbers. Remember, unless the number is a perfect square, the decimals produced are approximations.

The set of real numbers is said to be **dense** because between any two real numbers there is another real number.

EXAMPLE 4

TECHNOLOGY Find a real number between the value for π given by a calculator and the value for 22 divided by 7.

Solution

$$\pi \approx 3.141592654 \qquad \frac{22}{7} \approx 3.142857143$$

Let r represent the real number. Then

$$3.141592654 < r < 3.142857143$$

Any value between 3.141592654 and 3.142857143 is correct. A possible answer is 3.142. ◄

TRY THESE

List the subsets of the real numbers to which each number belongs.

1. -54

2. $\sqrt{100}$

3. $7.\overline{6}$

4. $-\sqrt{3}$

5. $2\frac{11}{46}$

6. $-4.\overline{589}$

7. $-\sqrt{170}$

8. $-\frac{85}{17}$

Show whether each square root is rational or irrational.

9. $-\sqrt{11}$

10. $\sqrt{\frac{144}{169}}$

11. $-\sqrt{\frac{24}{49}}$

12. $\sqrt{400}$

Use $<$, $=$, or $>$ to compare each pair of real numbers.

13. $\sqrt{14}$ ■ $\frac{15}{4}$

14. $0.\overline{6}$ ■ $\sqrt{0.38}$

15. $-\frac{41}{13}$ ■ $-\pi$

16. $3.505005000\ldots$ ■ $\frac{7}{2}$

Find a number between each pair of real numbers.

17. 9.326327328 and $\sqrt{87}$

18. $\frac{19}{22}$ and 0.87

19. $\frac{26}{33}$ and $\frac{35}{44}$

20. 18.1213141516 and $18.\overline{12}$

21. **WRITING MATHEMATICS** Is every integer a whole number? Explain why or why not.

22. **WRITING MATHEMATICS** Is every integer a rational number? Explain why or why not.

23. **HISTORY** Archimedes found two approximations for π: $3\frac{1}{7}$ and $3\frac{10}{71}$. Determine if these approximations are greater than or less than the value of π on a calculator.

24. **ESTIMATION** Justin estimates that the length of the sidewalk to his house is $\sqrt{148}$ ft, and Lester estimates it is 12.2 ft. If the actual length of the sidewalk is $12\frac{1}{6}$ ft, whose estimate is closer?

List the subsets of the real numbers to which each number belongs.

1. $-\dfrac{1}{3}$
2. 0.93774
3. $\sqrt{3.6}$
4. $0.7777\ldots$

5. $-\sqrt{\dfrac{81}{100}}$
6. $\sqrt{36}$
7. -1
8. $2.00\overline{32}$

Show whether each number is rational or irrational.

9. $-\sqrt{225}$
10. $\sqrt{\dfrac{94}{64}}$
11. 0.75
12. $-6.\overline{3}$

13. $5\dfrac{1}{25}$
14. $0.565656\ldots$
15. $-\sqrt{29}$
16. $-\dfrac{79}{119}$

Use $<$, $=$, or $>$ to compare each pair of real numbers.

17. $0.451451451\ldots$ ■ $\dfrac{451}{999}$
18. 6.090909 ■ $\dfrac{122}{20}$
19. $\sqrt{108}$ ■ $10.\overline{35}$

Find a number between each pair of real numbers.

20. $3.\overline{8}$ and $\sqrt{15}$
21. $-\dfrac{17}{21}$ and $-0.808080\ldots$
22. 26.94385073 and $\dfrac{539}{20}$

23. **WRITING MATHEMATICS** Is $\sqrt{45}$ between $6.\overline{6}$ and $6\dfrac{3}{4}$? How do you know?

24. **WRITING MATHEMATICS** Explain why $\sqrt{\dfrac{25}{36}}$ is not between $\dfrac{7}{8}$ and 0.9.

25. **HISTORY** About 1700 B.C. Egyptians used $\dfrac{256}{81}$ for π. About A.D. 125, Ch'ang Hong, a Chinese mathematician, used $\sqrt{10}$ for π. Is the value of π from your calculator greater than or less than each of these values?

26. **GOLDEN RATIO** Two common sizes for photographs are 8 in. × 10 in. and 5 in. × 7 in. Which set of dimensions forms a ratio that is closer to the golden ratio, ≈ 1.618?

Write these numbers in order from least to greatest.

27. $-2.6,\ \sqrt{5},\ 2.24,\ -\sqrt{10},\ -2\dfrac{2}{3},\ 2.\overline{23},$
    $-2.216215214,\ -\dfrac{215}{98},\ -\sqrt{5}$

28. $\dfrac{9}{10},\ 0.888\ldots,\ \sqrt{2},\ -1.1,\ 0.697989\ldots,\ \sqrt{\dfrac{16}{25}},\ 1.\overline{3}$

29. $\sqrt{\dfrac{49}{100}},\ -0.055,\ -0.5,\ 0.5,\ \dfrac{3}{4},\ 5.\overline{5},\ -5.678\ldots,\ \sqrt{11}$

30. **MODELING** Draw a diagram of the real number system. Show the relationships between the rational numbers, irrational numbers, integers, whole numbers, repeating decimals, terminating decimals, and nonterminating decimals. Explain your reasoning.

31. **WRITING MATHEMATICS** Do the set of rational and irrational numbers have any numbers in common? If so, describe these numbers. If not, explain why not.

32. **WRITING MATHEMATICS** Compare the heights of two students in your class using the trichotomy property. Are their heights real numbers? Explain.

33. **METALLURGY** The carbon content of steel, an alloy of mostly iron and carbon, varies from 0.1% to not more than 1.7%. If the carbon content is less than 0.1%, the iron is called *wrought iron*. If the carbon content is greater than 1.7%, the iron is called *cast iron*. Use the symbols < and > to show the relationship of carbon content between the three metals.

34. **PYRAMIDS** Each side of the base of the Great Pyramid in Egypt is $3\left(\sqrt{16}\right)\left(\sqrt{81}\right)\left(\sqrt{49}\right)$ ft. The height is 481 ft. Write an inequality comparing the length of a side and height.

THINK CRITICALLY

35. Is the set of whole numbers dense? Explain your answer.

36. Susan set the range values for the viewing window on her graphing utility as shown below.

$$\text{Xmin} = -5 \qquad \text{Ymin} = -1$$
$$\text{Xmax} = 5 \qquad \text{Ymax} = 5$$
$$\text{Xscl} = 1 \qquad \text{Yscl} = 1$$

She graphed $y_1 = \sqrt{\left(\frac{1}{4}\right)}, y_2 = 0.999999, y_3 = \sqrt{5}, y_4 = \sqrt{23},$

$y_5 = \pi, y_6 = \sqrt{17}, y_7 = \sqrt{2.25},$ and $y_8 = 30/11.$

Which line on the graph corresponds to each entry?

37. Given: $a = b, a > 0,$ and $b > 0.$ The following "proof" shows that $1 = 2.$ Find the mistake.

$a = b$	Given.
$ab = b^2$	Multiply both sides by b.
$ab - a^2 = b^2 - a^2$	Subtract a^2 from each side.
$a(b - a) = (b + a)(b - a)$	Factor each side.
$a = (b + a)$	Divide each side of the equation by $(b - a)$.
$a = a + a$	Substitute equal numbers.
$a = 2a$	Add.
$1 = 2$	Divide both sides of the equation by a.

MIXED REVIEW

TECHNOLOGY A calculator has 9 columns by 5 rows of equal size keys. Suppose you press one of the keys at random.

38. What are the odds that you will press one of the digits 0 to 9?

39. The ON and OFF keys are different. What is the probability you will press either the ON key or the OFF key?

STANDARDIZED TESTS Data for Exercises 40 and 41: 6.7, 2.8, 12.7, 26.8, 15.1, 2.8, 3.7, 22.1

40. The number 2.8 is the

    A. mean   **B.** median   **C.** mode   **D.** range

41. What is the median?

    A. 2.8   **B.** 9.7   **C.** 13.9   **D.** 11.58

GEOMETRY The formula for the volume of a rectangular solid is $V = lwh$.

42. Solve the formula for l.

43. Find the length of a rectangular solid with volume 160.875 m³, height 5.5 m, and width 4.5 m.

Show whether each number is rational or irrational.

44. -0.4 **45.** $\sqrt{89}$ **46.** $6.\overline{7}$ **47.** $0.118119120\ldots$

PROJECT Connection

In this activity you will build a balance scale using the materials you assembled in Getting Started.

1. Attach the paper clamp so its center is over the 50 cm mark on the meter stick. Slip the wire through the holes in the clamp. Cut off the tops of the cartons and weight them down. Make a small hole in the same spot near the top of each container and slip the wire through the holes.

2. Close to each end of the meter stick, place a piece of clay on the table. In each piece of clay, insert a pencil so that it is perpendicular to both the table and the balanced stick. Once the stick is still and balanced, mark its horizontal position on each pencil.

3. Open both paper clips and bend them to make hangers as shown. Cut two 6 cm squares from the closed ends of the sandwich bag. Attach one of these small bags to each clip.

4. Write a paragraph explaining how your balance can be used to compare the weights of a penny, nickel, dime, and quarter. Carry out your experiment and list the coins in increasing order of weight.

2.3 Properties of the Real Numbers

Explore

You are familiar with operations such as multiplication and have made multiplication tables to find the product of two integers. Suppose set $A = \{x, y, z\}$ has the operations $ and # as shown.

$	x	y	z
x	z	x	y
y	x	y	z
z	y	z	x

#	x	y	z
x	y	z	x
y	z	x	y
z	x	y	z

1. Is the operation $ commutative?

2. Is the operation # commutative?

3. Is there an identity element for the operation $? If so, what is it?

4. Describe the element z for the operation #?

5. Find the missing element.

$x \,\#\, \blacksquare = z$ $\qquad$ $y \,\#\, \blacksquare = z$ $\qquad$ $z \,\#\, \blacksquare = z$

6. Does $y \ \$ \ (x \ \# \ z) = (y \ \$ \ x) \ \# \ (y \ \$ \ z)$?

7. Does $y \ \# \ (x \ \$ \ z) = (y \ \# \ x) \ \$ \ (y \ \# \ z)$?

Build Understanding

In Explore, the parentheses are grouping symbols that show you the operation to perform first. The **order of operations** establishes the order in which mathematical operations are performed.

ORDER OF OPERATIONS

1. Perform operations within grouping symbols such as parentheses or brackets, beginning with the innermost.

2. Perform calculations involving exponents and roots.

3. Multiply and divide in order from left to right.

4. Add and subtract in order from left to right.

EXAMPLE 1

Evaluate: $[(7.2 \cdot 5)^2 + 3] \cdot 6 \div 3 - 1$

Solution

$$
\begin{aligned}
&[(7.2 \cdot 5)^2 + 3] \cdot 6 \div 3 - 1 && \text{Use the order of operations.} \\
={}&[36^2 + 3] \cdot 6 \div 3 - 1 && \text{Evaluate } 7.2 \cdot 5. \\
={}&[1296 + 3] \cdot 6 \div 3 - 1 && \text{Find } 36^2. \\
={}&1299 \cdot 6 \div 3 - 1 && \text{Do the operation inside brackets.} \\
={}&2598 - 1 && \text{Multiply and divide from left to right.} \\
={}&2597 && \text{Subtract.}
\end{aligned}
$$

COMMUNICATING ABOUT ALGEBRA

Enter the original expression in Example 1 into your graphing calculator. Do you get the same answer? Explain why or why not.

When using a graphing calculator to evaluate expressions, you may need to insert parentheses. For example, numerators or denominators of more than one term must be put in parentheses. Fractions as factors also must be put in parentheses.

Addition and multiplication in the set of real numbers have the following properties.

PROPERTIES OF ADDITION AND MULTIPLICATION

For all real numbers a, b, and c,

Closure Property
Addition	$a + b$ is a real number
Multiplication	$a \cdot b$ is a real number

Commutative Property
Addition	$a + b = b + a$
Multiplication	$a \cdot b = b \cdot a$

Associative Property
Addition	$a + (b + c) = (a + b) + c$
Multiplication	$a \cdot (b \cdot c) = (a \cdot b) \cdot c$

Identity Property
Addition	$a + 0 = 0 + a = a$
Multiplication	$a \cdot 1 = 1 \cdot a = a$

Inverse Property
Addition	$a + (-a) = 0$
Multiplication	$\dfrac{1}{a} \cdot a = a \cdot \dfrac{1}{a} = 1, \quad a \neq 0$

CHECK UNDERSTANDING

How would you enter the following in a graphing calculator?

$$\frac{14 - 5}{3} + \frac{21}{10 - 3} + \frac{3}{5}(6 - 2)$$

DISTRIBUTIVE PROPERTY OF MULTIPLICATION OVER ADDITION

For all real numbers a, b, and c,

$$
\begin{aligned}
a(b + c) &= ab + ac \\
(b + c)a &= ba + ca
\end{aligned}
$$

THINK BACK

Recall that $-a$ is the opposite, or *additive inverse*, of a, and that $\dfrac{1}{a}$ is the reciprocal, or multiplicative inverse, of a.

In Exercises 4–6 of
Explore, which of the
properties are
exhibited by set
$A = \{x, y, z\}$ and the
operations # and $?

Since the set of real numbers and the operations of addition and multiplication have the above properties, they form a mathematical system called a **field**. The field properties for real numbers can be used to prove other statements.

You can use your understanding of the order of operations and field properties to solve problems.

EXAMPLE 2

ENERGY COSTS Mr. Lopez usually keeps his thermostat set at 21°C both night and day. As a result of talking with a utility company representative, he learned that for each degree the temperature is lowered, he will save 4% of his heating expenses. What percent of his heating expenses will Mr. Lopez save if he lowers the temperature to 19°C from 6 A.M. to 11 P.M. and to 16°C from 11 P.M. to 6 A.M.?

Solution
There are 17 hours from 6 A.M. to 11 P.M.; the difference between 21°C and 19°C is 2°C. There are 7 hours from 11 P.M. to 6 A.M.; the difference between 21°C and 16°C is 5°C.

$$\text{percent savings} = (2 \cdot 17 \div 24 + 5 \cdot 7 \div 24) \cdot 4 \approx 11.5$$

Mr. Lopez would save about 11.5% of his heating expenses. ◀

TRY THESE

Evaluate each expression.

1. $2.5 \cdot 6 - 108 \div 6^2$

2. $\dfrac{\sqrt{(6 + 10)}}{2^3}$

3. $\dfrac{13^2 + 20}{(5 + 2)4}$

4. $\dfrac{72 - 12}{15} + \left(\dfrac{1}{5} \cdot 5^2 \cdot 3\right)$

5. $12 \div 3 + 3 \cdot 2 - 1$

6. $(7 - 0.5) - [(5) \cdot (4 - 9)]$

Which property is shown by each of the following?

7. $-9.67 \cdot 1 = -9.67$

8. $6(2 + 5) = 6(2) + 6(5)$

9. $3 \cdot (4 \cdot 5) = (3 \cdot 4) \cdot 5$

10. $161.5 \cdot 44.2 = 44.2 \cdot 161.5$

11. $17\left(\dfrac{1}{17}\right) = 1$

12. $\dfrac{12}{17} + 0 = \dfrac{12}{17}$

13. WRITING MATHEMATICS Is the commutative property true for subtraction? for division? Explain.

14. MODELING Write each of the following on an index card as shown below.

Using all of the cards, arrange them to form an expression which gives the greatest value possible (only single digit numbers are allowed). Can you find more than one way?

15. TECHNOLOGY Tamara entered the expression shown at the right into her graphing calculator and received an ERROR message. Determine the error and make the necessary change so that the expression results in a display of −3147.2.

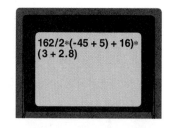

162/2*(-45 + 5) + 16)*(3 + 2.8)

16. SPACE TRAVEL The expression $\dfrac{\left(0.45 + \frac{1}{2}\right) \times 10^4}{1.7 \times 2}$ was used to calculate the weight of experimental equipment aboard a spacecraft. Determine the weight to the nearest kilogram.

PRACTICE

Evaluate each expression.

1. $\sqrt{49} + 3(4.5 + 2.75)$

2. $4.8 \div (1.44 \div 1.2) \cdot 2^4$

3. $(17 - 3)^2 + \dfrac{15}{16}$

4. $[192 \div (7 + 5)]^2 - 200$

5. $45 \div 5 + 7 \cdot \sqrt{9} - 6^2$

6. $2^4 + 10^2 \cdot (46 \div 23) - 0.77$

Which property is illustrated by each of the following?

7. $6.91 + 0 = 6.91$

8. $890.2 \cdot 1 = 890.2$

9. $\dfrac{1}{17} \cdot 17 = 1$

10. $59 + \dfrac{12}{13} = \dfrac{12}{13} + 59$

11. $\dfrac{1}{2}\left(\dfrac{1}{3} \cdot \dfrac{1}{4}\right) = \left(\dfrac{1}{2} \cdot \dfrac{1}{3}\right)\dfrac{1}{4}$

12. $1.9(2 + 1.5) = 1.9(2) + 1.9(.15)$

13. WRITING MATHEMATICS Is the associative property true for subtraction? for division? Explain.

WIND POWER The area swept by the blades of a wind turbine generator can be calculated using the formula $A = \pi d^2 \div 4$ where d is the diameter of the wind wheel.

14. Find the area swept by an 18-ft diameter wind wheel.

15. Find the area swept by a 36-ft diameter wheel.

16. WRITING MATHEMATICS Explain how you would find the diameter of a wind wheel that will have about twice the swept area of an 18-ft diameter wind wheel. What is the diameter?

EXTEND

Simplify each expression.

17. $9x + 4(2x^2 - 7x)$

18. $2\left(\dfrac{7a + 11a}{3}\right) - 22$

19. $3^3 + \left[\sqrt{25} + (5r - 3r) + 1\right]$

20. $56 - \dfrac{1}{2}(6c + 8d) + 22$

21. $\dfrac{a(9 + 5)}{7a} + 2a$

22. $-32h + 4^2 \cdot h - \sqrt{36}$

Kite Flying Use the kite pattern at the right for Exercises 23–25.

23. Write an expression that can be evaluated to determine the surface area of the kite. Measurements are in inches.

24. Find the kite's surface area in square inches.

25. What is the kite's surface area in square feet?

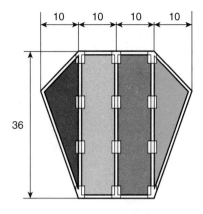

THINK CRITICALLY

Use the following expression for Exercises 26–28.

$$1 + 2^4 + 8 - 6\sqrt{2.2 + 1.8} \div 3 + 9$$

26. Evaluate the expression using the order of operations.

27. Place grouping symbols in the expression so that when it is evaluated the result is 88.

28. Place grouping symbols in the expression so that when it is evaluated the result is $27\frac{1}{3}$.

29. **Writing Mathematics** Examine the diameters and areas in Exercises 14–16. Explain why doubling the diameter does not result in twice the area swept by the wheel.

30. **Writing Mathematics** Consider the set $A = \{-1, 0, 1\}$. Explain why this set *is* closed under multiplication but *is not* closed under addition.

Box Design Use the box pattern. The dashed lines show folds. Measurements are in inches.

31. Write an expression that can be evaluated to approximate the surface area of the cardboard used for the box.

32. Evaluate the expression you wrote to estimate the surface area.

33. For the expression given, what expression would be evaluated to find the surface area of the cardboard if the dimensions were doubled?

34. Find the surface area if the dimensions of the box were doubled.

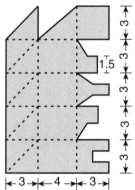

MIXED REVIEW

Add or subtract.

35. $\begin{bmatrix} -1 & 1 \\ 0 & 1 \end{bmatrix} - \begin{bmatrix} 4 & 4 \\ 0 & 3 \end{bmatrix}$

36. $\begin{bmatrix} -6 & 9 \\ -2 & 12 \end{bmatrix} + \begin{bmatrix} -6 & 5 \\ 10 & -17 \end{bmatrix}$

37. **Standardized Tests** Choose the letter of the property that is illustrated by the equation $(6 + 1.7)9 = 6(9) + 1.7(9)$.

    A. inverse property of multiplication   **B.** commutative property of addition
    C. identity property of multiplication   **D.** distributive property

List the subsets of the real number system to which each number belongs.

38. -12 39. 0.56 40. $\sqrt{15}$ 41. 777

Career
Environmental Engineer

Industries on the banks of rivers often empty their wastes into the river. The temperature of the waste is frequently far above the temperature of the river. Environmental engineers study how the discharged waste impacts the plant and animal life in the river. A formula for determining the final temperature of a river after waste has been discharged into it is

$$T_f = \frac{F_w T_w + F_r T_r}{F_w + F_r}$$

T_f = final temperature of the river in degrees Celsius
T_w = temperature of the waste in degrees Celsius
T_r = temperature of the river in degrees Celsius
F_w = flow of the waste in cubic meters per second (m³/s)
F_r = flow of the river in cubic meters per second (m³/s)

Decision Making

1. The maximum river temperature that can sustain trout is 22°C. A treatment plant discharges waste at 26°C flowing at 0.35 m³/s into a river at 21°C and 0.88 m³/s. Can the river sustain trout?

2. If the flow of the discharge remains the same, to what temperature must the waste be reduced to sustain trout?

3. If the temperature of the waste remains the same, how could the flow of the waste be altered to sustain trout?

4. Suppose that the temperature of the waste can vary. Let y represent the final water temperature of the river and x represent the temperature of the waste. Use the flow rates and river temperature from Question 1 to write an equation that illustrates the relationship between the final river temperature and the temperature of the waste.

5. What type of relationship is represented by the equation you wrote in Question 4?

6. Graph the equation you found in Question 4. Then graph $y = 22$. Where these lines intersect, what does the value of x represent? Use the TRACE feature to find this value of x.

2.4 Solving Equations and Inequalities in One Variable

Explore

SPOTLIGHT ON LEARNING

WHAT? In this lesson you will learn
- to use real number field properties and equality properties to solve equations and inequalities.

WHY? Equations and inequalities in one variable can help you solve problems involving personal finance, entertainment, and civil engineering.

Given: $S = \{47, 81, 34, 25, 70, 36, 54, 38, 18\}$

1. Sort the set into subsets based upon the relation "sum of digits are equal." For example, 34 and 70 are in the same subset because $3 + 4 = 7 + 0$.

2. Let @ represent the relation "sum of digits are equal." For any element x in S, is it true x @ x? Explain.

3. For any elements x and y in S, is it true if x @ y then y @ x? Explain.

4. For any elements x, y, z in S, is it true if x @ y and y @ z then x @ z? Explain.

Build Understanding

The properties you discovered for set S and relation @ in Explore are also true for the set of real numbers and the relation of equality.

PROPERTIES OF EQUALITY

For all real numbers a, b, and c,

$a = a$	Reflexive Property
If $a = b$, then $b = a$.	Symmetric Property
If $a = b$ and $b = c$, then $a = c$.	Transitive Property
If $a = b$, then a may replace b or b may replace a in any statement.	Substitution Property

---

You can use the addition, subtraction, multiplication, and division properties to create an equation *equivalent* to a given equation.

PROPERTIES OF EQUALITY

For all real numbers a, b, and c,

If $a = b$, then $a + c = b + c$.	Addition Property
If $a = b$, then $a - c = b - c$.	Subtraction Property
If $a = b$, then $ac = bc$.	Multiplication Property
If $a = b$, and $c \neq 0$, then $\frac{a}{c} = \frac{b}{c}$.	Division Property

---

You can apply the equality properties and the real number properties to solve equations in one variable.

EXAMPLE 1

Solve: $12x + 32 = 19 - 6\left(\frac{5}{3}x - 1\right)$

Solution

$$
\begin{aligned}
12x + 32 &= 19 - 6\left(\frac{5}{3}x - 1\right) \\
12x + 32 &= 19 - 10x + 6 &&\text{Distributive property.} \\
12x + 32 &= 25 - 10x &&\text{Combine like terms.} \\
10x + 12x + 32 &= 25 - 10x + 10x &&\text{Addition property.} \\
22x + 32 &= 25 &&\text{Combine like terms.} \\
22x + 32 - 32 &= 25 - 32 &&\text{Subtraction property.} \\
22x &= -7 &&\text{Combine like terms.} \\
x &= -\frac{7}{22} &&\text{Division property.}
\end{aligned}
$$

The solution is $-\frac{7}{22}$. Check by substituting in the original equation. ◄

You can also apply the equality properties to rewrite a formula in terms of a different variable.

EXAMPLE 2

PERSONAL FINANCE The formula $E = s + rt$ can be used to determine earnings E when you are given the monthly salary s, the commission rate r, and the total sales t. Solve this formula for t.

Solution

$$
\begin{aligned}
E &= s + rt \\
E - s &= s - s + rt &&\text{Subtraction property.} \\
E - s &= rt \\
\frac{E - s}{r} &= \frac{rt}{r} &&\text{Division property.} \\
\frac{E - s}{r} &= t
\end{aligned}
$$
◄

Just as you use equality properties to solve equations, you can use inequality properties to solve inequalities. You can replace $<$ with $\le$, $>$, or $\ge$ in these rules.

┌─ **PROPERTIES OF INEQUALITY** ──────────
│ **For all real numbers a, b, and c,**
│
│   If $a < b$, then $a + c < b + c$.   **Addition Property**
│   If $a < b$, then $a - c < b - c$.   **Subtraction Property**
└──────────────────────────────

ALGEBRA: WHO, WHERE, WHEN

In the sixteenth century, Francois Viète (1540–1603) of France used the word *aequalis* and the symbol $\sim$ to indicate equality. For Robert Recorde (1510–1558) of England, parallel lines showed equality. He was the first to use the symbol $=$.

THINK BACK

Recall that an inequality is two numbers or expressions separated by an inequality symbol.

COMMUNICATING ABOUT ALGEBRA

State the addition and subtraction properties of inequality if $a > b$, if $a \ge b$, and if $a \le b$.

THINK BACK

Recall that when graphing on a number line, a solid dot is used to show that the point is included in the solution; an open dot indicates that the point is not included.

EXAMPLE 3

Solve: $2.5x - 9 \leq 1.5x + 4$. Graph the solution.

Solution

$$2.5x - 9 \leq 1.5x + 4$$
$$2.5x - 1.5x - 9 \leq 1.5x - 1.5x + 4 \qquad \text{Subtraction property.}$$
$$x - 9 \leq 4 \qquad \text{Combine like terms.}$$
$$x - 9 + 9 \leq 4 + 9 \qquad \text{Addition property.}$$
$$x \leq 13 \qquad \text{Simplify.}$$

The solution is any real number less than or equal to 13.

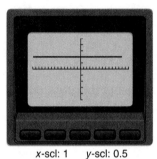

x-scl: 1   *y*-scl: 0.5

To solve this example on a graphing utility, first think of each side of the inequality separately.

$$y_1 = 2.5x - 9 \qquad y_2 = 1.5x + 4$$

You want to graph only $y_3 = y_1 \leq y_2$ so enter $y_3 = 2.5x - 9 \leq 1.5x + 4$. The graph shows $x \leq 13$. ◀

You can multiply or divide by a positive real number to produce an equivalent inequality. However, when you multiply or divide by a negative number, you must reverse the inequality sign to produce an equivalent inequality. You can replace $<$ with $\leq$, $>$, or $\geq$ in these rules.

CHECK UNDERSTANDING

What inequality symbol will make each statement true?

- If $a > b$, and c is positive, $ac \blacksquare bc$.
- If $a \geq b$, and c is negative, $ac \blacksquare bc$.
- If $a \geq b$, and c is positive, $\dfrac{a}{c} \blacksquare \dfrac{b}{c}$.
- If $a > b$, and c is negative, $\dfrac{a}{c} \blacksquare \dfrac{b}{c}$.

PROPERTIES OF INEQUALITY

For all real numbers a, b, and c,

If $a < b$, then $ac < bc$ if c is positive, and $ac > bc$ if c is negative.   **Multiplication Property**

If $a < b$, then $\dfrac{a}{c} < \dfrac{b}{c}$ if c is positive,   **Division Property**

and $\dfrac{a}{c} > \dfrac{b}{c}$ if c is negative.

EXAMPLE 4

Solve: $12 - 5x > 9x + 8$

Solution

$$12 - 5x > 9x + 8$$
$$-5x > 9x - 4 \qquad \text{Subtract 12 from both sides.}$$
$$-14x > -4 \qquad \text{Subtract } 9x \text{ from both sides.}$$
$$x < \frac{2}{7} \qquad \text{Divide by } -14. \text{ Reverse inequality symbol.}$$

Check by selecting a value of x where $x < \frac{2}{7}$. Try 0.

$$12 - 5x > 9x + 8$$
$$12 - 5(0) \overset{?}{>} 9(0) + 8 \qquad \text{Substitute 0 for } x.$$
$$12 > 8 \checkmark$$

Since 12 is greater than 8, the solution is all real numbers less than $\frac{2}{7}$. ◄

Two statements joined by *and* or by *or* form a compound statement. The temperature is below 40°F *and* above 32°F. I will go to lunch before 1 P.M. *or* after 3 P.M. Similarly, two inequalities joined by *and* or by *or* form a **compound inequality**.

$$t < 40 \quad and \quad t > 32 \qquad \text{conjunction}$$
$$g < 1 \quad or \quad g > 3 \qquad \text{disjunction}$$

A compound inequality joined by *and* is called a **conjunction**. Solutions that satisfy both parts of the compound inequality are the solution of the conjunction. A compound inequality joined by *or* is called a **disjunction**. Solutions that satisfy at least one part of the compound inequality are the solution of the disjunction.

To solve a compound inequality, solve each of the inequalities that form the compound inequality.

EXAMPLE 5

Solve and graph the solution of the compound inequality.

$$3\left(\frac{1}{2}x - 7\right) < -15 \quad \text{and} \quad -\frac{1}{3}(x + 7) \le -2$$

Solution

$$3\left(\frac{1}{2}x - 7\right) < -15 \qquad \text{Divide both sides by 3.}$$

$$\frac{1}{2}x - 7 < -5 \qquad \text{Add 7 to both sides.}$$

$$\frac{1}{2}x < 2 \qquad \text{Multiply both sides by 2.}$$

$$x < 4$$

$$-\frac{1}{3}(x + 7) \le -2 \qquad \text{Multiply both sides by } -3.$$
$$\qquad\qquad\qquad \text{Revserse inequality symbol.}$$
$$x + 7 \ge 6 \qquad \text{Subtract 7 from both sides.}$$

$$x \ge -1$$

$$-1 \le x \le 4$$

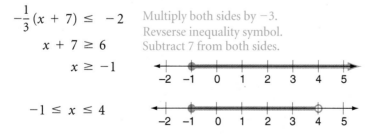

So, x is greater than or equal to -1 and less than 4. ◄

1. **WRITING MATHEMATICS** Look at the steps used to solve Example 5. Name the properties used.

Solve and check each equation.

2. $6x + 8.2 = -12.6 + 10x$

3. $2x - 4(3x - 5) = 5(-x + 2)$

4. $\frac{5}{8}x - 9 = 7(x + 3)$

5. $\frac{5t + 80}{3t} = 3$

6. $4\left(\frac{1}{2}x - 15\right) = 12x + 40$

7. $\frac{5(m + 9)}{4} = -3m$

Solve, graph, and check the solution of each inequality.

8. $9.3 - 7k < -6k + 11.2$

9. $2x + 5 \leq 5x - 4$ x

10. $9 + 5y \geq -1$ and $4y + 15 < 35$

11. $-c \geq \frac{c + 9}{4}$

12. $6.2x - 0.1 \geq \frac{x + 5.5}{0.2}$

13. $7n - 6 \geq 1$ or $\frac{5n}{3} < 1$

Is the number given a solution of the inequality?

14. 6, $\quad -6x + 5 \geq 4x - 2$

15. -7, $\quad \frac{5k}{12} \leq 3(k + 9)$

16. 0, $\quad \frac{2(t + 4)}{3} \leq t + 6$

Write each phrase as an inequality.

17. **TEMPERATURE** The temperature is at least 90°.

18. **PARKING** There are more than 200 cars in the parking lot.

19. **ENTERTAINMENT** There will be at least 40 people, but not more than 60, at the party.

20. **TRAVEL** The trip will cost a minimum of $90 and a maximum of $110.

21. **OFFICE SUPPLIES** André needs five identical ring binders and a stapler for his office. He does not want to spend more than $25. The stapler costs $6.95, and there are several styles of binder from which to choose. How much can he spend for each binder and stay within his budget? Disregard sales tax.

PRACTICE

Name the property illustrated by each equation or inequality.

1. $17.6x + 2.5 - 2.5 > 16.3x + 5.9 - 2.5$

2. $\frac{3}{2} \cdot \frac{2}{3}(x - 9) = \frac{3}{2} \cdot \frac{6}{7}x$

3. $x + 12.4 = 12.4 + x$

4. $4c + 12c < -12c + 12c - 148$

Solve and check each equation.

5. $4x + 9 = 2x - 1$

6. $3z + 1 = 4z - 4$

7. $\frac{2}{5}(3x + 9) = -74 + 2x$

8. $2(4m + 1) = 3(7m - 2)$

9. $\frac{1}{4}y - 1 = 2 + \frac{1}{5}y$

10. $0.2(4 - 3b) + 0.3b = 2.6$

Solve, graph, and check the solutions.

11. $0.7x - 0.1 < 2$

12. $-1.75x \leq 1.25x - 15$

13. $\dfrac{x + 3}{2} < 1$ or $9x + 2 > 11$

14. $-6(x + 2) \leq \dfrac{1}{3}(9x + 15)$

15. $\dfrac{1}{2}x - 2 > 6\left(-\dfrac{1}{4}x + 9\right)$

16. $\dfrac{9x}{3} \geq 12$ and $5(x - 3) < 2x$

17. WRITING MATHEMATICS Write a paragraph about the different types of compound inequalities. Explain how to solve them. Discuss what the solutions mean.

Match each compound inequality with the graph of its solution.

18. $-16 - 4x \leq 0$ and $11x - 5 \geq 61$

19. $\dfrac{22x - 6}{4} \leq -29$ or $4(3x + 7) \geq 64$

20. $\dfrac{3}{2}x - 9 \leq -3$ and $2x > -8$

21. $\dfrac{x}{2} \leq -0.5$ or $-2x < -10$

a.

b.

c.

d.

Write each phrase as an inequality.

22. TRANSPORTATION No more than 28 people ride on a bus.

23. MEDICINE At the most, 40 cases of the flu were reported.

24. SPORTS There were more than 450 people but fewer than 500 people at the baseball game.

Write and solve an inequality.

25. ENTERTAINMENT Joanna has $18 to spend on movies and snacks for herself and her brother. If movie tickets are $6.25 each, what is the most she can spend on snacks?

26. SAVING MONEY Stan has $45 saved toward a $389 bicycle. If he saves $\dfrac{1}{3}$ of the money he earns, what is the least Stan must earn to buy the bicycle?

EXTEND

27. WRITING MATHEMATICS Let R represent the relation "is paid the same hourly wage as" on a set of 30 people who work for a lawn maintenance company. Explain why this relation is reflexive, symmetric, and transitive.

Solve and check each compound inequality.

28. $\dfrac{3}{4}(2n + 9) > \dfrac{3}{8}$ and $\dfrac{15n}{6} + 1 \leq -9$

29. $15 < 7x + 9 < 26$

STOCK MARKET Write an inequality to show the range of each stock's price.

Stock	High	Low
IPP	16 3/4	14 5/8
BNN	56 1/2	55
HDD	35 1/8	28 1/8

30. Stock IPP

31. Stock BNN

32. Stock HDD

33. **FINANCE** Mr. Sanford has $5000 to invest. He wants to earn at least $380 a year to pay for his health club membership. He will invest some in bonds that pay 8% interest annually and the rest in a bank certificate that pays 6% annually. What is the least amount of money Mr. Sanford should invest in bonds?

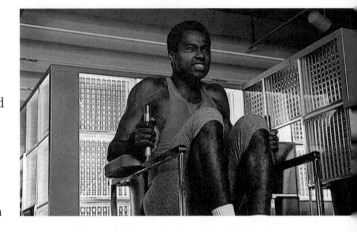

THINK CRITICALLY

Write *true* or *false* for each statement.

34. The solution of a disjunction is the intersection of the solutions of the two inequalities.

35. A number is a solution of a disjunction if it is the solution of both inequalities.

36. A number is a solution of a conjunction if it is the solution of both inequalities.

37. The solution of a disjunction must be the solution of one of the inequalities.

38. Write a compound inequality that has no solution. Explain why it has no solution.

39. Write a compound inequality that has all real numbers as the solution. Explain.

MIXED REVIEW

For the real numbers and the operations addition and multiplication, give each element.

40. the identity element for addition

41. the identity element for multiplication

42. the multiplicative inverse for a

43. the additive inverse for a

44. **STANDARDIZED TESTS** Which of the following is **not** a rational number?

   A. 1.321
   B. $\dfrac{1049}{6327}$
   C. 0.545454…
   D. 0.262262226…

Solve and check each equation.

45. $21 - 8x = 6x + 49$

46. $\dfrac{x + 5}{8} = x - 9$

47. $3(2x - 4) = -5x + 13$

48. Interpret this boxplot.

Hotel Prices on Palm Island, in dollars

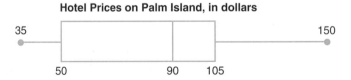

Have you ever felt a car's tires rhythmically roll over indentations in the surface of a road? The indentations in the road are expansion joints which allow the road surface to expand in the heat and contract in the cold without cracking. Civil engineers plan for the placement of these expansion joints during highway construction. Expansion is calculated by the following formula.

$E = kL(T - t)$, where

E = the amount of expansion in feet

T = the outside temperature in degrees Fahrenheit

t = the average temperature during highway construction in degrees Fahrenheit

L = the length of the section of highway under construction in feet

k = a constant, often 0.000012 for two-lane roads with units of $\dfrac{1}{\text{degrees Fahrenheit}}$

Decision Making

1. Suppose that a 3000-ft length of roadway was built when the average outside air temperature was 68°F. Weather records show that the maximum high temperature ever reported in the area was 105°F. If there were no expansion joints, what would be the maximum expansion of this section of roadway? How much is this as a percentage of the length of the section of roadway?

2. In August, the average high temperature is 97°F, and the average low temperature is 70°F along this stretch of roadway. Determine the expansion amounts at each of these temperatures.

3. Let E represent the expansion of the highway under the conditions listed in Question 2. Write an inequality for the expansion interval. Illustrate the expansion interval on a number line.

2.5 Solve Absolute Value Equations and Inequalities

Explore

SPOTLIGHT ON LEARNING

WHAT? In this lesson you will learn
• to solve absolute value equations and inequalities.

WHY? Absolute value inequalities help you solve problems in mechanical engineering and population characteristics.

● Recall that the absolute value of a number x, written $|x|$, is its distance from zero on the number line. Absolute value is a nonnegative number that shows distance, not direction.

Graph each pair of equations or inequalities. Determine if the pair is a conjunction or a disjunction. Then write a single absolute value equation or inequality for each graph.

1. $x = 5$ or $x = -5$

2. $x \le -6$ or $x \ge 6$

3. $x < 3$ and $x > -3$

4. $x - 7 = 2$ or $x - 7 = -2$

Build Understanding

● In Explore, you used the geometric idea of absolute value as a distance. The absolute value of a nonnegative real number is the number itself; for example, $|7| = 7$. The absolute value of a negative real number is the additive inverse of the number; for example, $|-10| = 10$. Absolute value can be defined algebraically as follows.

> **ABSOLUTE VALUE**
>
> For any real number a,
>
> $$|a| = a \text{ if } a \ge 0 \text{ and } |a| = -a \text{ if } a < 0.$$

To solve an absolute value equation, first write an equivalent disjunction. A solution may not satisfy the original absolute value equation. Values that do not check are called **extraneous solutions.**

CHECK UNDERSTANDING

Write an equivalent compound expression for each of the following.

$|E| < a$

$|E| \ge a$

EXAMPLE 1

Solve: $|x + 5| = -2x + 2$

Solution

$$x + 5 = -2x + 2 \quad \text{or} \quad x + 5 = -(-2x + 2)$$
$$3x = -3 \qquad \text{or} \quad x + 5 = 2x - 2$$
$$x = -1 \qquad \text{or} \qquad -x = -7$$
$$x = 7$$

Check

Try −1
$$|-1 + 5| \overset{?}{=} -2(-1) + 2$$
$$4 = 4 \checkmark$$

Try 7
$$|7 + 5| \overset{?}{=} -2(7) + 2$$
$$12 \ne -12$$

So, −1 is a solution and 7 is not a solution.   ◄

To solve an absolute value inequality, rewrite it as a compound inequality. When the inequality is "less than" or "less than or equal to," the solution is a *conjunction*. Write a *disjunction* when the inequality is "greater than" or "greater than or equal to."

EXAMPLE 2

Solve $|3x + 18| \leq 45$. Graph the solution.

Solution

$$-45 \leq 3x + 18 \leq 45 \quad \text{Write the conjunction.}$$
$$-63 \leq 3x \quad\quad \leq 27 \quad \text{Subtract 18 from each expression.}$$
$$-21 \leq \quad x \quad\quad \leq 9 \quad \text{Divide each expression by 3.}$$

The real numbers greater than or equal to -21 and less than or equal to 9 are the solution to the inequality $|3x + 18| \leq 45$.

Isolate the absolute value expression before rewriting an absolute value inequality as a compound inequality.

EXAMPLE 3

Solve $\frac{1}{2}|4x - 8| + 24 > 92$. Graph the solution.

Solution

$$\frac{1}{2}|4x - 8| + 24 > 92$$

$$\frac{1}{2}|4x - 8| > 68 \quad \text{Subtract 24 from each side.}$$
$$|4x - 8| > 136 \quad \text{Multiply both sides by 2.}$$
$$4x - 8 < -136 \text{ or } 4x - 8 > 136 \quad \text{Write the disjunction.}$$
$$4x < -128 \text{ or } 4x \quad\quad > 144 \quad \text{Add 8 to each expression.}$$
$$x < -32 \text{ or } \quad x \quad\quad > 36 \quad \text{Divide each expression by 4.}$$

The real numbers less than -32 or greater than 36 are the solutions to the inequality.

Some absolute value inequalities have no solution. For example, $|2x - 9| < -15$ has no solution because absolute value is not negative, so $|2x - 9|$ can never be less than or equal to any negative number.

Absolute value inequalities are often used to describe real world situations.

ALGEBRA: WHO, WHERE, WHEN

In Example 4, international standard paper dimensions are in approximately the ratio $\sqrt{2}$ to 1. Take half the longer dimension to create another size in the same ratio.

THINK BACK

Recall that *tolerance, margin of error*, and other indications of variance are often given in terms of $\pm$ (read *plus or minus*.)

EXAMPLE 4

INTERNATIONAL STANDARDS In the United States, paper is usually $8\frac{1}{2} \times 11$ in. The table below shows some international standard paper sizes. The measurements are given in millimeters.

	A1	A2	A3	A4	A5	A6	A7	A8
Width	594	420	297	210	148	105	74	52
Length	841	594	420	297	210	148	105	74

Tolerances of ± 1.5 mm are allowed on paper dimensions less than or equal to 150 mm. Tolerances of ± 2 mm are allowed on paper dimensions greater than 150 mm and less than or equal to 600 mm. Tolerances of ± 3 mm are allowed on paper dimensions greater than 600 mm. Write absolute value inequalities that describe the allowable length and width for paper size A5.

Solution
The standard size for A5 paper is 148 mm wide and 210 mm long.

The tolerance for 148 mm is ± 1.5 mm. The absolute value inequality describing allowable width is $|w - 148| \le 1.5$.

The tolerance for 210 mm is ± 2 mm. The absolute value inequality describing allowable length is $|l - 210| \le 2$.

Allowable dimensions for A5 paper are $|w - 148| \le 1.5$ and $|l - 210| \le 2$.

TRY THESE

Write each absolute value inequality as an equivalent compound inequality.

1. $|x + 4| < 9$ **2.** $|x - 3| > 6$ **3.** $|2.5x - 1| \ge 5.2$ **4.** $\left|\frac{1}{2}x - 5\right| \le 7$

Solve and check.

5. $|0.5x - 9| = 11$ **6.** $|3x + 1| = 2x + 3$ **7.** $|4x - 7| = 37$ **8.** $|3x + 7| = 2x - 4$

9. $|7x - 6| = -12$ **10.** $2|x + 2| = 4x - 6$ **11.** $4|3 - x| - 8 = 8$ **12.** $-|x + 2| = -4$

13. **WRITING MATHEMATICS** Compare and contrast solving absolute value equations with solving absolute value inequalities.

Solve and check. Graph the solution.

14. $|2x - 4| < 7$ **15.** $5|3 - x| < 30$ **16.** $|3x| \geq 21$ **17.** $0.5\,|3x + 5| < 5$

18. $|9x + 5| < 0$ **19.** $\left|4 - \frac{1}{2}x\right| \geq 2$ **20.** $\left|\frac{2x + 1}{3}\right| \geq 5$ **21.** $|2x + 1| + 9 \leq 14$

Match the absolute value inequality with its graph.

22. $|x + 2| > 1$ **23.** $|x + 2| < 1$ **24.** $|x + 2| = 1$

a.

b.

c.

25. INTERNATIONAL STANDARDS Use the international standard paper size information in Example 4. Write the absolute value inequalities that describe the allowable length and width for paper size A2.

26. HORTICULTURE A greenhouse manager determined that the amount of fertilizer used for each plant is about 925 mg. The amount used on any particular plant may vary by up to 70 mg. Write and solve an absolute value inequality describing this situation.

PRACTICE

Write each absolute value inequality as an equivalent compound inequality.

1. $\left|z + \frac{1}{2}\right| \leq 1\frac{1}{4}$ **2.** $|w + 9.3| < 12.3$ **3.** $|4p - 7| \geq 22$ **4.** $|c + 14| > 19$

Solve and check.

5. $|2x - 4| = 6$ **6.** $\left|\frac{1}{2}x - 8\right| = 12$ **7.** $|4x - 5| = x + 17$ **8.** $|5x + 4| = 9x - 12$

9. $|9x - 8| = -11x$ **10.** $|6.2x - 1| = -14$ **11.** $|15x| = 0$ **12.** $2|2x + 10| = 4x + 30$

Solve and graph. Check your solution.

13. $|5x - 10| < 15$ **14.** $|2x - 7| < 0$ **15.** $|5x| > 0$

16. $|-0.5k + 1| < 7$ **17.** $7|3x - 7| \geq 35$ **18.** $|5x - 4| - 8 > -3$

19. $|3p + 5| > 0$ **20.** $\left|-\frac{1}{2} + x\right| \geq 3$ **21.** $13 + |5c + 7| < 13$

22. $-12 + \left|\frac{3x + 2}{4}\right| < -9$ **23.** $-15 + 3|2n + 5| \leq 42$ **24.** $|-4x| > 20$

Match each inequality with its graph.

25. $|x + 6| < 2$ **26.** $|2x - 4| > 2$ **27.** $14 + |2x + 1| \leq 19$

a.

b.

c.

28. INTERNATIONAL STANDARDS Use the international standard paper size information in Example 4. Write absolute value inequalities that describe the allowable length and width for paper size A1.

29. ENTERTAINMENT A survey shows 19.4% of the viewing population watches a television drama. The margin of error is ±1.25%. Write and solve an absolute value inequality to describe this situation.

30. WRITING MATHEMATICS Describe the different types of graphs that can be the solution to an absolute value inequality. Give examples to illustrate your explanation.

EXTEND

Write an absolute value inequality for each compound inequality.

31. $x - 4 \leq -2$ or $x - 4 \geq 2$

32. $-6 < 2x$ and $2x < 6$

33. $5x - 6 < -5$ or $5x - 6 > 5$

34. $3(2x - 1) < 3$ and $3(2x - 1) > -3$

35. $-5 < 3x - 2$ and $3x - 2 < 5$

36. $5.6y - 9 \leq 17$ or $5.6y - 9 \geq 17$

Solve and check each inequality.

37. $|x + 3| \leq x + 3$

38. $|5 - x| > 5 - x$

39. $|12 - 3x| < 2x + 4$

40. $|4 - x| \geq 3x$

41. $|x - 6| < 2x$

42. $|2x + 1| \geq x + 2$

GRAPHING UTILITY Graph each absolute value inequality. Find the midpoint of the solution.

43. $|x - 4| \leq 5$

44. $|c - 22| \leq 14$

45. $|m - 12| \leq 30$

46. $|r - 9| \leq 3$

47. What conclusions can you draw about an equation in the form $|x - b| \leq c$?

48. What is the relationship between the equations $|x| = 3$ and $|x - 2| = 3$?

49. Are $|x - a| = b$ and $|x| - a = b$ equivalent equations? Justify your answer.

ENVIRONMENT An environmental engineer is monitoring the temperature of a river using a thermometer with an accuracy rating of ±0.2°C. In one 24-hour period, the high temperature is 12.4°C and the low temperature is 11.8°C.

50. What absolute value inequality can the engineer use to indicate the high temperature?

51. What absolute value inequality can the engineer use to indicate the low temperature?

52. SOIL CHEMISTRY A forester checks the pH of the soil in a forest using a pH meter with an accuracy rating of ±0.1. The pH reading for a soil sample is 7.4. What absolute value inequality can the forester record to describe the pH for the sample?

53. POPULATION CHARACTERISTICS In a certain population, the height of 75% of the people can be determined by the inequality $\left| \dfrac{h - 67.5}{2.6} \right| \leq 1$ where h is height measured in inches.

    a. Determine the interval of heights for this group.

    b. What can you conclude about the remaining 25% of the population? Explain.

THINK CRITICALLY

Write an absolute value inequality for each graph.

54.
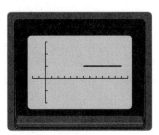
-8 -7 -6 -5 -4 -3 -2 -1 0 1 2 3 4 5 6 7

55.
-8 -7 -6 -5 -4 -3 -2 -1 0 1 2 3 4 5 6 7

56.
-8 -7 -6 -5 -4 -3 -2 -1 0 1 2 3 4 5 6 7

57.
-6 -5 -4 -3 -2 -1 0 1 2 3 4 5 6 7 8 9

If x-scale = 1, What absolute value inequality is shown on each screen?

58.
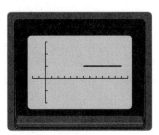

59.
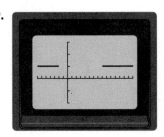

MIXED REVIEW

Find each number.

60. the additive identity of -7

61. the additive inverse of 62

62. the multiplicative identity of $\frac{12}{13}$

63. the multiplicative inverse of $\frac{17}{5}$

64. The distance traveled by a falling object is given by the formula $s = \frac{1}{2}gs = \frac{1}{2}gt^2$. Solve the formula for g.

PROJECT *Connection*   Use the balance you built in the Project Connection on page 67.

1. Make sure the stick is balanced. Insert three pennies in each small plastic bag. Hang the bags on the stick so that it balances.

2. Let d_1 represent the greater of the two centimeter markings and d_2 represent the smaller. Record d_1 and d_2. What is the centimeter marking at the balance point or fulcrum?

3. From the equation $|x - a| = b$, $x - a = b$ or $-(x - a) = b$. So $x = a + b$ or $x = a - b$. Let $a + b = d$, and $a - b = d_2$. Use these equations to write expressions for a and b in terms of d_1 and d_2.

4. Substitute the values you recorded for d_1 and d_2 to determine a and b. Considering the balancing experiment, explain what the absolute value equation $|x - a| = b$ represents.

5. Experiment to find another way to balance the two bags of pennies. Write an absolute value equation to describe the situation.

6. Suppose there is a balance when $|x - 50| = 18.5$. At what markings are the coin bags?

Career
Sound Engineer

Have you ever noticed protective sound barriers on either side of a highway? These barriers are constructed to lessen the traffic noise in the communities along the roadways. They are a product of research done by sound engineers. Sound engineers study sound and our tolerance to sound.

Sound waves are vibrations in the air. The number of vibrations per second, or *hertz*, is the frequency of the sound wave. Frequencies audible to the human ear are given by this absolute value inequality.

$$|f - 10,010| < 9990$$

Decision Making

1. Dogs are capable of hearing sounds with frequencies up to 25,000 hertz (Hz). Porpoises are capable of hearing sounds with frequencies up to 150,000 Hz. Illustrate how a sound engineer might visually represent the relationship between these animals' hearing and human hearing.

2. Hearing is also affected by the intensity of the sound. Intensity is measured in decibels. The absolute value inequality $|d - 60| < 60$ represents the range of intensities of sounds from barely audible to painfully loud. Express this intensity as a compound inequality.

3. At a rock concert, the sound reaches 115 to 120 decibels (dB). Based on this range, what can you conclude about the sound level at the concert?

Explore

As engineers plan for the future, they consider energy requirements and sources. The spreadsheet below presents projections for United States energy production, net imports, and energy consumption at five-year intervals from the years 1990 to 2010.

	A	B	C	D	E	F	G
1	ENERGY PRODUCTION 1990 – 2010, in quadrillion British thermal units (BTUs)						
2	YEAR		1990	1995	2000	2005	2010
3	PRODUCTION						
4	Petroleum		17.9	16.0	15.1	14.0	12.7
5	Natural Gas		17.8	19.4	21.2	20.9	20.4
6	Coal		21.6	23.5	25.5	29.9	35.2
7	Nuclear Power		5.9	6.1	6.2	6.5	6.5
8	Renewable Energy		6.8	7.7	8.3	9.2	10.1
9	TOTAL PRODUCTION		70.0	72.7	76.3	80.5	84.9
10	IMPORTS (+) and EXPORTS (–)						
11	Petroleum		16.1	19.3	21.0	23.4	26.0
12	Natural Gas		1.4	2.1	2.7	2.9	2.9
13	Coal and Other		–2.1	–2.1	–2.6	–3.5	–4.9
14	TOTAL NET IMPORTS		15.4	19.3	21.1	22.8	24.0
15	CONSUMPTION						
16	Petroleum Products		34.0	35.3	36.1	37.4	38.7
17	Natural Gas		19.2	21.5	23.9	23.8	23.3
18	Coal		19.5	21.4	22.9	26.4	30.3
19	Nuclear Power		5.9	6.1	6.2	6.5	6.5
20	Renewable Energy and Other		6.8	7.7	8.3	9.2	10.1
21	TOTAL CONSUMPTION		85.4	92.0	97.4	103.3	108.9

SPOTLIGHT ON LEARNING

WHAT? In this lesson you will learn
- to create an appropriate graph for presenting data from a spreadsheet.
- to interpret graphs created from spreadsheet data.

WHY? Graphing data can help show important relationships among data such as sources of energy in the United States.

ALGEBRA: WHO, WHERE, WHEN

The British thermal unit (BTU) is the amount of heat needed to raise the temperature of one pound of water 1°F. The term was first used about 1875–1880.

THINK BACK

Recall that a quadrillion is 10^{15} or 1,000,000,000,000,000.

An environmental engineer wants to show a client a comparison between petroleum production and natural gas production for the five-year intervals from 1990 to 2010. She uses the graphing capabilities in her spreadsheet program to make these two graphs.

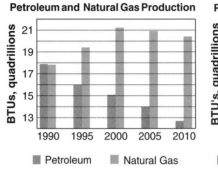

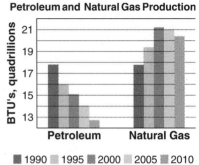

1. How are the two graphs alike?

2. How are the graphs different?

3. Give an example of a question that is answered by each graph.

4. Under what circumstances might the engineer select each graph to show her client?

CHECK UNDERSTANDING

About how many times as much petroleum is projected to be imported in 2010 as was imported in 1990?

About how many times greater is the projected export of coal and other energy in 2010 than it was in 1990?

Build Understanding

In a **bar graph** a series of bars is used to represent values. Bar graphs are often used to compare data at a given point in time.

EXAMPLE 1

Compare energy imported and exported in 1990, 2000, and 2010.

Solution
Use differently colored bars for each year. Group the bars for each energy source.

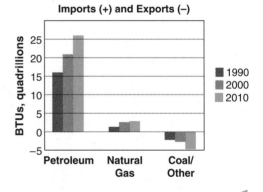

The graph illustrates that over time petroleum and natural gas imports increase. Exports of coal and other energy sources increase.

PROBLEM SOLVING TIP

In addition to selecting an appropriate graph to represent the data, it is important to select a suitable scale, label the scale and the axes, and title the graph.

A **line graph** is often used to show changes in data over time. It is particularly effective for showing trends and fluctuations in data. Double line graphs are useful for comparisons.

EXAMPLE 2

What is the expected trend for coal and petroleum consumption over the years 1990 to 2010?

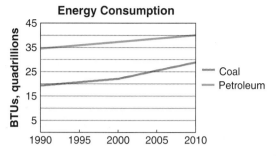

Energy Consumption

Solution

The demand for coal and petroleum will increase at about the same rate until approximately the year 2000. Then the demand for coal increases at a greater rate. ◄

A **circle graph** or **pie chart** represents data as portions, or sectors, of a circle. Each portion represents part of the whole.

EXAMPLE 3

Compare the relative amounts of energy consumption from each power source for 1995 and 2005.

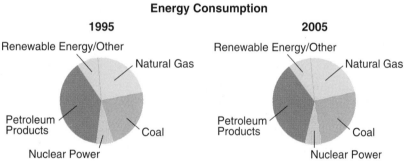

Energy Consumption

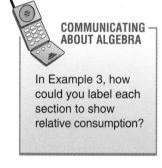

COMMUNICATING ABOUT ALGEBRA

In Example 3, how could you label each section to show relative consumption?

Solution

Comparing the sectors shows each is approximately the same for 1995 and 2005. Although the amount of energy consumed is greater in 2005, the relative energy consumption from each source is about the same. ◄

TRY THESE

Use the data in the spreadsheet on page 89 to make a graph for each presentation.

1. MANUFACTURING You are a manufacturer of solar panels. Show trends in production and consumption of renewable resources.

2. ARCHITECT You are an architect designing a factory. Your clients need to decide whether to power the factory with coal or with natural gas. Construct a graph to help them decide.

3. WRITING MATHEMATICS Select one of the situations above. What other information would you want to gather to help your client ?

PRACTICE

NEWS REPORTING You are a writer for *Energy News*. Use the spreadsheet on page 83 as your data source. Construct a graph to support each headline.

1. Renewable Energy Demand Outpaces Production!

2. Nuclear Power Production Remains Constant as Total Energy Production Increases

3. **WRITING MATHEMATICS** Write a short news article to accompany one of the headlines above.

4. **HOME HEATING** It is the year 2000. You are building a new home and selecting a heat supply. Using the spreadsheet information on page 83, which energy source would you select? Draw a graph to support your decision.

5. **WRITING MATHEMATICS** You are tutoring an eighth grade student. How would you explain when to use a circle graph and when to use a bar graph for presenting data?

EXTEND

6. What does this graph show?

7. For the year 2010, what inequality is illustrated?

8. What information is best illustrated by stacked bars?

9. Draw a stacked bar graph presenting a different set of information from the spreadsheet.

10. **WRITING MATHEMATICS** Explain how to match data with an appropriate graph. Illustrate your explanation with examples.

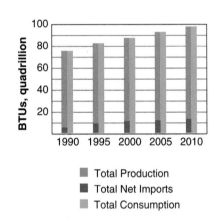

THINK CRITICALLY

11. Write a headline to accompany the graph below.

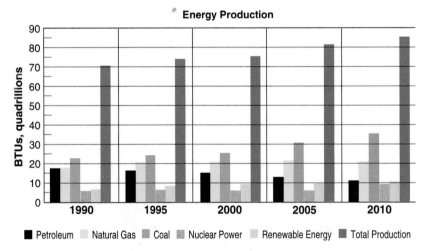

12. Draw a different graph of these data and suggest circumstances in which it might be used.

13. Find the data for the graph at the right in the spreadsheet on page 83. What do the data show?

Use the graph to project the following.

14. Consumption of petroleum products in 2015.

15. Consumption of coal in 2015.

16. Consumption of natural gas in 2015.

17. Create another graph that gives a different view of the data.

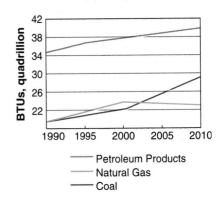

— Petroleum Products
— Natural Gas
— Coal

MIXED REVIEW

<small>STANDARDIZED TESTS</small> Simplify.

18. $[10^2 + (-8)] \cdot (-3) + 1$

    A. 235   **B.** -275 **C.** 2401   **D.** 325

19. $11^2 - 3 \cdot 5 + 1.49$

    A. 591.49   **B.** 101.53   **C.** 137.49   **D.** 107.49

The volume of a cylinder is given by the formula $V = \pi r^2 h$. Use this formula to solve Exercises 20 and 21.

20. Solve the formula for h.   **21.** Find h if $r = 7$ and $V = 1386$.

Complete each statement.

22. Two statements or inequalities joined by the word *or* is called a(n) __?__ .

23. Two statements or inequalities joined by the word *and* is called a(n) __?__ .

PROJECT *Connection*
    Use the balance you built in the Project Connection on page 67.

1. Insert 2 pennies N_1 in one small bag. Locate that bag 20 cm D_1 from the fulcrum.

2. Complete the table at the right. Insert the stated number of pennies in the second bag, hang it on the opposite side of the stick, and determine the distance from that bag to the fulcrum when you achieve a balance.

3. Determine the product $N_1 D_1$. Then determine $N_2 D_2$ for each case in the table. What do you notice?

4. Suppose that a bag with 6 pennies was placed at a distance of 11 cm from the fulcrum on the left side. Where might a bag with 5 pennies be placed to balance the stick?

Number of Pennies N_2	Distance D_2
1	
2	
3	
4	
5	
6	

2.7 Problem Solving File

Work Backward

Usually, when you solve a problem, you start at the beginning and proceed "forward," using a step-by-step method, until you obtain the solution. For some problems, however, an effective strategy is to begin at the end of the problem and work backward, using variables and equations as appropriate.

Problem

Marina had a supply of blank computer disks. Dan needed some disks, so Marina gave him half of her supply, plus one. Then Phyllis stopped by and took half of Marina's remaining disks, plus one. Next, Marina gave Roger half of the disks she had left, plus one. At this point, Marina saw that she had only one disk left. How many did she have originally?

Explore the Problem

1. To work backward, let x represent the number of disks Marina had *before* Roger took some. Represent what Roger took in terms of x. Then represent what Marina had left.

2. Since you know Marina ended up with one disk, what equation can you write?

3. Solve the equation you wrote in Question 2. Check that the value works in the problem.

4. Your solution to Question 3 tells you how many disks Marina had *after* Phyllis took some. You want to determine the number she had *before*. What do you notice about the description of the quantity Phyllis took?

5. How does this help you to write an equation for the number Marina had before Phyllis took some?

6. Solve the equation you wrote in Question 5.

7. Repeat the procedure once again to complete the solution. With how many disks did Marina begin? Work *forward* to check your answer.

Investigate Further

- When you use the work backward strategy, consider the relationships in the problem carefully so you choose the correct inverse operation.

Problem

A plant supervisor noted that one morning, workers produced ten times as many radios as were on hand when the day began. At 1:00 P.M. half of the total stock was shipped to McPherson's department store. During the rest of the afternoon, production amounted to three times what was left after the one o'clock shipment. At the end of the day, a shipment of 250 radios went to the Shanghai Hotel and another of 300 radios was sent to Save-a-Buck discount store. At this time, the supervisor noted that 1210 radios remained. How many radios were on hand at the beginning of the day?

8. Let x represent the number of radios before the shipments to the Shanghai Hotel and the Save-a-Buck discount store. Write and solve an equation to find x.

9. The number you determined in Question 6 tells you how many radios there were when afternoon production was completed. How does this number relate to the number (call it y) before afternoon production began? Determine y.

10. Write and solve an equation to determine the number of radios *before* the shipment to McPherson's.

11. The last step is to find the number of radios r before the morning production. What equation will you use? Remember to check your result by working forward.

12. WRITING MATHEMATICS For what types of problems might you try working backward?

Apply the Strategy

- 13. CAR PRICES The final price of Sharon's new car was $13,067. This price consisted of the base price plus $1170 in options; a 7% sales tax was then added. To this amount, a $175 dealer preparation fee and a $52 freight charge were added. What was the base price of the car?

14. AVIATION A plane gained altitude at a rate of 1000 ft/min, descended 1300 ft, then started to climb again for $2\frac{1}{4}$ min at a rate of 800 ft/min. The total gain in altitude was 6000 ft. How long did the plane climb at a rate of 1000 ft/min?

15. ACTIVITIES One half of the students in Mr. Miller's class worked on projects for the science fair. Of those who did not work on science projects, one third wrote short stories for a writing contest. One half of those who did not prepare science projects or write short stories acted in the school play. The other 5 students who did not work at a science, writing, or drama project created ceramic objects. How many students were in Mr. Miller's class?

16. AGES If you multiply Chan's age by 7, add 7, divide by 7 and then subtract 7, the result is 7. How old is Chan?

17. PRIZE VALUES Joseph won first prize in a chili cook off. His prize consists of kitchen equipment: the stove represents 50% of the total prize value, the refrigerator is worth 60% of the remaining value, the bread-making machine is worth $300 and the deluxe food processor is worth $220. What is the total value of the prize?

18. SYLIVIA'S SEASHELLS Sylvia took her four brothers—Lewis, Mark, Nico, and Oscar—to the beach. In the morning, Sylvia collected a total of 54 seashells. She gave each of her brothers some shells, leaving none for herself. In the afternoon, Lewis found two more shells, and Nico doubled the number of shells he already had. However, Mark and Oscar used their shells to dig in the sand and, as a result, Mark broke two of his shells and Oscar broke half of his shells. At this point, each of the boys had the same number of shells. How many shells had Sylvia given each of her brothers?

19. APPLIANCE PRICES Between 1985 and 1990 the price of a portable tape player decreased $30. From 1990 to 1993, the price decreased by 20%. In 1994 the price rose by $8, but by 1995 the price had fallen to $27, which was three-fourths of the 1994 price. Determine the price of the tape player in 1985.

20. SHARING APPLES Errol, Fran, and Gloria each picked some apples and then decided to share their harvest. First, Errol gave enough of his apples to Fran and Gloria to double the number each of them had to begin with. Then Fran gave to Errol and Gloria enough of the apples she now had to double the number Errol and Gloria each now had. Finally, Gloria gave enough of the apples she now had to double the number Errol and Fran each now had. At that point, each friend had 24 apples. With how many apples did each of them begin?

21. WRITING MATHEMATICS Create a problem of your own that can be solved by working backward. Your problem should have at least three steps. Exchange problems with some classmates.

REVIEW PROBLEM SOLVING STRATEGIES

WHAT'S FOR LUNCH?

1. Five students each select one sandwich from a tray containing two chicken sandwiches and three tuna sandwiches. Determine what each student ate if you know that Vernon and Laura each ate the same type of

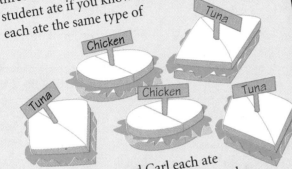

sandwich. Laura and Carl each ate a different type of sandwich. Carl and Janet each ate a different type of sandwich. Katherine is one of the students. Explain your reasoning.

ON TARGET

2. Suppose you can throw as many darts as you want at this board.
   a. What scores will be impossible for you to get?
   b. Make a conjecture about the completeness of your list above. Explain your reasoning.

STAMP COUNT

3. A new stamp was just issued with a picture of one of Vera's favorite rock-and-roll stars, so she bought 12 of them. The stamps are perforated and arranged in three rows and four columns as shown.

Vera promised to give her friend Tony four of the stamps. Tony requested that his four stamps all be joined together; that is, each stamp must share at least one full side with another stamp and cannot hang by just a corner. In how many different ways can Vera give Tony his stamps?

Work with two or three other students. Use the questions below to help you develop an organized method for counting and recording possibilities.
   a. In how many different ways can Vera give the stamps in a shape like the one shown at the top right?
   b. In how many different ways can Vera give the stamps in a shape like the one shown at the bottom right?
   c. What other shapes are possible for Tony's four stamps? How many different ways can each shape be made?
   d. What is the total number of different ways?

CHAPTER REVIEW

VOCABULARY

Choose the word from the list that completes each statement.

1. A(n) __?__ cannot be expressed as the ratio of two integers.

2. Following the __?__ , you multiply before you add.

3. According to the __?__ three numbers may be added in any group of two.

4. The __?__ states that if $a = b$ and $b = c$, then $a = c$.

5. A(n) __?__ does not check in the original equation.

a. Transitive property

b. Associative property

c. irrational number

d. extraneous solution

e. order of operations

Lesson 2.1   GRAPH OF THE NUMBER LINE   pages 59–61

• The tools of algebra and geometry may be used to locate square roots.

Right △ABC is constructed on the number line as shown.

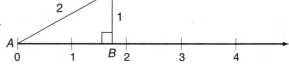

6. What is the length of $\overline{AB}$?

7. Tell how to locate $2\sqrt{3}$.

Lessons 2.2 THE STRUCTURE OF THE REAL NUMBERS   pages 62–67

• **Rational numbers** can be expressed as the ratio of two integers.

Tell whether the number is rational or irrational. If it is rational, express it as a ratio of two integers.

8. 0.125

9. $0.\overline{17}$

10. $\sqrt{625}$

11. π

Lessons 2.3   PROPERTIES OF THE REAL NUMBERS   pages 68–73

• The **field properties** include **closure, commutative, associative, distributive, identity,** and **inverse.**

Apply the order of operations to evaluate each expression.

12. $\dfrac{40 + 2}{10 - 3} + 15 \times 4$

13. $3 \cdot 4^2 + (3 + 4)^2$

Which property is illustrated by each of the following?

14. $8.1 + (3.2 + 5.4) = (8.1 + 3.2) + 5.4$

15. $7.9 + 15.6 = 15.6 + 7.9$

- Field properties and those of equality or inequality are used to solve equations and inequalities and prove other properties.

Solve and check each equation.

16. $14a + 1.3 = 2.7 + 7a$

17. $r - 2(4r - 5) = 1 - r\frac{3}{2}$

18. $\frac{4(z + 8)}{3} = -4z$

Solve, graph, and check the solution of each inequality.

19. $4x + 3 \geq 6x + 9$

20. $8 < -3y + 2 \leq 14$

21. $2t - 1 \leq -7 \text{ or } \frac{2t}{3} > 2$

- Write and solve an equivalent compound sentence. Check for **extraneous solutions**.

Solve and check.

22. $|3x - 2| = 5$

23. $|2y - 1| \leq 7$

24. $2z - |z + 2| = 11$

- Different graphs are appropriate to display the data of a spreadsheet.

The abbreviated spreadsheet at the right gives the grade point averages for the top three academic students at Faro High, who are being considered for valedictorian and salutatorian. Compare the students' grade averages at each of the three grade levels by constructing:

1	Grade Ledger				
2					
3	Grade Level	10	11	12	Average
4					
5	Student				
6	Blue	94.1	95.2	96.0	95.1
7	Green	97.0	96.0	92.3	95.1
8	Redd	92.1	98.1	95.1	95.1

25. a bar graph

26. a line graph

27. Comment on the relative effectiveness of each graph.

- A variety of problem situations can be solved by working backward.

Solve the problem.

28. Ruby paid $1113 including sales tax to buy a stereo system. She had saved the money from her part-time job in which she works 4 h per day and earns $5.25 per hour. While she was saving, her mother contributed enough money to cover the 6% sales tax. How many days did Ruby have to work to earn her portion of the cost?

CHAPTER ASSESSMENT

CHAPTER TEST

1. **STANDARDIZED TESTS** By constructing a right triangle with hypotenuse of 3 units, a value that can be easily located on the number line is

   A. $\sqrt{11}$ **B.** $\sqrt{10}$ **C.** $\sqrt{8}$ **D.** $\sqrt{7}$

Evaluate each expression.

2. $3 \cdot 5^3 - (3 - 5)^3$ 3. $\left(12 - 2\sqrt{121}\right)^2$

Evaluate each expression for $a = -1$, $b = -3$, $c = 4$.

4. $b^2 - 4ac$

5. $\dfrac{-b \pm \sqrt{b^2 - 4ac}}{2a}$

Solve and check each equation.

6. $3a - 7.2 = 2.7 + 6a$

7. $\dfrac{3(y - 7)}{5} = -6y$

8. $4(2x + 1) - 7 = 26 - 3(-2x - 5)$

9. $|4m - 1| = 7$

10. $|3 - 2x| = 11$

11. $|3z + 2| = 4z + 5$

12. $|3k + 8| + 4 = 0$

13. **WRITING MATHEMATICS** Explain why it is essential to check possible solutions for an equation that contains absolute value.

14. **STANDARDIZED TESTS** Which graph shows the solution to $-22 \le 7x - 1 < 13$?

    A.
    B.
    C.
    D.

Solve and graph the solution of each inequality.

15. $8t - 4 > 2t + 8$

16. $2x > -8$ and $x - 3 \le 0$

17. $9 \le -2z + 3 < 13$

18. $\dfrac{3y}{2} > 6$ or $y + 3 \le 1$

19. **STANDARDIZED TESTS** Which of the following values is in the solution set of $|2x - 1| < 9$?

    A. -5 **B.** 6 **C.** 5 **D.** 4

Solve and check each inequality.

20. $|2y - 3| < 7$

21. $5 + |2x + 1| > -4$

22. $x > |2x + 1|$

23. Use the spreadsheet on page 89 for the data. Compare the consumption of petroleum, natural gas, and coal over the years 1990 to 2010 by constructing a bar graph and a line graph. Comment on the significance of each graph.

24. Rent-a-Car charges $39 per day for a car, plus 35 cents per mile driven. If Mia rents a car for a day, what distance can she drive and have her total rental charge equal $60, not including tax?

25. The entertainment committee for the school dance has to make a decision about which band to hire. The Are-Nows will play for $300 plus 30% of the gate, and the Once-Agains will play for a flat fee of $540. The committee prefers the Are-Nows but has to be economical. If 200 students are expected to attend the dance, what is the most the committee can charge per ticket and still make the Are-Nows an economical choice?

PERFORMANCE ASSESSMENT

USE OPERATION TABLES For a finite mathematics system with a specific number of elements, a table can be used to define an operation. The table below defines the operation $\oplus$ for $\{0, 1, 2, 3, 4\}$.

$\oplus$	0	1	2	3	4
0	0	1	2	3	4
1	1	2	3	4	0
2	2	3	4	0	1
3	3	4	0	1	2
4	4	0	1	2	3

a. From the pattern shown, write a rule for performing addition in this system.
b. Applying this pattern, create a table that defines the operation of $\otimes$ for this system.
c. Determine whether or not the system $(\{0, 1, 2, 3, 4\}, \oplus, \otimes)$ forms a field. Explain how each of the field properties is satisfied or not.

FILL IN THE NUMBER LINE A compound inequality can produce two critical values on the number line, which separate the number line into the region between the two values and the two regions on either side. Depending on the nature of the inequality, its graph will occupy either the region between or the regions on either side. For example:

$$3 < 2x + 1 \leq 5 \qquad 2x + 1 < 3 \text{ or } 2x + 1 \geq 5$$

Create different inequalities varying the order of the symbols of inequality and also the algebraic expressions and values. Try to predict the two critical values and where the graph will fall before solving. Verify your predictions by solving.

Find some inequalities where this does not work and explain why.

CREATE A PROBLEM Write a problem that can be modeled by a linear inequality. Look through the problems of Lesson 2.7 for ideas. Solve your inequality and check the results in the words of your problem. When you are sure the problem is properly worded, have a partner try to solve it by writing a linear inequality.

DO RESEARCH Look through a textbook or other source to find a spreadsheet. Display two aspects of the data using different graphs. Draw some conclusions from the graphs. Show your graphs to a partner (without showing the spreadsheet) and ask your partner to interpret the graphs.

PROJECT *Connection*   As a class, work with your teacher to plan an Engineering Day. This event may include the following activities:

1. Invite some professional engineers to visit. Develop and have approved a list of questions that you wish these engineers to respond to. Provide the list in advance to each speaker so they may prepare.

2. Summarize the highlights of your research on different engineering fields. Design a display to include this information and illustrations of engineers "on the job." Your display should interest students in this career field, so be creative.

3. Write a report on your balance experiments and conclusions. Speculate how the laws of balance might be important in different fields of engineering.

• • • CUMULATIVE REVIEW • • •

1. **STANDARDIZED TESTS** If x is an integer that is located on the number line so that $2 < \sqrt{x} \le 3$, then the set of all possible values for x is

   A. no solution   **B.** $\{2, 3\}$
   C. $\{5, 6, 7, 8\}$ **D.** $\{5, 6, 7, 8, 9\}$

Tell whether the number is rational or irrational. If rational, write it as a ratio of integers.

2. $\sqrt{7}$ 3. 3.625

Evaluate each expression.

4. $5 \cdot 3^2 - (2 - 5)$ 5. $3(9 - 2)^3 - \dfrac{1 - 4}{-3}$

Which field property is shown by each of the following?

6. $-3.2 + 3.2 = 0$

7. $4 + (2 + 3) = 4 + (3 + 2)$

Solve and check.

8. $\dfrac{4(y - 2)}{3} = -2y$

9. $7x - 3(2x - 1) = 4 - x$

Solve, graph, and check the solution.

10. $6 \le 3z - 12 < 15$

11. $4x + 1 > 9$ or $\dfrac{2x}{5}$

Solve and check.

12. $2|4z + 1| = 18$ 13. $|2x - 3| \le 7$

14. **WRITING MATHEMATICS** How many elements are in the solution set of $3a - |a + 4| = 6$? Explain why this is so.

15. **WRITING MATHEMATICS** Name three types of graphs that might be appropriate in displaying data obtained from a spreadsheet, and tell the particular strength of each kind of graph.

Ken took a loan of $15,500 for 3 years at 8% to pay for a car.

16. Use the monthly payment formula to find his monthly payment M to the nearest dollar.

17. Calculate the total cost of the car over the period of the loan.

18. What is the total finance charge for the loan?

19. The February 1 inventories of jogging suits at Sam's Sports and at Tim's Togs, both owned by Lee Co., are summarized in the matrices below.

	Sam's Sports			Tim's Togs		
	S	M	L	S	M	L
red	12	6	10	2	0	3
white	0	15	2	5	15	12
blue	9	3	8	8	12	9

Write a matrix showing the combined inventory.

20. The center ring of Barney's Circus has a diameter of 100 ft. A rectangular region measuring 10 ft by 15 ft contains some equipment. The aerialist performing in the center ring drops a towel. Assuming that the towel lands at a random point in the ring, what is the probability that the towel lands in the equipment region to the nearest tenth of a percent?

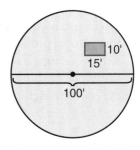

21. **WRITING MATHEMATICS** Calculate the mean for the data set that follows. Would the mean be a good measure of central tendency in this case? Explain. If the mean is not a good measure, tell which measure of central tendency would be more appropriate.

$7.7, 7.7, 7.7, 7.7, 7.7, 7.8, 8.0, 8.9, 9.0, 9.4$

• • • STANDARDIZED TEST • • •

STUDENT PRODUCED ANSWERS Solve each question and on the answer grid write your answer at the top and fill in the ovals.

Notes: Mixed numbers such as $1\frac{1}{2}$ must be gridded as 1.5 or 3/2. Grid only one answer per question. If your answer is a decimal, enter the most accurate value the grid will accommodate.

1. Consider a spiraled structure of right triangles in which the hypotenuse of one triangle becomes a leg of the next and the measure of the other leg is 1 unit. Given the measures shown, what is the length of the hypotenuse of the 15th triangle in the spiral?

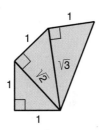

2. Express the rational number $2.\overline{7}$ as the ratio of two integers.

Use the operation table for addition shown at the right to answer Questions 3 and 4.

+	0	1	2	3
0	0	1	2	3
1	1	2	3	0
2	2	3	0	1
3	3	0	1	2

3. What is the inverse of 2?

4. Solve for x: $3 + x = 1 + 1$

5. Solve for $\frac{3x + 1}{4} = 10 - 6\left(\frac{1}{3}x - 1\right) + \frac{3}{4}$

6. Find the product of the integer solutions of $2 < 3x - 1 \le 8$.

Find the sum of the solutions for each of the following.

7. $|3z - 1| = 8$

8. $|2x + 1| = 7$

Find the sum of the integer solutions for each of the following.

9. $|x - 1| < 3$

10. $|2x + 7| \le 9$

11. Find the solution of $2x - |x + 1| = 7$

12. The line graph below compares the first-quarter sales of this year and last at Al's Used Cars.

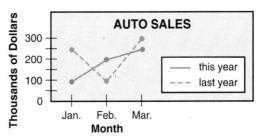

Grid the multiple of $1,000 that represents the difference between this year's total sales for the first quarter when compared to that of last year.

Use the Monthly Payment Formula to solve.

13. To buy a house, the Simons took a mortgage loan of $85,000 for 20 years at 8.1%. To the nearest dollar, how much would they have saved per month by obtaining a 7.8% mortgage for the same length of time? (Round both monthly payments to the nearest dollar before finding the difference.)

14. Of the 3 matrices below, $C = 2A - B$.

$$A = \begin{bmatrix} 3 & -3 & 2 \\ -1 & 4 & 0 \end{bmatrix} \qquad B = \begin{bmatrix} 3 & 3 & 0 \\ -1 & -4 & 2 \end{bmatrix}$$

$$C = \begin{bmatrix} 9 & -9 & x \\ -1 & 12 & y \end{bmatrix}$$

Find the value of $x + y$.

Use the following data set to answer Questions 15–16.

58	63	54	77	71	70
75	72	73	64	60	80
40	60	56	47	48	45
50	49	43	66	57	62

15. Find the mean and median of the data set.

16. Find the 1st and 3rd quartiles and the range of the data.

Functions and Graphs

Take a Look AHEAD

Make notes about things that look new.

• Find the page on which you will learn the meaning of each of the following terms: vertical line test; even function; odd function; inverse function.

• What special type of function is the focus of the Problem Solving File in this chapter?

Make notes about things that look familiar.

• The word "function" has many different meanings in everyday language. What are some of these meanings? Which do you think is closest to the mathematical meaning of "function"?

• Find the names of three common forms of linear functions.

• What four operations do you perform with real numbers?

DATA Activity

There's No Place Like Home

The housing market reflects the laws of supply and demand but also serves as an indicator of the general state of the economy. For example, more employment opportunities in a region will create an increased demand for housing that, in turn, may result in higher home prices. Higher prices may also result if an area restricts development. Demand may decrease if mortgage rates increase. The table presents data gathered by the U.S. Commerce Department for a region in the northeast.

SKILL FOCUS

▶ Read and interpret tables.

▶ Add, subtract, multiply, and divide real numbers.

▶ Determine the range of a set of data.

▶ Interpret the median of a set of data.

▶ Construct appropriate graphs.

Personal
FINANCE

In this chapter, you will learn how:

- **TAX ASSESSORS** use a property tax function to determine the assessed value of a home.
(Lesson 3.1, page 111)

- **STOCK BROKERS** can represent their fee schedules with a piecewise function.
(Lesson 3.4, page 131)

- **REAL ESTATE SALES AGENTS** use functions to determine their commissions and weekly earnings.
(Lesson 3.7, page 151)

New Single - Family Home Sales		
Year	New Homes Sold	Median Sale Price
1987	23,400	$195,100
1988	16,500	195,700
1989	15,000	202,200
1990	14,800	219,700
1991	12,700	202,000
1992	17,100	194,200
1993	17,900	191,400
1994	18,100	192,300

Use the table to answer the following questions.

1. During which period did new home sales show the sharpest increase?

2. What is the range of median sale prices for the years shown?

3. About how many new homes cost more than $219,700 in 1990? How do you know?

4. Describe the trend in new home sales?

5. Make a scatterplot of number of new homes sold and median sale price. What do you conclude about the relationship between these two variables? Explain.

6. **WORKING TOGETHER** Show both sets of data in a way that allows for quick visual comparisons.

PROJECT

A Place to Call Home

The most expensive purchase most people will ever make is of their own home. Purchasing a home can require a financial commitment for many years. How can you determine the costs involved in owning a home and the type of home you can afford depending on income level?

PROJECT GOAL

To research the costs involved in purchasing and maintaining a home.

PROJECT *Connections*

Lesson 3.1, page 110:
Determine statistics related to home prices.
Lesson 3.2, page 118:
Investigate the relationship between a home's square footage and the selling price.
Lesson 3.4, page 130:
Determine mortgage and closing costs, property taxes, and utility expenses.
Chapter Assessment, page 159:
Design a presentation to summarize monthly and annual costs for purchasing and maintaining a home.

Getting Started

Work in groups of four or five students.

1. Make a list of types of homes including single family houses, condominiums, prefabricated, or mobile homes. Select one or two types to research.

2. Discuss compromises you might make as a first-time buyer to reduce purchase cost. For example, would you be willing to commute longer distances or live in a house that may need repairs.

3. Collect real estate sections with home listings. Make a list of whom you could contact for information.

4. Try to find the simplest income-based rules used to provide an estimate of affordable home prices.

Internet Connection
www.swpco.com/
swpco/algebra2.html

3.1 Algebra Workshop
Symmetry

Think Back/Working Together

- Work with a partner. Trace figure *ABCDEF* onto a piece of paper. Fold it along the dashed line so that points *A* and *B* coincide.

  1. Name two other pairs of points that coincide.

  2. How many pairs of points coincide?

  3. Unfold the paper and refold it along the line connecting *F* and *C*. Name two pairs of points that coincide.

  4. How many sets of points coincide?

  5. What line segment can you fold the figure along to make points *A* and *C* coincide?

  6. What line segment can you fold the figure along to make points *B* and *F* coincide?

  7. How many line segments can this figure be folded along so that the two halves coincide?

Explore

- If a figure can be folded along a line so two halves coincide, the figure is said to be *symmetric about the line*, and the line is called a **line of symmetry** or **axis of symmetry.**

  8. Draw a square on a piece of paper. Draw the two diagonals about which the square is symmetric.

  9. What other axes of symmetry can you find for your square? Draw and describe them.

If a figure coincides with itself when it is rotated 180° about a point, the figure is said to be *symmetric about the point*. For example, rectangle *QRST* is symmetric about point *M*.

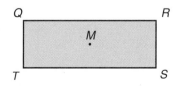

Trace each figure. Locate the point about which each figure is symmetric.

10. 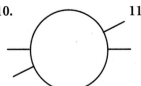 11.   12.

13. List the capital letters that have vertical line symmetry.

14. List the capital letters that have horizontal line symmetry.

15. List the capital letters that have point symmetry.

16. Examine a standard deck of 52 playing cards. Which cards have point symmetry?

17. Which cards in the standard deck have line symmetry?

Make Connections

Many graphs also have symmetries.

18. Which graphs have line symmetry about the y-axis?

a.   b.   c.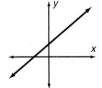

19. Which graphs have line symmetry about the x-axis?

a.   b.   c.

20. Which graphs have point symmetry about the origin?

a.   b. 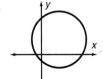 c.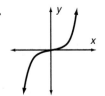

Make three copies of the graph at the right.

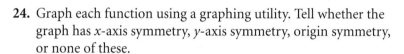

21. Extend the first graph to produce a figure that is symmetric about the *y*-axis.

22. Extend the second graph to produce a figure that is symmetric about the *x*-axis.

23. Extend the third graph to produce a figure that is symmetric about the origin.

24. Graph each function using a graphing utility. Tell whether the graph has *x*-axis symmetry, *y*-axis symmetry, origin symmetry, or none of these.

    a. $y = x^2$ **b.** $y = 12$ **c.** $y = x^3$ **d.** $y = 3x$

    e. $y = 3x + 1$ **f.** $y = x^2 + 5$ **g.** $y = x^3 + 2$ **h.** $y = (x + 2)^2$

PROBLEM SOLVING TIP

To decide if a figure is symmetric about a point, place the tip of your pencil at the point and rotate the page 180°. Determine if the figure you see is identical to the original figure.

25. Crossword puzzles are often designed to be symmetric. The top four rows of a crossword puzzle are displayed at the right. Copy the diagram and fill in the bottom three rows so that the puzzle has point symmetry about the center of the puzzle.

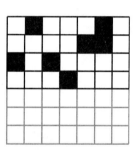

26. Draw a set of axes. Choose a point on the *x*-axis other than the origin and label it $(a, 0)$. Use a compass to construct a circle with center at $(a, 0)$. About which vertical line, horizontal line, and point is the circle symmetric? How many lines of symmetry does it have? How many points of symmetry does it have?

Summarize

27. **WRITING MATHEMATICS** Use the information from this lesson to write definitions of symmetry about the *x*-axis, symmetry about the *y*-axis, and symmetry about the origin.

28. **WRITING MATHEMATICS** Find pictures of logos, flags, or artwork that exhibit symmetry. Describe their symmetry or symmetries.

29. **MODELING** Graph $y = x^2$ in Quadrant I only. Select three points on the graph. How can you use a ruler to create a mirror image of the three points in Quadrant II.

30. MODELING Draw a figure that has at least two of the following symmetries: *y*-axis, *x*-axis, and origin. Discuss the symmetries of your figure.

31. THINKING CRITICALLY If you know that point (3, 4) is on a figure that has *y*-axis symmetry, what other point is also on the figure? In general, if a point (*a*, *b*) is on a figure that has *y*-axis symmetry, what other point is also on the figure?

32. THINKING CRITICALLY If you know that point (*a*, *b*) is on a figure that has *x*-axis symmetry, what other point is also on the figure?

33. THINKING CRITICALLY If you know that point (*a*, *b*) is on a figure that has origin symmetry, what other point is also on the figure?

34. GOING FURTHER Name a point about which the graph in Exercise 24g is symmetric.

35. GOING FURTHER Name a line about which the graph in Exercise 24h is symmetric.

PROJECT *Connection*

Explore prices for the type of homes your group selected.

1. Look through newspaper or other real estate advertisements. Circle about 20 listings of interest to your group. Assign each home a variable name and record the price.

2. Determine the mean, median, mode, and range of the home prices.

3. Let P_H represent the highest price of the homes and P_L represent the lowest price. Explain why the following function *cannot* be used to determine the mean price of all the homes you selected.

$$\text{mean} = \frac{1}{2}(P_H - P_L)$$

4. Graph your home prices on a number line. Label each point with its variable name.

5. Decide upon a price that best represents the home your group is interested in.

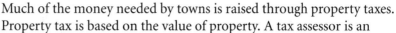

Much of the money needed by towns is raised through property taxes. Property tax is based on the value of property. A tax assessor is an official who estimates the value of property and computes property taxes. The estimated value of a property is its *assessed value*, which is not necessarily the same as its market value.

Decision Making

The assessment schedule for a town shows total property tax as the sum of flat rates and a percentage of structural rates.

Structural Rates, ft²		Flat Rates	
1st floor over basement	$3.00	Land	$1000 per acre
1st floor over slab	$2.25	1st bathroom	$0
2nd story	$2.00	Extra full bathroom	$100
Garage	$1.00	Half bathroom	$50
Dormer	$1.00	Fireplace	$125
Barn/stable	$0.70	Tennis court	$375
Deck	$0.50	Spa	$125
Shed	$0.70	Central air conditioning	$200
Screened porch	$1.10	Gas heat	$700
Vinyl-lined pool	$0.75		

Use the ad to find each assessed value.

> 2-story Colonial with 2 1/2 baths, frpl, full basement, CAC, 30 ft x 30 ft deck, 3/4 acre, 600 sq ft first fl, 1500 sq ft 2nd fl, 20 ft x 20 ft dormer, 12 ft x 21ft garage, 16 ft x 32 ft vinyl pool, gas heat, excellent cond. $178K.

1. The land

2. What is the total of the structural rate items?

3. What is the total of the flat rate items?

4. If the town has a tax rate of 95.5018% of assessed value for this year, determine the property tax on this house, to the nearest cent.

Relations, Functions, and Graphs

Explore/Working Together

● Work with a partner. Use a graphing utility to graph $y = x^2 + 2$. Set your calculator so integers are displayed. Begin tracing at the origin and trace to the right. Your partner should begin tracing at the origin and trace to the left. Trace one point at a time and record the coordinates.

1. Compare your results with your partner and identify any patterns.

2. What type of symmetry does the graph have?

3. Without doing any computations, explain why $x = 237$ and $x = -237$ will produce the same value of y in $y = 2x^2 - 17$.

Build Understanding

● A **relation** is a set of ordered pairs. The **domain** of the relation is the set of first coordinates of the ordered pairs. The **range** of the relation is the set of second coordinates of the ordered pairs.

Relations can be represented using tables or by graphing the ordered pairs on a Cartesian coordinate system. Recall that the coordinate plane has the x-axis and the y-axis meeting at the origin $(0, 0)$. The axes divide the plane into four quadrants which are numbered in a counterclockwise direction, as displayed at the right.

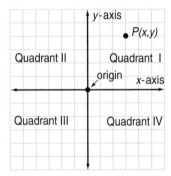

Relations can be described using a rule. You are familiar with relations expressed as $y = 2x + 4$. In this relation, $2x + 4$ is the rule and the domain and range are both all real numbers. A **mapping** demonstrates how each element of the domain is paired with the elements of the range.

A **function** is a relation in which each element in the domain is paired with one and only one element in the range.

The mapping at the right is a function. You can tell whether a relation represented by a graph is a function by applying the **vertical line test**. Draw or visualize all possible vertical lines that intersect the graph. If there is no vertical line that intersects the graph in more than one point, then the graph represents a function.

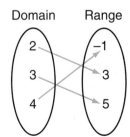

EXAMPLE 1

State the domain and the range. Tell whether the relation is a function.

a. $[(-4, 2), (-2, 1), (2, -1), (1, 4), (-4, -2)]$

b. 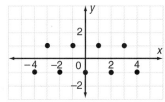   **c.**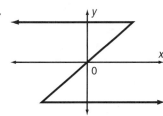

Solution

a. Domain: $\{-4, -2, 1, 2, 4]$ Range: $\{-2, -1, 1, 2, 4\}$

The relation is a function because each domain element is paired with only one range element.

b. Domain: $\{-4, -3, -2, -1, 0, 1, 2, 3, 4\}$ Range: $\{1, -1\}$

The relation is a function. Some vertical lines intersect the graph in only one point.

c. Domain: $\{$all real numbers$\}$ Range: $\{$all real numers$\}$

The relation is not a function. Any vertical line would intersect the graph in three points.   ◄

To decide whether a relation is a function, you may want to convert from one representation (mapping, table, graph, rule) to another.

EXAMPLE 2

TRAVEL A major international airline announced the following discounted one-way fares for travel from New York to Europe. Graph the relation on the coordinate axes, letting x represent the city. State its domain and range. Tell whether the relation is a function.

City	Athens	Berlin	Frankfurt	Lisbon	Milan	Paris	Rome
Fare	$299	$249	$249	$249	$299	$199	$199

COMMUNICATING ABOUT ALGEBRA

How would the solution to Example 2 differ if you let x represent the fare and y represent the city?

Solution

Domain: {Athens, Berlin, Frankfurt, Lisbon, Milan, Paris, Rome}
Range: {$199, $249, $299}

The relation is a function because each element in the domain is paired with one and only one element in the range.   ◄

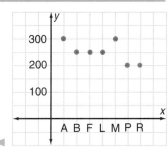

ALGEBRA: WHO, WHERE, WHEN

Coordinate geometry, the method of depicting the relationship between two variables by plotting points on a grid, was developed by the French mathematician and philosopher René Descartes (1596–1650).

THINK BACK

Recall that a function that is symmetric about the origin coincides with itself when it is rotated 180° about the origin.

When functions are written using **function notation**, the equation

$$y = 3x - 1 \qquad \text{is written as} \qquad f(x) = 3x - 1$$

The symbol $f(x)$ means "the value of f at x" and is read "f of x." The symbol $f(x)$ is another name for y and does **not** mean "f times x."

EXAMPLE 3

Let $f(x) = x^2 - 2x + 4$. Evaluate.

a. $f(-2)$ **b.** $f(0)$ **c.** $f(5)$

Solution

  a. $f(-2) = (-2)^2 - 2(-2) + 4 = 12$

  b. $f(0) = 0^2 - 2(0) + 4 = 4$

  c. $f(5) = 5^2 - 2(5) + 4 = 19$ ◄

As you discovered in Explore, the graph of $y = x^2 + 2$ is symmetric about the y-axis. You also discovered that x and $-x$ produced the same value of y. A function is an **even function** if and only if $f(x) = f(-x)$ for all values of x in its domain. All even functions are symmetric about the y-axis.

A function is an **odd function** if and only if $f(-x) = -f(x)$ for all values of x in its domain. All odd functions are symmetric about the origin. Functions can also be neither even nor odd.

> — PROPERTIES OF EQUALITY —
>
> A function is an *even function* if and only if $f(x) = f(-x)$.
>
> A function is an *odd function* if and only if $f(-x) = -f(x)$.

EXAMPLE 4

Graph $f(x) = x^3 + x$. Tell whether it appears to be *even*, *odd*, or *neither*. Justify your answer.

Solution

It appears to be symmetric about the origin and therefore appears to be an odd function.

$$\begin{aligned}
f(-x) &= (-x)^3 + (-x) \\
&= -x^3 - x \\
&= -(x^3 + x) \\
&= -f(x)
\end{aligned}$$

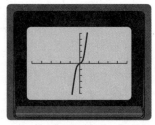

Xscl = 2 Yscl = 2

Since $f(-x) = -f(x)$ for all x, the function $f(x) = x^3 + x$ is odd.   ◄

Graph each relation. State its domain, range, and the quadrants in which it lies.

1. $\{(1, -2), (-3, 4), (-1, -2)\}$

2. $\{(-2, -2), (-2, 2), (2, 2)\}$

Determine whether the relation is a function.

3. $\{(3, 5), (4, 6), (1, 2), (3, 3)\}$

4. $\{(1, 6), (-1, 7), (0, -8)\}$

5. WRITING MATHEMATICS Explain the relationship between functions and relations in your own words.

Write a rule for each function. State the domain and range of the function.

6. $P = \{(2, 4), (3, 5), (-1, 1)\}$

7. $Q = \{(5, 2), (-2, -5), (0, -3)\}$

8. $R = \{(1, -3), (-4, 12), (5, -15)\}$

9. $S = \{(0, 0), (-2, -4), (4, 8)\}$

Use the vertical line test to determine which of the graphs represent functions.

10.

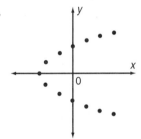

11.

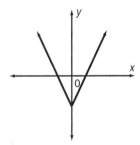

Let $f(x) = x^3 - 3x^2 + 7$. Evaluate.

12. $f(-3)$ **13.** $f(0)$ **14.** $f(5)$

15. COMMUTING The graph at the right shows the annual number of riders on three ferries that cross a river between two states. Make a mapping diagram for the relation, rounding the number of riders each year to the nearest quarter million. Is the relation a function?

16. GEOMETRY The volume of a sphere with radius r is given by $f(r) = \frac{4}{3}\pi r^3$. If $r = 5$ cm, find $f(5)$ and explain what it represents.

Ferry Ridership, in millions

Graph each function and determine whether it appears to be *odd*, *even*, or *neither*. Justify your answer.

17. $f(x) = x^4 - 2x^2$ **18.** $g(x) = x^3 - 2x$ **19.** $h(x) = 2x^2 + 2x$

Determine whether the relation is a function.

1. $\{(4, -7), (1, 6), (7, -7), (8, 1)\}$ **2.** $\{(1, 9), (9, 1), (0, -3), (2, -3)\}$

3. $\{(0, 0), (1, 1), (-2, 0), (1, 0)\}$ **4.** $\{(-2, -2), (3, -2), (1, -2), (3, 0)\}$

Write a rule for each function. State the domain and range of the function.

5. $P = \{(1, 4), (2, 5), (-2, 1)\}$ **6.** $Q = \{(4, 3), (-2, -3), (0, -1)\}$

7. $R = \left\{\left(1, \frac{1}{2}\right), \left(-\frac{1}{2}, -\frac{1}{4}\right), (-2, -1)\right\}$ **8.** $S = \left\{\left(16, 4\right), \left(-1, -\frac{1}{4}\right), \left(2, \frac{1}{2}\right)\right\}$

9. $T = \{(5, 24), (3, 8), (-1, 0)\}$ **10.** $U = \{(2, 5), (-2, -11), (4, 61)\}$

11. WRITING MATHEMATICS Find two graphs in the newspaper or a magazine. Tell whether the graphs represent functions. State the domain and the range. List five ordered pairs that satisfy each.

12. SALES Michelle sells personalized stationery. She makes a 20% commission on each sale. Write a rule T that represents her commission as a function. In which quadrant(s) will this function lie?

Use the vertical line test to determine which of the graphs represent functions.

13.
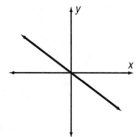

14.
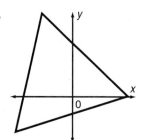

15.
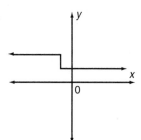

16.
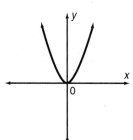

17.
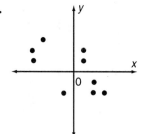

18.
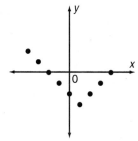

Let $f(x) = 2x^4 - x^2 - 10$. Evaluate.

19. $f(-5)$ **20.** $f(0)$ **21.** $f\left(\frac{1}{2}\right)$ **22.** $3f(6)$

Use a graphing utility to graph each function and determine whether it appears to be *odd*, *even*, or *neither*. Justify your answer.

23. $f(x) = 3x^4 - 2x^2 - 2x$ **24.** $g(x) = 2x^3 + 4x$ **25.** $h(x) = x^6 - 3x^2 - 1$

26. **TECHNOLOGY** Steve knows that his broken Celsius thermometer gives readings that are half the correct temperature for temperatures from −100°C to 100°C. Write a rule that represents the function Steve uses to find the correct temperature. Is the function *even*, *odd*, or *neither*? Explain your answer.

27. **SPORTS** The percentage of games won by the top team in each division of the National Basketball Association in the 1993–1994 season are listed at the right. If Division is the domain, is the relation a function? If Percentage is the domain, is the relation a function?

Division	Percentage
Atlantic	69.5
Central	69.5
Midwest	70.7
Pacific	76.8

Assume that q is a relation that includes the point $(-2, 1)$. What other point must be included in the relation if q is

28. even

29. symmetric about the *x*-axis

30. odd

31. **GEOGRAPHY** The table below lists the mean elevation in feet in several states. Graph the relation with the state as the domain and tell whether it is a function.

State	CT	MA	IA	PA	TX	WA	KS	ID	WY	CO
Elevation	500	500	1100	1100	1700	1700	2000	5000	6700	6800

EXTEND

A function f is said to be a **one-to-one function** if for any two elements in the domain of f, $f(x_1) = f(x_2)$ if and only if $x_1 = x_2$. You can use the **horizontal line test** to determine whether the graph of a function represents a one-to-one function. If no horizontal line can be drawn that intersects the graph in more than one point, then the graph represents a one-to-one function.

Determine whether each is a function and if it is one-to-one.

32.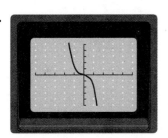

33.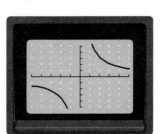

34.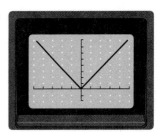

35. **GARDEN AREAS** Suppose 200 ft of fencing is to be used to fence in a rectangular garden. Let x represent the width of the garden.

   a. Write an expression for the length of the garden.
   b. Write an expression for the area A of the garden. Is A a function of x?
   c. If $A = A(x)$, find $A(5)$, $A(32)$, $A(50)$, and $A(95)$.
   d. What restrictions must be placed on the domain so the problem makes sense?
   e. Plot the expression for area A on a graphing calculator. Using ZOOM and TRACE find the maximum area that can be enclosed.
   f. Is A one-to-one?

36. **POSTAL REGULATIONS** If a box with a square cross-section is to be handled by the U.S. Postal Service, the sum of its length and cross-section perimeter cannot exceed 108 in. Let x represent the length (in inches) of each side of the square cross-section.

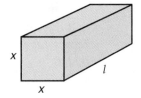

    a. Write an expression for the allowable length.

    b. Express its volume V in terms of x. Is V a function of x?

    c. Find $V(5)$ and $V(20)$.

    d. What restrictions must be placed on x so that the problem makes sense?

    e. Plot V on a graphing calculator. Use ZOOM and TRACE to find the maximum volume and value of x at that maximum volume.

THINK CRITICALLY

37. Explain why a graph that is symmetric about the x-axis cannot be the graph of a function.

38. Let $f(x) = 0.5x$. Does $f(8) + f(10) = f(8 + 10)$?

39. In general, does $f(a + b) = f(a) + f(b)$? Justify your answer.

40. For functions of the form $f(x) = Cx$, where C is a non-negative real number, does $f(a + b) = f(a) + f(b)$? Justify your answer.

41. Consider the maximum area of the rectangular garden in Exercise 35e. Using the same amount of fence how much area could be enclosed if the garden was circular? Round to nearest square foot.

MIXED REVIEW

42. **STANDARDIZED TESTS** Determine the mean of this set of data: 9, 39, 55, 56, 38, 10

    A. 47   **B.** 41.1   **C.** 34.5   **D.** 38.5

Write a rule for each function. State the domain and range of the function.

43. $P = \{(3, -9), (-1, 3), (-5, 15)\}$ 44. $Q = \{(12, 3), (0, 0), (20, 5)\}$

PROJECT *Connection* | Different-sized families have different needs with respect to home size. The "size" of a home is measured by computing the area of the floor space.

1. Use the home advertisements you circled in the Project Connection on page 118. For each home, record the number of square feet. If this information is not given, try to contact the seller or agent.

2. Use a graphing utility to determine a linear model of the form $y = a + bx$ that will allow you to predict the selling price of a home if you know the number of square feet.

3. Select a home that has either a much lower or higher price than your model would predict. Determine whether the home has other advantages or disadvantages to account for the price difference.

3.3 Linear Functions

Explore/Working Together

• Work with a partner.

1. Use a graphing utility to graph $y_1 = x$, $y_2 = x + 3$ and $y_3 = x - 2$ on the same set of axes.

2. Where is the graph of y_2 in relation to the graph of y_1?

3. Where is the graph of y_3 in relation to the graph of y_1?

4. Set your calculator so integers are displayed. Then trace along the graph to complete the table below.

	$x = 0$	$x = 1$	$x = 2$	$x = 3$	$x = 4$	$x = 5$
y_1	0					
y_2	3					
y_3	−2					

5. For a given value of x, what is the relationship between y_2 and y_1?

6. For a given value of x, what is the relationship between y_3 and y_1?

7. Predict where the graphs of $y_4 = x + 10$ and $y_5 = x - 9$ will be located in relation to y_1.

SPOTLIGHT ON LEARNING

WHAT? In this lesson you will learn
• to identify linear functions
• to write linear functions in slope-intercept, point-slope, and standard form.

WHY? Linear functions can help you solve problems involving housing, party costs, and life expectancy.

Build Understanding

• A **linear function** is a function which can be written in the form $f(x) = mx + b$, where m and b are constants. For a particular function such as $f(x) = 2x + 1$, you can input any value of x in the domain and find an output value. The input values are **independent variables**, and the output values are **dependent variables**. Thus, in the linear equation $y = 2x + 1$, the values of x are independent and the values of y are dependent.

THINK BACK

Recall that if (x_1, y_1) and (x_2, y_2) are two points on a line, then the slope of the line is given by $m = \dfrac{y_2 - y_1}{x_2 - x_1}$.

An equation of a line in **slope-intercept form** is written as $y = mx + b$, where m represents the slope and b represents the y-intercept. The line $y = 2x + 1$ is graphed at the right. Since the line crosses the y-axis at $(0, 1)$, its y-intercept is 1.

Use any two points on the line to obtain the values for the slope.

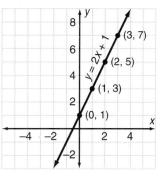

EXAMPLE 1

Write the equation of each line in slope-intercept form.

a.

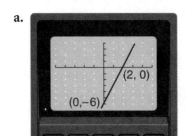

b.

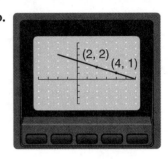

Solution

a. Since the graph crosses the y-axis at $(0, -6)$, its y-intercept is -6. Select any two points on the graph and calculate the slope m

$$m = \frac{y_1 - y_2}{x_1 - x_2} = \frac{0 - (-6)}{2 - 0} = \frac{6}{2} = 3$$

Thus, an equation of the line is $y = 3x - 6$.

b. Since the graph crosses the y-axis at $(0, 3)$, its y-intercept is 3. Select any two points on the graph and calculate the slope $m = \frac{-1}{2}$. Thus, an equation of the line is

$$y = -\frac{1}{2}x + 3$$

If you are given one point and the slope of a line or two points on a line, you can write the equation of the line in **point-slope form**. The point-slope of an equation of a line is

$$y - y_1 = m(x - x_1)$$

where m is the slope and (x_1, y_1) are the coordinates of a given point on the line.

EXAMPLE 2

Write the equation of the line that passes through the points $(1, 4)$ and $(3, -8)$ in point-slope form.

Solution

$$m = \frac{-8 - 4}{3 - 1} = \frac{-12}{2} = -6 \qquad \text{Find the slope.}$$

$$\begin{aligned} y - y_1 &= m(x - x_1) & \text{Use the point-slope form.} \\ y - 4 &= -6(x - 1) & m = -6; (x_1, y_1) = (1, 4) \end{aligned}$$

You can also express the equation of a line in **standard form**: $Ax + By = C$, where A, B, and C are integers and A and B are not both zero. The standard form of the equation of the line in Example 2 is $6x + y = 10$.

Two distinct lines are **parallel** if and only if their slopes are equal or are both undefined.

In Explore, you graphed $y_1 = x$, $y_2 = x + 3$ and $y_3 = x - 2$ on the same set of axes. Since the three graphs have the same slope but different y-intercepts, they are parallel. For each x-value, the y-values of y_2 are 3 more than the y-values of y_1 which causes the graph of y_2 to be shifted 3 units up. Similarly, for each x-value, the y-values of y_3 are 2 less than the y-values of y_1 which shifts the graph 2 units down. The graphs of y_2 and y_3 are **vertical translations** of the graph of y_1.

EXAMPLE 3

Without graphing, tell if each pair of equations have graphs that are vertical translations of each other. In which direction is the second graph shifted from the first graph?

a. $2x + 4y = 12$, $x = -2y + 4$ **b.** $x - 3y = 3$, $3x + y = -6$

Solution

a.
$$\begin{aligned} 2x + 4y &= 12 & x &= -2y + 4 \\ 4y &= -2x + 12 & 2y &= -x + 4 \\ y &= -\frac{1}{2}x + 3 & y &= -\frac{1}{2}x + 2 \end{aligned}$$
Write the equations in slope-intercept form.

Since the equations both have equal slopes, they are parallel lines and vertical translations of each other. The graph of the second equation is a vertical shift of the first graph by 1 unit downward.

b.
$$\begin{aligned} x - 3y &= 3 & 3x + y &= -6 \\ -3y &= -x + 3 & y &= -3x - 6 \\ y &= \frac{1}{3}x - 1 \end{aligned}$$
Write the equations in slope-intercept form.

Since the two equations have different slopes, they are not vertical translations of each other.   ◀

Two nonvertical lines are **perpendicular** if and only if their slopes are negative reciprocals of each other. Also, a vertical line and a horizontal line are perpendicular. For example, the lines in Example 3b have slopes of $\frac{1}{3}$ and -3. Since $\frac{1}{3}(-3) = -1$, the lines are perpendicular.

You can determine an equation of a line that is parallel or perpendicular to another line if you know its y-intercept.

EXAMPLE 4

Given the line $y = -5x + 10$, determine an equation of a line

a. parallel to the given line with y-intercept of 4

b. perpendicular to the given line with y-intercept of -2

CHECK UNDERSTANDING

In Example 3a, the description of the shift depends on which equation you choose as the reference line. Suppose you use $x = -2y + 4$ as the reference line. How will you describe the shift?

THINK BACK

Two numbers are negative reciprocals if their product is -1.

Solution

a. Since the slope of a parallel line is -5 and the y-intercept is 4, its equation is $y = -5x + 4$.

b. Since the product of the slope of the perpendicular line and -5 must equal -1, the slope of the perpendicular line is $\frac{1}{5}$. The y-intercept is -2, so the equation is $y = \frac{1}{5}x - 2$. ◄

You can use a graphing utility to find an equation of a line in slope-intercept form that models a given set of data. This is called **linear regression**. Your graphing utility will give you the equation of the line of best fit for your data. The slope and y-intercept are also given. The coefficient of correlation r, which is a measure of how well the line fits the data, will also be displayed.

THINK BACK

The closer the absolute value of r, the correlation coefficient is to 1, the better the correlation.

EXAMPLE 5

HOUSING The Garcia family is looking for a new home. They collected the information at the right about houses that are available in town.

Price (in thousands of dollars)	Area (in square feet)
325	4400
89	2250
101.8	2950
348	4800
219.9	3300
269.5	3800
125.8	2800
219.5	3500

a. Use a graphing utility to determine a linear model to use to predict the number of square feet in a house if they know the selling price. Round to the nearest hundredth.

b. Is there a strong correlation between price and area?

c. Use the model to predict the area of a house that is priced at $185,000. Round to the nearest square foot.

Solution

a. When you enter the prices as x-values and the areas as y-values, a graphing utility displays $y = 8.20x + 1734.30$ as the linear equation that best relates the price and the number of square feet.

b. Since the value of the correlation coefficient, 0.97, is close to 1, there is an excellent correlation between the price and the number of square feet in a house.

c. $y = 8.20x + 1734.30$
$= 8.20(185) + 1734.30$
$= 3251.30$

The area of the house should be approximately 3251 square feet. ◄

Lin Reg
y = ax + b
a = 8.198747824
b = 1734.303353
r = .9678085725

Write the equation of each line in slope-intercept form.

1.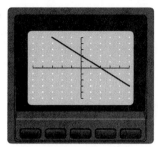

2.

3.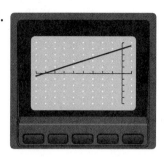

Write an equation of a line in point-slope, slope-intercept, and standard form that has slope *m* and passes through the given point.

4. $m = -2; (-1, 4)$

5. $m = 5; (3, 6)$

6. $m = -\frac{1}{4}; (4, -12)$

Write an equation of a line in slope-intercept form that passes through the given points.

7. $(-8, 5); (0, 0)$

8. $(6, 4); (-2, 8)$

9. $(-1, -5); (-6, 10)$

10. Write an equation of a line in standard form that passes through the points $(3, 6)$ and $(-4, 9)$.

11. WRITING MATHEMATICS There are three ways of writing linear functions: slope-intercept, point-slope and standard form. Write an example of each form for the same linear function.

12. PARTY COSTS A catering company will provide food for a party for $15 per guest plus a fixed charge of $200.

    a. Write a linear equation representing the relationship between the number of guests *x* and the total cost *y*.

    b. What is the slope of the line? What does it represent?

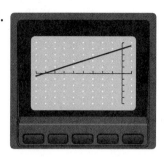

Given the line $y = 3x - 10$, determine an equation of a line.

13. parallel to the given line with *y*-intercept of –2.

14. perpendicular to the line with *y*-intercept of 6.

Tell whether the graph of the second equation is a vertical translation of the graph of the first equation. If so, in which direction and by how many units must the first graph be translated to obtain the graph of the second?

15. $-6x + 12y = 6, 8y = 4x - 8$

16. $5y - 10x = 25, -24 = 6x - 3y$

17. **WRITING MATHEMATICS** You know the y-intercept of a line is 5. Explain if this is enough information to determine the equation of a unique line. If it is, show how you find the equation. If it is not, make up the additional information you need and then show how to find the equation.

18. **LIFE EXPECTANCY** The table shows the life expectancy in years for people of a given age in 1991. Use a graphing utility to determine an equation that expresses life expectancy as a function of age in 1991. Is there a strong correlation between age and life expectancy?

Age	0	10	20	30	40	50	60	70	80
Life Expectancy	75.5	66.4	56.8	47.4	38.1	29.2	21.0	14.0	8.4

19. Use the equation you found in Exercise 18. What is a person's life expectancy at 35? at 45? at 55? Round to the nearest tenth.

PRACTICE

Determine the slope of the line that passes through each set of points.

1. $(2, 3), (5, 6)$ **2.** $(4, 6), (3, 7)$ **3.** $(7, -5), (2, -6)$ **4.** $(2, -4), (-6, 0)$

Write an equation of a line in point-slope, slope-intercept, and standard form that has slope m and passes through the given point.

5. $m = -3, (-2, 3)$ **6.** $m = 2, (1, 7)$ **7.** $m = -\frac{1}{2}, (6, -1)$

Write an equation of a line in standard form that passes through the given points.

8. $(6, 2), (9, 10)$ **9.** $(5, -2), (10, -4)$ **10.** $(-3, 9), (12, 6)$

11. **BUSINESS** Market research shows that the demand curve for a new computer game is linear. There is a demand of 300 units at \$40 and 400 units at \$30. Write the linear demand curve for this product where demand is a function of price and determine the number of games that will sell at a price of \$20.

Given the line $y = \frac{1}{4}x + 6$, determine an equation of a line

12. parallel to the given line with y-intercept of 0

13. perpendicular to the line with y-intercept of -7

Tell whether the graph of the second equation is a vertical translation of the graph of the first equation. If so, in which direction and by how many units must the first graph be translated to obtain the graph of the second?

14. $6x + 3y = 15, -7y = 14x - 49$ **15.** $y - 7 = 2x, 4y = -2x + 12$

16. $y + 5 = 4x, y + 4 = 5x$ **17.** $6y = 9x + 6, 10y = 15x$

18. **WRITING MATHEMATICS** Explain the concept of a vertical translation in your own words. Provide an original example of two lines that are vertical translations of each other.

19. MORTGAGES The first row of the chart below lists the approximate first-year rates on 1-year adjustable mortgages on the last day of each month. The second row lists the percentage of new mortgages that have adjustable interest rates. Determine a linear model to use to predict the percentage of new mortgages that have adjustable rates if you know the first-year interest rate. Round to the nearest hundredth. Is there a strong correlation?

	June	July	Aug.	Sept.	Oct.	Nov.	Dec.	Jan.	Feb.	Mar.
Mortgage Rate	5.3	5.4	5.4	5.7	5.9	6.4	6.8	6.7	6.4	6.3
Percent of Mortgages	43	41	43	46	49	55	55	59	53	44

EXTEND

20. GEOMETRY Demonstrate that a triangle with vertices at $A(2, 1)$, $B(4, 5)$, and $C(-2, 3)$ is a right triangle.

21. GEOMETRY Demonstrate that a triangle with vertices at $A(-1, -2)$, $B(-5, 2)$, and $C(4, 2)$ is not a right triangle.

22. GEOMETRY Demonstrate that the quadrilateral with vertices at $A(4, 5)$, $B(2, 6)$, $C(0, 2)$ and $D(2, 1)$ is a rectangle.

23. GEOMETRY Demonstrate that the quadrilateral with vertices at $A(2, 2)$, $B(5, 5)$, $C(4, 7)$ and $D(1, 4)$ is a parallelogram.

24. GEOMETRY Show that the quadrilateral in Exercise 28 is not a rectangle.

THINK CRITICALLY

25. The equation of a line is $Ax + By = C$ where A, B, and C are not equal to 0. Express each of the following in terms of A, B, and/or C.

    a. the slope   **b.** the y-intercept   **c.** the x-intercept

26. Write the intercept form of the equation of a line containing the point $(3, 2)$ and $(5, 8)$.

27. Write an equation of a line in standard form that passes through the origin and is perpendicular to the line $Ax + By = C$.

28. Find k such that the y-intercept of the line with equation $3x + 2ky + 9 = 0$ is -6.

MIXED REVIEW

Use the monthly payment formula to determine the monthly payment and deferred payment price for each car loan.

29. $10,000 at 8% for 4 years   **30.** $8950 at 9% for 3 years

31. STANDARDIZED TESTS Determine the probability of drawing a red 10 from a standard deck of 52 playing cards. Round to three decimal places.

    A. 0.384   **B.** 0.192   **C.** 0.038   **D.** 0.019

THINK BACK

Recall that two sentences joined by the word *or* form a disjunction. Two sentences joined by the word *and* form a conjunction.

Think Back

Recall that $|x|$, the absolute value of a number x, is the distance between x and 0 on a number line. To solve an absolute value equation, set up a disjunction and solve both parts for x.

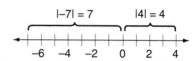

1. Express each absolute value equation as a disjunction.

   a. $|4x| = 10$ **b.** $|2x + 7| = 5$ **c.** $|3x - 1| = 11$

2. Solve each equation.

   a. $|3x| = 15$ **b.** $|4 - x| = 3$ **c.** $|2x - 3| = 7$

Recall that a figure is symmetric about a line if the figure can be folded along that line so the two parts coincide.

3. Use graph paper, draw the graph of $f(x) = x$ in Quadrant I only. Extend it to produce a graph symmetric about the x-axis.

4. Repeat Question 3, but extend the graph so it is symmetric about the y-axis.

Explore/Working Together

5. Work with a partner. Use a graphing utility to graph the functions $f(x) = x$ and $g(x) = |x|$ on the same coordinate plane.

6. Discuss the shapes of f and g. What letter of the alphabet best describes the shape of g?

7. At what point is the vertex of graph g located?

8. In which direction does the graph of g open?

9. Discuss the symmetries of each graph.

10. Why is the graph of g located only in Quadrants I and II?

11. Over what line can you "flip" the Quadrant III portion of the graph of f to obtain the Quadrant II portion of the graph of g?

The "flip" in Question 11 is called a **reflection**, and the line over which the graph is "flipped" is called the **line of reflection**. A reflection produces a mirror image of a figure.

12. If h is the reflection of $f(x) = x$ over the x-axis, then $h(x) = $ 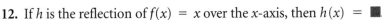.

A **translation** of a graph is a shift that produces the same graph in a new position. In Lesson 3.3, you saw that a **vertical shift** moves a graph upward or downward.

13. Graph $f(x) = |x|$, $g(x) = |x| + 4$, and $h(x) = |x| - 2$ on the same set of axes.

14. Determine the vertex of each function.

15. Compare the lines of symmetry and shape of the graphs above.

16. Where is the graph of g in relation to that of f?

17. Where is the graph of h in relation to that of f?

18. Graph several functions of the form $j(x) = |x| + c$, where c is a positive number. If a vertical translation of $f(x) = |x|$ is written as $j(x) = |x| + c$, write a general statement about the location of the graph of j in relation to the graph of f when c is a positive or negative number.

A **horizontal shift** moves a graph right or left.

19. Graph $f(x) = |x|$, $g(x) = |x + 3|$, and $h(x) = |x - 1|$ on the same set of axes.

20. Determine the vertex of each function in Question 19.

21. Compare the lines of symmetry and shapes of the graphs.

22. Where is the graph of g in relation to the graph of f?

23. Where is the graph of h in relation to the graph of f?

24. Graph several functions of the form $j(x) = |x - b|$, where b is a positive number and where b is a negative number. If a horizontal translation of $f(x) = |x|$ is written as $j(x) = |x - b|$, write a general statement about the location of the graph of j in relation to the graph of f when b is a positive number or negative number.

The graph of the function $g(x) = |x - b| + c$ will be shifted horizontally and vertically from the graph of $f(x) = |x|$.

25. Graph $g(x) = |x + 1| - 4$ on the same set of axes as $f(x) = |x|$. How is the graph of g similar to the graph of f? How is it different? Where is its vertex? To what line is it symmetric?

26. Graph a function of the form $g(x) = |x - b| + c$. Have your partner identify the function that you entered by studying the graph. Then switch roles with your partner.

Reflections and translations are transformations that produce graphs with the same size and shape as the original graph. **Dilations** produce graphs with slopes greater or less than those of the original graph.

27. Graph $f(x) = |x|$, $g(x) = 3|x|$, and $h(x) = \frac{1}{2}|x|$ on the same set of axes. Compare the graphs of g and h to the graph of f.

28. Graph other functions of the form $j(x) = a|x|$, where a is a positive real number. Use words "narrower" or "wider" to complete the statements about $j(x)$ compared to $f(x) = |x|$.

    a. If $a > 1$, the V-shaped graph is _____.
    b. If $0 < a < 1$, the V-shaped graph is _____.

29. Graph $f(x) = |x|$ and $g(x) = -|x|$ on the same set of axes. Is g a reflection, translation, or dilation of f? Explain.

30. How are the graphs of $f(x) = a|x|$ and $g(x) = -a|x|$ related?

You can determine the x-coordinate of the vertex of the graph of $g(x) = a|x - b| + c$ by setting $x - b$ equal to 0 and solving for b.

31. Find the x-coordinate of the vertex of the graph of $g(x) = 5|x - 3| + 2$.

32. How can you find the y-coordinate of the vertex?

33. How can you use the coordinates of the vertex to determine the equation of the line of symmetry?

34. Determine the vertex and the line of symmetry for the graph of each function.

    a. $f(x) = |x| + 6$
    b. $g(x) = |x - 8|$
    c. $h(x) = -|x - 1| + 4$

Make Connections

● Answer Exercises 35–36 without plotting the graphs.

35. Explain how each graph differs from the graph of $f(x) = |x|$.

    a. $g(x) = \frac{3}{5}|x|$ b. $h(x) = -3|x|$ c. $j(x) = 5.2|x|$

36. State the number of units and the direction in which the graph of each function is translated from the graph of $f(x) = |x|$.

    a. $g(x) = |x + 4| + 2$
    b. $h(x) = |x - 3| - 1$
    c. $i(x) = |x - 3|$

37. Match each function with its graph.

    a. $f(x) = |x + 3|$ **b.** $g(x) = |x| + 3$ **c.** $h(x) = |x - 2| + 1$

I.
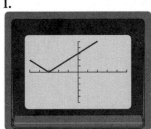

II.
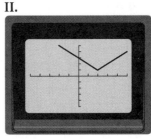

III.
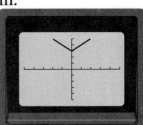

38. Match each function with its graph.

    a. $f(x) = \frac{1}{9}|x|$ **b.** $g(x) = 9|x|$ **c.** $h(x) = \frac{8}{9}|x|$

I.

II.
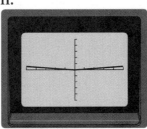

III.
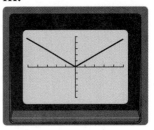

39. Write a function $g(x) = |x - b| + c$ whose graph represents

    a. a translation of 2 units down and 5 units left from the graph of $f(x) = |x|$

    b. a translation of 3 units up and 7 units right from the graph of $f(x) = |x|$

    c. a translation of 1 unit down and 6 units right from the graph of $f(x) = |x|$

    d. a translation of 4 units up and 8 units left from the graph of $f(x) = |x|$

40. Write a function of the form $f(x) = |x - b| + c$ that is represented by each graph. Assume Xscl and Yscl are 1.

a.
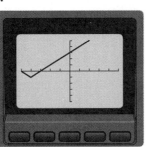

b.

c.
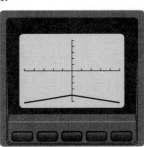

41. Use a graphing utility to graph $y = |4 - x|$ and $y = 3$ on the same set of axes. Find their points of intersection.

Algebra Workshop

Summarize

42. WRITING MATHEMATICS Explain how you can use a graphing utility to solve an absolute value equation with two variables.

43. WRITING MATHEMATICS If you are shown the graph of an absolute value function $y = |x - a| + b$ explain the steps you would use to determine the vertex and line of symmetry.

44. THINKING CRITICALLY The graph of $f(x) = |3x|$ is the same as the graph of $g(x) = 3|x|$. Is it also true that the graph of $h(x) = |-3x|$ is the same as the graph of $j(x) = -3|x|$? Explain.

45. THINKING CRITICALLY What portion of the graph of $f(x) = -x$ must be reflected to obtain the graph of $g(x) = |-x|$? What is the line of reflection?

46. MODELING Draw the graph of an absolute value equation that has only one solution. Do not use $x = 0$ as that solution.

47. GOING FURTHER A function defined differently over various parts of its domain is called a **piecewise function**. Graph each function.

a. $f(x) = \begin{cases} x^2, & x \le 3 \\ x, & x > 3 \end{cases}$ **b.** $g(x) = \begin{cases} 2x + 1, & x < -2 \\ 4, & x \ge -2 \end{cases}$

48. GOING FURTHER Express $f(x) = |x|$ as a piecewise function.

PROJECT *Connection*   Before purchasing a home, potential buyers use estimation to predict how much the home will cost per year and per month.

1. Find out about the procedures and responsibilities involved in getting a *mortgage*. Get data on monthly payments (or use the Monthly Payment Formula from Lesson 1.1) and down payments required to purchase a home in your group's price range. Find out what *points* and *closing costs* are and how they affect the cost of buying a home. Make one or more tables to compare and summarize your data.

2. Local governments collect *property taxes* from home owners to raise money to run the country. Contact the town hall or a real estate agent to find out property tax rates and to get an estimate of the annual property tax bill for the home you selected.

3. Find out how much homeowner's insurance costs, what it covers, and the payment schedule.

4. For utilities such as telephone, gas, oil, electric, and water, find an average and a range of each utility bill for your home. If the billing cycle is not monthly, convert the estimate into a monthly average and estimate a total monthly utility cost for your home.

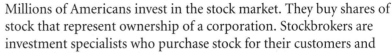

Millions of Americans invest in the stock market. They buy shares of stock that represent ownership of a corporation. Stockbrokers are investment specialists who purchase stock for their customers and provide investment advice. You cannot walk into a stock market and purchase stock. You can only purchase it through a broker.

Stockbrokers are paid a commission when they buy or when they sell shares of stock for customers. Commission rates vary, but they are often based on the value of the shares traded.

The commission function C, given in table form below, shows one stockbroker's fee schedule. Since it is defined differently over various parts of its domain, it is a **piecewise function**. It includes two flat fees for smaller transactions because the commission rate applied to transactions under $1000 is too small to make handling the transaction worthwhile for the broker.

Amount of Transaction (x)	Commission $C(x)$
$0 < x \le \$250$	$25
$\$250 < x \le \$1000$	$35
$x > \$1000$	2% of transaction value

1. Sketch the graph of the stockbroker's fee schedule.

2. Write an equation for the part of the graph that represents commissions on transactions over $1000. State its domain and range.

3. How much would a customer have to invest at a 2% commission rate for the commission to be equivalent to the flat rate on a $500 investment?

4. A stockbroker on this fee schedule makes the same amount of money if a client buys stock for $2000 and sells it for $1200 or buys it for $1200 and sells it for $2000. What is the value of this commission? Why would the broker prefer the client to make money?

Think Back

1. What line or point are even functions symmetric about?

2. What line or point are odd functions symmetric about?

Draw a graph that is

3. symmetric about the *x*-axis.

4. symmetric about the *y*-axis.

5. symmetric about the origin.

Explore/Working Together

You can predict the shape and location of many graphs by using some basic graphs as references. Work in groups of four.

6. Graph $f(x) = x^2$ and $g(x) = -x^2$ on the same set of axes.

7. Describe graph. Compare each graph's shape, vertex, and symmetries.

8. How can you reflect the graph of *f* to obtain the graph of *g*?

9. Have each member graph one pair of functions.

    a. $f(x) = x^3; -f(x) = -x^3$
    b. $f(x) = x^4; -f(x) = -x^4$
    c. $f(x) = x; -f(x) = -x$
    d. $f(x) = |x|; -f(x) = -|x|$

10. What is the relationship between the graphs of $f(x)$ and $-f(x)$?

Next, use the graph of $f(x) = x^2$ to investigate another change.

11. Use a graphing utility to graph $f(x) = x^2, g(x) = x^2 + 3$, and $h(x) = x^2 - 1$ on the same set of axes.

12. Find the vertex of the graph of each function.

13. Compare their lines of symmetry and shapes.

14. Where is the graph of *g* in relation to the graph of *f*?

15. Where is the graph of *h* in relation to the graph of *f*?

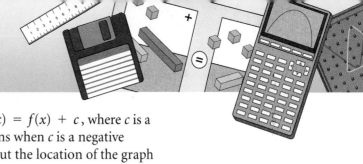

16. Graph several functions of the form $j(x) = f(x) + c$, where c is a positive number. Then see what happens when c is a negative number. Write a general statement about the location of the graph of j in relation to the graph of f.

Recall that another type of change causes a horizontal shift.

17. Use a graphing utility to graph $f(x) = x^3$, $g(x) = (x + 2)^3$, and $h(x) = (x - 4)^3$ on the same set of axes.

18. Find the point about which each graph is symmetric.

19. Where is the graph of g in relation to the graph of f?

20. Where is the graph of h in relation to the graph of f?

21. Graph several functions of the form $j(x) = f(x - b)$ where b is a positive number. Then see what happens when b is a negative number. Write a general statement about the location of the graph of j in relation to the graph of f.

22. Graph $g(x) = (x - 5)^2 - \dfrac{1}{2}$ on the same set of axes as $f(x) = x^2$. How is the graph of g similar to the graph of f? How is it different? What is its vertex? What line is g symmetric about?

23. Work in pairs. Graph a function of the form $g(x) = (x - b)^2 + c$. Have your partner identify the function that you entered by studying the graph. Then switch roles with your partner.

24. Work with a partner. Graph a function of the form $g(x) = (x - b)^3 + c$. Have your partner identify the function that you entered by studying the graph. Then switch roles with your partner.

You can observe patterns in the shape of graphs of even-degree functions and odd-degree functions.

25. Graph $f(x) = x^4$, $g(x) = 2x^4$, $h(x) = 3x^4$, and $j(x) = \dfrac{1}{2}x^4$ on the same set of axes. Tell how the graphs of g, h, and j differ from the graph of f.

26. In Question 25, you graphed several functions of the form $k(x) = af(x)$, where $a > 0$. If a dilation of f is written as $af(x)$, write a two-part general statement about the shape of the graph of k in relation to the graph of f for the case in which $a > 1$ and the case in which $0 < a < 1$.

27. Graph several functions of the form $f(x) = x^n$ where n is a positive even integer.

28. What is the general shape of these functions? Which line or point are they symmetrical about?

29. In which quadrants do the graphs of these functions lie?

30. Why are there no values of x for which $f(x)$ is negative?

31. When $x < 0$ are the y-values positive or negative? When $x > 0$ are the y-values positive or negative?

32. Graph several functions of the form $f(x) = x^n$ where n is a positive odd integer.

33. What is the general shape of these functions? Which line or point are they symmetrical about?

34. In which quadrants do the graphs of these functions lie?

35. When $x < 0$ are the y-values positive or negative? When $x > 0$ are the y-values positive or negative?

Make Connections

By applying the results of your explorations above, you can determine the equation of a function given some characteristics of its graph or make predictions about the appearance of the graph given its equation.

36. Write a function g that is the reflection of f about the x-axis.

    a. $f(x) = x^5$ **b.** $f(x) = -|x|$

    c. $f(x) = x - 1$ **d.** $f(x) = -x + 3$

    e. $f(x) = \sqrt{x}$ **f.** $f(x) = \dfrac{1}{x}$

37. State the direction(s) in which the graph of each function is translated from the graph of $f(x) = x^5$.

    a. $g(x) = (x - 2)^5 + 1$

    b. $h(x) = (x + 6)^5 - 3$

    c. $j(x) = \left(x + \dfrac{1}{2}\right)^5 + \dfrac{1}{4}$

38. Explain how each graph differs from the graph of $f(x) = x^2$.

    a. $g(x) = \dfrac{2}{3}x^2$ **b.** $h(x) = -2x^2$ **c.** $j(x) = 4x^2$

39. Match each function with its graph.

    a. $f(x) = (x - 1)^4$ **b.** $g(x) = x^4 + 1$ **c.** $h(x) = (x + 2)^3 - 3$

I.

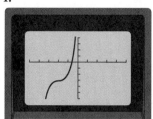

II.

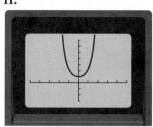

III.

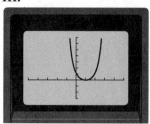

40. Match each function with its graph. Assume Xscl and Yscl are 1.

    a. $f(x) = 0.9x^2$ **b.** $g(x) = 0.1x^2$ **c.** $h(x) = 5x^2$

I.

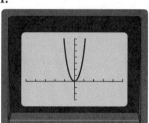

II.

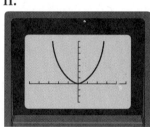

III.

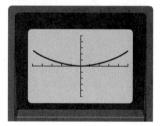

41. Match each function with its graph.

    a. $f(x) = x^3$ **b.** $g(x) = -x^5$ **c.** $h(x) = x^{15}$

I.

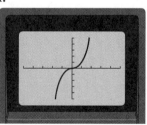

II.

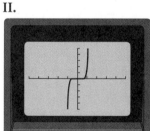

III.

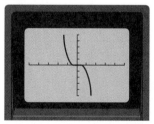

42. Write a function of the form $g(x) = (x - b)^{11} + c$ whose graph is a translation of

    a. 1 unit up and 8 units left from the graph of $f(x) = x^{11}$

    b. 4 units down and 5 units right from the graph of $f(x) = x^{11}$

Algebra Workshop

Summarize

43. **WRITING MATHEMATICS** Compare the effects that a, b, and c have on the graph of a function of the form $f(x) = a(x - b)^n + c$ with the effects that a, b, and c have on the graph of a function of the form $g(x) = a|x - b| + c$.

44. **THINKING CRITICALLY** Will the graphs of $f(x) = 3x^5 - 1$ and $g(x) = 3x^5$ ever intersect? Explain your answer.

45. **THINKING CRITICALLY** About what line is the graph of $g(x) = f(-x)$ a reflection of the graph of $f(x)$?

46. **MODELING** John launches a model rocket. Its height in feet is a function of time $h(t) = 200t - 16t^2$ where $t = 0$ at the time of launch and is measured in seconds.

   a. Graph this function. When will John's rocket hit the ground?

   b. Amy launches her rocket 3 seconds after John. Graph the height of Amy's rocket on the same axis as John's rocket using $j(t) = 200(t - 3) - 16(t - 3)^2$. Use the TRACE feature and find the height of Amy's rocket when John's hits the ground.

47. **GOING FURTHER** Graph each function. State the number of times that the graph of each function intersects the x-axis.

   a. $f(x) = (x + 3)^6$ b. $f(x) = x^2 + 2$

   c. $f(x) = (x - 2)^4 - 1$ d. $f(x) = -x^5 - 2$

   e. $f(x) = -x^9 + 3$ f. $f(x) = -(x - 3)^{11}$

48. **GOING FURTHER** Use the information from Question 47 to draw a conclusion about the maximum and minimum number of times that the graph of a function of the form $f(x) = a(x - b)^n + c$ where n is a positive even number can intersect the x-axis.

49. **GOING FURTHER** Graph several functions of the form $f(x) = a(x - b)^n + c$ where n is a positive odd number. What are the maximum and minimum number of times that a graph of this form can cross the x-axis?

50. **WRITING MATHEMATICS** Examine your results in Question 48 and 49. Compare the number of times a function of the form $f(x) = a(x - b)^n + c$ with an even exponent n intersects the x-axis to the number of times a function of the same form with an odd exponent intersects the x-axis.

3.6 Inverse Functions

Explore/Working Together

● Work with a partner. Use graph paper.

1. Make a list of ordered pairs for the function $f(x) = 2x + 8$. Your partner should make a list of ordered pairs for the function $g(x) = \dfrac{x - 8}{2}$.

2. Compile a list of the ordered pairs contributed by all teams in the class. Analyze the sets of points found. What pattern do you notice?

3. What operations does each function perform on an input value?

4. Rewrite g in slope-intercept form. Graph f and g on the same set of axes. Fold your graph along the line $y = x$. Where are f and g in relation to each other?

Build Understanding

● The **inverse** of a relation can be found by interchanging the first and second coordinates of each ordered pair in the relation. The domain of a relation is the range of its inverse relation, and the range of a relation is the domain of its inverse relation.

EXAMPLE 1

Let $F = \{(-2, 5), (-1, 4), (0, 3), (1, 4), (2, 5)\}$. Find the inverse of F.

Solution

Let G be the inverse of F. Find G by switching the first and second coordinates of each ordered pair in F.

$$G = \{(5, -2), (4, -1), (3, 0), (4, 1), (5, 2)\}$$

The graph of an inverse relation is a reflection of the original relation over the line $y = x$. The graphs of F and G are symmetric about the line $y = x$.

The inverse of every relation is a relation. However, the inverse of a function may or may not be a function.

In this case F is a function and G, the inverse of F, is not a function since 4 is paired with 1 and –1.

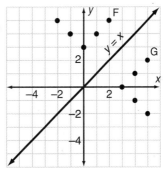

PROBLEM SOLVING TIP

Make a mapping to show if a relation is a function. Make a second mapping to show if its inverse is a function.

To find the inverse of a relation that is represented by an equation, interchange x and y in the equation. Then solve for y.

EXAMPLE 2

Find the inverse of the relation $y = 2x + 4$. Graph the relation, its inverse, and the line $y = x$ on the same set of axes.

Solution

$$y = 2x + 4$$
$$x = 2y + 4 \quad \text{Interchange } x \text{ and } y.$$
$$y = \frac{1}{2}x - 2 \quad \text{Solve for } y. \quad \blacktriangleleft$$

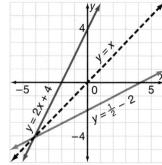

You can use the **horizontal line test** to determine whether or not a function f has an inverse that is also a function. If any horizontal line drawn on the graph of f passes through no more than one point on the graph of f, then the inverse of f is also a function. You can also apply the vertical line test to the graph of the inverse to determine whether it is a function.

If the inverse of a function f is a function, it is called the **inverse function** and is written f^{-1}. Note that the superscript is not an exponent.

EXAMPLE 3

Find the inverse of each function. Graph the function, its inverse, and the line $y = x$ on the same set of axes. Is the inverse a function?

a. $f(x) = 4x - 3$ **b.** $g(x) = x^2 - 1$

Solution

a. $f(x) = 4x - 3$
$$y = 4x - 3 \quad \text{Substitute } y \text{ for } f(x).$$
$$x = 4y - 3 \quad \text{Interchange } x \text{ and } y.$$
$$y = \frac{1}{4}x + \frac{3}{4} \quad \text{Solve for } y.$$

The inverse is a function:
$$f^{-1}(x) = \frac{1}{4}x + \frac{3}{4}$$

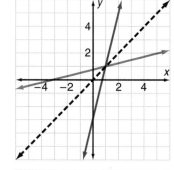

b. $g(x) = x^2 - 1$
$$y = x^2 - 1 \quad \text{Substitute } y \text{ for } f(x).$$
$$x = y^2 - 1 \quad \text{Interchange } x \text{ and } y.$$
$$x + 1 = y^2$$
$$y = \pm\sqrt{x + 1} \quad \text{Solve for } y.$$

For each value of x, there are two values of y, so the inverse is not a function. $\blacktriangleleft$

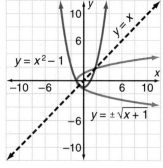

Inverse functions can be used to describe real world phenomena.

EXAMPLE 4

AREA The formula that allows you to convert square feet to acres is
$f(x) = \dfrac{x}{43{,}560}$ where x represents the number of square feet.

a. If the Kakutani family owns 80,000 square feet of property, how many acres do they own, to the nearest hundredth of an acre?

b. Determine the inverse function f^{-1}.

c. What does f^{-1} represent?

Solution

a. $f(80{,}000) = 1.84$ acres

b. $f(x) = \dfrac{x}{43{,}560}$

$y = \dfrac{x}{43{,}560}$ Substitute y for $f(x)$.

$x = \dfrac{y}{43{,}560}$ Interchange x and y.

$y = 43{,}560x$ Solve for y.

$f^{-1}(x) = 43{,}560x$

c. The inverse function $f^{-1}(x) = 43{,}560x$ converts acres to square feet, where x is the number of acres.

TRY THESE

Find the inverse of each relation or function.

1. $\{(1, 2), (3, 4), (5, 6), (7, 8)\}$

2. $\{(2, 4), (3, 6), (4, 8), (5, 10)\}$

3. $\{(2, 5), (3, 7), (-4, 5), (2, 9)\}$

4. $\{(3, -1), (4, -2), (3, 3), (6, -3)\}$

Write an equation for the inverse of each relation.

5. $y = x - 5$

6. $y = x + 4$

7. $y = -\dfrac{1}{2}x + 2$

8. $y = -\dfrac{1}{4}x + 4$

9. **WRITING MATHEMATICS** In your own words, explain how to find the inverse of a relation that is expressed by an equation. Provide an example.

Find f^{-1}. Graph f, f^{-1}, and the line $y = x$ on the same set of axes.

10. $f(x) = 5x - 8$ **11.** $f(x) = 3x + 7$ **12.** $f(x) = -2x + 6$ **13.** $f(x) = -3x + 3$

Graph each function f. Determine whether the inverse is also a function.

14. $f(x) = x^3$ **15.** $f(x) = x^5$ **16.** $f(x) = x^4$ **17.** $f(x) = x^6$

Match the graph of each function with the graph of its inverse.

18. 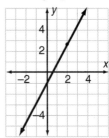 **19.** 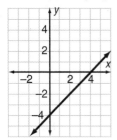 **20.** 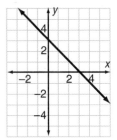 **21.**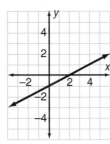

a. 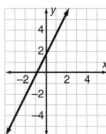 **b.** 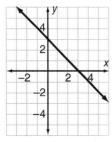 **c.** 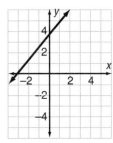 **d.**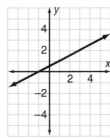

22. CRYPTOGRAPHY A simple cipher codes a number x using the function $f(x) = 4x - 9$. Find the inverse of this function and determine whether the inverse is also a function.

PRACTICE

Find the inverse of each relation or function.

1. $\{(4, 1), (5, 0), (6, 6), (7, 3)\}$

2. $\{(4, 3), (6, 5), (8, 7), (10, 9)\}$

3. $\{(-1, 4), (-3, 2), (-1, 1), (2, 2)\}$

4. $\{(4, -4), (5, -1), (4, 3), (3, -3)\}$

Write an equation for the inverse of each relation.

5. $y = x + 2$ **6.** $y = x - 7$ **7.** $y = -x + 2$ **8.** $y = -x - 8$

9. $y = 8x - 4$ **10.** $y = 6x - 3$ **11.** $y = 7x + 15$ **12.** $y = 3x - 10$

13. WRITING MATHEMATICS Explain two different methods for determining whether the inverse of a function is also a function. Use appropriate examples to illustrate your explanation. If you prefer one method, tell why.

14. TEMPERATURE SCALES The formula for converting temperatures from Fahrenheit to Celsius is $y = \frac{5}{9}(x - 32)$. Find the inverse of this function and determine whether it is also a function.

Find f^{-1}. Graph f, f^{-1}, and the line $y = x$ on the same set of axes.

15. $f(x) = 2x - 12$ **16.** $f(x) = 4x + 8$ **17.** $f(x) = -3x + 6$ **18.** $f(x) = -2x + 10$

Graph each function _f_. Use the horizontal line test to determine whether the inverse is also a function.

19. $f(x) = \frac{1}{2}x^8$ **20.** $f(x) = \frac{1}{4}x^{10}$ **21.** $f(x) = -x^7$ **22.** $f(x) = -x^3$

Match the graph of each function with the graph of its inverse.

23. 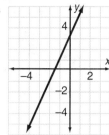 **24.** 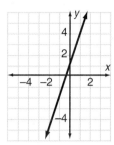 **25.** 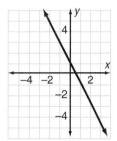 **26.**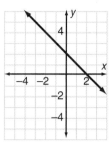

a. 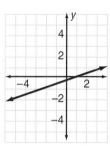 **b.** 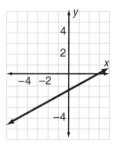 **c.** 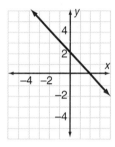 **d.**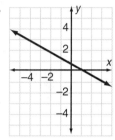

Sketch the graph of the inverse of each function. Then state whether or not the inverse is a function.

27. 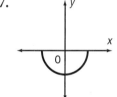 **28.** 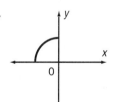 **29.** 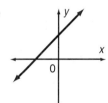 **30.**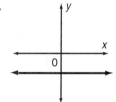

31. PHYSICS The formula that is used when matter is converted into energy is $E = mc^2$ where E is the energy in joules, m is the mass of the matter in kilograms, and c is the speed of light and is equal to 2.998×10^8 m/sec. Find the inverse of this function and determine whether the inverse is also a function.

32. PRINTING In typography, a point is a unit used to measure type size. One point is equivalent to 0.013837 inch. Write a formula that converts inches to points. Then find the inverse of this function. Round to the nearest hundredth. Is the inverse also a function?

33. GEOMETRY The volume of a sphere can be determined using the function $V = \dfrac{4\pi r^3}{3}$ where r represents the radius of the sphere. Write a formula to determine the radius, given the volume. Is this formula a function?

In the Extend section of Lesson 3.2, you learned that a function f is one-to-one if for any two elements in the domain of f, $f(x_1) = f(x_2)$ if and only if $x_1 = x_2$. If a function is one-to-one, then it will have an inverse function. It is often possible to restrict the domain of a function to make it one-to-one and ensure that it has an inverse function.

Refer to the graph of $h(x) = x^2 + 3$ in Example 3. If the domain is restricted to nonnegative real numbers, $h(x)$ will be one-to-one. The inverse of $h(x) = x^2 + 3$, $x \geq 0$, is $h^{-1}(x) = \sqrt{x - 3}$, $x \geq 3$.

Restrict the domain of f so that f^{-1} is also a function. State the restricted domain of f and the domain of f^{-1}.

34. $f(x) = x^2 + 5$ **35.** $f(x) = 3x^2$ **36.** $f(x) = \sqrt{x + 6}$ **37.** $f(x) = \sqrt{x + 3}$

38. GEOMETRY The circumference of a circle can be determined using the function $C = 2\pi r$ where r represents the radius of the circle. Write a formula to determine the radius, given the circumference. Is this formula a function?

39. Examine the functions and inverses in Exercises 11 and 12. What is the relation between these graphs and $y = x$?

40. Explain why the function $y = mx + b$ is a one-to-one function if $m \neq 0$, but is not a one-to-one function if $m = 0$.

Find f^{-1}.

41. $f(x) = \dfrac{x + 2}{x}$ **42.** $f(x) = \dfrac{x - 3}{x}$

43. Can the graphs of a linear function with slope $\neq 0$ and its inverse ever be perpendicular? Justify your answer.

44. In Try These Exercise 22, the inverse was a function. Explain why this is important.

Given the line $y = 5x + 2$, determine an equation of a line that is

45. parallel to the line with y-intercept of -3

46. perpendicular to the line with y-intercept of 4

Find the median of each set of data.

47. 3, 5, 9, 10, 12, 8, 7, 11, 25   **48.** $-1, 4, 7, 0, -3, -2, -4, 1, 9$

49. STANDARDIZED TESTS How much money can Rick borrow at 7% for 4 years if he can afford a monthly payment of $250?

     A. $12,000   **B.** $10,440.05   **C.** $840   **D.** $14,285.71

3.7 Algebra of Functions

Explore

1. Graph the function $h(x) = 1$. What happens to $h(x)$ as x increases?

2. Graph the Quadrant I portion of the function $f(x) = \frac{1}{x}$. What happens to $f(x)$ as x increases?

3. Graph the Quadrant I portion of $g(x) = \frac{1}{x} + 1$. What happens to $g(x)$ as x increases? What is the relationship between f and g?

4. Graph the function $j(x) = x$. What happens to $j(x)$ as x increases?

5. Graph the function $k(x) = 1^x$. What happens to $k(x)$ as x increases?

6. Graph the function $r(x) = x^x$. What happens to $r(x)$ as x increases?

7. Without graphing, explain why it is difficult to predict what will happen to $c(x) = \left(1 + \frac{1}{x}\right)^x$ as x increases.

8. Graph the Quadrant I portion of the function in Question 7. As x increases, what happens to $c(x)$?

Build Understanding

Given two functions f and g, you can form the functions $f + g$, $f - g$, $f \cdot g$, and $\frac{f}{g}$ by adding, subtracting, multiplying, and dividing, respectively.

Sum $(f + g)(x) = f(x) + g(x)$

Difference $(f - g)(x) = f(x) - g(x)$

Product $(f \cdot g)(x) = f(x) \cdot g(x)$

Quotient $\left(\frac{f}{g}\right)(x) = \frac{f(x)}{g(x)},\ g(x) \neq 0$

The domain of the sum, difference, and product functions is the set of all values of x that belong to both the domain of f and the domain of g. The domain of the quotient function is the set of all values of x that belong to the domain of f and the domain of g except values of x that make $g(x) = 0$.

EXAMPLE 1

Let $f(x) = 2x$ and $g(x) = x - 3$. Find each of the following and state its domain.

a. $f + g$ **b.** $f - g$ **c.** $f \cdot g$ **d.** $\dfrac{f}{g}$

Solution

a. $(f + g)(x) = f(x) + g(x) = 2x + x - 3 = 3x - 3;$

$$D_{f+g} = \{\text{all real numbers}\}$$

b. $(f - g)(x) = f(x) - g(x) = 2x - (x - 3) = x + 3;$

$$D_{f-g} = \{\text{all real numbers}\}$$

c. $(f \cdot g)(x) = f(x) \cdot g(x) = 2x(x - 3) = 2x^2 - 6x;$

$$D_{f \cdot g} = \{\text{all real numbers}\}$$

d. $\left(\dfrac{f}{g}\right)(x) = \dfrac{f(x)}{g(x)} = \dfrac{2x}{x - 3}$

$$D_{f/g} = \{x : x \neq 3\}$$

You can find $(f + g)(x)$ graphically by adding the values of the y-coordinates.

EXAMPLE 2

Graph $f(x) = x + 2$ and $g(x) = -2x - 1$ on the same set of axes. Make a table of values of $f(x)$ and $g(x)$ for $x = -3, -2, -1, 0, 1, 2,$ and 3. Add $f(x)$ and $g(x)$ to find $(f + g)(x)$. Then graph $(f + g)(x)$ on the same axes. Show that $f(2) + g(2) = (f + g)(2)$ on your graph.

Solution

x	-3	-2	-1	0	1	2	3
$f(x)$	-1	0	1	2	3	4	5
$g(x)$	5	3	1	-1	-3	-5	-7
$(f + g)(x)$	4	3	2	1	0	-1	-2

The graphs of $f(x) = x + 2$, $g(x) = -2x - 1$, and $(f + g)(x) = -x + 1$ are displayed at the right.

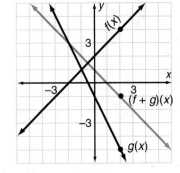

You can find the value of $(f + g)(2)$ by adding the values of $f(2)$ and $g(2)$ as shown on the graph. Since $f(2) = 4$ and $g(2) = -5$, $(f + g)(2) = 4 + (-5) = -1$.

COMMUNICATING ABOUT ALGEBRA

How can you find $(f - g)(x)$ graphically?

PROBLEM SOLVING TIP

When you find the composition of two functions, remember to work from the "inside out". For example, to find $f(g(x))$, first substitute for $g(x)$ and use $g(x)$ as the input for f.

A special way to combine two functions is by successive applications of the functions in a specific order. This operation is called **composition of functions**. Given two functions f and g, the composite function $(f \circ g)$ is written $(f \circ g)(x) = f(g(x))$ and is read "f of g of x." For example, if $f(x) = x - 2$ and $g(x) = x^2 + 1$, then

$$(f \circ g)(x) = f(x^2 + 1) = (x^2 + 1) - 2 = x^2 - 1$$

 CHECK UNDERSTANDING

EXAMPLE 3

Given $f(x) = \sqrt{x - 1}$ and $g(x) = 4x$, evaluate each function. State the domain of $f \circ g$ in part a and the domain of $g \circ f$ in part b and compare with the domain of f and the domain of g.

a. $(f \circ g)(x)$ **b.** $(g \circ f)(x)$ **c.** $(f \circ g)(5)$ **d.** $(g \circ f)(5)$

Solution

$D_f = x \geq 1$ D_g = all real numbers

a. $(f \circ g)(x) = f(g(x))$	**b.** $(g \circ f)(x) = g(f(x))$
$= f(4x)$	$= g\left(\sqrt{x - 1}\right)$
$= \sqrt{4x - 1}$	$= 4\sqrt{x - 1}$

$D_{f \circ g} = x : x \geq \dfrac{1}{4}$ since $D_{g \circ f} = x \geq 1$ since

$4x - 1$ must be non-negative $x - 1$ must be non-negative

c. $(f \circ g)(5) = f(g(5)) = f(20) = \sqrt{20 - 1} = \sqrt{19}$

d. $(g \circ f)(5) = g(f(5)) = g(2) = 4(2) = 8$ ◀

CHECK UNDERSTANDING

The domain of a $f(x)$ is denoted D_f. The domain of the sum of $f(x) + g(x)$ is denoted D_{f+g}. Write the notation for the domain of the difference, product, and quotient of $f(x)$ and $g(x)$.

You can use composition of functions to determine whether two functions are inverses of each other. Any two functions f and g are inverse functions if and only if $(f \circ g)(x) = x$ for all x in the domain of g and $(g \circ f)(x) = x$ for all x in the domain of f.

EXAMPLE 4

Determine whether $f(x) = \dfrac{1}{2}x + 3$ and $g(x) = 2x - 6$ are inverses of each other.

Solution

$(f \circ g)(x) = f(g(x))$	$(g \circ f)(x) = g(f(x))$
$= f(2x - 6)$	$= g\left(\dfrac{1}{2}x + 3\right)$
$= \dfrac{1}{2}(2x - 6) + 3$	$= 2\left(\dfrac{1}{2}x + 3\right) - 6$
$= x - 3 + 3$	$= x + 6 - 6$
$= x$	$= x$

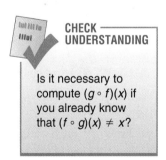 **CHECK UNDERSTANDING**

Is it necessary to compute $(g \circ f)(x)$ if you already know that $(f \circ g)(x) \neq x$?

Since $(f \circ g)(x) = x$ and $(g \circ f)(x) = x$, f and g are inverse functions. ◀

Composition of functions can be used in many real world applications.

EXAMPLE 5

ECONOMICS The wholesale price P for a package of blank videotapes is a function of the manufacturing cost per tape x in dollars. If there are five tapes per package and the manufacturer includes a $10 markup, the wholesale price is $P(x) = 5x + 10$. The retail price Q includes a 40% markup on the wholesale price. Thus, $Q(P) = P + 0.4P$.

a. Write a function that expresses the retail price of a package of five tapes in terms of x.

b. Determine the retail price of a package of five tapes that costs $0.52 per tape to manufacture.

Solution

a. $Q(P) = Q(5x + 10)$

$= (5x + 10) + 0.40(5x + 10)$

$= 7x + 14$

b. $Q = 7(0.52) + 14$

$= \$17.64$

TRY THESE

Find $(f + g)(x)$, $(f - g)(x)$, and $(f \cdot g)$ for each pair of functions.

1. $f(x) = 4x, g(x) = x + 6$

2. $f(x) = 5x, g(x) = x - 8$

3. $f(x) = 2x - 1, g(x) = 3x^2$

4. $f(x) = 3x + 2, g(x) = 2x^2$

Given the functions $f(x) = \sqrt{x}$ and $g(x) = x + 3$, state the domain of f and g. Then determine each of the following functions and state their domains.

5. $\left(\dfrac{f}{g}\right)(x)$

6. $\left(\dfrac{g}{f}\right)(x)$

Graph $f(x)$ and $g(x)$. Then find $(f + g)(x)$ graphically.

7. $f(x) = x - 1, g(x) = 2x - 2$

8. $f(x) = x^2 - 4, g(x) = x + 2$

Find $(f \circ g)(3)$ and $(g \circ f)(3)$ for each pair of functions.

9. $f(x) = x^2, g(x) = 3x$

10. $f(x) = 2x^2, g(x) = -2x$

Find $(f \circ g)(x)$ and $(g \circ f)(x)$ for each pair of functions.

11. $f(x) = 4x^2, g(x) = -x$

12. $f(x) = 5x^2, g(x) = -3x$

13. $f(x) = x + 1, g(x) = -x^2$

14. $f(x) = x - 2, g(x) = -2x^2$

15. WRITING MATHEMATICS Show that composition of functions is not commutative by giving a counterexample.

16. Consider the sets P, Q, and R at the right.

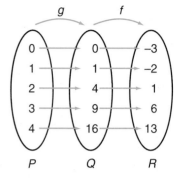

   a. Evaluate $(f \circ g)(2)$ **b.** Write a rule for g.

   c. Write a rule for f.   **d.** Write a rule for $f \circ g$.

Determine whether functions f and g are inverses of each other.

17. $f(x) = 2x - 4$ $g(x) = \frac{1}{2}x + 2$

18. $f(x) = \frac{1}{4}x - 6$ $g(x) = 4x + 6$

19. $f(x) = -x + 25$ $g(x) = -x + 25$

20. EMPLOYEE DISCOUNTS Greg works in a department store that provides its employees with a \$10 discount on all merchandise. During the store's January clearance sale, all men's suits are sold at 30% off the regular price.

   a. Let x represent the original price of a suit. Write a function $f(x)$ that represents the price that Greg pays for a suit before the sale.

   b. Write a function $g(x)$ that represents the price of a suit to a nonemployee during the sale.

   c. Write two composite functions that might represent the price that Greg pays for a suit during the sale.

   d. Which composite function would Greg rather the store use during the sale? Explain.

PRACTICE

Find $(f + g)(x)$, $(f - g)(x)$, and $(f \cdot g)(x)$ for each pair of functions.

1. $f(x) = 3x, g(x) = x - 4$ **2.** $f(x) = 6x, g(x) = x + 10$

3. $f(x) = 5x - 2, g(x) = 4x^2$ **4.** $f(x) = 4x - 4, g(x) = 5x^2$

Find $\left(\frac{f}{g}\right)(x)$ and $\left(\frac{g}{f}\right)(x)$ for each pair of functions. State their domains.

5. $f(x) = 3x + 1, g(x) = 2x + 1$ **6.** $f(x) = 4x - 3, g(x) = x + 2$

7. $f(x) = \sqrt{x - 3}, g(x) = x^2$ **8.** $f(x) = \sqrt{x + 2}, g(x) = 2x^2$

Graph f and g. Find $(f + g)(x)$ graphically.

9. $f(x) = x - 2, g(x) = 3x - 1$ **10.** $f(x) = 2x + 1, g(x) = x - 3$

11. $f(x) = x^2 - 2, g(x) = x + 1$ **12.** $f(x) = -x^2 + 1, g(x) = x + 2$

Graph f and g. Find $(f - g)(x)$ graphically.

13. $f(x) = 4x - 1, g(x) = -x - 3$ **14.** $f(x) = 3x - 2, g(x) = -2x + 4$

Find $(f \circ g)(4)$ and $(g \circ f)(4)$ for each pair of functions.

15. $f(x) = -2x^2, g(x) = 4x + 1$

16. $f(x) = -4x^2, g(x) = 2x - 6$

Find $(f \circ g)(x)$ and $(g \circ f)(x)$ for each pair of functions. State the domain.

17. $f(x) = 3x, g(x) = 2x - 4$

18. $f(x) = -x + 1, g(x) = 5x + 4$

19. $f(x) = 6x^2, g(x) = \frac{1}{2}x^3$

20. $f(x) = \frac{2}{x}, g(x) = x^2 - 3$

Find $(f \circ g)(x)$ and $(g \circ f)(x)$ for each pair of functions. State their domains.

21. $f(x) = 5x^2, g(x) = x + 3$

22. $f(x) = \sqrt{x + 4}, g(x) = 5x$

Determine whether functions f and g are inverses of each other.

23. $f(x) = 3x + 9$ and $g(x) = \frac{1}{3}x - 3$

24. $f(x) = 5x + 20$ and $g(x) = \frac{1}{5}x - 4$

25. $f(x) = -x + 16$ and $g(x) = -x + 16$

26. $f(x) = \frac{4}{x}, x \neq 0$ and $g(x) = -\frac{4}{x}, x \neq 0$

27. $f(x) = x^2 - 4, x \geq 0$ and $g(x) = \sqrt{x - 2}$

28. $f(x) = x^2 - 5, x \geq 0$ and $g(x) = \sqrt{x + 5}$

29. FINANCE Elisa is planning to invest 5% of her monthly net (not including the broker fee) earnings in stocks. Her broker's fee equals 8% of the dollar amount of the investment.

    a. Write an investment function $v(x)$ where x is Elisa's monthly net earnings.

    b. Write a broker's fee function $b(x)$ where x is the amount of the investment.

    c. Explain what $t(x) = b(v(x))$ represents. Write this function.

30. FINANCE Elisa's earnings in May were $4800. Use the function you wrote in Exercise 29 to determine the amount of her investment and the total cost of the investment.

31. CONSUMERISM A computer store offers a $200 rebate and a 15% discount off the price of new equipment.

    a. Write a cost-after-rebate function $r(x)$ where x is the price of the equipment.

    b. Write a discounted cost function $C(x)$ where x is the price of the equipment.

    c. Write a composite function for the final cost assuming you take the discount first.

    d. Write a composite function for the final cost assuming you take the rebate first.

    e. Determine the final cost of $5000 worth of equipment using each composite function. If you have a choice, should you take the discount or rebate first?

32. WRITING MATHEMATICS Explain how to use composition of functions to determine whether two functions are inverses of each other. Provide an example.

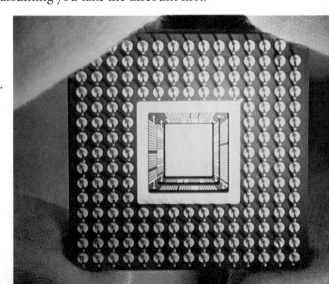

EXTEND

It is sometimes possible to "decompose" a composite function $f \circ g$ into two functions, f and g. Think of $g(x)$ as the "inner" function and $f(x)$ as the "outer" function in $(f \circ g)(x) = f(g(x))$. Use the following procedure to determine two functions f and g such that $(f \circ g)(x) = 4\sqrt{x} + 2$.

$$\text{Let } g(x) = \sqrt{x} \qquad \text{Inner function.}$$
$$\text{Let } f(x) = 4x + 2 \qquad \text{Outer function.}$$
$$(f \circ g)(x) = f(g(x)) \qquad \text{Check.}$$
$$= f(\sqrt{x}) \qquad \text{Apply } g.$$
$$= 4\sqrt{x} + 2 \qquad \text{Apply } f.$$

33. **WRITING MATHEMATICS** Are $f(x) = 4x + 2$ and $g(x) = \sqrt{x}$ the only functions such that $(f \circ g)(x) = 4\sqrt{x} + 2$? If your answer is yes, explain. If your answer is no, provide an example.

Determine two functions f and g such that $h(x) = (f \circ g)(x)$.

34. $h(x) = 3x^3 - 5$ 35. $h(x) = \sqrt{x} - 6$ 36. $h(x) = \sqrt{x} + 9$ 37. $h(x) = x^4 + x^2 + 1$

38. **CURRENCY EXCHANGE** On a particular day, one U.S. dollar was worth 87.17 Japanese yen. On the same day, one U.S. dollar was worth 0.6355 British pounds. Write a function that can be used to convert British pounds to yen on that day. Round to the nearest ten-thousandth.

39. **CURRENCY EXCHANGE** On the same day as that in Exercise 37, one French franc was worth 0.1957 U.S. dollars. Write a function that can be used to convert French francs to British pounds on that day. Round to the nearest ten-thousandth.

ECONOMICS The wholesale price W for a package of thumbtacks is a function of the manufacturing cost per tack x, in dollars. If there are 200 tacks per package and the manufacturer includes a $0.50 markup, the wholesale price is $W(x) = 200x + 0.50$. The retail price R includes a 25% markup on the wholesale price. Thus, $R(W) = W + 0.25W$.

40. Write a function that expresses the retail price of a package of 200 tacks in terms of x.

41. Determine the retail price of a package of 200 tacks that costs $0.0095 per tack to manufacture.

GEOMETRY A machine produces spherical soap bubbles that get larger and larger as time goes by. The surface area $S(r)$ of a sphere of radius r is given by $S(r) = 4\pi r^2$. Suppose the radius in millimeters of the bubble increases with time according to the formula $r(t) = \frac{3}{2}t^2$ for $t \geq 0$.

42. Find a formula for the surface area of the bubbles at any time $t \geq 0$.

43. Determine the surface area of a bubble when t is 10 seconds. Round to the nearest tenth.

THINK CRITICALLY

44. Show that the composition of any two linear functions is a linear function.

45. Show that the composition of any two constant functions is a constant function.

46. If f and g are one-to-one functions, is $f - g$ always a one-to-one function? Explain.

47. If $h(x) = (f \cdot g)(x)$, under what conditions will $h(x) = 0$?

Given $f(x) = x^2$ and $g(x) = 3x + 5$, find each value.

48. $(f \circ g)(x^2)$ **49.** $(g \circ f)(x^2)$ **50.** $(f \circ f)(x)$ **51.** $(g \circ g)(x)$

MIXED REVIEW

Solve each inequality.

52. $|x - 6| < 38$ **53.** $|x + 8| > 50$

54. Find the total financed price on a \$250,000 home for which you make a down payment of 30% and obtain a mortgage at 10% for 30 years.

55. Determine the probability of drawing a black face card from a pack of 52 playing cards to the nearest thousandth.

56. STANDARDIZED TESTS Determine $(f \circ g)(-2)$ and $(g \circ f)(-2)$ for the pair of functions $f(x) = -3x^2, g(x) = 2x + 8$

    A. $-48; -16$ **B.** $-12; 4$ **C.** $24; -32$ **D.** $432; 36$

In Questions 57–59 below, the graph of a function of the form $f(x) = a(x - b)^n + c$ is displayed. Tell whether a, b, and c are positive or negative and whether n is even or odd.

57. 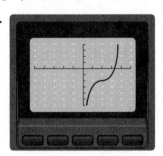 **58.**  **59.**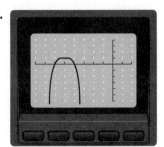

Career
Real Estate Sales Agent

Real estate sales agents are paid commission based on the selling price of the homes they sell. They may have to share their commission with the agency that employs them or another agent. Hopefully, their earnings are greater than their expenses.

1. If an agent receives a 5% commission on the sale of a $300,000 home, how much does she make?

2. Write a function *f* that an agent can use to compute a 3% commission on the selling price of a home.

3. If an agent sells an average of two $100,000 homes per month and receives a 3% commission, what is her annual salary?

4. A couple is advertising their house for $110,000. A family makes an offer of $102,000 for the house. The sellers do not want to sell the house for that low of a price. The real estate agent makes a rare decision to lower her 6% commission by 50% to cut the losses of the seller and to ensure that the sale finally takes place.

   a. Find the difference between the original and reduced commissions on the selling price of $102,000.
   b. At the original selling price and commission, how much would the sellers have left after paying the agent?
   c. At the offered price and original commission, how much would the sellers have left after paying the agent?
   d. At the offered price and reduced commission, how much would the sellers have left after paying the agent?

5. If a real estate agent receives a 5% commission and sells *M* houses at an average price of *D* dollars each month for an entire year, express his average weekly pay *W* as a function of *D* and *M*.

6. List the advantages and disadvantages of being paid in commission.

Direct Variation Functions

In many real world problems, the variables are related so that both directly increase or decrease together. For example, if you are eating peanuts, the total number of calories depends directly on the number of peanuts you eat. This type of function is called a **direct variation function** and is described by an equation of the form $y = kx$ where the constant of variation $k \neq 0$. Other examples of direct variation are distance when speed is constant ($d = st$) or the circumference of a circle ($c = 2\pi r$).

Problem

Sales tax varies directly with the purchase price of an item. The sales tax on an item that costs $25 is $1.75. Find the constant of variation and express the sales tax S as a function of purchase price P.

Explore the Problem

1. What is the general form of the equation for sales tax? Use S and P.

2. How can you use the given information to determine k?

3. Find k and write the equation. What does k represent?

4. Determine the sales tax on a purchase of $86. Compare the ratio S to P in this case with the ratio of S to P using the values from Question 2. What do you notice?

5. Find the sales tax on $135 by solving the proportion $\frac{1.75}{25} = \frac{S}{135}$. Show your work. Check using a different method.

6. If sales tax on an item was $74.13, what was the purchase price?

7. Look at the equation you wrote in Question 3. What type of function is a direct variation?

8. **WRITING MATHEMATICS** Explain how the equation of a direct variation $y = kx$ is a special case of the linear equation $y = mx + b$. When a direct variation is graphed, how does k relate to the slope of the graph? Through what point on the coordinate plane will the graph of a direct variation always pass?

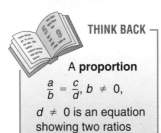

THINK BACK

A **proportion**

$\frac{a}{b} = \frac{c}{d}$, $b \neq 0$,

$d \neq 0$ is an equation showing two ratios are equal. In a proportion $ad = bc$.

Investigate Further

- In Explore the Problem, you found that for a direct variation function $y = kx$, the ratio $\frac{y}{x}$ is constant. Use this fact to determine if the relationship between two variables is a direct variation.

9. The graph at the right shows $y = x$. The region between the line $y = x$ and the x-axis from $x = 0$ to $x = 5$ is shaded. Use the formula for the area of a triangle $A = \frac{1}{2}bh$ to find the area of the shaded region.

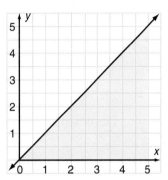

10. Copy and complete the table showing the area between the line $y = x$ and the x-axis from $x = 0$ to the given endpoint value of x.

11. If you double x, does the area double? If you triple x, does the area triple?

12. Is the relationship between x and the area a direct variation? Explain.

x	area	$\frac{\text{area}}{x}$
1		
2		
3		
4		
6		

13. The graph at the right shows the line $y = 4$. Copy and complete the table below showing the area between the line $y = 4$ and the x-axis from $x = 0$ to the given endpoint value of x.

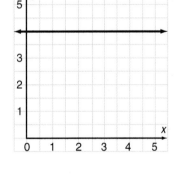

x	area	$\frac{\text{area}}{x}$
1		
2		
3		
4		
5		
6		

14. If you double x, does the area double? If you triple x, does the area triple?

15. Is the relationship between the x-value and the area a direct variation? Explain.

16. WRITING MATHEMATICS Write a brief report summarizing important ideas about direct variation. Discuss the form of the equation, how to determine if a relationship is a direct variation, and different strategies that you can use to solve problems. Include examples.

Apply the Strategy

● Assume that y varies directly as x.

17. When $x = 9$, $y = 30$. Determine y when $x = 24$.

18. When $x = 14$, $y = 103.6$. Determine y when $x = 10$.

19. When $x = 20$, $y = 6.8$. Determine x when $y = 17$.

20. FINANCE The annual simple interest on a loan varies directly with the amount of the loan. The annual interest on a loan of $750 is $60.

  a. Write an equation that expresses the amount of annual interest I as a function of the amount of the loan L.

  b. Determine the annual interest on a loan of $1600 at the same interest rate.

21. MEASUREMENT 55 in. is equivalent to 139.7 cm.

  a. Determine the constant of variation.

  b. How many centimeters are equivalent to 120 in.?

22. PHYSICS The distance a spring will stretch, S, varies directly with the weight, W, added to the spring. If a spring stretches 1.5 in. when 12 lb are added, how far will it stretch when 26 lb are attached?

23. BUSINESS Revenue R varies directly with items sold. If the revenue from the sale of 225 bracelets is $4488.75, how many bracelets must be sold for the revenue to be over $100,000?

24. DIRECT SQUARE VARIATION The relationship you explored in Questions 9–12 is an example of a **direct square variation**, a function that can be written in the form $y = kx^2$ where $k \neq 0$.

  a. Write an equation that shows the area between the line $y = x$ and the x-axis from $x = 0$ to a given value of x.

  b. Determine the area when $x = 14$.

25. THINKING CRITICALLY Assume the point (a, b) is on the graph of the function $y = kx$.

  a. Is $(2a, 2b)$ on the graph? Justify your answer.

  b. Is $(a + 2, b + 2)$ on the graph? Justify your answer.

Review Problem Solving Strategies

FLEA MARKET EXPEDITION

1. Karen and her friend Kate were shopping one day at a cash-only flea market. Karen spotted a denim jacket she liked that cost $20 but she didn't have enough money. Karen asked Kate, "Please lend me as much money as I have in my purse, so I can buy the jacket." Kate agreed. Karen then found a $20 pair of jeans and made the same request, to which Kate agreed. The same thing happened again when Karen wanted a $20 pair of sneakers. After buying the three items, Karen had no more money left. How much money did Karen have when she first arrived at the flea market? (Hint: Think about how much money Karen had when she saw the sneakers and what happened before that point.)

OLDER BUT WISER

2. The average age of a group of teachers is 50. The average age of a group of rock singers is 35. When combined, the average age of both teachers and rock singers is 40. What is the ratio of the number of teachers to the number of rock singers?

BLOCK BINGO

3. Show how to rearrange the blocks in a square array so that no letter nor number is repeated in one of the rows, columns, or diagonals.

A1	B1	C1	D1	E1
A2	B2	C2	D2	E2
A3	B3	C3	D3	E3
A4	B4	C4	D4	E4
A5	B5	C5	D5	E5

CHAPTER REVIEW

VOCABULARY

Choose the word from the list that completes each statement.

1. Two nonvertical lines are __?__ if and only if their slopes are negative reciprocals.   **a.** even

2. A function is __?__ if and only if $f(-x) = -f(x)$ for all values of x in its domain.   **b.** parallel

3. Two distinct lines are __?__ if and only if their slopes are equal or are both undefined.   **c.** odd

4. A function is __?__ if and only if $f(x) = f(-x)$ for all values of x in its domain.   **d.** perpendicular

Lesson 3.1   SYMMETRY   pages 107–111

● If a figure can be folded along a line such that the two parts coincide, then the figure is **symmetric about the line**. If a figure coincides with itself when it is rotated about a point, then the figure is **symmetric about the point**.

The point (2, 3) is on a figure. What other point is on the figure if it has

5. x-axis symmetry?   6. origin symmetry?   7. y-axis symmetry?

Lesson 3.2   RELATIONS, FUNCTIONS, AND GRAPHS   pages 112–118

● A **relation** is a set of ordered pairs. A **function** is a relation in which each element in the domain is paired with one and only one element in the range.

Determine whether each relation is a function.

8. $\{(0, 1), (1, 2), (4, 5), (1, 6)\}$ 9. $\{(3, -2), (-2, 3), (2, -3), (-3, 2)\}$

Evaluate $f(x) = 3x^3 - 2x^2 - x + 4$ for each value of x.

10. 0 11. -1 12. 2 13. -3

Lesson 3.3   LINEAR FUNCTIONS   pages 119–125

● A **linear function** is a function of the form $f(x) = mx + b$, where m and b are constants.

● Equations of lines can be expressed in **slope-intercept form:** $y = mx + b$, **point-slope form:** $(y - y_1) = m(x - x_1)$, and **standard form:** $Ax + By = C$.

Write an equation of a line in slope-intercept form that passes through the given points. Then rewrite the equation in standard form.

14. $(-4, 6), (0, 0)$ 15. $(1, -1), (7, -7)$ 16. $(5, 2), (1, 3)$

17. Given the line $y = 6x - 5$, determine an equation of a line that is perpendicular to the given line with y-intercept -7.

Lesson 3.4 EXPLORE ABSOLUTE VALUE pages 126–131

- The graph of an absolute value function of the form $f(x) = |x|$ is V-shaped and can be reflected, translated, and dilated.

Write a function of the form $g(x) = a|x - b| + c$ whose graph represents

18. a translation of 4 units down and 7 units right from the graph of $f(x) = |x|$

19. a reflection of $f(x) = |x|$ over the x-axis

Lesson 3.5 SYMMETRY IN OTHER FUNCTIONS pages 132–136

- The graphs of functions of the form $f(x) = x^n$, where n is a positive number can be reflected, translated, and dilated.

Explain how the graph of each function is translated from the graph of $f(x) = x^4$.

20. $g(x) = (x - 2)^4 + 9$ 21. $h(x) = (x + 5)^4 + 7$ 22. $j(x) = (x - 3)^2 - 6$

Lesson 3.6 INVERSE FUNCTIONS pages 137–142

- You can find the **inverse** of a relation represented by a set of ordered pairs by switching the first and second coordinates of each ordered pair.

Write an equation for the inverse of each function.

23. $y = x - 5$ 24. $y = 2x - 4$ 25. $y = -4x + 8$ 26. $y = -x + 4$

27. Find the inverse of $\{(9, 3), (1, 5), (-2, 6), (-3, -3)\}$. Is the inverse a function?

28. A simple cipher codes a number x using the function $f(x) = -3x + 5$. Find the inverse of this function and determine whether the inverse is also a function.

Lesson 3.7 ALGEBRA OF FUNCTIONS pages 143–151

- Given two functions f and g, you can form $f + g, f - g, f \cdot g$, and $\dfrac{f}{g}$ by adding, subtracting, multiplying, and dividing, respectively.

- **Composition of functions** is the successive application of the functions in a specific order.

Given $f(x) = 3x^2$ and $g(x) = x - 5$, evaluate each function.

29. $(f + g)(-1)$ 30. $(f \cdot g)(-1)$ 31. $(f \circ g)(-1)$ 32. $(g \circ f)(-1)$

Lesson 3.8 DIRECT VARIATION FUNCTIONS pages 152–154

- A **direct variation function** is described by an equation of the form $y = kx$, where the **constant of variation $k \neq 0$**.

33. If the sales tax on an item was \$58.50 in a state with a tax rate of 5%, what was the purchase price?

34. If y varies directly as x and $y = 70$ when $x = 20$ determine y when $x = 35$.

CHAPTER ASSESSMENT

CHAPTER TEST

1. **STANDARDIZED TESTS** If point (a, b) is on a figure that has x-axis symmetry, which of the following points is also on the figure?

   A. $(-a, b)$ **B.** $(a, -b)$
   C. (b, a) **D.** $(-a, -b)$

Determine whether the graph of each function has x-axis, y-axis, origin, or no symmetry.

2. $y = x^5$ 3. $y = x^2 - 1$

4. $y = 5x + 1$ 5. $y = 5x$

6. **WRITING MATHEMATICS** Write a paragraph that explains how to determine whether a function is odd, even, or neither.

Determine whether each relation is a function.

7. $\{(3, 4), (3, 5), (3, 6), (4, 4)\}$

8. $\{(6, 5), (5, 6), (9, 8), (8, 9)\}$

Evaluate $f(x) = 2x^3 - 3x^2 - x + 7$ for each value of x.

9. -5 10. 0 11. -2 12. 1

13. Write an equation of a line in slope-intercept form that passes through the points $(5, 4)$ and $(3, 10)$.

14. Given $y = -2x + 7$, determine an equation of a line that is parallel to the given line with y-intercept -4.

15. Explain how the graph of $h(x) = |x - 7| - 6$ is translated from the graph of $f(x) = |x|$.

Determine the vertex and the line of symmetry for the graph of each function.

16. $f(x) = |x + 3|$ 17. $g(x) = |x| + 2$

18. Write a function of the form $g(x) = |x - b| + c$ whose graph is a translation of 3 units down and 9 units left from the graph of $f(x) = |x|$.

Write a rule for a function g that is the reflection of f about the x-axis.

19. $f(x) = x^7$ 20. $f(x) = |2x|$

Explain how each graph differs from the graph of $f(x) = x^4$.

21. $g(x) = 4x^4$ 22. $f(x) = \frac{1}{2}x^4$

23. **WRITING MATHEMATICS** Explain how to use the horizontal line test to determine whether or not a function f has an inverse function.

24. Find the inverse of $\{(5, 5), (2, 1), (4, 2), (6, 1)\}$. Is the inverse a function?

25. **STANDARDIZED TESTS** If f and g are inverse functions, which of the following must be true?

     I. $(f \circ g)(x) = -1$
    II. $(g \circ f)(x) = x$
   III. $(f \circ g)(x) = x$

   A. I only   **B.** II only
   C. I and II   **D.** II and III

Let $f(x) = 3x$ and $g(x) = x + 2$. Determine each of the following and state their domains.

26. $f + g$ 27. $f - g$ 28. $f \cdot g$

29. f/g 30. $f \circ g$ 31. $g \circ f$

Assume that y varies directly as x.

32. When $x = 8$, $y = 36$. Find y when $x = 15$.

33. When $x = 18$, $y = 45$. Find x when $y = 90$.

PERFORMANCE ASSESSMENT

MEDIA MATH Look in old newspapers and magazines to find items that can be used to illustrate concepts in this chapter such as symmetric graphs, relations that are or are not functions, and direct variations. You may use graphs, diagrams, headlines, or portions of text. For each item, tell the concept it illustrates and provide a brief explanation. Arrange your materials in a poster or collage.

YOUR MIND: A GRAPHING UTILITY Sketch the graphs of $f(x) = x^2$, $g(x) = x^3$, and $h(x) = x^4$ on the same set of axes without using a graphing utility. Use a scale from 0 to 2 on both axes. Use three different colors for the graphs.

a. For which value(s) of x is $f(x) = g(x) = h(x)$?

b. For which value(s) of x is $h(x) < g(x) < f(x)$?

c. For which value(s) of x is $h(x) > g(x) > f(x)$?

USE FLOWCHARTS You can describe the process of finding the inverse of an equation using the flowchart below. Work with a partner to draw a flowchart that shows how to find the original equation if you know its inverse. Compare the flowcharts. Write three equations and ask your partner to find their inverses. Next, write three inverse equations and ask your partner to find the original equations. Then switch roles with your partner.

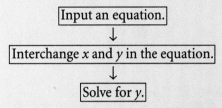

MAKE A PRESENTATION Write three equations whose graphs have x-axis, y-axis, and/or origin symmetries on the board. Present a handout showing the graphs of these equations. Demonstrate how to manipulate the paper to determine point and line symmetries.

PROJECT ASSESSMENT

PROJECT *Connection* Present all the information your group has collected on buying and maintaining a home.

1. Decide whether you will write a report, make a poster, or use a different format. Remember to include sources for all information.

2. Define new terms learned during the project.

3. Describe the type of home your group considered.

4. Identify the major costs involved in purchasing the home including an itemized list of closing costs.

5. Combine all of the financial information to determine the monthly cost of paying for and maintaining the home.

6. Make a list of other expenses that a homeowner could incur. Include periodic expenses (such as gardening or extermination) as well as unexpected expenses (such as a major appliance or structural breakdown).

7. Approximate the annual cost of owning the home.

8. As a class, discuss trends in home prices over the last ten years.

Copy the figure; draw all the lines of symmetry.

1.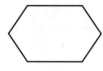

2.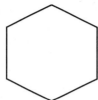

3. STANDARDIZED TESTS Which of the following mappings defines a relation that is not a function?

A. 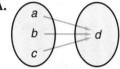   **B.**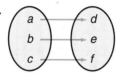

C. 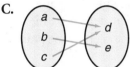   **D.**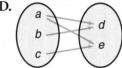

4. WRITING MATHEMATICS Explain how and why the vertical line test determines whether a relation represented by a graph is a function.

5. If $f(x) = |x^3 - 1|$, find $f(-1)$.

6. Write the equation of the line that passes through the points $(-2, 3)$ and $(0, 7)$ in point-slope form.

7. WRITING MATHEMATICS Without graphing, explain the relationship between the graphs of the equations $2y + 3x = 8$ and $6y = -9x - 12$.

8. Write a function of the form $g(x) = |x - b| + c$ whose graph represents a translation of 3 units down and 6 units left from the graph of $f(x) = |x|$.

9. Use a graphing utility to graph $y = |2x - 1|$ and $y = 9$ on the same set of axes. Find the points of intersection.

10. STANDARDIZED TESTS Which of the following graphs is symmetric with respect to origin?

A. 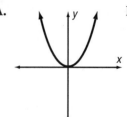   **B.**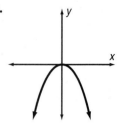

C. 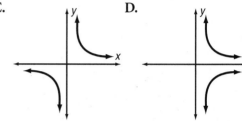   **D.**

11. Find the inverse of the function $f(x) = \dfrac{x + 1}{3}$.

12. If $f(x) = 2x^2$ and $g(x) = 3x + 1$, find $(f \circ g)(-3)$.

13. A mass of 48 kilograms causes a beam to bend 8 centimeters. If the amount of bending varies directly as the mass, what mass will cause the beam to bend 11 centimeters?

14. Rent-a-Cheapie charges \$39 per day for a car, plus 35 cents per mile driven. If Mia rents a car for a day, what distance can she drive and have her total rental charge equal \$60, not including tax?

15. Solve and check: $3x - |x + 1| = 7$

16. What is the theoretical probability of getting a red card in one draw from a standard deck of playing cards?

Let $A = \begin{bmatrix} 1 & 3 & 4 \\ 2 & -2 & -3 \end{bmatrix}$ $B = \begin{bmatrix} 0 & 9 & -5 \\ 3 & -4 & 6 \end{bmatrix}$

17. Find $A + B$.   **18.** Find $A - B$.

STANDARDIZED TEST

STANDARD FIVE-CHOICE Select the best choice for each question.

1. Which letter has both horizontal and vertical line symmetry?

   A. K   **B.** H   **C.** T   **D.** C   **E.** M

2. Which of the following relations is a function?

   A. $x = y^2$ **B.** $|y| = x$ **C.** $y = |x|$
   D. $xy^2 = 9$ **E.** $x^2 + y^2 = 25$

3. Which is true of the graph of $x = -1$?

      I. The slope is undefined
    II. It is parallel to the y-axis.
   III. It has a y-intercept of -1.

   A. I only   **B.** I and II   **C.** I and III
   D. II only   **E.** I, II, and III

4. The vertex of the graph of $y = 2|x + 4| + 5$ is

   A. $(-8, 5)$ **B.** $(8, 5)$ **C.** $(-8, 4)$
   D. $(-4, 5)$ **E.** $(4, 5)$

5. The function whose graph is a reflection of $y = |x - 2| + 4$ about the x-axis is

   A. $y = -|x - 2| + 4$
   B. $y = -|x - 2| - 4$
   C. $y = |2 - x| - 4$
   D. $y = |2 - x| + 4$
   E. $y = |x + 2| - 4$

6. The inverse of the function $y - 2 = 3x$ is

   A. $y = \dfrac{2 - x}{3}$ **B.** $y = \dfrac{2x}{3}$ **C.** $y = \dfrac{x - 2}{3}$

   D. $y = 3x - 2$ **E.** $y = \dfrac{-x - 2}{3}$

7. Which of the following statements is true about the data set $\{2, 5, 2, 15, 6\}$?

   A. median > mode   **B.** mode < median
   C. mean = mode   **D.** median = mode
   E. range = 10

8. The domain and range of the function $y = \sqrt{x - 1}$ are

   A. domain = real numbers
        range = real numbers
   B. domain = real numbers
        range = positive real numbers
   C. domain = real numbers
        range = nonnegative real numbers
   D. domain = real numbers ≥ 1
        range = real numbers
   E. domain = real numbers ≥ 1
        range = nonnegative real numbers

9. The area of a circle varies directly as the square of the radius. If the radius is doubled, the area is

   A. unchanged   **B.** doubled   **C.** quadrupled
   D. halved   **E.** divided by 4

10. Three consecutive steps in the solution of the equation $4x + 3 = 7x$ could be

    Step 1 $4x + (-4x) + 3 = 7x + (-4x)$
    Step 2 $0 + 3 = 7x + (-4x)$
    Step 3 $3 = 7x + (-4x)$

    The property that justifies Step 3 is

    A. additive inverse   **B.** additive identity
    C. addition   **D.** reflexive
    E. symmetric

11. **STANDARDIZED TESTS** Which of the graphs shown is symmetric about the x-axis?

    A. 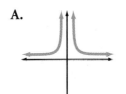   **B.**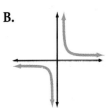

    C. 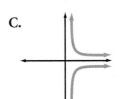   **D.**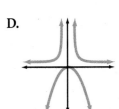

4 Systems of Linear Equations

Take a Look AHEAD

Make notes about things that look new.

- Copy the names of two different algebraic methods for solving systems of equations.
- What matrix operations did you learn about in Chapter 1? What new matrix operation will you learn about in this chapter?

Make notes about things that look familiar.

- Find a graph that shows two or more linear equations graphed on the same coordinate plane.
- Explain what is meant by a 2 × 3 matrix. Find an example.
- What is a square matrix? Find an example.

DATA Activity

Made in the U. S. A.

The word *manufacture* comes from the Latin words for hand (*manus*) and to make (*facere*). Manufactured items are usually classified as durable or nondurable. *Durable goods* are items with a life expectancy of three years or more (such as a tractor). *Nondurable goods* are items which last for three years or less (such as a bar of soap). All manufactured items are either *consumer goods*, sold in retail stores (such as groceries or clothing), or *producer goods*, (such as printing presses) used to make other products.

SKILL FOCUS

- Read and interpret tables.
- Add, subtract, multiply, and divide real numbers.
- Use estimation.
- Determine percent increase or decrease.
- Determine percent of a number.

Manufacturing

In this chapter, you will learn how:

- **CRAFTWORKERS** use systems of equations to determine the amount of time it takes one person to complete a job usually done by a team. (Lesson 4.2, page 175)

- **CRYPTOGRAPHIC TECHNICIANS** use matrix multiplication for coding and decoding messages. (Lesson 4.6, page 202)

- **SAFETY ENGINEERS** use matrix equations while planning budgets for safety awareness campaigns. (Lesson 4.7, page 209)

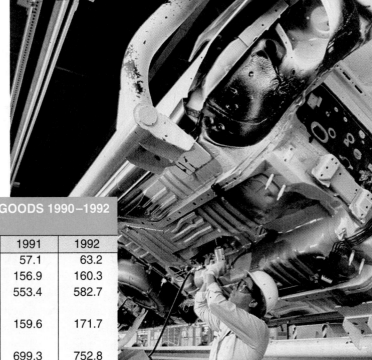

VALUE OF SHIPMENTS OF MANUFACTURED GOODS 1990–1992 (in billions of dollars)			
Durable goods industries	1990	1991	1992
Ceramic and glass products	63.5	57.1	63.2
Metal products	163.1	156.9	160.3
Industrial machinery and office equipment	562.5	553.4	582.7
Electronic equipment and household appliances	154.5	159.6	171.7
Transportation equipment	708.1	699.3	752.8
Instruments and measuring products	155.3	154.7	155.1
Nondurable goods industries	1990	1991	1992
Food and beverages	436.2	440.6	450.3
Textiles	66.0	67.8	72.5
Paper products	219.4	206.6	209.0
Chemicals and toiletries	540.7	543.3	560.1
Petroleum and coal products	172.6	160.4	158.0
Rubber and plastic products	101.4	103.6	106.7

1. What was the average value of transportation equipment shipped from 1990 to 1992? Round to the nearest tenth of a billion dollars.

2. Estimate the total value of shipments of nondurable goods for 1992. Explain your method.

3. To the nearest tenth of a percent, determine the percent change for shipment values from 1990 to 1992 for each of the following industries.

    a. ceramic and glass products
    b. food and beverages
    c. petroleum and coal products

4. In 1992, containers and boxes accounted for about 14.4% of paper product shipments. Determine the value of this category of paper products.

5. **WORKING TOGETHER** Make a list of 20 to 30 items around your school or home. Estimate the useful lifetime of each product, then classify it as durable or nondurable. Discuss any items you find difficult to classify.

A Manufacturing Simulation

When goods come out of the production area, they are packaged and shipped to warehouses, stores, offices, and homes. Packaging is an important part of the manufacturing process. Nobody wants soggy cereal or broken light bulbs. The manufacturer must select a cost-effective package that protects the items being shipped.

PROJECT GOAL

To simulate the design and manufacturing of a cereal package.

Getting Started

Work in groups of four or five students. Imagine that your company manufactures two different sizes of boxes and waxed inner bags used for packaging cereal.

1. Each group member should help to collect empty cereal boxes and waxed inner bags. Use only boxes and bags that are not torn and do not have pieces missing.

2. Select a large box, such as those used for flake-type cereals (typical dimensions are 19 cm long, 6.5 cm wide, and 24 cm high).

3. Obtain three sheets of white poster board and record the cost per sheet. Also obtain a roll of waxed paper and record the cost. Determine the area of the paper on the roll. You will use this information in the Project Connections.

PROJECT *Connections*

Lesson 4.2, page 174:
Design templates for cereal boxes and determine amounts of used and wasted materials.

Lesson 4.5, page 195:
Design templates for cereal box inner bags and determine amounts of used and wasted materials.

Lesson 4.7, page 208:
Create matrices to explore costs involved in producing cereal boxes and bags.

Chapter Assessment, page 223:
Experiment to collect data about production times for cereal boxes and bags and use this data to determine how many packages can be produced given certain conditions.

Internet Connection

www.swpco.com/
swpco/algebra2.html

Algebra Workshop
Exploring Systems of Equations

Think Back

● Use graph paper to answer each question.

1. Draw a line that has a positive slope. Identify and record a possible equation for the line.

2. Draw a line that has a negative slope. Identify and record a possible equation for the line.

3. Draw a line that has an undefined slope. Identify and record a possible equation for the line.

4. Draw a line that has zero slope. Identify and record a possible equation for the line.

5. Draw a graph that has both negative and positive slope. Identify and record a possible equation for the graph.

Explore/Working Together

● Work with a partner. Use a graphing utility.

6. Graph: $y = 2x + 3$

7. Have your partner choose a linear equation that will intersect the graph of the line $y = 2x + 3$. Graph the equation on the same coordinate plane as the first equation.

8. Use the TRACE feature, and if necessary the ZOOM feature, to find the point of intersection, if one exists. Record your partner's equation and the coordinates of the point of intersection. If the lines did not intersect, explain why.

9. If you found a point of intersection in Question 8, determine whether the ordered pair is a solution to both the original equation $y = 2x + 3$ and your partner's equation. Describe how you made the check.

10. Switch roles and repeat Questions 6–9.

11. As a result of your explorations, make a conjecture. How can you be sure that two equations will intersect at a single point? Test your conjecture by graphing several lines.

12. If two lines intersect at a single point, what must be true about the system of equations of those lines?

SPOTLIGHT ON LEARNING

WHAT? In this lesson you will learn
- to solve systems of linear equations by graphing.
- to determine whether a system of linear equations has one solution, no solution, or infinite solutions.

WHY? Knowing how to graph systems of linear equations will help you to solve problems that are modeled by more than one equation.

Make Connections

When you graph two linear equations, there are several possible relationships for the lines. Moreover, the relationship of the lines gives you information about points that are solutions to both equations.

13. Graph the two equations $y = 6x - 5$ and $y = 2x - 1$ on the same coordinate plane. At how many points do the lines intersect? If there is only one point of intersection, what are the coordinates of that point?

14. Is there a single ordered pair that satisfies both equations $y = 6x - 5$ and $y = 2x - 1$? If so, what is the ordered pair? If not, why not?

15. Graph $y = 6x - 5$ and $y = 6x - 10$ on the same coordinate plane. At how many points do the lines intersect? If there is only one point of intersection, what are the coordinates of that point?

16. Is there a single ordered pair that satisfies both equations $y = 6x - 5$ and $y = 6x - 10$? If so, what is the ordered pair? If not, why not?

17. Graph the two equations $y = 6x - 5$ and $2y = 12x - 10$ on the same coordinate plane. At how many points do the two lines intersect? If there is only one point of intersection, what are the coordinates of that point?

18. Is there a single ordered pair that satisfies both equations $y = 6x - 5$ and $2y = 12x - 10$? If so, what is the ordered pair? If not, why not?

19. Describe the graphs of two linear equations for each of the following situations.

   a. A single ordered pair satisfies both equations.

   b. An infinite number of ordered pairs satisfies both equations.

   c. No single ordered pair satisfies both equations.

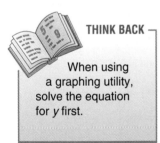

THINK BACK

When using a graphing utility, solve the equation for *y* first.

Summarize

20. WRITING MATHEMATICS Describe how you can use graphing to find the solution to a system of two linear equations.

21. Find the solution of $6x + y = 5$ and $5x + y = 3$.

22. Use a graphing utility to show that the ordered pair $(3.5, 4.75)$ is a solution of $3x + y = 15.25$ and $2x + 2y = 16.5$.

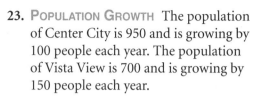

23. **POPULATION GROWTH** The population of Center City is 950 and is growing by 100 people each year. The population of Vista View is 700 and is growing by 150 people each year.

    a. Write an equation for each city that gives the total population y, x years from now.

    b. Find the ordered pair that satisfies both equations.

    c. Explain what the solution represents.

24. **WRITING MATHEMATICS** Explain how you can determine whether a pair of linear equations has a unique solution.

Match each system of equations with their graph. How many points represent solutions of each pair of equations?

25. $\begin{cases} 4x + 6y = 12 \\ 2x + 3y = -6 \end{cases}$
26. $\begin{cases} -2x + y = 2 \\ 3x + y = 2 \end{cases}$
27. $\begin{cases} -x + 2y = -8 \\ 3x - 6y = 24 \end{cases}$

a.

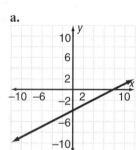

b.

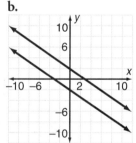

c.

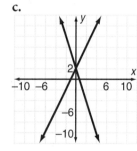

28. **THINKING CRITICALLY** Write a system of equations that has an infinite number of solutions. Explain how you found the equations.

29. **THINKING CRITICALLY** Is it possible to have a pair of equations with no solutions such that both lines have the same y-intercept? Explain your answer.

30. **GOING FURTHER** Think back to Questions 22 and 23. What might be some drawbacks of using the graphing method to find the solution to a system of linear equations?

31. **GOING FURTHER** Solve by graphing: $y - 4 = |x|$ and $y = 2x + 1$.

32. **GOING FURTHER** Solve by graphing: $y = x^2 - 2$ and $y - x = 0$.

Solving Systems of Linear Equations

Explore

When two or more linear equations are considered together, they are called a **system of linear equations,** or more simply, a **linear system.** Equivalent systems are systems that have the same solution.

1. Graph the systems of linear equations on the same coordinate plane. Use a different color to represent each system.

System 1
$$\begin{cases} x + 3y = -4 \\ x + y = 0 \end{cases}$$

System 2
$$\begin{cases} x = 2 \\ y = -2 \end{cases}$$

2. What do you notice about the intersections of the systems? What conclusion can you draw about the two systems?

3. How can you use the equation $x + y = 0$ from System 1 to express y in terms of x? Substitute this expression for y in $x + 3y = -4$. Then simplify to find a new equation for x.

4. Substitute the value you found for x into $x + y = 0$ to find a new equation for y.

5. What equations for x and y did you end up with? How do the equations in System 1 and System 2 relate to each other?

Build Understanding

The graphing method in Lesson 4.1 may provide only approximate solutions for some systems. For exact solutions, you can solve a system using an algebraic method called the **substitution method.**

In Explore, you found the solution to the system of equations $x + 3y = -4$ and $x + y = 0$ by solving one equation for one variable in terms of the other. Then you substituted that value into the other equation to find the second variable. This is the substitution method.

EXAMPLE 1

Solve the system using the substitution method.

$$\begin{cases} 2x + y = 6 & \textbf{Equation 1} \\ 3x + 4y = 4 & \textbf{Equation 2} \end{cases}$$

Solution

First solve one equation for one variable. Because the coefficient of y is 1 in Equation 1, it is easier to isolate y in that equation.

$2x + y = 6$	Equation 1
$y = 6 - 2x$	Solve for y.

Next, *substitute* the expression $(6 - 2x)$ for y in Equation 2 and solve for x.

$3x + 4y = 4$	Equation 2
$3x + 4(6 - 2x) = 4$	Substitute $(6 - 2x)$ for y.
$3x + 24 - 8x = 4$	Use the distributive property.
$24 - 5x = 4$	Combine like terms.
$-5x = -20$	Simplify.
$x = 4$	Divide both sides by -5.

Now substitute 4 for x in either of the original equations.

$2x + y = 6$	Equation 1
$2(4) + y = 6$	Substitute 4 for x.
$8 + y = 6$	Simplify.
$y = -2$	Solve for y.

Check

Check the solution in both original equations.

Equation 1	**Equation 2**
$2x + y = 6$	$3x + 4y = 4$
$2(4) + (-2) \stackrel{?}{=} 6$	$3(4) + 4(-2) \stackrel{?}{=} 4$
$6 = 6$ ✓	$4 = 4$ ✓

The solution to the system of equations is $(4, -2)$. You can also check the solution graphically. ◄

The substitution method of solving systems of linear equations works well when one of the variables in one of the equations has a coefficient of 1 or -1. With other coefficients, substitution can lead to complicated expressions. In such situations, it may be better to use the **elimination method** to solve the system.

The elimination method makes use of the addition and multiplication properties of equality. The equations in the system, or multiples of the equations, are added to or subtracted from each other to eliminate all but one of the variables. Once you have an equation with only one variable, solve this equation and use the result to solve the system.

PROBLEM SOLVING TIP

In Example 1, it is easy to solve for y in Equation 1 because it has a coefficient of 1. When you choose a variable to isolate, look for those with coefficients of 1 or -1.

COMMUNICATING ABOUT ALGEBRA

Explain why the system in Example 2 would be difficult to solve by graphing.

EXAMPLE 2

Solve the system using the elimination method.

$$\begin{cases} 2x - 3y = 0 & \textbf{Equation 1} \\ -4x - 1 = -3y & \textbf{Equation 2} \end{cases}$$

Solution

$$\begin{cases} 2x - 3y = 0 \\ -4x + 3y = 1 \end{cases}$$ Write Equation 2 in standard form so like terms are arranged in columns.

Because the coefficients of the y-terms are opposites, you can eliminate the y-term by adding the equations. Then solve for x.

$$\begin{array}{r} 2x - 3y = 0 \\ -4x + 3y = 1 \\ \hline -2x = 1 \end{array}$$ Add the equations.

$$x = -\frac{1}{2}$$ Solve for x.

To find y, substitute $-\frac{1}{2}$ for x in either of the original equations.

$$-4\left(-\frac{1}{2}\right) - 1 = -3y$$ Substitute for x in Equation 2.

$$1 = -3y$$ Simplify.

$$-\frac{1}{3} = y$$ Solve for y.

Check the solution $\left(-\frac{1}{2}, -\frac{1}{3}\right)$ in both original equations.   ◄

You may find that neither variable has coefficients with equal absolute values, so you cannot immediately add or subtract to eliminate a variable. You may be able to obtain coefficients you can add or subtract by multiplying both sides of one or both equations by a constant.

CHECK UNDERSTANDING

Explain how you can choose whether you will add or subtract the equations.

EXAMPLE 3

MANUFACTURING   Jay can assemble a doll in 3 minutes and a truck in 5 minutes during a 7 hour shift. Carol can package a doll in 4 minutes and a truck in 3 minutes during a 7 1/2 hour shift. How many of each toy should be produced so every doll and truck is assembled and packaged?

Solution

Let d represent the number of dolls and let t represent the number of trucks. Write a system of equations.

$$\begin{cases} 3d + 5t = 420 & \textbf{Equation 1} \text{ represents assembling.} \\ 4d + 3t = 450 & \textbf{Equation 2} \text{ represents packaging.} \end{cases}$$

You can obtain y-terms with opposite coefficients if you multiply both sides of Equation 1 by 4 and both sides of Equation 2 by 3.

$$
\begin{aligned}
3d + 5t &= 420 \rightarrow 12d + 20t = 1680 \quad \text{Multiply both sides by 4.} \\
4d + 3t &= 450 \rightarrow \underline{12d + 9t = 1350} \quad \text{Multiply both sides by 3.} \\
& 11t = 330 \quad\quad\ \text{Subtract the equations.} \\
& t = 30 \quad\quad\ \text{Solve for } t.
\end{aligned}
$$

$$
\begin{aligned}
3d + 5(30) &= 420 \quad \text{Substitute 30 for } t \text{ in the first equation.} \\
3d &= 270 \quad \text{Simplify.} \\
d &= 90 \quad\ \text{Solve for } d.
\end{aligned}
$$

They should produce 30 trucks and 90 dolls. ◄

COMMUNICATING ABOUT ALGEBRA

In Example 3, the variable d was eliminated. Explain to another student how you could eliminate the variable t instead of d.

As you saw in Lesson 4.1, a system can have one, none, or an infinite number of solutions. A system with *at least one* solution is called a **consistent system.** A consistent system with *exactly one* solution is **independent,** and a consistent system with an infinite number of solutions is **dependent.** A system with *no solutions* is called an **inconsistent system.**

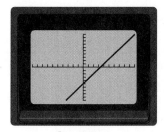

Consistent
Dependent

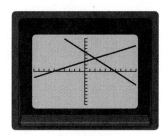

Consistent
Independent

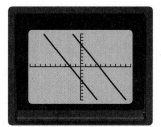

Inconsistent

EXAMPLE 4

Use the elimination method to solve the system. State whether the system is consistent or inconsistent. If the system is consistent, state whether it is independent or dependent.

$$
\begin{cases} -6x + 3y = -9 & \textbf{Equation 1} \\ -2x + y = -3 & \textbf{Equation 2} \end{cases}
$$

Solution

$$
\begin{aligned}
-6x + 3y = -9 &\rightarrow -6x + 3y = -9 \\
-2x + y = -3 &\rightarrow \underline{6x - 3y = 9} \quad \text{Multiply both sides by } -3. \\
& 0 = 0 \quad\ \text{Add the equations.}
\end{aligned}
$$

Adding the equations results in the true equation $0 = 0$. This means an infinite number of ordered pairs are solutions to the system. Since there is at least one solution, the system is consistent. There are an infinite number of solutions, so the system is dependent. Note that a false equation such as $4 = 0$ tells you a system is inconsistent. ◄

CHECK UNDERSTANDING

The system in Example 4 is a dependent system. What would a graph of this system look like? How could you have predicted the graph's appearance?

Use the substitution method to solve each system. Check your answer.

1. $\begin{cases} 4x + y = 9 \\ x - 2y = 0 \end{cases}$

2. $\begin{cases} 7x + 5y = 68 \\ x - 9y = -34 \end{cases}$

3. $\begin{cases} \dfrac{x}{4} - \dfrac{y}{5} = 9 \\ y = 5x \end{cases}$

Use the elimination method to solve each system. Check your answer.

4. $\begin{cases} 2x + 2y = 8 \\ 2x - 3y = 3 \end{cases}$

5. $\begin{cases} 3x + y = 15 \\ 2x + 2y = 16 \end{cases}$

6. $\begin{cases} 0.3x + 0.2y = 0.8 \\ 0.2x - 0.3y = 1.4 \end{cases}$

7. CONCERT TICKETS On the first night of a concert, Fish Ticket Outlet collected $67,200 on the sale of 1600 lawn seats and 2400 reserved seats. On the second night, the outlet collected $73,200 by selling 2000 lawn seats and 2400 reserved seats.

    a. Write a system of equations to find the cost of each type of seat, where x represents the cost of the lawn seats, and y represents the cost of the reserved seats.

    b. Solve the system to determine the cost of each type of seat.

Solve each system. State whether the system is *independent*, *dependent*, or *inconsistent*.

8. $\begin{cases} 2x - 3y = -7 \\ x + 2y = 7 \end{cases}$

9. $\begin{cases} 2x - 3y = -7 \\ 4x - 6y = -14 \end{cases}$

10. $\begin{cases} 2x - 3y = -7 \\ 4x - 6y = 8 \end{cases}$

11. WRITING MATHEMATICS Make a chart summarizing the possible solutions that can occur with a system of two linear equations. For each possibility, include a description, the number of solutions, the type of system, a graphical representation, and an example.

PRACTICE

Use the substitution method to solve each system. Check your answer.

1. $\begin{cases} x + y = 10 \\ x - y = 6 \end{cases}$

2. $\begin{cases} 3x - y = 5 \\ x + y = 0.5 \end{cases}$

3. $\begin{cases} x + y = 12 \\ 0.2x + 0.5y = 3.6 \end{cases}$

4. COFFEE BLENDS A gourmet shop wants to mix coffee beans that cost $3.00 per pound with coffee beans that cost $4.25 per pound to create 25 pounds of a new blend that costs $3.50 per pound. Find the number of pounds of each needed to produce the new blend.

Use the elimination method to solve each system. Check your answer.

5. $\begin{cases} x - 2y = 6 \\ -x + 3y = -4 \end{cases}$

6. $\begin{cases} 9x + 3y = -3 \\ 2x - 3y = -8 \end{cases}$

7. $\begin{cases} 3x + 4y = 4 \\ x - 2y = 2 \end{cases}$

8. $\begin{cases} 0.3x - 0.2y = 4 \\ -0.6x + 0.6y = 12 \end{cases}$

Describe two ways to transform each system so you can solve by elimination. Then solve the system and check your answer.

9. $\begin{cases} 2x + 3y = 17 \\ 5x + 7y = 29 \end{cases}$

10. $\begin{cases} 5x - 9y = 7 \\ 7y - 3x = -5 \end{cases}$

11. **WRITING MATHEMATICS** The solution to a system of equations is $x = 8$ and $y = 5$. Suppose that you multiply both sides of each equation by 2 to produce a new system of equations. Is the solution to the new system $x = 8$ and $y = 5$, or $x = 16$ and $y = 10$, or some other value? Explain your answer.

12. **GOLD ALLOYS** Pure gold is too soft for many applications. A gold alloy is a mixture of pure gold and other more durable materials. Pure gold is 24-carat, 18-carat gold is $\frac{18}{24}$ gold, 14-carat gold is $\frac{14}{24}$ gold, and 12-carat gold is $\frac{12}{24}$ gold. A jeweler has one bar of 12-carat gold alloy and one bar of 18-carat gold alloy. How many grams of each alloy must be mixed to obtain 15 grams of 14-carat gold?

Solve each system. State whether the system is *independent, dependent,* or *inconsistent.*

13. $\begin{cases} x - 3y = -3 \\ -2x + 6y = 12 \end{cases}$

14. $\begin{cases} 1 + y = x \\ y - 2x = 5 \end{cases}$

15. $\begin{cases} 4x - 2y = -19 \\ -6x - 3y = 1.5 \end{cases}$

16. $\begin{cases} 2x - 3y = 6 \\ 3y - 2x = -6 \end{cases}$

EXTEND

17. **GEOMETRY** The sum of the measures of the two acute angles in a right triangle is 90° and the difference is 16°. Find the measure of each angle.

18. **GEOMETRY** Two angles are supplementary. The measure of one angle is 6° less than 5 times the measure of the other angle. Find the measure of each angle.

19. **AVIATION** An airplane flies 3000 miles from New York to San Francisco with a tail wind. The trip takes 5 hours. On the return trip, the plane flies at the same speed against the same wind, and the trip takes 6 hours. If distance is equal to speed multiplied by time, find the speed of the plane and the speed of the wind.

Solve each system. Check your answer.

20. $\begin{cases} 2(x - y) - 4y = 2 \\ -6x - 8y = 2(-17 + y) \end{cases}$

21. $\begin{cases} \frac{1}{5}(x + y) = 6 - \frac{3}{2}y \\ \frac{3}{5}(x + 2) = \frac{1}{2}y + \frac{16}{5} \end{cases}$

22. **CHECKING ACCOUNTS** Two banks offer special checking plans. At Standard Bank, a checking account has a monthly service charge of $13.00 and a charge of 10¢, for every check written during the month. At Guarantee Trust, each check costs 10¢, and the monthly service charge is $14.25. How many checks have to be written in one month for the checking plan at Guarantee Trust to be cheaper than the plan at Standard Bank? Explain your answer.

THINK CRITICALLY

23. Can a system of two linear equations have exactly two solutions? Explain.

Solve each system. Assume a and b are constants.

24. $\begin{cases} ax + by = 2 \\ -ax + 2by = 1 \end{cases}$

25. $\begin{cases} x + y = 2a \\ x - y = 2b \end{cases}$

26. $\begin{cases} 3x + y = a \\ x - 3y = b \end{cases}$

What value of c will make each of the following a dependent system?

27. $\begin{cases} 12x - 18y = 3 \\ 4x - 6y = c \end{cases}$

28. $\begin{cases} cx - 14y = -6 \\ -15x + 21y = 9 \end{cases}$

What value of c will make each of the following an inconsistent system?

29. $\begin{cases} x + 3y = 4 \\ -2x - cy = 3 \end{cases}$

30. $\begin{cases} x - 2y = c \\ -0.5x + y = -2 \end{cases}$

MIXED REVIEW

Find the slope and y-intercept of each line

31. $y = 2x + 3$

32. $2y = -6x + 10$

33. $3x - 4y = 12$

34. $3y + 10 = 0$

35. STANDARDIZED TESTS Determine which ordered pair is a solution of the system
$3x + y = 2.5$ and $2x - y = 0$.

    A. $(1, 0.5)$ **B.** $(1, 2)$ **C.** $(0.5, 1)$ **D.** $(-0.5, 1)$

PROJECT *Connection* You will need two sheets of poster board, a centimeter ruler, scissors, and the two cereal boxes you obtained in Getting Started.

1. Carefully open and flatten each cereal box. Be sure to keep all tabs that are necessary for joining sides of the box.

2. On one of the poster boards, draw a pattern or *template* for each box. Mark the fold lines and dimensions. Cut out each template and discard the waste.

3. As a group, plan how to cut out as many boxes as possible from the other sheet of poster board using the templates to trace the box outlines. Your plan must include at least one of each size box. You may not cut any of the boxes into smaller pieces.

4. After all the outlines are drawn, estimate the percent of the poster board that will be wasted W and the percent that will be used for boxes U.

5. Let C represent the cost of one sheet of the poster board. Determine $U\%$ of C and $W\%$ of C. Explain what these amounts represent.

6. Devise a method to estimate the unit cost for each size box.

Career
Craft Worker

As an alternative to mass-produced goods, many companies specialize in providing consumers with handcrafted one-of-a-kind goods. These businesses rely on skilled artisans such as woodworkers, weavers, potters, quilters, and painters who may spend many hours completing a single item.

Daria and Frank work as partners in a company that makes dollhouses. Their job is to paint the completed dollhouses that come from the carpentry group. Daria and Frank wonder how long it would take each of them to paint a dollhouse alone.

Decision Making

1. When Daria and Frank work together, it takes them 2 hours to paint a single dollhouse. What fractional part of the job is complete in 1 hour if they work together?

2. If d represents the number of hours it would take for Daria to paint a dollhouse alone, then $\frac{1}{d}$ represents the portion of the job that Daria could complete in 1 hour. If f represents the number of hours it would take for Frank to paint a dollhouse alone, then what does $\frac{1}{f}$ represent?

3. Using $\frac{1}{d}$ and $\frac{1}{f}$, write an equation to represent the portion of the job that is completed after 1 hour if Daria and Frank work together.

4. One afternoon, after they had been painting a dollhouse for 1 hour, Daria was called away. It took Frank another 3 hours to finish painting the doll-house. Using $\frac{1}{d}$ and $\frac{1}{f}$, write an equation to represent this situation.

5. Let $x = \frac{1}{d}$ and $y = \frac{1}{f}$. Rewrite the equations from Question 3 and Question 4 in terms of x and y.

6. Solve this system of equations. Then use the solutions to find d and f. Check your answers.

4.3 Solving Systems in Three Variables

Explore

A nut producer ships gift packs of different nuts. Flat cardboard dividers are used to separate a box or tin into equal sections and each type of nut is packed into a different section. Work with a partner, and use paper, scissors, and tape to model each of the following sets of dividers. You can cut a notch in each sheet of paper so they will slide together easily.

1. Use two dividers intersecting in a common line to package four types of nuts in a square box.

2. Use three dividers intersecting in a common line to package six types of nuts in a round tin.

3. Use three dividers to package six types of nuts in a rectangular box. Two of the dividers are parallel, while the third is perpendicular to the parallel pieces. Do all three dividers have a common point or line of intersection?

4. Use two dividers to package three types of nuts in a box. The dividers separate the nuts into three layers. Do the two dividers have a common point or line of intersection?

5. Use the four-pack divider in Question 1 and another divider to package eight types of nuts in a square box. The four-pack divider is intersected horizontally by the third divider to separate the box into two levels. Do all three dividers have a common point or line of intersection?

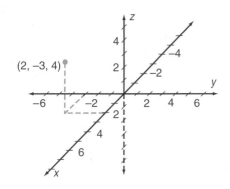

Build Understanding

Just as an ordered pair (x, y) is a solution of a linear equation in two variables, an **ordered triple** (x, y, z) is a solution of a **linear equation in three variables** $Ax + By + Cz = D$ where A, B, and C are not all zero.

SPOTLIGHT ON LEARNING

WHAT? In this lesson you will learn
- to use the substitution and elimination methods to solve systems of linear equations in three variables.
- to interpret systems of linear equations in three variables geometrically.

WHY?
Understanding how to solve systems/ equations in three variables can help you to solve problems in mixing paint pigments, lumber cutting, and animal nutrition.

The graph of the point $(2, -3, 4)$ is shown on a three-dimensional coordinate system at the bottom of page 176. The x-axis, y-axis, and z-axis are perpendicular to each other.

To graph the plane $2x + 4y + 6z = 12$, determine the points at which it intersects the x-axis (y and z are zero), the y-axis (x and z are zero), and the z-axis (x and y are zero). Connect the points with line segments and shade the resulting triangular region of the plane.

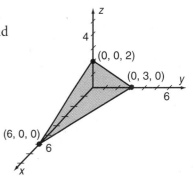

A **system of linear equations in three variables** usually consists of three equations. In Explore, you used paper to model different ways planes can intersect. The system has exactly one solution when the three planes intersect in a single point as shown in Question 5.

You can solve systems of linear equations in three variables. First use elimination to reduce the system to two equations in two variables.

EXAMPLE 1

Solve the system using the elimination method.

$$\begin{cases} 4x - 2y - 3z = 5 & \textbf{Equation 1} \\ -8x - y + z = -5 & \textbf{Equation 2} \\ 2x + y + 2z = 5 & \textbf{Equation 3} \end{cases}$$

Solution

Use two of the equations to eliminate one of the variables.

$$\begin{array}{ll} -8x - y + z = -5 & \text{Add to eliminate } y\text{-term.} \\ \underline{2x + y + 2z = 5} & \\ -6x + 3z = 0 & \text{First new equation.} \end{array}$$

You need to get a second equation with the same variables, x and z.

$$\begin{array}{lll} 2x + y + 2z = 5 \rightarrow & 4x + 2y + 4z = 10 & \text{Multiply Equation 3 by 2.} \\ 4x - 2y - 3z = 5 \rightarrow & \underline{4x - 2y - 3z = 5} & \text{Add to eliminate } y\text{-term.} \\ & 8x + z = 15 & \text{Second new equation.} \end{array}$$

$$\begin{cases} 8x + x = 15 \\ -6x + 3z = 0 \end{cases} \quad \text{New system.}$$

Use elimination again to solve the new system of equations for x.

$$\begin{array}{lll} 8x + z = 15 \rightarrow & -24x - 3z = -45 & \text{Multiply by } -3. \\ -6x + 3z = 0 \rightarrow & \underline{-6x + 3z = 0} & \text{Add to eliminate the } z\text{-term.} \\ & -30x = -45 & \text{Solve for } x. \\ & x = \dfrac{3}{2} \end{array}$$

Next, use the value of x to solve for z.

$$\begin{array}{ll} 8\left(\dfrac{3}{2}\right) + z = 15 & \text{Substitute } \dfrac{3}{2} \text{ for } x \text{ in second new equation.} \\ 12 + z = 15 & \text{Simplify.} \\ z = 3 & \text{Solve for } z. \end{array}$$

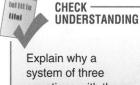

CHECK
UNDERSTANDING

Explain why a system of three equations with three unknowns could have 0, 1, or infinitely many solutions, but couldn't have exactly two solutions.

CHECK UNDERSTANDING

In Example 1, the y-terms were eliminated first to produce a system that had only x- and z-terms. Could you eliminate the x- or z-terms first, instead of the y-terms? Explain.

Finally, substitute the values for x and z into any of the three original equations to solve for y.

$$2x + y + 2z = 5 \quad \text{Equation 3}$$
$$2\left(\frac{3}{2}\right) + y + 2(3) = 5 \quad \text{Substitute } \frac{3}{2} \text{ for } x \text{ and } 3 \text{ for } z.$$
$$3 + y + 6 = 5 \quad \text{Simplify.}$$
$$y = -4 \quad \text{Solve for } y.$$

The solution to the system of equations is the ordered triple $\left(\frac{3}{2}, -4, 3\right)$. Check the solution in all three original equations. ◀

When one or more of the equations has a missing variable, substitution can be an effective method for solving the system.

EXAMPLE 2

PAINT MIXTURES An artist can create colors by mixing pigments. To mix a particular color using red, blue, and yellow pigments, the artist uses three times as much yellow pigment as red pigment. The amount of yellow pigment plus twice the amount of blue pigment is 80% of the mixture. If the sum of the amounts of red, yellow, and blue pigments is 100% of the mixture, what percent of each color is in the mixture?

Solution

Let r represent the percent of red pigment, let b represent the percent of blue pigment, and let y represent the percent of yellow pigment. Then you can model the problem with the linear system:

$$\begin{cases} r + b + y = 100 & \textbf{Equation 1} \\ 3r = y & \textbf{Equation 2} \\ 2b + y = 80 & \textbf{Equation 3} \end{cases}$$

Both Equation 2 and Equation 3 are missing a variable, but include a y-term.

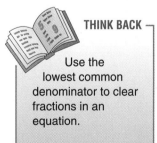

THINK BACK

Use the lowest common denominator to clear fractions in an equation.

$$r = \frac{y}{3} \qquad\qquad \text{Solve Equation 2 for } r.$$

$$b = \frac{1}{2}(80 - y) \qquad\qquad \text{Solve Equation 3 for } b.$$

$$b = 40 - \frac{y}{2}$$

Now substitute the expressions for r and b into Equation 1.

$$r + b + y = 100 \quad \text{Equation 1}$$
$$\left(\frac{y}{3}\right) + \left(40 - \frac{y}{2}\right) + y = 100 \quad \text{Substitute } \frac{y}{3} \text{ for } r, \text{ and } 40 - \frac{y}{2} \text{ for } b.$$
$$2y + 240 - 3y + 6y = 600 \quad \text{Multiply both sides by 6.}$$
$$5y = 360 \quad \text{Simplify.}$$
$$y = 72 \quad \text{Solve for } y.$$

To find r, substitute 72 for y in the equation $r = \dfrac{y}{3}$:

$$r = \frac{y}{3} = \frac{1}{3}(72) = 24$$

To find b, substitute 72 for y in the equation $b = 40 - \dfrac{y}{2}$:

$$b = 40 - \frac{y}{2} = 40 - \frac{1}{2}(72) = 4$$

The solution is the ordered triple (24, 4, 72). The paint mixture contains 24% red pigment, 4% blue pigment, 72% yellow pigment. Check these values in all three of the original equations. ◄

COMMUNICATING ABOUT ALGEBRA

Explain to another student a different way you could have eliminated the y-terms in Example 2 to reduce the system to a system in two variables.

Like systems of linear equations with two variables, systems of linear equations with three variables can be consistent and independent, consistent and dependent, or inconsistent.

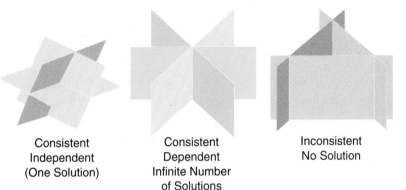

Consistent Independent (One Solution)

Consistent Dependent Infinite Number of Solutions

Inconsistent No Solution

The systems in Examples 1 and 2 are both consistent and independent since they have one solution. When solving a system, if you eliminate the variables and obtain a false statement, such as $0 = 3$, then the system has no solution and is inconsistent. If you eliminate the variables and obtain a true statement, such as $0 = 0$, then the system is dependent. A dependent system has an infinite number of solutions.

TRY THESE

Determine whether the given ordered triple is a solution of the system.

1. $\begin{cases} 4x - y + z = 6 \\ 2x + y + 2z = 3 \\ 3x - 2y + z = 3 \end{cases}$ (2, 1, −1)

2. $\begin{cases} 4x + 2y + 5z = 6 \\ 2x - y + z = 5 \\ x + 2y - z = 0 \end{cases}$ (1, 2, 3)

3. $\begin{cases} 4x + 2y + 5z = 6 \\ 2x - y + z = 5 \\ x + 2y - z = 2 \end{cases}$ (2, −1, 0)

Use either the elimination method or the substitution method to solve each system.

4. $\begin{cases} x - y + 2z = -3 \\ x + 2y + 3z = 4 \\ 2x + y + z = -3 \end{cases}$

5. $\begin{cases} 2x + 3y + 12z = 4 \\ 4x - 6y + 6z = 1 \\ x + y + z = 1 \end{cases}$

6. $\begin{cases} 5x - 3y + 2z = 13 \\ 2x + 4y - 3z = -9 \\ 4x - 2y + 5z = 13 \end{cases}$

7. **LUMBER CUTTING** At a lumber company, there are three sawmills, A, B, and C. When all three of them are working, 5700 board-feet of lumber can be produced in one day. Mills A and B together can produce 3400 board-feet in a day, while mills B and C together can produce 4200 board-feet.

   a. Write a system of equations to find the number of board-feet each mill can produce by itself in a day, where a represents the board-feet from mill A, b represents the board-feet from mill B, and c represents the board-feet from mill C.

   b. Determine the number of board-feet each mill can produce in one day.

8. **WRITING MATHEMATICS** Discuss why it is necessary to eliminate the same variable when you reduce a system of three equations in three variables to a system of two equations.

Solve and check each system. State whether the system is *independent, dependent,* or *inconsistent.*

9. $\begin{cases} -x + 4y - 3z = 2 \\ 2x - 8y + 6z = 1 \\ 3x - y + z = 0 \end{cases}$
10. $\begin{cases} 2x - 3y + 3z = -15 \\ 3x + 2y - 5z = 19 \\ 5x - 4y - 2z = -2 \end{cases}$
11. $\begin{cases} x + y = 9 \\ y + z = 7 \\ x - z = 2 \end{cases}$

PRACTICE

1. **WRITING MATHEMATICS** Write a real world problem that involves a system of linear equations in three variables. Interpret what the solution means in the situation.

Solve and check each system.

2. $\begin{cases} x + y + z = 6 \\ 2x - y + 3z = 9 \\ x + 2y + 2z = 9 \end{cases}$
3. $\begin{cases} 0.2x + 0.3y - 0.1z = 0.1 \\ 0.1x + 0.2y + 0.2z = 0.5 \\ 0.1x - 0.1y + 0.1z = 0.6 \end{cases}$
4. $\begin{cases} 2x - 3y + z = 5 \\ x + 3y + 8z = 22 \\ 2x - y + 2z = 9 \end{cases}$

5. $\begin{cases} 2x - y + 3z = 4 \\ x + 2y - z = -3 \\ 4x + 3y + 2z = -5 \end{cases}$
6. $\begin{cases} 2x + 5y + 2z = 9 \\ 4x - 7y - 3z = 7 \\ 3x - 8y - 2z = 9 \end{cases}$
7. $\begin{cases} \dfrac{1}{2}x + \dfrac{1}{2}y + z = \dfrac{1}{2} \\ \dfrac{1}{2}x - \dfrac{1}{4}y - \dfrac{1}{4}z = 0 \\ \dfrac{1}{4}x + \dfrac{1}{12}y + \dfrac{1}{6}z = \dfrac{1}{6} \end{cases}$

8. $\begin{cases} 4x + 2y - 3z = 6 \\ x - 4y + z = -4 \\ -x + 2z = 2 \end{cases}$
9. $\begin{cases} 4y - z = -13 \\ 3y + 2z = 4 \\ 6x - 5y - 2z = 0 \end{cases}$
10. $\begin{cases} x + y = 7 \\ 3y + 2z = 9 \\ 4x + z = 5 \end{cases}$

11. **ANIMAL NUTRITION** A veterinarian wants to make a food mix for guinea pigs that contains 23 g of protein, 6.2 g of fat, and 16 g of moisture. He has three mixes available, the compositions of which are shown in the table. How many grams of each mix should be used to get the desired new mix?

Mix	Protein (grams)	Fat (grams)	Moisture (grams)
A	0.2	0.02	0.15
B	0.1	0.06	0.10
C	0.15	0.05	0.05

Solve and check each system. State whether the system is *independent*, *dependent*, or *inconsistent*.

12. $\begin{cases} x + 2y + z = 1 \\ x - y + z = 1 \\ 2x + y + 2z = 2 \end{cases}$ 13. $\begin{cases} x + z = 0 \\ x + y + 2z = 3 \\ y + z = 2 \end{cases}$ 14. $\begin{cases} 4x + 9y = 8 \\ 8x + 6z = -1 \\ 6y - 6z = -1 \end{cases}$

EXTEND

15. BERRY PICKING A strawberry grower picked a total of 87 quarts of strawberries in 3 days. On Friday, she picked 15 more quarts than on Thursday. On Saturday, she picked 3 quarts fewer than on Friday. How many quarts of strawberries did she pick each day?

16. TELEVISION PRODUCTION A company produces three color television sets—models L, M, and N. Each model L set requires 2 h electronics work, 2 h assembly time, and 1 h finishing time. Each model M set requires 1 h electronics work, 3 h assembly time, and 1 h finishing work. Each model N set requires 3 h of electronics work, 2 h assembly time, and 2 h finishing time. Every week there are 100 h available for electronics work, 100 h available for assembly, and 65 h available for finishing. Write and solve a system of linear equations to find the number of each model produced each week.

17. WRITING MATHEMATICS Explain why the following system is an inconsistent system.

$$\begin{cases} 2x + 3y - z = 5 \\ 4x + 6y - 2z = 15 \\ x - 4y + 3z = 5 \end{cases}$$

18. WRITING MATHEMATICS What can you deduce about this system without solving it?

$$\begin{cases} x + 2y - 3z = -12 \\ -4x + 8y + z = -7 \\ -5x + 6y + 4z = 5 \end{cases}$$

THINK CRITICALLY

19. ANIMAL NUTRITION A veterinarian wants to make a food mix for hamsters that contains 18.5 g of protein, 4.9 g of fat, and 13 g of moisture. She has three mixes available. The table shows the amount of protein, fat, and moisture in each mix as a *percent* of the total weight (in grams) of each mix. How many grams of each mix should she use to get the desired new mix?

Mix	Protein (percent)	Fat (percent)	Moisture (percent)
A	20	2	15
B	10	6	10
C	15	5	5

20. WRITING MATHEMATICS Suppose you solved a linear system in three variables, and two of the equations were equivalent, while the third was different. Is this system independent, dependent, or inconsistent? Describe the intersection of the graph of this system.

21. The points (1, 1, 2), (3, 2, –6), and (6, 0, –4) are solutions to the equation $z = C - Ax - By$. Find A, B, and C.

22. MODELING Model two ways three planes can intersect and be a dependent system.

23. MODELING Model three ways three planes can produce an inconsistent system.

24. Solve the system. $\left(\text{Hint: Let } l \text{ represent } \dfrac{1}{x}, \text{ let } m \text{ represent } \dfrac{1}{y}, \text{ and let } n \text{ represent } \dfrac{1}{z}.\right)$

$$\begin{cases} \dfrac{2}{x} - \dfrac{1}{y} - \dfrac{3}{z} = -1 \\[2mm] \dfrac{2}{x} - \dfrac{1}{y} + \dfrac{1}{z} = -9 \\[2mm] \dfrac{1}{x} + \dfrac{2}{y} - \dfrac{4}{z} = 17 \end{cases}$$

25. THREE-DIMENSIONAL COORDINATE SYSTEM You can use three mutually perpendicular axes as shown on p. 176. A point $(2, -5, 4)$ is located by moving 2 units on the x-axis, -5 units on the y-axis, and 4 units on the z-axis. Pairs of axes determine three *coordinate planes*; the xy-plane, the yz-plane, and the xz-plane.

    a. Graph the points $(-3, 2, 4)$, $(0, 4, -2)$, and $(5, -1, 3)$.

    b. Find the points at which the plane $2x + 4y + 6z = 12$ intersects the x-axis, the y-axis, and the z-axis.

    c. Graph the points you determined in part b. Connect them with line segments and shade the resulting triangular region that is the graph of $2x + 4y + 6z = 12$.

MIXED REVIEW

Solve for y.

26. $2x - 2y = 6$ **27.** $2x - 5y = 7x - 40$ **28.** $(y - 2) = 7x + 25$ **29.** $10(x + y) = 2(5x - 5)$

Indicate the dimensions of each matrix.

30. $\begin{bmatrix} 3 & -2 & 5 \end{bmatrix}$ **31.** $\begin{bmatrix} 3 \\ -2 \\ 5 \end{bmatrix}$ **32.** $\begin{bmatrix} 4 & 1 \\ 0 & 3 \end{bmatrix}$ **33.** $\begin{bmatrix} 4 & -7 & 12 \\ 10 & -5 & 0 \end{bmatrix}$

Determine whether the following systems are *independent*, *dependent*, or *inconsistent*.

34. $\begin{cases} x - 2y + 4z = -10 \\ -3x + 6y - 12z = 20 \\ 2x + 5y + z = 12 \end{cases}$ **35.** $\begin{cases} 2x - 8y + 2z = -10 \\ -x + 4y - z = 5 \\ 3x - 12y + 3z = -15 \end{cases}$

36. STANDARDIZED TESTS The largest angle of a triangle is twice the measure of the smallest, and the third angle is $20°$ less than the largest. Find the measure of the smallest angle.

    A. $20°$ **B.** $40°$ **C.** $60°$ **D.** $80°$

Let $f(x) = 2x + 3$ and $g(x) = 4x^2$. Determine the following.

37. $(f \circ g)(x)$ **38.** $(g \circ f)(x)$ **39.** $f^{-1}(x)$ **40.** $f(x) + g(x)$

Compare the graph of each of the following with the graph of $f(x) = |x|$

41. $f(x) = |x - 2| + 6$ **42.** $f(x) = 3|x|$ **43.** $f(x) = |x + 8| - 1$

4.4 Augmented Matrices

Explore

Recall from Lesson 1.3, a **matrix** is a rectangular arrangement of numbers in rows and columns enclosed with brackets. Each number is called an **element** of the matrix.

$$\begin{bmatrix} 2 & 1 & -5 \\ 1 & -3 & 1 \end{bmatrix}$$

1. What are the elements of Row 1 in the above matrix? of Row 2?

2. Compare the elements of the matrix with the coefficients and constants in this system of equations. What do you notice?

$$\begin{cases} 2x + y = -5 \\ x - 3y = 1 \end{cases}$$

3. Use any method to solve the system in Question 2.

4. Interchange Row 1 and Row 2 of the matrix, so that Row 2 becomes Row 1 and Row 1 becomes Row 2. Rewrite the matrix to show these changes.

5. Multiply each element in Row 1 by -2. Then, for each column in Row 2 add the new value of the element in the corresponding column in Row 1. (For example, you first multiply the 1 in Row 1, Column 1 by -2. Then you add -2 to the 2 in Row 2, Column 1 to get 0). Rewrite the matrix using the new values in Row 2.

6. Multiply each element in Row 2 by $\frac{1}{7}$. Rewrite the matrix using the new values in Row 2.

7. How could you write a system of equations that corresponds to the matrix you just created? Write and solve the system.

8. Compare the system you wrote in Question 7 with the system given in Question 2. What do you notice about the two systems? How are the matrices that correspond to the systems related?

Build Understanding

In Explore you saw that you can write a special matrix that represents a system of equations. This type of matrix is called an **augmented matrix.** Augmented matrices are sometimes written with a line to separate the coefficients of the variables from the constants.

$$\begin{bmatrix} 3 & 2 & | & 15 \\ -1 & -2 & | & -7 \end{bmatrix} \quad \begin{bmatrix} 1 & 0 & -2 & | & -5 \\ 0 & -1 & 3 & | & 3 \\ -2 & 0 & 1 & | & 4 \end{bmatrix}$$

Represent any variable missing from an equation with 0.

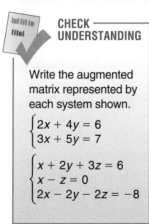

THINK BACK

To find the sum of two rows in a matrix, you add each element from one row to the corresponding element in the second row. To multiply a row by a scalar, you multiply each element in the row by the scalar.

COMMUNICATING ABOUT ALGEBRA

Suppose in Example 1 you perform more row operations on the matrix

$$\begin{bmatrix} 1 & 1 & 2 & | & 13 \\ 0 & 1 & -1 & | & 4 \\ 0 & 0 & 1 & | & 2 \end{bmatrix}$$

Add Row 3 to Row 2. Then subtract new Row 2 from Row 1. Finally add -2 (Row 3) to Row 1. The new matrix is

$$\begin{bmatrix} 1 & 0 & 0 & | & 3 \\ 0 & 1 & 0 & | & 6 \\ 0 & 0 & 1 & | & 2 \end{bmatrix}$$

How does this shorten the work of solving the solution?

PROBLEM SOLVING TIP

Some graphing calculators and computer software allow you to perform row operations on matrices.

You used several operations to transform the original matrix in Explore. These **row operations** produce new matrices that lead to systems having the same solution as the original system. The row operations are:

- Interchange any two rows.

- Replace any row with a nonzero multiple of that row.

- Replace any row with the sum of that row and a multiple of another row.

You can use augmented matrices and row operations to solve systems of equations. First write an augmented matrix for the system. Next, use row operations to transform the matrix into the reduced form

$$\begin{bmatrix} 1 & \blacksquare & \blacksquare & | & \blacksquare \\ 0 & 1 & \blacksquare & | & \blacksquare \\ 0 & 0 & 1 & | & \blacksquare \end{bmatrix}$$

where $\blacksquare$ represents any real number. Then rewrite the reduced matrix as a system of equations and use substitution to solve the system.

EXAMPLE 1

PACKAGING Fruit-in-a-Basket sells three types of fruit samplers. The 13 oz sample has 1 apple, 1 grapefruit and 2 plums. The 17 oz sampler has 1 apple, 3 grapefruit, and 1 plum. The 32 oz sampler has 2 apples, 3 grapefruit, and 4 plums. How much does each type of fruit weigh?

Solution

Write a system of equations.

$$\begin{cases} x + y + 2z = 13 & \textbf{Equation 1} \\ x + 2y + z = 17 & \textbf{Equation 2} \\ 2x + 3y + 4z = 32 & \textbf{Equation 3} \end{cases}$$

$$\begin{bmatrix} 1 & 1 & 2 & | & 13 \\ 1 & 2 & 1 & | & 17 \\ 2 & 3 & 4 & | & 32 \end{bmatrix} \quad \text{Write the augmented matrix.}$$

$$\begin{bmatrix} 1 & 1 & 2 & | & 13 \\ 0 & 1 & -1 & | & 4 \\ 2 & 3 & 4 & | & 32 \end{bmatrix} \quad \text{New Row 2} = -1(\text{Row 1}) + \text{Row 2}$$

$$\begin{bmatrix} 1 & 1 & 2 & | & 13 \\ 0 & 1 & -1 & | & 4 \\ 0 & 1 & 0 & | & 6 \end{bmatrix} \quad \text{New Row 3} = -2(\text{Row 1}) + \text{Row 3}$$

$$\begin{bmatrix} 1 & 1 & 2 & | & 13 \\ 0 & 1 & -1 & | & 4 \\ 0 & 0 & 1 & | & 2 \end{bmatrix} \quad \text{New Row 3} = -1(\text{Row 2}) + \text{Row 3}$$

Now the matrix is in reduced form, and you can write the system of equations represented by the matrix.

$$\begin{cases} x + y + 2z = 13 \\ y - z = 4 \\ z = 2 \end{cases}$$

Use substitution to find the solution is $(3, 6, 2)$. Check this solution in the original system. Therefore, 1 apple weighs 3 oz, 1 grapefruit weighs 6 oz, and 1 plum weighs 2 oz. ◄

You can determine if a system is dependent or inconsistent from the reduced matrix. Look at the reduced matrix for the following system.

$$\begin{cases} 2x + 6y = -3 \\ x + 3y = 2 \end{cases} \rightarrow \begin{bmatrix} 2 & 6 & | & -3 \\ 1 & 3 & | & 2 \end{bmatrix} \rightarrow \begin{bmatrix} 1 & 3 & | & -\frac{3}{2} \\ 0 & 0 & | & \frac{7}{2} \end{bmatrix}$$

Notice that elements in Row 2 represent the equation $0 = \frac{7}{2}$. Since this is untrue, the system is inconsistent and there is no solution.

As another example, look at the reduced matrix for this system.

$$\begin{cases} 2x - y + z = 4 \\ -6x + 3y - 3z = -12 \\ 8x - 4y + 4z = 16 \end{cases} \rightarrow \begin{bmatrix} 2 & -1 & 1 & | & 4 \\ -6 & 3 & -3 & | & -12 \\ 8 & -4 & 4 & | & 16 \end{bmatrix} \rightarrow \begin{bmatrix} 1 & -\frac{1}{2} & \frac{1}{2} & | & 2 \\ 0 & 0 & 0 & | & 0 \\ 0 & 0 & 0 & | & 0 \end{bmatrix}$$

Either row of 0s tells you that the equation represented by that row was derived from Row 1. Therefore this is a dependent system, and there is no unique solution.

EXAMPLE 2

State whether the system of equations represented by each reduced matrix is consistent and independent, consistent and dependent, or inconsistent. Find the solution, if one exists.

a. $\begin{bmatrix} 1 & 2 & | & 11 \\ 0 & 1 & | & 5 \end{bmatrix}$
b. $\begin{bmatrix} 1 & 3 & 4 & | & 10 \\ 0 & 1 & 3 & | & -\frac{31}{4} \\ 0 & 0 & 0 & | & \frac{23}{4} \end{bmatrix}$
c. $\begin{bmatrix} 1 & -1.5 & | & 2 \\ 0 & 0 & | & 0 \end{bmatrix}$

Solution

a. The reduced matrix represents a consistent and independent system. The solution is $(1, 5)$.

b. The reduced matrix represents an inconsistent system. There is no solution.

c. The reduced matrix represents a consistent and dependent system. There are infinitely many solutions. ◄

PROBLEM SOLVING TIP

When transforming augmented matrices into reduced form, work only on the first column until the first row has a 1 and all other rows are zero. Then start on column 2.

ALGEBRA: WHO, WHERE, WHEN

The first recorded use of matrices to solve linear systems appears in an ancient Chinese text called *Nine Chapters on the Mathematical Art*. This book, which was written at least 1700 years ago, also contains the first use of negative numbers.

CHECK UNDERSTANDING

What would the graph of the system in Example 2b look like? What would the graph of the system for 2c look like?

Write the augmented matrix for each system of linear equations. Do not solve.

1. $\begin{cases} x + 2y = 7 \\ x - y = -2 \end{cases}$

2. $\begin{cases} 2x + 4y - 3z = -18 \\ 3x + y - z = -5 \\ x - 2y + 4z = 14 \end{cases}$

3. $\begin{cases} 2x - 3y = 12 \\ 3y + z = -12 \\ 5x - 3z = 3 \end{cases}$

Write the linear system represented by each reduced matrix. Solve the system.

4. $\begin{bmatrix} 1 & -2 & | & 7 \\ 0 & 1 & | & -2 \end{bmatrix}$

5. $\begin{bmatrix} 1 & 0 & 0 & | & 0.65 \\ 0 & 1 & 0 & | & 1.2 \\ 0 & 0 & 1 & | & 0.43 \end{bmatrix}$

6. $\begin{bmatrix} 1 & 0 & -4 & | & 5 \\ 0 & 1 & -12 & | & 13 \\ 0 & 0 & 1 & | & -\frac{1}{2} \end{bmatrix}$

Begin with the following matrix each time. Perform the indicated row operations.

$$\begin{bmatrix} 6 & -3 & | & -9 \\ 2 & 4 & | & 22 \end{bmatrix}$$

7. Interchange Row 1 and Row 2.

8. New Row 2 $= \dfrac{1}{2}$ (Row 2)

9. New Row 2 $= -\dfrac{1}{3}$ (Row 1) + Row 2

10. WRITING MATHEMATICS Explain how solving a system of equations with an augmented matrix is similar to solving by elimination.

Use augmented matrices to solve each system, if possible. State whether the system is independent, dependent, inconsistent, or consistent.

11. $\begin{cases} 2x + 4y - 10z = -2 \\ 3x + 9y - 21z = 0 \\ x + 5y - 12z = 1 \end{cases}$

12. $\begin{cases} x - 2y + z = 4 \\ 3x - 6y + 3z = 12 \\ -2x + 4y - 2z = -8 \end{cases}$

13. $\begin{cases} x + 5y - 3z = 4 \\ 4x - 3y + 2z = 4 \\ 8x - 6y + 4z = 14 \end{cases}$

14. $\begin{cases} 2x + 7y + 15z = -12 \\ 4x + 7y + 13z = -10 \\ 3x + 6y + 12z = -9 \end{cases}$

15. INVESTMENTS Jason has three investments totaling $2500 and receives $212 per year in simple interest. Part of the money is invested at 7%, part is invested at 8%, and part is invested at 9%. There is $1100 less invested at 8% than at 9%.

a. Write a system of equations that would allow you to find the amount of money invested at each rate.

b. Write an augmented matrix to represent the system in Part a.

c. Solve the system to find the amount invested at each rate.

PRACTICE

Write the augmented matrix for each system of linear equations. Do not solve.

1. $\begin{cases} 3x + 4y = 7 \\ -5x + 2y = 10 \end{cases}$

2. $\begin{cases} \frac{1}{3}x + \frac{1}{2}y + z = -1 \\ x - y + \frac{1}{5}z = 1 \\ x + y + z = 5 \end{cases}$

3. $\begin{cases} x - 3y = 6 \\ y + 2z = 2 \\ 7x - 3y - 5z = 14 \end{cases}$

Write the linear system represented by each reduced matrix. Solve the system.

4. $\begin{bmatrix} 1 & 0 & \frac{5}{7} \\ 0 & 1 & \frac{8}{7} \end{bmatrix}$

5. $\begin{bmatrix} 1 & 2 & 3 & 4 \\ 0 & 1 & 2 & 4 \\ 0 & 0 & 1 & 2 \end{bmatrix}$

6. $\begin{bmatrix} 1 & -3 & 2 & 0 \\ 0 & 1 & 0.7 & 0.7 \\ 0 & 0 & 1 & 1 \end{bmatrix}$

7. WRITING MATHEMATICS Explain how you can recognize an inconsistent system when you are using augmented matrices to solve a linear system. How can you recognize a dependent system?

Use augmented matrices to solve each system, if possible. State whether the system is independent, dependent, inconsistent, or consistent.

8. $\begin{cases} x + 4y = 8 \\ 3x + 5y = 3 \end{cases}$

9. $\begin{cases} -4x + 12y = 36 \\ x - 3y = 9 \end{cases}$

10. $\begin{cases} \frac{1}{3}x + \frac{1}{5}y = 2 \\ \frac{1}{3}x - \frac{1}{2}y = -\frac{1}{3} \end{cases}$

11. $\begin{cases} x + y + z = 1 \\ x + 2y + 3z = 4 \\ 4x + 5y + 6z = 7 \end{cases}$

12. $\begin{cases} 4x - y - 3z = 1 \\ 8x + y - z = 5 \\ 2x + y + 2z = 5 \end{cases}$

13. $\begin{cases} x - y + 5z = -6 \\ 6y - 16z = 28 \\ x + 3y + 2z = 5 \end{cases}$

14. CHEMISTRY How many ounces of 5% hydrochloric acid and of 20% hydrochloric acid must be combined to get 10 oz of a solution that is 12.5% hydrochloric acid? Write a system of linear equations, then use augmented matrices to find the solution.

15. FOOD PRICES Rosa bought apples, cheddar cheese, and tomatoes at the grocery store. The apples cost $0.70 a pound, the cheese cost $1.50 a pound, and the tomatoes cost $0.80 a pound. Rosa bought twice as many pounds of apples as cheese, and one more pound of tomatoes than cheese. Her total bill was $8.20. Write a system of equations that allows you to find the number of pounds of each item Rosa bought. Then use augmented matrices to solve the system.

Solve each system.

16. $\begin{cases} 3x = 2y \\ x + y = 5 \end{cases}$

17. $\begin{cases} x - 3y = -4z \\ 2x + z = 4 - 2y \\ 3x - 4y + 2z = 25 \end{cases}$

18. Write an augmented matrix for this system. Do not solve.

$$\begin{cases} 2x + 2y + 2z = -10 + 2w \\ w + x + y + z = -5 \\ x + 4z + 3w = y - 2 \\ w + 3x - 2y + 2z = -6 \end{cases}$$

19. WRITING MATHEMATICS Solve the system at the right using each of the three methods—substitution, elimination, and augmented matrices. Which method was the easiest? Which method was the most difficult? Explain why you might use different methods to solve different types of systems.

$$\begin{cases} 4x - y - 3z = 1 \\ 8x + y - z = 5 \\ 2x + y + 2z = 5 \end{cases}$$

20. ELECTRIC CURRENTS An electric circuit is made of three resistors, R_1, R_2, and R_3, a 6-volt battery, and a 12-volt battery. I_1, I_2, and I_3 represent the amount of current (in amperes) that flows across each resistor. The system of linear equations for the three currents is shown at the right. Find the three currents if:

$$\begin{cases} I_1 - I_2 + I_3 = 0 \\ R_1 I_1 + R_2 I_2 = 6 \\ R_2 I_2 + R_3 I_3 = 12 \end{cases}$$

   a. $R_1 = R_2 = R_3 = 3$ ohms
   b. $R_1 = R_3 = 4$ ohms, and $R_2 = 1$ ohm

21. GEOMETRY The perimeter of a triangle is 56 cm. The longest side measures 12 cm less than the sum of the other two sides. The longest side is also 26 cm less than three times the shortest side. Find the lengths of the three sides.

22. Use an augmented matrix to solve.

$$\begin{cases} w + x + y = 6 - 2z \\ w + 2x + z = -2 \\ 3y + w + x = 2z + 12 \\ w + x + 5z = 4y - 16 \end{cases}$$

23. Are these two matrices equivalent? Justify your answer.

$$\begin{bmatrix} 2 & 5 & 4 & | & 4 \\ 1 & 4 & 3 & | & 1 \\ 1 & -3 & -2 & | & 5 \end{bmatrix} \text{ and } \begin{bmatrix} 1 & 0 & 0 & | & 3 \\ 0 & 1 & 0 & | & -2 \\ 0 & 0 & 1 & | & 2 \end{bmatrix}$$

24. Harold was using the augmented matrix method to solve the system on the right. After he used a row operation to perform the first step, he had an augmented matrix that looked like the one on the right below. Which of the row operations below do you think is best for Harold to do next? Explain.

$$\begin{cases} 2x + y + 2z = 11 \\ 2x - y - z = -3 \\ 3x + 2y + z = 9 \end{cases}$$

   a. Add Row 1 to Row 2.
   c. Multiply Row 2 by -1 and add it to Row 3.

   b. Add Row 3 to Row 2.
   d. Multiply Row 1 by -2 and add it to Row 2.

$$\begin{bmatrix} 1 & \frac{1}{2} & 1 & | & \frac{11}{2} \\ 2 & -1 & -1 & | & -3 \\ 3 & 2 & 1 & | & 9 \end{bmatrix}$$

25. COMPOUND INTEREST An investment of $2560 is made at interest rate i, compounded annually. In 2 years, it grows to $3240. If the formula for compound interest is $A = P(1 + i)^t$, where A is the amount grown, P is the principal, and t is the number of years, find the interest rate.

26. STANDARDIZED TESTS Choose the equation with the greatest slope.

   A. $y = \frac{2}{3}x - 1$ **B.** $y = -\frac{3}{2}x + 1$ **C.** $y = -x + 1$ **D.** $y = -\frac{5}{3}x - \frac{2}{3}x + 1$

Multiplying Matrices

Explore

Video Game Works sells several types of video games. The table on the left gives the number of each type of game sold during a two-week period. The table on the right gives the price of each game.

Number of Games Sold			
	Strategy	**Simulation**	**Arcade**
Week 1	62	74	123
Week 2	57	76	135

Game Prices	
Game Type	**Price, \$**
Strategy	19
Simulation	25
Arcade	19

1. Write the information in the tables in two matrices. What are the dimensions of each matrix?

2. Determine the amount spent on each video game during Week 1. Determine the amount spent during Week 2. Explain your method.

3. Determine the *total* amount of money spent on video games for each week. Explain your method.

4. Create a matrix for the totals each week. What are the dimensions?

5. Compare the number of rows in the totals matrix and in the number of games sold matrix. Compare the number of columns in the totals matrix and the price matrix. What do you notice?

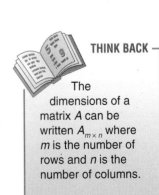

Build Understanding

When you used two matrices in Explore to find a third matrix showing the totals for each week, you performed **matrix multiplication.** You could multiply the matrices because the number of columns in the first matrix is equal to the number of rows in the second matrix.

To multiply two matrices, multiply corresponding elements and then add those products. The product matrix has the same number of rows as the first matrix and the same number of columns as the second matrix.

$$A_{m \times n} \cdot B_{n \times p} = AB_{m \times p}$$

equal

EXAMPLE 1

Use the given matrices to find each product, if possible.

$$A = [2 \quad 4 \quad 6] \qquad B = \begin{bmatrix} 1 \\ 3 \\ 5 \end{bmatrix} \qquad C = \begin{bmatrix} 1 & 2 \\ 3 & 4 \\ 5 & 6 \end{bmatrix} \qquad D = \begin{bmatrix} -2 & 4 \\ 3 & -5 \end{bmatrix}$$

a. AB **b.** BA **c.** AC **d.** BC **e.** CD

Solution

a. The number of columns in A equals the number of rows in B, so the product AB is defined. Since A is a 1×3 matrix and B is a 3×1 matrix, AB is a 1×1 matrix.

$$AB = [2 \quad 4 \quad 6] \begin{bmatrix} 1 \\ 3 \\ 5 \end{bmatrix} = [2(1) + 4(3) + 6(5)] = [44]$$

b. The number of columns in B equals the number of rows in A, so the product BA is defined and is a 3×3 matrix.

$$BA = \begin{bmatrix} 1 \\ 3 \\ 5 \end{bmatrix} [2 \quad 4 \quad 6] = \begin{bmatrix} 1(2) & 1(4) & 1(6) \\ 3(2) & 3(4) & 3(6) \\ 5(2) & 5(4) & 5(6) \end{bmatrix} = \begin{bmatrix} 2 & 4 & 6 \\ 6 & 12 & 18 \\ 10 & 20 & 30 \end{bmatrix}$$

c. The number of columns in A equals the number of rows in C, the product AC is defined and is a 2×2 matrix.

$$AC = [2 \quad 4 \quad 6] \begin{bmatrix} 1 & 2 \\ 3 & 4 \\ 5 & 6 \end{bmatrix}$$

$$= [2(1) + 4(3) + 6(5) \quad\quad 2(2) + 4(4) + 6(6)]$$

$$= [44 \quad 56]$$

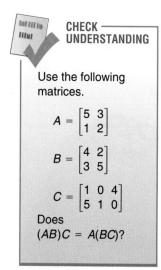
d. Because the number of columns in B does not equal the number of rows in C, the product BC is not defined.

e. The number of columns in C equals the number of rows in D, so product CD is defined and is a 3×2 matrix.

$$\begin{bmatrix} 1 & 2 \\ 3 & 4 \\ 5 & 6 \end{bmatrix} \begin{bmatrix} -2 & 4 \\ 3 & -5 \end{bmatrix} = \begin{bmatrix} 1(-2) + 2(3) & 1(4) + 2(-5) \\ 3(-2) + 4(3) & 3(4) + 4(-5) \\ 5(-2) + 6(3) & 5(4) + 6(-5) \end{bmatrix} = \begin{bmatrix} 4 & -6 \\ 6 & -8 \\ 8 & -10 \end{bmatrix} \blacktriangleleft$$

In parts a and b of Example 1 the products AB and BA are not the same. For most matrices, matrix multiplication is *not commutative*. Matrix multiplication is *associative* and *distributive*. Therefore, if A, B, and C have the correct number of rows and columns, then

$A(BC) = (AB)C$	Associative Property.
$A(B + C) = AB + AC$	Left Distributive Property.
$(B + C)A = BA + CA$	Right Distributive Property.

The *zero product property* does not hold for matrix multiplication. That is, the product of two nonzero matrices can equal a zero matrix.

Matrix multiplication can require a large number of calculations, so it is often easier, and more accurate, to use a graphing calculator or computer software to find the product.

EXAMPLE 2

FOOD PRODUCTION Milk, butter, and cheese are produced throughout the world. The matrix on the left shows the percentage of the total world production of these dairy products for four continents—Africa, Asia, North America, and Europe. The matrix on the right gives the total world production in metric tons for each dairy product. Find the total amount (in metric tons) of dairy products produced in each of the four continents.

Percentage of World Production			
	Milk	**Butter**	**Cheese**
Africa	2.9	2.3	3.3
Asia	10.5	21.5	5.1
North America	18.2	9.8	24.1
Europe	36.4	36.3	47.2

Total World Production	
In Metric Tons	
Milk	474,020
Butter	7,611.826
Cheese	14,475,276

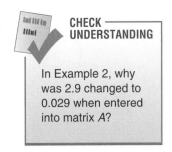

CHECK UNDERSTANDING

In Example 2, why was 2.9 changed to 0.029 when entered into matrix *A*?

Solution

Enter percentage of world production as matrix *A* and total world production as matrix *B*. Multiply the two matrices.

```
MATRIX [A] 4 x 3
[ .029    .023    .033 ]
[ .105    .215    .051 ]
[ .182    .098    .241 ]
[ .364    .363    .472 ]
```

```
MATRIX [B]  3 x 1
[ 474020          ]
[ 7.61E6          ]
[ 1.45E7          ]
```

```
[A]*[B]
[ [ 666502.686   ]
  [ 2424553.766  ]
  [ 4320772.104  ]
  [ 9767966.39   ] ]
```

This product matrix shows that Africa produced a total of 666502.686 metric tons of dairy products, Asia produced 2424553.766 metric tons, North America produced 4320772.104 metric tons, and Europe produced 9767966.39 metric tons.

Find the dimensions of each product AB, if it is defined.

1. $A_{2\times3}B_{3\times2}$　　　　**2.** $A_{5\times1}B_{1\times2}$　　　　**3.** $A_{1\times3}B_{1\times3}$　　　　**4.** $A_{2\times3}B_{3\times5}$

For each pair of matrices, find AB and BA.

5. $A = \begin{bmatrix} -3 & 7 & 2 \end{bmatrix}$　$B = \begin{bmatrix} 1 \\ 4 \\ -5 \end{bmatrix}$　　　　**6.** $A = \begin{bmatrix} -8 & 3 \\ -4 & 4 \end{bmatrix}$　$B = \begin{bmatrix} 1 & -4 \\ 2 & 0 \end{bmatrix}$

7. $A = \begin{bmatrix} 4 & -3 & 1 \\ -5 & 2 & 2 \end{bmatrix}$　$B = \begin{bmatrix} 2 & 1 \\ 0 & 1 \\ -4 & 7 \end{bmatrix}$　**8.** $A = \begin{bmatrix} 1 & 0 & 2 \\ 1 & 0 & 2 \\ 1 & 0 & 2 \end{bmatrix}$　$B = \begin{bmatrix} 2 & 2 & 2 \\ 0 & 0 & 0 \\ -1 & -1 & -1 \end{bmatrix}$

9. WRITING MATHEMATICS Suppose you are given two matrices, A and B. If both the products AB and BA are defined, what must be true about the dimensions of A and B? Give examples to support your answer.

Use the matrices A, B, and C to perform the indicated operations, if possible.

$$A = \begin{bmatrix} -1 & 2 & 1 \\ 2 & 4 & -3 \end{bmatrix} \quad B = \begin{bmatrix} 2 & -2 & 2 \\ 4 & 3 & 0 \\ 1 & -3 & 4 \end{bmatrix} \quad C = \begin{bmatrix} 2 & 3 & 1 \\ -1 & 2 & -1 \\ 1 & 3 & 2 \end{bmatrix}$$

10. AB　　　　**11.** BA　　　　**12.** BC　　　　**13.** CB

14. AC　　　　**15.** CA　　　　**16.** $A(B + C)$　　　　**17.** $AB + AC$

18. FOOTBALL SCORES One Sunday, four teams played football. The table on the right shows the number of times each team scored. The table on the left shows the number of points each type of score is worth. Use matrices to find the total points scored by each team.

	Points
Touchdown	6
Field Goal	3
Point-After Kick	1
Point-After Run/Pass	2
Safety	2

Number of Times Scored					
	Touchdown	Field Goal	Point-After Kick	Point-After Run/Pass	Safety
Bears	3	2	3	0	0
Lions	1	2	0	1	0
Tigers	2	3	2	0	1
Falcons	3	1	2	1	0

PRACTICE

Find the dimensions of each product, if it is defined.

1. $A_{1\times3}B_{2\times1}$　　　　**2.** $B_{2\times1}A_{1\times3}$　　　　**3.** $A_{3\times4}B_{4\times5}$　　　　**4.** $B_{4\times5}A_{3\times4}$

5. WRITING MATHEMATICS Describe how the conditions for multiplying two matrices differ from the conditions for adding and subtracting two matrices.

Multiply the matrices, if possible.

6. $[4 \ -2 \ 3] \begin{bmatrix} 2 \\ 3 \\ -5 \end{bmatrix}$

7. $[5 \ 2] \begin{bmatrix} 2 \\ -5 \end{bmatrix}$

8. $\begin{bmatrix} 1 \\ 2 \\ 3 \end{bmatrix} \begin{bmatrix} -2 & 3 \\ 4 & 1 \end{bmatrix}$

9. $\begin{bmatrix} 4 & -2 \\ 0 & 3 \\ -7 & 5 \end{bmatrix} \begin{bmatrix} 3 \\ 4 \end{bmatrix}$

10. $\begin{bmatrix} -1 & -2 & -3 \\ 0 & 1 & 0 \\ 4 & 5 & 6 \end{bmatrix} \begin{bmatrix} -2 & 4 & 0 \\ -3 & 0 & -8 \end{bmatrix}$

11. $\begin{bmatrix} 1 & -5 & 0 \\ 4 & 1 & -2 \\ 0 & -1 & 3 \end{bmatrix} \begin{bmatrix} 3 & 0 & -1 \\ 0 & 4 & 2 \\ 5 & -3 & 1 \end{bmatrix}$

Use the following matrices to determine if the expressions in each pair are equal.

$$A = \begin{bmatrix} 1 & 2 \\ 0 & -3 \end{bmatrix} \quad B = \begin{bmatrix} 2 & -1 \\ 3 & 1 \end{bmatrix} \quad C = \begin{bmatrix} 3 & 1 \\ -2 & 0 \end{bmatrix}$$

12. $A(B + C)$ and $AB + AC$

13. $(B + C)A$ and $BA + CA$

14. RETAILING Beachware stores record the prices of items in the matrix on the left. The chain also records the daily sales at three stores in the matrix on the right. Find the product of the two matrices. Explain what the product represents.

Beach Towel	Suntan Lotion	Sunglasses
$[\$19.95$	$\$5.50$	$\$11.25]$

	Store 1	Store 2	Store 3
Beach Towel	$[10$	23	$16]$
Suntan Lotion	32	31	55
Sunglasses	$[20$	47	$51]$

EXTEND

15. WRITING MATHEMATICS Suppose you are given the product matrices $AB_{2 \times 2}$, $AB_{3 \times 1}$, and $AB_{2 \times 4}$. For each product matrix, create two different examples that show the dimensions of possible matrices that could be used to give that product matrix.

16. Calculate: $[1 \ 2 \ 0] \begin{bmatrix} 1 & 4 & 3 \\ 0 & 1 & 2 \\ 3 & 2 & 1 \end{bmatrix} \begin{bmatrix} 1 \\ 2 \\ -1 \end{bmatrix}$

Use the given matrices to determine if the expressions in each pair are equal. $A^2 = AA$ **and** $B^2 = BB.$

$$A = \begin{bmatrix} -1 & 0 \\ 2 & 1 \end{bmatrix} \quad B = \begin{bmatrix} 1 & -1 \\ 0 & 2 \end{bmatrix}$$

17. $(A + B)(A - B)$ and $A^2 - B^2$

18. $(A + B)(A - B)$ and $A^2 + BA - AB - B^2$

19. GEOMETRY Matrix multiplication is used to find the coordinates of the vertices of a geometrical figure that has been rotated 90° counterclockwise about the origin. Use a matrix which gives the coordinates of the vertices of a figure in each column, with the x-values in the first row, and the y-values in the second row. Multiply the *rotation* matrix by the coordinate matrix. The resulting matrix will give the coordinates of the rotated figure.

For a triangle, use the rotation matrix $\begin{bmatrix} 0 & -1 \\ 1 & 0 \end{bmatrix}$ and the coordinate matrix $\begin{bmatrix} -1 & 0 & 4 \\ 1 & -3 & 2 \end{bmatrix}$.

a. Determine the product of the rotation matrix times the coordinate matrix.

b. Determine the coordinates of the vertices of the rotated triangle.

c. Draw the original triangle and the rotated triangle.

20. College Costs The number of undergraduate students and graduate students at a small college and the costs of tuition, room, and board are given below. Compute the product of the top matrix times the bottom matrix and describe what the entries in the product matrix mean.

	Undergraduate	Graduate
In / State	2000	1000
Out-of-State	1800	800

$$\begin{bmatrix} 2000 & 1000 \\ 1800 & 800 \end{bmatrix}$$

	Tuition	Room	Board
Undergraduate	8000	1600	2400
Graduate	9500	2500	2400

$$\begin{bmatrix} 8000 & 1600 & 2400 \\ 9500 & 2500 & 2400 \end{bmatrix}$$

THINK CRITICALLY

Solve for x and y.

21. $\begin{bmatrix} 5 & -2 \\ 1 & 7 \end{bmatrix}\begin{bmatrix} 2 & 6 \\ 3 & -4 \end{bmatrix} = \begin{bmatrix} 4 & x \\ y & -22 \end{bmatrix}$

22. $\begin{bmatrix} 1 & 2 \\ 3 & 4 \end{bmatrix}\begin{bmatrix} -3 & x \\ 2 & -1 \end{bmatrix} = \begin{bmatrix} 1 & 3 \\ y & 11 \end{bmatrix}$

23. $\begin{bmatrix} -8 & 3 \\ -4x & y \end{bmatrix}\begin{bmatrix} 1 & -4 \\ 2 & 0 \end{bmatrix} = \begin{bmatrix} -2 & 32 \\ 4 & 16 \end{bmatrix}$

24. $\begin{bmatrix} 4 & 2y \\ 2x & 2 \end{bmatrix}\begin{bmatrix} 2 & 2 \\ -4 & -4 \end{bmatrix} = \begin{bmatrix} 0 & 0 \\ 0 & 0 \end{bmatrix}$

25. **Writing Mathematics** Suppose you are given three matrices, A, B, and C. You know the product $(AB)C$ is defined. If A is a 2×2 matrix and C is a 4×4 matrix, what are the dimensions of B? Explain your answer.

26. **Ore Production** The matrix shows the *percentage* of the total world production of copper, lead, and tin for four countries in one year. For that year, the total world production of copper, lead, and tin was 8791.8 t, 3455.6 t, and 203.8 t, respectively.

Percentage of World Production			
	Copper	Lead	Tin
China	4.2	9.0	14.7
Canada	8.6	10.6	1.9
Peru	3.4	4.3	2.1
Russia	11.3	15.0	7.4

a. Determine the total amount in thousands of metric tons of ore produced in each of the four countries to the nearest tenth.

b. Suppose the next year the total world production of copper, lead, and tin was 9231.1 t, 4001.0 t, and 183.5 t, respectively. Use the same percentages of total world production for each country. Determine the change in thousands of metric tons of the total amount of ore produced in each of the four countries to the nearest tenth.

27. Let $A = \begin{bmatrix} 1 & 2 \\ 0 & 0 \end{bmatrix}$

    a. Determine A^2. What do you notice?

    b. What will the product matrix be for A^5? A^n? Explain.

    c. Try to create another matrix B with the property $B^2 = B$.

MIXED REVIEW

Identify the property illustrated by each of the following.

28. $4 + 0 = 4$

29. $5\left(\frac{1}{5}\right) = 1$

30. $-a + a = a + (-a) = 0$

31. $a(1) = (1)a = a$

Solve and graph on a number line.

32. $|x| = 3$ **33.** $|x| < 3$ **34.** $|x| \geq 3$ **35.** $|x + 2| < 3$

36. GEOMETRY A 10-foot ladder leans against a wall. The bottom of the ladder is 6 ft from the wall.

    a. If the bottom of the ladder is then pulled out 1 ft farther, how much to the nearest hundredth does the top of the ladder move down the wall?

    b. If the bottom of the ladder is then pulled out another foot, how much to the nearest hundredth does the top move down the wall?

    c. If the bottom of the ladder is then pulled out another foot, how much to the nearest hundredth does the top move down the wall?

37. STANDARDIZED TESTS Which of the following products is not defined?

    A. $\begin{bmatrix} -2 & -5 \\ 6 & -3 \end{bmatrix}\begin{bmatrix} 1 & 0 \\ 0 & 1 \end{bmatrix}$ **B.** $\begin{bmatrix} -2 & -5 \\ 6 & -3 \end{bmatrix}\begin{bmatrix} 4 & 0 \\ 1 & -1 \\ 2 & -3 \end{bmatrix}$ **C.** $\begin{bmatrix} -2 & 1 & 0 \\ 3 & 2 & 5 \end{bmatrix}\begin{bmatrix} 4 & 0 \\ 1 & -1 \\ 2 & -3 \end{bmatrix}$ **D.** $\begin{bmatrix} 2 \\ 7 \end{bmatrix}\begin{bmatrix} 1 & 2 & 6 \end{bmatrix}$

PROJECT *Connection*   For this activity, you will need the inner cereal bags collected in Getting Started, a sheet of poster board, and scissors.

1. Carefully open and flatten each bag. On the poster board, draw a template for each bag. Mark the fold lines and dimensions, then cut out the templates.

2. As a group, design a plan to cut out as many bags as possible with as little waste as possible. Your plan must include at least one of each size bag. You may not cut any of the bags into smaller pieces.

3. You do not have to actually cut out all the bags. Use your plan to estimate the percent of the total roll of waxed paper that would be used (U) and the percent of the roll that would be wasted (W).

4. Let C represent the cost of one roll of waxed paper. Determine $U\%$ of C and $W\%$ of C. Write a summary of your results.

Identity and Inverse Matrices

Explore

SPOTLIGHT ON LEARNING

WHAT? In this lesson you will learn
• to write the identity matrix for a matrix.
• to find the multiplicative inverse of a matrix.
• to use inverse matrices to solve a matrix equation.

WHY? Matrices and their inverses have many real life applications such as coding and decoding messages.

Use the following matrices for Questions 1–4.

$$A = \begin{bmatrix} 11 & 3 \\ 7 & 2 \end{bmatrix} \quad B = \begin{bmatrix} 2 & -3 \\ -7 & 11 \end{bmatrix} \quad I = \begin{bmatrix} 1 & 0 \\ 0 & 1 \end{bmatrix}$$

1. Find the products of AI and IA. What do you notice about the two products and the original matrices?

2. Find the products of BI and IB. What do you notice about the two products and the original matrices?

3. Recall from Lesson 2.3 that for any real number a, the *multiplicative identity* of a is 1. Based on what you saw in Questions 1 and 2, describe how matrix I is similar to the multiplicative identity.

4. Find the products of AB and BA. What do you notice about the two products?

Build Understanding

CHECK UNDERSTANDING

Does $DI = D$? Is I the identity matrix for D? Explain.

$$D = \begin{bmatrix} 1 & 2 & 3 \\ 4 & 5 & 6 \end{bmatrix}$$

$$I = \begin{bmatrix} 1 & 0 & 0 \\ 0 & 1 & 0 \\ 0 & 0 & 1 \end{bmatrix}$$

In Explore, you used a matrix called the **identity matrix**. The identity matrix is a square matrix with 1s along the diagonal and 0s everywhere else.

$$I_{2 \times 2} = \begin{bmatrix} 1 & 0 \\ 0 & 1 \end{bmatrix} \qquad I_{3 \times 3} = \begin{bmatrix} 1 & 0 & 0 \\ 0 & 1 & 0 \\ 0 & 0 & 1 \end{bmatrix}$$

When you multiply a matrix A by its identity matrix I, then the product is A,

$$AI = IA = A$$

When one of the matrices being multiplied is an identity matrix, matrix multiplication is commutative.

In Explore, the product AB was the identity matrix. When the product of two matrices is the identity matrix, the matrices are **multiplicative inverses,** or **inverses,** of each other. The inverse of square matrix A is written as A^{-1}. Only square matrices have inverses.

To determine the inverse of a square matrix, first form an augmented matrix with matrix A on the left of the vertical line, and the corresponding identity matrix I on the right. Perform row operations to obtain an equivalent matrix with the identity matrix I on the left. The matrix on the right will be A^{-1}.

Find A^{-1} if $A = \begin{bmatrix} 4 & -3 \\ 1 & 2 \end{bmatrix}$.

Solution

First form an augmented matrix with matrix A on the left of the vertical line, and the corresponding identity matrix I on the right of the vertical line.

$$\left[\begin{array}{cc|cc} 4 & -3 & 1 & 0 \\ 1 & 2 & 0 & 1 \end{array}\right]$$

Now use the appropriate row operations to try to obtain an equivalent matrix that has the identity matrix I on the left. The resulting matrix on the right will be A^{-1}.

$$\left[\begin{array}{cc|cc} 1 & 2 & 0 & 1 \\ 4 & -3 & 1 & 0 \end{array}\right] \qquad \text{Interchange Row 1 and Row 2.}$$

$$\left[\begin{array}{cc|cc} 1 & 2 & 0 & 1 \\ 0 & -11 & 1 & -4 \end{array}\right] \qquad \text{New Row 2} = -4(\text{Row 1}) + (\text{Row 2}).$$

$$\left[\begin{array}{cc|cc} 1 & 2 & 0 & 1 \\ 0 & 1 & -\dfrac{1}{11} & \dfrac{4}{11} \end{array}\right] \qquad \text{New Row 2} = -\dfrac{1}{11}(\text{Row 2}).$$

$$\left[\begin{array}{cc|cc} 1 & 0 & \dfrac{2}{11} & \dfrac{3}{11} \\ 0 & 1 & -\dfrac{1}{11} & \dfrac{4}{1} \end{array}\right] \qquad \text{New Row 1} = -2(\text{Row 2}) + (\text{Row 1}).$$

Notice that the identity matrix I is now on the left of the augmented matrix. The matrix on the right is the inverse matrix

$$A^{-1} = \begin{bmatrix} \dfrac{2}{11} & \dfrac{3}{11} \\ -\dfrac{1}{11} & \dfrac{4}{11} \end{bmatrix}$$

Check this solution by showing that $AA^{-1} = A^{-1}A = I$. ◄

Some square matrices do not have inverses. You can examine the reduced augmented matrix you create to determine whether the matrix A has an inverse. If the augmented matrix has one or more rows that contain all zero elements to the left of the vertical line, then A has no multiplicative inverse.

For example, if you had obtained $\left[\begin{array}{cc|cc} 1 & 0 & \dfrac{2}{11} & \dfrac{3}{11} \\ 0 & 0 & -\dfrac{1}{11} & \dfrac{4}{11} \end{array}\right]$ in Example 1,

this would have meant the original matrix A had no inverse.

EXAMPLE 2

If $A = \begin{bmatrix} 1 & 0 & 1 \\ 2 & 1 & 0 \\ 1 & -1 & 1 \end{bmatrix}$, find A^{-1} if it exists.

Solution

$\begin{bmatrix} 1 & 0 & 1 & | & 1 & 0 & 0 \\ 2 & 1 & 0 & | & 0 & 1 & 0 \\ 1 & -1 & 1 & | & 0 & 0 & 1 \end{bmatrix}$ Write the augmented matrix.

$\begin{bmatrix} 1 & 0 & 1 & | & 1 & 0 & 0 \\ 2 & 1 & 0 & | & 0 & 1 & 0 \\ 1 & -1 & 1 & | & 0 & 0 & 1 \end{bmatrix}$ New Row 2 $= -2(\text{Row 1}) + (\text{Row 2})$.

$\begin{bmatrix} 1 & 0 & 1 & | & 1 & 0 & 0 \\ 0 & 1 & -2 & | & -2 & 1 & 0 \\ 1 & -1 & 1 & | & 0 & 0 & 1 \end{bmatrix}$ New Row 3 $= -1(\text{Row 1}) + (\text{Row 3})$.

$\begin{bmatrix} 1 & 0 & 1 & | & 1 & 0 & 0 \\ 0 & 1 & -2 & | & -2 & 1 & 0 \\ 0 & 0 & -2 & | & -3 & 1 & 1 \end{bmatrix}$ New Row 3 $= (\text{Row 2}) + (\text{Row 3})$.

$\begin{bmatrix} 1 & 0 & 1 & | & 1 & 0 & 0 \\ 0 & 1 & -2 & | & -2 & 1 & 0 \\ 0 & 0 & 1 & | & \frac{3}{2} & -\frac{1}{2} & -\frac{1}{2} \end{bmatrix}$ New Row 3 $= -\frac{1}{2}(\text{Row 3})$.

$\begin{bmatrix} 1 & 0 & 1 & | & 1 & 0 & 0 \\ 0 & 1 & 0 & | & 1 & 0 & -1 \\ 0 & 0 & 1 & | & \frac{3}{2} & -\frac{1}{2} & -\frac{1}{2} \end{bmatrix}$ New Row 2 $= 2(\text{Row 3}) + (\text{Row 2})$.

$\begin{bmatrix} 1 & 0 & 0 & | & -\frac{1}{2} & \frac{1}{2} & \frac{1}{2} \\ 0 & 1 & 0 & | & 0 & 0 & -1 \\ 0 & 0 & 1 & | & \frac{3}{2} & -\frac{1}{2} & -\frac{1}{2} \end{bmatrix}$ New Row 1 $= -1(\text{Row 3}) + (\text{Row 1})$.

Since the identity matrix is on the left side, matrix A does have an inverse which is the matrix on the right side.

$$A^{-1} = \begin{bmatrix} -\frac{1}{2} & \frac{1}{2} & \frac{1}{2} \\ 1 & 0 & -1 \\ \frac{3}{2} & -\frac{1}{2} & -\frac{1}{2} \end{bmatrix}$$

Check the solution by showing that $AA^{-1} = A^{-1}A = I$. ◀

If a matrix has an inverse, the inverse can be used to solve a **matrix equation** of the form $AX = B$, where A, X, and B are matrices.

PROBLEM SOLVING TIP

Graphing calculators can be used to find the inverse of a matrix.

THINK BACK

In general, matrix multiplication is not commutative. Therefore, $A^{-1}AX = A^{-1}B$ may not be the same as $AXA^{-1} = BA^{-1}$.

EXAMPLE 3

TECHNOLOGY Solve the matrix equation in the form $AX = B$ for X.

$$\begin{bmatrix} 3 & 3 & -1 \\ -5 & -4 & 2 \\ -2 & -2 & 1 \end{bmatrix} X = \begin{bmatrix} 3 \\ -7 \\ -3 \end{bmatrix}$$

Solution

Use a graphing calculator or computer software that will compute inverse matrices to compute A^{-1}.

$$A^{-1} = \begin{bmatrix} 0 & -1 & 2 \\ 1 & 1 & -1 \\ 2 & 0 & 3 \end{bmatrix} \quad \text{Calculate } A^{-1}.$$

Multiply both sides by A^{-1}. Be sure to put A^{-1} to the left on each side. Then simplify the equation.

$$\begin{bmatrix} 0 & -1 & 2 \\ 1 & 1 & -1 \\ 2 & 0 & 3 \end{bmatrix} \begin{bmatrix} 3 & 3 & -1 \\ -5 & -4 & 2 \\ -2 & -2 & 1 \end{bmatrix} X = \begin{bmatrix} 0 & -1 & 2 \\ 1 & 1 & -1 \\ 2 & 0 & 3 \end{bmatrix} \begin{bmatrix} 3 \\ -7 \\ -3 \end{bmatrix}$$

$$\begin{bmatrix} 1 & 0 & 0 \\ 0 & 1 & 0 \\ 0 & 0 & 1 \end{bmatrix} X = \begin{bmatrix} 0 & -1 & 2 \\ 1 & 1 & -1 \\ 2 & 0 & 3 \end{bmatrix} \begin{bmatrix} 3 \\ -7 \\ -3 \end{bmatrix} \quad A^{-1}A = I$$

$$X = \begin{bmatrix} 0 & -1 & 2 \\ 1 & 1 & -1 \\ 2 & 0 & 3 \end{bmatrix} \begin{bmatrix} 3 \\ -7 \\ -3 \end{bmatrix} \quad IX = X$$

> **CHECK UNDERSTANDING**
>
> What must the dimensions of X be? Explain.

This leaves you with the equation $X = A^{-1}B$, which you can solve using matrix multiplication.

$$X = \begin{bmatrix} 1 \\ -1 \\ -3 \end{bmatrix} \quad \text{Use matrix multiplication.}$$

Multiply to check that $AX = B$.   ◀

TRY THESE

Determine whether each statement is true or false.

1. $\begin{bmatrix} 5 & 3 \\ 3 & 2 \end{bmatrix} \begin{bmatrix} 2 & -3 \\ -3 & 5 \end{bmatrix} = \begin{bmatrix} 1 & 0 \\ 0 & 1 \end{bmatrix}$

2. $\begin{bmatrix} -11 & 1 \\ 5 & -13 \end{bmatrix} \begin{bmatrix} 1 & 0 & 0 \\ 0 & 1 & 0 \\ 0 & 0 & 1 \end{bmatrix} = \begin{bmatrix} -11 & 1 \\ 5 & -13 \end{bmatrix}$

3. $\begin{bmatrix} 2 & -1 & 1 \\ 1 & -2 & 3 \\ 4 & 1 & 2 \end{bmatrix} \begin{bmatrix} \dfrac{7}{3} & -1 & \dfrac{1}{5} \\ -\dfrac{10}{3} & 0 & \dfrac{5}{3} \\ -3 & 2 & 1 \end{bmatrix} = \begin{bmatrix} 1 & 0 & 0 \\ 0 & 1 & 0 \\ 0 & 0 & 1 \end{bmatrix}$

4. $\begin{bmatrix} \dfrac{1}{2} & -\dfrac{2}{3} \\ \dfrac{5}{6} & \dfrac{1}{3} \end{bmatrix} \begin{bmatrix} 1 & 0 \\ 0 & 1 \end{bmatrix} = \begin{bmatrix} 1 & 0 \\ 0 & 1 \end{bmatrix} \begin{bmatrix} \dfrac{1}{2} & -\dfrac{2}{3} \\ \dfrac{5}{6} & \dfrac{1}{3} \end{bmatrix}$

Find the inverse of each matrix, if it exists.

5. $A = \begin{bmatrix} 8 & 5 \\ 5 & 3 \end{bmatrix}$ **6.** $A = \begin{bmatrix} 2 & 4 \\ 4 & 8 \end{bmatrix}$ **7.** $A = \begin{bmatrix} 2 & -4 \\ 1 & 3 \end{bmatrix}$

8. $A = \begin{bmatrix} 0 & 0 & 3 \\ 0 & -2 & 0 \\ 4 & 0 & 0 \end{bmatrix}$ **9.** $A = \begin{bmatrix} 2 & -1 & 0 \\ 3 & 0 & 1 \\ -2 & 4 & 0 \end{bmatrix}$ **10.** $A = \begin{bmatrix} 1 & -4 & 8 \\ 1 & -3 & 2 \\ 2 & -7 & 10 \end{bmatrix}$

Solve each matrix equation for X.

11. $\begin{bmatrix} 4 & -2 \\ 1 & 5 \end{bmatrix} X = \begin{bmatrix} -1 \\ 1 \end{bmatrix}$ **12.** $\begin{bmatrix} 1 & 0 & 1 \\ 2 & 1 & 0 \\ 1 & -1 & 1 \end{bmatrix} X = \begin{bmatrix} 1 \\ 3 \\ 4 \end{bmatrix}$

PRACTICE

1. What is the product when a matrix A is multiplied by the identity matrix?

2. What is the product when a matrix A is multiplied by its inverse A^{-1}?

Determine whether the matrices are inverses of each other.

3. $A = \begin{bmatrix} 0 & -1 \\ 1 & 0 \end{bmatrix}$ $B = \begin{bmatrix} 0 & 1 \\ -1 & 0 \end{bmatrix}$ **4.** $A = \begin{bmatrix} 2 & -4 \\ 1 & 3 \end{bmatrix}$ $B = \begin{bmatrix} -0.3 & -0.4 \\ 0.1 & 0.2 \end{bmatrix}$

5. $A = \begin{bmatrix} 3 & 1 & 0 \\ -1 & -3 & 6 \\ -2 & 2 & 4 \end{bmatrix}$ $B = \begin{bmatrix} \frac{3}{8} & \frac{1}{8} & -\frac{1}{4} \\ -\frac{1}{8} & -\frac{3}{8} & \frac{3}{4} \\ -\frac{1}{4} & \frac{1}{4} & \frac{1}{2} \end{bmatrix}$ **6.** $A = \begin{bmatrix} -1 & 18 & 6 \\ -7 & 136 & 45 \\ 1 & -21 & -7 \end{bmatrix}$ $B = \begin{bmatrix} -7 & 0 & -6 \\ -4 & 1 & 3 \\ 11 & -3 & -10 \end{bmatrix}$

Find the inverse of each matrix, if it exists.

7. $\begin{bmatrix} -2 & 6 \\ -1 & 3 \end{bmatrix}$ **8.** $\begin{bmatrix} 1 & 1 \\ 3 & 4 \end{bmatrix}$ **9.** $\begin{bmatrix} 14 & 8 \\ 6 & 4 \end{bmatrix}$

10. $\begin{bmatrix} 1 & 0 & 0 \\ 0 & 4 & 7 \\ 0 & 1 & 2 \end{bmatrix}$ **11.** $\begin{bmatrix} -2 & 5 & 3 \\ 4 & -1 & 3 \\ 4 & -10 & -6 \end{bmatrix}$ **12.** $\begin{bmatrix} -2 & 2 & 3 \\ 1 & -1 & 0 \\ 0 & 1 & 4 \end{bmatrix}$

13. Explain why the matrix $\begin{bmatrix} 1 & 2 & 3 \\ 4 & 5 & 10 \end{bmatrix}$ does not have a multiplicative inverse.

14. WRITING MATHEMATICS Suppose you were given a matrix equation of the form $AX = B$. Describe the steps you would take to find A^{-1} if you did not have a graphing calculator. Then describe the steps you would take to solve the equation for X.

Solve each matrix equation for X.

15. $\begin{bmatrix} 5 & 4 \\ -3 & 2 \end{bmatrix} X = \begin{bmatrix} 10 \\ -16 \end{bmatrix}$ **16.** $\begin{bmatrix} 2 & -1 & 0 \\ 1 & 4 & 2 \\ 3 & -2 & 1 \end{bmatrix} X = \begin{bmatrix} -5 \\ 15 \\ -7 \end{bmatrix}$

The formula below can be used to find the inverse of a 2 × 2 matrix provided $ad \neq bc$. If $ad = bc$, A^{-1} does not exist.

If $A = \begin{bmatrix} a & b \\ c & d \end{bmatrix}$, then $A^{-1} = \dfrac{1}{ad - bc} \begin{bmatrix} d & -b \\ -c & a \end{bmatrix}$

Find the inverse, if it exists, for each of the following.

17. $\begin{bmatrix} 4 & 7 \\ 1 & 2 \end{bmatrix}$ **18.** $\begin{bmatrix} 6 & 12 \\ 2 & 4 \end{bmatrix}$ **19.** $\begin{bmatrix} 5 & 9 \\ 6 & 11 \end{bmatrix}$

20. HOURLY WAGES The matrix equation describes the hours worked by two part-time employees on two days and the total wages paid for those two days. Matrix A gives the hours worked by each employee on each day. Matrix X describes the hourly wage in dollars for each employee. Matrix B gives the total wages paid in dollars for both employees on each day. Find the inverse for matrix A. Use A^{-1} to find matrix X. What are the values for x and y?

$$\begin{array}{c} \text{Day 1} \\ \text{Day 2} \end{array} \underset{A}{\begin{bmatrix} 3 & 2 \\ 5 & 3 \end{bmatrix}} \underset{X}{\begin{bmatrix} x \\ y \end{bmatrix}} = \underset{B}{\begin{bmatrix} 25 \\ 40 \end{bmatrix}}$$

(columns labeled: Employee 1, Employee 2, Hourly Wage, Total Wages)

THINK CRITICALLY

21. WRITING MATHEMATICS Explain how an inverse matrix in matrix multiplication is like the multiplicative inverse in multiplication.

For each matrix, find a formula for A^{-1}.

22. $A = [a]$, $a \neq 0$ **23.** $A = \begin{bmatrix} a & 0 \\ 0 & b \end{bmatrix}$, $ab \neq 0$ **24.** $A = \begin{bmatrix} 0 & 0 & a \\ 0 & b & 0 \\ c & 0 & 0 \end{bmatrix}$, $ab \neq 0$

25. State the property or definition used in each step.

$$\begin{aligned} AX &= B \\ A^{-1}(AX) &= A^{-1}B \\ (A^{-1}A)X &= A^{-1}B \\ IX &= A^{-1}B \\ X &= A^{-1}B \end{aligned}$$

26. Solve for X: $\begin{bmatrix} 3 & 2 \\ 5 & 4 \end{bmatrix} X + \begin{bmatrix} 1 \\ 4 \end{bmatrix} = \begin{bmatrix} -2 \\ 1 \end{bmatrix}$

MIXED REVIEW

Determine whether each of the following is rational or irrational.

27. -9.032 **28.** $\sqrt{6}$ **29.** $-\sqrt{49}$ **30.** $5.30300300030000\ldots$

31. STANDARDIZED TESTS Determine the inverse of the function $g(x) = 2x + 5$.

     A. $g^{-1}(x) = \dfrac{1}{2x + 5}$ **B.** $g^{-1}(x) = \dfrac{x - 5}{2}$ **C.** $g^{-1}(x) = -2x - 5$ **D.** $g^{-1}(x) = 10 - \dfrac{x}{2}$

32. CAR SALES Jorge sold 20 new cars. He sold 4 more subcompact cars than compact cars, and one more full-size car than compact cars. How many of each type of car did he sell?

Career
Cryptographic Technician

Secret messages might seem like kid's stuff, but to the military and many businesses coded messages are important. Cryptographic technicians operate equipment for encoding, decoding, and sending secret messages. A machine converts the message into code form and then sends it on. When the technician receives a coded message, a decoding machine translates it back into usable form.

Matrices and inverse matrices can be used to code and decode messages. First, assign a number to each letter of the alphabet, starting with 1 for A and going to 26 for Z. Use 0 to indicate a space. Next, write out the message using the numbers and group each pair of numbers into a 1×2 matrix, as shown:

$$\begin{array}{cccc} \text{G E T} & \text{H} & \text{E L} & \text{P} \\ [7 \quad 5] \quad [20 \quad 0] & [8 & 5] & [12 \quad 16] \end{array}$$

To encode a message, choose a 2×2 matrix A that has an inverse and multiply the uncoded matrices by A. The result will be a series of coded matrices. To decode the message, multiply the coded matrices by A^{-1}. Then translate the numbers back into letters.

Decision Making

Suppose you work as a cryptographic technician. Use matrix A for Questions 1–5.

$$A = \begin{bmatrix} 1 & 2 \\ 2 & 5 \end{bmatrix}$$

1. Convert the password HI MOM into numbers and group the numbers into 1×2 matrices.

2. Multiply each matrix by the matrix A to code the password. The matrix should be on the left, and matrix A should be on the right. What is the coded message?

3. After you send the password, you get a message back that uses the same coding system. How should you decode the message if the message reads

   $[39 \quad 86]\ [15 \quad 30][47 \quad 113]\ [43 \quad 101]\ [13 \quad 26].$

4. Translate the message from Question 3.

5. Use a 2×2 matrix that has an inverse to code a short message. Then give the coded message and the matrix to a friend to decode.

4.7 Solve Systems Using Inverse Matrices

Explore

Use the linear system $\begin{cases} 11x + 3y = -4 \\ 7x + 2y = 5 \end{cases}$ to answer Questions 1–7.

1. Write a 2 × 2 matrix that contains the coefficients of x in the first column and the coefficients of y in the second column.

2. Write a 2 × 1 matrix that shows the variables in the system, with x in the first row and y in the second row.

3. Let the coefficients matrix be matrix A, and the variables matrix be matrix X. Find the product AX. Compare the product and the linear system. What does this tell you about the two matrices and the linear system?

4. Now write a 2 × 1 matrix that shows the constants in the system.

5. Let the constants matrix be matrix B. Use the matrices to write the matrix equation $AX = B$.

6. How do you solve a matrix equation of the form $AX = B$ for X?

7. Use the matrix equation from Question 5 to determine x and y.

Build Understanding

You can write a system of linear equations as a matrix equation of the form $AX = B$. The **coefficients matrix** A contains the coefficients of the system, the **variables matrix** X contains the variables, and the **constants matrix** B contains the constants. If the inverse of the coefficients matrix exists, you use an inverse matrix to solve the system.

EXAMPLE 1

Use an inverse matrix to solve the system.

$$\begin{cases} -2x + 2y + 3z = 1 \\ x - y = 3 \\ y + 4z = -2 \end{cases}$$

Solution

$$\begin{bmatrix} -2 & 2 & 3 \\ 1 & -1 & 0 \\ 0 & 1 & 4 \end{bmatrix}\begin{bmatrix} x \\ y \\ z \end{bmatrix} = \begin{bmatrix} 1 \\ 3 \\ -2 \end{bmatrix}$$

Remember to place zeros in the matrix where coefficients are zero.

CHECK UNDERSTANDING

For Example 1, describe how the matrices in the matrix equation correspond to the original system of equations.

Use a graphing calculator or augmented matrices to find A^{-1}. Then multiply both sides of the equation by A^{-1}.

$$\begin{bmatrix} x \\ y \\ z \end{bmatrix} = \begin{bmatrix} -\dfrac{4}{3} & -\dfrac{5}{3} & 1 \\ -\dfrac{4}{3} & -\dfrac{8}{3} & 1 \\ \dfrac{1}{3} & \dfrac{2}{3} & 0 \end{bmatrix} \begin{bmatrix} 1 \\ 3 \\ -2 \end{bmatrix}$$

$$\begin{bmatrix} x \\ y \\ z \end{bmatrix} = \begin{bmatrix} \dfrac{25}{3} \\ -\dfrac{34}{3} \\ \dfrac{7}{3} \end{bmatrix}$$

The solution is $x = -\dfrac{25}{3}, y = -\dfrac{34}{3}, z = \dfrac{7}{3}$. ◄

One advantage of using inverses to solve systems of equations is that the method can save you time when you are solving several systems that have the same coefficients and variables matrices but different constants matrices.

EXAMPLE 2

MIXTURES A grocery store carries two different types of canned orange drinks. One drink contains 20% pure orange juice, the other contains 10% pure orange juice.

a. Suppose you wanted to mix the two drinks together to produce 30 liters of a drink that is 15% pure orange juice. How many liters of each juice would you use?

b. Suppose you wanted to mix the two drinks together to produce 40 liters of a drink that is 16% pure orange juice. How many liters of each juice would you use?

Solution

a. If x represents the amount in liters of the 20% drink, and y represents the amount in liters of the 10% drink, then

$$x + y = 30$$

The amount of pure juice in the 20% drink is $0.20x$ and the amount of pure juice in the 10% drink is $0.10y$. When the two parts are combined, there will be $0.15(30)$ pure juice in the mixture, so

$$0.20x + 0.10y = 0.15(30)$$

$$\begin{cases} x + y = 30 \\ 2x + y = 45 \end{cases}$$ Multiply 2nd equation by 10.

$$\begin{bmatrix} 1 & 1 \\ 2 & 1 \end{bmatrix} \begin{bmatrix} x \\ y \end{bmatrix} = \begin{bmatrix} 30 \\ 45 \end{bmatrix}$$ Write the matrix equation.

$$\begin{bmatrix} x \\ y \end{bmatrix} = \begin{bmatrix} -1 & \dfrac{1}{10} \\ 2 & -\dfrac{1}{10} \end{bmatrix} \begin{bmatrix} 30 \\ 45 \end{bmatrix}$$ Multiply both sides of the inverse.

$$\begin{bmatrix} x \\ y \end{bmatrix} = \begin{bmatrix} 15 \\ 15 \end{bmatrix}$$ Multiply.

Therefore, you would need 15 liters of the 20% mixture and 15 liters of the 10% mixture to make 30 liters of a 15% mixture.

b. $\begin{cases} x + y = 40 \\ 2x + 1y = 64 \end{cases}$ Write the system of equations.

$$\begin{bmatrix} 1 & 1 \\ 2 & 1 \end{bmatrix}\begin{bmatrix} x \\ y \end{bmatrix} = \begin{bmatrix} 40 \\ 64 \end{bmatrix}$$ Write the matrix equation.

$$\begin{bmatrix} x \\ y \end{bmatrix} = \begin{bmatrix} -1 & \dfrac{1}{10} \\ 2 & -\dfrac{1}{10} \end{bmatrix} \begin{bmatrix} 40 \\ 64 \end{bmatrix}$$ The coefficients matrix and its inverse are the same as in part a.

$$\begin{bmatrix} x \\ y \end{bmatrix} = \begin{bmatrix} 24 \\ 16 \end{bmatrix}$$

This means you would use 24 liters of the 20% mixture and 16 liters of the 10% mixture to make 40 liters of a 16% mixture. ◄

PROBLEM SOLVING TIP

Sometimes when you use a graphing calculator to find the inverse of a matrix, you may get an element that contains a very small number in scientific notation. An example is 1E − 14, which represents 1×10^{-14}. This is due to a rounding error in the calculator program. If this happens, you may assume that entry should be 0, but it is a good idea to use another method to check the inverse matrix.

TRY THESE

Determine whether the matrix equation is equivalent to the linear system.

1. $\begin{bmatrix} 4 & 5 \\ 1 & -2 \end{bmatrix}\begin{bmatrix} x \\ y \end{bmatrix} = \begin{bmatrix} 2 \\ 3 \end{bmatrix}$ and $\begin{cases} 4x + 5y = 2 \\ x - 2y = 3 \end{cases}$ **2.** $\begin{bmatrix} 7 & 1 \\ 1 & 7 \end{bmatrix}\begin{bmatrix} x \\ y \end{bmatrix} = \begin{bmatrix} 4 \\ 8 \end{bmatrix}$ and $\begin{cases} 7x + y = 4 \\ x - 7y = 8 \end{cases}$

3. $\begin{bmatrix} 1 & 13 & 5 \\ 20 & -2 & -6 \\ -4 & -1 & -11 \end{bmatrix}\begin{bmatrix} x \\ y \\ z \end{bmatrix} = \begin{bmatrix} 3 \\ -3 \\ 2 \end{bmatrix}$ and $\begin{cases} x + 13y + 5z = 3 \\ 20x - 2y - 6z = -3 \\ -4x - y - 11z = 2 \end{cases}$

4. $\begin{bmatrix} 0 & 1 & -1 \\ 2 & 3 & 0 \\ 1 & 0 & -2 \end{bmatrix}\begin{bmatrix} x \\ y \\ z \end{bmatrix} = \begin{bmatrix} 4 \\ 5 \\ 6 \end{bmatrix}$ and $\begin{cases} 2x + 3y = 5 \\ x - 2z = 6 \\ y - z = 4 \end{cases}$

Write a matrix equation for each system of linear equations.

5. $\begin{cases} 3x - 7y = 7 \\ 7x + 3y = 3 \end{cases}$ **6.** $\begin{cases} x + 5y - 10z = 13 \\ 2x - y + 3z = 18 \\ -4x + 6y + 12z = 7 \end{cases}$ **7.** $\begin{cases} -x + z = 6 \\ 4y + 3z = -1 \\ x - y = 0 \end{cases}$

8. WRITING MATHEMATICS When you want to solve a linear system using an inverse matrix, for which matrix in the matrix equation do you need to find the inverse? Describe how you use that inverse to solve the system.

Solve each linear system.

9. $\begin{cases} x + 3y = 10 \\ 2x + 5y = 2 \end{cases}$

10. $\begin{cases} 2x - 3y = 6 \\ -x + 2y = -4 \end{cases}$

11. $\begin{cases} x + 2y = 5 \\ 2x - 5y = -8 \end{cases}$

12. $\begin{cases} x + 3y + 3z = 14 \\ x + 3y + 4z = 17 \\ x + 4y + 3z = 15 \end{cases}$

13. $\begin{cases} -x + 3y + z = 1 \\ 2x + 5y = 3 \\ 3x + y - 2z = -2 \end{cases}$

14. $\begin{cases} y - z = -4 \\ 4x + y = -3 \\ 3x - y + 3z = 1 \end{cases}$

15. FLOWER FARMING A flower farmer wants to plant three types of bulbs: gladiolas, irises, and tulips. The gladiolas cost \$75 per acre to plant, the irises cost \$100 per acre, and the tulips cost \$50 per acre.

a. The farmer wants to plant 200 acres of bulbs and spend a total of \$15,000. The farmer decides to plant twice as many gladiolas as irises. Write a system of equations and use a matrix equation to find the total number of acres of each type of flower.

b. The farmer decides to plant 250 acres instead, using the same amount of money. The farmer wants to keep the ratio of gladiolas to irises the same as in part a. Write the matrix equation and find the total number of acres planted of each type of flower.

PRACTICE

1. WRITING MATHEMATICS Describe the steps you would take to write the system
$\begin{cases} 0.135x + 0.405y = 1.000 \\ 0.040x + 0.877y = 3.150 \end{cases}$ as a matrix equation. Then write the matrix equation.

Write the linear system represented by each matrix equation.

2. $\begin{bmatrix} -1 & -5 \\ 3 & -3 \end{bmatrix}\begin{bmatrix} x \\ y \end{bmatrix} = \begin{bmatrix} 10 \\ 3 \end{bmatrix}$

3. $\begin{bmatrix} 1 & 0 & 5 \\ -8 & 4 & 0 \\ 0 & 0 & 2 \end{bmatrix}\begin{bmatrix} x \\ y \\ z \end{bmatrix} = \begin{bmatrix} 12 \\ 0 \\ 4 \end{bmatrix}$

4. $\begin{bmatrix} 0.4 & 0.8 & -0.2 \\ 1.2 & -0.4 & 0.4 \\ 0.6 & 0.4 & -0.4 \end{bmatrix}\begin{bmatrix} x \\ y \\ z \end{bmatrix} = \begin{bmatrix} 7 \\ 9 \\ 4 \end{bmatrix}$

Solve each linear system.

5. $\begin{cases} 8x + 5y = -6 \\ 5x + 3y = 2 \end{cases}$

6. $\begin{cases} x + 2y + 3z = -1 \\ 2x - 3y + 4z = 2 \\ -3x + 5y - 6z = 4 \end{cases}$

7. $\begin{cases} x + 2y - 4z = 16 \\ y - z = 4 \\ x - y = 1 \end{cases}$

8. ACID SOLUTIONS A chemist has one solution of acid and water that is 15% acid and a second solution that is 75% acid. The chemist wants to mix the two solutions to get 20 gallons of a solution that is 39% acid. Use a matrix equation to determine the number of gallons of each solution the chemist should use to make the new solution.

Solve each linear system.

9. $\begin{cases} 3x + 5y = -4 \\ 2x + 4y = -2 \end{cases}$

10. $\begin{cases} -x - 3y = 14 \\ 5y - 2z = 2 \\ 2x + 2z = -2 \end{cases}$

11. $\begin{cases} 0.1x + 0.3y + 0.1z = 1.4 \\ 0.1x + 0.5y + 0.2z = 1.8 \\ 0.2x + 0.6y + 0.3z = 0.8 \end{cases}$

12. STOCK INVESTMENTS Shelly has inherited $80,000. She invests part of her money in a certificate of deposit that produces a return of 7% per year. She divides the rest equally between a money market fund at 12% per year and a tax-free bond at 6% per year.

    a. Use an inverse matrix to determine the amount of money invested in each account if Shelly receives an annual return of $6800 on all three investments.

    b. Shelly decides to invest only three-fourths of her money in the three accounts, and she receives an annual return of $5000. Determine the amount invested in each account.

EXTEND

Solve the system of equations using the given inverse matrix.

13. $\begin{cases} 2x - 4y = 10 \\ x + 3y = 20 \end{cases}$ $\quad A^{-1} = \dfrac{1}{10}\begin{bmatrix} 3 & 4 \\ -1 & 2 \end{bmatrix}$

14. $\begin{cases} 3x + y = 2 \\ 2x - y + 2z = -5 \\ x + y + z = 5 \end{cases}$ $A^{-1} = \dfrac{1}{9}\begin{bmatrix} 3 & 1 & -2 \\ 0 & -3 & 6 \\ -3 & 2 & 5 \end{bmatrix}$

15. Write the linear system equivalent to the following matrix equation.

$$\begin{bmatrix} 2 & 4 & -5 & 12 \\ 4 & -1 & 12 & -1 \\ -1 & 4 & 0 & 2 \\ 2 & 10 & 1 & 0 \end{bmatrix}\begin{bmatrix} w \\ x \\ y \\ z \end{bmatrix} = \begin{bmatrix} 2 \\ 5 \\ 13 \\ 5 \end{bmatrix}$$

Use an inverse matrix and the given constant matrix to solve the following linear system.

$$\begin{cases} -8x + 3y = c \\ -4x + 4y = d \end{cases}$$

16. $\begin{bmatrix} c \\ d \end{bmatrix} = \begin{bmatrix} -16 \\ -8 \end{bmatrix}$

17. $\begin{bmatrix} c \\ d \end{bmatrix} = \begin{bmatrix} 2 \\ 4 \end{bmatrix}$

18. $\begin{bmatrix} c \\ d \end{bmatrix} = \begin{bmatrix} \dfrac{1}{6} \\ \dfrac{1}{2} \end{bmatrix}$

19. WRITING MATHEMATICS Write three problems similar to Exercises 17, 18, and 19 for a linear system in three variables. Also give the inverse of the coefficient matrix that can be used to solve the problems.

20. GEOMETRY In a triangle, the measure of the largest angle is 70° greater than the measure of the smallest angle. The measure of the largest angle is twice as large as the measure of the remaining angle. Use an inverse matrix to find the measure of each angle.

THINK CRITICALLY

21. WRITING MATHEMATICS You can use inverse matrices to solve linear systems only when the coefficients matrix has an inverse. If the coefficients matrix does *not* have an inverse, what might this tell you about the linear system you are trying to solve?

Use an inverse matrix to solve each linear system for *x* and *y*.

22. $\begin{cases} x + ay = 1 \\ 2x + ay = 0 \end{cases}$

23. $\begin{cases} bx + ay = 2 \\ bx - ay = 10 \end{cases}$

24. $\begin{cases} 4y + bx = 6 \\ 3y + bx = 4 \end{cases}$

25. Solve the following linear system using the given inverse of the coefficient matrix.

$$\begin{cases} x + 2y + 3z + w = 18 \\ x + 3y + 3z + 2w = 24 \\ 2x + 4y + 3z + 3w = 31 \\ x + y + z + w = 10 \end{cases} \quad A^{-1} = \begin{bmatrix} 1 & -2 & 1 & 0 \\ 1 & -2 & 2 & -3 \\ 0 & 1 & -1 & 1 \\ -2 & 3 & -2 & 3 \end{bmatrix}$$

26. **PRODUCTION TIMES** Three companies produce three products—tables, lamps, and chairs. In one hour, the first company produces 1 table, 3 lamps, and 2 chairs. In the same amount of time, the second company produces 2 tables, 1 lamp, and 1 chair. The hourly production for the third company is 2 tables, 3 lamps and 2 chairs.

a. Find the number of hours that each company needs to operate so their combined production fills an order for 60 tables, 78 lamps, and 56 chairs.

b. Find the number of hours that each company needs to operate so their combined production fills an order for 42 tables, 52 lamps, and 39 chairs.

MIXED REVIEW

Solve the following for y.

27. $2x - 5y = 7x - 40$

28. $5(y - 2) = 7x + 25$

29. $10(x + y) = 2(5x - 5)$

Solve the following linear systems.

30. $\begin{cases} x - 3y = 2 \\ 6x + 5y = -34 \end{cases}$

31. $\begin{cases} 0.3x + 0.2y = -0.9 \\ 0.2x - 0.3y = -0.6 \end{cases}$

32. $\begin{cases} \dfrac{1}{5}x + \dfrac{1}{2}y = 6 \\ \dfrac{3}{5}x - \dfrac{1}{2}y = 2 \end{cases}$

PROJECT *Connection*   In this activity, you will use matrices to explore costs involved in producing your cereal boxes and bags.

1. Estimate the percent of the roll of waxed paper that will be needed to provide enough bags for the boxes that can be cut from one sheet of poster board. Let Z represent this percent. To determine the waxed bag cost for these boxes, find $Z\%$ of the cost of the whole roll of waxed paper.

2. Estimate the percent of this length of waxed paper that will be used for bags X and the percent that will be wasted Y.

3. Set up a 1×2 matrix C. Enter the cost of one sheet of poster board in C_{11} and the cost of the waxed paper needed for the boxes from one poster board in C_{12}.

4. Set up a 2×2 matrix Q. Enter the percent used of the poster board in Q_{11}, the percent of board wasted in Q_{12}, the values for X in Q_{21}, and the value for Y in Q_{22}.

5. Determine the product $C \times Q$. What do the entries in this matrix represent?

Career
Safety Engineer

If you walk through a manufacturing plant, you may see posters illustrating safe working procedures, notices describing safety seminars, and inspectors checking machinery. These are all signs that a safety engineer has been at work. Safety engineers are concerned with all aspects of a business that relate to the safety of the workers and the consumers.

Decision Making

A safety engineer employed by Chemix Manufacturing oversees all safety issues at three different sites. The engineer has developed a three-part plan that includes awareness posters, safety seminars, and site inspections. The budget is $7,100 for the posters, $14,300 for the seminars, and $52,600 for the operations inspections. The budget breakdown by plant is shown in the following table.

	Western Plant	Southern Plant	Eastern Plant
Posters	15%	10%	5%
Seminars	25%	10%	25%
Inspections	60%	80%	70%

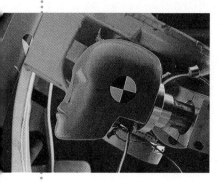

1. Write the budget amounts for each plant as matrix A.

2. Write the totals for posters, seminars, and inspections as matrix B.

3. Let x represent the total amount budgeted for the Western plant, let y represent the total amount budgeted for the Southern plant, and let z represent the total amount budgeted for the Eastern plant. Use the matrices you wrote in Questions 1 and 2 to write a matrix equation of the form $AX = B$.

4. Solve the matrix equation. Explain what the solution means.

5. Suppose that the safety engineer changes the budget amounts for the posters, seminars, and inspections to $6,200, $13,100, and $54,700, respectively. What would the totals per plant then be?

4.8 Determinants and Cramer's Rule

Explore

SPOTLIGHT ON LEARNING

WHAT? In this lesson you will learn
- to evaluate the determinant of a 2×2 or a 3×3 matrix.
- to use Cramer's rule to solve systems of linear equations.

WHY? Cramer's rule can help you solve problems in the retail industry.

● Use the following system for Questions 1–5.

$$\begin{cases} a_1x + b_1y = c_1 & \textbf{Equation 1} \\ a_2x + b_2y = c_2 & \textbf{Equation 2} \end{cases}$$

1. Multiply both sides of Equation 1 by b_2, and multiply both sides of Equation 2 by $-b_1$.

2. Add the two equations from Question 1, then solve for x.

3. Now, multiply both sides of Equation 1 by $-a_2$, and multiply both sides of Equation 2 by a_1.

4. Add the two equations from Question 3, then solve for y.

5. Notice the expressions for x and y have the same denominator, $a_1b_2 - a_2b_1$. You can represent this denominator as a square array of numbers enclosed by two vertical lines: $D = \begin{vmatrix} a_1 & b_1 \\ a_2 & b_2 \end{vmatrix}$. The vertical. bars symbolize the determinant of a matrix. Go back to the original system of equations and write a coefficients matrix for the system. What do you notice about the coefficients matrix and the determinant?

Build Understanding

● Associated with every square matrix A is a real number called the **determinant of A** and written **det A**. The determinant of a 2×2 matrix

$$A = \begin{bmatrix} a_1 & b_1 \\ a_2 & b_2 \end{bmatrix} \quad \text{is defined as} \quad \det A = \begin{vmatrix} a_1 & b_1 \\ a_2 & b_2 \end{vmatrix} = a_1b_2 - a_2b_1$$

As shown in the diagram, the value of a 2×2 determinant is the difference of the products of the entries on the two diagonals.

EXAMPLE 1

Evaluate each determinant.

a. $\begin{vmatrix} 1 & -2 \\ 3 & 4 \end{vmatrix}$ 　　　　**b.** $\begin{vmatrix} 5 & -2 \\ -1 & -1 \end{vmatrix}$ 　　　　**c.** $\begin{vmatrix} -1 & 2 \\ -3 & 6 \end{vmatrix}$

Solution

a. $a_1b_2 - a_2b_1 = 1(4) - 3(-2) = 4 + 6 = 10$

b. $a_1b_2 - a_2b_1 = 5(-1) - (-1)(-2) = -5 - 2 = -7$

c. $a_1b_2 - a_2b_1 = -1(6) - (-3)(2) = -6 + 6 = 0$ ◀

The determinant of a 3×3 matrix $\begin{bmatrix} a_1 & b_1 & c_1 \\ a_2 & b_2 & c_2 \\ a_3 & b_3 & c_3 \end{bmatrix}$ is defined as

$$\begin{vmatrix} a_1 & b_1 & c_1 \\ a_2 & b_2 & c_2 \\ a_3 & b_3 & c_3 \end{vmatrix} = a_1b_2c_3 + a_2b_3c_1 + a_3b_1c_2 - a_1b_3c_2 - a_2b_1c_3 - a_3b_2c_1$$

This rule for evaluating a 3×3 determinant can be hard to remember, so two other methods are commonly used.

The first method is called the **diagonals method**. Start by copying the first two columns to the right of the third column. Then subtract the sum of the products along the three bottom-to-top diagonals from the sum of the products along the three top-to-bottom diagonals.

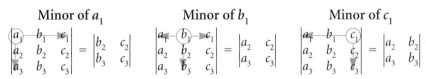

$$= a_1b_2c_3 + b_1c_2a_3 + c_1a_2b_3 - a_3b_2c_1 - b_3c_2a_1 - c_3a_2b_1$$

The second method is called **expansion by minors** and expresses the determinant in terms of three 2×2 determinants. For a 3×3 determinant, the **minor** of an element is a 2×2 determinant that is formed when the row and column containing that element are deleted.

Minor of a_1
$$\begin{vmatrix} a_1 & b_1 & c_1 \\ a_2 & b_2 & c_2 \\ a_3 & b_3 & c_3 \end{vmatrix} = \begin{vmatrix} b_2 & c_2 \\ b_3 & c_3 \end{vmatrix}$$

Minor of b_1
$$\begin{vmatrix} a_1 & b_1 & c_1 \\ a_2 & b_2 & c_2 \\ a_3 & b_3 & c_3 \end{vmatrix} = \begin{vmatrix} a_2 & c_2 \\ a_3 & c_3 \end{vmatrix}$$

Minor of c_1
$$\begin{vmatrix} a_1 & b_1 & c_1 \\ a_2 & b_2 & c_2 \\ a_3 & b_3 & c_3 \end{vmatrix} = \begin{vmatrix} a_2 & b_2 \\ a_3 & b_3 \end{vmatrix}$$

Multiply each element in one row or column by its minor. If the element is in a position marked with $+$, the product is added. If the element is in a position marked by a $-$, the product is subtracted.

$$\begin{vmatrix} + & - & + \\ - & + & - \\ + & - & + \end{vmatrix} \qquad \begin{vmatrix} a_1 & b_1 & c_1 \\ a_2 & b_2 & c_2 \\ a_3 & b_3 & c_3 \end{vmatrix} = a_1 \begin{vmatrix} b_2 & c_2 \\ b_3 & c_3 \end{vmatrix} - b_1 \begin{vmatrix} a_2 & c_2 \\ a_3 & c_3 \end{vmatrix} + c_1 \begin{vmatrix} a_2 & b_2 \\ a_3 & b_3 \end{vmatrix}$$

ALGEBRA: WHO, WHERE, WHEN

The discovery of determinants is attributed to the Japanese mathematician Seki Takakazu (1642–1708). Gottfried Wilhelm von Leibnitz (1646–1716) developed the modern theory of determinants.

EXAMPLE 2

Evaluate: $\begin{vmatrix} 1 & 3 & -2 \\ -1 & -2 & -3 \\ 1 & 1 & 2 \end{vmatrix}$

a. Using the diagonals method.
b. Using expansion by minors about the first column.

Solution

a. Rewrite the first and second columns of the determinant to the right of the third column. Then add the top-to-bottom diagonal products, and subtract the bottom-to-top products.

$$\begin{vmatrix} 1 & 3 & -2 \\ -1 & -2 & -3 \\ 1 & 1 & 2 \end{vmatrix} \begin{matrix} 1 & 3 \\ -1 & -2 \\ 1 & 1 \end{matrix} = -4 + (-9) + 2 - 4 - (-3) - (-6) = -6$$

b. Find the minors for the first column.

Minor of a_1
$$\begin{vmatrix} 1 & 3 & 2 \\ -1 & -2 & -3 \\ 1 & 1 & 2 \end{vmatrix} = \begin{vmatrix} -2 & -3 \\ 1 & 2 \end{vmatrix}$$

Minor of a_2
$$\begin{vmatrix} 1 & 3 & -2 \\ -1 & -2 & 3 \\ 1 & 1 & 2 \end{vmatrix} = \begin{vmatrix} 3 & -2 \\ 1 & 2 \end{vmatrix}$$

Minor of a_3
$$\begin{vmatrix} 1 & 3 & -2 \\ -1 & -2 & -3 \\ 1 & 1 & 2 \end{vmatrix} = \begin{vmatrix} 3 & -2 \\ -2 & -3 \end{vmatrix}$$

Next, multiply each element in the column by its minor, and find the product using the appropriate sign pattern.

$$\begin{vmatrix} 1 & 3 & -2 \\ -1 & -2 & -3 \\ 1 & 1 & 2 \end{vmatrix} = 1\begin{vmatrix} -2 & -3 \\ 1 & 2 \end{vmatrix} - (-1)\begin{vmatrix} 3 & -2 \\ 1 & 2 \end{vmatrix} + 1\begin{vmatrix} 3 & -2 \\ -2 & -3 \end{vmatrix} = -6$$

Determinants can be used to solve systems of equations. Recall from Explore that you can write the denominator of the expression for x and y as a determinant D. You can also write the numerators of the expressions for x and y as determinants D_x and D_y.

$$D_x = c_1 b_2 - c_2 b_1 = \begin{vmatrix} c_1 & b_1 \\ c_2 & b_2 \end{vmatrix} \qquad D_y = a_1 c_2 - a_2 c_1 = \begin{vmatrix} a_1 & c_1 \\ a_2 & c_2 \end{vmatrix}$$

Replacing the numerators and denominators with the appropriate determinants gives the solution of $\begin{cases} a_1 x + b_1 y = c_1 \\ a_2 x + b_2 y = c_2 \end{cases}$

$$x = \frac{D_x}{D} = \frac{\begin{vmatrix} c_1 & b_1 \\ c_2 & b_2 \end{vmatrix}}{\begin{vmatrix} a_1 & b_1 \\ a_2 & b_2 \end{vmatrix}} \qquad y = \frac{D_y}{D} = \frac{\begin{vmatrix} a_1 & c_1 \\ a_2 & c_2 \end{vmatrix}}{\begin{vmatrix} a_1 & b_1 \\ a_2 & b_2 \end{vmatrix}} \qquad \text{where } D \neq 0$$

This method for solving systems of equations is called **Cramer's rule.**

EXAMPLE 3

Use Cramer's rule to solve the system: $\begin{cases} 5x + 7y = -1 \\ 6x + 8y = 1 \end{cases}$

Solution

Begin by evaluating the determinants D, D_x, and D_y.

$$D = \begin{vmatrix} a_1 & b_1 \\ a_2 & b_2 \end{vmatrix} = \begin{vmatrix} 5 & 7 \\ 6 & 8 \end{vmatrix} = 5(8) - 6(7) = -2$$

$$D_x = \begin{vmatrix} c_1 & b_1 \\ c_2 & b_2 \end{vmatrix} = \begin{vmatrix} -1 & 7 \\ 1 & 8 \end{vmatrix} = (-1)8 - 7(1) = -15$$

$$D_y = \begin{vmatrix} a_1 & c_1 \\ a_2 & c_2 \end{vmatrix} = \begin{vmatrix} 5 & -1 \\ 6 & 1 \end{vmatrix} = 5(1) - (-1)6 = 11$$

Now you can use Cramer's rule to find the solution

$$x = \frac{D_x}{D} = \frac{-15}{-2} = \frac{15}{2} \qquad y = \frac{D_y}{D} = \frac{11}{-2} = -\frac{11}{2}$$

The solution to the system is $x = \frac{15}{2}$ and $y = -\frac{11}{2}$. You can check this solution in the original system. ◄

Cramer's rule can also be used with systems of three equations in three variables. The solution of

$$\begin{cases} a_1x + b_1y + c_1z = d_1 \\ a_2x + b_2y + c_2z = d_2 \\ a_3x + b_3y + c_3z = d_3 \end{cases} \text{ is } \quad x = \frac{D_x}{D} \qquad y = \frac{D_y}{D} \qquad z = \frac{D_z}{D}$$

$$D = \begin{vmatrix} a_1 & b_1 & c_1 \\ a_2 & b_2 & c_2 \\ a_3 & b_3 & c_3 \end{vmatrix} \qquad D_x = \begin{vmatrix} d_1 & b_1 & c_1 \\ d_2 & b_2 & c_2 \\ d_3 & b_3 & c_3 \end{vmatrix} \qquad D_y = \begin{vmatrix} a_1 & d_1 & c_1 \\ a_2 & d_2 & c_2 \\ a_3 & d_3 & c_3 \end{vmatrix} \qquad D_z = \begin{vmatrix} a_1 & b_1 & d_1 \\ a_2 & b_2 & d_2 \\ a_3 & b_3 & d_3 \end{vmatrix}$$

TRY THESE

Evaluate the determinant.

1. $\begin{bmatrix} 2 & 4 \\ 3 & -1 \end{bmatrix}$

2. $\begin{bmatrix} -3 & -1 \\ 4 & -2 \end{bmatrix}$

3. $\begin{bmatrix} -3 & 2 \\ 6 & -4 \end{bmatrix}$

Evaluate the determinant.

4. $\begin{bmatrix} 1 & 2 & 0 \\ 0 & 2 & 1 \\ 1 & 1 & 1 \end{bmatrix}$

5. $\begin{bmatrix} -1 & -2 & -3 \\ 3 & 4 & 2 \\ 0 & 1 & 2 \end{bmatrix}$

6. $\begin{bmatrix} 1 & 1 & 2 \\ 5 & 5 & 7 \\ 3 & 3 & 1 \end{bmatrix}$

7. WRITING MATHEMATICS What is a minor of an element in a 3×3 matrix? Create a chart to show the minor of each element in the given matrix. $\begin{bmatrix} 2 & -3 & 5 \\ 1 & 0 & -1 \\ 5 & 3 & 14 \end{bmatrix}$

Evaluate each determinant.

8. $\begin{bmatrix} 4 & -4 & 6 \\ 2 & 8 & -3 \\ 0 & -5 & 0 \end{bmatrix}$

9. $\begin{bmatrix} 3 & 2 & -2 \\ -2 & 1 & 4 \\ -4 & -3 & 3 \end{bmatrix}$

10. $\begin{bmatrix} 2 & \frac{1}{2} & 2 \\ -1 & 4 & -3 \\ \frac{1}{2} & 1 & 1 \end{bmatrix}$

11. Show that the given determinant equation is another way to write the slope-intercept form of the equation of a line. $\begin{vmatrix} y & x \\ m & 1 \end{vmatrix} = b$

Solve using Cramer's rule.

12. $\begin{cases} 2x + y = 1 \\ 5x + 3y = 2 \end{cases}$

13. $\begin{cases} x + y = 0 \\ x - y = 0 \end{cases}$

14. $\begin{cases} 3x - 2y = 7 \\ 3x + 2y = 9 \end{cases}$

15. $\begin{cases} 2x + 4y + 3z = 6 \\ x - 3y + 2z = -7 \\ -x + 2y - z = 5 \end{cases}$

16. $\begin{cases} x + y - z = 2 \\ -x + y + z = 3 \\ x + y + z = 4 \end{cases}$

17. $\begin{cases} 3x + 5z = 0 \\ 2x + 3y = 1 \\ y - 2z = 11 \end{cases}$

18. RAINFALL Mawsynram, India is the wettest place in the world. In an average year, Mawsynram receives 6.5 times more rain than Beijing, China. In an average year the number of inches of rain that falls in Beijing is three times more than the amount that falls in Paris, France. The combined total rainfall in all three cities averages 564 inches.

a. Write a system of equations to find the total average amount of rain that falls in all three cities. Then write Cramer's rule for the three equations.

b. Use Cramer's rule to determine the yearly average rainfall in inches for each city.

PRACTICE

Evaluate the determinant.

1. $\begin{vmatrix} 6 & 0 \\ -3 & 4 \end{vmatrix}$

2. $\begin{vmatrix} -3 & -1 \\ -5 & \frac{1}{2} \end{vmatrix}$

3. $\begin{vmatrix} 1 & 0 \\ 0 & 1 \end{vmatrix}$

Evaluate the determinant for each matrix.

4. $\begin{bmatrix} 1 & 0 & 0 \\ -2 & 4 & 3 \\ 5 & -2 & 1 \end{bmatrix}$

5. $\begin{bmatrix} 1 & 3 & 7 \\ -2 & 6 & 4 \\ 3 & 7 & -1 \end{bmatrix}$

6. $\begin{bmatrix} 1 & 4 & 3 \\ 2 & 1 & 6 \\ 3 & -2 & 9 \end{bmatrix}$

7. WRITING MATHEMATICS Solve the given system using Cramer's rule and describe each step.

$\begin{cases} 2x - 3y + 5z = 27 \\ x + 2y - z = -4 \\ 5x - y + 4z = 27 \end{cases}$

8. WORKING HOURS An electrician, a carpenter, and a plumber are hired to work on a house. The electrician earns \$32.50 an hour, the carpenter earns \$40.75 an hour, and the plumber earns \$62 an hour. Together they worked a total of 21.5 hours and earned a total of \$1059.75. The carpenter worked two fewer hours than the plumber. Write a system of equations then use Cramer's rule to determine the number of hours the electrician worked.

Solve using Cramer's rule, if possible.

9. $\begin{cases} 4x + 3y = 0 \\ 3x - 4y = 25 \end{cases}$

10. $\begin{cases} -2x + 4y = 3 \\ 3x - 7y = 1 \end{cases}$

11. $\begin{cases} 4x + 2y = 3 \\ -8x - 4y = 9 \end{cases}$

12. $\begin{cases} y + 4z = 6 \\ 3x + z = 7 \\ 5y - z = 9 \end{cases}$

13. $\begin{cases} 2x + 2y + z = 1 \\ x + 3y + 2z = 5 \\ x - y - z = 6 \end{cases}$

14. $\begin{cases} 3x + 2y - z = 4 \\ 3x - 2y + z = 5 \\ 4x - 5y - z = -1 \end{cases}$

EXTEND

15. WRITING MATHEMATICS Can two different 3×3 matrices have the same determinant? If so, give an example.

Solve each determinant equation for x.

16. $\begin{vmatrix} x & -2 \\ 5 & 3 \end{vmatrix} = 1$

17. $\begin{vmatrix} x + 3 & 4 \\ x - 3 & 5 \end{vmatrix} = -7$

18. $\begin{vmatrix} 2 & x & -1 \\ -1 & 3 & 2 \\ -2 & 1 & 1 \end{vmatrix} = -12$

19. **Writing Mathematics** Write a 3×3 matrix with a row or column of zeros. What is the value of the determinant? What type of system does the matrix represent?

20. **Writing Mathematics** Write a 3×3 matrix with two identical rows or columns. What is the value of the determinant? What type of system does the matrix represent?

21. **Geometry** The area of a triangle with the vertices (x_1, y_1), (x_2, y_2), and (x_3, y_3) is the absolute value of the given determinant. Find the area of a triangle with the vertices $(-1, 4)$, $(4, 8)$, and $(1, 1)$.

$$\frac{1}{2}\begin{vmatrix} x_1 & y_1 & 1 \\ x_2 & y_2 & 1 \\ x_3 & y_3 & 1 \end{vmatrix}$$

22. **Retailing** A gift shop has helium-filled balloons for all occasions priced according to size—small, medium, and large. Alice spent $17 for one large, two medium, and four small balloons. It cost Bryan $22 for six small, one medium, and two large balloons. For $25, Lucille got two large, three medium, and four small balloons. How much did Ken pay for three of each size? Write a system of equations, then solve using Cramer's rule.

THINK CRITICALLY

23. Evaluate the given determinant by examining and comparing rows.

$$\begin{vmatrix} 3 & 5 & 1 & 9 \\ 0 & 5 & 2 & 0 \\ 2 & -1 & -1 & -1 \\ -4 & 2 & 2 & 2 \end{vmatrix}$$

24. Show that an equation of a line that passes through the points (x_1, y_1) and (x_2, y_2) can be written as shown.

$$\begin{vmatrix} x & y & 1 \\ x_1 & y_1 & 1 \\ x_2 & y_2 & 1 \end{vmatrix} = 0$$

25. Use the result from Exercise 24 to find the equation of the line through $(6, 2)$ and $(3, -2)$. Write in standard form.

26. For what system of equations does Cramer's rule yield the following determinants.

$$D = \begin{vmatrix} 1 & 2 \\ 3 & 4 \end{vmatrix} \qquad D_x = \begin{vmatrix} 1 & 2 \\ 0 & 4 \end{vmatrix}$$

Solve for x and y using Cramer's rule.

27. $\begin{cases} ax + by = -1 \\ bx + ay = 1 \end{cases}$

28. $\begin{cases} ax + by = \dfrac{b}{a} \\ x + y = \dfrac{1}{b} \end{cases}$

MIXED REVIEW

29. **Standardized Tests** Solve for x: $|3 - 4x| > 2$

   A. $x < \dfrac{1}{4}$ or $x > \dfrac{5}{4}$ B. $\dfrac{1}{4} < x < \dfrac{5}{4}$ C. $x = \dfrac{1}{4}$ or $x = \dfrac{5}{4}$ D. $-\dfrac{5}{4} < x < \dfrac{1}{4}$

30. **Dress Sizes** Sizes of clothing are not the same in different countries. For example, a size 8 dress in the United States is a size 40 in France. The function that will convert dress sizes in the United States to those in France is $f(x) = x + 32$. Use this function to find the dress sizes in France that correspond to sizes 6, 10, 14, and 18 in the United States. Then determine if the function has an inverse. If so, find a formula for the inverse.

Problem Solving File

Using Matrices with Directed Graphs

If you look at a road map, you will see a collection of locations connected in different ways by roads. A **directed graph** is a geometrical representation of a map. It shows locations as points, and roads—or other forms of connections—as lines. Connections may be possible in only one direction on some of the paths and in both directions on other paths. You can analyze directed graph problems using matrices.

Problem

SummerSun Company has five plants. The directed graph at the right shows how supplies are transferred between plants. Find the total number of ways in which supplies can be transferred between two plants without being routed through another plant.

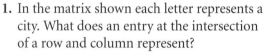

Explore the Problem

1. In the matrix shown each letter represents a city. What does an entry at the intersection of a row and column represent?

2. Use the directed graph to complete the matrix. Write 0 if there are no ways in which supplies can be directly transferred from one plant to another. Use 1 if there is one way.

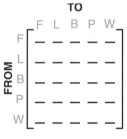

3. What is the sum of the elements in the first row? What does this number represent?

4. What is the sum of the elements for the other four rows? Explain what each sum represents.

5. Add the sums of the rows. What does the total represent?

6. Find the number of ways in which supplies can be transferred directly to Boston from another city. How did you use the matrix to find this number?

Investigate Further

● You can also use matrices to solve problems that involve *indirect* connections between two locations.

7. The manager at SummerSun Company wants to find the total number of ways in which supplies can be shipped between any two plants if there is one stop between the two plants. Why did the manager use a 2 to show the number of connections between Fitchburg and Providence?

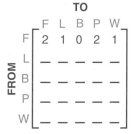

8. Copy and complete the matrix shown in Question 7 for the other plants. Then find the total number of ways in which supplies can be transferred between two plants if there is one stop in-between.

9. Square the matrix in Question 2 and compare it to the matrix in Question 8. What do you notice? How could this information help you solve directed graph problems that involve stopping at one point between two locations?

PROBLEM SOLVING PLAN

• Understand
• Plan
• Solve
• Examine

Apply the Strategy

● The directed graphs below show the connections between stations for two subway systems. Use the graphs for questions 10–12.

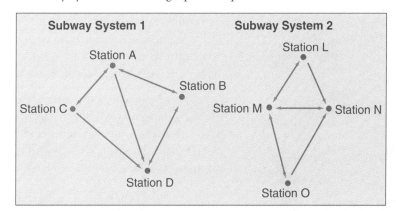

10. Determine a matrix for each system that shows the number of direct connections from one station to another. Does one system have more direct connections? If so, which one?

11. Determine a matrix for each system that shows the number of ways someone could travel between two stations with one stop. Explain your method. Does one system have more one-stop connections? If so, which one?

12. Find the cube of each of the matrices in Question 10. What would each resulting matrix represent?

VACATIONING Two friends, Kevin and Angela, are visiting a theme park. The directed graph shows five locations in the park and the one-way roads connecting them.

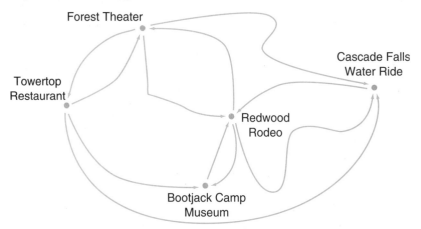

Forest Theater

Cascade Falls
Water Ride

Towertop
Restaurant

Redwood
Rodeo

Bootjack Camp
Museum

13. Determine a matrix for the number of roads that lead directly from one location to another location.

14. If Kevin and Angela start at Forest Theater, how many other locations can they travel to directly? From how many locations could they get to Forest Theater?

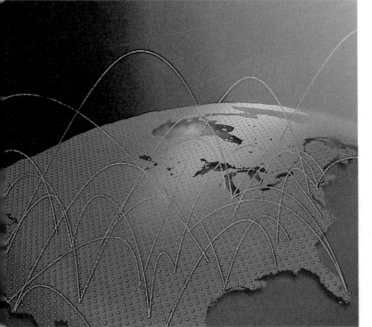

15. Determine a matrix for the number of ways in which Kevin and Angela could travel between two locations with one stop on the way. How many different choices do the friends have?

16. Kevin and Angela are only interested in traveling between two *different* locations with one stop on the way. How could the matrix from Question 15 help find the total number of trails?

17. **MODELING** The matrix at the right represents the number of nonstop flights between cities A, B, C, D, and E. Draw a directed graph that corresponds to the matrix.

$$
\begin{array}{c}
\begin{array}{ccccc} & A & B & C & D & E \end{array} \\
\begin{array}{c} A \\ B \\ C \\ D \\ E \end{array}
\left[
\begin{array}{ccccc}
0 & 1 & 1 & 1 & 0 \\
1 & 0 & 0 & 1 & 0 \\
1 & 0 & 0 & 1 & 0 \\
1 & 1 & 1 & 0 & 1 \\
0 & 0 & 0 & 1 & 0
\end{array}
\right]
\end{array}
$$

Review Problem Solving Strategies

SUMTHING IS UP

4	5
5	6

6	7	8
7	8	9
8	9	10

1. The 2 by 2 number square has a sum of 20 or 4 × 5.

The 3 by 3 number square has a sum of 72 or 9 × 8.

a. Write a similar 4 by 4 number square starting with 7. Find the sum and express it using the same pattern of factors as above. Explain how you determined the factor.

b. What will be the sum of a similar 50 by 50 number square starting with 50?

c. A similar 100 by 100 number square has a sum of 3,000,000. With what number did the square begin?

ON THE FACE OF IT

2. At a factory that produces children's blocks, wood is received in cubes measuring 3 inches on an edge. Workers then cut each large cube into twenty-seven 1-inch cubes.

a. If a clamp holds the large block together while it is cut, how many cuts are needed to produce the twenty-seven small cubes?

b. The factory manager wonders if it is possible to do the job with fewer cuts if the pieces are rearranged after each cut. For example, if the pieces are stacked as shown after the first cut, the second cut will cut through more wood than if the piece had not been moved.

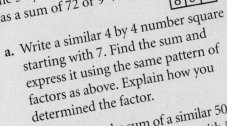

Now, the number of pieces increase after each cut and then many arrangements are possible. Will the manager have to try all the ways or can the problem be solved with logical reasoning? Try thinking about the faces of the small cubes formed in Part a in relation to the faces of the original cube.

TRADING UP

3. Rhonda bought a mathematical software package at a computer store. However, after speaking with her teacher, she realized she needed a different package, one that would cost twice as much. Rhonda went back to the store, returned the first software package and walked out carrying the more expensive software. The store owner stopped her and said that she still owed the store money. Rhonda explained that she did not owe anything as she had already paid for the first software, which is half the cost of the new software, plus she had returned the first software for the other half of the cost. Was Rhonda's argument correct? Explain.

CHAPTER REVIEW

VOCABULARY

Choose the word from the list that completes each statement.

1. A system with at least one solution is __?__.

2. A consistent system with exactly one solution is __?__.

3. The matrix used to represent a system of equations is the __?__.

a. independent

b. consistent

c. augmented matrix

Lesson 4.1 SYSTEMS OF EQUATIONS   pages 165–167

- The graphs of a pair of linear equations *intersect* (have a unique solution), are *parallel*, (have no solution), or *coincide* (have infinitely many solutions)

Solve the system graphically and check.

4. $2x - y = 3$
   $y = -x$

5. $y = x + y$
   $2x + y = 1$

6. $y = 3x + 3$
   $y = 3x + 6$

Lessons 4.2 and 4.3 ALGEBRAIC SOLUTIONS OF SYSTEMS   pages 168–175, 176–182

- To solve of system of linear equations in two variables, use the *substitution method* when an equation has a coefficient of 1 or -1 and the *elimination method* with other coefficients.

- To solve a system of linear equations in three variables, work with two pairs of equations to eliminate one variable, producing a system with two equations in two variables.

Solve the systems algebraically and check.

7. $7x + 2y = 37$
   $y = 3x - 1$

8. $5x - 2y = 20$
   $2x + 3y = 27$

9. $3x + y + 2z = 6$
   $x + y + 4z = 3$
   $2x + 3y + 2z = 2$

Lesson 4.4 AUGMENTED MATRICES   pages 183–188

- The augmented matrix $\begin{bmatrix} a_1 & b_1 & | & c_1 \\ a_2 & b_2 & | & c_2 \end{bmatrix}$ represents the system $\begin{cases} a_1 x + b_1 y = c_1 \\ a_2 x + b_2 y = c_2 \end{cases}$. Use *row*
operations to transform the augmented matrix to $\begin{bmatrix} 1 & 0 & | & r \\ 0 & 1 & | & s \end{bmatrix}$. The solution is $x = r, y = s$.

Use an augmented matrix to solve the system.

10. $\begin{cases} 3x + y = 4 \\ 2x + y = 2 \end{cases}$

11. $\begin{cases} y = 1 - 4x \\ 3x + 2y = 2 \end{cases}$

12. $\begin{cases} x - 3z = 2 \\ 3x + y2z = 5 \\ 2x + 2y + z = 4 \end{cases}$

- You can multiply two matrices when the number of columns in the first matrix equals the number of rows in the second matrix.

Use the matrices A, B, C to perform the indicated operations, if possible.

$$A = \begin{bmatrix} 1 & 2 & 3 \end{bmatrix} \qquad B = \begin{bmatrix} 4 \\ 5 \\ 6 \end{bmatrix} \qquad C = \begin{bmatrix} 1 & 4 \\ 2 & 5 \\ 3 & 6 \end{bmatrix}$$

13. AB **14.** BA **15.** AC **16.** BC

Lessons 4.6 and 4.7   IDENTITY AND INVERSE MATRICES   pages 196–202, 203–209

- An *identity matrix* is the square matrix I whose main diagonal contains only 1's. All other elements are 0. The product of a matrix A and its **inverse A^{-1}** is an identity matrix I.

- If the inverse of the coefficient matrix of a system of linear equations exists, you can solve the system using inverse matrices.

Determine the inverse of each matrix, if it exists, and verify that $A \cdot A^{-1} = I$.

17. $A = \begin{bmatrix} 4 & -4 \\ 3 & 2 \end{bmatrix}$ **18.** $A = \begin{bmatrix} 6 & 3 \\ 8 & 4 \end{bmatrix}$ **19.** $A = \begin{bmatrix} -1 & 1 & 0 \\ -1 & 0 & 1 \\ 6 & 2 & -3 \end{bmatrix}$

Use an inverse matrix to solve each system.

20. $\begin{cases} 2x - 5y = 2 \\ 3x - 7y = 1 \end{cases}$ **21.** $\begin{cases} 4x = 3y + 11 \\ 5x - 6y = 9 \end{cases}$ **22.** $\begin{cases} 6x - 2y + z = 16 \\ 2x - y + 5z = 2 \\ 2x - 3z = 8 \end{cases}$

Lesson 4.8   DETERMINANTS AND CRAMER'S RULE   pages 210–215

- You can use the *diagonals method* or *expansion by minors* to evaluate a determinant.

Evaluate each determinant.   **Use Cramer's Rule to solve the system.**

23. $\begin{vmatrix} 3 & 7 \\ 4 & 2 \end{vmatrix}$ **24.** $\begin{vmatrix} 3 & 4 & 8 \\ 7 & 1 & 6 \\ 2 & 9 & 5 \end{vmatrix}$ **25.** $\begin{cases} x + y = 4 \\ 5x + 2y = 11 \end{cases}$ **26.** $\begin{cases} -5x + 16y = -7 \\ 3x - 9y = 4 \end{cases}$

Lesson 4.9   USING MATRICES WITH DIRECTED GRAPHS   pages 216–219

- You can analyze directed graph problems using matrices.

  The directed graph shows the Air America flight network.

27. Use the graph to create a matrix and determine the total number of nonstop routes in the network.

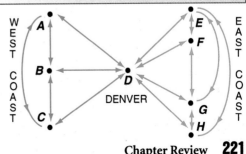

CHAPTER ASSESSMENT

CHAPTER TEST

1. **WRITING MATHEMATICS** Explain why square matrices are the only matrices that have multiplicative inverses.

2. **STANDARDIZED TESTS** Select the solution of the system $x + y = 7$ and $y = 2$.

   A. $(2, 5)$ **B.** $(5, 2)$ **C.** $(3, 4)$ **D.** $(4, 3)$

Solve graphically and check.

3. $\begin{cases} y - x = 1 \\ x + 2y = -4 \end{cases}$

4. $\begin{cases} 3x + 2y = 6 \\ x = 4 \end{cases}$

Use the substitution method to solve the system. State whether the system is consistent, inconsistent, independent or dependent.

5. $\begin{cases} 2x + 2y = 8 \\ 3x + 3y = 12 \end{cases}$

6. $\begin{cases} x + y = 8 \\ 2x + 2y = 3 \end{cases}$

Use the elimination method to solve each system.

7. $\begin{cases} x + 3y = 10 \\ 4x - 3y = 5 \end{cases}$

8. $\begin{cases} 2x + y = 6 \\ x = 3y + 10 \end{cases}$

9. Solve graphically.
   $\begin{cases} 2y = -3x \\ 3x + y = -3 \\ y - x = 5 \end{cases}$

10. Solve algebraically.
    $\begin{cases} 3x - 2y - 3z = -1 \\ 6x + y + 2z - 7 \\ 9x + 3y + 4z = 9 \end{cases}$

11. Use an augmented matrix to solve the system.
    $\begin{cases} x + 2y = 4 \\ 3x - y = 5 \end{cases}$

Use matrices A, B, and C to find each product, if possible.

$A = \begin{bmatrix} 1 \\ 4 \\ 5 \end{bmatrix}$ $B = \begin{bmatrix} 2 & 3 & 6 \end{bmatrix}$ $C = \begin{bmatrix} 2 & 4 & 0 & 6 \\ 1 & 5 & -2 & 3 \end{bmatrix}$

12. AB 13. BA 14. BC 15. $2A \cdot B$

Find the inverse of each matrix, if it exists.

16. $\begin{bmatrix} 2 & 8 \\ 1 & 4 \end{bmatrix}$

17. $\begin{bmatrix} 3 & 2 & 1 \\ 1 & 1 & -1 \\ 4 & 3 & 1 \end{bmatrix}$

18. Use an inverse matrix to solve the system.
    $\begin{cases} x - 2y + 3z = 3 \\ 2x + y + 5z = 8 \\ 3x - y - 3z = -22 \end{cases}$

Evaluate each determinant.

19. $\begin{vmatrix} 1 & -2 & 5 \\ 1 & 4 & -1 \\ 2 & 6 & 3 \end{vmatrix}$

20. $\begin{vmatrix} 0 & 4 & 1 \\ 2 & 3 & -2 \\ 4 & -1 & -3 \end{vmatrix}$

Use Cramer's Rule to solve each system.

21. $\begin{cases} x = 2y + 1 \\ x + 2y = 4 \end{cases}$

22. $\begin{cases} x + y + z = 3 \\ 2x - y + 2z = 6 \\ 5x + 2y + 2z = 0 \end{cases}$

23. The table shows prices of 3 "penny stocks" (A, B, C) over a 3-month period and the amount Ms. Li invested when she bought the same number of shares in each stock in each month. How many shares did she buy each month?

	A	B	C	Cost
March	0.02	0.02	0.03	65.00
April	0.03	0.03	0.02	85.00
May	0.03	0.02	0.01	65.00

24. If A-G are students, and the arrows in the directed graph indicate that one person knows the other's phone number, create a matrix and find the number of direct calls that can be made.

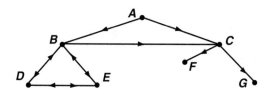

PERFORMANCE ASSESSMENT

BUILD POLYGONS Write a system of linear equations in two variables that will form a polygon. State the coordinates of the vertices in the polygon. Find the area of the polygon.

For example, the lines $y + x = 2$, $y = x$, and $y = -3$ form $\triangle ABC$ with vertices $A(-3, -3)$, $B(5, -3)$ and $C(1, 1)$.

$$\text{Area}_{\triangle ABC} = \frac{1}{2}(AB)(CH) = \frac{1}{2}(8)(4) = 16$$

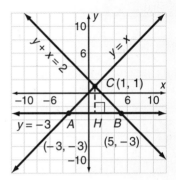

TIC-TAC-TOE Work in groups of three. Each group member should make a list of nine different questions about the concepts in this chapter and an answer key. The questions should be varied and include definitions, descriptions of procedures, or a specific problem to be solved. Each group member should create a Tic-Tac-Toe board by writing one question in each of the gameboard squares. Play three rounds so that each of you is the judge for the board she or he created and a player for the other two rounds.

DESIGN A NETWORK Create a directed graph to illustrate a real-life situation such as an airline network. Use the graph to create a matrix that represents the situation. Tell what the sums of the various rows of the matrix represent and what the total of the sums of the rows represents.

CREATE MATRICES Cut out eight squares of paper. Label each one with a number from 1–8. Arrange your squares into two matrices for which the product exists. Calculate the product. Repeat this procedure for all pairs of matrices that can be constructed with eight elements for which the product exists.

PROJECT ASSESSMENT

PROJECT *Connection*

For this final activity, you will need all your templates, a sheet of poster board, waxed paper, scissors, tape, and a timer.

1. Using the large box template, one student should trace the outline on paper board, cut it out, and assemble it using tape. A second student should do the same for the smaller box. A third student should record the time (in minutes) for each box to be "manufactured". Let A represent the time for the large box and B represent the time for the small box.

2. Follow similar steps to trace, assemble, and time the large and small inner bags. Let C represent the time for the large bad and D represent the time for the small bag.

3. Solve the following problem using the data collected above. Your company manufactures both sizes of bag and box. You have one cutter/folder/assembler for boxes and another for bags. The large sizes requires A minutes for the box and C minutes for the bag. The small size requires B minutes for the box and D minutes for the bag. If your plant operates 12 hours a day, how many complete packages of each size can be produced?

CUMULATIVE REVIEW

1. **WRITING MATHEMATICS** Explain how the graph of the system $3y - 2x = 12$ and $2x - 6 = 3y$ relates to the solution of the system.

Solve the system algebraically and check.

2. $\begin{cases} \dfrac{y}{2} = x + 1 \\ 4x - y = 6 \end{cases}$

3. $\begin{cases} 3x - 4y = 12 \\ 4x - y = -10 \end{cases}$

Write and solve a system of equations for the problem.

4. The combination to Arnie's locker is a 3-digit number, the sum of whose digits is 14. The units digit is equal to the sum of the hundreds digit and the tens digit. The number with the digits reversed exceeds the original number by 297. Find the original number.

5. **STANDARDIZED TESTS** The augmented matrix for the linear system $\begin{cases} 3x - y - 1 = 0 \\ 7x + 2y = 37 \end{cases}$ is

A. $\begin{bmatrix} 3 & -1 & | & -1 \\ 7 & 2 & | & 37 \end{bmatrix}$ B. $\begin{bmatrix} 1 & 0 & | & 3 \\ 0 & 1 & | & 8 \end{bmatrix}$

C. $\begin{bmatrix} 3 & -1 & | & 1 \\ 7 & 2 & | & 37 \end{bmatrix}$ D. $\begin{bmatrix} 0 & 1 & | & 3 \\ 1 & 0 & | & 8 \end{bmatrix}$

Let $A = \begin{bmatrix} 4 & 2 \\ 3 & 1 \end{bmatrix}$ and $B = \begin{bmatrix} 5 & 6 \\ 0 & -7 \end{bmatrix}$.

6. Find AB.

7. Find A^{-1}.

8. Use an inverse matrix to solve the system
$\begin{cases} x - 3y = -2 \\ 3x - 2y + z = 5 \\ 2x + y + 2z = 4 \end{cases}$

Evaluate each determinant.

9. $\begin{vmatrix} 7 & -4 \\ -3 & -2 \end{vmatrix}$

10. $\begin{vmatrix} 1 & 2 & 8 \\ 3 & -1 & 6 \\ 5 & 4 & 7 \end{vmatrix}$

11. Write an equation of the line that is parallel to the line $2y = 3x - 1$ and passes through the point $(-4, 5)$.

12. The Pregolya River flows through the Russian city that was known in the 1700's as Königsberg.

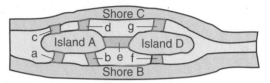

There are two islands, A and D, in the river. Seven bridges, a, b, c, d, e, f, and g, connect the islands to each other and to the shores, B and C. Draw a network to model the situation and list the ways to get from B to C by going only one way on any bridge.

13. **STANDARDIZED TESTS** Which of the following statements are true if the relation $(a, 1)$, $(2, 3)$, $(b, 3)$ is to be a function?

   I. It cannot be a function since there are two second values that are the same.
   II. The values of a and b cannot be the same.
   III. Neither a nor b can be 2.
   A. I only   **B.** II only
   C. III only   **D.** II and III

14. Express $1.\overline{17}$ as a ratio of two integers.

15. **WRITING MATHEMATICS** Explain why or why not the systems ({integers}, $+$, $\times$) is a field.

16. In one roll of a number cube, what is the theoretical probability of getting a number that is less than 5?

17. Light travels about 186,000 miles per second. Express, in scientific notation, the number of miles light travels in a year. (This distance is called a *light-year*.)

Find the inverse of each function.

18. $f(x) = 9x + 11$ 19. $g(x) = -4$

QUANTITATIVE COMPARISON In each question, compare the quantity in Column I with the quantity in Column II. Select the letter of the correct answer from these choices:

A. The quantity in Column I is greater.
B. The quantity in Column II is greater.
C. The two quantities are equal.
D. The relationship cannot be determined by the information given.

Notes: In some questions, information which refers to one or both columns is centered over both columns. A symbol used in both columns has the same meaning in each column. All variables represent real numbers. Most figures are not drawn to scale.

Column I	Column II

1.
$$y = 2x - 1$$
$$x = y$$

x	1

2.
$$2m + n = -1$$
$$3n + 4p = 9$$
$$3m - p = 0$$

$p \div m$	$p + n$

3. $A = \begin{bmatrix} 1 & 2 & 3 \end{bmatrix}$ $\qquad B = \begin{bmatrix} 1 & 4 \\ 2 & 5 \\ 3 & 6 \end{bmatrix}$

The element in row 1 column 2 of AB	The element in row 1 column 2 of $A + B$

4. The element in row 2 column 1 of the 2 × 2 identity matrix | The element in row 2 column 2 of the 3 × 3 identity matrix

5. $A = \begin{bmatrix} 2 & -1 \\ 4 & 4 \end{bmatrix}$

The element in row 2 column 1 of A^{-1}	The element in row 2 column 2 of A^{-1}

Column I	Column II

6. $\begin{vmatrix} -4 & 0.5 \\ 16 & 2 \end{vmatrix}$ | $\begin{vmatrix} 2 & -1 & 3 \\ 1 & 2 & 3 \\ 3 & -2 & 1 \end{vmatrix}$

7. The number of lines of symmetry in | The number of lines of symmetry in

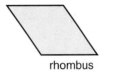
rhombus

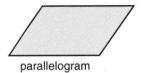
parallelogram

8. line l is $y = -x - 8$

The slope of a line that is parallel to line l	The slope of a line that is perpendicular to line l

9. $f(x) = 3|2x - 3| - 4$

The x-coordinate of the vertex of the graph	The y-coordinate of the vertex of the graph

10. $f(x) = x^2$ and $g(x) = x - 1$

$[f \circ g](3)$	$[g \circ f](3)$

11. y varies directly as x^3

x	0	1	2
y	0	7	56

The value of y when $x = -3$	The value of x when $y = 189$

12. $\sqrt{5} + \sqrt{11}$ | $\sqrt{16}$

13. $\dfrac{x^3}{3}$ | $\left(\dfrac{x}{3}\right)^3$

14. The probability of 7 in one roll of a number cube | The probability of an even number in one random selection from $\{-2, 0, 2, 4, 6\}$

5 Polynomials and Factoring

Take a Look
AHEAD

Make notes about things that look new.
- Locate and copy an example of the difference of two squares and an example of a perfect square trinomial.
- What is another name for the x-intercepts of a polynomial function?
- What property will be applied when you solve polynomial equations by factoring?

Make notes about things that look familiar.
- What operations can be performed with polynomials? What operations are inverses of each other?
- Give an example of a number written in exponent form. Identify the exponent and the base. Explain how to find the standard form of the number.
- Copy an example that shows the use of the distributive property.

DATA Activity

Product Changes

To stimulate sales, most businesses regularly change their products or introduce new ones. A *product line* is any group of closely related products while a *product mix* is the mix of all products offered for sale by a company. For example, cereals are a product line that might be part of a product mix of cereals, cookies, and crackers. Existing products might be modified by changes in packaging, or composition, or the line might be extended to include a totally new variety.

SKILL FOCUS

- ▶ Read and interpret a table.
- ▶ Add, subtract, multiply, and divide real numbers.
- ▶ Solve percent problems.
- ▶ Design a survey and analyze the results.

PROFIT & LOSS

In this chapter, you will see how:

- **MOVIE THEATER MANAGERS** use polynomials to analyze their costs and profits. (Lesson 5.2, page 241)

- **MANUFACTURERS** use polynomial functions to model aspects of business such as production levels and profits. (Lesson 5.5, page 261)

ITEM	Food	Beverages	Health and Beauty	Household Products	Pet Products
NEW PRODUCT INTRODUCTIONS, 1994					
NUMBER	3883	807	2655	378	161
CATEGORY %					
New brand	21.1	32.6	19.3	17.2	31.7
Brand extensions	1.4	1.1	1.0	2.9	0.6
Line extensions	77.5	66.3	79.7	79.9	67.7
CHANGES %					
Formulation	56.5	50.8	47.8	50.0	51.8
New markets	0.9	—	3.4	11.1	3.4
Packaging	13.3	17.5	8.6	—	3.4
Positioning	29.3	31.7	39.1	33.3	34.5
Technology	—	—	1.1	5.6	6.9

1. What was the total number of new product introductions for the categories shown?

2. What percent of new products were beverages? Round to the nearest tenth of a percent.

3. How many new health and beauty products represented line extensions?

4. The term "positioning" refers to a new product presented for new users or uses compared to existing products in its category. In which item category were approximately 1138 products the result of a change in positioning?

5. **WORKING TOGETHER** Design a survey to determine consumer attitudes about changing products to increase sales. Do people feel changes provide an improved or more useful product? Do they think changes are of little value and only serve to increase price? If possible, carry out your survey near a supermarket or shopping mall.

The Business Game

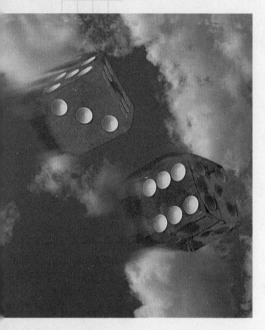

Perhaps you have heard it said that the world of business is like a game — strategy and luck influence the outcomes, there are ups and downs, winners and losers. Board games themselves are a big business. A game that achieves long-term popularity will earn millions of dollars for its manufacturer. Like any other product, developing a marketable new game is a lengthy process during which an initial idea may undergo many changes from start to finish. For this project, imagine you have been hired by a toy company to create a new board game about business and making money!

PROJECT GOAL

To design a board game which centers around the idea of profit and loss in business.

Getting Started

Work in groups of three to five students.

1. Make a list of board games that group members have played. Briefly explain and discuss each game. Which games are most similar to the type of game you will design in this project?

2. Brainstorm a list of general features that all games have in common. Also think about why you enjoy playing some games more than others.

PROJECT Connections

Lesson 5.1, page 233:
Investigate profit functions and choose a format for the game board.
Lesson 5.2, page 240:
Plan key features of the game and the mathematical skills that will be involved.
Lesson 5.5, page 260:
Prepare the final game board and instructions for playing the game.
Chapter Assessment, page 269:
Prepare descriptions and advertisements for the game. Play and evaluate games.

3. Decide your group's five favorite games. For each of these top five games, discuss the set-up of the board, how players move around, and how a winner is determined.

4. Find out the meaning of key business terms such as revenue, fixed costs, variable costs, and profit.

✉ Internet Connection

www.swpco.com/
swpco/algebra2.html

5.1 Algebra Workshop
Polynomials

Think Back

- A **monomial** is a real number, a variable, or a product of a real number and one or more variables. A **polynomial** is a monomial, or a sum or difference of monomials. You can use Algeblocks to model a polynomial.

  1. What polynomial is modeled on the Basic mat below?

  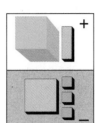

  2. How can you simplify the expression modeled below?

  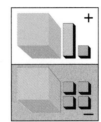

SPOTLIGHT ON LEARNING

WHAT? In this lesson you will learn to use Algeblocks
- to model operations with polynomials.
- to find the greatest common factor of a polynomial.
- to find the binomial factors of a trinomial.

WHY? Polynomials can help you solve geometric problems about areas, volumes, and missing dimensions.

Explore

- Work with a partner and Algeblocks.

  3. **a.** Model $2x^3 + 3x^2 - 4x + 2$ on a Basic mat.

  b. Add $x^3 - 4x^2 + 3x - 5$ to the mat.

  c. Simplify the expression that results.

  d. Read the sum of the two polynomials from the mat.

Use Algeblocks to find each sum.

4. $(x^3 - 4x^2 + 7) + (-2x^3 + 3x^2 - 5x + 1)$

5. $(-3x^3 + 2x^2 - 5x) + (x^3 + 4x + 4)$

6. **a.** Model $3x^3 + 2x + 1$ on a mat.

b. Add the *opposite* of $2x^3 - x^2 - 2x + 3$ to the mat.

c. What operation have you modeled?

d. Simplify. Then read the result from the mat.

Use Algeblocks to find each difference.

7. $(2x^3 - x + 5) - (x^3 - 2x + 1)$

8. $(-x^3 + 3x^2 - 2x) - (-x^3 - 3x^2 + 2x + 1)$

THINK BACK

Recall that finding the opposite of the expression changes the sign of each term. You can place the expression on the mat and then move each piece to the opposite region.

To use Algeblocks in multiplying two polynomials, place the blocks for one factor along the horizontal axis and place the blocks for the other factor along the vertical axis of a Quadrant Mat.

9. a. What multiplication is shown on the mat?

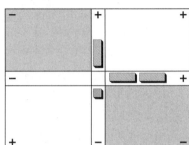

b. Form rectangular areas with Algeblocks in the quadrants bounded by the blocks that represent the factors.

c. The product is the sum of the rectangular areas. Write a polynomial for the product.

10. a. Place the factors $(2x - 1)$ and $(x - 3)$ on the axes of a Quadrant Mat.

b. Form rectangular areas in the quadrants bounded by the blocks that represent the factors.

c. What does the sum of the rectangular areas represent?

d. What is the product $(2x - 1)(x - 3)$?

Use Algeblocks to find each product.

11. $(2x - 3)(x + 2)$

12. $(x - 2)(x + 2)$

13. $(x + 2)^2$

A **binomial** is a polynomial with two terms. A **trinomial** is a polynomial with three terms.

14. Examine Questions 11–13. Is the product of two binomials always a trinomial? If not, when is the product a binomial?

15. a. Model each of the monomials $2x^2$ and $6x$ with Algeblocks.

b. For each monomial in part a, form rectangles that have the length of one side in common. What is the greatest common length of the two rectangles?

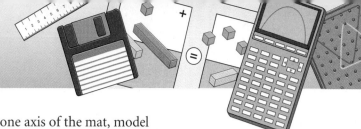

c. Model $2x^2 + 6x$ in Quadrant I. On one axis of the mat, model the greatest common length.

d. Show the other length of the rectangle on the other axis. Read the dimensions from the mat.

e. Write $2x^2 + 6x$ as the product of a monomial factor and a binomial factor.

THINK BACK

To model in a quadrant means to form rectangular regions.

For each model, write the polynomial and its factored form.

16. 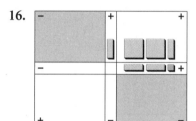　**17.**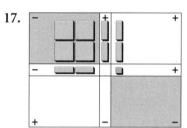

18. a. On a Quadrant Mat, place Algeblocks for $x^2 + 4x + 3$ in Quadrant I. Arrange the blocks to form a rectangle.

b. Along the horizontal and vertical axes, place blocks to represent the dimensions of the rectangle you formed in part a.

c. From the axes, read the two binomial factors of $x^2 + 4x + 3$.

19. a. Using all four quadrants, place Algeblocks for $x^2 - 6x + 5$ on a Quadrant Mat. Arrange the blocks to form rectangles. Place blocks for the dimensions of the rectangles on the horizontal and vertical axes. Draw a diagram to show your work.

b. From the mat, read the factors of $x^2 - 6x + 5$.

c. Compare your diagram for part a with those of your classmates. Describe any differences you see. Did the differences affect the answer to part b?

20. a. Using appropriate quadrants, place Algeblocks for $x^2 + x - 12$ on a Quadrant mat. What do you discover when you try to arrange the blocks into rectangles?

b. What zero pair can you add in order to arrange the Algeblocks into rectangles? Show your arrangement.

c. From the axes, read the binomial factors.

PROBLEM SOLVING TIP

When factoring trinomials, you may need to try several arrangements until you find one that works. Noting the sign of each term will help you decide on the quadrants to use. You can add blocks that form zero pairs to complete rectangles.

Use Algeblocks to find the factors of each trinomial.

21. $x^2 + 2x - 8$　　　　　**22.** $2x^2 - 5x - 3$

Make Connections

Each Algeblocks model below represents an operation on two polynomials. For each model, write two different expressions and their results.

23.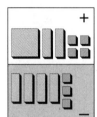

24.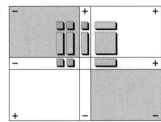

25.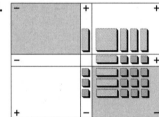

26.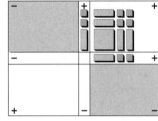

The product of two binomials is usually a trinomial. The products of some binomials have special patterns.

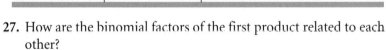

Binomial	Product	Example
$(x + c)(x - c)$	$x^2 - c^2$	$(x + 3)(x - 3) = x^2 - 9$
$(x + c)(x + c)$	$x^2 + 2cx + c^2$	$(x + 3)(x + 3) = x^2 + 6x + 9$
$(x - c)(x - c)$	$x^2 - 2cx + c^2$	$(x - 3)(x - 3) = x^2 - 6x + 9$

27. How are the binomial factors of the first product related to each other?

28. Describe the first product.

29. How are the binomial factors of the second and third products related to each other?

30. Describe the second product.

31. Describe the third product.

Match each product with its binomial factors.

32. $x^2 - 25$

33. $x^2 + 10x + 25$

34. $x^2 - 10x + 25$

a. $(x + 5)(x + 5)$

b. $(x - 5)(x - 5)$

c. $(x + 5)(x - 5)$

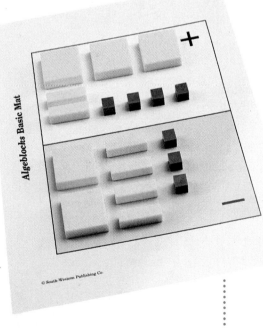

Algeblocks Basic Mat

© South-Western Publishing Co.

Summarize

35. **WRITING MATHEMATICS** Explain how you would use Algeblocks to find the factored form of $4x^2 - 12x$. Describe each step.

36. **MODELING** Use Algeblocks to find the factored form of $x^2 - x - 6$.

37. **THINKING CRITICALLY** Write the binomial factors whose product is shown on the mat.

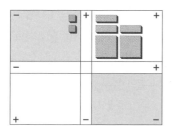

38. **GOING FURTHER** Use Algeblocks to make the shape shown. Write an expression for the perimeter and the area.

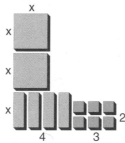

PROJECT *Connection*

Your first important decision will be the configuration of the game board. The game will center around the mathematics of profit and loss. Below is a list of some profit functions, where x is the number of units produced.

$P(x) = -0.001x^2 + 60x + 150,000$
$P(x) = -0.001x^2 + 125x - 250,000$
$P(x) = -135x^3 + 6,400x^2 - 200,000$
$P(x) = -35x^3 + 2,000x^2 - 27,000$

1. Assign each function to at least one group member. Have each group member graph the function on a graphing utility. Sketch each graph noting x-intercepts and places where the function changes direction. Note where the graph indicates a profit and where it indicates a loss.

2. Discuss the format of your game. Will players move around a board with squares or along a path that looks like the graph of a profit function? Will the function used be the same for each player or vary depending on a chance feature such as drawing a card? Sketch one or more possible game boards.

5.2 Operations With Polynomials

SPOTLIGHT ON LEARNING

WHAT? In this lesson you will learn
- to use the properties of exponents.
- to add, subtract, and multiply polynomials.

WHY? Polynomial expressions can help you model and solve problems about the dimensions of a painting, the area of a lawn, or the volume of a box.

THINK BACK

The expression a^n is called a **power**. An **exponent** tells you how many times a number or variable, called the **base**, is used as a factor.
$$2^3 = 2 \cdot 2 \cdot 2$$
$$a^4 = a \cdot a \cdot a \cdot a$$

Explore

1. **a.** What is the meaning of $a^6 \cdot a^4$?

   b. Write the result using an exponent.

   c. What is the meaning of $b^3 \cdot b^2$?

   d. Write the result using an exponent.

   e. Write a rule for multiplying terms with like bases.

2. **a.** Write a statement to explain the meaning of $(a^6)^4$.

   b. Write an expression to explain the meaning of $(a^6)^4$.

   c. To evaluate this expression, apply the rule in Question 1e.

   d. Write a rule for raising a power to a power.

3. **a.** What is the meaning of $(ab)^5$?

   b. Simplify the expression you wrote in part a and write the result using exponents. What properties did you use?

   c. Write a rule for finding a power of a product.

4. **a.** What is the meaning of $\left(\dfrac{a}{b}\right)^4$?

   b. Simplify the expression you wrote in part a.

   c. Write a rule for finding a power of a quotient.

5. **a.** What is the meaning of $\dfrac{a^6}{a^4}$?

   b. Divide and write the result using an exponent.

   c. Write a rule for dividing terms with like bases.

6. **a.** Simplify the quotient $\dfrac{a^4}{a^4}$.

   b. Use the rule in Question 5c to evaluate $\dfrac{a^4}{a^4}$.

   c. Draw a conclusion from 6a and 6b. $a^0 = 1$

7. **a.** Divide common factors to simplify the quotient $\dfrac{a^4}{a^6}$.

   b. Use the rule in Question 5c to evaluate $\dfrac{a^4}{a^6}$.

   c. Draw a conclusion from 7a and 7b.

Build Understanding

The rules you discovered in Explore apply for all positive integers m and n and all real numbers a and b.

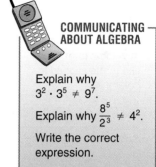
> **PROPERTIES OF EXPONENTS**
>
> For any real numbers a and b and positive integers m and n:
>
> | | | |
> |---|---|---|
> | Product of Powers | $a^m \cdot a^n = a^{m+n}$ | |
> | Power of a Power | $(a^m)^n = a^{mn}$ | |
> | Power of a Product | $(ab)^m = a^m b^m$ | |
> | Power of a Quotient | $\left(\dfrac{a}{b}\right)^m = \dfrac{a^m}{b^m}$ | $b \neq 0$ |
> | Quotient of Powers | $\dfrac{a^m}{a^n} = a^{m-n}$ | $a \neq 0$ |
> | Zero Exponent | $a^0 = 1$ | $a \neq 0$ |
> | Negative Exponent | $a^{-n} = \dfrac{1}{a^n}; \ \dfrac{1}{a^{-n}} = a^n$ | $a \neq 0$ |

EXAMPLE 1

Simplify each expression.

a. $3k^2(-2k^3)$ **b.** $(y^5)^3$ **c.** $(-3a^2)^4$

d. $\left(\dfrac{t}{2}\right)^3$ **e.** $\dfrac{16x^6}{8x^2}$ **f.** 5^{-2}

Solution

a. $3k^2(-2k^3) = 3(-2)k^{2+3} = -6k^5$

b. $(y^5)^3 = y^{5 \cdot 3} = y^{15}$

c. $(-3a^2)^4 = (-3)^4(a^2)^4 = 81a^8$

d. $\left(\dfrac{t}{2}\right)^3 = \dfrac{t^3}{2^3} = \dfrac{t^3}{8}$

e. $\dfrac{16x^6}{8x^2} = \dfrac{16}{8}(x^{6-2}) = 2x^4$

f. $5^{-2} = \dfrac{1}{5^2} = \dfrac{1}{25}$

> **THINK BACK**
>
> A polynomial is usually arranged in **descending order** (highest to lowest degree) or **ascending order** (lowest to highest degree) of one of its variables. The **degree** of a polynomial is the highest exponent to which a variable is raised.

The properties of exponents are used to simplify polynomials. A polynomial is in *simplest form* when all like terms are combined. When you combine polynomials by addition or subtraction, the result has the same exponent as the terms being simplified.

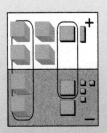

EXAMPLE 2

Add: $(4x^3 - 3x + 2x^2) + (4 + 2x^3 - 2x)$

Solution

$\qquad (4x^3 + 2x^2 - 3x) + (2x^3 - 2x + 4)$ Put in descending order.

$= (4x^3 + 2x^3) + 2x^2 + (-3x - 2x) + 4$ Group like terms.

$= (4 + 2)x^3 + 2x^2 + [-3 + (-2)]x + 4$ Use distributive property.

$= 6x^3 + 2x^2 - 5x + 4$

The sum of the two polynomials is $6x^3 + 2x^2 - 5x + 4$.   ◀

Recall that you can subtract a number by adding its opposite.

$$a - b = a + (-b)$$

The opposite of a polynomial is the opposite of each of its terms.

$$-(3x^2 - 7x + 15) = -3x^2 + 7x - 15$$

EXAMPLE 3

Subtract: $(4x^3 - 2x^2 + x - 3) - (2x^3 - x^2 + 1)$

Solution

$\qquad (4x^3 - 2x^2 + x - 3) - (2x^3 - x^2 + 1)$

$= (4x^3 - 2x^2 + x - 3) + (-2x^3 + x^2 - 1)$ Add the opposite.

$= (4x^3 - 2x^3) + (-2x^2 + x^2) + x + (-3 - 1)$ Group like terms.

$= 2x^3 - x^2 + x - 4$

The difference of the two polynomials is $2x^3 - x^2 + x - 4$.   ◀

Sometimes, you will need to multiply by a polynomial by a monomial to solve an equation that you have used to model a problem. Use the distributive property.

EXAMPLE 4

ART A painting 30 in. long is in a frame 4-in. wide. The area of the frame alone is 448 in.^{2} Find the width of the framed painting.

Solution

Let $x = $ the width of the framed painting. The overall length is $(30 + 8)$ in. and the width of the painting is $(x - 8)$ in.

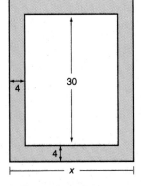

| Area of Framed Painting | $-$ | Area of Painting | $=$ | Area of Frame |

$$x(30 + 8) - 30(x - 8) = 448 \qquad \text{Area} = \text{length} \times \text{width}$$
$$38x - (30x - 240) = 448 \qquad \text{Distributive property.}$$
$$38x + (-30x + 240) = 448 \qquad \text{Add the opposite.}$$
$$8x + 240 = 448$$
$$8x = 208 \qquad \text{Subtract 240 from both sides.}$$
$$x = 26 \qquad \text{Divide both sides by 8.}$$

Check

If $x = 26$, the dimensions of the framed painting are 26 in. $\times$ 38 in., resulting in an area of 988 in.^{2} The width of the painting is $(26 - 8)$ or 18 in. Thus, the area of the painting is 30 in. $\times$ 18 in. or 540 in.^{2} The area of the frame is $(988 - 540)$ or 448 in.^{2} The overall width of the framed painting is 26 in. ◄

To multiply two binomials, use the distributive property twice.

EXAMPLE 5

Multiply: $(2x + 1)(x - 3)$

Solution

$$(2x + 1)(x - 3)$$
$$= 2x(x - 3) + 1(x - 3) \qquad \text{Distribute the terms of the first binomial.}$$
$$= 2x^2 - 6x + x - 3 \qquad \text{Distributive property.}$$
$$= 2x^2 - 5x - 3 \qquad \text{Collect like terms.} \qquad ◄$$

You can also use the mnenomic device FOIL to help you remember which terms to multiply when multiplying two binomials. FOIL stand for First terms, Outside terms, Inside terms, and Last terms.

You can multiply two polynomials of any number of terms using the distributive property.

EXAMPLE 6

Multiply: $(3x^2 + 4x - 2)$ by $(2x - 3)$

Solution

$$(2x - 3)(3x^2 + 4x - 2) \qquad \text{Distribute the terms of the first factor.}$$
$$= 2x(3x^2 + 4x - 2) - 3(3x^2 + 4x - 2)$$
$$= 6x^3 + 8x^2 - 4x - 9x^2 - 12x + 6 \qquad \text{Distributive property.}$$
$$= 6x^3 - x^2 - 16x + 6 \qquad \text{Combine like terms.}$$

You can also use a vertical arrangement.

$$\begin{array}{r} 3x^2 + 4x - 2 \\ 2x - 3 \\ \hline -9x^2 - 12x + 6 \\ 6x^3 + 8x^2 - 4x \\ \hline 6x^3 - x^2 - 16x + 6 \end{array}$$

◄

COMMUNICATING ABOUT ALGEBRA

How is algebraic multiplication by a polynomial similar to arithmetic multiplication by a multi-digit number? How is it different?

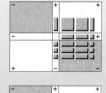

ALGEBLOCKS MODEL

Place $2x + 1$ in the horizontal axis, and place $x - 3$ in the vertical axis. Form rectangular areas in all quadrants bounded by pieces.

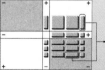

Remove zero pairs. Read the product:
$2x^2 - 5x - 3$

Simplify.

1. $3x^2 \cdot 9x^6$

2. $(3x^6)^2$

3. $9x^6 \div 3x^2$

4. $\left(\dfrac{1}{6}\right)^2$

Rewrite the expression using only positive exponents.

5. y^{-4}

6. $6x^{-2}$

7. $\dfrac{3}{x^{-3}}$

8. $\dfrac{m^{-2}}{n^{-3}}$

Simplify.

9. $2x^0$

10. $(2x)^0$

11. $2^0 x$

12. $(2+x)^0$

13. MODELING Show the product $(2x - 1)(x - 2)$ using Algeblocks.

14. GEOMETRY Tania is mowing a rectangular lawn that is 40 ft wide. She has already mowed a 5-foot strip around the entire lawn. If the area of the strip Tania has mowed is 900 ft², what is the overall length of the lawn?

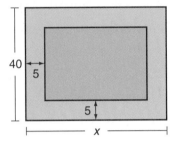

Simplify.

15. $(3x^2 + 5x - 1) + (-2x + 3)$

16. $(-2x^2 + 4x^3 - 3) + (x^3 - 2)$

17. $(7y^2 - 4y + 6) - (3y^2 - 2y - 7)$

18. $3z^2(2z^2 - 4z + 5)$

19. $(3t + 2)(2t - 3)$

20. $(2w^2 - 3w + 4)(w - 3)$

21. WRITING MATHEMATICS Explain how subtraction of polynomials is related to addition of polynomials.

22. GEOMETRY Write a polynomial to represent the area of the right triangle shown below.

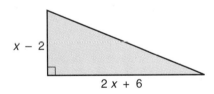

23. GEOMETRY Write a polynomial to represent the area of the trapezoid.

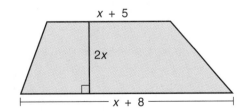

PRACTICE

Simplify.

1. $-3^2(3)^2$

2. $(a + b)^2(a + b)^3$

3. $(4m^2n^3)^2$

4. $\dfrac{16x^8}{4x^4}$

Rewrite the expression using only positive exponents.

5. 2^{-3}

6. $\dfrac{1}{2}y^{-2}$

7. $\dfrac{-4}{x^{-4}}$

8. $\dfrac{b^{-3}}{\frac{1}{a^{-2}}}$

Simplify.

9. $-3x^0$

10. $(-3x)^0$

11. $(-3)^0x$

12. $(x - 3)^0$

13. **GEOMETRY** A rectangle is formed from square $ABCD$ by doubling the length of side AB and decreasing the length of side BC by 4 units. Write a polynomial to represent the area of the rectangle.

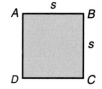

14. **MODELING** Show $(3x^3 - 2x^2 + 1) - (x^3 - 2x^2 - 2)$ using Algeblocks.

15. **MODELING** Show the product $(3x - 2)(x - 3)$ using Algeblocks.

Simplify.

16. $(4y^2 - 2y + 5) + (-2y^2 + y - 3)$

17. $(-2 + z^2) + (4z^3 - 3z^2 + 6)$

18. $(2k^2 + 3k + 1) - (3k^2 - 2k - 5)$

19. $(-8m^2 - 3m + 3) - (-2m^2 - 5)$

20. $-4h^2(h^3 - 2h^2 + h - 3)$

21. $3ab(2a^2 + 4ab - b^2)$

22. $(3y - 4)(4y + 3)$

23. $(4m + 3)(2m - 1)$

24. $(2 - h)(3 + 2h)$

25. $(1 + r)(2 - 4r)$

26. $(5k + 2)(5k - 2)$

27. $(3z - 4)(3z + 4)$

28. $(a + b)(a + b)$

29. $(2m + 3)(2m + 3)$

30. $(x^2 + 3x - 4)(2x - 3)$

31. $(r^2 + rt - t^2)(r - 2t)$

32. **WRITING MATHEMATICS** Is the product of two binomials always a trinomial? Explain.

EXTEND

GEOMETRY Use the diagram at the right for Exercises 33–36.

33. Write a polynomial for the area of the shaded region shown.

34. If $x = 5$, evaluate the area of each rectangle and of the shaded region.

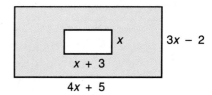

35. Suppose that the larger rectangle represents a sheet of paper and that a 1-unit square is cut from each corner of the paper. Then the paper is folded up to form an open box. Write a polynomial to represent the volume of the open box.

36. If $x = 5$, find the volume of the open box.

Determine the exponents that will make each expression equal to 1. Explain your reasoning.

37. $\left(\dfrac{x^2y^4}{x^5y^2}\right)\left(\dfrac{x^{\blacksquare}y^3}{x^4y^{\blacksquare}}\right)$

38. $\left(\dfrac{x^3y^{-2}}{x^{-4}y^{-1}}\right)\left(\dfrac{x^{-8}y^{\blacksquare}}{x^{\blacksquare}y^7}\right)$

39. $\left(\dfrac{x^{-2}y^3}{x^{-4}y^{-1}}\right)^2\left(\dfrac{x^{\blacksquare}y^{\blacksquare}}{x^6y^{-5}}\right)$

40. WRITING MATHEMATICS Could you find the product $(3x + 2)(2x + 4)$ using the order FILO or LOIF? Explain. (Assume the letters have the same meaning as in the lesson.)

THINK CRITICALLY

41. Use $a = 3$ and $b = 2$ to show that $(a + b)^3 \neq a^3 + b^3$.

42. Square *MATH* of side length 5 overlaps rectangle *HELP* with dimensions t and $(10 - t)$. Determine which of the following areas is larger. Explain your reasoning.

the area of *HELP* that is not in the shaded overlap
the area of *MATH* that is not in the shaded overlap

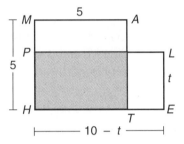

MIXED REVIEW

Determine the slope of the line.

43. $y = 3x - 2$

44. $2y = 8x + 9$

45. $y + 2x = 7$

Determine the inverse of each function.

46. $f(x) = -\dfrac{2}{3}x$

47. $f(x) = \dfrac{x + 1}{2}$

48. $y = 3x + 2$

Use a graphing utility to determine the number of points of intersection of the graphs of the system of equations.

49. $\begin{cases} x^2 + y^2 = 25 \\ y = x \end{cases}$

50. $\begin{cases} xy = 8 \\ y + x = 0 \end{cases}$

51. $\begin{cases} y = x^2 \\ xy = -4 \end{cases}$

52. STANDARDIZED TESTS The value of -3^{-2} is

A. $\dfrac{1}{9}$ **B.** $-\dfrac{1}{9}$ **C.** 9 **D.** -9

PROJECT *Connection* In this activity, you will plan some other features of your game that will make it interesting.

1. Discuss how players will move around the board. Will they roll dice, draw cards, or use some other method to determine moves? What physical items will your game require?

2. When, where, and what kinds of problems will players solve in your game? Consider such problems as substituting values into a polynomial, multiplying binomials, or factoring polynomials. Try to make your game involve a combination of luck and skill.

3. How will a winner be decided? Will a winner have to reach a destination, accumulate a certain number of points or an amount of money? Is it necessary to have a scoring method?

Career
Movie Theater Manager

Running a movie theater may seem to be a relatively simple business to an outsider. However, for the manager, there are many decisions that need to be made. What kinds of movies to show? How many employees will be needed? How much food needs to be ordered on a weekly basis?

The manager for Triple Cinema is currently reviewing costs and examines the following information:

Expenses (per weekend)
Advertising...........................$300.00
Employee Pay.....................$800.00
Heat....................................$700.00
Tape Rental.......................$3000.00
Food...................................$500.00
TOTAL.............................$5300.00

Total seating capacity for the three theaters is 900 people.
Price of a ticket is $5.00.
Food profit averages $0.50 per sale.
Each movie is shown five times throughout the weekend.

Decision Making

1. Let x represent total movie attendance. Write a monomial to represent the income from ticket sales.

2. Write a binomial to represent the profits from one weekend, based upon selling tickets only.

3. Write a monomial to represent the food profit, if 30% of the people who came to the cinema bought food.

4. Write a polynomial to represent the total profits from the weekend, using the information you obtained in Questions 2 and 3.

5. Use a graphing utility to graph the polynomial in Question 4.

   a. What meaning does the y-intercept have?

   b. What meaning does the x-intercept have?

5.3 Factoring

SPOTLIGHT ON LEARNING

What? In this lesson you will learn
- to remove the greatest common factor from a polynomial.
- to factor the difference of two squares.
- to factor a perfect square trinomial and other factorable trinomials.

Why? Factoring is a key step in solving problems about genetics, number theory, and geometry.

Explore

Like an integer, a monomial has prime factors.

1. Write the monomial whose prime factors are $2 \cdot 2 \cdot 2 \cdot x \cdot x$.

2. Write the monomial whose prime factors are $2 \cdot 2 \cdot 3 \cdot x \cdot x \cdot x \cdot x \cdot x$.

3. What is the greatest common factor of the monomials you wrote in Questions 1 and 2?

4. What is the greatest common factor of $30x^4$, $15x^2$, and $10x^5$?

5. Multiply each pair of binomials.

   a. $(x + 2)(x - 2)$ **b.** $(2x + 3)(2x - 3)$

6. Describe the ways in which the products of Question 5 are similar.

7. Describe the ways in which the factors given in each part of Question 5 are related.

8. Multiply each pair of binomials.

   a. $(x + 2)(x + 2)$ **b.** $(2x - 3)(2x - 3)$

9. Describe the ways in which the products of Question 8 are similar.

10. Describe the ways in which the factors given in each part of Question 8 are related.

THINK BACK

In the product ab, the numbers a and b are the **factors**. The **prime factorization** of an integer shows the integer as a product of prime factors:

$12 = 2 \cdot 2 \cdot 3$
$\quad = 2^2 \cdot 3$
$8 = 2 \cdot 2 \cdot 2 = 2^3$

There is only one prime factorization of an integer.

Build Understanding

When a polynomial is expressed as the product of its prime factors, the polynomial is in **factored form**. One type of factoring is to factor the **greatest common factor (GCF)** from each term of the polynomial.

EXAMPLE 1

Factor each polynomial.

a. $4x^3 + 16x^2 - 8x$ **b.** $36a^2b^4 - 18a^3b^3$

Solution

a. $4x^3 + 16x^2 - 8x$

$= 2 \cdot 2 \cdot x \cdot x \cdot x + 2 \cdot 2 \cdot 2 \cdot 2 \cdot x \cdot x - 2 \cdot 2 \cdot 2 \cdot x$

$= 4x(x^2) + 4x(4x) - 4x(2)$ The GCF is $4x$.

$= 4x(x^2 + 4x - 2)$ Distributive property.

b. $36a^2b^2 - 18a^3b^3$

$\qquad = 3 \cdot 3 \cdot 2 \cdot 2 \cdot a \cdot a \cdot b \cdot b - 3 \cdot 3 \cdot 2 \cdot a \cdot a \cdot a \cdot b \cdot b \cdot b$

$\qquad = 18a^2b^2(2) - 18a^2b^2(ab) \qquad$ The GCF is $18b^2$.

$\qquad = 18a^2b^2(2 - ab) \qquad\qquad$ Distributive property.

To check that you factored correctly, multiply the factors. ◀

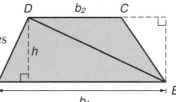

CHECK UNDERSTANDING

Is $2x$ a factor of $8x^2 + 6x + 2$? Explain.

By removing the greatest common factor of a polynomial, you can present some geometric formulas in compact form.

EXAMPLE 2

GEOMETRY Use the area of the two triangles in the diagram to write the formula for the area of the trapezoid.

Solution

Trapezoid $ABCD$ is the sum of the areas of $\triangle ABD$ and $\triangle CBD$.

The height of both triangles is h.

$$\text{Area}_{ABCD} = \text{Area}_{\triangle ABD} + \text{Area}_{\triangle CBD}$$

$$\qquad\quad = \frac{1}{2} \cdot b_1 \cdot h + \frac{1}{2} \cdot b_2 \cdot h \qquad \text{The GCF is } \frac{1}{2}h.$$

$$\qquad\quad = \frac{1}{2}h(b_1 + b_2) \qquad\qquad ◀$$

When two binomials have first terms that are equal and last terms that are opposites they are called **conjugates**. The product of conjugates is called the **difference of two squares**.

> **DIFFERENCE OF TWO SQUARES**
>
> $a^2 - b^2 = (a + b)(a - b)$

EXAMPLE 3

Factor each polynomial.

a. $y^2 - 16$ **b.** $4x^2 - y^6$ **c.** $0.49a^2 - 0.04$

Solution

a. $y^2 - 16 = (y + 4)(y - 4)$

b. $4x^2 - y^6 = (2x + y^3)(2x - y^3)$

c. If the terms do not have integer coefficients, factor out a rational number first.

$\quad 0.49a^2 - 0.04 = 0.01(49a^2 - 4) = 0.01(7a + 2)(7a - 2)$ ◀

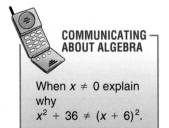
The product of two identical factors is called a **perfect square trinomial**. Notice that the middle term in each perfect square trinomial is twice the product of the square roots of the first and last terms of the trinomial.

> **PERFECT SQUARE TRINOMIAL**
>
> $$(a + b)^2 = a^2 + 2ab + b^2$$
> $$(a - b)^2 = a^2 - 2ab + b^2$$

EXAMPLE 4

Factor as a perfect square trinomial, if possible.

   a. $z^2 + 10z + 25$ **b.** $y^2 - 6y + 9$ **c.** $x^2 + 5x + 4$

Solution

   a. $z^2 + 10z + 25 = (z + 5)^2$ $2 \cdot \sqrt{z^2} \cdot \sqrt{25} = 10z$

   b. $y^2 - 6y + 9 = (y - 3)^2$ $2 \cdot \sqrt{y^2} \cdot \sqrt{9} = 6y$

   c. $x^2 + 5x + 4$ is not factorable $2 \cdot \sqrt{x^2} \cdot \sqrt{4} \neq 5x$
    as a perfect square trinomial.

Example 4c did not factor as a perfect square trinomial, but can be factored into two binomials. Factoring a trinomial into two binomials is the reverse of multiplying two binomials whose product is a trinomial. List all the factors of the coefficient of the last term, then choose the factors that have a sum equal to the coefficient of the middle term.

EXAMPLE 5

Factor each trinomial.

   a. $x^2 + 5x + 4$ **b.** $x^2 - 7x + 12$

   c. $x^2 + x - 6$ **d.** $x^2 - 3x - 10$

Solution

   a. $x^2 + 5x + 4 = (x + \blacksquare)(x + \blacksquare)$
    List all factors of 4. Find the pair whose sum is 5.

Factors of 4	Sum of the Factors	
1, 4	5	Factors 1 and 4 have a sum of 5.
2, 2	4	
−1, −4	−5	
−2, −2	−4	

   So, $x^2 + 5x + 4$ factored is $(x + 4)(x + 1)$.

To check that you have the correct factors, multiply the binomials.

$$(x + 1)(x + 4) = x^2 + 4x + x + 4 = x^2 + 5x + 4$$

b. $x^2 - 7x + 12 = (x + \blacksquare)(x + \blacksquare)$ List all factors of 12.
$$= (x - 3)(x - 4)$$ Factors -3 and -4 have a

So, $x^2 - 7x + 12$ factored is $(x - 3)(x - 4)$.   sum of -7.

c. $x^2 + x - 6 = (x + \blacksquare)(x + \blacksquare)$ List all factors of -6.
$$= (x + 3)(x - 2)$$ Factors 3 and -2 have a sum of 1.

So, $x^2 + x - 6$ factored is $(x + 3)(x - 2)$.

d. $x^2 - 3x - 10 = (x + \blacksquare)(x + \blacksquare)$ List all factors of -10.
$$= (x - 5)(x + 2)$$ Factors -5 and 2 have

So, $x^2 - 3x - 10$ factored is $(x - 5)(x + 2)$.   a sum of -3.   ◀

When the coefficient of the x^2 term of the trinomial is greater than 1, you will have to try various combinations of factors of the first and third terms to determine the correct factors. Remember to factor a GCF first, if possible.

EXAMPLE 6

Factor each trinomial.

a. $2x^2 + 7x + 3$ **b.** $6x^2 + x - 5$

Solution

a. $2x^2 + 7x + 3 = (2x + \blacksquare)(x + \blacksquare)$

The factors of 3 are 1, 3 and -1, -3. You must test the different possibilities to see which produces the original trinomial.

Trial Factors	Product
$(2x + 3)(x + 1)$	$2x^2 + 5x + 3$
$(2x - 3)(x - 1)$	$2x^2 - 5x + 3$
$(2x + 1)(x + 3)$	$2x^2 + 7x + 3$ This is the product you want.
$(2x - 1)(x - 3)$	$2x^2 - 7x + 3$

So, $2x^2 + 7x + 3 = (2x + 1)(x + 3)$.

b. Since the coefficient of the x^2-term is not prime, list all the positive factors. Factors of $6x^2$ are $3x$, $2x$ and $1x$, $6x$. Factors of -5 are -1, 5 and 1, -5. List the product of every possible combination of binomial factors.

Trial Factors	Product
$(6x + 1)(x - 5)$	$6x^2 - 29x - 5$
$(6x - 1)(x + 5)$	$6x^2 + 29x - 5$
$(6x + 5)(x - 1)$	$6x^2 - x - 5$
$(6x - 5)(x + 1)$	$6x^2 + x - 5$ This is the product you want.
$(2x + 1)(3x - 5)$	$6x^2 - 7x - 5$
$(2x - 1)(3x + 5)$	$6x^2 + 7x - 5$
$(2x + 5)(3x - 1)$	$6x^2 + 13x - 5$
$(2x - 5)(3x + 1)$	$6x^2 - 13x - 5$

So, $6x^2 + x - 5$ is factored $(6x - 5)(x + 1)$.   ◀

CHECK UNDERSTANDING

Can a given trinomial have different sets of binomial factors? Explain.

Factor. Check by multiplying.

1. $3y^2 - 6y$

2. $21a^3b^2 - 14a^2b$

3. $3c(r - t) + 2d(r - t)$

Use factoring to evaluate the expression mentally.

4. $35 \cdot 49 + 35 \cdot 51$

5. $\frac{22}{7} \cdot 1600 - \frac{22}{7} \cdot 900$

Factor.

6. $4h^2 - 25 \ (2h + 5)(2h - 5)$

7. $0.16m^2 - 0.25$

8. $(a + b)^2 - c^2$

9. WRITING MATHEMATICS Explain how the special product $(a + b)(a - b) = a^2 - b^2$ can be used to mentally evaluate the product $25 \cdot 15$. (Hint: Think of 25 as $20 + 5$.)

10. MODELING Use Algeblocks to model the product $x^2 + 6x + 9$, and find the factors of the trinomial.

Factor each trinomial. Check by multiplying.

11. $x^2 + 8x + 16$

12. $y^2 + 10y + 25$

13. $z^2 - 4z + 4$

14. MODELING Use Algeblocks to model the product $x^2 - x - 6$, and find the factors of the trinomial.

Factor each trinomial. Check by multiplying.

15. $x^2 + 4x + 3$

16. $a^2 - 2a - 8$

17. $m^2 + m - 12$

18. MODELING Use Algeblocks to model the product $2x^2 - 5x - 12$, and find the factors of the trinomial.

19. GEOMETRY Express the shaded area as a polynomial in factored form.

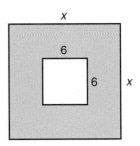

Factor each trinomial. Check by multiplying.

20. $3x^2 + 7x + 2$

21. $2z^2 + 17z + 8$

22. $4m^2 + 4m - 15$

PRACTICE

Factor.

1. $p + prt$

2. $28m^4n^3 - 70m^2n^4$

3. $4r(y + z) + 7s(y + z)$

4. $\pi r^2 + 2\pi rh$

5. $9ab^2 - 6ab + 3a$

6. $a(p - q) - b(p - q)$

Use factoring to evaluate the expression mentally.

7. $\frac{1}{2} \cdot 153 + \frac{1}{2} \cdot 47$

8. $\frac{1}{2} \cdot 7 \cdot 6.3 + \frac{1}{2} \cdot 7 \cdot 1.7$

Factor.

9. $p^2 - 100$

10. $\frac{1}{9}t^2 - \frac{1}{81}$

11. $(u + v)^2 - 169$

12. $225 - y^2$

13. $1.44r^2 - 0.36s^4$

14. $25 - (m + n)^2$

15. WRITING MATHEMATICS Explain how the special product $(a + b)^2 = a^2 + 2ab + b^2$ can be used to mentally evaluate the number 23^2.

16. MODELING Use Algeblocks to model the product $x^2 - 8x + 16$. Find the factors.

Factor each trinomial. Check by multiplying.

17. $t^2 + 2t + 1$

18. $x^2 + 16x + 64$

19. $w^2 - 10w + 25$

20. $x^2 - 4x + 4$

21. $16 + 8m + m^2$

22. $49 - 14s + s^2$

23. $4x^2 + 4x + 1$

24. $9a^2 - 12a + 4$

25. $1 + 6z + 9z^2$

26. MODELING Use Algeblocks to model the product $12 - x - x^2$. Find the factors.

Factor each trinomial. Check by multiplying.

27. $z^2 + 7z + 12$

28. $t^2 - 4t - 5$

29. $n^2 - 5n - 6$

30. $2 + 3m + m^2$

31. $15 - 8n + n^2$

32. $12 - r - r^2$

33. MODELING Use Algeblocks to model the product $5 + 8x - 4x^2$. Find the factors.

Factor each trinomial. Check by multiplying.

34. $2x^2 + 5x + 2$

35. $2a^2 + 7a + 6$

36. $4x^2 - 4x - 15$

37. $9 + 9x + 2x^2$

38. $8 + 14r + 3r^2$

39. $12 - 13a - 4a^2$

40. GEOMETRY Express the surface area of the rectangular prism as a polynomial in factored form.

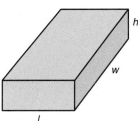

41. GEOMETRY Express the surface area of the cylinder as a polynomial in factored form.

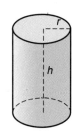

EXTEND

Factor.

42. $9x^2y^2 + 24xy + 16$

43. $4m^2n^2 - 20mn + 25$

44. $25a^4b^2 + 10a^2b + 1$

45. $7x^4 - 6x^2y - y^2$

46. $10x^2y^2 + 11xyz + 3z^2$

47. $15c^2d^2 + 13cdf + 2f^2$

GENETICS In human reproduction, certain genes are *dominant* (such as brown eyes) and other genes are *recessive* (such as blue eyes). Suppose p represents the ratio of dominant gene B in a population and q represents the ratio of recessive gene b. The next generation is the result of $(p + q)^2$, namely,

> p^2 members with *pure dominant* genes BB
> q^2 members with *pure recessive* genes bb
> $2pq$ members with *heterozygous* genes Bb

In the village of Chelmus, the recessive blue-eyed gene b has a ratio of 1:4, and the dominant brown-eyed gene B has a ratio of 3:4. Note that $p + q = 1$.

48. For the next generation, what ratios would you predict for

    a. pure brown-eyed members?

    b. heterozygous brown-eyed members?

    c. pure blue-eyed members?

49. If Chelmus expects its next generation to number approximately 32,000, about how many brown-eyed citizens would you predict?

THINK CRITICALLY

Replace each n so that each polynomial is a perfect square trinomial.

50. $x^2 + 16x + n$ **51.** $y^2 - 10y + n$

52. $4z^2 + 12z + n$

Factor. Check by multiplying.

53. $3k^{2a} - 10^2 + 3$

54. $(2x + 1)^2 - 3(2x + 1) - 10$

55. $(x^2 + x - 2)^2 - x^4$

MIXED REVIEW

Solve each equation for the indicated letter.

56. $ax + b = d - cx$ for x **57.** $S = \pi a(b + c)$ for c

58. STANDARDIZED TESTS The solution of $|2x + 1| = -3$ is

    A. ± 2 **B.** 2   **C.** -2 **D.** no real number

59. STANDARDIZED TESTS If $f(x) = \dfrac{x + 3}{x - 1}$, then $f(a + 1)$ is equal to

    A. $\dfrac{a + 3}{a - 1} + 1$ **B.** $\dfrac{a + 3}{a - 1}$ **C.** $\dfrac{a + 4}{a}$ **D.** 4

5.4 Factoring Completely

Explore

1. Use the distributive property to multiply each pair of polynomials.

   a. $(x - 1)(x^2 + x + 1)$ **b.** $(x - 2)(x^2 + 2x + 4)$

2. Describe any patterns you notice in each part of Question 1.

3. For each binomial factor, use the pattern you observed above to create a trinomial factor and guess the product.

   a. $(x - 3)$ **b.** $(x - 4)$

4. Use the distributive property to multiply each pair of polynomials.

   a. $(x + 1)(x^2 - x + 1)$ **b.** $(x + 2)(x^2 - 2x + 4)$

5. Describe any patterns you notice in each part of Question 4.

6. For each binomial factor, use the pattern you observed in Question 4 to create a trinomial factor and guess the product.

   a. $(x + 3)$ **b.** $(x + 4)$

Build Understanding

As you saw in Explore, if a polynomial is the difference or sum of two cubes, the factors are a binomial and a related trinomial.

DIFFERENCE OR SUM OF TWO CUBES

$$a^3 - b^3 = (a - b)(a^2 + ab + b^2)$$
$$a^3 + b^3 = (a + b)(a^2 - ab + b^2)$$

EXAMPLE 1

Factor.

a. $x^3 - 1000$ **b.** $y^3 + 125$ **c.** $z^9 + \dfrac{1}{8}$

Solution

a. $x^3 - 1000 = x^3 - 10^3 = (x - 10)(x^2 + 10x + 100)$

b. $y^3 + 125 = y^3 + 5^3 = (y + 5)(y^2 - 5y + 25)$

c. $z^9 + \dfrac{1}{8} = \dfrac{1}{8}(8z^9 + 1)$ Factor out $\dfrac{1}{8}$.

$\qquad = \dfrac{1}{8}((2z^3)^3 + 1^3)$

$\qquad = \dfrac{1}{8}(2z^3 + 1)(4z^6 - 2z^3 + 1)$

PROBLEM SOLVING TIP

Here is a brief list of factoring methods you can apply:
- Greatest common factor
- Difference of two squares
- Perfect square trinomial
- Two binomial factors
- Grouping
- Sum or difference of two cubes

A polynomial is not considered factored until it is factored completely. A polynomial is factored completely when it is expressed as the product of prime factors, the prime factorization. Factoring a polynomial completely may take several steps. For example, $2x^2 + 14x + 24$ has a GCF of 2 and is factored $2(x^2 + 7x + 12)$. The trinomial $x^2 + 7x + 12$ is factored $(x + 3)(x + 4)$. So, $2x^2 + 14x + 24$ is factored completely as $2(x + 3)(x + 4)$.

EXAMPLE 2

Factor completely.

a. $2x^2 - 18$ **b.** $6x^2 + 10x + 4$

Solution

a. $2x^2 - 18 = 2(x^2 - 9)$ GCF = 2 Factor.
$= 2(x + 3)(x - 3)$ Difference of two squares.
Check by multiplying the factors.

b. $6x^2 + 10x + 4 = 2(3x^2 + 5x + 2)$ GCF = 2 Factor.
$= 2(3x + 2)(x + 1)$ Factor again.
Check by multiplying the factors. ◄

Even if there is no common factor, it may still be necessary to perform more than one factoring. Completely factoring polynomials can help you solve real world problems.

EXAMPLE 3

MANUFACTURING The volume of a box is $x^4 - 16$. Factor to find the dimensions of the box in terms of x.

Solution

The expression $x^4 - 16$ is a difference of two squares. One of its factors, $x^2 - 4$, is also a difference of two squares. So you must factor again.

$x^4 - 16 = (x^2 + 4)(x^2 - 4)$ Factor.
$= (x^2 + 4)(x + 2)(x - 2)$ Factor again.

The dimensions of the box are $x^2 + 4$, $x + 2$, and $x - 2$. ◄

ALGEBLOCKS MODEL

Place the product $xy + y - 2x - 2$ on the mat in the arrangement shown, and find its factors.

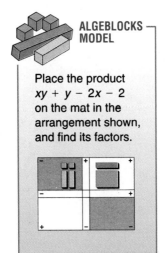

The common factor may be a monomial. In the polynomial $2ax^2 + 4ax + 10a$, the monomial $2a$ is a factor of all the terms.

$2ax^2 + 4ax + 10a = 2a(x^2 + 2x + 5)$

The common factor may be a binomial. In $a(m + n) + b(m + n) + c(m + n)$, the binomial $(m + n)$ is a factor of all the terms.

$a(m + n) + b(m + n) + c(m + n) = (m + n)(a + b + c)$

Some terms may have one common factor and other terms may have a different common factor. In this case, you **factor by grouping.**

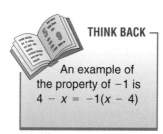

EXAMPLE 4

Factor.

    a. $x^2 - 4x + 12y - 3xy$ **b.** $x^2 - 6x + 9 - z^2$

Solution

   a. $x^2 - 4x + 12y - 3xy$

$$= (x^2 - 4x) + (12y - 3xy) \quad \text{Group the terms.}$$
$$= x(x - 4) + 3y(4 - x) \quad \text{Factor out GCFs.}$$
$$= x(x - 4) + 3y(-1)(x - 4) \quad \text{Use the property of } -1.$$
$$= x(x - 4) - 3y(x - 4) \quad \text{Factor out the common}$$
$$= (x - 4)(x - 3y) \quad \text{factor } (x - 4).$$

   b. The first three terms form a perfect square trinomial.

$$x^2 - 6x + 9 - z^2$$
$$= (x^2 - 6x + 9) - z^2 \quad \text{Group the terms.}$$
$$= (x - 3)^2 - z^2 \quad \text{Factor.}$$
$$= (x - 3 + z)(x - 3 - z) \quad \text{Difference of two squares.} \quad ◄$$

THINK BACK

An example of the property of -1 is
$4 - x = -1(x - 4)$

PROBLEM SOLVING TIP

In 4b, you might have begun with $(x^2 - z^2) + (-6x + 9)$ and factored getting $(x + z)(x - z) + 3(-2x + 3)$, whichcannot be factored. You may need to try several approaches to find one that works.

TRY THESE

Factor completely. Check by multiplying.

  1. $3x^2 - 27$ **2.** $2y^2 - 4y - 30$ **3.** $6z^2 - 9z - 6$

  4. GEOMETRY In the diagram, the smaller of the two circles has a radius of length r, and the larger circle has a radius of length R.

    a. Does the expression $\pi(R - r)^2$ represent the area of the shaded region? Explain using a numerical example.

    b. Express the area of the shaded region as a polynomial in factored form.

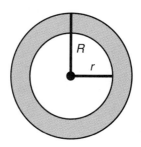

Factor completely. Check by multiplying.

  5. $b^4 - 625$ **6.** $y^8 - 256$

  7. MODELING Use Algeblocks to find the factors of $2xy + 2y - 3x - 3$.

Factor by grouping. Check by multiplying.

  8. $4xy + 8xz + 6y + 12z$ **9.** $x^2 - 5x + 15y - 3xy$

10. $y^3 + 4y^2 - 9y - 36$ **11.** $z^2 + 8z + 16 - x^2$

12. WRITING MATHEMATICS Describe two different ways in which the polynomial $3xy + 6xz + 4y + 8z$ can be grouped and factored. Compare the results.

Factor. Check by multiplying.

13. $r^3 - s^3$ **14.** $m^3 + n^3$

15. $64x^3 - 1$ **16.** $y^6 + .001$

Factor completely. Check by multiplying.

1. $5y^2 - 80$

2. $175x^2 - 343$

3. $3x^2 - 15x + 18$

4. $5y^2 - 25y - 30$

5. $4x^2 + 16x + 16$

6. $3z^2 - 24z + 48$

7. WRITING MATHEMATICS Explain the factorization of $4x^2 - 36$ if you do not first factor out the GCF. Compare the results to those you get if you do first factor out the GCF.

8. DESIGN A corporate logo consists of an "S" in a red rectangle surrounded by a blue border of equal width on all sides.

    a. If the area of the red rectangle is $(187 - 56x + 4x^2)$ square inches, write expressions that represent the dimensions of the red rectangle.

    b. How do the expressions you wrote in part a relate to the dimensions of the whole logo and the size of the border?

    c. If the area of the red rectangle is 91 in.^{2}, what is its length and width? What is the width of the border?

Factor. It will be easier to factor polynomials with integer coefficients. Check by multiplying.

9. $t^4 - 81$

10. $1296 - x^4$

11. $a^8 - 16$

12. $\dfrac{1}{27} - b^{12}$

13. MODELING Use Algeblocks to find the factors of $3y + 3xy - 2x - 2$.

Factor by grouping. Check by multiplying.

14. $5ab + 10ac - 3b - 6c$

15. $2ax + 3 + x + 6a$

16. $y^2 - 3y + 6x - 2xy$

17. $10y - 5xy - 6x + 3x^2$

18. $z^2 + 8z + 16 - q^2$

19. $t^2 - 10t + 25 - u^2$

20. $36 - a^2 - 2a - 1$

21. $25 - r^2 - 4r - 4$

22. WRITING MATHEMATICS Explain the possible methods that may be used to factor a polynomial of two terms.

23. GEOMETRY The area of a rectangle is $2ax - bx + cx + 2ay - by + cy$ and its width is $x + y$. Determine the length.

EXTEND

24. Below are the patterns for factoring the sum of two cubes and the sum of two fifth powers. Write the pattern for factoring the sum of two seventh powers.

$$a^3 + b^3 = (a + b)(a^2 - ab + b^2)$$
$$a^5 + b^5 = (a + b)(a^4 - a^3b + a^2b^2 - ab^3 + b^4)$$

25. WRITING MATHEMATICS In your own words, describe the factors for the sum of any two like odd powers, $a^n + b^n$, where n is an odd integer.

26. The expression $x^4 + x^2y^2 + y^4$ would be a perfect square trinomial if the middle term were $2x^2y^2$. Use the technique of adding a zero pair that you use in Algeblock models. By adding the zero pair x^2y^2 and $-x^2y^2$, the original nonfactorable expression becomes the factorable expression $x^4 + 2x^2y^2 + y^4 - x^2y^2$. Complete the factoring.

Use the zero pair technique to factor. Check by multiplying.

27. $9x^4 + 2x^2y^2 + y^4$

28. $x^4 + 64$

THINK CRITICALLY

Factor. Check by multiplying.

29. $(x - 2)^3 + 64$

30. $x^4 + x^2 - 20$

31. $(3x - 1)^2 - 3(x - 1) - 10$

32. $(y^2 + y - 2)^2 - (y^2 - y - 6)^2$

33. $(x - y)^3 - 27$

MIXED REVIEW

Use the alphabetical list of the world's eight highest waterfalls.

Falls	Height, ft
Angel	3212
Cuquenán	2000
Great	1600
Mardalsfoss	2149
Ribbon	1612
Sutherland	1904
Tugela	2014
Yosemite	2425

34. Find the range.

35. Find the median.

36. Find the first quartile.

37. STANDARDIZED TESTS The point whose coordinates are $(4, -2)$ lies on a line whose slope is $\frac{2}{3}$. Find the coordinates of another point on this line.

A. $(1, 0)$
B. $(2, 1)$
C. $(6, 1)$
D. $(7, 0)$

Given the line k whose equation is $2y = 5x + 3$.

38. Write an equation of the line that is parallel to k and passes through $(1, 5)$.

39. Write an equation of the line that is perpendicular to k and passes through $(-2, 3)$.

40. STANDARDIZED TESTS If an equation of the axis of symmetry of a parabola is $x = 1$, which could be an equation of the parabola?

A. $y = x^2 + 2x + 1$
B. $y = x^2 - 2x + 1$
C. $y = 2x^2 + x - 2$
D. $y = 3x^2 + 6x + 2$

5.5 Solving Polynomial Equations by Factoring

Explore

1. a. Use a graphing utility to graph the polynomial function $y = x^3 + 2x^2 - 15x$. Use range values that will show you all the features of the graph.

   b. How many times does the graph cross the x-axis? At what values of x?

   c. Factor the polynomial $x^3 + 2x^2 - 15x$.

   d. What relationship do you see between the x-intercepts of the graph and the factors of the polynomial?

2. a. Repeat all the parts of Question 1 using the polynomial function $y = x^4 - 5x^2 + 4$.

   b. Is the relationship from Question 1d the same for the polynomial in Question 2a?

3. The graph of a polynomial function and the range for the viewing window is shown.

   a. What are the x-intercepts of the graph?

   b. Write the factors of the related polynomial.

   c. Write the polynomial.

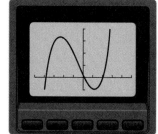

Xmin = −5   Ymin = −5
Xmax = 5   Ymax = 20
Xscl = 1   Yscl = 5

Build Understanding

As you observed in Explore, the factors of a polynomial function are related to the x-intercepts of its graph, that is, to the zeros of the function. The zeros of a function are the solutions or *roots* of the related polynomial equation when $y = 0$.

You know that when you multiply a number by zero, the product is zero. Conversely, if a product is zero, then at least one of its factors is zero.

> **THE ZERO PRODUCT PROPERTY**
>
> For any real numbers a and b, if $ab = 0$, then $a = 0$, or $b = 0$, or both $a = 0$ and $b = 0$.

The zero product property can be used in solving a polynomial equation.

EXAMPLE 1

Find all the real solutions to each equation.

a. $x(2x - 5) = 0$

b. $(y + 3)(y - 5) = 0$

Solution

Use the zero product property.

a. $x(2x - 5) = 0$

$$x = 0 \quad \text{or} \quad 2x - 5 = 0$$
$$x = 0 \quad \text{or} \quad 2x = 5$$
$$x = 0 \quad \text{or} \quad x = \frac{5}{2}$$

Check to show that the solutions are 0 and $\frac{5}{2}$.

b. $(y + 3)(y - 5) = 0$

$$y + 3 = 0 \quad \text{or} \quad y - 5 = 0$$
$$y = -3 \quad \text{or} \quad y = 5$$

Check to show that the solutions are -3 and 5. ◀

A polynomial equation of degree 2 is called a **quadratic** equation. Its general form is $ax^2 + bx + c = 0$. You can solve some quadratic equations by factoring and applying the zero product property. To apply the zero product property, the factored polynomial must equal zero.

PROBLEM SOLVING TIP

When solving an equation whose highest term has a negative coefficient, make it positive by multiplying both sides of the equation by -1. This will simplify the factoring process.

EXAMPLE 2

Solve each equation.

a. $-x^2 + 9x - 18 = 0$ **b.** $r^2 = 9r$

Solution

a.
$$-x^2 + 9x - 18 = 0$$
$$-1(-x^2 + 9x - 18) = -1 \cdot 0 \qquad \text{Multiply both sides by } -1.$$
$$x^2 - 9x + 18 = 0$$
$$(x - 6)(x - 3) = 0 \qquad \text{Factor the trinomial.}$$
$$x - 6 = 0 \quad \text{or} \quad x - 3 = 0 \quad \text{Zero product property.}$$
$$x = 6 \quad \text{or} \quad x = 3$$

Check to show that the solutions are 3 and 6.

b.
$$r^2 = 9r$$
$$r^2 - 9r = 0 \quad \text{Make one side of the equation equal to zero.}$$
$$r(r - 9) = 0 \quad \text{Factor out the GCF.}$$
$$r = 0 \qquad \text{or} \quad r - 9 = 0 \quad \text{Zero product property.}$$
$$r = 0 \qquad \text{or} \quad r = 9$$

Check to show that the solutions are 0 and 9. ◀

COMMUNICATING ABOUT ALGEBRA

In Example 2b why was the equation not solved like this?
$$r^2 = 9r$$
$$\frac{r^2}{r} = \frac{9r}{r}$$
$$r = 9$$

You will often use quadratic equations to model real world problems.

EXAMPLE 3

ART A mural is to be painted on a museum wall that is 40 ft by 50 ft. A mosaicist laid a uniform strip of colored tiles around the edge. How wide is the strip of mosaic tile if 60% of the wall is left for the mural?

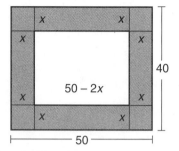

Solution

Let x represent the uniform width of the tiled border.
Then $50 - 2x$ represents the base of the rectangle to be painted, and $40 - 2x$ represents the height of the rectangle to be painted.

The area of the mural to be painted is 60% of the wall area.

$$
\begin{aligned}
(50 - 2x)(40 - 2x) &= 0.6(50 \cdot 40) && \text{Change 60\% to decimal.}\\
2000 - 180x + 4x^2 &= 1200 && \text{Multiply each side.}\\
4x^2 - 180x + 800 &= 0 && \text{Subtract 1200 from both sides.}\\
x^2 - 45x + 200 &= 0 && \text{Divide both sides by 4.}\\
(x - 5)(x - 40) &= 0 && \text{Factor.}\\
x - 5 = 0 \quad \text{or} \quad x - 40 &= 0 && \text{Zero product property}\\
x = 5 \quad \text{or} \qquad\qquad x &= 40
\end{aligned}
$$

Since $x = 40$ is not a possible solution, $x = 5$.

Check

If $x = 5$, the dimensions of the mural to be painted are 40 ft by 30 ft. The area is 1200 ft², which is 60% of the original area.

The tiled border is 5 ft wide. ◄

CHECK UNDERSTANDING

In Example 3 why is $x = 40$ not a possible solution?

The zero product property can be used to solve a polynomial equation of any degree, when the polynomial is factorable.

EXAMPLE 4

Solve: $x^3 + x^2 - 9x - 9 = 0$

Solution

$$
\begin{aligned}
x^3 + x^2 - 9x - 9 &= 0 \\
(x^3 + x^2) + (-9x - 9) &= 0 && \text{Group terms.}\\
x^2(x + 1) - 9(x + 1) &= 0 && \text{Factor by grouping.}\\
(x + 1)(x^2 - 9) &= 0 && \text{Factor out the GCF, } (x + 1).\\
(x + 1)(x + 3)(x - 3) &= 0 && \text{Factor the difference of squares.}\\
x + 1 = 0 \quad \text{or} \quad x + 3 = 0 \quad \text{or} \quad x - 3 &= 0 && \text{Zero product}\\
x = -1 \quad \text{or} \qquad x = -3 \quad \text{or} \qquad x &= 3 && \text{property}
\end{aligned}
$$

Check to show that the solutions are -1, -3, and 3. ◄

Sometimes a factor is repeated in the factorization of a polynomial. For example, 5 is a **double solution** of $x^2 - 10x + 25 = 0$ because the factored form is $(x - 5)^2$. Solutions that are obtained from repeated factors are called **multiple solutions**. For the cubic equation $(x - 5)^3 = 0$, 5 is a **triple solution**, or a solution with **multiplicity** 3.

EXAMPLE 5

Solve: $4x^3 + 24x^2 + 36x = 0$

Solution

$$4x^3 + 24x^2 + 36x = 0$$
$$4x(x^2 + 6x + 9) = 0 \quad \text{Factor out the GCF.}$$
$$4x(x + 3)^2 = 0 \quad \text{Factor the perfect square trinomial.}$$
$$4x = 0 \quad \text{or} \quad x + 3 = 0 \quad \text{or} \quad x + 3 = 0 \quad \text{Zero product}$$
$$x = 0 \quad \text{or} \quad x = -3 \quad \text{or} \quad x = -3 \quad \text{property.}$$

So, -3 is a double solution.

Check to show that the solutions are 0, -3, and -3. ◄

You can write a polynomial equation if you know its solutions.

EXAMPLE 6

Write a polynomial equation for which 2 is a solution and -3 is a double solution.

Solution

If the solutions are $x = 2$ and $x = -3$ and $x = -3$, then the factors are $(x - 2)(x + 3)(x + 3)$.

$$(x - 2)(x + 3)(x + 3) = 0$$
$$(x - 2)(x^2 + 6x + 9) = 0 \quad \text{Multiply.}$$
$$x^3 + 6x^2 + 9x - 2x^2 - 12x - 18 = 0 \quad \text{Multiply again.}$$
$$x^3 + 4x^2 - 3x - 18 = 0 \quad \text{Combine like terms.}$$

The solutions 2, -3, -3 are for $x^3 + 4x^2 - 3x - 18 = 0$ ◄

TRY THESE

Solve each equation. Check by substituting.

1. $x^2 - 7x = 0$ **2.** $m^2 + 5m = 0$ **3.** $k^2 - 4 = 0$

4. $x^2 - 7x + 10 = 0$ **5.** $a^2 + 7a + 12 = 0$ **6.** $-c^2 + 2c + 15 = 0$

7. $t^2 = 8t$ **8.** $35 = z^2 + 2z$ **9.** $r^2 - 5r - 20 = 4$

10. **WRITING MATHEMATICS** To solve the equation $x(x - 4) = 12$, Joe said to set each factor equal to 12. Margo said to set one factor equal to 4 and the other to 3. Cynthia disagreed with both Joe and Margo. With whom do you agree? Why?

11. GEOMETRY A 3 in. wide by 5 in. high photo is surrounded by a frame made of four strips of wood. The top and bottom strips are twice the width of the side strips. If the total area of the framed photo is 45 in.^{2} what is the width of a side strip of the frame?

Solve and check each equation.

12. $x^3 + x^2 - 16x - 16 = 0$

13. $x^3 - x^2 - 25x + 25 = 0$

14. $3x^3 + 30x^2 + 75x = 0$

15. $-2x^3 + 2x^2 + 12x = 0$

For each set of numbers, write a polynomial equation that has the given numbers as solutions. Answer in standard form, not factored form.

16. 0, 1

17. 2, 3

18. 4, −7

19. 0, 3, 3

20. 0, 1, −2

21. 1, −2, −4

PRACTICE

Solve and check each equation.

1. $-k^2 + 8k = 0$

2. $2m^2 - 12m = 0$

3. $6z^2 + 18z = 0$

4. $y^2 - 121 = 0$

5. $4p^2 - 36 = 0$

6. $6g^2 - 96 = 0$

7. $x^2 - 7x + 12 = 0$

8. $y^2 - 10y + 21 = 0$

9. $-z^2 + 6z + 16 = 0$

10. $a^2 + 4a + 3 = 0$

11. $2b^2 - b - 3 = 0$

12. $-6c^2 + c + 1 = 0$

13. $6x^2 + 5x - 6 = 0$

14. $m^2 = 9m$

15. $10 = k^2 + 3k$

16. $15 = r^2 + 2r$

17. $x^2 - 3x - 25 = 3$

18. $z^2 - 11z + 28 = 10$

19. WRITING MATHEMATICS Explain why the quadratic equation $x^2 + 4 = 0$ does not have real solutions.

20. NUMBER THEORY Two positive integers are in the ratio 1:2. If their product is added to their sum, the result is 119. Find the integers.

21. INTERIOR DESIGN A floor is covered with 1728 square tiles. If tiles 2 in. longer on each side are used, only 432 tiles are needed. Find the length of each side of the smaller tile.

22. GEOMETRY Rick has a rectangular piece of cardboard that is twice as long as it is wide. From each of the four corners of the cardboard, Rick cuts a 2 in. square and turns up the flaps to form an uncovered box. If the volume of the box is 896 in.^{3}, what were the dimensions of the original piece of cardboard?

Solve and check each equation.

23. $x^3 + x^2 - 36x - 36 = 0$ 24. $d^3 - d^2 - 9d + 9 = 0$ 25. $p^3 + 12p^2 + 36p = 0$

26. $x^3 - 14x^2 + 49x = 0$ 27. $-3x^3 + 3x^2 + 36x = 0$ 28. $5n^3 - 20n^2 - 25n = 0$

29. **CONSTRUCTION** A piece of wire 52 in. long is cut into two pieces, and then each piece is formed into a square. If the sum of the areas of the two squares is 97 in.^{2}, how long are the pieces of wire?

30. **BUSINESS** An importer of jade figurines has a cost function given by $C(x) = 2x^2 + 10x - 420$ and a revenue function given by $R(x) = x^2 + 60x - 945$ where x is the number of figurines in thousands.

   a. Determine the profit function $f(x)$ by subtracting cost from revenue.

   b. When will the importer just break even?

For each set of numbers, write a polynomial equation that has given numbers as solutions. Answer in standard form, not factored form.

31. $0, 3$ 32. $2, 5$ 33. $7, -2$

34. $-3, -5$ 35. $4, 4$ 36. $-5, -5$

37. $0, 6, 6$ 38. $0, 3, -4$ 39. $2, 5, 5$

40. $3, -2, -2$ 41. $2, 5, 8$ 42. $-3, 4, -6$

EXTEND

Solve and check each equation.

43. $(3x + 1)^2 + 5(3x + 1) = 0$ 44. $(x - 2)^2 - 4(x - 2) - 12 = 0$

45. $x^4 - 13x^2 + 36 = 0$ 46. $x^4 - 29x^2 + 100 = 0$

47. **GEOMETRY** The difference between the lengths of the legs of a right triangle is 1 in. The difference between the lengths of the hypotenuse and the longer leg is 1 in. Determine the lengths of all the sides of the triangle.

48. **NUMBER THEORY** The square of the units digit of a two-digit number is 2 more than the number. The difference between the units digit and the tens' digit is 3. Determine all such numbers.

49. **NUMBER THEORY** The sum of the squares of two consecutive positive odd integers is 1570. Determine the integers.

THINK CRITICALLY

Solve each system of equations.

50. $x^2 + y^2 = 18$
    $x = y$

51. $x^2 + 3 = y$
    $y = 5 - x$

52. $y + 3 = 2x$
    $3x^2 = 4 + xy$

53. TECHNOLOGY Use a graphing utility to graph the polynomial function $y = x(x - 5)^2$. How does the multiple solution to the polynomial equation $x(x - 5)^2 = 0$ show up on the graph?

MIXED REVIEW

Perform the indicated matrix operations.

54. $\begin{bmatrix} 4 & 3 \\ -2 & 0 \\ 1 & 5 \end{bmatrix} + \begin{bmatrix} 0 & -7 \\ 1 & 8 \\ 6 & -4 \end{bmatrix}$

55. $\begin{bmatrix} 3 & -8 \\ 1 & -1 \end{bmatrix}\begin{bmatrix} 2 & 0 \\ 3 & -5 \end{bmatrix}$

56. STANDARDIZED TESTS For which set of data do the mean, median, and mode all have the same value?

A. 2, 2, 2, 4, 10   **B.** 2, 2, 4, 10, 12   **C.** 2, 6, 6, 6, 10   **D.** 2, 2, 6, 10, 20

Tell if the given relation for the set shown satisfies the reflexive property. If not, give a counterexample.

57. "is greater than" {real numbers}

58. "is the square of" {0, 1}

59. STANDARDIZED TESTS Which mapping represents a one-to-one function?

A. 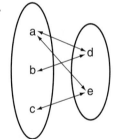   **B.** 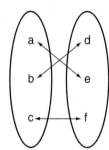   **C.** 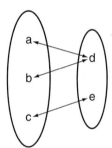   **D.**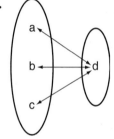

PROJECT *Connection*

In this activity, you will complete the last stages of development for your game. Test play your game as needed to refine it.

1. Make a final drawing of your game board or boards. Remember, the board should be attractive as well as functional.

2. Prepare any kind of problems that players will need to solve. Consider how other players will know if the answer is correct.

3. Decide and write the instructions that must be provided with your game. Make sure they are complete and accurate. Include such items as the number of players, how to begin, how to move, variations of play, and what determines a winner.

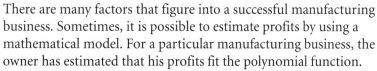

There are many factors that figure into a successful manufacturing business. Sometimes, it is possible to estimate profits by using a mathematical model. For a particular manufacturing business, the owner has estimated that his profits fit the polynomial function.

$$f(x) = 0.0028x^3 - 1.4x^2 + 200x - 1500$$

The polynomial is based on current production levels. Due to space limitations and supplies, the number of items produced cannot exceed 325 items.

Decision Making

1. Graph the polynomial using a graphing utility.

2. Describe the graph when the level of production is between 0 and 150 items. When is maximum profit achieved in this interval? What is the profit?

3. Describe the graph when the level of production is between 150 and 250 items. When is the profit a minimum in this interval? What is the profit?

4. What is the most profit the owner can expect to make? How many items must be produced?

5. What strategy should the owner plan to keep profits high?

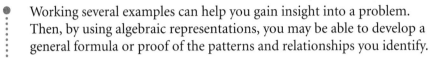
Use Algebra to Generalize

Working several examples can help you gain insight into a problem. Then, by using algebraic representations, you may be able to develop a general formula or proof of the patterns and relationships you identify.

Problem

Identify a pattern in the following set of examples, then make a conjecture and prove it.

$$0 + 1 + 2 = 3$$
$$1 + 2 + 3 = 6$$
$$7 + 8 + 9 = 24$$
$$13 + 14 + 15 = 42$$
$$95 + 96 + 97 = 288$$

Explore the Problem

1. Describe each expression on the left.

2. What does each number on the right have in common?

3. Try a few more examples. Then make a conjecture about what might be true for all cases.

4. How can you represent a sum of three consecutive whole numbers? Simplify your expression.

5. Use factoring to prove your conjecture. What else can you conclude about the actual value of a sum generated by three consecutive integers?

6. Predict the sum of $1023 + 1024 + 1025$. Check your answer.

7. **WRITING MATHEMATICS** Write some other generalizations that occur to you as you think about the problem above. Try to prove or disprove your conjectures.

Investigate Further

Here is another problem that can be used as a number trick once you understand how it works.

8. Choose any two numbers to be the first two elements in a sequence, such as 3 and 8. Form a sequence of 10 numbers where each new item is the sum of the two preceding numbers:

3, 8, 11, 19, 30, ■ , ■ , ■ , ■ , ■

Determine the missing numbers. Circle the seventh number in the sequence.

9. Find the sum of all 10 numbers.

10. What is the relationship between the seventh number and this sum?

11. Try the steps using other starting numbers. Is the result the same?

12. Let *m* represent the first number and *n* represent the second number. Express all 10 numbers of the sequence using *m* and *n*. What is the seventh number?

13. Write and simplify an expression using *m* and *n* for the sum of all the numbers. Explain the relationship you observe.

14. Suppose your beginning numbers for the same type of sequence are 5 and 9. Predict the sum of the 10 numbers. Check your prediction.

15. WRITING MATHEMATICS Find a number trick in a book or magazine that features games and puzzles. Try to analyze and explain the trick using algebra.

> **PROBLEM SOLVING PLAN**
>
> • Understand
> • Plan
> • Solve
> • Examine

Apply the Strategy

LANDSCAPING The patio below consists of five 1-foot slate squares.

16. What is the perimeter of the patio? What is the area?

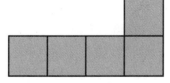

17. If only the vertical dimensions of the patio are doubled, how it will look is shown at the right. What is the new perimeter? What is the new area?

18. Draw the new patio that is obtained if only the horizontal dimension of the figure is doubled. What is the new perimeter? What is the new area?

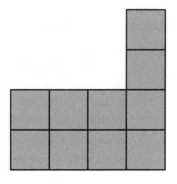

19. Draw the new patio obtained if both the horizontal and vertical dimensions of the original are doubled. What are the perimeter and area of this new patio?

20. Which method(s) doubled the original perimeter? Which methods doubled the original area?

21. Let p represent the perimeter of the original patio, z represent the total length of the horizontal segments and r represent the total length of the vertical segments. Express p in terms of z and r. What are the numerical values of z and r?

22. Express the perimeter that resulted from each of the doubling methods in terms of z and r. Simplify each expression. Verify the numerical values.

23. Express the sum of the perimeters of the patios from Parts b and c in terms of z and r. What is the relationship to the original patio? Verify the numerical values.

24. Predict how the perimeter of the original patio could change if each dimension was tripled. If each dimension was n-tripled.

25. How did the area change when you doubled both dimensions? Predict how the area would change if you tripled both dimensions. What if both dimensions were n-tripled?

26. NUMBER THEORY Consider the following numbers of the form abc, abc.

$$217,217 \div 11 = 19,747$$
$$308,308 \div 11 = 28,028$$
$$596,596 \div 11 = 54,236$$

Try some other examples using numbers of the same form. Then show that 11 is always a factor of these numbers. What other two numbers are also always factors of these numbers?

27. NUMBER THEORY Take *any* two real numbers whose sum is one. Square the larger number and add the result to the smaller number. Then square the smaller number and add the result to the larger number. What do you discover? Show why this is true.

28. FOUR SQUARE Multiply any four consecutive whole numbers and add one to the product. What do you notice? Make a conjecture and prove it. (Hint: Factor by grouping. Remember, you can break a monomial into a sum to get more convenient terms.)

REVIEW PROBLEM SOLVING STRATEGIES

OFFICE SUPPLIES

1. Glenda had a balance scale and some other items on her desk. She noticed the following:
   I. A calculator and a stapler balance a pencil sharpener.
   II. The stapler alone balances a calculator and a tape dispenser.
   III. Two pencil sharpeners balance three tape dispensers.
   So, Glenda wondered, how many calculators would balance a stapler?

   a. If you try to solve this problem, what equations can you write? How many variables are there?

   b. Consider situation II above. Suppose you add a calculator to each side of the balance. Write a new equation.

   c. Think about how the new equation can help you solve the problem. Then explain the steps you use.

PAGE BY PAGE

2. A publisher asked an author how many pages were in her manuscript. The author replied that she used 1563 digits to renumber the pages of the manuscript. How many pages were in the manuscript? Explain your strategy.

HAVE A SEAT

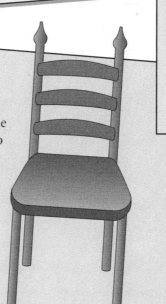

3. Kyra had 14 chairs. Can she arrange the chairs along the four walls of a room so that each wall has the same number of chairs? Explain.

· · · CHAPTER REVIEW · · ·

VOCABULARY

Choose the word from the list that completes each statement.

1. Addition and subtraction, and multiplication and division are __?__ .

2. __?__ binomials are alike in their first terms and opposite in their second.

3. Removing a __?__ is the inverse of distributing a multiplier.

4. The special product with two identical binomial factors is a __?__ trinomial.

5. The __?__ states that if $ab = 0$, then at least one of a, b, is 0.

a. greatest common factor

b. zero product property

c. inverse operations

d. conjugate

e. perfect square

Lesson 5.1   POLYNOMIALS   pages 229–233

● Algeblocks may be used to model operations with polynomials.

Interpret the models.

6. What polynomial is modeled on the mat below?

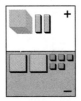

7. What polynomial has been added to the mat of Question 1?

8. Simplify the sum that is modeled in Question 2.

9. Write a difference of two polynomials that could also be represented by the model of Question 2.

10. Explain why a single model can represent both a sum and a difference.

11. What binomials are modeled on the mat below?

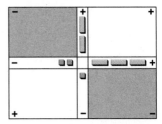

12. What is the product of the binomials in Question 6?

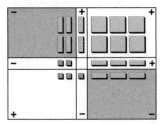

13. Write two divisions that are also represented by the model in Question 7.

14. What other operation with polynomials can the model in Question 7 represent?

Lesson 5.2   OPERATIONS WITH POLYNOMIALS pages 234–241

- You use properties of exponents when you perform operations with polynomials.
- When adding or subtracting polynomials, only like terms can be combined. When multiplying polynomials, apply the distributive property.

Perform the indicated operations.

15. $2x^3(-3x^4)$ **16.** $(4y^3)^2$ **17.** $12m^4 \div 3m$ **18.** $3x^0$ **19.** 2^{-2}

20. $(4x^2 - 3x - 7) + (-3x - 2)$ **21.** $(-x^3 - 2x^2 - 3) - (x^3 + x^2 - x + 1)$

22. $(4x - 1)(2x + 3)$ **23.** $3k^2(2k^2 - 5k + 2)$ **24.** $(1 - 2x)(1 + 2x)$

Lesson 5.3   FACTORING pages 242–248

- The *greatest common factor* (GCF) of a polynomial is determined from the GCF of all the coefficients and the highest power of variables that appear in every term.
- Special products have factors that are binomial conjugates or identical binomials. A trinomial may be factorable into two unrelated binomials.

Factor.

25. $24x^3y - 8xy^4$ **26.** $4x^2 - 25$ **27.** $y^2 - 10y + 25$ **28.** $2x^2 + x - 3$

Lesson 5.4   FACTORING COMPLETELY pages 249–253

- Repeated factoring may be necessary. Always examine for a common factor.
- The difference and sum of two cubes are special products.

Factor.

29. $6ax^2 - 15ax - 9a$ **30.** $z^4 - 16$ **31.** $ab + b - 2a - 2$ **32.** $x^3 - 64$

Lesson 5.5   SOLVING POLYNOMIAL EQUATIONS BY FACTORING pages 254–260

- To solve some polynomial equations, factor and set each factor equal to 0.

Use factoring to solve.

33. $x^2 - 5x = 0$ **34.** $4a^2 - 64 = 0$ **35.** $z^2 - 6z + 9 = 0$ **36.** $r^2 - 3r + 6 = 4$

37. $3x^2 - 10x + 8 = 0$ **38.** $2x^3 - 6x^2 - 20x = 0$

39. $x^3 + x^2 - 4x - 4 = 0$ **40.** $y^5 - 4y^3 = 0$

Lesson 5.6   USE ALGEBRA TO GENERALIZE pages 262–264

- By using algebraic representations and methods, you can develop formulas to describe patterns or prove general statements.

41. The sum of any seven consecutive whole numbers always has what number as a factor? Prove your conjecture.

CHAPTER ASSESSMENT

CHAPTER TEST

1. **WRITING MATHEMATICS** Explain how the algeblocks model shown below can be used to represent different operations.

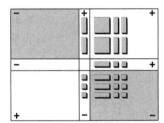

Apply the appropriate property of exponents.

2. $y \cdot y^3$ 3. $2m^3(-m^4)$ 4. $(3x^2)^3$

5. $4d^0$ 6. $\dfrac{50z^5}{5z}$ 7. $2(4^{-2})$

8. **STANDARDIZED TESTS** The value of $2(1 + x)^0$ when $x = 2$ is

   A. 6^0 **B.** 1   **C.** 2   **D.** 6

Perform the indicated operations.

9. $(4x^3 - 3x^2 + x - 2) + (3 - 2x - 5x^3)$

10. $(-2y^2 - y - 4) - (3y^2 + y - 2)$

11. $4w^4z(3z^2 - 2wz + 2w^2)$

12. $(2x + 4)(3x - 2)$

13. $(3q^2 - 5q + 1)(2q - 3)$

Write a trinomial to represent the area.

14.

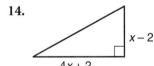

$x - 2$

$4x + 2$

15.

$x + 3$

$4x$

$x + 7$

Factor.

16. $21a^2b - 7ab$ 17. $25 - 9x^2$

18. $y^2 - 4y - 12$ 19. $4m^2 + 20m + 25$

20. $6z^2 - 13z + 6$ 21. $12ax^2 - 26ax - 10a$

22. $2m^2 - 50$ 23. $x^2 - 2x + 8y - 4xy$

24. $a^3 + 2a^2 - 4a - 8$

25. $m^3 - 125$

26. $r^3 + 64$

27. **WRITING MATHEMATICS** Write a paragraph to explain how you would solve and check $x^2 + x + 7 = 9$.

Solve and check.

28. $x^2 + 7x = 0$

29. $3y^2 - 27 = 0$

30. $3a^2 - 5a + 2 = 0$

31. $2y^2 + 5y + 5 = 8$

32. $x^2 - 14x + 49 = 0$

33. $x^3 - x^2 - 12x = 0$

34. $m^3 + m^2 - 9m - 9 = 0$

35. $z^4 - 5z^2 = 0$

Write and solve an equation for each problem.

36. Find three consecutive odd integers such that twice the square of the first is one less than the product of the second and third.

37. The measure of the altitude of a triangle is 4 inches greater than the measure of the base. If the area of the triangle is 30 square inches, find the dimensions of the triangle.

PERFORMANCE ASSESSMENT

USE ALGEBLOCKS Use Algeblocks to show division of a trinomial by a binomial. Follow the example, which is a model of

$$(6x^2 + 7x + 2) \div (3x + 2) = 2x + 1$$

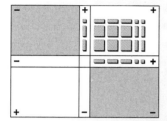

Have a partner determine a suitable trinomial and a binomial divisor by multiplying two binomials.

DO RESEARCH Look through textbooks in other subject areas such as science or economics to find formulas that are factorable. For example, the total distance traveled by an object during uniformly accelerated motion can be expressed as a function of initial velocity, acceleration, and time.

$$\Delta s = v_i \Delta t + \frac{1}{2}a(\Delta t)^2$$

$$\Delta s = \Delta t \left(v_i + \frac{1}{2}a \cdot \Delta t \right)$$

USE ALGEBLOCKS Use Algeblocks to show factoring the difference of two squares. Follow the example, which is a model of

$$x^2 - 4 = (x + 2)(x - 2)$$

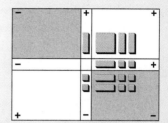

Also, illustrate factoring a trinomial. Have a partner determine suitable trinomials by multiplying two binomials.

POLYNOMIAL POOL Work in groups of four or five students. Each student writes on an index card a polynomial of at least three terms. Student 1 shows the polynomial to the group. Each group members factors the polynomial or states that it cannot be factored. The group should compare work, agree on the correct answer, and record each polynomial and its factors. Continue until each group member has displayed a polynomial for the others to factor.

PROJECT ASSESSMENT

PROJECT *Connection*   Design a name and package for your game. Prepare a description to convince the toy company that the game is a marketable product. Also create an advertisement to attract people to buy your game.

1. As a class, brainstorm ideas for and design an evaluation form that groups can complete when trying each other's games. Consider evaluating the directions, the level of challenge, and the enjoyment.

2. Play other games with your group and decide which games are successful and why.

3. Decide which games would be adaptable to a computer format. What changes could be made? Which games would make the best learning tools for students?

CUMULATIVE REVIEW

1. **STANDARDIZED TESTS** Which of the statements below are true of this Algeblocks model?

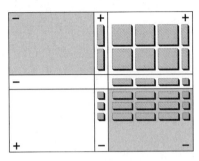

I. The model displays the product $(6x^2 + 2x)(-9x - 3)$

II. The model displays the factorization of $6x^2 - 7x - 3$

III. The model displays the quotient $(6x^2 - 7x - 3) \div (3x + 1)$

A. I, II, and III   **B.** I and II

C. II only   **D.** II and III

Perform the indicated operations.

2. $(3a^2 - 5a - 7) + (-5a^2 - 3a + 7)$

3. $(6x^2 = 3x + 1) - (-2x^2 - 4x - 8)$

4. $4m^3(-2m^4 - 7m^2 + 1)$

5. $(4z - 3)(3z + 4)$

6. **WRITING MATHEMATICS** Explain how to find the greatest common factor of a polynomial.

Factor.

7. $3m^3 - 12m$

8. $9z^2 - 6z + 1$

9. $xy + 3x - 2y - 6$

10. $y^3 - 27$

11. $6a^2 + 33a - 63$

12. $36x^3y^2 + 66x^2y^2 - 210xy^4$

Solve and check.

13. $3q^2 - 243 = 0$

14. $3x^2 + 10x - 8 = 0$

15. $z^3 + 2z^2 - 25z - 50 = 0$

16. $2x^2 + 40x + 192$

17. $3y^2 - 24y + 48$

18. Prove that all four-digit numbers of the form *abba* (such as 3443 or 8118) have a factor of 11.

19. Mr. Provera has three types of vending machines (*A*, *B*, *C*) that offer snacks, juices, and fruits. The table shows the number of items of each type held by each machine and today's total sales. If each machine was sold out at the end of the day, how many of each type of machine does he own?

Item	A	B	C	Mon. Sales
Snacks	10	12	15	327
Juices	24	18	20	576
Fruits	16	0	14	318

20. **WRITING MATHEMATICS** Explain why it is not always possible to use Cramer's rule to solve a system of equations.

21. The horsepower developed by a steam engine varies directly as the number of revolutions per minute at which it is operated. If an engine that is operated at 160 rpm develops 1280 horsepower, what horsepower will develop if the engine operates at 220 rpm?

22. For simple interest, the amount *A* accumulated after investing a principal sum *p* at a rate of *r*% over a period of *t* years is given by $A = p + prt$.

   a. Solve the formula for *t*.

   b. Solve the formula for *p*.

··· STANDARDIZED TEST ···

STANDARD FIVE-CHOICE Select the best choice for each question.

1. The expression $(2x^2)^3$ is equivalent to

   A. $6x^5$ **B.** $6x^6$ **C.** $8x^5$
   D. $8x^6$ **E.** $5x^5$

2. In factored form, $49 - 14x - x^2$ is

   A. $7(1 - x)^2$ **B.** $7(x - 1)^2$ **C.** $(7 - x)^2$
   D. $(7 + x)^2$ **E.** $(7 - x)(7 + x)$

3. When factored completely, $3x^2 - 48$ is

   A. $3(x^2 - 16)$ **B.** $3(x - 4)^2$
   C. $(3x + 12)(x - 4)$ **D.** $(3x - 12)(x + 4)$
   E. $3(x + 4)(x - 4)$

4. To solve the equation $x^2 - 4 = 5$, which of the following statements apply?

   I. Factor the left side into $(x + 2)(x - 2)$ and set each factor equal to 5.
   II. The factors to use are $(x + 3)(x - 3)$
   III. Set each factor equal to 0.

   A. I only   **B.** II only   **C.** III only
   D. II and III   **E.** I and II

5. If $\frac{1}{2}x + 2 = y$, then $2x + 2$ equals

   A. $2y$ **B.** $2y - 2$ **C.** $4y$
   D. $4y - 2$ **E.** $4y - 6$

6. When a 3×3 matrix A is multiplied by the 3×3 identity matrix I, the result is

   A. I **B.** A^{-1} **C.** A
   D. A^2 **E.** $-A$

7. If $D = \begin{vmatrix} 12 & -4 \\ -8 & -2 \end{vmatrix}$, which of the following determinants has the same value as D?

   A. $\begin{vmatrix} -4 & 12 \\ -2 & -8 \end{vmatrix}$ **B.** $\begin{vmatrix} -8 & -2 \\ 12 & -4 \end{vmatrix}$

   C. $\begin{vmatrix} 8 & 2 \\ -12 & 4 \end{vmatrix}$ **D.** $\begin{vmatrix} -12 & 4 \\ 8 & 2 \end{vmatrix}$

   E. $\begin{vmatrix} 8 & 1 \\ -1 & 7 \end{vmatrix}$

8. Which of the following shows the graphic solution of the system $y - x = 1$ and $y = 1$?

   A. 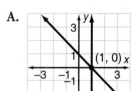   **B.**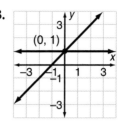

   C. 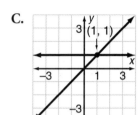   **D.**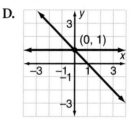

   E.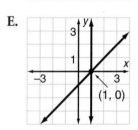

9. If $f(x) = \dfrac{x - 2}{x + 1}$, then $f(n + 1)$ is equal to

   A. $f(n) + 1$ **B.** $\dfrac{n + 1}{n - 2}$ **C.** $\dfrac{n + 2}{n - 1}$

   D. $\dfrac{n - 2}{n + 1}$ **E.** $\dfrac{n - 1}{n + 2}$

10. An equation of the line that is perpendicular to $2y = 6x + 1$ and passes through $(1, -3)$ is

    A. $x + 3y - 8 = 0$
    B. $x - 3y - 8 = 0$
    C. $3x + y - 8 = 0$
    D. $3x - y - 8 = 0$
    E. $3x - y + 8 = 0$

11. If the mean of ten consecutive integers that are arranged in increasing order is 25.5, then the mean of the first five of these integers is

    A. 25.5   **B.** 23   **C.** 20.5
    D. 19   **E.** 17.75

6 Quadratic Functions and Equations

Take a Look AHEAD

Make notes about things that look new.
- Describe the general appearance of graphs in this chapter. How are they different than graphs you studied earlier?
- What methods have you already used to solve quadratic equations? What new methods will be explained in this chapter?

Make notes about things that look familiar.
- Copy three examples of quadratic equations. What is the degree of a quadratic equation?
- Try to find at least two formulas or equations that you have worked with before. To what types of applications do these formulas relate?

DATA Activity

Now Boarding

In 1903, a small gasoline-powered airplane built by two American bicycle makers, Wilbur and Orville Wright, became the first to lift a passenger into the air and fly successfully. During the 1920s and 1930s, commercial airlines relied on propeller aircraft. The first commercial jet airliners began service in the 1950s, making long-distance travel routine. The table on the next page summarizes traffic at the nation's ten busiest airports.

SKILL FOCUS

- ▶ Use estimation.
- ▶ Add, subtract, multiply, and divide real numbers.
- ▶ Solve percentage problems.
- ▶ Compare and order real numbers.
- ▶ Choose an appropriate graph.

In this chapter, you will see how:

- **ENGINE MECHANICS** use a formula to adjust the operation of an automobile's internal combustion engine. *(Lesson 6.1, page 281)*

- **CARTOGRAPHERS** use quadratic equations as they map the Earth. *(Lesson 6.2, page 286)*

- **BRIDGE ENGINEERS** work with quadratic models that represent suspension cables. *(Lesson 6.3, page 295)*

TRANSPORTATION

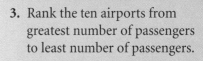

Busiest U.S. Airports, 1994
(in thousands)

	Departures	Passengers Boarded
Atlanta, Hartsfield International	273	22,295
Chicago, O'Hare	384	29,134
Dallas / Ft. Worth International	357	24,656
Denver, Stapleton International	183	14,328
Detroit, Metro Wayne	146	11,027
Los Angeles International	192	18,457
Minneapolis / St. Paul	137	10,378
Newark International	141	10,965
Phoenix, Sky Harbor International	148	11,295
San Francisco International	152	14,003

Use the table to answer the following questions.

1. Estimate the total number of aircraft departures for the airports shown. Explain your method.

2. The total number of departures for all U.S. airports is about 7,200,000. About what percent of this total is represented by the ten busiest airports?

3. Rank the ten airports from greatest number of passengers to least number of passengers.

4. What type of graph would you use to show the number of passengers at each airport? Explain your choice.

5. The number of passengers shown for Phoenix represents a 234% increase over the 1988 figures. How many passengers boarded planes in Phoenix in 1988? Round to the nearest thousand.

6. **WORKING TOGETHER** Research other data about air travel, such as on-time arrivals and departures. Display your data graphically.

Everybody's on the Move

When you hear the word transportation, you probably first think of cars, trains, and airplanes. Next, you might think of buses and boats. How about skateboards, monorails, gliders, in-line skates, skis, helicopters, golf carts, cable cars, horses, rockets, and hot-air balloons? These are all modes of transportation, too. Did you forget to include two of the earliest forms of transportation, swimming and walking? In this project, you will explore another form of transportation helpful to many people the world over — the wheelchair.

PROJECT GOAL

To gather data about the physically challenged resulting in increased awareness.

Getting Started

You will work both as a class and in small groups.

1. As a class, list local pharmacies, hospitals, surgical supply stores, doctors, and other sources that may have wheelchairs available.

2. Divide the class into groups and assign each group some of the listed sources to contact regarding their willingness to loan your class a wheelchair.

3. Discuss all plans and progress with your teacher to ensure that you are in compliance with all applicable school regulations.

4. Arrange to pick up the wheelchair from the donor, enlisting the help of school or parent volunteers if necessary.

PROJECT Connections

Lesson 6.1, page 280:
Plan questions and data recording sheets for the wheelchair study.

Lesson 6.2, page 285:
Use data to analyze speeds and distances relating to wheelchair transportation.

Lesson 6.3, page 294:
Design a safety device for wheelchairs.

Chapter Assessment, page 315:
Summarize project results using graphs and statistics; apply what has been learned to designing improved equipment or accessibility for wheelchair users.

✉ **Internet Connection**

www.swpco.com/ swpco/algebra2.html

Explore

● In Lesson 5.5, you learned to solve quadratic equations by factoring. However, you will not be able to solve most quadratic equations using this method. One method that you can use to solve *any* quadratic equation is called **completing the square**. Use Algeblocks and consider the quadratic equation $x^2 + 6x + 3 = 0$.

1. Write the equation so that the terms containing a variable are all on one side of the equation and the constant is on the other side.

2. **a.** Use Algeblocks to model the equation you found in Question 1.

   b. Divide the x-blocks into two equal groups and place them on adjacent sides of the x^2-block.

3. Complete the square on the left side by adding 9 unit blocks. What else must you do? Why? Sketch how the mat looks now.

4. Remove zero pairs. Write the equation it shows.

5. One side of the equation in Question 4 should be a perfect square trinomial. Write the equation with this trinomial in factored form. Then take the square root of both sides and solve for x.

Solve each equation using Algeblocks to complete the square.

6. $x^2 + 4x + 1 = 0$ 7. $x^2 + 2x - 6 = -4$ 8. $x^2 + 8x + 4 = 0$

Look for a pattern in your solutions to Questions 6–8 to answer the following question.

9. If you use Algeblocks to model an equation of the form $x^2 + bx = c$, how many unit blocks are necessary to complete the square on the left side of the equation? Express your answer in terms of b.

SPOTLIGHT ON LEARNING

WHAT? In this lesson you will learn
• to solve quadratic equations by completing the square.

WHY? Solving quadratic equations can help you solve problems about landscaping, travel, and geometry.

ALGEBRA: WHO, WHERE, WHEN

A geometric demonstration of completing the square is in writings of the ninth-century Moslem scholar, al-Kwarismi.

Build Understanding

● When you complete the square, you obtain a perfect square trinomial on one side of the equation. In Explore Question 9, you observed that the constant term of the trinomial can be obtained by dividing b by 2 and then squaring the result. Thus, you can make the expression $x^2 + bx$ into a perfect square trinomial by adding $\left(\dfrac{b}{2}\right)^2$.

When you complete the square to solve a quadratic equation, you must maintain equality. If you add a constant to one side of the equation, you must add the same constant to the other side of the equation.

EXAMPLE 1

Solve $x^2 + 16x + 25 = 0$ by completing the square.

Solution

$$x^2 + 16x + 25 = 0 \qquad \text{Add } -25 \text{ to both sides}$$

$$x^2 + 16x = -25$$

$$x^2 + 16x + 64 = -25 + 64 \qquad \left(\frac{16}{2}\right)^2 = 64; \text{ Add 64 to both sides.}$$

$$(x + 8)^2 = 39 \qquad \text{Factor the left side.}$$

$$x + 8 = \pm\sqrt{39} \qquad \text{Take the square root of both sides.}$$

$$x = -8 \pm \sqrt{39} \qquad \text{Solve for } x.$$

Check your solutions in the original equation.

$$\left(-8 + \sqrt{39}\right)^2 + 16\left(-8 + \sqrt{39}\right) + 25 \stackrel{?}{=} 0 \qquad 0 = 0 \checkmark$$

$$\left(-8 - \sqrt{39}\right)^2 + 16\left(-8 - \sqrt{39}\right) + 25 \stackrel{?}{=} 0 \qquad 0 = 0 \checkmark$$

Thus, $-8 + \sqrt{39}$ and $-8 - \sqrt{39}$ are both solutions to the equation. ◀

When completing the square, the coefficient of x^2 must be 1. If the coefficient of x^2 is not 1, begin by dividing both sides of the equation by the coefficient.

EXAMPLE 2

Solve $2x^2 - 5x - 9 = 0$ by completing the square.

Solution

$$2x^2 - 5x - 9 = 0$$

$$x^2 - \frac{5}{2}x - \frac{9}{2} = 0 \qquad \text{Divide both sides by 2.}$$

$$x^2 - \frac{5}{2}x = \frac{9}{2} \qquad \text{Add } \frac{9}{2} \text{ to both sides.}$$

$$x^2 - \frac{5}{2}x + \frac{25}{16} = \frac{9}{2} + \frac{25}{16} \qquad \text{Add } \left(\frac{5}{4}\right)^2 \text{ to both sides.}$$

$$\left(x - \frac{5}{4}\right)^2 = \frac{97}{16} \qquad \text{Factor left side. Simplify right side.}$$

$$x - \frac{5}{4} = \pm\sqrt{\frac{97}{16}} \qquad \text{Take the square root of both sides.}$$

$$x = \frac{5 \pm \sqrt{97}}{4} \qquad \text{Solve for } x.$$

Use your calculator to check that $\frac{5 + \sqrt{97}}{4}$ and $\frac{5 - \sqrt{97}}{4}$ are solutions. ◀

You can use quadratic equations to model many real world problems.

EXAMPLE 3

LANDSCAPE ARCHITECTURE A landscape architect designed a garden such that a grass path of uniform width surrounds a rectangular fountain. The fountain is 60 m long and 40 m wide, and the area of the grass path is 616 m². Determine the width of the grass path to the nearest hundredth of a meter.

Solution

Let x represent the width of the path.

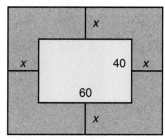

Area of Path	=	Area of Garden	−	Area of Fountain

$$616 = (60 + 2x)(40 + 2x) - (60 \cdot 40)$$
$$616 = 2400 + 120x + 80x + 4x^2 - 2400$$
$$616 = 4x^2 + 200x$$
$$154 = x^2 + 50x \qquad \text{Divide both sides by 4.}$$
$$154 + 625 = x^2 + 50x + 625 \qquad \text{Add } \left(\frac{50}{2}\right)^2 \text{ to both sides.}$$
$$779 = (x + 25)^2 \qquad \text{Factor right side.}$$
$$\pm \sqrt{779} = x + 25 \qquad \text{Take the square root of both sides.}$$
$$-25 \pm \sqrt{779} = x \qquad \text{Solve for } x.$$

The width of the path is $-25 + \sqrt{779} \approx 2.91$ m. ◄

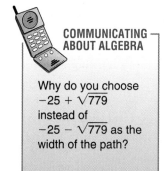

COMMUNICATING ABOUT ALGEBRA

Why do you choose $-25 + \sqrt{779}$ instead of $-25 - \sqrt{779}$ as the width of the path?

TRY THESE

Convert each expression to a perfect square trinomial.

1. $x^2 + 10x$ **2.** $x^2 - 20x$ **3.** $x^2 - 13x$

4. NUMBER THEORY Solve by completing the square. Determine a number whose square is 120 more than twice the number.

Find the value of b that makes the left side of each equation a perfect square trinomial.

5. $x^2 + bx + 36 = 8$ **6.** $x^2 - bx + 16 = 0$ **7.** $x^2 - bx + 49 = \frac{11}{3}$

8. WRITING MATHEMATICS Explain how to solve a quadratic equation by completing the square. Provide an example.

Solve each equation by completing the square. Check your answer.

9. $x^2 - 8x = -15$

10. $x^2 + 14x + 3 = 0$

11. $x^2 + 8x - 2 = 0$

12. $x^2 - 12x - 4 = 5$

13. $x^2 - 18x - 7 = 3$

14. $2x^2 + 5x + 6 = 8$

15. $\dfrac{x^2}{2} + 4x = 10$

16. $\dfrac{x^2}{2} + 5x = -1$

17. $\dfrac{x^2}{2} - 4x - 5 = 0$

18. HOME FURNISHINGS The length of a rectangular carpet is 2 ft less than twice its width. Determine the dimensions of the carpet if its area is 180 ft².

19. GEOMETRY The length of a rectangle is 2 yd more than its width. The area is 15 yd². Find the dimensions of the rectangle.

PRACTICE

Convert each to a perfect square trinomial.

1. $x^2 + 40x$

2. $x^2 - 2x$

3. $x^2 - 11x$

4. $x^2 + 5x$

5. $x^2 - \dfrac{10}{3}x$

6. $x^2 + \dfrac{5}{2}x$

7. NUMBER THEORY Solve by completing the square. The product of two consecutive positive odd integers is 143. What is the sum of the integers?

Find the value of b that makes the left side of each equation a perfect square trinomial.

8. $x^2 + bx + 81 = 2$

9. $x^2 - bx + 144 = 0$

10. $x^2 - bx + 225 = -\dfrac{1}{4}$

11. WRITING MATHEMATICS To solve $-x^2 + 4x = 5$ by completing the square, the first step Manuel wrote was $-x^2 + 4x + 4 = 5 + 4$. Is Manuel's work correct? Explain.

Solve each equation by completing the square. Check your answer.

12. $x^2 + 10x + 9 = 0$

13. $x^2 + 4x - 8 = 0$

14. $-x^2 + 14x + 7 = -9$

15. $x^2 - 24x - 2 = 1$

16. $4x^2 - 10x = 6$

17. $6x + 6 = -x^2$

18. $\dfrac{x^2}{4} + 2x = 6$

19. $\dfrac{x^2}{5} + 2x = -2$

20. $\dfrac{x^2}{2} - 6x - 8 = 0$

21. BOOK PRODUCTION A page of a book contains 80 square inches of print. It has margins of $\dfrac{1}{2}$ in. on both sides, and margins of 1 in. on the top and bottom. Determine the dimensions of the page if the width across the page is three-fourths of the length.

22. **TRAVEL** The FLY AWAY travel agency offers a vacation package to South America. They have calculated that the equation $y = -x^2 + 68x$ where x is the number of people that buy the package models their profit per person.

a. If their profit was $900 per person, how many people went to South America?

b. If their profit was $1075 per person, how many people went to South America?

23. Here is an example of solving a quadratic equation by completing the square. There is an error. Find it and solve the equation correctly.

$$3x^2 - 6x - 15 = 0$$
$$x^2 - 2x - 5 = 0$$
$$x^2 - 2x = 5$$
$$x^2 - 2x + 1 = 5$$
$$(x - 1)^2 = 5$$
$$x - 1 = \pm\sqrt{5}$$
$$x = 1 \pm \sqrt{5}$$

EXTEND

The solutions to quadratic equations may contain variables. For example, to solve the equation $x^2 + ax - 20a^2 = 0$ by completing the square, you can use the following steps.

$$x^2 + ax - 20a^2 = 0$$
$$x^2 + ax = 20a^2 \qquad \text{Add } 20a^2 \text{ to both sides.}$$
$$x^2 + ax + \frac{a^2}{4} = 20a^2 + \frac{a^2}{4} \qquad \text{Add } \left(\frac{a}{2}\right)^2 = \frac{a^2}{4} \text{ to both sides.}$$
$$\left(x + \frac{a}{2}\right)^2 = \frac{81a^2}{4} \qquad \text{Factor the left side.}$$
$$x + \frac{1a}{2} = \pm\frac{9a}{2} \qquad \text{Take the square root of both sides.}$$
$$x = -\frac{1a}{2} \pm \frac{9a}{2} \qquad \text{Solve for } x.$$
$$x = 4a \quad \text{or} \quad x = -5a$$

Solve each equation for x by completing the square. Check your answer.

24. $x^2 - 4ax - 21a^2 = 0$

25. $x^2 + 7ax - 18a^2 = 0$

26. $x^2 - \frac{8a}{3}x = a^2$

27. $x^2 + \frac{15a}{4}x = a^2$

28. **GEOMETRY** Determine the length of a side of a square whose diagonal is 12 ft longer than the length of a side. Round your answer to the nearest hundredth of a foot.

Write each equation in the form $a(x - h)^2 + b(y - k)^2 = c$ by completing the square first for the x-values and then for the y-values.

29. $x^2 + 4x + y^2 + 6y = 5$

30. $x^2 + y^2 + 6x - 8y + 10 = 0$

THINK CRITICALLY

31. WRITING MATHEMATICS Write a quadratic equation that you can solve by completing the square or by factoring.

32. WRITING MATHEMATICS Write a quadratic equation that you can solve by completing the square, but not by factoring.

33. WRITING MATHEMATICS If you can solve a particular quadratic equation by completing the square or by factoring, which method would you prefer to use? Explain.

34. Can you obtain a real number solution to the quadratic equation $x^2 - 2x + 6 = 0$? If you can, provide the solution. If you cannot, explain why not.

35. Solve the quadratic equation $4x^2 + \dfrac{4a}{3}x - \dfrac{4a^2}{3} = 0$ for x in terms of a.

MIXED REVIEW

36. STANDARDIZED TESTS The system of equations consisting of $x = 3$ and $3y = 2x + 1$ is

A. consistent and dependent   **B.** inconsistent

C. consistent and independent   **D.** none of the above

Factor each expression.

37. $36x^2 - 100$

38. $2x^2 - 88x$

Determine $f^{-1}(x)$ for each function.

39. $f(x) = \dfrac{1}{2}x - 8$

40. $f(x) = \dfrac{1}{4}x + 2$

PROJECT *Connection*   Plan your study, taking into account wheelchair availability.

1. Decide on the number of hours each participant will spend in the wheelchair. During that time, the participant must use the wheelchair to perform the regular daily routine.

2. Select pairs of student volunteers. One student will use the wheelchair and the other student will serve as an assistant. Set up a schedule for wheelchair use. Decide if weekends will be included in the study.

3. Prepare a log sheet that volunteers will use to record all times and activities.

4. Complete a list of items for all volunteers to monitor. For example, you might have participants indicate the time needed to travel a standard distance when they first begin the experiment and then again at the end. Create appropriate data sheets as needed.

5. Prepare a list of safety precautions. If possible, obtain advice from an experienced wheelchair user.

Many people are keeping their automobiles for longer periods of time. Engine mechanics repair or rebuild the internal combustion engines of automobiles to make them last longer.

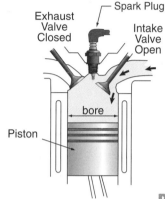

Exhaust Valve Closed

Spark Plug

Intake Valve Open

bore

Piston

The power generated by an internal combustion engine starts in the cylinders of the engine block. Larger cylinders can create more power. The diameter of the cylinder is called the **bore** (b) and the height traveled by the piston is called the **stroke** (s). A diagram of a piston that compresses an air-fuel mixture in the cylinder is shown.

The displacement, or volume V, of the engine is often given in cubic inches or liters. It is calculated by finding the volume of one cylinder and then multiplying by the number of cylinders.

Decision Making

1. Express the radius of a cylinder of an engine in terms of its bore.

2. Express the volume of a cylinder of an engine as a function of its bore and stroke.

3. Express the displacement of an eight-cylinder engine as a function of its bore and stroke.

4. Determine the engine displacement of a six-cylinder engine with a bore of 3.026 in. and a stroke of 4.245 in. Round your answer to the nearest cubic inch.

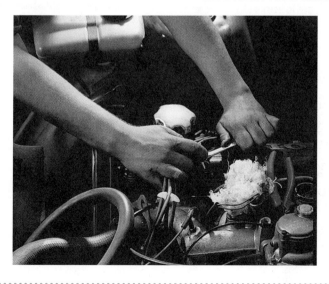

5. An engine mechanic may need to increase the diameter of an engine's cylinders. This is called boring out the engine. A mechanic is going to bore out the engine in Question 4 by x in. so that it has a displacement of 184 in.[3]

   a. Write an equation that models this situation.

   b. Solve the equation. Round to the nearest thousandth.

   c. What is the resulting bore?

Think Back

● Use the word *reflection*, *translation*, or *dilation* to make each of the following statements true.

1. The graph of the function $h(x) = -|x|$ is a ___?___ of the graph of the function $f(x) = |x|$ over the x-axis.

2. The graph of the function $j(x) = 2|x|$ is a ___?___ of the graph of the function $f(x) = |x|$.

3. The graph of the function $k(x) = |x + 1|$ is a ___?___ of the graph of the function $f(x) = |x|$.

Explain what you must do to the graph of $f(x) = |x|$ to produce each graph of g.

4. $g(x) = |x| + 4$ 5. $g(x) = |x + 2|$ 6. $g(x) = |x - 5| - 6$

Explore

● The quadratic function $f(x) = x^2$ is graphed at the right.

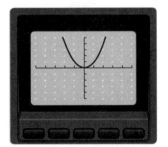

7. In which direction does the graph open?

8. Why is the graph located in Quadrants I and II?

9. Graph the function $g(x) = -x^2$.

10. In which direction does the graph open?

11. In which quadrants is the graph located?

12. What must you do to the graph of f to obtain the graph of g?

You can translate the graph of a quadratic equation to produce the same graph in a new position. Recall that a vertical translation moves a graph up or down.

13. Graph $f(x) = x^2$, $p(x) = x^2 + 1$, and $q(x) = x^2 - 2$ on the same set of axes.

14. Where is the graph of p in relation to the graph of f?

15. Where is the graph of q in relation to the graph of f?

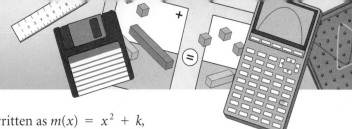

16. If a vertical translation of $f(x) = x^2$ is written as $m(x) = x^2 + k$, describe the location of the graph of m in relation to the graph of f when k is positive and when k is negative.

Recall that a horizontal translation moves a graph left or right.

17. Graph $f(x) = x^2$, $p(x) = (x + 3)^2$, and $q(x) = (x - 4)^2$ on the same set of axes.

18. Where is the graph of p in relation to the graph of f?

19. Where is the graph of q in relation to the graph of f?

20. If a horizontal translation of $f(x) = x^2$ is written as $n(x) = (x - h)^2$, describe the location of the graph of n in relation to the graph of f when h is positive and when h is negative.

The graph of $g(x) = (x - h)^2 + k$ will be translated horizontally and vertically from the graph of $f(x) = x^2$.

21. Work with a partner. Graph a function of the form $g(x) = (x - h)^2 + k$. Have your partner identify the function by studying the graph. Then switch roles.

The graph $g(x) = ax^2$ is a dilation of the graph of $f(x) = x^2$.

22. Graph $f(x) = x^2$, $p(x) = \frac{1}{2}x^2$, and $q(x) = 2x^2$ on the same set of axes.

23. Discuss similarities and differences of the graphs of f and p.

24. Discuss similarities and differences of the graphs of f and q.

25. Discuss the relationship of the graphs $p(x) = \frac{1}{2}x^2$ and $t(x) = -\frac{1}{2}x^2$.

You have solved quadratic equations by factoring and by completing the square. Another way to solve quadratic equations is by graphing both sides of the equation and finding their points of intersection.

26. Solve the quadratic equation $x^2 + x - 10 = 2$ by factoring.

27. Solve the equation in Question 26 by completing the square.

28. Use a graphing utility to graph $y_1 = x^2 + x - 10$ and $y_2 = 2$ on the same set of axes. Use the ZOOM and TRACE features to determine the points of intersection. The x-values of the points of intersection are the solutions. Compare your results with those of Questions 26 and 27.

Make Connections

29. Write a rule for a function g that is the reflection of f over the x-axis.

   a. $f(x) = 3x^2$ **b.** $f(x) = -2x^2$ **c.** $f(x) = -\frac{1}{4}x^2$ **d.** $f(x) = -2.4x^2$

30. State the direction(s) and number of units in which the graph of each function is translated from the graph of $f(x) = x^2$.

   a. $j(x) = (x + 5)^2 + \frac{1}{2}$ **b.** $k(x) = x^2 - 8.8$

   c. $m(x) = (x - 9)^2 - 3.33$ **d.** $n(x) = \left(x - \frac{1}{2}\right)^2 + \frac{1}{4}$

31. Compare the widths of the graphs of each of the functions. List the functions from narrowest to widest.

   a. $j(x) = 2.5x^2$ **b.** $k(x) = 0.9x^2$ **c.** $m(x) = \frac{3}{7}x^2$ **d.** $n(x) = \frac{9}{8}x^2$

32. Match each function with its graph.

   a. $p(x) = (x - 1)^2 - 2$ **b.** $q(x) = x^2 + 3$ **c.** $r(x) = (x + 1)^2 + 2$

   I. 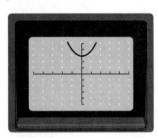   **II.** 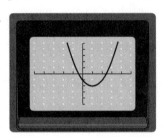   **III.**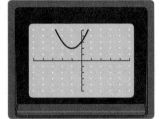

33. Match each function with its graph.

   a. $p(x) = \frac{1}{3}x^2$ **b.** $q(x) = \frac{4}{3}x^2$ **c.** $r(x) = \frac{2}{3}x^2$

   I. 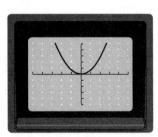   **II.** 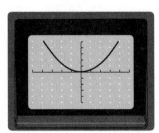   **III.**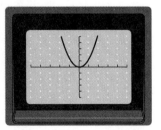

34. Write a function of the form $g(x) = (x - h)^2 + k$ whose graph represents the translation of $f(x) = x^2$.

   a. 1 unit down and 5 units left   **b.** 2 units up and 6 units right

   c. 3 units down and 7 units right   **d.** 4 units up and 8 units left

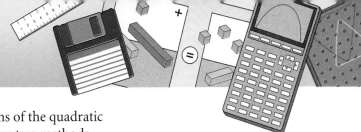

35. Use a graphing utility to find the solutions of the quadratic equation $2x^2 - 6x - 16 = 4$ using these two methods.

 a. Graph $y_1 = 2x^2 - 6x - 16$ and $y_2 = 4$ on the same set of axes and determine the x-values of their points of intersection.

 b. Graph $y_1 = 2x^2 - 6x - 20$ and determine the x-values of the points at which the graph intersects the x-axis.

36. Did you obtain the same answer for Question 35 a and b? Explain.

37. At what point or points on the graph of a quadratic equation of the form $ax^2 + bx + c = 0$ are the real solutions located?

Summarize

38. **WRITING MATHEMATICS** Explain how the graph of $g(x) = (x - h)^2 + k$ differs from that of $f(x) = x^2$.

39. **WRITING MATHEMATICS** Explain how the graph of $g(x) = -ax^2$ differs from that of $f(x) = ax^2$.

40. **WRITING MATHEMATICS** Explain how to use a graphing utility to determine the solutions of a quadratic equation in the form $ax^2 + bx + c = d$.

41. **THINKING CRITICALLY** Jacob says that he can obtain the graph of $f(x) = -(x + 2)^2 + 1$ by graphing $j(x) = x^2$, translating it up 1 unit, translating it left 2 units, and then reflecting it over the x-axis. Stacey says that she can obtain the graph of f by reflecting the graph of j over the x-axis, translating it up 1 unit, and then translating it left 2 units. Which student is correct? Explain.

42. **GOING FURTHER** Graph a quadratic equation that has no real number solutions.

PROJECT *Connection*   Use your data to analyze wheelchair travel.

1. Measure the radius of your chair's wheel. Determine its circumference and area.

2. How long does it take you to go 100 m? How many revolutions does your wheel make?

3. Road races sponsored by Wheelchair Athletics USA typically include a 1500-meter course for men and an 800-meter course for women. Based on your answer to Question 2, what would your time be for these races?

4. In 1995, the world record for a 26.2-mile wheelchair marathon was 1 h 21 min 23 s. At your present rate, how long would it take you to complete a marathon course?

Career
Cartographer

Cartographers create maps that represent areas as small as your neighborhood and as large as the Earth itself. It is difficult to create an accurate map of the Earth because you must represent a curved surface on a flat piece of paper.

One method early cartographers used is called a Mercator projection. They projected the Earth onto a cylinder wrapped around a globe touching at the equator. Distortion in the projection results when the lines of longitude are projected as parallel on the cylinder.

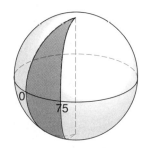

Decision Making

1. The diameter of the Earth at the equator is approximately 7926 miles. Determine the radius at the equator.

2. The surface area S of the Earth can be represented by the quadratic equation $S = 4\pi r^2$. Use your result from Question 1 to determine the surface area to the nearest square mile.

Any point on the Earth's surface can be located using latitude and longitude. Longitude is measured east and west from the Prime Meridian, 180° in each direction for a total of 360°.

3. What percentage of the Earth's surface is represented by the shaded sector displayed at the right?

4. Determine the area of the shaded sector to the nearest square mile.

5. Determine the length of the segment of the equator in the shaded sector to the nearest mile.

6. Assume the Earth is a perfect sphere. Determine the length of one of the lines of longitude that borders the shaded sector to the nearest mile.

7. If this portion of the Earth's surface were represented by a rectangle, how many square miles greater would the area be than that of the sector in Question 4?

Explore/Working Together

- You have studied quadratic functions of the form $g(x) = a(x - h)^2 + k$. Recall a function of this form is a transformation of the quadratic function $f(x) = x^2$.

    1. Express the function $j(x) = 3(x - 1)^2 + 2$ without parentheses.

    2. Graph both forms of the equation in Question 1 on the same set of axes. Do you obtain the same graph? Explain.

    3. Work with a partner. Use a graphing utility to graph the function $f(x) = 2x^2 + 3$. Set your calculator so integers are displayed when using the trace feature. One of you trace to the right of $x = 0$ and record the x- and y-coordinates. The other trace to the left of $x = 0$ and record the x- and y-coordinates.

    4. Compare results. What pattern do you notice?

    5. Which of the following is true for $f(x) = 2x^2 + 3$?

        a. $f(x) = -f(x)$ **b.** $f(x) = f(-x)$ **c.** $f(-x) = -f(x)$

    6. Is $f(x) = 2x^2 + 3$ an odd function or an even function?

SPOTLIGHT ON LEARNING

WHAT? In this lesson you will learn
- to determine the vertex and axis of symmetry of the graph of a quadratic function.
- to graph a quadratic function.
- to determine the equation of a parabola from its graph.

WHY? Graphing quadratic functions can help you solve problems about construction, urban planning, and business.

Build Understanding

- A **quadratic function** is a function that can be written in the form $f(x) = ax^2 + bx + c$, where a, b, and c are real numbers and $a \neq 0$.

    The U-shaped graph of a quadratic function is called a **parabola**. For $f(x) = ax^2 + bx + c$, the graph opens upward if $a > 0$ and downward if $a < 0$.

    The graph of a quadratic function has either a **minimum** (lowest) point or a **maximum** (highest) point called the **vertex** of a parabola. The vertical line that passes through the vertex is called the **axis of symmetry**.

    The graph of $f(x) = x^2 - 8x + 10$ is displayed at the right. It opens upward because $a > 0$. It has a vertex at $(4, -6)$ and an axis of symmetry at $x = 4$.

    When a quadratic equation is in the form $y = ax^2 + bx + c$, the x-coordinate of the vertex is $x = \frac{-b}{2a}$. To find the y-coordinate of the vertex, substitute the x-value into the original equation and solve for y. The axis of symmetry is the vertical line $x = \frac{-b}{2a}$.

THINK BACK

For an odd function, $f(-x) = -f(x)$.

For an even function, $f(-x) = f(x)$.

CHECK UNDERSTANDING

If $a = 0$ in a function of the form $f(x) = ax^2 + bx + c$, what type of function is it?

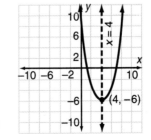

EXAMPLE 1

Determine the vertex and axis of symmetry of each parabola. Then state whether the parabola opens upward or downward.

a. $f(x) = -2x^2 + 8x - 5$ **b.** $g(x) = \frac{1}{2}x^2 - 3x - 4$

Solution

a. For $f(x) = -2x^2 + 8x - 5$, $a = -2$ and $b = 8$.

$$\frac{-b}{2a} = \frac{-8}{2(-2)} = 2 \qquad \textit{x-coordinate of vertex}$$

$$f(2) = -2(2)^2 + 8(2) - 5 = 3 \qquad \textit{y-coordinate of vertex}$$

The vertex is $(2, 3)$ and the axis of symmetry is $x = 2$. Since a is negative, the parabola opens downward.

b. For $g(x) = \frac{1}{2}x^2 - 3x - 4$, $a = \frac{1}{2}$ and $b = -3$.

$$\frac{-b}{2a} = \frac{-(-3)}{2\left(\frac{1}{2}\right)} = 3 \qquad \textit{x-coordinate of vertex}$$

$$f(3) = \frac{1}{2}(3)^2 - 3(3) - 4 = -8.5 \qquad \textit{y-coordinate of vertex}$$

The vertex is $(3, -8.5)$ and the axis of symmetry is $x = 3$. Since a is positive, the parabola opens upward. ◄

You can reflect points on the graph of a parabola about the axis of symmetry to determine other points on the graph.

EXAMPLE 2

The points $(0, -6)$, $(3, -9)$, and $(-2, 6)$ are on the graph of $f(x) = x^2 - 4x - 6$. For each point, find its point of reflection.

Solution

For $f(x) = x^2 - 4x - 6$, $a = 1$ and $b = -4$.

The axis of symmetry is $x = \frac{-(-4)}{2(1)}$, or $x = 2$.

The reflection of $(0, -6)$ is $(4, -6)$.

The reflection of $(3, -9)$ is $(1, -9)$.

The reflection of $(-2, 6)$ is $(6, 6)$. ◄

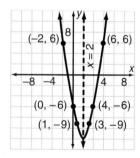

Use your knowledge about the direction in which a parabola opens, its axis of symmetry, vertex, x-intercepts, y-intercept, and the reflection of the y-intercept to help you graph the parabola.

EXAMPLE 3

Graph: $y = x^2 + 6x + 8$

Solution

The graph opens upward because $a > 0$. $1 > 0$

Axis of symmetry: $x = \dfrac{-b}{2a} = \dfrac{-6}{2(1)}$ $x = -3$

Vertex: $f(-3) = (-3)^2 + 6(-3) + 8$
The vertex is $(-3, -1)$.

x-intercepts: $0 = x^2 + 6x + 8$ Solve equation.
$\phantom{x\text{-intercepts: }} 0 = (x + 4)(x + 2)$
$\phantom{x\text{-intercepts: }} x = 4 \quad \text{and} \quad x = 2$
The graph crosses the x-axis at $(-4, 0)$ and $(-2, 0)$.

y-intercept: $f(0) = 0^2 + 6(0) + 8$
The graph crosses the y-axis at $(0, 8)$.

Since the point $(0, 8)$ is 3 units to the right of the axis of symmetry, its reflection $(-6, 8)$ is 3 units to the left of the axis of symmetry. ◀

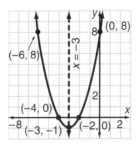

You can solve many real world problems by finding the maximum or minimum value of a quadratic equation.

THINK BACK

The solutions of a quadratic equation are the points at which its graph crosses the x-axis.

EXAMPLE 4

CONSTRUCTION The Berger family wants to build a rectangular dog run in their backyard using 240 ft of fencing. If their goal is to maximize the area of the pen, what should the dimensions of the run be? What is the maximum area of the pen?

Solution

Let x represent the width of the pen. Since the perimeter of the pen is 240 ft, the length of the pen must be $\dfrac{240 - 2x}{2} = 120 - x$ ft. The area of the pen is $x(120 - x) = -x^2 + 120x$.

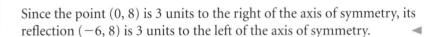

120 − x

x x

120 − x

Graph $y = -x^2 + 120x$. Use the ZOOM and TRACE features of your graphing utility to find the maximum point is $(60, 3600)$. Since the width of the pen is 60 ft, the length is 120 ft − 60 ft = 60 ft.

The maximum area is 3600 ft². ◀

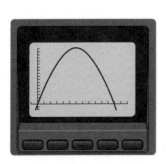

You can determine the equation of a parabola if you know three points on its graph.

EXAMPLE 5

Determine the equation of the graph shown at the right.

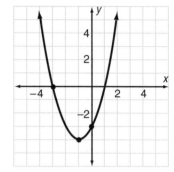

Solution

The equation should be in the form $y = ax^2 + bx + c$. Substitute the x-values and y-values for each point into the equation to create a system of three equations.

For $(0, -3)$: $c = -3$
For $(-3, 0)$: $9a - 3b + c = 0$
For $(-1, -4)$: $a - b + c = -4$

Use $c = -3$ on the other two equations. Solve for a.

$$9a - 3b - 3 = 0 \quad \rightarrow \quad 9a - 3b - 3 = 0$$
$$a - b - 3 = -4 \quad \rightarrow \quad \underline{3a - 3b - 9 = -12} \quad \text{Multiply by 3.}$$
$$6a \qquad\quad + 6 = 12 \quad \text{Subtract.}$$

Since $6a + 6 = 12$, $a = 1$. Substitute $a = 1$ and $c = -3$ into the second equation to obtain $9 - 3b - 3 = 0$. Thus, $b = 2$. Check these values in the third equation.

The equation of the parabola is $y = x^2 + 2x - 3$.   ◄

PROBLEM SOLVING TIP

In Example 5, you can use any of the methods you learned in Chapter 4 for solving systems of equations.

COMMUNICATING ABOUT ALGEBRA

In Example 5, will you obtain the equation $y = x^2 + 2x - 3$ if you select three points on the graph other than $(0, -3)$, $(-3, 0)$ and $(-1, -4)$? Explain.

TRY THESE

Determine the vertex and axis of symmetry of each parabola. Then state whether the parabola opens upward or downward.

1. $f(x) = -x^2 + 12x - 2$ **2.** $f(x) = -2x^2 - 4x + 5$ **3.** $f(x) = 4x^2 + 10$

4. $f(x) = 7x^2 + 14$ **5.** $f(x) = 6x - 1 - 6x^2$ **6.** $f(x) = -10x + 9 + 5x^2$

7. WRITING MATHEMATICS   Explain how to determine the point of reflection of a given point (x, y) on the graph of a quadratic function of the form $f(x) = ax^2 + bx + c$. Provide an example.

For Exercises 8–11, the graph of the quadratic equation contains the three given points. For each point, determine its point of reflection on the graph.

8. $y = x^2 + 2x - 5$; $(-5, 10)$, $(0, -5)$, $(2, 3)$

9. $y = x^2 - 4x + 6$; $(-1, 11)$, $(0, 6)$, $(3, 3)$

10. $y = -3x^2 - 18x - 1$; $(-4, 23)$, $(0, -1)$, $(2, -49)$

11. $y = -2x^2 + 12x + 2$; $(-1, -12)$, $(0, 2)$, $(4, 18)$

12. **PHYSICS** The height h of an object t seconds after being released can be modeled by the equation $h(t) = -\frac{1}{2}at^2 + vt + s$ where a is the acceleration due to gravity, v is the upward speed of the object upon release, and s is the starting height of the object. At the surface of the Earth, a is approximately 32 ft/s². A model rocket is launched from the Earth with an upward speed of 160 ft/s.

   a. Write a function that represents the height of the rocket as a function of time.

   b. Use a graphing utility to determine the maximum height attained by the rocket to the nearest foot.

   c. How many seconds after launch does the rocket reach maximum height?

   d. How many seconds after launch does the rocket hit the ground?

Graph each equation.

13. $y = x^2 - 4x - 5$ **14.** $y = 2x^2 + 8x + 6$ **15.** $y = -x^2 - 4x + 12$

Use a system of equations to determine the equation of each parabola.

16.
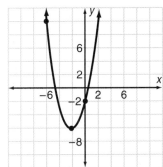

17.
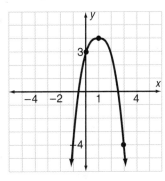

18.
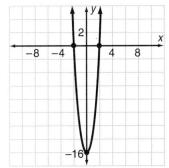

19.
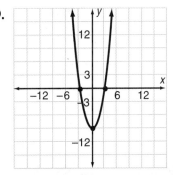

20. **URBAN PLANNING** The transit authority in Stoneville is considering lowering the fare to ride the subways. A study determined that the daily revenue from people riding the subway can be represented by $R(x) = -5x^2 + 370x + 19{,}800$ where x represents the number of $0.01 decreases from the current fare of $1.10. What fare will yield the highest revenue?

Determine the vertex and axis of symmetry of each parabola. Then state whether the parabola opens upward or downward.

1. $f(x) = -x^2 + 4x + 1$ **2.** $f(x) = -2x^2 - 8x + 3$ **3.** $f(x) = 6x^2 + 18$

4. $f(x) = 5x^2 + 15$ **5.** $f(x) = 2x - 1 - 3x^2$ **6.** $f(x) = -5x + 3 + 2x^2$

7. WRITING MATHEMATICS Explain how to determine if the vertex of a quadratic function in the form $f(x) = ax^2 + bx + c$ is a minimum or maximum without graphing.

For Exercises 8–11, the graph of the quadratic equation contains the three given points. For each point, find its point of reflection on the graph.

8. $y = x^2 + 8x - 2$; $(-6, -14)$, $(0, -2)$, $(1, 7)$

9. $y = x^2 - 12x + 1$; $(-1, 14)$, $(0, 1)$, $(4, -31)$

10. $y = -5x^2 - 20x - 3$; $(-5, -28)$, $(0, -3)$, $(2, -63)$

11. $y = -6x^2 + 24x + 2$; $(-1, -28)$, $(0, 2)$, $(3, 20)$

12. MANUFACTURING BOXES ARE US! wants to manufacture a box for which the bottom has a perimeter of 64 cm, the height is 12 cm, and the volume is a maximum. What should the dimensions of the bottom be? What is the maximum volume of the box?

Graph each equation.

13. $y = x^2 - 2x - 3$ **14.** $y = x^2 - 6x - 16$ **15.** $y = 4x^2 + 18x + 8$

16. $y = 2x^2 - 8x + 12$ **17.** $y = -x^2 - 4x + 3$ **18.** $y = -x^2 + 2x + 8$

Use a system of equations to determine the equation of each parabola.

19.

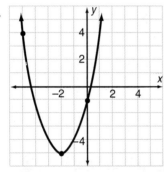

20.

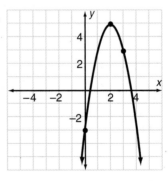

21.

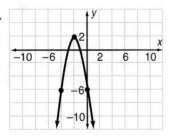

22.

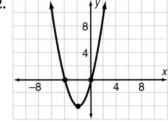

23. STUDY TIME Judging by his past performance on mathematics exams, Ross can estimate the grade that he will receive on a mathematics exam using the function $G(t) = -t^2 + 8t + 78$, where t represents the number of hours that he spends studying. How many hours should he study to achieve his maximum grade? What is the maximum grade that he can achieve?

EXTEND

You can use the quadratic regression feature on your graphing utility to determine the equation of a parabola if you are given three points on its graph. Your graphing utility will provide you with an equation of the form $y = ax^2 + bx + c$. For example, enter the points $(0, -3)$, $(-3, 0)$, and $(-1, -4)$ from Example 5 into your graphing utility. Use the quadratic regression feature to obtain the equation $y = x^2 + 2x - 3$.

Use quadratic regression to determine the equation of the parabola that contains each set of points.

24. $(-1, -42)$, $(2, -30)$, $(5, 0)$

25. $(-3, 4)$, $(1, 16)$, $(2, 24)$

26. $(-5, 16)$, $(0, 6)$, $(4, 70)$

27. $(-1, -8)$, $(0, -7)$, $(3, 32)$

28. PHYSICS Elliott threw a ball straight up with an upward speed of 40 ft/s. His hand was 8 ft above the ground when he released the ball. Use the information provided in Exercise 12 to write the function that represents the height of the ball as a function of time. Determine the maximum height attained by the ball.

29. PHYSICS If Elliott released the ball in Exercise 28 when his hand was 7 ft above the ground, would the maximum height attained by the ball be 1 ft less than your answer to Exercise 28? Explain.

30. BUSINESS A coffee shop sells special blends of coffee at $1.50/mug. They sell approximately 200 mugs per day. For each $0.05 decrease in the price, the shop will sell approximately 10 more mugs per day. Determine the revenue function for the coffee shop's revenue on coffee. Determine the maximum daily revenue that the shop can earn on coffee and the price at which they should sell the coffee.

31. If a quadratic function of the form $f(x) = ax^2 + bx + c$ is an even function, what is the equation of its axis of symmetry?

32. For the function in Exercise 31 to be an even function, what must the value of b be?

33. If you enter two points into your graphing utility, will it provide you with a quadratic equation that models the data? Explain.

34. If you enter four points into your graphing utility, will it provide you with a quadratic equation that models the data? Explain.

35. For what values of a and b is the vertex of the function $f(x) = ax^2 + bx - 7$ at $(-4, -39)$?

36. For what values of a and b is the vertex of the function $f(x) = ax^2 + bx + 6$ at $(1, 7)$?

MIXED REVIEW

Determine the mode of each set of data.

37. 3, 4, 5, 4, 6, 4, 5, 7, 3, 7

38. 0, -1, 1, 0, -1, 0, 1, 2

39. STANDARDIZED TESTS The equation of a line in slope-intercept form that has a slope of -3 and passes through the point $(2, 4)$ is

 A. $-3x + y = 10$ **B.** $-y = 3x - 10$ **C.** $y = -3x + 10$ **D.** $y = 3x - 10$

Approximate each number to the nearest ten-thousandth.

40. $\sqrt{139}$

41. $\sqrt{555}$

Determine the vertex and axis of symmetry of each parabola.

42. $y = 3x^2 - 24x + 2$

43. $y = -2x^2 - 16x - 10$

Solve for x.

44. $|x| = 3$

45. $|x - 6| = 5$

46. $5|2x - 1| < -2$

47. $|x + 3| > 1$

Find $f^{-1}(x)$.

48. $f(x) = 4x$

49. $f(x) = 3x + 6$

50. $f(x) = x^2$

51. $f(x) = \dfrac{1}{x + 1}$

PROJECT *Connection*   As you may have discovered, wheelchair travel has certain special safety hazards associated with it. Because riders are sitting only 2 to 4 feet above the road surface, they can be difficult to see.

1. Make a list of safety hazards for wheelchair riders. Suggest ways safety can be improved.

2. Design a reflective device that could make using a wheelchair safer at night. Indicate the materials used and dimensions of the device. Remember the device will have to withstand wet weather.

Civil engineers design bridges to connect the vast network of roads and railroads throughout the world. They use mathematics, physics, and engineering to create bridges that will withstand the weight of the vehicles that pass over them, their own weight, the wind, and so on. Some bridges are supported from below, while others are suspended from towers above.

When the main cables of a suspension bridge are hung without any weight on them, they form a curve called a *catenary*. If weights are attached to them, their shape resembles a parabola and can be approximated by a quadratic equation.

If you model the roadway of San Francisco's Golden Gate Bridge with the x-axis, the shape of each main suspension cable can be approximated by the quadratic equation $y = \frac{1}{9000}x^2 + 5$. In this model, the y-axis is the axis of symmetry.

Decision Making

1. What are the coordinates of the lowest point on the cable?

2. How long is a vertical hanger that hangs from the main cable at a point that is 1600 ft from the axis of symmetry? Round your answer to the nearest foot.

3. How many hangers on this bridge would be 1600 ft from the axis of symmetry?

4. If the span between the two towers is 4200 ft, how high above the roadway are the cables connected to the towers? Round your answer to the nearest foot.

5. The straight horizontal overpass shown in the figure is supported by a parabolic arch that can be modeled by the equation $f(x) = -0.05x^2 + 5.25x - 47.5$. Determine the width of the arch at road level.

6. If the overpass is 12 ft above the top of the arch, what equation can you use to model it?

6.4 Use the Quadratic Formula

Explore

SPOTLIGHT ON LEARNING

WHAT? In this lesson you will learn
- to solve quadratic equations using the quadratic formula.
- to use the discriminant to determine the nature of the solution(s) of a quadratic equation.
- to determine a quadratic equation given its solutions.

WHY? Using the quadratic formula can help you solve problems in geometry, framing, and business.

THINK BACK

$$\sqrt{\frac{a}{b}} = \frac{\sqrt{a}}{\sqrt{b}}$$

In Lesson 6.1, you solved quadratic equations by completing the square. By completing the square for the general form of a quadratic equation $ax^2 + bx + c = 0$, a formula is derived that can be used to solve any quadratic equation. Replace each ■ to complete each step. Provide a reason for each step.

$$ax^2 + bx + c = 0$$

$$x^2 + \frac{b}{a}x + \frac{c}{a} = 0$$

When completing the square the coefficient of x^2 must be 1.

1. $x^2 + \dfrac{b}{a}x = $ ■

2. $x^2 + \dfrac{b}{a}x + $ ■ $= -\dfrac{c}{a} + $ ■

3. $x^2 + \dfrac{b}{a}x + \left(\dfrac{b}{2a}\right)^2 = -\dfrac{c}{a} + $ ■

4. $\left(x + ■\right)^2 = -\dfrac{c}{a} + \dfrac{b^2}{4a^2}$

5. $\left(x + \dfrac{b}{2a}\right)^2 = $ ■

6. $x + \dfrac{b}{2a} = $ ■ $\sqrt{\dfrac{b^2 - 4ac}{4a^2}}$

7. $x = $ ■ $+ \dfrac{\pm\sqrt{b^2 - 4ac}}{\sqrt{4a^2}}$

8. $x = \dfrac{-b}{2a} + \dfrac{\pm\sqrt{b^2 - 4ac}}{■}$

9. $x = \dfrac{-b \pm \sqrt{b^2 - 4ac}}{■}$

Build Understanding

The result of your work in Explore is called the **quadratic formula**. You can use this formula to solve any quadratic equation in the standard form $x^2 + bx + c = 0$.

THE QUADRATIC FORMULA

For a quadratic equation of the form $ax^2 + bx + c = 0$ where a, b, and c are real numbers and $a \neq 0$,

$$x = \frac{-b \pm \sqrt{b^2 - 4ac}}{2a}$$

EXAMPLE 1

Use the quadratic formula to solve $2x^2 + 3x - 20 = 0$.

Solution

In $2x^2 + 3x - 20 = 0$, $a = 2$, $b = 3$, and $c = 20$.

$$x = \frac{-b \pm \sqrt{b^2 - 4ac}}{2a}$$ Write the quadratic formula.

$$x = \frac{-3 \pm \sqrt{3^2 - 4(2)(-20)}}{2(2)}$$ Substitute values of a, b and c.

$$x = \frac{-3 \pm \sqrt{169}}{4}$$ Simplify under the radical symbol.

$$x = \frac{-3 \pm 13}{4}$$ Evaluate $\sqrt{169}$.

The solutions are $x = \dfrac{-3 + 13}{4} = 2.5$ and $x = \dfrac{-3 - 13}{4} = -4$.

Check your solutions in the original equation.

$$2x^2 + 3x - 20 = 0 \qquad\qquad 2x^2 + 3x - 20 = 0$$
$$2(2.5)^2 + 3(2.5) - 20 \overset{?}{=} 0 \qquad 2(-4)^2 + 3(-4) - 20 \overset{?}{=} 0$$
$$0 = 0 \checkmark \qquad\qquad\qquad 0 = 0 \checkmark$$

The solutions are 2.5 and -4. ◀

The expression $b^2 - 4ac$ under the radical symbol in the quadratic formula is called the **discriminant**. You can use the discriminant to determine the nature of the solutions of a quadratic equation.

> **For a quadratic equation of the form $ax^2 + bx + c = 0$, the discriminant is $b^2 - 4ac$.**
>
> - If $b^2 - 4ac > 0$, the equation has two real solutions.
> - If $b^2 - 4ac = 0$, the equation has one real solution.
> - If $b^2 - 4ac < 0$, the equation has no real solutions.

EXAMPLE 2

Use the discriminant to determine the number of solutions and the nature of the solutions of each quadratic equation.

a. $x^2 - 6x + 9 = 0$ **b.** $x^2 - 6x - 2 = 0$ **c.** $x^2 - 6x + 12 = 0$

Solution

a. $b^2 - 4ac = (-6)^2 - 4(1)(9) = 0$; one real, rational solution

b. $b^2 - 4ac = (-6)^2 - 4(1)(-2) = 44$; two real, irrational solutions

c. $b^2 - 4ac = (-6)^2 - 4(1)(12) = -12$; no real solutions ◀

COMMUNICATING ABOUT ALGEBRA

Christine asserts that although Example 2a has only one *real* solution, it is a double solution. Explain why she is correct.

You can also graph a quadratic equation to determine how many real solutions it has. If its graph intersects the x-axis in two points, then the quadratic equation has two real solutions. If its graph intersects the x-axis in only one point, then the quadratic equation has one real solution. If its graph does not intersect the x-axis, then the quadratic equation has no real solutions. The equations from Example 2 are graphed below.

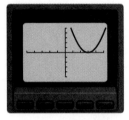  

one real solution   **two real solutions**   **no real solutions**

The sum and product of the solutions of a quadratic equation are related to the coefficients of the equation. If you know the sum and product, you can use them to determine the quadratic equation.

> **SUM AND PRODUCT OF SOLUTIONS OF A QUADRATIC EQUATION**
>
> If s_1 and s_2 are the solutions of a quadratic equation of the form $ax^2 + bx + c = 0$, $a \neq 0$,
>
> then $s_1 + s_2 = -\dfrac{b}{a}$ and $s_1 s_2 = \dfrac{c}{a}$.

EXAMPLE 3

Determine the sum and product of the solutions to the equation $2x^2 - 4x + 2 = 0$.

Solution

In $2x^2 - 4x + 2 = 0$, $a = 2, b = -4$, and $c = 2$.

The sum: $-\dfrac{b}{a} = -\left(\dfrac{-4}{2}\right) = 2$

The product: $\dfrac{c}{a} = \dfrac{2}{2} = 1$

If you solve this equation, you find it has a double solution of $x = 1$. Since $x = 1$, $s_1 = 1$ and $s_2 = 1$.

So, $s_1 + s_2 = 2$ and $s_1 s_2 = 1$.   ◄

> **QUADRATIC EQUATION WITH SOLUTIONS S_1 AND S_2**
>
> If the solutions of a quadratic equation are s_1 and s_2 then its equation is
> $$x^2 - (s_1 + s_2)x + s_1 s_2 = 0$$

EXAMPLE 4

Write a quadratic equation that has the solutions 4 and -5.

Solution

Let $s_1 = 4$ and $s_2 = -5$.

$$s_1 + s_2 = 4 + (-5) = -1$$
$$s_1 s_2 = 4(-5) = -20$$

$$x^2 - (-1)x + (-20) = 0 \qquad x^2 - (s_1 + s_2)x + s_1 s_2 = 0$$
$$x^2 + x - 20 = 0$$

The equation is $x^2 + x - 20 = 0$. ◀

You can use the quadratic formula to solve real world problems.

EXAMPLE 5

GEOMETRY The number of square feet in the area of a square banquet room is 357 greater than the number of feet in its perimeter. Determine the dimensions of the room.

Solution

Let x represent the length of the room.

$$\text{Area} - \text{Perimeter} = 357$$
$$x^2 - 4x = 357$$
$$x^2 - 4x - 357 = 0$$

In $x^2 - 4x - 357 = 0$, $\quad a = 1, b = -4,$ and $c = -357$.

$$x = \frac{-b \pm \sqrt{b^2 - 4ac}}{2a} \qquad \text{Write the quadratic formula.}$$

$$x = \frac{-(-4) \pm \sqrt{(-4)^2 - 4(1)(-357)}}{2(1)} \qquad \text{Substitute for } a, b \text{ and } c.$$

$$x = \frac{4 \pm \sqrt{1444}}{2} \qquad \text{Simplify under the radical symbol.}$$

$$x = \frac{4 \pm 38}{2} \qquad \text{Evaluate } \sqrt{1444}.$$

The solutions are $x = \dfrac{4 + 38}{2} = 21$ and $x = \dfrac{4 - 38}{2} = -17$.

Check

$$s_1 + s_2 = 21 + -17 = 4 \qquad -\frac{b}{a} = -\left(\frac{-4}{1}\right) = 4 \checkmark$$

$$s_1 s_2 = 21(-17) = -357 \qquad \frac{c}{a} = \frac{-357}{1} = -357 \checkmark$$

The dimensions of the room are 21 ft by 21 ft. ◀

CHECK UNDERSTANDING

In Example 5, why did you ignore the solution $x = -17$ when you determined the dimensions?

Use the quadratic formula to solve each equation. Check your answers.

1. $2x^2 + 4x - 1 = 0$

2. $4y^2 + 2y - 8 = 0$

3. $9z^2 - 30z + 25 = 0$

4. **WRITING MATHEMATICS** Jeffrey claims that in order to determine the quadratic equation with solutions 3 and 6, he will write the equation $(x - 3)(x - 6) = 0$ and then multiply the left side to obtain $x^2 - 9x + 18 = 0$. Does his method provide the correct answer? If so, write a general rule for finding the quadratic equation with solutions s_1 and s_2. If not, explain why not.

Use the discriminant to determine the number of solutions and the nature of the solutions of each quadratic equation.

5. $2x^2 + 6x + 8 = 0$

6. $4y^2 + 12y + 9 = 0$

7. $3z^2 - 2z - 10 = 0$

Use a graphing utility. Determine the number of times that the graph of each quadratic equation intersects the x-axis.

8. $16p^2 - 8p + 1 = 0$

9. $3q^2 - 5q + 6 = 0$

10. $-2r^2 + 9r + 18 = 0$

Determine the value(s) of k for which each equation will have exactly one solution.

11. $kx^2 + 20x + 25 = 0$

12. $36x^2 - kx + 1 = 0$

13. $9x^2 + 24x + k = 0$

14. **BUSINESS** Irma's Items makes custom hand-painted T-shirts. The company's projected annual revenue can be modeled by the function $R(x) = 12x^2 + 120x + 111$, where x is the number of items produced, in hundreds. The cost to produce these items can be modeled by the function $C(x) = 4x^2 + 35x + 91$. If the company sells every item that it produces, it will make a profit of $222. Write a function that models profit. How many items did the company sell?

Determine the sum and the product of the solutions of each equation.

15. $f^2 - 12f + 4 = 0$

16. $2g^2 - 8g = 4$

17. $3h^2 = 4h + 7$

Write a quadratic equation that has the given solutions.

18. 4 and $-\dfrac{1}{2}$

19. -6 and -7

20. -5 and $-\dfrac{1}{4}$

21. **NUMBER THEORY** Find a number whose square is 135 greater than six times the number.

22. **WRITING MATHEMATICS** If the discriminant of a quadratic equation is less than zero, explain why the equation has no real solutions.

PRACTICE

Use the quadratic formula to solve each equation. Check your answers.

1. $3x^2 - 9x + 1 = 0$

2. $4y^2 - 7y + 2 = 0$

3. $3z^2 + 16z + 5 = 0$

4. $5p^2 - 2p - 6 = 0$

5. $-3q^2 - 5q + 8 = 0$

6. $-16r^2 - 24r + 9 = 0$

7. AMUSEMENT PARKS Amusement park rides can simulate free fall for the rider. Designers of these rides use the equation $d = -16t^2$ to find the distance traveled by the ride as a function of time. If the ride drops 48 ft, how many seconds is it in free fall?

Use the discriminant to determine the number of solutions and the nature of the solutions of each quadratic equation.

8. $4x^2 + 2x + 7 = 0$

9. $2y^2 - 5y - 5 = 0$

10. $-36z^2 + 12z - 1 = 0$

11. $16p^2 + 24p + 9 = 0$

12. $2q^2 + 7q - 30 = 0$

13. $5r^2 + 24r - 5 = 0$

Use a graphing utility. Determine the number of times that the graph of each quadratic equation intersects the x-axis.

14. $4x^2 - 4x + 1 = 0$

15. $9y^2 + 9y + 1 = 0$

16. $3z^2 - 5z + 6 = 0$

17. $-2p^2 + 3p - 8 = 0$

18. $3q^2 + 2q - 10 = 0$

19. $6x^2 - 3x - 1 = 0$

Determine the value(s) of k for which each equation will have exactly one solution.

20. $kx^2 - 40x + 25 = 0$

21. $25x^2 + kx + 9 = 0$

22. $9x^2 - 42x + k = 0$

23. WRITING MATHEMATICS Describe the graph of a quadratic equation that has no real solutions. Give examples of real world situations where quadratic equations with no real solutions can be used to model the situation. Identify what the x-axis models.

Determine the sum and the product of the solutions of each equation.

24. $f^2 - 15f + 5 = 0$

25. $g^2 + 8g - 12 = 0$

26. $3h^2 - 7h = -7$

27. $4m^2 = 5m + 9$

28. $\frac{1}{2}n^2 = \frac{1}{4}n + 3$

29. $\frac{1}{2}p^2 - \frac{1}{2}p = -3$

Write a quadratic equation that has the given solutions.

30. 5 and -12

31. -10 and $\frac{9}{11}$

32. -8 and $\frac{11}{12}$

33. $-\frac{2}{3}$ and $-\frac{4}{9}$

34. FRAMING ARTWORK Elena just completed a needlepoint that is 24 in. by 12 in. Before framing it, she will add a mat that will increase the dimensions of the piece by an equal amount on all four sides. The mat will add 160 in.² to the area of the needlepoint. Determine the dimensions of the needlepoint plus the mat.

NUMBER THEORY For each exercise, use the sum and the product of the numbers to write a quadratic equation. Then use the quadratic formula to solve for the numbers.

35. The sum of the numbers is -3 and their product is -40.

36. The sum of the numbers is 11 and their product is -152.

37. The sum of the numbers is 35 and their product is -750.

You can use the quadratic formula to solve for x in terms of another variable.

In $x^2 + 2qx - 8q^2 = 0$, $a = 1$, $b = 2q$, and $c = -8q^2$.

$$x = \frac{-b \pm \sqrt{b^2 - 4ac}}{2a}$$

Write the quadratic formula.

$$x = \frac{-2q \pm \sqrt{(2q)^2 - 4(1)(-8q^2)}}{2(1)}$$

Substitute values of a, b and c.

$$x = \frac{-2q \pm \sqrt{36q^2}}{2}$$

Simplify under the radical symbol.

$$x = \frac{-2q \pm 6q}{2}$$

Evaluate $\sqrt{36q^2}$.

$$x = \frac{-2q + 6q}{2} \text{ and } x = \frac{-2q - 6q}{2}$$

The solutions are $x = 2q$ and $x = -4q$

Solve each quadratic equation for x in terms of q.

38. $x^2 + qx - 30q^2 = 0$ **39.** $6x^2 + qx - 2q^2 = 0$ **40.** $5x^2 + 14qx - 3q^2 = 0$

Write a quadratic equation that has the given solutions.

41. $6f$ and $2f$ **42.** $5g$ and $-8g$ **43.** $-7h$ and $-3h$

THINK CRITICALLY

44. Which of these methods can be used to solve any quadratic equation with real roots?

     a. quadratic formula   **b.** factoring   **c.** completing the square   **d.** graphing

45. For what values of k are there no real solutions to the quadratic equation $-x^2 - 8x + k = 0$?

MIXED REVIEW

Solve each system of equations. Check your answer.

46. $\begin{cases} 4b + 6c = 16 \\ -2a - b = 4 \\ a + b + c = 1 \end{cases}$ **47.** $\begin{cases} 2a + b - c = -4 \\ a - 2b + c = 13 \\ a + b + c = 4 \end{cases}$

Use factoring to solve each quadratic equation. Check your answer.

48. $x^2 + x - 20 = 0$ **49.** $2x^2 - 7x - 30 = 0$

50. STANDARDIZED TESTS The nature of the solutions of $5x^2 + x - 9 = 0$ is

     A. one real, rational solution   **B.** two real, irrational solutions

     C. two real, rational solutions   **D.** no real solutions

Explore

1. Let $y = x^2$ in the equation $x^4 - 7x^2 + 10 = 0$. Rewrite the equation in terms of y.

2. What type of equation is your answer to Question 1?

3. Let $z = \sqrt{x}$ in the equation $x - 9\sqrt{x} - 18 = 0$. Rewrite the equation in terms of z.

4. What type of equation is your answer to Question 3?

5. Factor $x^3 + 11x^2 - 6x = 0$. Is one of the factors quadratic?

┌─ SPOTLIGHT
ON LEARNING

WHAT? In this lesson you will learn
- to write equations that are not quadratic in quadratic form.
- to solve equations that are not quadratic using quadratic techniques.

WHY? Solving equations in quadratic form can help you solve problems about astronomy and space planning.

Build Understanding

Some equations that are not quadratic can be written in **quadratic form**. For example, $x^4 - 17x^2 + 72 = 0$ is $(x^2)^2 - 17(x^2) + 72 = 0$. If you can express an equation in quadratic form, you can solve it using the methods you use to solve quadratic equations.

EXAMPLE 1

Solve: $x^4 - 29x^2 + 100 = 0$

Solution

$$x^4 - 29x^2 + 100 = 0 \qquad \text{Write as a quadratic in } x^2.$$
$$(x^2)^2 - 29(x^2) + 100 = 0$$
$$(x^2 - 4)(x^2 - 25) = 0 \qquad \text{Factor.}$$
$$x^2 = 4 \quad \text{or} \quad x^2 = 25 \qquad \text{Zero product property.}$$
$$x = \pm 2 \quad \text{or} \quad x = \pm 5 \qquad \text{Solve for } x.$$

THINK BACK

Recall that a quadratic equation can be written in the form $ax^2 + bx + c = 0$.

Check

$$x^4 - 29x^2 + 100 = 0$$
$$(\pm 2)^4 - 29(\pm 2)^2 + 100 \overset{?}{=} 0$$
$$16 - 116 + 100 \overset{?}{=} 0$$
$$0 = 0 \checkmark$$
$$x^4 - 29x^2 + 100 = 0$$
$$(\pm 5)^4 - 29(\pm 5)^2 + 100 \overset{?}{=} 0$$
$$625 - 725 + 100 \overset{?}{=} 0$$
$$0 = 0 \checkmark$$

The solutions are -2, 2, -5, and 5. ◄

When you solve an equation in quadratic form, you may obtain a solution that does not satisfy the original equation.

EXAMPLE 2

Solve: $x - 2\sqrt{x} - 8 = 0$

Solution

$$x - 2\sqrt{x} - 8 = 0$$
$$(\sqrt{x})^2 - 2(\sqrt{x}) - 8 = 0 \qquad \text{Write as a quadratic in } \sqrt{x}.$$
$$(\sqrt{x} - 4)(\sqrt{x} + 2) = 0 \qquad \text{Factor.}$$
$$\sqrt{x} = 4 \text{ or } \sqrt{x} = -2 \qquad \text{Zero product property.}$$
$$x = 16 \qquad \text{Solve for } x.$$

Since $\sqrt{x}$ cannot be a negative number, eliminate the solution $\sqrt{x} = -2$.

Check

$$x - 2\sqrt{x} - 8 = 0$$
$$16 - 2\sqrt{16} - 8 \overset{?}{=} 0$$
$$16 - 2(4) - 8 \overset{?}{=} 0$$
$$0 = 0 \checkmark$$

◀

An equation can be quadratic in a binomial. You can solve such an equation by substituting a variable for the binomial.

EXAMPLE 3

Solve: $(x - 3)^2 - 4(x - 3) - 21 = 0$

Solution

$$(x - 3)^2 - 4(x - 3) - 21 = 0 \qquad \text{The equation is quadratic in } (x - 3).$$
$$p^2 - 4p - 21 = 0 \qquad \text{Substitute } p \text{ for } (x - 3).$$
$$(p - 7)(p + 3) = 0 \qquad \text{Factor.}$$
$$p = 7 \quad \text{or} \quad p = -3 \qquad \text{Zero product property.}$$
$$x - 3 = 7 \quad \text{or } x - 3 = -3 \qquad \text{Substitute } (x - 3) \text{ for } p.$$
$$x = 10 \quad \text{or} \qquad x = 0$$

The solutions are 0 and 10. Check both solutions. ◀

Recall other methods for solving quadratic equations. Include completing the square, the quadratic formula, and graphing.

EXAMPLE 4

Solve graphically: $x^3 - 2x^2 - 29x = -30$

Solution

Enter $y_1 = x^3 - 2x^2 - 29x$ and $y_2 = -30$ into your graphing utility. The solutions to this equation are the x-values of the points of intersection. Use the ZOOM and TRACE features of your graphing utility to find the solutions of $x = -5$, 1, and 6. ◀

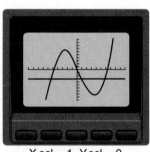

X scl = 1 Y scl = 0

You can solve certain cubic equations that contain quadratic factors using quadratic techniques.

EXAMPLE 5

OFFICE SPACE Wolk and Company provides cubicles for their data entry employees. The length of each cubicle is 3 ft more than the width, the height is 1 ft less than the width, and the volume is 96 times the width. Determine the dimensions of the cubicle.

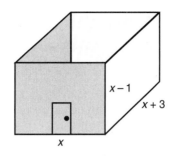

Solution

Let x = the width of the cubicle.

$$96x = (x + 3)(x)(x - 1) \qquad V = lwh$$
$$96x = x^3 + 2x^2 - 3x \qquad \text{Multiply.}$$
$$0 = x^3 + 2x^2 - 99x \qquad \text{Factor out common } x.$$
$$0 = x(x^2 + 2x - 99)$$
$$0 = x(x + 11)(x - 9) \qquad \text{Factor.}$$
$$0 = x \text{ or } 0 = x + 11 \text{ or } 0 = x - 9$$
$$0 = x \text{ or } -11 = x \quad \text{ or } 9 = x$$

Since the width of the cubicle must be positive, you can eliminate 0 and -11 as possible solutions. So, the width is 9 ft, the length is 9 ft + 3 ft = 12 ft, and the height is 9 ft − 1 ft = 8 ft. ◄

TRY THESE

Determine whether each equation can be expressed in quadratic form. If so, write it in quadratic form. If not, write *no*. Do not solve.

1. $x^4 - x^3 + 12 = 0$ **2.** $x^4 + x^2 - x = 0$

3. $x^4 + 4x^2 - 5 = 0$ **4.** $x^4 - 8x^2 + 12 = 0$

5. $x - 7\sqrt{x} + 10 = 0$ **6.** $x - 6\sqrt{x} - 72 = 0$

7. **WRITING MATHEMATICS** Miguel solved the equation $(m + 3)^2 - 10(m + 3) + 24 = 0$ and obtained the answers $m = 4$ and $m = 6$. What mistake did he make? Explain.

Factor each equation to determine whether it contains a quadratic factor. If so, write the quadratic factor. If not, write *no*. Do not solve.

8. $x^3 - 7x = 0$ **9.** $x^3 + 5x = 0$

10. $x^5 - 4x^4 + 9x^3 = 0$ **11.** $x^5 + 8x^4 - 9x^3 = 0$

12. $x^3 - 2x^2 - 5 = 0$ **13.** $x^3 + 3x^2 - 10 = 0$

14. **GEOMETRY** The longer leg of a right triangle is three times the square root of the shorter leg. Determine the length of the shorter leg if the length of the hypotenuse is 6 cm.

Solve each equation. Check your answer.

15. $x^4 - 14x^2 + 45 = 0$ **16.** $x^4 - 15x^2 + 36 = 0$ **17.** $x - 4\sqrt{x} - 5 = 0$

18. $(x - 5)^2 - 5(x - 5) + 6 = 0$ **19.** $(x - 4)^2 - 10(x - 4) + 9 = 0$

20. NUMBER THEORY The fourth power of a certain number is 36 less than 12 times the square of the number. Determine the number.

PRACTICE

Determine whether each equation can be expressed in quadratic form. If so, write it in quadratic form. If not, write *no*. Do not solve.

1. $x^4 - 5x^2 + 15 = 0$ **2.** $x^4 + 6x^2 - 9 = 0$ **3.** $x^4 + 7x^3 - 5 = 0$

4. $x^4 - 2x + 2 = 0$ **5.** $x - 3\sqrt{x} + 1 = 0$ **6.** $x - 9\sqrt{x} + 3 = 0$

7. WRITING MATHEMATICS Write three equations that are not quadratic but can be written in quadratic form.

Factor each equation to determine whether it contains a quadratic factor. If so, write the quadratic factor. If not, write *no*. Do not solve.

8. $3x^3 + 6x = 0$ **9.** $4x^3 + 2x = 0$ **10.** $x^5 - 12x^4 + 2x^3 = 0$

11. $x^5 + 6x^4 - 3x^3 = 0$ **12.** $2x^5 - 2x - 5 = 0$ **13.** $5x^3 + 5x^2 - 5 = 0$

14. GEOMETRY The length of a rectangle is twice the square root of its width. If each diagonal measures 10 in. determine the width of the rectangle.

Write each equation in quadratic form and use any method to solve. Check your answer.

15. $x^4 - 3x^2 - 40 = 0$ **16.** $x^4 - 14x^2 - 32 = 0$

17. $9x^4 - 18x^2 + 8 = 0$ **18.** $12x^4 + 5x^2 - 2 = 0$

19. $x - 3\sqrt{x} - 4 = 0$ **20.** $x - 8\sqrt{x} + 15 = 0$

21. $x - 12\sqrt{x} + 32 = 0$ **22.** $x - 14\sqrt{x} + 24 = 0$

23. $(x - 6)^2 - 17(x - 6) + 70 = 0$ **24.** $(x - 1)^2 - 16(x - 1) + 60 = 0$

25. $(x + 2)^2 - 4(x + 2) - 32 = 0$ **26.** $(x + 1)^2 - (x + 1) - 20 = 0$

27. BUSINESS Box-It-Up sells storage boxes. The length of one popular size box is 12 in. longer than the width, the height is 6 in. longer than the width, and the volume is 2016 times the width. Determine the dimensions of the box.

EXTEND

Solve and check each equation.

28. $\left(\dfrac{x + 1}{x - 1}\right)^2 + \left(\dfrac{x + 1}{x - 1}\right) - 6 = 0$ **29.** $2\left(\dfrac{x + 3}{x - 3}\right)^2 - 9\left(\dfrac{x + 3}{x - 3}\right) + 10 = 0$

30. **PHYSICS** One method for determining the distance d in feet from ground level to the water level in a well is by dropping a stone into the well and measuring the time t in seconds until the splash is heard. The formula that relates t and d is

$$t = \frac{\sqrt{d}}{4} + \frac{d}{1100}$$

a. Determine the distance if the time between the drop and when the sound in heard is $5\frac{4}{11}$ s.

b. Use a calculator to approximate to the nearest foot the distance if the time between the drop and when the sound is heard is 10 s.

31. **ASTRONOMY** Astronomers have found that the diameter and depth of craters on the moon are related by the equation

$$1000D = 108d^2 + 803d + 620$$

where D is the diameter in meters and d is the depth in meters. If the diameter of a crater is 115 m, what is its depth to the nearest meter? (*Hint:* Solve by graphing.)

THINK CRITICALLY

Solve each equation. Check your answer.

32. $|a - 2|^2 - 9|a - 2| = -18$

33. $|b + 1|^2 - 10|b + 1| = -16$

34. $2 - \frac{11}{x} - \frac{6}{x^2} = 0$

35. $3 - \frac{7}{x} - \frac{20}{x^2} = 0$

36. Why does the equation $x^2 + 3x - 4 = 0$ have two solutions when the equation $x + 3\sqrt{x} - 4 = 0$ has only one solution?

MIXED REVIEW

Solve each equation. Check your answer.

37. $|x - 5| = 10$

38. $|x + 8| = 10$

39. **STANDARDIZED TESTS** The probability of obtaining three tails when you flip three coins is

A. $\frac{1}{3}$ B. $\frac{1}{6}$ C. $\frac{1}{8}$ D. $\frac{1}{9}$

Factor.

40. $x^3 - 125$

41. $x^3 - 1000$

Solve each equation. Check your answer.

42. $x^4 - 15x^2 + 54 = 0$

43. $x - 3\sqrt{x} - 40 = 0$

Problem Solving File

Uniform and Nonuniform Motion

Suppose the motion of a car along a straight road is described by a function $y = f(t)$ where t represents time, measured in seconds from some designated starting position, and y represents the car's position, measured in feet from the starting position. Then the *velocity* or speed of the car is measured in feet per second. You could measure the velocity at any given moment with a speedometer on the car or a radar device fixed on the road.

Problem

Given a function that describes the position of the car, how can you determine the velocity of the car at a specific moment

    a. if the car's motion is uniform?

    b. if the car's motion is nonuniform?

Explore the Problem

Suppose the position of the car is measured at time t_0 and again at a later time t.

1. What are the car's positions at t_0 and t?

2. Represent the elapsed time $t - t_0$ as Δt (Δ is the capital Greek letter "delta" which stands for difference). So, $\Delta t = t - t_0$ and $t = t_0 + \Delta t$. Represent the difference in the car's positions, or distance traveled, in terms of t_0 and Δt. Call this difference Δy.

3. What does the following ratio represent?

$$\frac{\Delta y}{\Delta t} = \frac{f(t_0 + \Delta t) - f(t_0)}{t - t_0}$$

4. If the ratio in Question 3 is the same for all choices of t_0 and Δt, then the motion is called *uniform* and $\frac{\Delta y}{\Delta t}$ is the velocity v of the car. Show that uniform motion is described by a linear function.

5. In the equation you wrote in Question 4, let $m = v$ and $b = y_0 - vt_0$. Write the new equation and describe its form.

6. **WRITING MATHEMATICS** Show that if $y = mt + b$, then $\frac{\Delta y}{\Delta t} = m$ for any t_0 and Δt.

7. The position of a car moving on a straight road is described by $y = 4t - 3$. What is its velocity at $t = 5$? Explain.

PROBLEM SOLVING TIP

The notation Δt, read "delta t," is not the product of Δ and t, but a single expression denoting the difference between two values of t.

Investigate Further

PROBLEM
SOLVING PLAN

• Understand
• Plan
• Solve
• Examine

Now suppose the motion of the car is *not* uniform. Then the ratio of the distance traveled to elapsed time depends on which t_0 and t are chosen.

Therefore the equation

$$\frac{\Delta y}{\Delta t} = \frac{f(t_0 + \Delta t) - f(t_0)}{t - t_0}$$

is the *average velocity* during the time interval Δt. In the graph at the right, the average velocity is the slope of the line through the points

$$(t_0, f(t_0)) \quad \text{and} \quad (t, f(t))$$

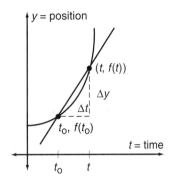

If Δt is very small, then you would expect the average velocity $\frac{\Delta y}{\Delta t}$ to be close to the instantaneous velocity v. The error $\frac{\Delta y}{\Delta t} - v$ should approach 0 as Δt approaches 0.

8. Why can't Δt be allowed to equal 0?

9. The position of a car moving along a straight road is described by $y = t^2 + 1$. To determine its instantaneous velocity at $t = 3$, begin by calculating the average velocity for a time interval Δt starting at time $t_0 = 3$. Complete each step.

$$\frac{\Delta y}{\Delta t} = \frac{f(3 + \Delta t) - f(3)}{\Delta t}$$

$$= \frac{[(3 + \Delta t)^2 + 1] - \blacksquare}{\Delta t}$$

$$= \frac{\blacksquare + \blacksquare + \blacksquare + \blacksquare - \blacksquare}{\Delta t}$$

$$= \frac{\blacksquare + \blacksquare}{\Delta t}$$

$$= \blacksquare + \blacksquare$$

10. If Δt gets arbitrarily small, does the last expression in Question 9 approach some number? Explain.

11. The number you determined in Question 10 is the instantaneous velocity at $t = 3$. What is the velocity?

Apply the Strategy

In the following exercises, y represents the position in feet from a fixed reference point of a car traveling along a straight road and t represents the time in seconds from a fixed starting time. Determine whether the motion is uniform or nonuniform and determine the velocity at the specified time t.

12. $y = 4t - 2; t = 3$

13. $y = 7t + 8; t = 5$

14. $y = 2t^2 + 3; t = 2$

15. $y = 3t^2 - 1; t = 4$

16. $y = t^2 + 2t - 6; t = 1$

17. WRITING MATHEMATICS Can you imagine a situation where you would not want to use the method of this lesson to determine an instantaneous velocity? Explain.

18. Suppose the motion equation for a particle moving on a line is

$$y = 4t^2$$

The instantaneous velocities v at different times t are as follows:

$t = 3 \qquad y = 24 \text{ ft/s}$

$t = 5 \qquad y = 40 \text{ ft/s}$

$t = 8 \qquad y = 64 \text{ ft/s}$

Look for a pattern. Can you write an expression that gives the velocity at any time t? Use your expression to predict the velocity at $t = 11$ sec.

Review Problem Solving Strategies

TILE TRICKS

1. The tray below consists of 15 yellow tiles surrounded by 20 green tiles.

a. Make a rectangular design consisting of 8 yellow tiles and 22 green tiles.

b. Make a rectangular design consisting of 8 yellow tiles and 16 green tiles.

c. Find two other rectangular designs using 16 green tiles. How many yellow tiles are inside each?

d. Can you discover a relationship between the dimensions of the yellow rectangle and the number of green tiles? How do the dimensions relate to the number of yellow tiles?

e. Can you make a rectangular design where the number of outer green tiles equals the number of inner yellow tiles? Show your design or explain why no design is possible.

THE ROAD HOME

2. Each evening, Mr. Lee meets his wife's train at 5:30 P.M. and drives her home. One Friday, Mrs. Lee caught an earlier train that pulled into her station at 4:30 P.M. To get some exercise, she decided to walk along the road her husband always takes. After some time, she spotted his car, got in and rode the rest of the way home. As they entered the house, Mrs. Lee noticed that it was 30 minutes earlier than they usually arrive. Assume Mr. Lee would have arrived at the station punctually, that the ride takes the same amount of time each way, and that the rate of travel was the same as every other day.

a. How much less time than usual did Mr. Lee spend driving? How much less time did his trip take in each direction?

b. How long did Mrs. Lee walk?

c. If Mrs. Lee had walked another 5 minutes before meeting her husband, how much earlier than usual would they have arrived home?

AN AGE-OLD QUESTION

3. For the first quarter of her life, Roy's aunt lived in Mexico. Then she spent one sixth of her life in Texas and one half of her life in Illinois. For the final seven years of her life, her home was in Maryland. How old was Roy's aunt when she died?

CHAPTER REVIEW

· · · · · ·

VOCABULARY

Choose the word from the list that completes each statement.

1. The graph of $y = x^2 + 1$ is a ___?___ of the graph of $y = x^2$.

2. The expression $b^2 - 4ac$ is called the ___?___.

3. The graph of a quadratic function is a ___?___.

4. The ___?___ is the vertical line $x = -\dfrac{b}{2a}$.

5. The maximum or minimum point is at the ___?___ of the parabola.

a. parabola

b. vertex

c. axis of symmetry

d. discriminant

e. translation

Lesson 6.1   SOLVE QUADRATIC EQUATIONS BY COMPLETING THE SQUARE   pages 275–281

- To solve quadratic equations you can write the equation so that one side is a perfect square trinomial.

- To make the expression $x^2 + bx$ into a perfect square trinomial, add $\left(\dfrac{b}{2}\right)^2$.

Find the value of b that makes the left side of each equation a perfect square trinomial.

6. $x^2 + bx + 256 = 0$ **7.** $x^2 - bx + 400 = 100$ **8.** $x^2 - bx - \dfrac{25}{36} = 0$

Solve each equation by completing the square.

9. $x^2 + 9x + 18 = 0$ **10.** $x^2 - 10x = 2$ **11.** $4x^2 - 6x = 5$

Lesson 6.2   ALGEBRA WORKSHOP: EXPLORE QUADRATIC FUNCTIONS   pages 282–286

- The graph of $f(x) = x^2$ is translated vertically when a constant is added or subtracted to the squared term.

- The graph of $f(x) = x^2$ is translated horizontally when a constant is added to x before squaring the quantity.

- The width of $f(x) = x^2$ changes when the squared term has a coefficient. When the coefficient is less than 1, the graph is wider than the graph of $f(x) = x^2$. When the coefficient is greater than 1, the graph is narrower than the graph of $f(x) = x^2$.

Explain how the graph of each function is translated from the graph of $f(x) = x^2$.

12. $g(x) = (x - 3)^2 - 5$ **13.** $h(x) = (x + 6)^2$ **14.** $j(x) = (x + 4)^2 - 8$

15. List the functions from narrowest to widest.

    a. $f(x) = 3.3x^2$ **b.** $g(x) = \dfrac{6}{5}x^2$ **c.** $h(x) = 3.6x^2$ **d.** $j(x) = \dfrac{5}{6}x^2$

- A **quadratic function** is a function that can be written in the form $f(x) = ax^2 + bx + c$, where a, b, and c are real numbers and $a \neq 0$. The x-value of the vertex is $\frac{-b}{2a}$.

- The graph opens upward when $a > 0$ and downward when $a < 0$.

Find the vertex and axis of symmetry. Graph each function.

16. $f(x) = x^2 + 2x - 3$ **17.** $g(x) = x^2 - 4x + 4$ **18.** $h(x) = -x^2 - 4x + 5$

19. A rocket is launched from Earth with an upward speed of 144 ft/sec. Use a graphing utility to determine the maximum height attained by the rocket to the nearest foot.

- For a quadratic equation of the form $ax^2 + bx + c = 0$ where a, b, and c are real numbers and $a \neq 0$, $x = \dfrac{-b \pm \sqrt{b^2 - 4ac}}{2a}$

- If the solutions of a quadratic equation are s_1 and s_2, then its equation is $x^2 - (s_1 + s_2)x + s_1 s_2 = 0$.

Use the quadratic formula to solve each equation.

20. $2x^2 - 4x - 5 = 0$ **21.** $5y^2 + y - 7 = 0$ **22.** $2z^2 - 24z + 10 = 0$

Write a quadratic equation that has the given solutions.

23. -8 and 6 **24.** 3 and $\dfrac{1}{4}$ **25.** 4 and $-\dfrac{1}{2}$

- If you can express an equation in quadratic form, you can solve it with the methods you use to solve quadratic equations.

Solve each equation.

26. $x^4 - 16x^2 + 48 = 0$ **27.** $x + 11\sqrt{x} - 26 = 0$ **28.** $(x - 2)^2 + 7(x - 2) + 12 = 0$

- Uniform motion is described by a linear distance function where the slope represents the velocity at any time t. When the motion is not uniform, you can determine the instantaneous velocity by examining what happens to the average velocity as the time interval gets very small.

Assume y represents the distance traveled by a car in feet and t represents the time in seconds. Tell if the motion is uniform or nonuniform and determine the velocity at t.

29. $y = 6t - 2; t = 4$ **30.** $y = 2t^2 + 1; t = 4$

CHAPTER ASSESSMENT

CHAPTER TEST

Determine the value of b that would make the left side of each equation a perfect square trinomial.

1. $x^2 + bx + 169 = 6$

2. $x^2 - bx - \dfrac{121}{196} = 9$

3. **WRITING MATHEMATICS** Write a paragraph that explains how the equation $x^2 - 8x - 50 = 0$ can be solved by completing the square and how it can be solved using the quadratic formula.

Solve each equation by completing the square.

4. $x^2 + 12x + 11 = 0$ 5. $x^2 - 4x - 1 = 7$

6. $2x^2 + 8x - 16 = 0$ 7. $\dfrac{x^2}{5} - 4x + 3 = 0$

8. The sum of a number and its reciprocal is $\dfrac{130}{63}$. What is the number?

Write a function of the form $g(x) = (x - h)^2 + k$ whose graph represents:

9. A translation of 9 units left and 3 units up from the graph of $f(x) = x^2$.

10. A translation of 1 unit left and 10 units down from the graph of $f(x) = x^2$.

11. List the functions from narrowest to widest.

    a. $f(x) = 1.4x^2$ **b.** $f(x) = \dfrac{1}{2}x$

    c. $f(x) = 0.55x^2$ **d.** $f(x) = 0.29x^2$

Determine the vertex and axis of symmetry of each parabola. Then state whether the parabola opens upward or downward.

12. $f(x) = x^2 - 10x + 6$

13. $f(x) = -\dfrac{1}{4}x^2 + 5x - 7$

Use the quadratic formula to solve each equation.

14. $3x^2 - 6x + 2 = 0$

15. $25x^2 - 60x + 36 = 0$

16. $7x^2 - 3x - 1 = 0$

Use the discriminant to determine the number of and the nature of the solutions of each quadratic equation.

17. $x^2 + 5x + 6 = 0$

18. $x^2 - 12x + 36 = 0$

19. $x^2 + 5x + 7 = 0$

20. $x^2 + 4x + 2 = 0$

Determine the sum and the product of the solutions of each equation.

21. $3x^2 + 4x - 9 = 0$

22. $\dfrac{1}{4}x^2 - \dfrac{1}{2}x - 8 = 0$

Write a quadratic equation that has the given solutions.

23. 6 and -3 24. 11 and -12

25. **STANDARDIZED TESTS** Which expression cannot be expressed in quadratic form?

    A. $x^4 - x^2 + 8$ **B.** $x - 8\sqrt{x} + 6$
    C. $x^4 - 3x^2 - 8$ **D.** $x^4 - x^3 - 2$

Factor each equation to determine whether it contains a quadratic factor. If so, write the quadratic factor. If not, write "no."

26. $x^3 + 5x + 3 = 0$ 27. $x^3 - 6x^2 - 4x = 0$

28. Solve: $x^4 + x^2 - 30 = 0$

29. Solve: $x + 3\sqrt{x} - 108 = 0$

PERFORMANCE ASSESSMENT

USE ALGEBLOCKS Show how to model the equation $x^2 - 6x = 7$ with Algeblocks. Explain to a partner how to solve the equation by completing the square. Then have your partner show the steps for solving $x^2 + 4x = 5$.

DESIGN A DEMONSTRATION Use an overhead projector. Draw a coordinate grid on a transparency. Draw a second grid on a piece of paper. Graph the parabola with the equation $f(x) = x^2$ on the second grid, cut it out, and place it on the coordinate axes of the first grid.

a. Show how the parabola can be translated and reflected by moving it around the grid. State the equation of the parabola after each transformation.

b. Demonstrate how to determine whether a quadratic equation has zero, one, or two real solutions from examining its graph.

PIZZA PIES Obtain the prices of at least four different sizes of plain cheese pizza. Write an equation that relates the diameter and the area of a pizza. What type of equation is it? Plot your data on a graph with radius on the x-axis and price on the y-axis. Then plot your data on a different graph with area on the x-axis and price on the y-axis. Connect all points with a smooth line or curve on both graphs. Comment on the shapes of the graphs. Why do you think that many pizza stores should price pizza by the area rather than by the diameter?

EQUATION ART Parabolic curves appear in art and architecture. Find examples of parabolic designs in books or magazines. Lay a coordinate grid over each design. Try to determine the equations of the parabola(s) in each design. Then state the vertex and axis of symmetry for each parabola. Tell whether the parabola opens upward or downward.

PROJECT ASSESSMENT

PROJECT *Connection*   Summarize your findings and apply what you have learned.

1. Review the data logs from your group and discuss how you can use graphs, tables, and statistical measures to effectively represent your data.

2. Plan your presentation. It should include a commentary about students' experiences using the wheelchair. What was the most difficult part about being in the chair? Did anything surprise you as being easier to do than you expected? Which areas of the school and community are "wheelchair friendly"? Which are not?

3. Choose one of the following applications to work on.
   a. Design a wheelchair of the future.
   b. Design a room in the home (such as a kitchen) or a public space (such as a library) to accommodate wheelchair users.

CUMULATIVE REVIEW

1. **WRITING MATHEMATICS** Explain how you would convert the expression $2x^2 - 60x$ to a perfect square trinomial. Write the result in factored form.

State the direction(s) and number of units in which the graph of each function is translated from the graph of $f(x) = x^2$.

2. $g(x) = (x + 3)^2$ 3. $h(x) = x^2 + 3$

Determine the vertex and axis of symmetry of each parabola, and state whether the parabola opens upward or downward.

4. $y = -x^2 + 4x - 6$ 5. $y = 3 - 8x + 2x^2$

Solve and check.

6. $x^4 - 13x^2 + 36 = 0$

7. $(x - 3)^2 + 2(x - 3) - 8 = 0$

8. The base of a triangle is 6 inches greater than its height. The area of the triangle is 80 in.^{2}. Determine the base and height of the triangle.

Factor.

9. $3x^4 - 48$ 10. $y^3 + 125$

Multiply.

11. $(2x - 3)(2x^2 + 3x - 2)$

12. $(2m + 1)(m^2 - 3m + 5)$

13. **STANDARDIZED TESTS** The result of subtracting $6 + 2x$ from the sum of $2 - x$ and $3x^2 + 3x + 4$ is

   A. $-3x^2$ **B.** $3x^2 - 2x + 4$

   C. $3x^2$ **D.** $3x^2 + 2x + 6$

14. The owner of a shoe store is ordering stock. Which measure of central tendency is useful in determining the stock needed? Explain.

Write matrices to model the problem and solve.

15. For her spring black/white collection, the designer has used 5 yd of fabric, 4 yd of braiding, and 3 packets of beads for a white design. A black design used 6 yd of fabric, 5 yd of braiding, and 2 packets of beads. The fabric used cost $40/yd, the braiding $20/yd, and the beads $15 per packet. Determine the total cost for each design.

Write a system of equations to model the problem and use matrices to solve the system.

16. Ms. Paul invested $25,000, part in CDs at 4% and the rest in bonds at 6%. If the total interest on these investments last year was $1300, how much did she place in each type of investment?

17. **STANDARDIZED TESTS** The inverse of the function $\{(1, 2), (3, 4), (5, 6), (7, 8)\}$ is

   A. $\{(-1, 2), (-3, 4), (-5, 6), (-7, 8)\}$

   B. $\{(1, -2), (3, -4), (5, -6), (7, -8)\}$

   C. $\{(-1, -2), (-3, -4), (-5, -6), (-7, -8)\}$

   D. $\{(2, 1), (4, 3), (6, 5), (8, 7)\}$

18. The market research department of a company planning to introduce a new model ceiling fan gave the management the following demand/price forecast:

Price	Demand
$80	7200
$120	4800
$150	3000
$200	0

   Write the equation that models the relationship of demand d as a function of price p.

19. Solve the inequality $-3 \le \dfrac{2x + 3}{4} < 7$.

• • • STANDARDIZED TEST • • •

STUDENT PRODUCED ANSWERS Solve each question and on the answer grid write your answer at the top and fill in the ovals.

Notes: Mixed numbers such as $1\frac{1}{2}$ must be gridded as 1.5 or 3/2. Grid only one answer per question. If your answer is a decimal, enter the most accurate value the grid will accommodate.

1. If the expression $x^2 - 9x + k$ is a perfect square trinomial, what is the value of k?

2. By how many units to the left is the graph of $f(x) = x^2$ translated to produce the graph of $g(x) = 2(x + 3)^2$?

Use the following information to answer Questions 3 and 4.

The profit function for a company that manufactures sunglasses is $P(x) = 350x - 15{,}000$ where x is the number of units produced weekly.

3. Determine the level of production that yields the maximum profit.

4. What is the greatest number of items the company can produce and just break even?

5. If the equation $ax^2 + bx + c = 0$ has integral coefficients and its solutions are $\left\{\frac{1}{2}, \frac{1}{2}\right\}$, determine the value of a.

6. Find the product of the solutions of the equation $(x - 2)^2 - 4(x - 2) - 5 = 0$.

7. The expression $(4x^2)^3$ may be written in the form $64x^n$. Determine the value of n.

8. The expression $4y^2 - 12y + 9$ may be written in the form $(ay + b)^2$. Determine the value of b.

9. Determine the constant term of a polynomial equation whose solutions are $-2, 3, -1$.

10. If $P(-1, 3)$ is reflected about the line $y = -x$, what is the x-coordinate of P?

11. The health club's swimming pool can be filled by water pipes A, B, and C according to the three methods listed in the table.

	A	B	C
Method 1	closed	open for 4 hours	open for 6 hours
Method 2	open for 5 hours	open for 2 hours	open for 3 hours
Method 3	open for 5 hours	closed	open for 6 hours

How many hours would it take pipe A to fill the pool by itself?

12. How many lines of symmetry does a regular pentagon have?

13. Solve for x:
$$\frac{3(x - 1)}{4} = -12x.$$

14. The Hurleys run a small mail-order business. Of their total monthly expenses, 40% goes for rent, one-third of what is left is for cleaning and repairs, 25% of what is left after that is for advertising, and 80% of what is left after that is for utilities. The Hurleys spend the remaining $96 for office supplies. How much are their monthly expenses?

To answer Questions 15 and 16, use the box plot below, which is a summary of data obtained by asking 30 children to rate the taste of a new flavor of soft drink on a scale of 1 (terrible) through 10 (terrific).

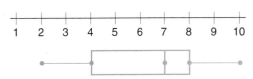

15. What is the interquartile range?

16. How many of the scores were between 4 and 8?

7 Inequalities and Linear Programming

Take a Look AHEAD

Make notes about things that look new.

• Find and copy an example of a quadratic inequality. Think about real-life situations in which you might need to solve this type of problem.

• Linear programming is the main focus of Lessons 7.5, 7.6 and 7.7. List all the new terms in these lessons. Can you guess the meaning of any of them?

Make notes about things that look familiar.

• How do you think solving and graphing a linear inequality in two variables may be similar to your work with inequalities in one variable? What are some possible differences?

• Prepare a chart of all the inequality symbols. Next to each symbol, write the different word phrases the symbol can represent.

DATA Activity

Can We Count on You?

During election periods, candidates want to know how popular they are with different populations or about what issues people are most concerned. So, data is collected and used to influence the policies of the political parties and the strategies a candidate adopts. One fundamental question asked is which political party an individual identifies with. The Democratic Party and the Republican Party are the two main political parties in the U. S. *Independents* may switch between the two parties or support a third party candidate.

SKILL FOCUS

▶ Read and interpret a table.

▶ Add, subtract, multiply, and divide real numbers.

▶ Determine the experimental probability of an event.

▶ Determine the odds in favor of an event.

▶ Solve percent problems.

▶ Construct a divided bar graph.

Government

In this chapter, you will see how:

- **IRS Customer Service Representatives** use inequalities and functions to represent parts of the tax code. (Lesson 7.2, page 331)

- **Members of Congress** use inequalities to establish regulations for utility companies. (Lesson 7.5, page 349)

- **Environmental Protection Agents** use linear programming to help companies comply with federal laws. (Lesson 7.6, page 357)

Political Party Identification Percent				
	Characteristic	**Democrat**	**Republican**	**Independent**
Gender	Male	42	46	12
	Female	52	37	11
Race	White	43	46	11
	African-American	81	10	9
Education	Grade school	59	24	17
	High school	52	34	14
	College	43	50	7

Use the table to answer the following questions.

1. Of the groups shown, which has the greatest percent of individuals identifying themselves as Independent?

2. What is the probability that a randomly selected female will identify with the Republican Party?

3. What are the odds that a randomly selected African-American responded *Democrat*?

4. If 1500 randomly selected college graduates were asked their political identification, about how many more would answer Republican then Democrat?

5. **Working Together** Work in groups of three students. Have each member of the group work with one of the characteristics — gender, race, or education — and construct a horizontally divided bar graph (the whole bar represents 100%) for each line of data. Use the same scale for all graphs so that the graphs can be displayed together to provide an at-a-glance representation of the information in the table.

PROJECT

Taxing Facts

If you have ever received a paycheck, or know somebody who has, you might already know that your take-home pay is less than your actual, or gross pay, because money is taken out for *income tax*. You are often required to pay *sales tax* on purchases. Homeowners usually pay *property taxes*. Travelers often have to pay *hotel taxes*, while car owners are charged *federal excise taxes*. Nobody likes to pay taxes, but they understand why the taxes are being collected. In this project, you will become an informed taxpayer!

PROJECT GOAL

To research, analyze, and present information about how the government collects and spends tax money.

Getting Started

Work in groups of six students. Each project connection should be completed by two students.

1. Read each of the project connections and decide how your group is going to divide the work. Plan to access information using varied sources such as library books, government publications, and personal interviews. It will take time to set up an interview, or to write a letter and receive a response, so begin immediately.

2. Define key terms associated with principles of taxation such as direct and indirect taxes, progressive and regressive taxes.

3. Think of all the services provided by federal, state, and city governments. List some questions you have about these services. Where does the money come from? Who decides how it is spent? What services do you feel are worthwhile? What services directly effect you?

PROJECT *Connections*

Lesson 7.2, page 330:
Research Social Security and Medicare to find out what taxpayers pay and what the benefits are.

Lesson 7.3, page 338:
Find out where federal and state income tax dollars come from, and how they are spent.

Lesson 7.4, page 345:
Find out what items are subject to sales tax and taxed at what rate. Explore property taxes.

Chapter Assessment, page 365:
Compile your information for a presentation.

Internet Connection

www.swpco.com/
swpco/algebra2.html

320

Algebra Workshop
Graph Linear Inequalities

Think Back

● Copy the table and complete the table.

	Inequality	Solution	Graph of Solution
	$x + 5 > 8$	$x > 3$	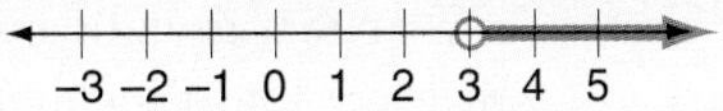
1.	$2x - 1 \leq -11$		
2.	$-3x < 15$		
3.	$3x + 2 \geq 8$		

4. For each inequality in the table, pick a point in the graph of the solution and show that it satisfies the inequality.

5. For each inequality in the table, pick a point that is *not* in the solution and show that it does *not* satisfy the inequality.

Explore/Working Together

● Work with a partner. You will each need graph paper and a straight edge. Do not let your partner see your work.

6. Write a linear equation of the form $y = mx + b$ and graph it.

7. Create a linear inequality by replacing the equal symbol in your equation with one of the symbols $<$, $>$, $\leq$ or $\geq$.

8. Find four points that are solutions of your inequality. Do not pick points on the line you graphed. Show algebraically that each point you chose satisfies the inequality.

9. Label the coordinates of each solution you determined in Question 8 at its location on your graph. Write the word "yes" next to these four points to signify that they are solutions.

10. Find four points that are *not* solutions of your inequality. Do not pick points on the line you graphed. Show algebraically that each point you chose does not satisfy the inequality.

11. Label the coordinates of each solution you determined in Question 10 at its location on your graph. Write the word "no" next to these four points to signify that they are not solutions.

SPOTLIGHT ON LEARNING

WHAT? In this lesson you will learn
- to determine the ordered pairs that are solutions of linear inequalities.
- to graph linear inequalities in two variables.

WHY? Determining the solution to linear inequalities by graphing can help you solve problems about finance and exercise.

THINK BACK

An open circle is used to denote $<$ or $>$. A closed circle is used for $\geq$ or $\leq$.

THINK BACK

When multiplying or dividing both sides of an inequality by a negative number, remember to reverse the inequality sign.

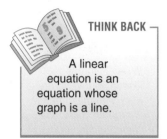

THINK BACK

A linear equation is an equation whose graph is a line.

12. Pick two points on the line you graphed. Show algebraically that these points are solutions to your original equation and may or may not be solutions to your inequality.

13. Compile a list of the ten points you identified. Next to each set of ordered pairs, write "yes," "no," "yes and on boundary line," or "no and on boundary line."

14. Exchange lists with your partner. Use this list to determine your partner's linear inequality. Check each other's work and discuss the strategies you used.

15. What relationship did you notice between the points that are solutions and the line you graphed? What did you notice about points that are not solutions and the line?

CHECK UNDERSTANDING

Write four inequalities that are related to the corresponding equation $y = 3x - 4$.

Making Connections

The graph of a linear inequality in two variables shows all the solutions of the inequality. The graph of a linear inequality divides the coordinate plane into two **half-planes**. The line itself is the boundary of each half plane.

The graph of a linear inequality includes all points in a half-plane. Shaded half-planes show where the solutions can be found. When the inequality symbol is $\leq$ or $\geq$, draw the boundary as a solid line. This denotes a **closed half-plane**, or a half-plane that includes the boundary. When the inequality symbol is $<$ or $>$, draw the boundary as a dashed line. This denotes an **open half-plane**, or a half-plane that does not include the boundary.

Graph each of the inequalities in the table on the next page. First, write the boundary line and graph it. Indicate whether the line is dashed or solid. Then, test a point in each half-plane to see which half-plane is the graph of the inequality. The first one is done for you.

Inequality	Boundary Line	Point in Half-Plane	Point in Other Half-Plane
$2x + y < 9$	$2x + y = 9$ dashed	$(0, 0);$ $2(0) + 0 \overset{?}{<} 9$ YES	$(6, 4);$ $2(6) + 4 \overset{?}{<} 9$ NO
16. $y \geq 5 + 4x$			
17. $2y < x - 1$			
18. $3x \leq y - 6$			

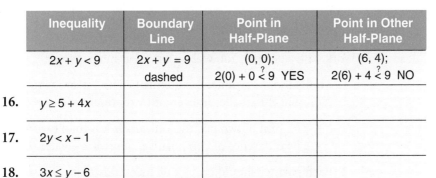

19. For which inequalities in the table did you shade the half-plane above the boundary line? For which inequalities did you shade the half-plane below the line? What generalizations can you make?

Summarize

20. WRITING MATHEMATICS Graph the linear inequality $y > 3x + 1$. Explain how you decide whether or not points on the boundary line are solutions to the inequality and which half-plane to shade. Label specific points you chose to decide which half-plane to shade.

21. MODELING How many pairs of positive integers have a sum less than 9? Write a linear inequality that models this situation. Graph your inequality and use the graph to determine the number of different pairs of positive integers that have a sum of less than 9. The pair $(2, 1)$ is the equivalent to $(1, 2)$ in this problem.

22. THINKING CRITICALLY Graph $y = |x|$. Consider $y \leq |x|$. Pick ten points that are solutions of the inequality and label them Y_1 through Y_{10}. Then pick ten points that are not solutions and label these points N_1 through N_{10}. Describe what you know about the location of the solutions to this absolute value inequality.

23. THINKING CRITICALLY All of the linear inequalities you have worked with in this lesson have an infinite number of solutions. How many solutions does the inequality $|x + y| < -5$ have? Explain your answer.

24. GOING FURTHER Graph the absolute value inequality $y > |x + 4|$.

25. GOING FURTHER There are inequality symbols other than $<, \leq, >,$ and $\geq$. Describe the meaning of the two inequality symbols used in the following two linear inequalities.

$$y \neq -2x - 1 \quad \text{and} \quad y \not< x + 2$$

7.2 Graph Systems of Linear Inequalities

Explore/Working Together

- Work with a partner. You will each need graph paper and a straightedge.

  1. Model the following two statements with linear inequalities. Let x represent the smaller integer and y represent the larger integer.

     - The sum of two positive integers is less than twelve.
     - The difference of two positive integers is greater than three.

  2. Each partner should graph one inequality from Question 1. Make sure you use the same x and y scales.

  3. Hold the two graphs you and your partner drew up to the light and move the graph papers until the axes coincide. Locate the region in which the solutions overlap. Determine four sets of ordered pairs that are located in the overlapping region and record them.

  4. Verify that the sum of the numbers is less than 12 and the difference between them is greater than 3.

  5. How many pairs of integers satisfy both statements? How can you tell?

  6. Find an ordered pair that satisfies the first statement, but not the second.

  7. Find an ordered pair that satisfies the second statement, but not the first.

  8. Find an ordered pair that does not satisfy either statement.

COMMUNICATING ABOUT ALGEBRA

Why is $(-1, 6)$ not a solution to the system of inequalities in Question 1?

Build Understanding

- The two inequalities you wrote in Explore form a **system of linear inequalities**. To solve a system of inequalities, you must determine the ordered pairs that satisfy both inequalities. Graph both inequalities in the system on the same coordinate plane. Use a dashed line when the inequality symbol is $<$ or $>$ to indicate the line is not included in the solution. If the symbol $\leq$ or $\geq$, use a solid line to include the line in the solution.

  The region in which the graphs overlap contains all the solutions of the system. There are infinitely many solutions in the overlapping region, so graphing is the most efficient way to report them. Most graphing utilities allow you to graph systems of inequalities.

EXAMPLE 1

Solve the system of linear inequalities by graphing.

$$\begin{cases} y < 2x + 5 \\ x + 2y \geq 8 \end{cases}$$

Solution

Graph $y = 2x + 5$ with a dashed line to show that points on the line $y = 2x + 5$ do not satisfy the inequality $y < 2x + 5$. Test points on the line to verify this. Test points on either side of the line, such as $(0, 0)$ and $(-4, 0)$ to see which half-plane is the solution to the inequality.

Shade the half-plane that contains $(0, 0)$, because $(0, 0)$ makes $y < 2x + 5$ a true statement.

Next, graph $x + 2y = 8$ on the same set of coordinates axes. Use a solid line since points on the line $x + 2y = 8$ satisfy the inequality $x + 2y \geq 8$. Testing points above and below this line shows that the half-plane above the line should be shaded.

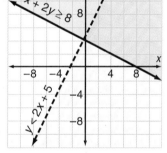

The region with the double shading is where the solutions to each inequality overlap. Choose a point in the double-shaded region such as $(2, 5)$ and check it in both of the original inequalities.

$$y < 2x + 5 \qquad\qquad x + 2y \geq 8$$
$$5 \overset{?}{<} 2(2) + 5 \qquad 2 + 2(5) \overset{?}{\geq} 8$$
$$5 < 9 \checkmark \qquad\qquad 12 \geq 8 \checkmark$$

The double-shaded region is the solution to the system of inequalities. The boundary line $x + 2y = 8$ is part of the solution, but the boundary line $y = 2x + 5$ is not. ◄

Recall that to solve an absolute value inequality, you restated it as two inequalities using *and* or *or*. You can graph an absolute value inequality as a system of inequalities.

EXAMPLE 2

Graph: $|y| \leq 5$

Solution

Rewrite $|y| \leq 5$ as a system of two inequalities.

$$\begin{cases} y \leq 5 \\ y \geq -5 \end{cases}$$

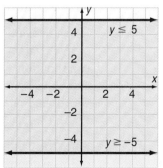

Graph the system. The double-shaded region represents the solution to the inequality. ◄

PROBLEM SOLVING TIP

When testing points in half-planes to solve a linear inequality, use the origin when possible since substituting $(0, 0)$ into an inequality can usually be done mentally.

CHECK UNDERSTANDING

What part of the graph shown in Example 1 contains the points that do not satisfy either inequality?

THINK BACK

Graphs of equations of the form $y = a$, where a is a constant, are lines parallel to the x-axis. Graphs of equations of the form $x = b$ are parallel to the y-axis.

Systems of linear inequalities can be used to solve real world problems.

EXAMPLE 3

BUDGETING Alison and Kareem are planning the entertainment for their wedding reception. The party will last at least 4 hours. They want to hire a band that charges $300 per hour and a disc jockey who charges $150 per hour. They want to keep the music expenses under $1200. The band and the disc jockey charge by the whole hour only.

a. Write a system of linear inequalities that models this situation.

b. Solve the system of inequalities by graphing.

c. What possible combinations of band and disc jockey hours can Alison and Kareem use?

d. What solutions do not apply? Explain.

Solution

a. Let x represent the number of hours the disc jockey works. Let y represent the number of hours the band works.

The number of hours Alison and Kareem need music can be modeled by the inequality

$$x + y \geq 4$$

The amount they plan to spend can be modeled by the inequality

$$150x + 300y < 1200$$

b. Graph the inequalities, on the same coordinate axes. Test points in all half-planes to determine the solution to each inequality. The solution to the system is represented by the double-shaded region.

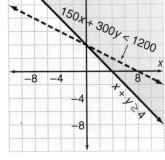

c. Since the band and disc jockey charge by the whole hour, the possible combinations of disc jockey hours and band hours are represented by the points with integer-valued coordinates that are in the solution of the system. These are $(1, 3)$, $(2, 2)$, $(3, 1)$, $(3, 2)$, $(4, 1)$, and $(5, 1)$.

d. Since the band and disc jockey are paid for whole hours only, points without integer-valued coordinates are not suitable solutions. Also, points with negative coordinates do not apply. ◄

1. Determine without graphing if the ordered pairs are solutions to the system.

$$\begin{cases} 2x + y > 10 \\ 3x \le 5y - 1 \end{cases}$$

   a. $(1, 10)$ **b.** $(0, 0)$ **c.** $(5, 3)$

   d. $(0, 2)$ **e.** $(-3, 13)$ **f.** $(-3, 14)$

Solve each system of linear inequalities by graphing.

2. $\begin{cases} y \ge 2x - 4 \\ y > -3x + 1 \end{cases}$ 3. $\begin{cases} 2y > x \\ x + y < 10 \end{cases}$ 4. $\begin{cases} y \ge 5 \\ y \le \frac{1}{2}x + 1 \end{cases}$

5. $\begin{cases} x + y \le -8 \\ 2x - 3y > 12 \end{cases}$ 6. $\begin{cases} y > 2x \\ |x| \le 6 \end{cases}$ 7. $\begin{cases} 2x + 3y < 6 \\ 3x - 4y \ge 0 \end{cases}$

8. $\begin{cases} y \ge -x - 3 \\ 2x + 5y < 10 \end{cases}$ 9. $\begin{cases} y < 2x + 3 \\ y \ge 2x + 1 \end{cases}$ 10. $\begin{cases} 2x + y \le 3 \\ y < 6 + x \end{cases}$

EMPLOYMENT Joelle needs to earn at least $100 a week for a ski vacation she plans to take next month. She earns $6 per hour working at a local golf course and $4 per hour as a mother's helper. She does not want to work more than 18 hours per week.

11. Write and graph a system of linear inequalities that models the weekly number of hours Joelle can work at each job and how much money she needs to earn.

12. Graph the system you wrote in Exercise 11.

13. If Joelle's employers pay her in half-hour increments only, name five possible combinations of hours Joelle can work at each job per week to have enough money for the vacation.

14. Write the system of linear inequalities whose solution is represented by the graph at the right.

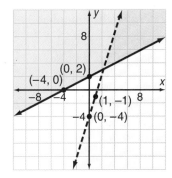

15. **WRITING MATHEMATICS** The two boundary lines for the system of linear inequalities below divide the coordinate plane in four regions.

$$\begin{cases} y < 3x + 2 \\ y \ge x - 1 \end{cases}$$

Graph the system. On your paper, in each region, write an explanation of which inequality is satisfied by the ordered pairs in that region.

PRACTICE

1. For the system of linear inequalities below, state whether the indicated region of the coordinate plane is a solution of the system, of one inequality, or of neither inequality. If the region is a solution to only one inequality, state which inequality.

$$\begin{cases} y < 3x - 5 \\ x + 2y \geq 7 \end{cases}$$

   a. A   **b.** B   **c.** C   **d.** D

Solve each system of linear inequalities by graphing.

2. $\begin{cases} y \leq x - 3 \\ y > 2x \end{cases}$

3. $\begin{cases} 2x + y > 7 \\ x - 2y < 1 \end{cases}$

4. $\begin{cases} 2x - y \geq 3 \\ \frac{1}{2}x + y \leq 2 \end{cases}$

5. $\begin{cases} 3y + 2x < 24 \\ y \geq \frac{2}{3}x \end{cases}$

6. $\begin{cases} 5x + 2y > 10 \\ 2x - 3y < 6 \end{cases}$

7. $\begin{cases} 19 - x + y < 1 \\ x + 2y \leq 5 \end{cases}$

8. $\begin{cases} |y| > 4 \\ x \leq -1 \end{cases}$

9. $\begin{cases} y < 3x - 2 \\ y \geq -2 \end{cases}$

10. **NUMBER THEORY** The sum of two positive numbers is less than 8 and the difference between the numbers is greater than 6.

    a. Write a system of linear inequalities that models this situation and state the restrictions on x and y.

    b. Find five pairs of rational numbers that are not integers that are solutions to the system. Check that your answers satisfy the original conditions of the problem.

11. **HEALTH** Many people who do aerobic exercise monitor their pulse rates during exercise. A formula is used to compute ideal pulse rates, based on your age.

    - Subtract your age from 220. Then find 72% of that difference. An ideal pulse rate should be greater than this number.

    - Subtract your age from 220 and find 87% of that difference. An ideal pulse rate should be less than this number.

    a. If x represents age and y represents pulse rate, write a system of inequalities that models the recommended pulse rates.

    b. Use a graphing utility to graph the system.

    c. Express the range of ideal pulse rates for a 40-year-old during exercise as an inequality. Round to the nearest integer.

12. **WRITING MATHEMATICS** Compare and contrast solving systems of linear equations and solving systems of linear inequalities.

13. **POLITICS** A congressional candidate wishes to use a combination of radio and television advertising. Research has shown that each 1-min television spot reaches 0.2 million voters and each 1-min radio spot reaches 0.05 million voters. The candidate wants to reach at least 2 million voters and plans to buy no more than 30 min of advertising.

   a. Write the system of inequalities that model the candidate's needs. Let x represent minutes of television and y represent minutes of radio.

   b. Graph the system.

   c. What are two time combinations the candidate can use?

   d. Could the candidate use all radio or all TV advertising to reach 2 million voters?

   e. How many voters would be reached with 20 min of TV and 10 min of radio? 30 min of TV and no radio?

EXTEND

Write a system of inequalities whose solution is represented by each graph.

14.
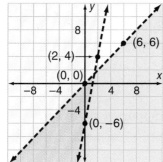

15.
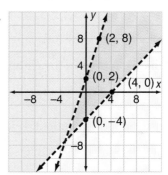

Solve each of the following systems of inequalities by graphing.

16. $\begin{cases} y \geq |x| \\ y < 2x + 5 \end{cases}$

17. $\begin{cases} |x + 2| < 3 \\ |y| > 4 \end{cases}$

Graph the solution to each of the following systems of inequalities.

18. $\begin{cases} x \geq 0 \\ y \geq 0 \\ y \leq x \end{cases}$

19. $\begin{cases} x > 5 \\ y < 3x + 2 \\ y \leq 10 \end{cases}$

20. $\begin{cases} y \geq x + 4 \\ x + 5y < 15 \\ 2y < x \end{cases}$

21. $\begin{cases} x \geq 1 \\ y \geq -5 \\ y \leq 1 \\ 2x + y \leq 4 \end{cases}$

22. Graph the following equation-inequality system and find four points that are solutions. Explain why you selected those four points.

$$\begin{cases} y < -4x - 1 \\ y = 2x + 3 \end{cases}$$

THINK CRITICALLY

23. Describe the region that is the solution of the inequality $|3y - 4x| \neq -4$. Explain.

24. Can you solve $\begin{cases} x + 2y > -15 \\ x + 2y < -25 \end{cases}$? Explain.

25. WRITING MATHEMATICS Nick says that the system $\begin{cases} y \leq 2x + 11 \\ y \geq 2x - 5 \end{cases}$ has no solutions because the boundary lines are parallel and do not intersect. Marilyn says he is incorrect, that it is possible for systems of inequalities with parallel boundary lines to have solutions. Graph the system and explain who is correct.

26. Answer true or false. If false, give a reason or example that supports your answer.

"For all real numbers x, $\sqrt{x} \leq x$."

MIXED REVIEW

27. Which of the following numbers are rational?

$$-\sqrt{2}, \ 17, \ 1.\overline{32}, \ 1.3, \ \sqrt{17}, \ \sqrt{64}, \ \frac{3}{8}, \ \sqrt{\frac{5}{2}}, \ 0$$

28. STANDARDIZED TESTS If $x > 1$, which statement is always true?

   A. $x > 2$ **B.** $x^2 > 1$ **C.** $x^3 \geq 8$ **D.** $-x > 0$

29. Graph the solution set to the inequality $2x - 7 > 5$ on the number line.

30. Determine the x-intercept of the graph of $y = 4x - 9$.

31. Determine the roots of the quadratic equation $x^2 - 8x + 15 = 0$.

Write true or false for each statement. If false, give an example to support your answer.

32. The graph of $y = -x^2 - 7$ lies in Quadrants I and II.

33. $(2x + 5)^2 = 4x^2 + 25$.

PROJECT *Connection*

1. Write or visit your local Social Security office and get brochures on benefits and costs. Interview a worker. Express the amount taken out of a paycheck for Social Security as an inequality.

2. Find out the maximum amount a person could have paid into Social Security for each of the past five years. Explain how these numbers are computed and express the Social Security tax schedule for each year using an inequality.

3. List some forms used by the Social Security office and give the use of each form. What form is used to obtain personal records of all Social Security contributions?

4. Find information on Medicare, including its history, benefits to taxpayers, and how much taxpayers contribute to Medicare each year. Obtain data on the Medicare tax for each of the past five years. Express the formula for computing the Medicare tax for each year using equations or inequalities.

Most people who earn money in the United States must file an income tax return with the Internal Revenue Service (IRS). There are hundreds of IRS tax forms and many pages of regulations. IRS customer service representatives receive thousands of phone calls annually from people with tax questions.

Use if Your Filing Status is Married Filing Jointly			
If Your Taxable Income Is Over—	But Not Over—	Your Tax Is—	Of the Amount Over—
$ 0	$ 38,000	............15%	$ 0
38,000	91,850	$ 5,700.00 + 28%	38,000
91,850	140,000	20,778.00 + 31%	91,850
140,000	250,000	35,704.50 + 36%	140,000
250,000	....	75,304.50 + 39.6%	250,000

Decision Making

1. The following functions express the tax function $Y = t(x)$ using inequalities. Y_1 represents the first line of the table, Y_2 the second line, and so on. Use the table to write functions for Y_4 and Y_5.

$$Y_1 = 0.15x \qquad\qquad\qquad\quad \text{if } 0 < x \le 38,000$$
$$Y_2 = 0.28(x - 38,000) + 5,700 \quad \text{if } 38,000 < x \le 91,850$$
$$Y_3 = 0.31(x - 91,850) + 20,778 \quad \text{if } 91,850 < x \le 140,000$$

2. Determine the tax on five income amounts at which the tax function changes. Graph and connect your points.

3. One way to simplify the tax process is with a flat tax—one tax rate, and one tax function, for everybody. An example of a flat tax is "24% of all income over $9000". Write a function $f(x)$ for this flat tax.

4. Graph $f(x)$ on the same graph used in Question 2. Determine the points where the flat tax function intersects the tax function.

5. Shade the area that represents where $f(x) > t(x)$. Explain whether people in this income range would most likely be for the flat tax function $f(x)$.

7.3 Solve Quadratic Inequalities in One Variable

Explore

SPOTLIGHT ON LEARNING

WHAT? In this lesson you will learn
- to solve quadratic inequalities in one variable.

WHY? Solving quadratic inequalities can help you solve problems in carpentry, computer design, and physics.

THINK BACK

A quadratic expression can be written in the form $ax^2 + bx + c$ where a, b, and c are real numbers and $a \neq 0$.

Examine the expression $x^2 + 6x - 7$.

1. Find two values of x for which $x^2 + 6x - 7$ is greater than zero.

2. Find two values of x for which $x^2 + 6x - 7$ is less than zero.

3. Find two values of x for which $x^2 + 6x - 7$ is equal to zero.

Examine the expression $(x + 7)(x - 1)$.

4. Find two values of x for which $(x + 7)(x - 1)$ is greater than zero. Explain any strategies you use.

5. Find two values of x for which $(x + 7)(x - 1)$ is less than zero.

6. Find two values of x for which $(x + 7)(x - 1)$ is equal to zero.

7. Graph $y_1 = x^2 + 6x - 7$ and $y_2 = (x + 7)(x - 1)$. What do you notice about these two functions?

8. If you were asked to determine mentally the values of $x^2 + 6x - 7$ and $(x + 7)(x - 1)$ at $x = 11$, which expression would you prefer to evaluate? Why?

Build Understanding

In Explore, you found that the factored form of a quadratic expression may be more convenient to use when you need to evaluate the expression mentally. In the quadratic equation $ax^2 + bx + c = 0$, if you replace the equal symbol with $\leq$, $\geq$, $<$, or $>$, you create a **quadratic inequality**. To determine whether a value is a solution to a quadratic inequality, you can substitute values for x into the inequality and see if the resulting statement is true. If the expression can be factored, you may want to factor it and substitute into the factored form.

THINK BACK

The expression $a < 0$ can be interpreted as "a is a negative number" and $a > 0$ can be interpreted as "a is a positive number."

EXAMPLE 1

Determine whether each value of x is a solution of $x^2 - 3x - 5 < 0$.

a. 4   **b.** −2

Solution

a.
$$x^2 - 3x - 5 < 0$$
$$(4)^2 - 3(4) - 5 \overset{?}{<} 0$$
$$-1 < 0 \text{ True}$$

b.
$$x^2 - 3x - 5 < 0$$
$$(-2)^2 - 3(-2) - 5 \overset{?}{<} 0$$
$$5 < 0 \text{ False}$$

So, $x = 4$ is a solution of $x^2 - 3x - 5 < 0$. ◄

Notice that the quadratic expression in Example 1 was not easily factorable, so substitution was used. When a quadratic expression is factorable, you can substitute into the standard quadratic form or use **number line analysis** on the factored form.

EXAMPLE 2

Solve the quadratic inequality $2x^2 - 7x - 4 > 0$. Graph the solution set on a number line.

Solution

$$2x^2 - 7x - 4 = 0 \qquad \text{Write related equation.}$$
$$(2x + 1)(x - 4) = 0 \qquad \text{Factor.}$$
$$2x + 1 = 0 \quad \text{or} \quad x - 4 = 0 \qquad \text{Zero product property.}$$
$$x = -\frac{1}{2} \quad \text{or} \quad x = 4 \qquad \text{Solve.}$$

These solutions separate the number line into three intervals. To determine which of these intervals hold the solutions find the signs of the values of the factors $(2x + 1)$ and $(x - 4)$ and their product in each interval. Make the **sign graph** shown below.

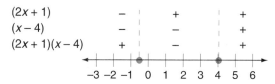

THINK BACK

If $ab < 0$ then a and b have opposite signs. If $ab > 0$ then a and b have the same signs.

Since, $2x^2 - 7x - 4 > 0$, the intervals where both factors are positive or both are negative are the solutions. Since $0 \not> 0$, $-\frac{1}{2}$ and 4 are not included in the solution. Graph the solutions $x < -\frac{1}{2}$ and $x > 4$.

Quadratic inequalities model many real world situations.

EXAMPLE 3

HOME CONSTRUCTION The Grudman family has a rectangular deck that measures 20 ft by 40 ft. They plan to increase the deck's area by increasing the length and the width by the same amount. The town building code restricts decks to a maximum of 1500 ft². How much could the existing length and width be increased so that the deck's area does not exceed 1500 ft²?

PROBLEM SOLVING TIP

Remember, in problems involving geometry, a helpful strategy is to draw a diagram.

Solution

Let x represent the number of feet added to the length and width. The area of the new deck can be expressed as $(x + 40)(x + 20)$. Now solve the following quadratic inequality.

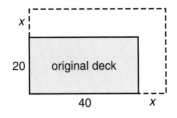

$$(x + 40)(x + 20) \leq 1500$$
$$x^2 + 60x + 800 \leq 1500 \qquad \text{Multiply.}$$
$$x^2 + 60x - 700 \leq 0 \qquad \text{Write in standard form.}$$
$$(x + 70)(x - 10) \leq 0 \qquad \text{Factor.}$$

Since this inequality is "less than or equal to," -70 and 10 are possible solutions. The three intervals $x < -70$, $-70 < x < 10$, and $x > 10$ can be examined using number line analysis.

$(x + 70)$	$-$ $+$	$+$
$(x - 10)$	$-$ $-$	$+$
$(x + 70)(x - 10)$	$+$ $-$	$+$

$-90 \ -70 \ -50 \ -30 \ -10 \quad 0 \quad 10 \quad 30$

Since $(x + 70)(x - 10) \leq 0$, the solutions must be in the interval from -70 to 10. While $-70 \geq x \geq 10$ solves the quadratic inequality, negative values of x make no sense as linear measurements. Therefore, the Grudmans can add up to and including 10 ft to both the length and the width and still comply with the town ordinance. ◀

Recall that you can solve linear equations in one variable by thinking of the expression on each side of the equal symbol as a function having the form $y = f(x)$. This is also true of inequalities.

If two equations are graphed on the same coordinate plane, their solutions can be compared.

EXAMPLE 4

Solve: $(x + 3)(x - 2) \leq (x + 1)(x - 4)$

a. Solve by graphing   **b.** Solve algebraically

Solution

a. $(x + 3)(x - 2) \leq (x + 1)(x - 4)$
$y_1 = (x + 3)(x - 2)$ Write two equations in the form $y = f(x)$.
$y_2 = (x + 1)(x - 4)$

Graph the two equations.

The graph of $y_1 = (x + 3)(x - 2)$ is less than or equal to the graph of $y_2 = (x + 1)(x - 4)$ for all values of x less than or equal to $\frac{1}{2}$.

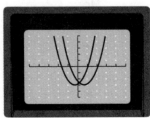

b. $(x + 3)(x - 2) \leq (x + 1)(x - 4)$
$$x^2 + x - 6 \leq x^2 - 3x - 4 \qquad \text{Multiply.}$$
$$x - 6 \leq -3x - 4 \qquad \text{Subtract } x^2 \text{ from each side.}$$
$$4x \leq 2 \qquad \text{Solve for } x.$$
$$x \leq \frac{1}{2}$$

CHECK UNDERSTANDING

In Example 5, verify that the inequality is not true if $x > \frac{1}{2}$.

To check, select some values of $x \leq \frac{1}{2}$ and verify that the original inequality is true.

The quadratic inequality $(x + 3)(x - 2) \leq (x + 1)(x - 4)$ is true when $x \leq \frac{1}{2}$.

TRY THESE

Determine whether each given value of x is a solution of the quadratic inequality.

1. $x^2 - 4x + 6 < 0; x = 3$
2. $x^2 + 5x - 3 > 0; x = 6$
3. $x^2 + 1.5x - 6.25 < 0; x = -4.5$
4. $2x^2 - 8x \geq 0; x = 4$

State the signs of the factors of each quadratic inequality.

5. $(x - 3)(x + 5) \geq 0$
6. $(x + 4)(x - 2) \leq 0$
7. $(x + 10)(x + 3) < 0$
8. $x^2 - 10x + 21 > 0$

Solve each inequality. Graph the solution on a number line. Check your answer by selecting values and verifying the inequality is true.

9. $x^2 - 2x - 24 < 0$
10. $x^2 + x - 12 \leq 0$
11. $x^2 - 4x - 5 > 0$
12. $2x^2 - 5x - 3 \geq 3$
13. $3x^2 + 2x - 5 < 0$
14. $x^2 - 6x + 9 > 0$

ROCKETRY The height h in feet of a rocket above the ground at time t seconds is given by $h = -16t^2 + vt + s$ where v is the initial velocity and s is the initial height. A toy rocket is initially 80 ft above the ground and is launched straight up with a velocity of 48 ft/s.

15. During what time interval will the rocket be at least 16 ft above the ground?

16. During what time interval will the rocket be at most 112 ft above the ground?

17. **WRITING MATHEMATICS** Suppose you are working a problem where x represents dollars spent. You find that you must solve the quadratic inequality $(x + 3)(x - 1) > 0$. The algebraic solution is $x < -3$ or $x > 1$. What is the solution to the application? Explain.

Determine whether each given value of x is a solution of the quadratic inequality.

1. $x^2 + 8x - 1 < 0; x = 0$

2. $3x^2 - 2x + 5 < 0; x = -1$

3. $x^2 + 0.75x - 4 \leq 0; x = 1.5$

4. $x^2 + \frac{3}{2} < 0; x = -\frac{2}{3}$

Solve each inequality. Graph the solution on a number line. Check your answer by selecting values and verifying the inequality is true.

5. $x^2 - 4x + 3 < 0$

6. $x^2 - 6x < 0$

7. $x^2 - 5x + 4 > 0$

8. $x^2 - 3x - 10 < 0$

9. $2x^2 + 3x - 2 \geq 0$

10. $3x^2 - 27 > 0$

11. $x^2 + 8x + 16 \leq 0$

12. $3x^2 - 7x + 2 \leq 0$

13. $-2x^2 - 5x + 3 > 0$

14. PHYSICS A projectile is fired straight up from the ground. Its initial velocity is 208 ft/sec. At any time t its height in feet is given by $h = -16t^2 + 208t$. When will the projectile be at least 480 ft off the ground?

15. CARPENTRY The length of a desktop is to be made 30 in. longer than the width. What are possible widths of this desktop that would result in an area of at least 1800 in.²?

Use the graph below for Exercises 16-19.

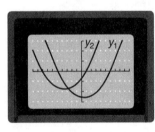

16. On what interval is $y_1 < y_2$?

17. On what interval is $y_1 > y_2$?

18. On what interval is $y_1 \leq y_2$?

19. On what interval is $y_1 \geq y_2$?

20. Replace y_1 with the expression $x^2 - x - 30$ and y_2 with the expression $x^2 + 7x - 8$. Solve Exercises 18 and 19 above algebraically. How do the algebraic results compare with the graphic results?

Use the graph below for Exercises 21-24.

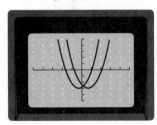

21. On what interval is $y_1 < y_2$?

22. On what interval is $y_1 > y_2$?

23. On what interval is $y_1 \leq y_2$?

24. On what interval is $y_1 \geq y_2$?

25. Replace y_1 with the expression $x^2 - 4x + 4$ and y_2 with the expression $-x^2 + 5x$. Solve Exercises 23 and 24 above algebraically. How do the algebraic results compare with the graphic results?

Solve each of the following algebraically and graphically.

26. $x^2 - x - 12 \leq x^2 + 6x$

27. $x^2 + 4x - 5 > x^2 - x - 6$

28. **ARCHITECTURE** The architect for a home wants the length of the living room to be 10 ft less than twice its width. What could the width measure so that the area of the room is at least 300 ft²?

29. **JEWELRY** A jeweler makes silver bracelets. She has determined that if she sells the bracelets at $60 - x$ dollars where x is the number of bracelets produced each week, then she is able to sell all of the bracelets made. How many bracelets must she make so that her revenue is at least $500 per week?

30. **WRITING MATHEMATICS** Give an example of a quadratic inequality that has no solution. Describe the appearance of the graph of the quadratic expression in your inequality. Compare your observations with other students. What generalization can you make about the graphs of quadratic inequalities with no solution?

EXTEND

31. Determine the solutions that satisfy

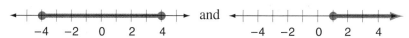

32. Determine the solutions that satisfy

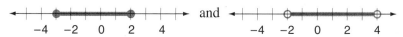

33. Determine the solutions that satisfy both $x^2 - x \geq 0$ and $x^2 - 1 \geq 0$.

34. Determine the solutions that satisfy both $x^2 + x - 6 \leq 0$ and $x^2 - 5x + 4 \geq 0$.

35. Solve: $-4 \leq x^2 - 4x \leq 12$

36. Solve: $-7 < x^2 + 8x < 20$

THINK CRITICALLY

Determine a possible quadratic inequality whose solutions are graphed.

37.
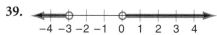

38.

39.

40.

41. COMPUTER DESIGN Originally, a computer chip was designed as a square. In the upgrade, one side of the chip is to be 2 cm longer than the other side. The total area of the chip must be between 15 and 24 cm² inclusive. Find an interval for the measure of the shortest side of the new computer chip.

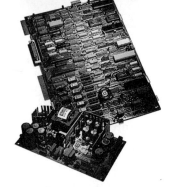

MIXED REVIEW

42. Simplify: $\left|(-4)^3\right|\dfrac{-(36+8)}{2}$

43. Simplify: $-5t^3 + 4t - 8t^2 + t - 8$

44. MAP SCALE On a map, 1.5 cm represents 30 km. What is the distance between two cities on the map that are 9 cm apart?

45. Solve for y: $-2 < 2y + 8 < 18$

46. Solve for k: $|-4 + 3k| \le 5$

47. GEOMETRY The length of a rectangular prism is twice the square of its width. The depth is 4 less than the length. Write an expression for the volume of this prism.

48. STANDARDIZED TESTS Determine k such that the solution of the quadratic inequality $2x^2 + kx - 12 \le 0$ is $-1.5 \le x \le 4$.

   A. 10   **B.** 5   **C.** −5   **D.** −11

PROJECT *Connection*

1. Write to the Internal Revenue Service for a copy of Publication 17, *Your Federal Income Tax*. This publication may be available in your local library or federal tax office.

2. Obtain a copy of the most recent set of Internal Revenue Service tax tables. Pick five lines from different parts of the table and express these lines using inequalities.

3. Research the different sources from which the government receives money. Construct a circle graph to display this information.

4. Research how the government spends the money it collects. Construct a circle graph to display this information. List some services provided by federal income tax money.

5. Find out if your state has a state income tax. If it does, get a copy of the state income tax tables and express five different lines in the table using inequalities. Contact the appropriate state income tax office and request statistical information on the collection and spending of state tax money. If your city also has an income tax, complete Question 5 using the city tax tables.

6. Find out what a tax deduction is. Illustrate, using a sample numerical example, how a tax deduction can save a taxpayer money. Consult reliable sources and make a list of ten items that are tax deductible, and ten that are not. Prepare a poster or transparency displaying all of this information.

Explore/Working Together

• Work with a partner. Use the inequality $y < x^2 - 8x + 7$.

1. Use substitution to test each of the following ordered pairs (x, y) and determine if they satisfy the inequality.

   a. $(0, 8)$ b. $(-3, 1)$ c. $(4, -10)$

   d. $(8, 9)$ e. $(10, 2)$ f. $(5, 0)$

2. Use trial and error to identify three other ordered pairs that satisfy the inequality. Name these ordered pairs G, H and I.

3. Use trial and error to identify three other ordered pairs that do *not* satisfy the inequality. Name these ordered pairs J, K and L.

4. Graph the equation $y = x^2 - 8x + 7$ on graph paper.

5. Plot and label the points A–L on the graph. Next to each point, write *no* or *yes* to indicate whether or not the ordered pair represented by the point satisfies the inequality.

6. What conclusion can you make about the points that satisfy the inequality and the points that do not satisfy the inequality?

SPOTLIGHT ON LEARNING

WHAT? In this lesson you will learn
• to graph quadratic inequalities.

WHY? Graphing quadratic inequalities can help you visualize the solutions to problems about transportation, recreation, and pharmacology.

Build Understanding

• Consider the quadratic equation $y = ax^2 + bx + c$.

If you replace the equal symbol in any quadratic equation in two variables with $<$, $>$, $\leq$, $\geq$, or $\neq$, you create a **quadratic inequality in two variables**. Solutions to quadratic inequalities in two variables can be represented on the coordinate plane. The graph of a quadratic equation in two variables is a parabola that divides the coordinate plane into two regions. The parabola is the boundary of each of these two regions.

Shaded regions show where solutions can be found. When the inequality symbol is $\leq$ or $\geq$, the parabolic boundary also contains solutions. A region that includes the boundary is called a **closed set**. The parabola is drawn as a solid curve.

CHECK UNDERSTANDING

In Explore, you can use your graphing utility to substitute each x-value into the equation $y = x^2 - 8x + 7$ and find a y-value. How can this y-value be used to determine if the inequality is true?

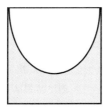
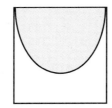

The symbol $<$, $>$, or $\neq$ means that the parabola itself contains no solutions to the inequality. A region that does not include the boundary is an **open set**. The parabola is drawn as a dashed curve.

Substitute to determine if a point is a solution to a quadratic inequality.

EXAMPLE 1

Determine if each ordered pair is a solution of $y \leq x^2 + 3x - 4$.

a. $(1, 2)$ **b.** $(5, 30)$

Solution

a. $y \leq x^2 + 3x - 4$

$2 \overset{?}{\leq} 1^2 + 3(1) - 4$

$2 \leq 0$ **False**

b. $y \leq x^2 + 3x - 4$

$30 \overset{?}{\leq} 5^2 + 3(5) - 4$

$30 \leq 36$ **True**

The ordered pair $(5, 30)$ is a solution of $y \leq x + 3x - 4$. ◄

The solutions to a quadratic inequality can be displayed in a graph.

EXAMPLE 2

Find the solutions of $y > x^2 - 2x - 5$.

Solution

Graph the corresponding quadratic equation $y = x^2 - 2x - 5$.

The axis of symmetry is $x = 1$, the vertex is $(1, -6)$ and the parabola opens upward. The *y*-intercept is -5 and the *x*-intercepts are -1.45 and 3.45.

The inequality symbol is $>$, so the parabola is not part of the solution. Test points in both regions. Choose points easy to evaluate.

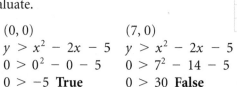

$(0, 0)$ $(7, 0)$

$y > x^2 - 2x - 5$ $y > x^2 - 2x - 5$

$0 > 0^2 - 0 - 5$ $0 > 7^2 - 14 - 5$

$0 > -5$ **True** $0 > 30$ **False**

The point $(0, 0)$ is a solution of the inequality, so all other points in the region containing $(0, 0)$ are also solutions. Shade that region. ◄

Quadratic inequalities have many real world applications.

EXAMPLE 3

HOME IMPROVEMENT The square footage covered by one gallon of paint is printed on the label of the can. An employee of the paint store wants to make a graphic display to help customers determine how much paint to buy.

a. The graph will be for square rooms, with 8-ft ceilings, in which the ceiling as well as the walls will be painted. Let w represent the width of the room, and let c represent paint coverage in square feet. Write an inequality for the coverage which must exceed the area.

b. Graph the inequality using w- and c-axes.

c. Bob has a square room with an 8-ft ceiling and 16-ft width. How many square feet of paint coverage must he purchase?

Solution

a. The area of the ceiling is w^2. The area of each wall is $8w$, and there are four walls, so $4(8w) = 32w$. The total area to be painted is $w^2 + 32w$, ignoring the area of windows and doors. The amount of coverage c must be greater than the area to be painted.

$$c \geq w^2 + 32w$$

b. Let w be the horizontal axis and let c be the vertical axis. Graph $c = w^2 + 32w$. The axis of symmetry is $w = -16$. The vertex is $(16, -256)$. The parabola opens up. The c-intercept is 0 and the w-intercepts are 0 and -32.

The boundary parabola is a solid curve. Test points in both regions.

$(0, 1)$ $(2, 0)$
$c \geq w^2 + 32w$ $c \geq w^2 + 32w$
$1 \overset{?}{\geq} 0^2 + 32(0)$ $0 \overset{?}{\geq} 2^2 + 32(2)$
$1 \geq 0$ **True** $0 \geq 68$ **False**

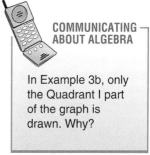

Shade the region containing $(0, 1)$.

c. $c \geq w^2 + 32w$
$c \geq 16^2 + 32(16)$
$c \geq 768$

The customer must purchase enough paint to cover at least 768 ft². The purchase can be represented by any point with w-coordinate equal to 16 that is in the shaded region. The customer may have to buy more than necessary because the paint is sold by the gallon. ◀

COMMUNICATING ABOUT ALGEBRA

In Example 3b, only the Quadrant I part of the graph is drawn. Why?

Determine whether each ordered pair is a solution of $y \le x^2 - 8x + 2$.

1. $(0, 0)$ **2.** $(5, 3)$ **3.** $(-3, 10)$ **4.** $(1, -5)$

5. $(-2, -2)$ **6.** $(-1, 0)$ **7.** $(-10, 192)$ **8.** $(0, 2)$

Graph each inequality. Indicate the intercepts and vertex of the boundary curve.

9. $y > x^2 + 1$ **10.** $y \le x^2 - 7x + 12$ **11.** $y < -x^2 + x$

12. $y \ge x^2 + 9x + 8$ **13.** $y > (x - 7)^2$ **14.** $y \le x^2 + x + 5$

15. WRITING MATHEMATICS Write a quadratic inequality whose solution set does not include the origin. Use a parabola that has axis of symmetry $x = 2$. Graph the inequality and explain how you determined it.

16. TRANSPORTATION A car's braking distance under ideal conditions can be approximated by the formula $b = 0.05s^2$ where s is the speed in miles per hour and b represents the number of feet it takes to bring the car to a full stop once brakes are applied. Ideal conditions include a dry, clean road, good brakes and suspension on car, and good tires. Many accidents happen because drivers do not make sufficient adjustments for less than ideal conditions, which increase braking distance. Express the formula for braking distance as an inequality that allows for changes in driving conditions.

PRACTICE

Determine whether each ordered pair is a solution of $y > 2x^2 - x - 3$.

1. $(1, 5)$ **2.** $(-1, 5)$ **3.** $(12, 0)$ **4.** $(0, -4)$ **5.** $(2, 3)$

Graph each inequality. Indicate the intercepts and vertex of the boundary curve.

6. $y \le 2x^2 + 7$ **7.** $y > -4x^2 + x$ **8.** $y < x^2 + 9x + 10$

9. $y \ge 2x^2 + 7x + 3$ **10.** $y \le -0.5x^2$ **11.** $y \le 16 - x^2$

12. $y > -x^2 + 2x + 1$ **13.** $y > (x - 5)^2$ **14.** $y + 3 > -x^2 + 4x - 1$

15. RECREATION Humphrey plans to install a rectangular pool and patio in a 50-by 60-foot area of his backyard. The perimeter of the pool will be a patio of uniform width which is made of concrete bordered by flagstones where the patio meets the edge of the pool. The patio's width is represented by x.

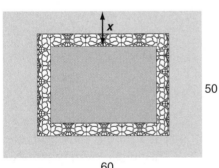

a. Express the length of the pool in terms of x.

b. Express the width of the pool in terms of x.

c. Determine the area of the pool and surrounding patio.

d. Express the area of the pool in terms of x.

e. The natural stones placed around the pool will reduce the amount of concrete required for the patio. Since they are irregularly shaped, it is difficult to determine exactly how much concrete is saved. Write an inequality that expresses the fact that y, the amount of concrete needed in square feet, is less than what the patio area would be if there were no surrounding stones.

f. If x, the width of the patio including the stones, is 6 ft, and the stones cover 200 ft^2, is 900 ft^2 of concrete enough to cover the concrete part of the patio? Explain.

16. **LANDSCAPE ARCHITECTURE** Marti installs in-ground sprinkler systems used to water lawns. Water is sprayed out of sprinkler heads that rotate 360°. The radius of the water spray is determined by the water pressure. The heads are set up so that the circular areas they cover overlap to insure complete watering.

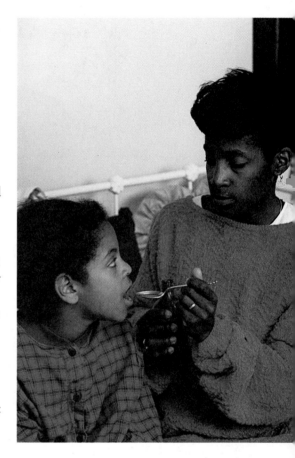

   a. Write a quadratic inequality that models the area 10 of these heads could cover.

   b. Graph the inequality you wrote in Part a.

   c. If $r \leq 14$ ft, determine the maximum area that could be watered by the 10 heads. Round to the nearest square foot.

17. **PHARMACOLOGY** The maximum daily dosage of a certain type of children's medication depends on the child's body weight according to the function $d \leq 0.05w^2 + 2w$ where d represents dosage in milligrams and w represents the child's weight in pounds. If you have 3000 mg of the medication, what is the range of weights that a child who needs a 5-day treatment can have? Round to the nearest pound.

18. **WRITING MATHEMATICS** List two ways graphing quadratic inequalities in two variables is similar to graphing linear inequalities in two variables. List two ways it is different.

EXTEND

19. Given the system of inequalities

$$\begin{cases} y \leq 6 \\ y > x^2 + 2 \end{cases}$$

   a. Graph the inequalities on the same coordinate plane.

   b. At what points do the boundary lines intersect?

   c. Explain whether or not the point (1, 2) is a solution to the system.

   d. Find three points that are solutions of the system. Verify that your points are solutions to both inequalities.

20. Below is the graph of the boundaries of the system of inequalities

$$\begin{cases} y < x^2 - 8x + 15 \\ y \geq x + 7 \end{cases}$$

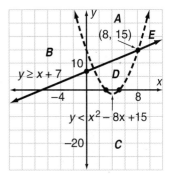

a. Which region(s) have solutions to the first inequality?

b. Which region(s) have solutions to both inequalities?

c. Which region(s) have solutions to the second inequality, but not the first inequality?

d. Which region(s) have solutions to the first inequality, but not the second inequality?

21. Graph: $\begin{cases} y < -x^2 + 4 \\ y > |x| \end{cases}$

22. **a.** Graph $y = x^2 - 9x + 8$. Label the x- and y-intercepts and the vertex.

b. Graph $y > x^2 - 9x + 8$. Shade the solution set.

c. Use your knowledge of absolute value and graphing to graph $y = |x^2 - 9x + 8|$. Label all intercepts. Explain your thinking.

d. Use your answers to Parts a, b and c above and your knowledge of inequalities to graph $y > |x^2 - 9x + 8|$.

THINK CRITICALLY

23. If $x > 0$, which of the following could not represent sides of a triangle? Explain.

a. $x + 3,\ x + 4,\ x + 6$

b. $x + 5,\ 3x + 6,\ x + 1$

c. $x + 5,\ x + 4,\ x + 8$

24. WRITING MATHEMATICS Graph the solution to the inequality $x^2 > 0$. Describe the solution in words. How would you display the solution set on the coordinate axes?

25. Use the quadratic equations $y = x^2$ and $y = x^2 + 10$ to create a system of quadratic inequalities whose intersection contains

a. no points

b. infinitely many points

26. Graph the solution of the inequality $y \leq |x^2 - 9|$.

27. Graph the solution of the inequality $y \neq x^2 - 4$.

28. WRITING MATHEMATICS If the discriminant of the quadratic inequality $y < x^2 + bx + c$ is negative, all points on the x-axis are solutions to the inequality. Explain why, using words and illustrations.

29. **STANDARDIZED TESTS** Which of the following numbers is the largest?

    A. 1.7295   **B.** 1.7329   **C.** $\sqrt{3}$ **D.** 1.732

30. For what values of x is $\dfrac{3x}{x^2 - 5x + 4}$ undefined?

31. Determine the multiplicative inverse of $\dfrac{1}{x - 7}$.

32. For what values of x is $\dfrac{\sqrt{x - 4}}{3}$ defined?

33. Graph the system of inequalities.

$$\begin{cases} y < 2x + 1 \\ y \geq -x + 4 \end{cases}$$

34. A circle is inscribed in a square whose side is 10 ft. Determine the area of the circle. (Use $\pi = 3.14$).

35. At what point do the graphs of $x = 5$ and $y = 2x + 3$ intersect?

36. Express 0.71 as the quotient of two integers.

37. Express $0.\overline{71}$ as the quotient of two integers.

38. Graph $f(x) = |x + 3|$ without the aid of a graphing utility.

39. Graph $g(x) = |x| + 3$ without the aid of a graphing utility.

40. Find all real numbers that are greater than their squares.

PROJECT *Connection*

1. Find out if your state has a sales tax, and if your county has its own sales tax. List the services provided by the money that is collected in sales tax.

2. When shopping in an area that has a sales tax, you must be sure you have enough money, including the tax, to purchase an item. If x represents the "sticker price" of an item, t represents that sales tax rate, and M represents the amount of money you have to spend, create an inequality that can be used to compute the maximum sticker price you can afford, given the sales tax rate and the amount of money you have to spend. Make a poster that illustrates the use of the inequality using the sales tax rate in your area or a nearby area.

3. Interview a travel agent about taxes on hotel rooms (in your locality and elsewhere) and airline tickets. Interview a gas station owner about the different taxes on one gallon of gasoline. Interview a supermarket manager and/or a pharmacist about what items in their stores are subject to sales tax, and what items are not. Summarize the information you obtain.

4. Interview a local tax assessor or homeowner about property taxes. Write an explanation of the various parts of a property tax bill. List the services paid for by our property taxes. Find out the range of property taxes paid in your school district and express this as an inequality.

Algebra Workshop
Systems of Inequalities: Polygonal Regions

Think Back/Working Together

● A **polygon** is a closed plane figure formed by at least three line segments.

 1. Which of the following shapes are polygons? Explain.

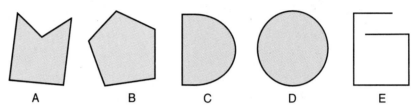

Work with a partner. Follow the directions below using separate sheets of graph paper for each activity.

 2. On a coordinate plane, sketch a line in the first quadrant so that it forms a polygon with the *x*- and *y*-axes. Compare sketches. What generalizations can you make about this polygon?

 3. On a coordinate plane, sketch two lines in the first quadrant so that they form a polygon with the *x*- and *y*-axes. Compare sketches. What generalizations can you make about this polygon?

Explore

● A closed **polygonal region** is formed by the intersection of a system of linear inequalities.

 4. Examine the graphs below. What is the difference between the solutions pictured in each graph?

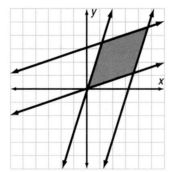 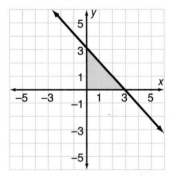

The polygonal region on the left is the solution to a system of four linear inequalities. The polygonal region on the right is the solution to the linear system $x \geq 0, y \geq 0$, and $x + y \leq 3$.

Work with your partner. You each need graph paper, scissors, and clear tape.

5. Draw a coordinate plane on one sheet of graph paper. On the other sheet, shade in an isosceles right triangle, each leg measuring 5 units. Cut out this right triangle. Tape the triangle onto the coordinate plane so that it lies entirely within the first quadrant and its legs are parallel to the x- and y- axes. Place the vertices on points with integer coordinates. Draw the lines that form the sides of the triangle. Determine the three inequalities that define the triangular polygonal region. Trade graphs with your partner and determine the inequalities for your partner's triangular polygonal region. Compare results.

6. Draw a coordinate plane on a sheet of graph paper. On another sheet, shade in an isosceles right triangle, each leg measuring 5 units. Cut out this right triangle. Tape the triangle onto the coordinate plane so that its sides are not parallel to either of the axes, but its vertices lie on points with integer coordinates. Draw the lines that form the sides of the triangle. Determine the three inequalities that define the triangular polygonal region. Trade graphs with your partner and determine the inequalities for your partner's triangular polygonal region. Compare results.

7. Draw a coordinate plane on one sheet of graph paper. On another other sheet, shade a square with sides 5 units and cut it out. Tape the square onto the coordinate plane so that it lies entirely within the first quadrant with its sides parallel to the x- and y-axes. Place the vertices on points with integer coordinates. Draw the lines that form the sides of the square. Determine the four inequalities that define the square polygonal region. Trade graphs with your partner and determine the inequalities for your partner's square polygonal region. Compare results.

8. Draw a coordinate plane on a sheet of graph paper. On another sheet, shade a square with sides 5 units and cut it out. Tape the square onto the coordinate plane so that its sides are not parallel to either of the axes, but its vertices lie on points with integer coordinates. Draw the lines that form the sides of the square. Determine the four inequalities that define the square polygonal region. Trade graphs with your partner and determine the inequalities for your partner's square polygonal region. Compare results.

THINK BACK

If you know the coordinates of two points on a line, you can use the slope-intercept or point-slope forms to determine the equation of the line.

Algebra Workshop

Making Connections

Consider the four inequalities labeled L_1, L_2, L_3, and L_4. Examine regions A, B, and C.

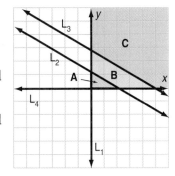

9. Which of the inequalities are satisfied by any point in region A? Explain.

10. Which of the inequalities are satisfied by any point in region B? Explain.

11. Which of the inequalities are satisfied by any point in region C? Explain your reasoning.

12. Write the four possible inequalities L_1, L_2, L_3, and L_4 in the graph pictured above. Which of these inequalities confine the regions under consideration to the first quadrant only?

Summarize

13. WRITING MATHEMATICS Consider the graph of the system of these linear inequalities.

$$\begin{cases} x \geq 0 \\ y \geq 0 \\ -2x + y \leq 10 \\ x - 2y \leq 6 \\ y + 5x \leq 52 \end{cases}$$

Do the coordinates of the vertices of the polygonal region satisfy all of the inequalities that form the polygonal region? Explain.

14. MODELING Write the system of linear inequalities represented by the graph at the right.

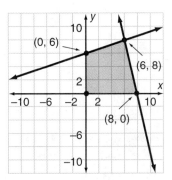

15. THINKING CRITICALLY A polygonal region has vertices $(0, 0)$, $(0, 8)$, $(9, 5)$ and $(3, 0)$. Determine the system of linear inequalities that defines this polygonal region.

16. GOING FURTHER Define a system of linear inequalities that would form a pentagonal polygonal region in the first quadrant, two of whose sides are on the x- and y-axes.

Members of Congress must be aware of the concerns of the voters they represent. They must also be knowledgeable about controversial issues, such as the use of nuclear energy. Federal and state governments have very strict regulations on construction and operation of nuclear power plants. How much of our electricity is generated by nuclear power plants? Examine the data in the table. The percent represents the percent of electricity generated by nuclear energy.

Year	1980	1981	1982	1983	1984	1985	1986	1987	1988	1989	1990
Percent	11.0	11.9	12.6	12.7	13.6	15.5	16.6	17.7	19.5	19.0	20.5

Decision Making

1. Represent the years by their last two digits. For example, 1998 would be entered as 98, and 2012 would be entered as 112. Why can't 2012 be entered simply as 12?

2. The data above appears to be positively correlated. Use a graphing utility to determine the equation of the regression line and the correlation coefficient. Explain whether or not it is a good predictor.

3. If this trend continues, what percent of our electricity will be generated by nuclear energy in the year 2003?

4. At this rate, in what year will 50% of our power come from nuclear energy?

5. Congresswoman A has a graph in which the area below the regression equation is shaded. She claims this area represents a cutback in nuclear dependency. Write this area as an inequality.

6. Congressman B argues the percent of electricity from nuclear power should remain at or below the 1990 level and shades in this area on Congresswoman A's graph. Write this area as an inequality.

7. Congresswoman A says the point (120, 39) represents a cutback in nuclear dependency. What do these coordinates represent?

8. Graph the two inequalities on the same coordinate axes and show the location of (120, 39). Explain why Congresswoman A describes this point as a cutback and Congressman B describes the same point as an increase in nuclear energy dependency.

The Objective Function

Explore

1. On a sheet of graph paper, construct the polygonal region defined by the following inequalities.

$$x \geq 0 \qquad y \geq 0 \qquad y \leq 5 \qquad y + x \leq 10$$

2. On the same coordinate axes, draw the line $y = -\frac{1}{2}x$.

3. Place a pencil over the line $y = -\frac{1}{2}x$. Slowly slide the pencil over the polygonal region keeping it parallel to the line $y = -\frac{1}{2}x$. What are the coordinates of the last point in the polygonal region that the pencil passes over?

4. Position the pencil anywhere outside of the polygonal region but not parallel to any of its sides. Slowly slide the pencil over the region keeping it parallel to its original position. What are the coordinates of the last point in the polygonal region that the pencil passes over?

5. What generalization can you make about the last point in the polygonal region that the pencil passes over?

Build Understanding

Linear programming is a method used by business and government to help manage resources and time. **Constraints** are conditions that limit available resources. In linear programming, such constraints are represented by inequalities. The intersection of the graphs of a system of constraints includes all possible solutions and is known as a **feasible region**.

In Explore, you determined that the last point in the polygonal region that the pencil passed over was at a vertex. The line represented by the pencil is known as the **objective function.** The equation of this line can represent quantities such as revenue, profit, or cost. In business the objective function is used to determine how to make maximum profit with minimum cost.

By substituting values within the constraints in an objective function, you can determine the values that will **maximize** or **minimize** the objective function. The maximum or minimum values of the objective function will occur when the function passes through a vertex of the feasible region. Solutions to linear programming problems are often restricted to the first quadrant of the coordinate plane. If the feasible region is not completely within the first quadrant and it should be, include $x \geq 0$ and $y \geq 0$ as constraints, if necessary.

EXAMPLE 1

MANUFACTURING Universe Corporation makes two types of athletic shoes; an aerobics shoe and a high-top basketball shoe. The shoes are assembled by machine and then finished by hand. It takes 0.25 h for the machine assembly and 0.1 h by hand to make an aerobics shoe. It takes 0.15 h on the machine and 0.2 h by hand to make the basketball shoe. At their manufacturing plant, Universe can allocate no more than 900 machine hours and no more than 500 hand hours per day. The profit is $10 on each aerobic shoe and $15 on each basketball shoe. Determine how many of each type of shoe should be made to maximize the profit.

Solution

If x represents the number of aerobic shoes made and y represents the number of basketball shoes made, then the profit objective function is $P = 10x + 15y$.

Write an inequality to represent each constraint.

Machine hours: $0.25x + 0.15y \leq 900$
Hand hours: $0.1x + 0.2y \leq 500$

Include $x \geq 0$ and $y \geq 0$ so feasible region is in the first quadrant.

Graph the system of inequalities.

The vertices of the feasible region are at $(0, 0)$, $(3600, 0)$, $(0, 2500)$ and $(3000, 1000)$.

Evaluate the objective function at each of the vertices of the feasible region.

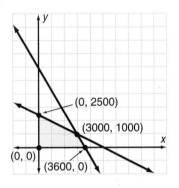

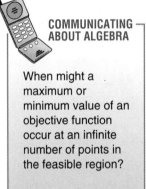

COMMUNICATING ABOUT ALGEBRA

When might a maximum or minimum value of an objective function occur at an infinite number of points in the feasible region?

Vertex	$10x$ $+ 15y$	Profit P, dollars
(0, 0)	10(0) + 15(0)	0
(3600, 0)	10(3600) + 15(0)	36,000
(0, 2500)	10(0) + 15(2500)	37,500
(3000, 1000)	10(3000) + 15(1000)	45,000 (maximum)

Under the given daily constraints, the maximum daily profit Universe should expect to make is $45,000. To do this, they would have to produce and sell 3000 aerobic shoes and 1000 basketball shoes per day.   ◄

You will not always get integer coordinates at the vertices of a feasible solution. If the maximizing or minimizing vertex has decimal coordinates, round the values to the nearest whole number so they make sense in the context of the problem.

7.6   **The Objective Function**   **351**

Explain which
constraints in
Example 2 intersect
to form each of the
vertices.

PROBLEM
SOLVING TIP

If you use a
graphing utility, the
vertices of the
feasible region may
not be easy to locate
in the original
window. It might be
necessary to use the
TRACE and ZOOM
or INTERSECT
features in order to
determine the
coordinates.

EXAMPLE 2

MANUFACTURING Carpet City manufactures wool and nylon fiber rugs.
The warehouse has 915 lots of wool fiber and 1120 lots of nylon fiber
readily available for manufacturing of their WearRite and UltraGuard
carpets for the week of July 5. A roll of WearRite uses 15 lots of wool
fiber and 35 lots of nylon fiber. A roll of UltraGuard uses 40 lots of
wool fiber and 45 lots of nylon fiber. Carpet City makes a profit of
$180 per roll on WearRite and $300 per roll on UltraGuard.

It takes 4 hours to manufacture a roll of WearRite and 3 hours to
manufacture a roll of UltraGuard. Carpet City has a maximum of
98 manufacturing hours per week and can only work on one type of
carpet at a time. How many rolls of each type carpet should they
manufacture during this week in order to maximize the profit?

Solution

Let x represent the number of rolls of WearRite and y represent the
number of rolls of UltraGuard manufactured. Make a table to organize
the data.

Carpet (Number of Rolls)	Number of Lots of Wool	Number of Lots of Nylon	Labor Hours	Profit, dollars
WearRite	$15x$	$35x$	$4x$	$180x$
UltraGuard	$40y$	$45y$	$3y$	$300y$
Totals	915	1120	98	P

The objective function for total profit P is
$P = 180x + 300y$

Write the constraints as inequalities.
Graph the feasible region.

$$15x + 40y \leq 915$$
$$35x + 45y \leq 1120$$
$$4x + 3y \leq 98$$
$$x \geq 0$$
$$y \geq 0$$

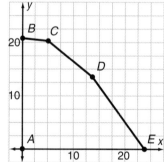

Evaluate $P = 180x + 300y$ at each vertex.

A $(0, 0)$ $P = 180(0) + 300(0) = 0$
B $(0, 22.875)$ $P = 180(0) + 300(22.875) = 6862.50$
C $(5, 21)$ $P = 180(5) + 300(21) = 7200$ (maximum)
D $(14, 14)$ $P = 180(14) + 300(14) = 6720$
E $(24.5, 0)$ $P = 180(24.5) + 300(0) = 4410$

The maximum profit that Carpet City can make for the week of July 5
is $7200. This will be achieved by manufacturing 5 rolls of WearRite
and 21 rolls of UltraGuard.   ◄

Determine if each point is within the feasible region $x \geq 0$, $y \geq 0$, and $5x + 2y \leq 30$.

1. $(10, 7)$ **2.** $(1, 6)$ **3.** $(2.75, 6.1)$ **4.** $\left(4\frac{3}{4}, 7\frac{1}{2}\right)$ **5.** $(5, 2.5)$

Determine the maximum value of $P = 15x + 12y$ for each feasible region.

6.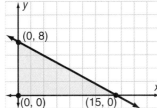

7.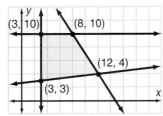

Determine the minimum value of the objective function $C = 3x + 2y$ for the graph of each feasible region.

8.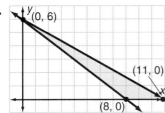

9.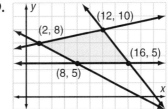

For the given vertices of each feasible region in Exercises 10–13, determine the minimum and maximum value of the objective function and identify the coordinates at which they occur.

10. $P = 10x + 6y$ $(0, 10)$ $(5, 15)$ $(8, 8)$ $(12, 0)$

11. $P = 4x + 5y$ $(2, 5)$ $(2, 9)$ $(6, 11)$ $(8, 5)$

12. $C = 1.25x + 0.75y$ $(0, 4)$ $(9, 15)$ $(20, 2)$

13. $C = 120x + 180y$ $(6, 6)$ $(6, 10)$ $(10, 12)$ $(13, 11)$ $(13, 6)$

SMALL BUSINESS A group of students are making and selling short-sleeved and long-sleeved tie-dyed shirts. Their costs are $3.00 for the short-sleeved shirts and $5.00 for the long-sleeved shirts. A local store owner has agreed to sell their shirts but will only take up to a total of 50 short- and long-sleeved shirts combined. In addition, the store owner said they must sell at least 15 short-sleeved shirts and 10 long-sleeved shirts to continue selling at the store. Let x equal the number of short-sleeved shirts and y equal the number of long-sleeved shirts.

14. Write the objective function for cost.

15. Write the inequalities that express the constraints.

16. Graph the inequalities and determine coordinates of the vertices of the feasible region.

17. How many of each type must they sell to minimize cost?

18. MODELING Write the system of linear inequalities whose feasible region is represented by the graph at the right.

19. WRITING MATHEMATICS The constraints $x \geq 0$ and $y \geq 0$ are sometimes known as "reality" or "real world" constraints. Why might they be given such a name?

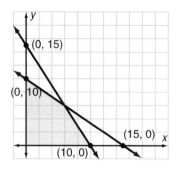

(0, 15)

(0, 10)

(15, 0) x

(10, 0)

PRACTICE

Identify the vertices of the feasible region defined by the constraints.

1. $x \geq 0;$ $y \geq 0;$ $7x + 9y \leq 63$

2. $x \geq 0;$ $y \geq 0;$ $y + 2x \geq 8;$ $y + x \leq 8$

3. $x \geq 0;$ $y \geq 0;$ $8y - 4x \leq 48;$ $y + x \leq 10$

For the given vertices of each feasible region in Exercises 4–7, find the minimum and maximum value of the objective function and identify the coordinates at which each occurs.

4. $A = 2x + 3y$ $(0, 0)$ $(12, 0)$ $(0, 16)$

5. $A = 3x - 2y$ $(0, 0)$ $(12, 0)$ $(0, 16)$

6. $A = 0.5x + 0.75y$ $(6, 6)$ $(6, 12)$ $(10, 12)$ $(15, 6)$

7. $A = 10x + 12y$ $(4, 9)$ $(6, 14)$ $(12, 12)$ $(14, 2)$

8. Determine the maximum value of the objective function $P = -4x + 5y$ under the following constraints. Identify the point at which it occurs.

$$x \geq 0;\quad y \geq 0;\quad y + x \geq 5;\quad 2y + x \leq 28;\quad 15y - 4x \geq 20$$

9. Determine the minimum value of the objective function $C = 5x + 2y$ under the following constraints. Identify the point at which it occurs.

$$x \geq 2;\quad 4 \leq y \leq 8;\quad 3y + x \leq 30$$

10. Determine two different objective functions P and R such that point A of the feasible region yields the maximum value for P, and point B of the feasible region yields the maximum value of R.

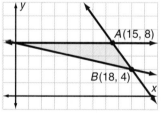

A(15, 8)

B(18, 4)

x

Find the maximum of each objective function under the given constraints. Identify the point at which the maximum value occurs.

11. $P = 5x + 6y$ Constraints: $y + 3x \leq 6; y + 2x \leq 5; x \geq 0; y \geq 0$

12. $P = 2x + y$ Constraints: $y \geq 1; 3x - 9y \leq 0; x + y \leq 16$

13. $P = -2x + 3y$ Constraints: $1 \leq x \leq 4; x + y \leq 10; x - y \geq 0$

14. $P = -2x + 4y$ Constraints: $x \geq 0; y \geq 0; x + y \leq 4; x - y \leq 0$

Find the minimum of each of the following objective functions under the given constraints. Identify the point in the feasible region at which the minimum value occurs.

15. $C = x + 7y$ Constraints: $x \geq 0$; $y \geq 0$; $x - y \geq 0$; $x - 4y \geq -9$

16. $C = 1.5x + 2.5y$ Constraints: $9 \leq x \leq 15$; $5 \leq y \leq 15$; $x + y \leq 24$

17. $C = 3x + 4y$ Constraints: $y + x \geq 8$; $y - x \leq 4$; $y - 2x \geq -10$

18. $C = -x + 3y$ Constraints: $x \geq 0$; $y \geq 0$; $x \leq 6$; $y \leq 8$; $x + 2y \geq 10$

19. **AGRICULTURE** Montez Farms owns a 3600-acre field. The farmers want to plant Iceberg lettuce which yields $200 per acre and Romaine lettuce which yields $250 per acre. To prevent loss due to disease, the farmers should plant no more than 3000 acres of Iceberg and no more than 2500 acres of Romaine. How many acres of each crop should Montez plant in order to maximize profits? What is the maximum profit?

20. **SMALL BUSINESS** Sasha owns and operates the Stand In-Line Skate Shop. She makes a profit of $40 on each pair of adult skates and $20 on each pair of child skates sold. Sasha can stock at most 80 pairs of skates on her shelves. Sasha orders skates once every 6 weeks and can order up to 50 pairs of each type of skate. How many of each type of skate must Sasha stock and sell in a 6-week period in order to maximize her profits? What is her maximum profit?

21. **MEDIA RESEARCH** A media research company is planning a week long survey of people's television viewing habits. The company knows they will need statisticians and interviewers for the survey. Each statistician receives $900 per week, whereas each interviewer receives $550 per week. The data collection phase of the survey will require at least 210 h and the data analysis phase will require at least 150 h. A statistician spends 10 h per week collecting data and 30 h per week analyzing data. An interviewer spends 30 h per week collecting data and 10 h per week analyzing data. How many of each type of worker should be assigned to minimize the cost of the survey?

22. **RETAILING** Fifty pounds of regular coffee and 90 pounds of decaffeinated coffee are available to make special holiday 5 lb gift packages. The Morning Mix Package contains four 1 lb bags of regular coffee and one 1 lb bag of decaffeinated coffee. The Coffee Break Mix contains four 1 lb bags of decaffeinated coffee and one 1 lb bag of regular coffee. The company makes $30 on the Morning Mix coffees and $40 for the Coffee Break coffees.

    a. If only Coffee Break packages were made, how many could be produced?

    b. If only Morning Mix packages were made, how many could be produced?

    c. How many of each type package should be produced to maximize the profits?

    d. Will all the available coffee be used? Explain.

23. WRITING MATHEMATICS Examine the graph at the right. Explain why it is impossible to find the maximum of function $P = 2x + 3y$ on the feasible region pictured.

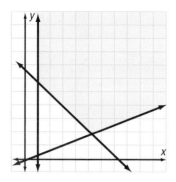

EXTEND

METALLURGY A metals manufacturer needs 20 tons of iron, 15 tons of copper, and 8 tons of zinc for a particular order. Two types of ore are available. Grade A ore contains 15% iron, 5% copper, and 2% zinc and costs $30 per ton. Grade B ore contains 10% iron, 10% copper, and 8% zinc, and costs $40 per ton. Let x equal the number of tons of ore A and y equal the number of tons of ore B.

24. Write an inequality that compares the total iron content that can be extracted from ores A and B with the manufacturer's iron needs.

25. Write an inequality that compares the total copper content that can be extracted from ores A and B with the manufacturer's copper needs.

26. Write an inequality that compares the total zinc content that can be extracted from ores A and B with the manufacturer's zinc needs.

27. Graph the three inequalities you have just identified.

28. Determine the vertices of the feasible region.

29. Write the objective function that is to be minimized.

30. Determine the minimum value of the objective function subject to the given constraints. What are the coordinates of the vertices of the feasible region at which the minimum occurs? What do these coordinates represent?

MIXED REVIEW

31. STANDARDIZED TESTS Jim scored 98, 89, 88, and 80 on the first four math tests. What must he score on the fifth exam to achieve an average of at least 90?

A. 90 **B.** 95 **C.** 88.75 **D.** 93

Solve each of the following.

32. $2x + 8 = -x - 1$ **33.** $-3y + 4 \le 25$ **34.** $|2x + 3| = 9$

Solve each of the following systems.

35. $\begin{cases} 3x + 2y = 16 \\ -4x + y = -3 \end{cases}$ **36.** $\begin{cases} -2x + 3y = 11 \\ x - y = -4 \end{cases}$

37. Determine the inverse of $y = 5 - 2x$.

The Environmental Protection Agency (EPA) was formed in 1970. EPA personnel are responsible for enforcing environmental protection laws regarding water quality, air pollution, toxic substances, and solid and hazardous waste. The EPA posts strict guidelines to which both the private and public sectors must adhere in order to maintain a safe and clean environment.

Decision Making

The chemical company near a river produces chemicals and generates waste. EPA regulations no longer permit disposal of this waste into the river, so the company must find a means of dealing with it. One method is to convert the waste so it can be sold as products.

One conversion formula calls for 2 lb of A and 3 lb of B to be added to 3 lb of waste to produce product C. A second conversion method is to add 4 lb of A and 1 lb of B to 4 lb of waste to produce product D. For inventory control, the company will limit the use of A and B to 1 lb each for each batch of waste converted. Product C sells for $1/lb and D sells for $3/lb. The two methods can be summarized in a table.

	A	B
Product C	2	3
Product D	4	1

1. What objective function can be used to maximize the income from products C and D? Explain this function.

2. The inequality $2C + 4D \leq 1$ models the limitations on the amount of product A to be used. Explain this constraint.

3. The inequality $3C + D \leq 1$ models the limitations on the amount of product B to be used. Explain this constraint.

4. Are there any other constraints that need to be considered? If so, state them and explain their function in this situation.

5. Graph the feasible region and find the solution that maximizes the objective function.

6. In what ratio should C and D be produced in order to maximize the income?

Linear Programming in Investment Planning

People invest to make money. Investing in stocks, bonds, and mutual funds can be risky, but smart investments can yield substantial returns.

One type of investment is bonds. A **bond** is a loan to a corporation or a government. States, cities, towns, and villages often buy and sell bonds. Bonds are sold to raise money for important municipal projects such as highways, airports, and schools. Bonds are purchased with the hope that the investment will make money. The investor is interested in maximizing returns on the investment while reducing the risk of losing money. Bonds, which are riskier investments, must pay a higher interest rate in order to attract investors.

Problem

A financial manager for a small town plans to invest in two bonds. Bond A pays an annual interest rate of 9.5% and Bond B pays an annual interest rate of 11%. The voters have approved a plan for an investment of up to $200,000. To reduce risk, the voters and the financial manager agree that the investment in Bond B should be limited to at most 35% of the total. The financial manager thinks that to get the best return, the total investment should be no less than $100,000. How much should be invested in each bond to maximize the return?

Explore the Problem

Let x represent the amount invested in Bond A, and let y represent the amount invested in Bond B.

1. Copy and complete the table.

Investment	Amount Invested	Interest Rate
Bond A		
Bond B		

2. What constraints represent limitations on the total amount invested for Bonds A and B?

3. What constraint represents a limitation on the amount invested in Bond B?

4. Write this constraint so it can be entered in a graphing utility.

5. Write a decimal expression in terms of *x* to represent the return on investing in Bond A.

6. Write a decimal expression in terms of *y* to represent the return on the investment in Bond B.

7. Let *R* represent the return objective function that is to be maximized. Write this linear objective function *R* in terms of *x* and *y*.

8. Explain the significance of this function.

9. To graph the constraints on a graphing utility, what window might you use? Why?

10. Graph the constraints. Determine the vertices of the feasible region.

11. At which vertex is the return function a maximum?

12. What investment combination of Bonds A and B should the town financial manager purchase?

13. How much will the city earn?

Investigate Further

Given the constraints in the problem, assume that there is a fee of $\frac{3}{4}$% of the Bond A purchase price and $\frac{1}{2}$% of the Bond B purchase price.

14. Write a decimal expression for the fee on the purchase of *x* dollars of Bond A.

15. Write a decimal expression for the fee on the purchase of *y* dollars of Bond B.

16. Write the linear objective function *F* which represents the total fees.

17. At which vertex of the feasible region will the fees be at a minimum?

18. What are the fees for the purchase of those bonds?

19. WRITING MATHEMATICS In this situation, is it possible for the town to maximize returns and minimize fees at the same time? Explain.

20. WRITING MATHEMATICS If you were the financial manager, what investment decision would you make? Explain.

PROBLEM SOLVING PLAN

- Understand
- Plan
- Solve
- Examine

ALGEBRA: WHO, WHERE, WHEN

To solve a linear programming problem with three variables, you would need to graph a feasible region that is the boundary and interior of a polyhedron. If the problem has four or more variables, a graphical solution is not possible and the approach must be entirely algebraic. The *simplex method*, developed by George Dantzig, uses matrices to narrow the choice of vertices that need to be tested.

Apply the Strategy

21. **COLLEGE INVESTMENTS** Rhea has $31,500 that her parents want her to invest for the greatest return before she leaves for college in one year. They agree that bonds and money market funds are good short-term investments. Rhea decides to invest no less than $10,000 and no more than $20,000 in the money market fund. She also will invest up to 75% of the money market amount in bonds. The money market funds have an annual return of 7.5% while the bonds pay 10.5%.

    a. What are the constraints on the total amount invested?

    b. What are the constraints on each type of investment?

    c. Write the return objective function that is to be maximized.

    d. What combination of investments will yield the greatest return on Rhea's college fund? How much will Rhea earn?

22. **RETIREMENT INVESTMENTS** The Allisons recently retired. They need at least $25,000 per year to maintain a comfortable lifestyle. Their Social Security and pension income is $19,800 per year. Their savings total $100,000. They can invest in a safe bank account at 4% per year or a riskier bond fund at 10% per year or both. They decide to invest no more than 40% in the bond fund. How much should they invest in each to maximize their return?

23. **BONDS** An investor has $60,000 to invest in two types of bonds. She expects an annual yield of 5% on Bond A and 10% on Bond B. She decides to invest at least one-fourth of her money in Bond B and no less than one-half in Bond A. She also decides that the total investment should be at least $30,000.

    a. How much should she invest in each type of bond to maximize her return?

    b. She must pay a 1% fee on the total Bond A investment and a 0.8% fee on the total Bond B investment. What combination of investments would minimize her fees and yet adhere to the constraints outlined above?

24. **WRITING MATHEMATICS** Determine the annual yields on two types of real world investments. You can look at newspaper advertisements, go to a bank, and/or interview a financial planner. Assume you have at most $50,000 to invest. Identify constraints that might apply to your investment plan. Explain how you would invest based upon a linear programming analysis.

REVIEW PROBLEM SOLVING STRATEGIES

A SPECIAL NUMBER

1. There is a five-digit number N with this interesting property: If you write a 1 after N, the new number is three times as large as N with a 1 written before it. Can you discover what N is? Use these questions to help you think about the problem.

a. If N is a five-digit number and you write a 1 after it, how many digits does the new number have? How can you express the value of the new number in terms of N?

b. If you write 1 before N, how can you express the value of the new number in terms of N?

c. What equation relates the two new numbers?

d. What is N?

DIGITALLY CORRECT

2. A certain number has four digits, the sum of which is 14. If you switch the first and the last digits, the new number is 2997 greater than the original. If you switch the two middle digits of the original number, the new number is 90 greater than the original. The sum of this second changed number plus the original is 4780.

a. What is the original number?

b. Analyze the steps that were used in the problem. Try making up a problem of your own.

ONE, TWO, THREE, KICK

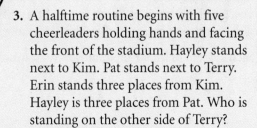

3. A halftime routine begins with five cheerleaders holding hands and facing the front of the stadium. Hayley stands next to Kim. Pat stands next to Terry. Erin stands three places from Kim. Hayley is three places from Pat. Who is standing on the other side of Terry?

CHAPTER REVIEW

VOCABULARY

Choose the word from the list that completes each statement.

1. A shaded region that does not include the boundary is a(n) __?__.

2. A(n) __?__ is a simple closed shape formed by line segments.

3. In linear programming, the linear inequalities are called __?__.

4. The solution of a system of linear inequalities is called the __?__.

5. When the solution of an inequality includes the boundary it is a(n) __?__.

6. When the solution of an inequality does not include the boundary it is a(n) __?__.

a. closed half-plane

b. constraints

c. feasible region

d. open half-plane

e. open set

f. polygon

Lesson 7.1   GRAPH LINEAR INEQUALITIES   pages 321–323

- The graph of a linear equation separates the coordinate plane into three parts: points on the line, points above the line, and points below the line. Points above the line and below the line form two **half-planes**. The line forms the **boundary** of each of the half-planes.

- The graph of a linear inequality in two variables includes all points in a half-plane. For inequalities with $\leq$ or $\geq$, the half-plane includes the boundary line and is a **closed half-plane**. For inequalities with $<$ or $>$, the half-plane does not include the boundary line and is an **open half-plane**.

Tell whether the ordered pair is a solution of the linear inequality.

7. $y \leq 2x + 6;\ (1, 10)$

8. $4y \geq 3x;\ (0, 0)$

9. $5x - y > 3;\ (1, 3)$

Graph each linear inequality.

10. $y < x - 1$

11. $5x + 10y \geq 30$

12. $4y - 3x \leq 24$

Lesson 7.2   GRAPH SYSTEMS OF LINEAR INEQUALITIES   pages 324–331

- If you graph a **system of linear inequalities** on a coordinate plane, the region in which the graphs overlap contains all the solutions of the system.

Solve each system of linear inequalities by graphing.

13. $\begin{cases} y \geq 2x \\ x + y < 8 \end{cases}$

14. $\begin{cases} y < 5 \\ y > x - 2 \end{cases}$

15. $\begin{cases} y \leq 2x + 3 \\ 2x + y > 0 \end{cases}$

16. The sum of two positive integers is less than ten and the difference between the integers is greater than four. Name four ordered pairs of integers that are solutions to the system. Let x represent the larger integer and y represent the smaller integer.

- To determine whether a value is a solution to a **quadratic inequality**, substitute it into the inequality and see if the resulting statement is true. If the inequality can be factored, factor it and substitute the value into the factored form.

Determine whether each value of x is a solution of the quadratic inequality.

17. $x^2 + 2x - 80 \geq 0; x = 9$

18. $x^2 + 6x - 2 > 0; x = 0$

19. $4x^2 - 8x - 2.25 < 0; x = -\dfrac{1}{4}$

20. $3x^2 + 2x - 5 \leq 0; x = \dfrac{1}{2}$

- The graph of a **quadratic inequality in two variables** includes all points in one of the regions created by the graph of the corresponding quadratic equation.

Determine whether each ordered pair is a solution of $y > 2x^2 - 7x - 3$.

21. $(3, 1)$ **22.** $(0, -3)$ **23.** $(5, 15)$ **24.** $(8, -1)$

Graph each inequality. Indicate the vertex and intercepts of the boundary curve.

25. $y \leq x^2 - 4x + 1$ **26.** $y > -0.25x^2$ **27.** $y < 9 - 16x^2$

- The intersection of a system of linear inequalities can define a polygonal region.

28. Determine the vertices of the polygonal region defined by the following system of inequalities: $x \geq 0, y \geq 0, y \geq x - 4,$ and $7y \leq -2x + 35$.

- **Linear programming** is a technique in which graphs of inequalities are used to find a maximum or minimum value of a linear function called the **objective function**.

- A maximum or minimum value always occurs at a vertex of the feasible region.

29. Determine the maximum value of the objective function $P = -3x + 4y$ given $x \geq 0, y \geq 0,$ $x + 3y \geq 18, x + y \leq 12, y - x \leq 4$. Identify the point at which it occurs.

- A **bond** is a loan issued to a corporation or a government.

30. Joe has $90,000 to invest in two types of bonds. He expects an annual yield of 6% on Bond A and 12% on Bond B. He wants to invest at least one-third of his money in Bond B and no more than one-half of his money in Bond A. He also wants his total investment to be at least $50,000. How much should he invest in each type of bond to maximize his return?

CHAPTER TEST

Tell whether the ordered pair is a solution of the linear inequality.

1. $y > 3x + 1; (3, 5)$ **2.** $y < 2x - 5; (2, -4)$

3. How many solutions does the inequality $|x + y| < -2$ have?

Graph each system of linear inequalities.

4. $\begin{cases} y \le 3x - 1 \\ x + y \ge 3 \end{cases}$ **5.** $\begin{cases} y + 2x \le 4 \\ 4x + 3y \ge 12 \end{cases}$ **6.** $\begin{cases} y \le 4x + 1 \\ y > 2x \end{cases}$

7. **STANDARDIZED TESTS** Which value(s) of x are solutions of $x^2 + 2x - 1 < 0$?

    I. 0   **II.** -1 **III.** -3

    A. II only   **B.** III only
    C. I and II   **D.** I and III

Solve each inequality algebraically.

8. $(x + 4)(x - 3) \le (x + 2)(x - 5)$

9. $(x + 6)(x - 2) \ge (x + 3)(x - 1)$

10. Write the system of linear inequalities whose feasible region is represented by the graph at the right.

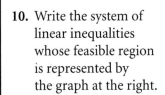
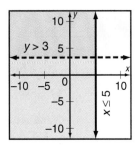

Solve each inequality. Graph the solution.

11. $x^2 + 3x - 40 \le 0$ **12.** $x^2 - 4x - 45 \ge 0$

Graph each inequality. Indicate the vertex and intercepts of the boundary.

13. $y \ge x^2 + 2x - 3$ **14.** $y \le 6x^2 - 5x - 6$

15. Determine the vertices of the polygonal region defined by the following system of inequalities: $x \ge 0$, $y \ge 0$, $10y \le -3x + 80$, and $4y \ge 5x - 30$.

16. **WRITING MATHEMATICS** Write a paragraph that outlines the steps to solving a linear program.

Find the values of x and y for the objective function under the given constraints.

17. Maximize $P = 3x + 2y$
    $2x + y \le 12; y \le 5 ; x \ge 0; y \ge 0;$

18. Minimize $C = 20x + 30y$
    $x \ge 1; x + y \ge 7; 3x + y \le 15;$

19. **WRITING MATHEMATICS** Explain why the feasible region graphed at the right does not yield a maximum value for its objective function.

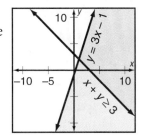

20. Stacey has $12,000 to invest. AAA bonds yield 6%, while B bonds yield 9%. Her policy is to invest at least twice as much in AAA bonds as in B bonds. How much should she invest in each type of bond to maximize her return? What is her annual return?

21. The Pep Club can buy bumper stickers for $1.00 each and "Homer Hankies" for $1.25 each. It can buy no more than 300 bumper stickers and no more than 150 hankies. The members must sell at least 200 items. They want to sell the bumper stickers for $2.00 each and the hankies for $2.50 each.

How many of each item should they sell to minimize costs? maximize profit?

PERFORMANCE ASSESSMENT

COMMUNITY INEQUALITIES Research a situation in your school, local community, or state for which linear programming can be used to allocate limited resources. State the objective function and the constraints.

FLASH YOUR GRAPH Work with a partner to make a set of five pairs of flashcards with linear and quadratic inequalities. Write the inequality on one card in each pair and graph the inequality on the other card in each pair. Exchange your cards with another group and practice matching the inequalities with their graphs.

SHADE YOUR AREA Draw a set of coordinate axes on graph paper. Shade a right triangle that has an area of 12 square units and does not have a vertex of $(0, 0)$. Write a system of linear inequalities that describes the shaded area.

TILE MODELS The designs on kitchen and bathroom ties are often based on simple shapes such as squares and triangles. Find a tile pattern that you like. Copy the design onto a set of coordinate axes on graph paper. Describe the regions in the design using systems of inequalities.

PROJECT *Connection*

Your group can focus on the following six topics from the Project Connections to make a presentation about taxes to the class.

- Social Security
- Medicare
- Federal Income Tax
- State Income Tax
- Sales Tax
- Property Tax

1. Plan your presentation by creating an outline. Each person should present the material they specialized in. Use the graphics completed for the Project Connections, and original material, to help your group convey the necessary information. Rehearse your presentation and time it. Make sure it fits in the allotted time.

2. As a group, prepare a pretest on taxes called "Your Tax I.Q." Have your classmates take this quiz before your presentation. Make sure it includes questions that will be answered in your presentation.

3. Give your presentation. Follow it up by handing out a posttest — another copy of "Your Tax I.Q." — to the class. Let them take the quiz again. Go over the quiz. Ask your audience to compare their grades on the pretest and posttest.

4. Have a class discussion on this question: What do you think would happen if paying income tax became an option?

CUMULATIVE REVIEW

1. Show the ratio of the area of the shaded region of the circle to the area of the rectangle is π.

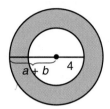 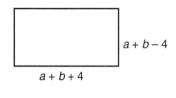

Solve each system by graphing.

2. $\begin{cases} |y| < 2 \\ x \geq -2 \end{cases}$ 3. $\begin{cases} x + y > 6 \\ y \leq 2x - 5 \end{cases}$

4. The following matrices show the number of miles driven in a year by each of a four-member household and the cost per mile of driving each of their two cars.

	Miles Driven		Cost per Mile
	Car 1	Car 2	(in dollars)
Sam	5412	2546	Car 1 [0.23]
Avi	4543	3756	Car 2 [0.31]
Jan	1790	1928	
Tim	3578	1578	

How much would the household have saved if Jan had always driven Car 1?

5. **WRITING MATHEMATICS** Explain how to graph the solution of $y > 3x + 7$.

6. Argon refinery produces at most 30,000 barrels of crude oil and diesel oil in a week. At least 10,000 barrels must be crude oil and at least 5,000 barrels must be diesel oil.

   a. Write and graph the inequalities for the constraints. Determine the feasible region.

   b. If crude oil sells for $18.50 per barrel and diesel oil for $20.75 per barrel, determine the production output that will yield a maximum revenue for Argon.

7. The number of gallons that a car uses varies directly with the number of miles driven. If Meosha's car uses 11.5 gal to go 150 mi, how many gallons does it need to go 375 mi?

8. Solve the inequality $x^2 - 5x \geq 6$ and graph the solution set on the number line.

9. Determine a system of inequalities whose solution is the triangle and its interior region shown below.

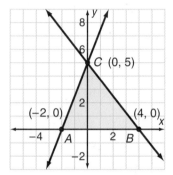

10. **STANDARDIZED TESTS** Which of the following statements are true for $4x^2 - 12x + 9 = 0$?

   I.   The discriminant is 0.
   II. The sum of the roots is 12.
   III. The graph has one x-intercept.

   A. I, II, and III   **B.** I only
   C. I and II   **D.** I and III

11. In a bowling tournament, Jon had the following scores for his first 8 games.

   250 219 245 224 215 232 274 217

   Jon estimates that he needs a mean of 240 for 12 games in order to be a prize winner. What must he average in his last 4 games?

12. Graph the inequality $y \leq -x^2 + 5x - 4$. Indicate the intercepts and vertex of the boundary curve.

STANDARDIZED TEST

STANDARD FIVE-CHOICE Select the best choice for each question.

1. A box contains marbles of four colors: red, white, blue, and yellow. There are 8 more red marbles than white, and twice as many white marbles as each of the blue and yellow. If a marble is drawn at random from the 56 that are in the box, the probability that it will be blue is

   A. 1:56   **B.** 1:7   **C.** 1:8
   D. 2:7   **E.** 3:7

2. The value of the determinant shown at the right is $\begin{vmatrix} -21 & 12 \\ 7 & 4 \end{vmatrix}$

   A. 196   **B.** −196   **C.** 108
   D. −108   **E.** 0

3. Which of the following systems is represented by the shaded region of the graph?

   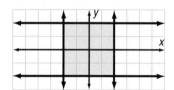

   A. $\begin{cases} |y| \geq 1 \\ |x| \leq 1 \end{cases}$ **B.** $\begin{cases} |y| \leq 1 \\ |x| \geq 1 \end{cases}$ **C.** $\begin{cases} |y| \leq 1 \\ |x| \leq 1 \end{cases}$

   D. $\begin{cases} |y| \geq 1 \\ |x| \geq 1 \end{cases}$ **E.** $\begin{cases} |y| < 1 \\ |x| < 1 \end{cases}$

4. If $\dfrac{\text{area of } ABEF}{\text{area of } FECD} = \dfrac{4}{1}$, then the dimensions of $ABCD$ are

   A ———————— B
   (with $x+1$ on upper right, F———E, x, D——$x+3$——C)

   A. $3\frac{1}{3} \times 1\frac{1}{3}$ **B.** 7×9 **C.** $3\frac{1}{2} \times 1\frac{2}{3}$
   D. 6×7 **E.** $3\frac{1}{4} \times 3\frac{1}{2}$

5. Which of the following statements are true for the function $f(x) = x^3$?

   I. $f(x)$ is one-to-one.
   II. $f^{-1}(x)$ exists and is a function.
   III. $f(x)$ has point symmetry.

   A. I and II   **B.** I and III   **C.** II and III
   D. I only   **E.** I, II, and III

6. Which of the following ordered pairs does not satisfy the greatest integer function $y = [x]$?

   A. $(2, 2)$ **B.** $(2.1, 2)$ **C.** $(2, 2.1)$
   D. $(2.99, 2)$ **E.** $(-3.1, -4)$

7. The solution of $2x^2 + 5x - 3 \leq 0$ is

   A. $-3 \leq x \leq \frac{1}{2}$

   B. $-\frac{1}{2} \leq x \leq 3$

   C. $(x \leq -3) \text{ or } \left(x \geq \frac{1}{2}\right)$

   D. $\left(x \leq -\frac{1}{2} \text{ or } (x \geq 3)\right)$

   E. $(x \leq -3) \text{ or } \left(x \geq -\frac{1}{2}\right)$

8. The height h of a projectile at any time t can be modeled by the quadratic function $h = -16t^2 + v_0 t + h_0$, where v_0 is the initial upward velocity and h_0 is the initial height of the object. If a soccer ball resting on the ground is kicked with an initial upward velocity of 48 ft/s, the maximum height the ball will reach is

   A. 48 ft   **B.** 36 ft   **C.** 30 ft
   D. 24 ft   **E.** 18 ft

8. The solution set of $(x - 1)8x^2 + (x - 1)(-2x) + (x - 1)(-3)$ is

   A. $\left\{-\frac{1}{2}, \frac{3}{4}\right\}$ **B.** $\left\{\frac{1}{2}, -\frac{3}{4}\right\}$ **C.** $\left\{-\frac{1}{2}, -\frac{3}{4}\right\}$

   D. $\left\{1, -\frac{1}{2}, \frac{3}{4}\right\}$ **E.** $\left\{-1, -\frac{1}{2}, \frac{3}{4}\right\}$

8 Exponents and Radicals

Take a Look
AHEAD

Make notes about things that look new.
- What type of exponents will you work with for the first time in this chapter?
- What exponent is used to denote a square root?
- Locate and copy an example of a pure imaginary number and a complex number.

Make notes about things that look familiar.
- What symbol is used to denote a square root?
- Copy at least one example that shows multiplication of expressions containing exponents and at least one that shows division of expressions containing exponents. Explain the multiplication and division properties for integral exponents.

DATA Activity

Fly Me to the Moon
The aerospace industry includes the manufacturing of aircraft, missiles, space vehicles and propulsion systems, related products such as flight simulators, electronics, and ground support systems, as well as the operation of facilities needed for air and space travel. Although some parts of this industry have experienced a downturn in recent years, there are still many employment opportunities for engineers, computer specialists, electricians, mechanics, and radar technicians. The table on the next page gives some key figures about aerospace industry sales.

SKILL FOCUS
- Add, subtract, multiply, and divide real numbers.
- Use scientific notation.
- Solve percent problems.
- Determine average rate of change.
- Solve proportions.
- Display data graphically.

In this chapter you will learn how:

- **METEOROLOGISTS** collect data and use formulas with radicals and exponents to predict weather conditions for space missions.
  (Lesson 8.2, page 380)

- **AEROSPACE ENGINEERS** use algebraic methods to determine the velocity and energy requirements for space missions.
  (Lesson 8.4, page 391)

AEROSPACE

Aerospace Industry Sales (In billions of Dollars)			
	1993	**1994**	**1995**
Total Aircraft	66.5	58.2	56.7
Civil	33.8	26.2	25.8
Military	32.7	32.0	30.9
Missiles	8.1	7.3	6.6
Space Vehicles and Parts	28.9	28.5	27.8
Related Products and Services	20.7	18.8	18.2

Use the table to answer the following questions.

1. What were the total aerospace sales for each of 1993, 1994, and 1995? Write answers in scientific notation.

2. In 1994, what percent of total aircraft sales went to the military? Round to the nearest tenth of a percent.

3. What was the average yearly change in missile sales for 1993–1995?

4. When expressed in terms of 1987 dollars, the value of the 1995 related products and services category would be 14.5 billion dollars. What would the value of the 1995 space vehicles and parts category be in terms of 1987 dollars?

5. **WORKING TOGETHER** Research data about the number of people employed in the aerospace industry for the last three to five years. Display your data graphically and write two questions that can be answered using your graph.

Blast Off!

In simplest terms, Newton's Third Law of Motion states that for every action there is an equal and opposite reaction. This law is used in the propulsion system of a rocket. Escaping gases exert pressure on the rocket and cause it to move in a forward direction. This forward movement is known as the rocket's *forward thrust*. The forward thrust of the rocket acts in the opposite direction of the propellants that are ejected from the rocket.

Although the study of rocketry is quite complex, this project will introduce you to Newton's law and give you an opportunity to experiment with a balloon rocket.

PROJECT GOAL

To explore relationships present in the thrust and altitude capabilities of a balloon rocket.

Getting Started

Work in groups of four students.

1. In preparation for the Project Connections, gather the following materials: some oblong balloons, plastic drinking straws, tape, about 10 meters of string, twist ties, an expandable meter stick or metal tape measure, and graph paper.

2. Two group members should research topics that are involved in rocketry including *engine thrust* and *jet propulsion*. The other two group members should compile a chart showing the formulas for determining the volume of a sphere, the volume of a cylinder, and the circumference of a circle. These group members should be prepared to identify the variables used and explain the relationships among the formulas.

PROJECT *Connections*

Lesson 8.2, page 379:
   Develop a method for approximating the volume of an oblong balloon.

Lesson 8.4, page 390:
   Build and launch a balloon rocket; determine relationships between volume and distance traveled.

Lesson 8.5, page 398:
   Compare results for a double balloon rocket system; investigate relationships between volume and maximum altitude.

Project Assessment, page 417:
   Present and discuss project results; design experiments to test other motion relationships.

Algebra Workshop

Explore Exponents and Radicals

Think Back

● Simplify each expression. You may want to review the properties of exponents in Lesson 5.1.

1. $y^4 \cdot y^5$

2. $\dfrac{n^6}{n}$

3. $\dfrac{k^7 \cdot k^9}{k^2}$

4. $x^2(x+3)^6(x+3)^4$ **5.** $(5m-7)^0$

6. $\dfrac{14p^3q^5}{2n^3pq^2}$

Simplify, writing each expression with positive exponents.

7. b^{-7}

8. $\dfrac{4x^{-3}y^4}{6x^{-5}y^{10}}$

9. $\dfrac{a^5(a+1)^{-9}}{(a+1)^{-3}}$

Write each expression with positive exponents and without parentheses.

10. $(2h)^4$

11. $5(e^3)^4$

12. $(a^5b^3)^4$

13. $\left(\dfrac{d^3z^0}{w^4}\right)^5$

14. $\left(\dfrac{x^{-3}}{x^2}\right)^4\left(\dfrac{x^6}{x^{-5}}\right)^3$

15. $\left(\dfrac{a^{-2}}{b^{-3}}\right)^{-4}$

SPOTLIGHT ON LEARNING

WHAT? In this lesson you will learn
• to use the properties of exponents.
• to explore the properties of radicals and rational exponents.

WHY? Exponents and radicals can help you solve problems such as finding the area of a rectangle or the volume of a rectangular prism.

Explore

● **16.** Use the Pythagorean theorem in $\triangle ADE$ to determine the length of $\overline{AD}$.

17. Use the Pythagorean theorem in $\triangle CDF$ to determine the length of $\overline{CD}$.

18. By counting squares, determine the area of rectangle $ABCD$.

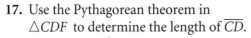
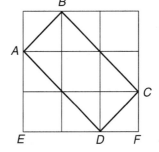

19. How does this exploration show that $\sqrt{8} \cdot \sqrt{2} = \sqrt{16}$?

20. Make a conjecture about the value of $\sqrt{8} \cdot \sqrt{18}$. Then use a geometrical demonstration like the one you did above to verify your conjecture.

21. Make a conjecture about the value of $\sqrt{5} \cdot \sqrt{5}$. Then use a calculator to verify your conjecture. Explain your method.

22. Use $\triangle AEG$ and $\triangle ECF$ to show that $AC = \sqrt{5} + \sqrt{5}$. Explain your method.

23. Use $\triangle ABC$ to determine AC. Is $\sqrt{5} + \sqrt{5} = \sqrt{10}$? Why or why not?

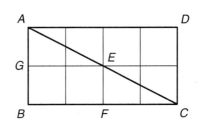

24. Make a conjecture about the value of $\sqrt{8} + \sqrt{8}$. Then use a geometrical demonstration like you did in Question 20 to verify your conjecture.

Decide whether each of the following is true or false. Use a calculator to verify.

25. $\sqrt{6} \cdot \sqrt{5} = \sqrt{30}$

26. $\sqrt{6} + \sqrt{6} = \sqrt{12}$

27. $\sqrt{11} \cdot \sqrt{11} = 11$

28. $\sqrt{11} + \sqrt{11} = \sqrt{44}$

29. $\sqrt{9} + \sqrt{16} = \sqrt{9 + 16}$

30. $\sqrt{9} \cdot \sqrt{16} = \sqrt{9 \cdot 16}$

31. Write the positive square root of each of the following: 1, 4, 9, 16, 25, 36.

32. Using a calculator, raise each of the numbers in Question 31 to the $\frac{1}{2}$ power.

33. Make a conjecture about what it means to raise a number to the $\frac{1}{2}$ power.

34. Test your conjecture with other numbers. Does it appear to be true?

35. Write the cube root of each of the following: 1, 8, 27, 64, 125.

36. Use a calculator to raise each of the numbers in Question 35 to the $\frac{1}{3}$ power.

37. Make a conjecture about what it means to raise a number to the $\frac{1}{3}$ power.

38. Test your conjecture with other numbers. Does it appear to be true?

39. Predict the $\frac{1}{4}$ power of 81 and the $\frac{1}{5}$ power of 32. Then evaluate each power using a calculator.

40. What does it mean to raise a number to the $\frac{1}{4}$ power? to the $\frac{1}{5}$ power?

41. Use a calculator to evaluate each expression. Look for patterns.

   a. $9^{\frac{3}{2}}$ $(9^3)^{\frac{1}{2}}$ $(9^{\frac{1}{2}})^3$

   b. $8^{\frac{5}{3}}$ $(8^5)^{\frac{1}{3}}$ $(8^{\frac{1}{3}})^5$

   c. $256^{\frac{3}{4}}$ $(256^3)^{\frac{1}{4}}$ $(256^{\frac{1}{4}})^3$

42. Use the patterns in Question 41 to write two expressions equivalent to $128^{\frac{5}{7}}$. Evaluate each expression.

Make Connections

State whether each equation is true or false. Give an example or counterexample to support your answer.

43. $\sqrt{a} + \sqrt{b} = \sqrt{a + b}$ **44.** $\sqrt{a} - \sqrt{b} = \sqrt{a - b}$

45. $\sqrt{a}\,\sqrt{b} = \sqrt{ab}$ **46.** $\dfrac{\sqrt{a}}{\sqrt{b}} = \sqrt{\dfrac{a}{b}}$

47. Write an equation of the form $a^x = b$ expressing each of the following relationships.

    a. a is the square root of b **b.** a is the cube root of b

    c. a is the fourth root of b **d.** a is the nth root of b

48. Write an equation of the form $a = b^x$ expressing each of the relationships in Question 47.

Write each of the following expressions using an exponent.

49. the square root of k **50.** the cube root of 7

51. the sixth root of 5   **52.** the ninth root of 11.2

53. the cube root of x squared   **54.** the eleventh root of 15 raised to the fifth power

Summarize

55. **MODELING** Show that the volume of the rectangular prism equals 12.

56. **THINKING CRITICALLY** Is $\sqrt{a + b}$ ever equal to $\sqrt{a} + \sqrt{b}$? Give an example.

57. **GOING FURTHER** Solve for x:
$$\left(x^{\frac{1}{2}}\right)^{\frac{1}{3}} = 2$$

58. **GOING FURTHER** Write $(n^{\frac{1}{4}})^{\frac{1}{5}}$ using a single radical.

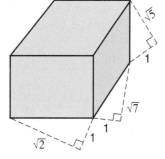

59. **MODELING** Use a graphing utility to evaluate $2^{\frac{1}{5}}$. Explain your method.

60. **WRITING MATHEMATICS** The domain of the function $f(x) = x^{\frac{1}{2}}$ is restricted to nonnegative values of x. Explain why.

61. **WRITING MATHEMATICS** Are any values excluded from the domain of the function $f(x) = x^{\frac{1}{3}}$? Explain.

8.2 Rational Exponents and Radicals

Explore

SPOTLIGHT ON LEARNING

WHAT? In this lesson you will learn
- to evaluate rational expressions.
- to simplify rational expressions.

WHY? Evaluating expressions with rational exponents and radicals can help you solve problems in commerce, aerospace, and physics.

1. Evaluate 2^n for $n = 0$, 1, 2, 3, and 4. Would you expect $2^{2.5}$ to be midway between $2^{2.0}$ and $2^{3.0}$? Explain.

2. Graph $y = 2^x$. Use the TRACE feature to approximate $2^{2.5}$. Compare with a calculator value for $2^{2.5}$.

3. Do your graphical results confirm or prove false what you expected in Question 1? Explain.

4. Use a calculator to evaluate $\sqrt{2^5}$ and $\left(\sqrt{2}\right)^5$. Compare your results with the value of $2^{2.5}$. What do you notice?

5. Using the patterns suggested by Question 4, write and evaluate two radical expressions equal to $3^{1.5}$. Then approximate $3^{1.5}$ graphically. Did you get the same results each time? If not, why might they have been different?

Build Understanding

ALGEBRA: WHO, WHERE, WHEN

Historians of mathematics agree that German mathematician Christoff Rudolff introduced the use of the radical symbol in 1525. Leonard Euler claimed that Rudolff modified the letter *r*, the first letter in the Latin word *radix* (root), to obtain the symbol. Other authorities believe the symbol evolved from a dot.

Recall that if a and b are real numbers and $b^2 = a$, then b is a **square root** of a.

$$5^2 = 25$$
$$(-5)^2 = 25$$

5 and -5 are square roots of 25

When a number has more than one square root, the nonnegative square root is called the **principal root**. $\sqrt{25}$ indicates the principal root of 25.

$$\sqrt{25} = 5 \qquad \text{principal square root}$$
$$-\sqrt{25} = -5$$

These ideas can be extended to roots other than square roots.

Let n be an integer greater than 1. If a and b are real numbers with $b^n = a$, then b is the **principal nth root** of a, $b = \sqrt[n]{a}$. The integer n is the **index** of the radical. The index 2 is normally not written. The real number a is the **radicand**.

$$\text{index} \rightarrow \sqrt[n]{a} \leftarrow \text{radical symbol}$$
$$\nwarrow \text{radicand}$$

For every real number a, $a^n \geq 0$ for all *even* values of n.

$$2^8 = 256 \qquad (-2)^8 = 256 \qquad 3^6 = 729 \qquad (-3)^6 = 729$$

Therefore, a negative number cannot have a real nth root for even values of n.

Notice that with odd values of n, there is only one root.

$$\sqrt[5]{32} = 2 \qquad \sqrt[5]{-32} = -2$$

Odd roots exist for both negative and nonnegative values of a radicand.

> - If n is odd, then $\sqrt[n]{a}$ is the nth root of a.
> - If n is even and $a \geq 0$, then $\sqrt[n]{a}$ is the nonnegative nth root of a.
> - If n is even and $a \leq 0$, then $\sqrt[n]{a}$ is not a real number.

EXAMPLE 1

Evaluate each radical expression, if possible.

a. $\sqrt{-36}$ **b.** $\sqrt[4]{0.0081}$ **c.** $\sqrt[3]{-125}$

Solution

a. Since the square of any number is positive, a negative number cannot have a real square root. Therefore, $\sqrt{-36}$ has no real solution.

b. The radicand is positive, so the expression has a real fourth root.

$$(0.3)^4 = 0.0081 \qquad (-0.3)^4 = 0.0081$$

So, $\sqrt[4]{0.0081} = 0.3$

c. $(-5)^3 = -125$, so $\sqrt[3]{-125} = -5$ ◄

CHECK UNDERSTANDING

How can you change Example 1b so that the solution is -0.3?

A square root can be simplified by looking for pairs of equal factors in the prime factorization of the radicand.

$$\sqrt{54} = \sqrt{3 \cdot 3 \cdot 3 \cdot 2} = \sqrt{3 \cdot 3} \cdot \sqrt{3 \cdot 2} = 3\sqrt{3 \cdot 2} = 3\sqrt{6}$$

Apply a similar method to simplify an nth root. Look for groups of n equal factors in the prime factorization of the radicand. You can simplify because the nth root of each such group is the factor itself. When the index is even, you must indicate the principal root of a variable as the absolute value of the variable.

COMMUNICATING ABOUT ALGEBRA

Explain why $\sqrt[n]{x^n}$ equals x for odd values of x.

EXAMPLE 2

Simplify.

a. $\sqrt[3]{80}$ **b.** $-\sqrt[4]{1250k^4}$

Solution

a. $\sqrt[3]{80} = \sqrt[3]{2 \cdot 2 \cdot 2 \cdot 2 \cdot 5} = \sqrt[3]{2 \cdot 2 \cdot 2} \cdot \sqrt[3]{2 \cdot 5} = 2\sqrt[3]{10}$

b. $-\sqrt[4]{1250k^4} = -\sqrt[4]{2 \cdot 5^4 \cdot k^4} = -5\sqrt[4]{2} \cdot \sqrt[4]{k^4} = -5|k|\sqrt[4]{2}$ ◄

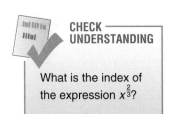

PROBLEM
SOLVING TIP

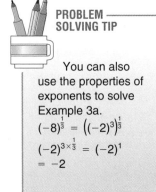

You can also
use the properties of
exponents to solve
Example 3a.
$(-8)^{\frac{1}{3}} = ((-2)^3)^{\frac{1}{3}}$
$(-2)^{3 \times \frac{1}{3}} = (-2)^1$
$= -2$

Rational exponents are defined as follows.

If m and n are positive integers and $\sqrt[n]{a}$ is a real number, then

$$a^{\frac{1}{n}} = \sqrt[n]{a} \qquad a^{\frac{m}{n}} = \left(\sqrt[n]{a}\right)^m = \sqrt[n]{a^m}$$

The properties of integral exponents are also true for rational exponents.

EXAMPLE 3

Evaluate each expression.

   a. $(-8)^{\frac{1}{3}}$ **b.** $49^{-\frac{1}{2}}$ **c.** $6.3^{\frac{9}{5}}$ **d.** $(10,000)^{-\frac{3}{4}}$

Solution

   a. $(-8)^{\frac{1}{3}} = \sqrt[3]{-8} = \sqrt[3]{(-2)^3} = -2$

   b. $49^{-\frac{1}{2}} = \dfrac{1}{49^{\frac{1}{2}}} = \dfrac{1}{\sqrt{49}} = \dfrac{1}{7}$

   c. $6.3^{\frac{9}{5}} \approx 27.467111$ Use a calculator to approximate.

   d. $(10,000)^{-\frac{3}{4}} = \dfrac{1}{(10,000)^{\frac{3}{4}}} = \dfrac{1}{\left(\sqrt[4]{10,000}\right)^3} = \dfrac{1}{10^3} = \dfrac{1}{1000}$ ◄

A radical expression is in **simplest form** when it is written using the least possible index.

EXAMPLE 4

Write each expression in simplest radical form.

   a. $\sqrt[8]{16}$ **b.** $3^{\frac{1}{3}} m^{\frac{5}{6}} n^{\frac{3}{4}}$

Solution

   a. $\sqrt[8]{16} = \sqrt[8]{2^4} = 2^{\frac{4}{8}} = 2^{\frac{1}{2}} = \sqrt{2}$

   b. Rewrite exponents using the LCD.

$$3^{\frac{1}{3}} m^{\frac{5}{6}} n^{\frac{3}{4}} = 3^{\frac{4}{12}} m^{\frac{10}{12}} n^{\frac{9}{12}} = (3^4 m^{10} n^9)^{\frac{1}{12}} = \sqrt[12]{3^4 m^{10} n^9}$$ ◄

Real world formulas often contain radical or exponential expressions that can be evaluated using a calculator.

EXAMPLE 5

AEROSPACE The atmospheric pressure p in lb/in.², at an altitude of h feet above sea level can be approximated by

$$p = 14.7\left(\sqrt[25,000]{2.718}\right)^{-h}$$

   a. Write the formula in exponential form, removing the radical.

   b. Find the atmospheric pressure at sea level.

   c. Jet aircraft must be pressurized at high altitudes. If the supersonic Concorde is cruising at 60,000 ft, what is the pressure?

Solution

a. $p = 14.7\left(\sqrt[25,000]{2.718}\right)^{-h}$

$\quad p = 14.7(2.718)^{\frac{-h}{25,000}}$

b. At sea level, $h = 0$.

$\quad p = 14.7(2.718)^{\frac{-0}{25,000}} = 14.7(2.718)^0 = 14.7(1) = 14.7$

The atmospheric pressure at sea level is 14.7 lb/in.²

c. $h = 60,000$

$\quad p = 14.7(2.718)^{-\frac{60,000}{25,000}} \approx 1.33$

The atmospheric pressure at 60,000 ft is about 1.33 lb/in.² ◀

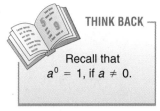

THINK BACK

Recall that
$a^0 = 1$, if $a \neq 0$.

TRY THESE

Evaluate each radical expression, if possible.

1. $-\sqrt{81}$

2. $\sqrt[3]{-8}$

3. $\sqrt[4]{-16}$

4. $\sqrt[7]{-1}$

Simplify.

5. $\sqrt{45}$

6. $-2\sqrt{200}$

7. $-\sqrt{24x^2}$

8. $\sqrt[3]{-108x^3}$

Evaluate each expression. Use a calculator for decimal approximations.

9. $16^{\frac{1}{2}}$

10. $125^{-\frac{1}{3}}$

11. $7.23^{\frac{5}{4}}$

12. $-(64^{\frac{5}{6}})$

Write each expression in simplest radical form.

13. $\sqrt[4]{4}$

14. $\sqrt[9]{(-20)^3}$

15. $\sqrt{5}\sqrt[3]{5}$

16. $a^{\frac{5}{6}}b^{\frac{1}{2}}$

17. **AEROSPACE** A donut-shaped space station of radius r ft must spin at an angular velocity of $\sqrt{gr}$ ft/s in order to produce artificial gravity equal to Earth's gravitational pull where g is the acceleration of gravity (32 ft/s²). The radius of the space station is 192 ft. Determine and write in simplest radical form the spin velocity that will produce artificial gravity equal to that on Earth.

18. **WRITING MATHEMATICS** If $x^2 = y^2$, does $x = y$? Explain.

PRACTICE

Write each expression using a radical symbol.

1. $3^{\frac{1}{2}}$

2. $8^{\frac{5}{3}}$

3. $(x + y)^{\frac{3}{10}}$

4. $p^{-\frac{2}{5}}$

Write each expression using an exponent.

5. $\sqrt[4]{15}$

6. $\sqrt[8]{e^9}$

7. $\dfrac{1}{\sqrt[5]{n}}$

8. $\sqrt[9]{(3x^5 - y)^2}$

Evaluate each radical expression, if possible.

9. $-\sqrt{121}$

10. $\sqrt[4]{1}$

11. $\sqrt{0.09}$

12. $\sqrt[5]{-243}$

13. $\sqrt[7]{0}$

14. $\sqrt{-0.25}$

15. $\sqrt[5]{6^{10}}$

16. $\sqrt[8]{(-10)^{24}}$

Simplify.

17. $\sqrt{98}$

18. $-3\sqrt{375}$

19. $\sqrt{2h^2}$

20. $-\sqrt[4]{32}$

21. $\sqrt{0.64x^6}$

22. $\sqrt[4]{10,000x^8}$

23. $\sqrt{16a^2b^4}$

24. $-\sqrt[3]{-64k^{12}}$

25. $\sqrt[5]{32x^{25}}$

26. $\sqrt{(c-2)^2}$

27. $\sqrt[7]{(x+4)^7}$

28. $\sqrt{y^2-4y+4}$

29. $1024^{0.1}$

30. $(2^{k-2})(2^{2-k})$

31. $\sqrt[3]{b^{12n-15}}$

32. WRITING MATHEMATICS Will a radical expression with an even index and a real number as the radicand always be a real number? Justify your answer.

Evaluate each expression. Use a calculator for decimal approximations.

33. $4^{\frac{1}{2}}$

34. $27^{-\frac{1}{3}}$

35. $16^{\frac{3}{4}}$

36. $64^{-\frac{2}{3}}$

37. $49^{\frac{3}{2}}$

38. $9^{\frac{5}{2}}$

39. $-125^{\frac{4}{3}}$

40. $32^{\frac{2}{5}}$

41. $(9^{\frac{4}{3}})^{\frac{9}{8}}$

42. $81^{\frac{3}{8}}81^{\frac{7}{8}}$

43. $12.08^{\frac{7}{2}}$

44. $\left(\frac{2}{3}\right)^{\frac{2}{3}}$

Write each expression in simplest radical form.

45. $\sqrt[6]{7^3}$

46. $36^{\frac{1}{4}}$

47. $\sqrt[16]{3^4}$

48. $\sqrt[6]{8}$

49. $\sqrt[3]{2}\sqrt[4]{2^3}$

50. $\sqrt[20]{x^8}\sqrt[5]{x^2}$

51. $m^{\frac{3}{5}}n^{\frac{2}{3}}$

52. $\sqrt{\sqrt[3]{5}}$

53. PHYSICS The distance d in feet that an object falls in t seconds is given by $d = 16t^2$.

    a. Solve the formula for t using an exponent to write your answer.

    b. Write a radical expression for t. Then determine how long it takes an object to fall 864 ft, expressing your answer in simplest radical form.

54. GEOMETRY The volume V of a regular octahedron with edge e is given by $V = \frac{e^3\sqrt{2}}{3}$. The edge of one octahedron is $\sqrt{24}$ cm. Write the volume in simplest radical form.

55. COMMERCE An item sold for d dollars in 1967. The price P of the item t years after 1967 can be approximated using the *consumer price index* formula $P = d\left(\sqrt[16]{2.7}\right)^t$.

    a. Write the formula in exponential form.

    b. The annual tuition at Euclid University was $2400 in 1967. Approximate the 1995 tuition at the university.

56. WRITING MATHEMATICS Write a paragraph explaining why it is possible to find a real number that is the cube root of -64 but not the square root of -64.

EXTEND

Determine the domain of each function.

57. $f(x) = x^{\frac{1}{2}}$

58. $f(x) = x^{\frac{1}{7}}$

59. $f(x) = (2x + 6)^{\frac{1}{4}}$

60. $f(x) = (-5x^2)^{\frac{1}{2}}$

61. WRITING MATHEMATICS For what values of x will $\sqrt{(x + 5)^2} = x + 5$? Explain how you determined your answer.

Solve for x.

62. $2x^5 - 10 = 0$

63. $6x^{\frac{2}{3}} = 150$

64. $\sqrt[4]{x^5} = 32$

65. $x^9 + x^2 = 0$

THINK CRITICALLY

66. Determine the domain of the function $f(x) = \dfrac{\sqrt{x + 5}}{x^2 - 5x + 6}$.

Write each expression using a single radical symbol.

67. $\dfrac{1}{a^{-\frac{1}{n}}}$

68. $\sqrt[n]{\sqrt[m]{a}}$

69. $a^{\frac{x}{z}} \cdot a^{\frac{y}{z}}$

70. $(a^{\frac{w}{x}})^{\frac{y}{z}}$

71. Find two values of x that satisfy this equation: $\sqrt[3]{x} = \sqrt{x}$

MIXED REVIEW

Factor completely.

72. $y^2 - 81$

73. $9a^2 - 4$

74. $2c^4d - 32d$

75. $x^3 - 1$

76. STANDARDIZED TESTS Choose the letter of the value equal to $16^{-\frac{3}{4}}$.

    A. $\sqrt[3]{16^{-4}}$ **B.** $\dfrac{1}{8}$ **C.** -8 **D.** $\dfrac{1}{12}$

PROJECT *Connection*   Begin by inflating an oblong balloon. Notice that the shape of the balloon is similar to that of a cylinder with a hemisphere at either end.

1. Instead of trying to measure the radius of the inflated balloon, you can find its measure indirectly by solving the circumference formula for r: $r = \dfrac{2\pi}{c}$. Write a formula for the volume of the balloon (volume of cylinder + volume of sphere).

2. Inflate the balloon to its fullest capacity. Measure the length of the cylindrical portion. Use a piece of string to measure the circumference. Determine the radius and volume.

3. Deflate the balloon a bit and determine the radius and volume at this size. Repeat this procedure three more times (ignore the "knob" at the balloon end).

4. Graph the relationship between the radius of the balloon and the volume. Let volume be the independent variable graphed on the horizontal axis and the radius be the dependent variable on the vertical axis. Describe the graph. Use your graphing utility to determine a function that might approximate such a relationship.

Career
Meteorologist

Although scientists have gone a long way toward conquering outer space, they have yet to conquer the weather here on Earth. Space missions can be delayed by clouds and rain. For that reason, the work of meteorologists and their ability to predict weather for launch and recovery are critical to the success of any space mission.

On March 16, 1926, the first liquid fuel rocket rose 41 ft into the air over Auburn, Massachusetts, and traveled a distance of 184 ft. In 1986, *Pioneer 10* crossed the orbit of Pluto 3.67 billion miles from Earth.

Decision Making

1. When *Pioneer 10* crossed the orbit of Pluto, how many times as far as the distance traveled by the first liquid fuel rocket had it traveled?

2. Weather satellites can measure the diameter of a storm. Using the formula $t = \left(\dfrac{d}{6}\right)^{\frac{3}{2}}$, where d represents storm diameter in miles, meteorologists approximate the duration t of a storm in hours.

   a. Write the formula for t using a radical symbol.

   b. Determine the duration of a storm with a diameter of 24 mi.

   c. Use a graphing utility to graph $t = \left(\dfrac{d}{6}\right)^{\frac{3}{2}}$. Trace to approximate the diameter of a storm that lasted 35 minutes.

3. The January–June average daily temperature t at Kennedy Space Center in Florida can be approximated by $t = 49.6 \sqrt[25]{D^6}$ where D is the date expressed in months (for example, January 1 = 1.0; April 15 = 4.5).

   a. Write the formula in exponential form.

   Approximate the average temperature on each date.

   b. June 1 **c.** March 9

4. The formula

   $$T_{WC} = 0.045(5.27\sqrt{V} + 10.45 - 0.28V)(T - 33) + 33$$

   gives the wind chill temperature T_{WC} in degrees Celsius at temperature T (degrees Celsius) and wind velocity V (mi/h). Determine the wind chill temperature at 0°C and wind velocity 36 mi/h.

HURRICANE ANDREW
24 AUGUST 1992
5 AM EDT 928 MB

Algebra Workshop
Graphing Radical Functions

Think Back

1. Use a graphing utility to graph $y = 2x + 6$.

2. Explain how you can tell from the graph that $y = 2x + 6$ represents a function.

3. Explain how you can tell from the equation that $y = 2x + 6$ represents a function.

4. Let $f(x) = 2x + 6$. Determine $f^{-1}(x)$, the inverse of $f(x)$.

5. Use a graphing utility to graph $f(x)$ and $f^{-1}(x)$ on the same coordinate axes.

6. The line $f(x) = 2x + 6$ contains the point $(3, 12)$. Without looking at the graph of $f^{-1}(x)$, how do you know that the graph contains the point $(12, 3)$?

7. Describe the graph of $f^{-1}(x)$ as a *reflection* of the graph of $f(x) = 2x + 6$.

8. Describe the graph of $y = x^2$.

9. Tell how the graph of each equation differs from the graph of $y = x^2$.

   a. $y = x^2 + 5$ **b.** $y = x^2 - 7$ **c.** $y = (x + 3)^2$

   d. $y = (x - 4)^2$ **e.** $y = 4x^2$ **f.** $y = -\frac{1}{2}x^2$

> **SPOTLIGHT ON LEARNING**
>
> **WHAT?** In this lesson you will learn
> • to graph radical functions.
> • to identify graphs of radical functions.
>
> **WHY?** Graphing radical functions can help you solve problems about geometry and navigation.

Explore

10. Complete the table of ordered pairs for the relation $y = x^2$.

x	-4	-3	-2	-1	0	1	2	3	4
y									

11. Is the relation $y = x^2$ a function? Explain.

12. Complete the table of ordered pairs for the relation that is the *inverse* of the relation $y = x^2$.

x	0	1	4	9	16
y					

13. Is the inverse of $y = x^2$ a function? Explain.

14. On the same coordinate axes, plot the ordered pairs in both of the tables in Questions 10 and 12. Draw a curve connecting each set of points. Do your results verify your answers to Questions 11 and 13? Explain.

15. Graph the line $y = x$ on the coordinate axes with the two curves. How are the curves related to the line? Explain.

16. Give reasons to support the following statement: "The relation $y = \pm\sqrt{x}$ is the inverse of the relation $y = x^2$."

17. Graph $y = \sqrt{x}$ using a graphing utility. What expression did you enter for y?

18. The graph that you obtained in Question 7 is not identical to the graph of the inverse of the relation $y = x^2$ that you drew in Question 14. Explain the difference.

19. Devise a method for plotting the entire graph of the inverse of $y = x^2$ using a graphing utility. Describe your method and graph the inverse relation.

PROBLEM SOLVING TIP

In Question 19 if you are using a graphing utility, learn how it graphs inverses.

For Questions 20–24, graph the given pair of equations on the same screen with the graph of $y = \sqrt{x}$. Note how the constant appears to affect the position and shape of the graph of $y = \sqrt{x}$.

20. **a.** $y = \sqrt{x} + 5$ **b.** $y = \sqrt{x} - 5$

21. **a.** $y = \sqrt{x + 3}$ **b.** $y = \sqrt{x - 3}$

22. **a.** $y = 2\sqrt{x}$ **b.** $y = \frac{1}{2}\sqrt{x}$

23. **a.** $y = 2\sqrt{-x}$ **b.** $y = \frac{1}{2}\sqrt{-x}$

24. **a.** $y = 3\sqrt{x}$ **b.** $y - -3\sqrt{x}$

Make Connections

For Questions 25–29, describe how the inclusion of a constant c into the equation $y = \sqrt{x}$ affects the graph of the equation.

25. $y = \sqrt{x} + c$

26. $y = \sqrt{x + c}$

27. $y = c\sqrt{x}$ $(c > 0)$

28. $y = c\sqrt{-x}$ $(x < 0)$

29. $y = c\sqrt{x}$ $(c < 0)$

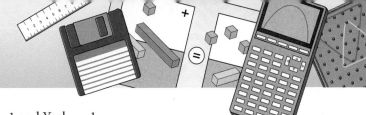

Match each graph with its equation. $Xscl = 1$ and $Yscl = 1$.

30.

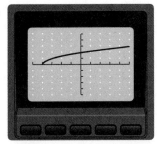

31.

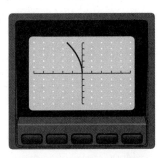

CHECK UNDERSTANDING

In Questions 30–35, why are there large intervals of the domains that produce no points on the graphs?

32.

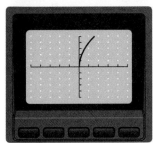

33.

34.

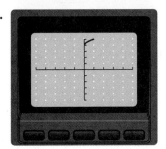

35.

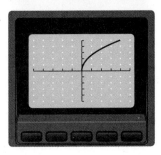

a. $y = \sqrt{x} + 4$ **b.** $y = \sqrt{x + 4}$ **c.** $y = 4\sqrt{x}$

d. $y = \frac{1}{4}\sqrt{x}$ **e.** $y = -4\sqrt{x}$ **f.** $y = 4\sqrt{-x}$

36. The screen shows the graph of the equation $y = x^3$. Compare the graph with the graph of $y = x^2$.

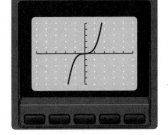

37. Give reasons to support the following statement: "The relation $y = \sqrt[3]{x}$ is the inverse of the relation $y = x^3$."

38. Without graphing the equation $y = \sqrt[3]{x}$, describe in detail how you think the graph will look. Give reasons for your conjectures.

39. Graph $y = \sqrt[3]{x}$ using a graphing utility or graph paper. If your graph differs from the one you described in Question 38, give possible reasons for the differences.

40. Compare and contrast the graphs of $y = \sqrt{x}$ and $y = \sqrt[3]{x}$.

Without graphing the equation, make a conjecture about how the graph of each equation differs from the graph of $y = \sqrt[3]{x}$.

41. $y = 5\sqrt[3]{x}$ **42.** $y = \sqrt[3]{x} + 9$ **43.** $y = \sqrt[3]{x - 4}$

44. $y = -\sqrt[3]{x}$ **45.** $y = \frac{2}{3}\sqrt[3]{x}$ **46.** $y = \sqrt[3]{-x}$

Summarize

CHECK UNDERSTANDING

Let $x = 64$. Evaluate y in each part of Question 47. What do your results tell you about whether or not y is a function?

47. WRITING MATHEMATICS Which of the following relations is a function? Explain how you know.

     a. $y = \pm\sqrt{x}$ **b.** $y = \sqrt{x}$ **c.** $y = \sqrt[3]{x}$

48. THINKING CRITICALLY Without drawing the graph, describe how you could obtain the graph of $y = -3\sqrt{x + 6} - 9$ from the graph of $y = \sqrt{x}$.

49. THINKING CRITICALLY Solve for x. Explain how you determined the solutions.

     a. $\sqrt{x} = \pm\sqrt[3]{x}$ **b.** $\sqrt[3]{x} = \sqrt[5]{x}$

50. GOING FURTHER Without graphing, describe any similarities you would expect to find between the graphs in each pair of equations.

     a. $y = \sqrt{x}$ and $y = \sqrt[4]{x}$ **b.** $y = \sqrt[3]{x}$ and $y = \sqrt[5]{x}$

51. THINKING CRITICALLY The equations $y = \sqrt[4]{x}$ and $y = \sqrt[5]{x}$ are graphed in the viewing window with

Xmin = 50   Ymin = 2
Xmax = 51   Ymax = 3
Xscl = 1   Yscl = 1

Which graph is which?

PROBLEM SOLVING TIP

In Question 52, remember to use parentheses when entering expressions into your graphing utility.

52. GOING FURTHER Solve graphically.

     a. $\sqrt[4]{x - 1} = \sqrt[5]{x + 3}$

     b. $2.1\sqrt[4]{x + 61.79} = \sqrt[6]{-x + 15.43} + 1.983$

     c. $\sqrt[4]{x - 23.3375} - 2 = \frac{1}{2}\sqrt[5]{x - 61.4}$

8.4 Exponential and Radical Expressions

Explore

- Decide whether each equation is true or false. Look for patterns.

  1. Square roots

  a. $\sqrt{36} + \sqrt{9} = \sqrt{36 + 9}$ **b.** $\sqrt{36} - \sqrt{9} = \sqrt{36 - 9}$

  c. $\sqrt{36} \cdot \sqrt{9} = \sqrt{36 \cdot 9}$ **d.** $\sqrt{36} \div \sqrt{9} = \sqrt{36 \div 9}$

  2. Cube roots

  a. $\sqrt[3]{27} + \sqrt[3]{8} = \sqrt[3]{27 + 8}$ **b.** $\sqrt[3]{27} - \sqrt[3]{8} = \sqrt[3]{27 - 8}$

  c. $\sqrt[3]{27} \cdot \sqrt[3]{8} = \sqrt[3]{27 \cdot 8}$ **d.** $\sqrt[3]{27} \div \sqrt[3]{8} = \sqrt[3]{27 \div 8}$

  3. Fourth roots

  a. $\sqrt[4]{81} + \sqrt[4]{16} = \sqrt[4]{81 + 16}$ **b.** $\sqrt[4]{81} - \sqrt[4]{16} = \sqrt[4]{81 - 16}$

  c. $\sqrt[4]{81} \cdot \sqrt[4]{16} = \sqrt[4]{81 \cdot 16}$ **d.** $\sqrt[4]{81} \div \sqrt[4]{16} = \sqrt[4]{81 \div 16}$

  4. Use your results to conjecture whether each of the following is true.

  a. $\sqrt[n]{a} + \sqrt[n]{b} = \sqrt[n]{a + b}$ **b.** $\sqrt[n]{a} - \sqrt[n]{b} = \sqrt[n]{a - b}$

  c. $\sqrt[n]{a} \cdot \sqrt[n]{b} = \sqrt[n]{ab}$ **d.** $\sqrt[n]{a} \div \sqrt[n]{b} = \sqrt[n]{a \div b}$

SPOTLIGHT ON LEARNING

WHAT? In this lesson you will learn
- to simplify radical and exponential expressions.
- to add, subtract, multiply, and divide radicals.

WHY? Simplifying radical and exponential expressions can help you solve real world problems about sports and aerospace.

Build Understanding

- In Lesson 8.2, you learned to simplify a radical by using the least possible index. A radical expression is in *simplest form* when

  - The radicand contains no fractions.
  - No denominator contains a radical.

You can simplify radical expressions using these properties.

PROPERTIES OF RADICALS

For real numbers a and b and integer $n > 1$,

$$\sqrt[n]{a} \cdot \sqrt[n]{b} = \sqrt[n]{ab} \qquad \text{PRODUCT PROPERTY}$$

If n is even, both a and b must be nonnegative.
For real numbers a and b, $b \neq 0$, and integer $n > 1$,

$$\frac{\sqrt[n]{a}}{\sqrt[n]{b}} = \sqrt[n]{\frac{a}{b}} \qquad \text{QUOTIENT PROPERTY}$$

If n is even, both a and b must be nonnegative.

EXAMPLE 1

Simplify.

    a. $3\sqrt[3]{49n^2} \cdot 2\sqrt[3]{35n^2}$ **b.** $\sqrt[5]{\dfrac{3}{4}} \cdot \sqrt[5]{\dfrac{5}{8}}$

Solution

    a. $3\sqrt[3]{49n^2} \cdot 2\sqrt[3]{35n^2}$

$$= 3 \cdot 2\sqrt[3]{49n^2 \cdot 35n^2} \qquad \text{Use the product property.}$$

$$= 6\sqrt[3]{7 \cdot 7 \cdot 7 \cdot 5 \cdot n^2 \cdot n^2} \qquad \text{Factor.}$$

$$= 6 \cdot 7 \cdot n\sqrt[3]{5n} \qquad \text{Simplify.}$$

$$= 42n\sqrt[3]{5n} \qquad \text{Multiply.}$$

    b. $\sqrt[5]{\dfrac{3}{4}} \cdot \sqrt[5]{\dfrac{5}{8}} = \sqrt[5]{\dfrac{3}{4} \cdot \dfrac{5}{8}} \qquad \text{Use the product property.}$

$$= \sqrt[5]{\dfrac{15}{32}} \qquad \text{Multiply.}$$

$$= \dfrac{\sqrt[5]{15}}{\sqrt[5]{32}} \qquad \text{Use the quotient property.}$$

$$= \dfrac{\sqrt[5]{15}}{2} \qquad \text{Simplify.} \qquad \blacktriangleleft$$

CHECK UNDERSTANDING

In the solution to Example 1a, why does $3 \cdot 2\sqrt[3]{7^3 \cdot 5 \cdot n^3 \cdot n}$ simplify to $3 \cdot 2 \cdot 7 \cdot n\sqrt[3]{5n}$?

An exponential expression is in *simplest form* when:

- Negative exponents have been eliminated.
- Fractional exponents have been removed from the denominator.
- Fractional exponents are in lowest terms.

EXAMPLE 2

Simplify: $25^{-\frac{6}{4}}$

Solution

$$25^{-\frac{6}{4}} = 25^{-\frac{3}{2}} = (5^2)^{-\frac{3}{2}} = 5^{2 \cdot -\frac{3}{2}} = 5^{-3} = \dfrac{1}{5^3}, \text{ or } \dfrac{1}{125} \qquad \blacktriangleleft$$

When comparing radical expressions, it is easier if there are no radicals in the denominator. To remove a radical from a denominator, use the process of **rationalizing the denominator**. Multiply by 1 in the form $\dfrac{r}{r}$ where r is an expression that will produce a perfect nth power when it is multiplied by the denominator.

THINK BACK

To multiply two binomials, find and add the products of the **F**irst, **O**uter, **I**nner, and **L**ast terms. When conjugates are multiplied, the outer and inner terms sum to zero.

The binomials $a\sqrt{b} + c\sqrt{d}$ and $a\sqrt{b} - c\sqrt{d}$, where a, b, c, and d are real numbers are called **conjugates**. Use the conjugate to rationalize a binomial denominator containing a radical.

EXAMPLE 3

Simplify.

a. $\dfrac{6}{5\sqrt{7}}$
  b. $\sqrt[3]{\dfrac{3}{4x^7}}$
  c. $\dfrac{1 + \sqrt{3}}{4 - \sqrt{3}}$

Solution

a. $\dfrac{6}{5\sqrt{7}} \cdot \dfrac{\sqrt{7}}{\sqrt{7}} = \dfrac{6\sqrt{7}}{5\sqrt{7^2}}$ Multiply by $\dfrac{\sqrt{7}}{\sqrt{7}} = 1$.

$\phantom{\dfrac{6}{5\sqrt{7}} \cdot \dfrac{\sqrt{7}}{\sqrt{7}}} = \dfrac{6\sqrt{7}}{5 \cdot 7} = \dfrac{6\sqrt{7}}{35}$

b. $\sqrt[3]{\dfrac{3}{4x^7}} = \dfrac{\sqrt[3]{3}}{\sqrt[3]{4x^7}}$ Quotient property.

$\phantom{\sqrt[3]{\dfrac{3}{4x^7}}} = \dfrac{\sqrt[3]{3}}{\sqrt[3]{2^2\, x^7}} \cdot \dfrac{\sqrt[3]{2x^2}}{\sqrt[3]{2x^2}}$ Rationalize the denominator.

$\phantom{\sqrt[3]{\dfrac{3}{4x^7}}} = \dfrac{\sqrt[3]{6x^2}}{\sqrt[3]{2^3\, x^9}}$ Product property.

$\phantom{\sqrt[3]{\dfrac{3}{4x^7}}} = \dfrac{\sqrt[3]{6x^2}}{2x^3}$ Simplify.

c. $\dfrac{1 + \sqrt{3}}{4 - \sqrt{3}} \cdot \dfrac{4 + \sqrt{3}}{4 + \sqrt{3}}$

$ = \dfrac{(1 + \sqrt{3})(4 + \sqrt{3})}{(4 - \sqrt{3})(4 + \sqrt{3})}$ Multiply by the conjugate of the denominator.

$ = \dfrac{4 + \sqrt{3} + 4\sqrt{3} + 3}{16 + 4\sqrt{3} - 4\sqrt{3} - 3}$ Multiply binomials.

$ = \dfrac{7 + 5\sqrt{3}}{13}$ Simplify. ◀

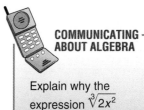

COMMUNICATING ABOUT ALGEBRA

Explain why the expression $\sqrt[3]{2x^2}$ was chosen to rationalize the denominator in Example 3b.

CHECK UNDERSTANDING

Show how to use the formula for the difference of two squares to multiply $\left(2 + \sqrt{3}\right)\left(2 - \sqrt{3}\right)$.

Radicals with the same index and the same radicand are **like radicals**. Like radicals can be added or subtracted.

EXAMPLE 4

Simplify.

a. $3\sqrt{5} + 8\sqrt{5}$
  b. $\sqrt[3]{250} - \sqrt[3]{54}$

Solution

a. $3\sqrt{5} + 8\sqrt{5} = 11\sqrt{5}$

b. $\sqrt[3]{250} - \sqrt[3]{54} = \sqrt[3]{5^3 \cdot 2} - \sqrt[3]{3^3 \cdot 2}$ Simplify.

$\phantom{\sqrt[3]{250} - \sqrt[3]{54}} = 5\sqrt[3]{2} - 3\sqrt[3]{2}$ Subtract like radicals.

$\phantom{\sqrt[3]{250} - \sqrt[3]{54}} = 2\sqrt[3]{2}$ ◀

Use the rules for simplification when you solve real world problems involving radical and exponential expressions.

EXAMPLE 5

SPORTS A volleyball has a volume of 864π cm³. A tennis ball has a volume of 32π cm³. By how much does the radius of the volleyball exceed that of the tennis ball?

Solution

The formula for the volume of a sphere is $V = \frac{4}{3}\pi r^3$.

Volleyball

$$\frac{4}{3}\pi r^3 = 864\pi$$

$$r = \sqrt[3]{\frac{3 \cdot 864\pi}{4\pi}}$$

$$r = \sqrt[3]{\frac{3 \cdot 4 \cdot 6 \cdot 6 \cdot 6}{4}}$$

$$r = 6\sqrt[3]{3}$$

Tennis Ball

$$\frac{4}{3}\pi r^3 = 32\pi$$

$$r = \sqrt[3]{\frac{3 \cdot 32\pi}{4\pi}}$$

$$r = \sqrt[3]{\frac{3 \cdot 2 \cdot 2 \cdot 2 \cdot 2 \cdot 2}{2 \cdot 2}}$$

$$r = 2\sqrt[3]{3}$$

difference in radii: $6\sqrt[3]{3} - 2\sqrt[3]{3} = 4\sqrt[3]{3}$

The radius of the volleyball exceeds that of the tennis ball by $4\sqrt[3]{3}$ cm. ◄

TRY THESE

Give the conjugate of each expression.

1. $2 + \sqrt{6}$

2. $3\sqrt{11} - 8\sqrt{16}$

3. $-\sqrt{3.5} + 5$

4. $7 - \sqrt{21}$

Give the expression you would multiply by to rationalize the denominator.

5. $\dfrac{\sqrt{5}}{\sqrt{6}}$

6. $\dfrac{8 - \sqrt{10}}{\sqrt{7} - 3\sqrt{3}}$

7. $\dfrac{5\sqrt[3]{12n}}{14\sqrt[3]{10n^5}}$

8. $\dfrac{3}{\sqrt[5]{72x^6}}$

Simplify.

9. $\sqrt{3} \cdot \sqrt{5}$

10. $2\sqrt[3]{9} \cdot \sqrt[3]{3}$

11. $8^{-\frac{2}{3}}$

12. $\dfrac{\sqrt{5}}{\sqrt{7}}$

13. $\sqrt[3]{\dfrac{8}{9n}}$

14. $\dfrac{3x}{\sqrt{20x}}$

15. $\sqrt{10} + \sqrt{40} - 3\sqrt{250}$

16. $\sqrt[4]{2k} - \sqrt[4]{162k}$

17. $\dfrac{1 + \sqrt{2}}{1 - \sqrt{2}}$

18. **GEOMETRY** A rectangular picture frame measures $\sqrt{600}$ in. $\times$ $\sqrt{384}$ in. Determine the perimeter and area of the frame.

19. **WRITING MATHEMATICS** Explain what it means to say that a radical expression is in simplest form.

Simplify.

1. $\sqrt{11} \cdot \sqrt{7}$

2. $\sqrt[3]{4} \cdot 5\sqrt[3]{6}$

3. $\sqrt[6]{3.2} \cdot \sqrt[6]{20}$

4. $\sqrt[3]{9} \cdot \sqrt[3]{-9}$

5. $\sqrt{18x^2y} \cdot \sqrt{2yz}$

6. $\sqrt{5ab} \cdot \sqrt{10ab}$

7. $\sqrt{147} \div \sqrt{3}$

8. $8\sqrt[3]{56} \div 4\sqrt[3]{7}$

9. $64^{-\frac{1}{2}}$

10. $243^{-\frac{4}{5}}$

11. $\left(\dfrac{1}{8}\right)^{-\frac{3}{4}}$

12. $\dfrac{1}{x^{-\frac{8}{10}}}$

13. $\dfrac{\sqrt{3}}{\sqrt{5}}$

14. $\dfrac{6\sqrt[3]{32}}{\sqrt[3]{2}}$

15. $\dfrac{\sqrt[5]{-24x^{20}}}{\sqrt[5]{12x^5}}$

16. $\sqrt{\dfrac{7}{12b}}$

17. $\sqrt{\dfrac{25x^3}{y^2}}$

18. $\sqrt[3]{\dfrac{162x^5y^7}{6x^2y^2}}$

19. $\dfrac{\sqrt{10a^2b}}{\sqrt{18a^5b^3}}$

20. $\dfrac{12h}{\sqrt[4]{18h}}$

21. **PHYSICS** The velocity v in meters per second of a subatomic particle is given by $y = \sqrt{\dfrac{2K}{m}}$ where K is the kinetic energy in Newton-meters of the particle and m is its mass in kilograms. Write the formula in simplified form.

Simplify.

22. $6\sqrt{3} + 8\sqrt{3}$

23. $8\sqrt{45} + 7\sqrt{20}$

24. $\sqrt[3]{54} - \sqrt[3]{128}$

25. $\sqrt[4]{48m^5} - \sqrt[4]{768m}$

26. $\left(\sqrt{2} + \sqrt{3}\right)\left(\sqrt{2} + \sqrt{3}\right)$

27. $\left(\sqrt{10} - \sqrt{6}\right)^2$

28. $\left(\sqrt{7} - \sqrt{5}\right)\left(\sqrt{7} + \sqrt{5}\right)$

29. $\left(2\sqrt{x} - 6\sqrt{y}\right)^2$

30. $\dfrac{3}{3 + \sqrt{3}}$

31. $\dfrac{5 - \sqrt{8}}{\sqrt{2}}$

32. $\dfrac{\sqrt{k} + 1}{\sqrt{k} - 1}$

33. $\dfrac{2\sqrt{5} - 3\sqrt{2}}{3\sqrt{6} - 4\sqrt{3}}$

34. **NAVIGATION** The primary radar system at La Guardia airport guides all aircraft within a radius of r_1 mi of the airport and covers an area of 90π mi². Secondary radar extends to a radius of r_2 miles and covers 1440π mi². Determine the difference in the radii of the circles.

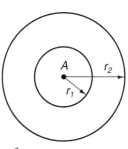

35. **AEROSPACE** In 1915, Einstein proved the mass of an object increases as its velocity increases. At a velocity of v, an object with a mass of m at rest will have a mass m_v where $m_v = m\left(1 - \dfrac{v^2}{c^2}\right)^{-\frac{1}{2}}$ where c is the velocity of light.

   a. An astronaut has a mass of 100 kg at rest. In a rocket traveling at a velocity of 299,000 km/s, what is the astronaut's mass? The velocity of light is 299,793 km/s.

   b. Write the expression for mass in simplest exponential form.

36. **WRITING MATHEMATICS** Explain why the product of two conjugates always produces an expression without a radical.

Simplify.

37. $\sqrt{\dfrac{2}{3}} + \sqrt{\dfrac{3}{2}}$

38. $\sqrt[4]{9} - \sqrt{\dfrac{1}{3}}$

39. $\sqrt[4]{144} + 3\sqrt[4]{9} - 5\sqrt{48}$

40. $\left(\dfrac{x^{-2}y^6}{9}\right)^{-\frac{1}{2}}$

41. $\dfrac{x^{\frac{1}{2}}}{x^{-\frac{1}{2}} + x^{-\frac{3}{2}}}$

42. $\dfrac{m^{-\frac{2}{3}}n^{\frac{1}{2}}}{n^{-\frac{3}{2}}\sqrt[3]{m}}$

43. WRITING MATHEMATICS Explain how you can determine without a calculator which is greater, $\dfrac{3}{\sqrt{3}}$ or $\dfrac{5}{\sqrt{5}}$.

44. Verify that $\sqrt{7} - \sqrt{2}$ is a square root of $9 - 2\sqrt{14}$. [*Hint:* What is $\left(\sqrt{7} - \sqrt{2}\right)^2$?]

Verify each equation.

45. $\sqrt{5} + \sqrt{3} = \sqrt{8 + 2\sqrt{15}}$

46. $\sqrt{11} - \sqrt{3} = \sqrt{14 - 2\sqrt{33}}$

Use the patterns in Exercises 45–46 to determine the square root of each expression.

47. $11 + 2\sqrt{30}$

48. $17 - 2\sqrt{70}$

49. $13 - 2\sqrt{42}$

50. Rationalize the denominator of $\dfrac{1}{\sqrt[3]{n} - 1}$. (*Hint:* Factor $n - 1$ as the difference of two cubes.)

MIXED REVIEW

Solve each system.

51. $\begin{cases} x + y + z = 5 \\ 2x + 3y - z = 9 \\ -4x - y + 6z = -4 \end{cases}$

52. $\begin{cases} 2x + y + 2z = 0 \\ 3x - 2y - 3z = 1 \\ -x + 3y + z = -5 \end{cases}$

Solve for x. Check your answer.

53. $x^2 - 2x - 15 = 0$

54. $2x^2 + 5x + 1 = 0$

55. $2x^2 = 5$

56. STANDARDIZED TESTS Which represents $\sqrt{\dfrac{3}{8}}$ in simplest form?

A. $\dfrac{\sqrt{24}}{8}$

B. $\sqrt{3}$

C. $\left(\dfrac{3}{8}\right)^{\frac{1}{2}}$

D. $\dfrac{\sqrt{6}}{4}$

PROJECT *Connection*   In this activity, you will build and launch your balloon rocket.

1. Inflate the balloon and close it tightly with the twist tie. Tape the drinking straw to the balloon as shown.

2. Thread the 10-meter length of string through the straw. Stretch the string across a long room or hallway and fasten each end to the back of a chair.

3. Launch your rocket several times by removing the twist tie. Begin by inflating the balloon to capacity. Take measurements to determine the volume, release the balloon at one end of the string, and measure the distance. Repeat each time inflating the balloon a bit less.

4. Graph and discuss your results.

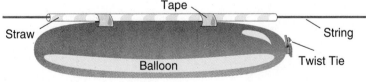

Career
Aerospace Engineer

Early in the twentieth century, most eyes on the skies were focused on the newly invented airplane. However, physicist Robert H. Goddard (1882–1945) was looking far into the future, to a time when rockets would transport humans to the moon and beyond. Turned down by the Army in his request for financial support, and forbidden by the Massachusetts fire marshall from conducting his dangerous experiments anywhere in the state, Goddard traveled to New Mexico to find wide open spaces for his research. He settled in Roswell, where he constructed a launch tower and continued his experiments. Because of Goddard's vision and work, today's aerospace engineers design and build spacecraft.

Decision Making

1. A rocket must attain *escape velocity* before it can escape the gravitational pull of a planet. The escape velocity v (km/s) for a planet of mass m (kg) and radius r (km) is given by

$$v = (3.7 \times 10^{-10})\sqrt{\frac{m}{r}}$$

   Find Earth's escape velocity ($m = 5.98 \times 10^{24}$ kg, $r = 6376$ km).

2. Much of the energy for a mission is generated by solar cells attached to the outside of the spacecraft. Each square centimeter of a cell generates $\sqrt{0.12}$ watts of electric power. An equilateral triangular solar cell measures 24 cm on a side.

   a. Determine s, the *semiperimeter* of the cell (semiperimeter is one-half of the perimeter).

   b. Use *Heron's Formula*, $A = \sqrt{s(s-a)(s-b)(s-c)}$, to determine A, the area of a triangle with semiperimeter s and sides measuring a, b, and c. Find the area of the solar cell.

   c. How much energy can be delivered by one cell?

3. An aerospace engineer designing a satellite communications system wants to know the maximum distance d to Earth from satellite S at altitude h. Find d in terms of h and r the radius of Earth.

8.5 Solve Radical Equations

Explore

- Use the linear equation $x + 2 = -2x - 1$.

  1. Solve the equation algebraically.

  2. Solve the equation graphically. Describe your method.

  3. Create a quadratic equation by squaring both sides of the original equation.

  4. Solve the quadratic equation algebraically.

  5. Solve the quadratic equation graphically.

  6. Compare the solutions of the linear and the quadratic equations. How does squaring an equation appear to affect the solutions of the equation?

Build Understanding

- An equation containing a variable in a radicand is called a **radical equation**. Some examples of radical equations are $\sqrt{x} = 5$, $\sqrt[3]{z + 1} = 9$, $\sqrt[4]{x + 8} = \sqrt[4]{2x}$. Equations that contain rational exponents are equivalent to radical equations. They are radical equations in exponential form. You can solve these types of equations graphically.

EXAMPLE 1

Solve graphically.

    a. $\sqrt{x + 5} = \sqrt{x} + 1$ **b.** $2 + x^{\frac{1}{3}} = x^{\frac{2}{3}}$

Solution

    a. The screen shows the graphs of the equations $y_1 = \sqrt{x + 5}$ and $y_2 = \sqrt{x} + 1$. The x-coordinate 4 of the point of intersection $(4, 3)$ represents the solution of the equation, $\sqrt{x + 5} = \sqrt{x} + 1$.

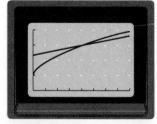

    b. The graphs of the equations $y_1 = 2 + x^{\frac{1}{3}}$ and $y_2 = x^{\frac{2}{3}}$ intersect at $(8, 4)$. The solution of the equation is $x = 8$. ◀

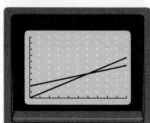

To solve radical equations algebraically, use inverse operations. The following property assures that you can raise both sides of an equation to the same power.

$$\text{If } a = b, \text{ then } a^n = b^n.$$

When solving a radical equation, *isolate* the radical on one side of the equation before raising both sides to the same power. An invalid solution, called an **extraneous solution,** is sometimes introduced when you raise both sides of an equation to a power. Because of the possibility of extraneous solutions, you must always check solutions by substituting in *the original equation.* Reject any solutions that do not check.

COMMUNICATING ABOUT ALGEBRA

What is the converse of the statement if $a = b$, then $a^n = b^n$? Is the converse true? Explain.

EXAMPLE 2

Solve for x.

a. $\sqrt[3]{x + 12} = -5$

b. $2 + \sqrt{2x - 1} = x$

Solution

a.
$$\sqrt[3]{x + 12} = -5$$
$$\left(\sqrt[3]{x + 12}\right)^3 = (-5)^3 \qquad \text{Cube both sides of the equation.}$$
$$x + 12 = -125$$
$$x = -137$$

Check
$$\sqrt[3]{-137 + 12} \overset{?}{=} -5$$
$$\sqrt[3]{-125} = -5 \checkmark$$

The solution is -5.

b.
$$2 + \sqrt{2x - 1} = x$$
$$\sqrt{2x - 1} = x - 2 \qquad \text{Isolate the radical.}$$
$$\left(\sqrt{2x - 1}\right)^2 = (x - 2)^2 \qquad \text{Square both sides of the equation.}$$
$$2x - 1 = x^2 - 4x + 4$$
$$0 = x^2 - 6x + 5 \qquad \text{Simplify.}$$
$$0 = (x - 5)(x - 1) \qquad \text{Factor.}$$
$$x = 5 \quad \text{or} \quad x = 1 \qquad \text{Solve for } x.$$

Check
$$2 + \sqrt{2(5) - 1} \overset{?}{=} 5 \qquad\qquad 2 + \sqrt{2(1) - 1} \overset{?}{=} 1$$
$$2 + \sqrt{9} \overset{?}{=} 5 \qquad\qquad 2 + \sqrt{1} \overset{?}{=} 1$$
$$2 + 3 \overset{?}{=} 5 \qquad\qquad 2 + 1 \overset{?}{=} 1$$
$$5 = 5 \checkmark \qquad\qquad 3 \neq 1$$

The solution is 5. ◄

You may need to raise both sides of the equation to a given power twice before obtaining a solution. First, however, you need to isolate a radical term.

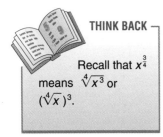

EXAMPLE 3

Solve: $\sqrt{2x + 7} - \sqrt{x - 5} = 3$

Solution

$$\sqrt{2x + 7} - \sqrt{x - 5} = 3$$
$$\sqrt{2x + 7} = \sqrt{x - 5} + 3 \qquad \text{Isolate the radical } \sqrt{2x + 7}.$$
$$\left(\sqrt{2x + 7}\right)^2 = \left(\sqrt{x - 5} + 3\right)^2 \qquad \text{Square both sides.}$$
$$2x + 7 = \left(\sqrt{x - 5} + 3\right)\left(\sqrt{x - 5} + 3\right)$$
$$2x + 7 = x - 5 + 6\sqrt{x - 5} + 9 \qquad \text{Simplify.}$$
$$x + 3 = 6\sqrt{x - 5} \qquad \text{Isolate the radical.}$$
$$(x + 3)^2 = \left(6\sqrt{x - 5}\right)^2 \qquad \text{Square both sides.}$$
$$x^2 + 6x + 9 = 36(x - 5)$$
$$x^2 + 6x + 9 = 36x - 180$$
$$x^2 - 30x + 189 = 0 \qquad \text{Simplify.}$$
$$(x - 21)(x - 9) = 0 \qquad \text{Factor.}$$
$$x = 21 \quad \text{or} \quad x = 9$$

Check both solutions in the original equation. The solutions are 21 and 9.

To solve an equation with rational exponents raise both sides of the equation to the same power. Choose the power so the resulting exponent of the variable is 1.

EXAMPLE 4

Solve for x: $x^{\frac{3}{4}} = 15.625$

Solution

$$x^{\frac{3}{4}} = 15.625$$
$$\left(x^{\frac{3}{4}}\right)^{\frac{4}{3}} = 15.625^{\frac{4}{3}} \qquad \text{Raise both sides to the } \frac{4}{3} \text{ power.}$$
$$x = 39.0625$$

The solution is 39.0625.

Many of the formulas used by physicists and aerospace engineers involve radicals and exponents.

EXAMPLE 5

AEROSPACE The orbital radius r in meters of an Earth satellite is given by $r = \sqrt[3]{\dfrac{GMt^2}{4\pi^2}}$ where G is the universal gravitation constant, 6.67×10^{-11} N-m²/kg², M is the mass of Earth, 5.98×10^{24} kg, and t is the orbital period in seconds. A satellite is in orbit 22,260 mi above Earth's surface. What is unusual about the period of the orbit?

Solution

You can determine the orbital period, the length of time it takes the satellite to make one complete revolution around Earth. To find the orbital radius add the radius of Earth to the distance above Earth's surface.

$$r = 3{,}963 + 22{,}260 = 26{,}223 \text{ mi}$$

To express the orbital radius in meters, use 1 mi $= 1609$ m.

$$26{,}223 \text{ mi} \times 1{,}609 \text{ m/mi} \approx 4.22 \times 10^7 \text{ m}$$

Solve for the formula t.

$$\frac{GMt^2}{4\pi^2} = r^3$$

$$GMt^2 = 4\pi^2 r^3$$

$$t^2 = \frac{4\pi^2 r^3}{GM}$$

$$t = \sqrt{\frac{4\pi^2 r^3}{GM}}$$

$$t = \sqrt{\frac{4\pi^2 (4.22 \times 10^7)^3}{(6.67 \times 10^{-11})5.98 \times 10^{24}}} \quad \text{Substitute values.}$$

$$t = 8.63 \times 10^4 = 86{,}300 \text{ s} \div 3600 \text{ s/h} \approx 24.0 \text{ h}$$

Since Earth's rotational period is 24 hours, the satellite is in *geosynchronous* orbit. That means the satellite makes one complete revolution around Earth in exactly the time it takes the planet to rotate once. The satellite always stays above the same point on Earth's surface and, to an observer on the ground, appears not to be moving at all.

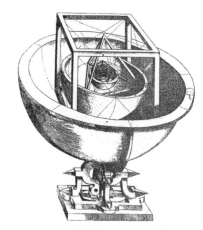

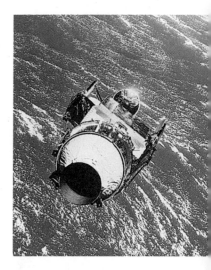

TRY THESE

Solve and check.

1. $\sqrt{x - 4} = 7$

2. $3 = \sqrt{2x - 7} - 2$

3. $\sqrt[3]{x} = -4$

4. $\sqrt[4]{3x - 5} + 11 = 18$

5. $x^{\frac{1}{7}} = -3$

6. $x^{\frac{3}{5}} = 1.728$

7. $\sqrt{x - 5} = \sqrt{4x - 29}$

8. $\sqrt{x + 7} - x = 1$

9. $\sqrt{x + 12} - \sqrt{x} = 2$

10. **AEROSPACE** Solve the formula $r = \sqrt[3]{\dfrac{GMt^2}{4\pi^2}}$ for M.

11. **MANUFACTURING** The radius r of a sphere with volume V is given by $r = \left(\dfrac{3V}{4\pi}\right)^{\frac{1}{3}}$. Write and solve an equation for finding the volume of a spherical ball bearing with a radius of 0.145 cm.

12. **WRITING MATHEMATICS** Explain how the following pair of equations illustrates the creation of extraneous solutions: $x = 3$ and $x^2 = 9$.

PRACTICE

Solve and check.

1. $\sqrt{x} = 11$

2. $-6 = -\sqrt{x - 9}$

3. $\sqrt{x + 11} - 7 = 3$

4. $-\sqrt{3x - 2} + 6 = 1$

5. $8 - \sqrt{\dfrac{x}{4}} = -1$

6. $16 = 9 + \sqrt{\dfrac{9x - 1}{2}}$

7. $-3 = \sqrt[3]{x}$

8. $\sqrt[3]{-x - 6} - 3 = -9$

9. $\sqrt[4]{5x + 16} + 11 = 14$

10. $2 - \sqrt[4]{\dfrac{5x + 8}{3}} = 0$

11. $\sqrt[5]{\dfrac{1}{x}} - 1 = -0.8$

12. $\sqrt[7]{x^7} = 3.62$

13. $x^{\frac{1}{2}} - 8 = 0$

14. $x^{\frac{1}{5}} + 3 = 1$

15. $1000 = x^{\frac{3}{2}}$

16. $(3x + 5)^{\frac{2}{3}} = 25$

17. $(4x + 3.713)^{\frac{3}{4}} = 6.859$

18. $\left(\dfrac{1}{x} - 0.00007\right)^{0.8} = 0.0081$

19. **PHYSICS** When two forces F_1 and F_2 pull at right angles to each other, the resultant or effective force R is given by the formula $R = \sqrt{F_1{}^2 + F_2{}^2}$.

   a. Solve the formula for F_1.

   b. Suppose Leroy and Dana are helping Fred get his car out of a ditch. If 1000 lb of force is needed, and Leroy's truck is exerting a force of 800 lb, how much force will Dana's car have to exert?

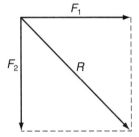

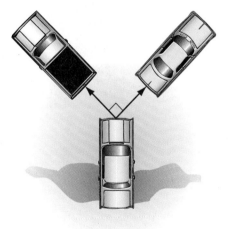

20. **WRITING MATHEMATICS** How can you tell mentally that the equation $\sqrt{x - 4} = -\sqrt{3x + 1}$ has no real solution? Explain.

Solve and check.

21. $\sqrt{4x + 4} = \sqrt{5x - 4}$

22. $\sqrt{\dfrac{1}{x + 1}} = \sqrt{\dfrac{x - 1}{8}}$

23. $3\sqrt{2x + 3} = \sqrt{x + 10}$

24. $\sqrt{x + 4} = x - 2$

25. $\sqrt{x + 16} - 4 = x + 6$

26. $x - \sqrt{3x + 1} = 3$

27. $\sqrt{x + 10} = 8 - \sqrt{x - 6}$

28. $\sqrt{3x + 3} - \sqrt{6x + 7} = -1$

29. $\sqrt{x + 5} + \sqrt{1 - 2x} = 4$

30. $1 - \sqrt{3x + 1} + \sqrt{x + 4} = 0$

Solve each formula for the given variable.

31. $f = \sqrt{\dfrac{\pi F^2}{4A}}$, for F (a formula for the f-stop of a camera)

32. $r = \sqrt[3]{\dfrac{A}{P}} - 1$, for A (a formula for interest rate in banking)

33. $a = \sqrt[3]{p^2}$, for p (a formula for the distance of a satellite from the body it is orbiting)

34. $E = Z\dfrac{\sigma}{\sqrt{n}}$, for n (a formula in statistics for the maximum error of estimation)

35. **AEROSPACE** The distance d in miles from a spacecraft to Earth's horizon is approximately $\left(\dfrac{3h}{2}\right)^{\frac{1}{2}}$ where h is the height of the craft in feet. An astronaut in the space shuttle Columbia spots a mountain range on the horizon 1260 mi away. Find the shuttle height in miles.

36. **GEOMETRY** The distance d in the coordinate plane from (x_1, y_1) to (x_2, y_2) is given by $d = \sqrt{(x_2 - x_1)^2 + (y_2 - y_1)^2}$.

   a. The distance from $(4, -2)$ to $(-1, y)$ is 13. Find y.

   b. The distance from $(-15, 24)$ to $(x, -11)$ is 37. Find x.

   c. Solve the equation for x_2.

EXTEND

37. **AUTO SAFETY** A driver whose eyes are h meters above the road is d meters from lettering on the highway. For maximum legibility, the letters should be length $L = \dfrac{d^{\frac{9}{4}}}{400h}$.
A driver's eye height is 1.2 m. The ideal length of letters in a STOP message is 2.8m. Find the distance from the driver to the message for maximum legibility.

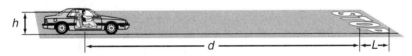

38. **PHYSICS** The period of a pendulum T in seconds is the length of time it takes for the pendulum to make one complete swing back and forth. The formula $T = 2\pi\sqrt{\dfrac{L}{32}}$ gives the period T for a pendulum of length L in feet.

   a. Solve the formula for L.

   b. If you wanted to build a grandfather clock with a pendulum that swings back and forth once every 3 seconds, how long would you make the pendulum? Round to the nearest hundredth.

Solve and check.

39. $x - 7\sqrt{x} + 12 = 0$ (*Hint:* How would you factor $x^2 - 7x + 12 = 0$? Try a similar method.)

40. $x - 11\sqrt{x} + 30 = 0$ **41.** $x - 18 = -7\sqrt{x}$ **42.** $2x - 5\sqrt{x} - 12 = 0$

Solve and check.

43. $\sqrt{x + 1} - \sqrt{x - 2} - \sqrt{4x - 11} = 0$

44. $\sqrt{\sqrt{x - 4} + x} = 2$

45. WRITING MATHEMATICS Explain how you can use critical thinking to solve the equation $\sqrt[5]{x^2} = -\sqrt[5]{x^2}$.

46. AEROSPACE Use the formula you found in Exercise 10 in Try These and the information in Example 5 to find the mass of the sun. Use the fact that Earth revolves around the sun in 365 days at a distance of 93 million miles. Be sure to use units of meters and seconds. About how many times as massive as Earth is the sun?

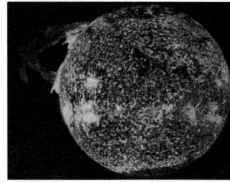

MIXED REVIEW

47. STANDARDIZED TESTS A cube numbered 1 to 6 is thrown. Choose the letter giving the probability that a number other than 3 comes up.

   A. 3   **B.** $\frac{1}{6}$ **C.** $\frac{1}{2}$ **D.** $\frac{5}{6}$

Solve for the given variable.

48. $P = 2l + 2w$, for w **49.** $C = \frac{5}{9}(F - 32)$, for F **50.** $A = \frac{1}{2}h(a + b)$, for a

Find $f(g(x))$.

51. $f(x) = 2x, g(x) = -3x + 2$ **52.** $f(x) = x^2, g(x) = x + 1$

53. $f(x) = \frac{1}{x}, g(x) = \sqrt{x}$ **54.** $f(x) = 16^x, g(x) = -\frac{3}{4}$

55. Solve: $\sqrt{x + 3} - \sqrt{x - 2} = 1$

PROJECT *Connection*   Adapt the experiment you used in the Project Connection on page 390 to explore some other relationships.

1. Use the same two-chair set-up as before, but this time make two balloon rockets. Tape the two balloons onto the straw so that the first can act as a "booster" rocket for the second. Gather data for the two-balloon system and make a graph. Compare the data to that from a single balloon. For example, do two balloons, each with volume V, travel approximately twice as far as a single balloon with volume V?

2. Use one balloon-and-straw rocket, but this time fasten one end of the string to the top of a ladder, bookcase, wall, or any high location. Walk away from this location until you run out of string and can make a triangle with the ground. Fasten the other end at this spot. Launch the rocket several times from ground level, each time inflating it a bit less. Calculate each volume as before and measure the maximum height achieved each time. Graph and discuss your results.

8.6 Complex Numbers and the Quadratic Equation

Explore

1. Solve the quadratic equation $x^2 - x - 6 = 0$ algebraically.

2. Graph $y = x^2 - x - 6$. Explain how you can use the graph to find the solutions of the equation $x^2 - x - 6 = 0$.

3. Graph each of the following.
   a. $y = x^2 + x + 1$ **b.** $y = -x^2 + 3x - 4$

4. What happens when you attempt to use the method you described in your answer to Question 2 to solve the related equations? $x^2 + x + 1 = 0$ and $-x^2 + 3x - 4 = 0$?

5. Find the values of the discriminants of the equations in Questions 2 and 4. What can you conclude about the relationship between the value of the discriminant and the solutions of a quadratic equation?

Build Understanding

As the human understanding of numbers has evolved, so has our number system. At first, people used only whole numbers.

$$x + 3 = 5$$

Rational numbers and negative numbers were introduced when the concepts of fractions and numbers less than zero began to make sense.

$$x + 4 = 2$$
$$5x = 3$$

The Pythagoreans of the sixth century B.C. were the first to recognize the need for irrational numbers.

$$x^2 = 5$$

With the inclusion of the irrationals, the number line was complete. Still, not all equations could be solved, and during the sixteenth century mathematicians began to look at equations like $x^2 = -3$ and to ask how they could be solved. Once again, the answer required the creation of a new class of numbers, the **imaginary numbers**.

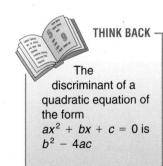

THINK BACK

The discriminant of a quadratic equation of the form $ax^2 + bx + c = 0$ is $b^2 - 4ac$

ALGEBRA: WHO, WHERE, WHEN

Imaginary numbers were first proposed around 1545, when Italian physician and mathematician Girolamo Cardano found himself searching for two numbers with a sum of 10 and a product of 40. Cardano was led to the solutions $5 + \sqrt{-15}$ and $5 - \sqrt{-15}$. Because of the curious nature of such numbers, mathematician René Descartes called them "imaginary."

CHECK UNDERSTANDING

Describe the pattern in the powers of i.

PROBLEM SOLVING TIP

The product and quotient properties of radicals hold only for real numbers. If you try to use the product property in Example 3b, you will get
$\sqrt{-3} \cdot \sqrt{-12} = \sqrt{36} = 6$ rather than the correct answer, -6.

> **IMAGINARY NUMBERS**
>
> The imaginary unit i is defined by $i = \sqrt{-1}$.
> The number i satisfies the equation $i^2 = -1$.
> A pure imaginary number is a number of the form bi, where b is a real number and $b \neq 0$.

Square roots of negative numbers can be expressed in terms of i. A simplified radical expression has no negative signs inside a radical.

EXAMPLE 1

Simplify.

  a. $\sqrt{-3}$ **b.** $\sqrt{-64}$ **c.** $\sqrt{-20}$

Solution

  a. $\sqrt{-3} = \sqrt{3}\sqrt{-1} = i\sqrt{3}$ **b.** $\sqrt{-64} = \sqrt{64}\sqrt{-1} = 8i$

  c. $\sqrt{-20} = \sqrt{4}\sqrt{5}\sqrt{-1} = 2i\sqrt{5}$ ◄

Successive powers of i produce a pattern.

$i^1 = i$ $i^5 = i^4 \cdot i = 1 \cdot i = i$

$i^2 = -1$ $i^6 = i^4 \cdot i^2 = 1 \cdot (-1) = -1$

$i^3 = i^2 \cdot i = -1 \cdot i = -i$ $i^7 = i^4 \cdot i^3 = 1 \cdot (-i) = -i$

$i^4 = i^2 \cdot i^2 = -1(-1) = 1$ $i^8 = i^4 \cdot i^4 = 1 \cdot 1 = 1$

EXAMPLE 2

Simplify: **a.** i^{27} **b.** $12i^{48}$

Solution

  a. $i^{27} = (i^4)^6 \cdot i^3 = 1 \cdot (-i) = -i$

  b. $12i^{48} = 12(i^4)^{12} = 12 \cdot 1 = 12$ ◄

Always express the square roots of negative numbers in terms of i before adding, subtracting, multiplying, or dividing them.

EXAMPLE 3

Simplify.

  a. $\sqrt{-9} + \sqrt{-16}$ **b.** $\sqrt{-3} \cdot \sqrt{-12}$ **c.** $\dfrac{8}{2i}$

Solution

  a. $\sqrt{-9} + \sqrt{-16} = 3i + 4i = (3 + 4)i = 7i$

  b. $\sqrt{-3} \cdot \sqrt{-12} = i\sqrt{3} \cdot i\sqrt{12} = i^2\sqrt{36} = 6i^2 = -6$

  c. $\dfrac{8}{2i} = \dfrac{8}{2i} \cdot \dfrac{i}{i} = \dfrac{8i}{2i^2} = \dfrac{4i}{-1} = -4i$ ◄

With imaginary numbers you can solve quadratic equations that do not have real solutions.

EXAMPLE 4

Solve: $2x^2 + 5x + 6 = 0$

Solution

By the quadratic formula,

$$x = \frac{-5 \pm \sqrt{5^2 - 4(2)(6)}}{2(2)} = \frac{-5 \pm \sqrt{-23}}{4} = -\frac{5}{4} \pm \frac{i\sqrt{23}}{4} \blacktriangleleft$$

Recall that the x-intercepts of the graph of a quadratic equation in the form $ax^2 + bx + c = 0$ represent the real number solutions of the equation. Because the solutions of the equation $2x^2 + 5x + 6 = 0$ are imaginary, the graph does not intersect the x-axis. Recall from Lesson 6.4 that you can determine the nature of the solutions of a quadratic equation by checking the discriminant.

> ── NATURE OF THE SOLUTIONS OF A QUADRATIC EQUATION ──
>
> The solutions of the equation $ax^2 + bx + c = 0$ are
> - **real and unequal if $b^2 - 4ac > 0$.**
> - **rational if $b^2 - 4ac$ is a perfect square**
> - **real and equal if $b^2 - 4ac = 0$.**
> - **imaginary if $b^2 - 4ac < 0$.**

PROBLEM SOLVING TIP

Example 4 could also be solved by completing the square.

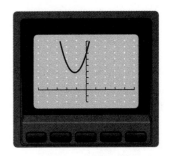

COMMUNICATING ABOUT ALGEBRA

Explain why the solutions of a quadratic equation are imaginary when the discriminant is negative.

EXAMPLE 5

Without solving, determine the nature of the solutions.

a. $x^2 - 5x - 7 = 0$ **b.** $-3x^2 + 2x - 1 = 0$

c. $4x^2 + 4x + 1 = 0$

Solution

a. $a = 1, b = -5, c = -7$

$b^2 - 4ac = (-5)^2 - 4(1)(-7) = 25 + 28 = 53$

Since $53 > 0$, the solutions are real and unequal.

b. $b^2 - 4ac = -8$, so the solutions are imaginary.

c. $b^2 - 4ac = 0$, so the solutions are real and equal. $\blacktriangleleft$

The solutions to Example 4, $-\frac{5}{4} + \frac{i\sqrt{23}}{4}$ and $-\frac{5}{4} - \frac{i\sqrt{23}}{4}$, each

consist of the sum of a real number, $-\frac{5}{4}$, and an imaginary number,

$\pm \frac{i\sqrt{23}}{4}$. Such a combination is called a *complex number*.

CHECK UNDERSTANDING

Draw a diagram of the relationships among all the sets of numbers you have studied.

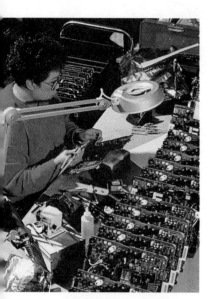

A **complex number** is a number of the form $a + bi$ where a and b are real numbers and i is the imaginary unit. The real part of a complex number $a + bi$ is a. The imaginary part is bi.

Two complex numbers are equal if and only if their real parts are equal and their imaginary parts are equal. That is, $a + bi = c + di$ if and only if $a = c$ and $b = d$.

Every real number a is a complex number because $a = a + 0i$. Similarly, every imaginary number bi is a complex number because $bi = 0 + bi$. Therefore, the real numbers and the imaginary numbers are subsets of the complex number system.

Despite the vagueness of the term "imaginary," both complex and imaginary numbers have widespread real world applications.

EXAMPLE 6

ELECTRICAL ENGINEERING In a circuit, opposition to the electrical current that does not involve a loss of energy is called *reactance*. The total reactance X_T and two types of reactance, *inductive reactance X_L* and *capacitive reactance X_C*, are related by the formula $X_T = X_L - X_C$. Each variable can be expressed as a pure imaginary number. Determine the capacitive reactance in a circuit with an inductive reactance of $27i$ ohms and a total reactance of $19i$ ohms.

Solution

$$X_T = X_L - X_C$$
$$19i = 27i - X_C$$
$$X_C = 27i - 19i$$
$$X_C = 8i$$

The capacitive reactance is $8i$ ohms. ◀

TRY THESE

Simplify.

1. $\sqrt{-25}$

2. $3\sqrt{-200}$

3. $\sqrt{-\dfrac{1}{2}}$

4. $\left(\sqrt{-15}\right)^2$

5. i^{17}

6. $-5i^{46}$

7. $14i - 9i + 3i$

8. $10i \cdot (-3i)$

9. $20 \div 5i$

10. $\sqrt{-2} + \sqrt{-18}$

11. $\sqrt{-3}\sqrt{-5}$

12. $\dfrac{10i}{\sqrt{-5}}$

Solve for x. Check your answer.

13. $x^2 + 81 = 0$

14. $6x^2 + 4 = 0$

15. $3x^2 - 6x + 4 = 0$

Without solving, determine the nature of the solutions of the equation.

16. $4x^2 - 12x + 9 = 0$

17. $3x^2 + 1 = 2x$

18. $x(5x + 2) = 3$

19. **ELECTRICAL ENGINEERING** An electrical circuit for a music amplification system has an inductive reactance of $11.61i$ ohms and a capacitive reactance of $9.45i$ ohms. Determine the total reactance of the circuit.

20. **WRITING MATHEMATICS** In which of the following sets of numbers—complex, imaginary, pure imaginary, real, rational, irrational—do each of the numbers $\sqrt[3]{64}$, $\sqrt{-64}$, and $2 + i\sqrt{2}$ belong? Explain.

PRACTICE

Simplify.

1. $\sqrt{-98}$
2. $\sqrt{-800}$
3. $-n\sqrt{-36}$
4. $i\sqrt{-45}$

5. $3i\left(2i\sqrt{-16}\right)$
6. $-i(8i)\sqrt{-243}$
7. $\left(\sqrt{-11}\right)^2$
8. $i\left(\sqrt{-18}\right)^2$

9. $\sqrt{-\dfrac{4}{9}}$
10. $\sqrt{-\dfrac{15}{16}}$
11. $\dfrac{2\sqrt{5}}{\sqrt{-3}}$
12. $\dfrac{3\sqrt{-6}}{4\sqrt{-12}}$

13. i^{13}
14. i^{104}
15. i^{67}
16. i^{50}

17. $3i(i^9)$
18. $-9i(i^{19})$
19. $-5i + 14i$
20. $6.9i + 8.3i + 0.47i$

21. $778i + 697i$
22. $0.091i - 0.77i$
23. $23i^{87} + 16i^{33}$
24. $-8.1i^{31} + 11.5i^{55}$

25. $(-8i)(9i)$
26. $2.7i(-4.1i)$
27. $3i^{13} \cdot 4i^{23}$
28. $6 \cdot 4i^{34}$

29. **WRITING MATHEMATICS** Franco says that a real number is an imaginary number and an imaginary number can be a real number. Is Franco correct? Explain. If he is not correct, how can you change his statement to make it correct?

Simplify.

30. $2\sqrt{-3} + 5\sqrt{-3}$
31. $\sqrt{-18} - \sqrt{-32}$
32. $5i - \sqrt{-5}$
33. $i\sqrt{2} + 3\sqrt{-200}$

34. $\sqrt{-6}\sqrt{-6}$
35. $-4\sqrt{-8}\sqrt{-4.5}$
36. $\sqrt{-5}\left(2\sqrt{-9}\right)$
37. $\sqrt{-10}\sqrt{-10}\sqrt{-10}$

38. $\dfrac{8}{i}$
39. $\dfrac{15}{9i}$
40. $\dfrac{3}{\sqrt{-3}}$
41. $\dfrac{12i}{3\sqrt{-32}}$

Solve for x. Check your answer.

42. $x^2 + 16 = 0$
43. $x^2 + 11 = 0$
44. $4x^2 + 100 = -700$

45. $x^2 + x + 1 = 0$
46. $5x^2 - 6x + 5 = 0$
47. $3(x - 1) = 2x^2 + 1$

Without solving, determine the nature of the solutions of the equation.

48. $x^2 + 5x + 7 = 0$
49. $2x^2 - 3x - 4 = 0$
50. $6x^2 + 5x + 3 = 0$

51. $9x^2 - 6x + 1 = 0$
52. $4x^2 + 20x + 25 = 0$
53. $-3(x^2 - 2) = 4$

54. **ELECTRICAL ENGINEERING** The capacitive reactance of a circuit is $17.85i$ ohms. The total reactance of the circuit is $8.08i$ ohms. Find the inductive reactance of the circuit.

EXTEND

55. **WRITING MATHEMATICS** Give an example of each of the following or explain why it is not possible: a number that is *not* a rational number; a number that is *not* an imaginary number; a number that is *not* a complex number.

Simplify.

56. i^{-1}

57. i^{-2}

58. i^{-3}

59. i^{-4}

60. What is unusual about the values of i^{-2} and i^{-4}?

61. Is there a pattern in the successive negative powers of i? If there is, describe it.

Determine the values of x and y for which each equation is true.

62. $4x + 2yi = 20 + 6i$

63. $5x + 3yi = 35 - 18i$

THINK CRITICALLY

Let ai and bi represent pure imaginary numbers where a and b are real numbers. Decide whether the set of pure imaginary numbers is closed under each operation. Give reasons.

64. addition

65. subtraction

66. multiplication

67. division

68. Explain the fallacy in the following "proof."

a. $\sqrt{-1} = \sqrt{-1}$

b. $\sqrt{\dfrac{1}{-1}} = \sqrt{\dfrac{-1}{1}}$

c. $\dfrac{\sqrt{1}}{\sqrt{-1}} = \dfrac{\sqrt{-1}}{\sqrt{1}}$

d. $\sqrt{1} \cdot \sqrt{1} = \sqrt{-1} \cdot \sqrt{-1}$

e. $1 = -1$

MIXED REVIEW

Determine each product.

$$A = \begin{bmatrix} 1 & 2 \\ 4 & 3 \end{bmatrix} \qquad B = \begin{bmatrix} -3 & 5 \\ 2 & -1 \end{bmatrix} \qquad C = \begin{bmatrix} 1 & -1 \\ -1 & 1 \end{bmatrix}$$

69. $5A$

70. $-3C$

71. AB

72. BC

Evaluate each expression if $M = 2 + \sqrt{2}$, $N = -3 - 2\sqrt{2}$, and $P = 2 - \sqrt{2}$

73. MN

74. P^2

75. MP

76. $M + N + P$

77. **STANDARDIZED TESTS** Simplify $6 \div \sqrt{-2}$. Choose the letter of the correct answer.

A. $-3\sqrt{2}$

B. $-3i\sqrt{2}$

C. -6

D. $6i\sqrt{2}$

8.7 Operations with Complex Numbers

Explore

1. Name the property of real numbers that justifies each step.

   a. $(4 + 2x) + (3 - 5x) = (2x + 4) + (3 - 5x)$

   b. $ = 2x + (4 + 3) - 5x$

   c. $ = (4 + 3) + 2x - 5x$

   d. $ = (4 + 3) + (2x - 5x)$

   e. $ = (4 + 3) + x(2 - 5)$

   $ = 7 - 3x$

2. Follow the above steps to find the sum of the complex numbers $4 + 2i$ and $3 - 5i$.

3. Make conjectures about the results of performing the following operations on the complex numbers $a + bi$ and $c + di$.

   a. $(a + bi) + (c + di)$ **b.** $(a + bi) - (c + di)$

> **SPOTLIGHT ON LEARNING**
>
> **WHAT?** In this lesson you will learn
> • to add, subtract, multiply, divide, and graph complex numbers.
>
> **WHY?** Operations with complex numbers are used in problems about biological, physical, and social sciences.

Build Understanding

Like the set of real numbers, the set of complex numbers is a field. So the commutative, associative, and distributive properties can be applied to operations with complex numbers. As the Explore activity suggests, this means that you can add and subtract complex numbers by combining their real parts and combining their imaginary parts.

> **THINK BACK**
>
> A *field* is a set of elements for which two operations are defined. The set is closed, and has commutative, associative, identity, and inverse properties, and is distributive.

EXAMPLE 1

Simplify.

   a. $(6 + 4i) + (7 - 9i)$ **b.** $(1 - 2i) - (6 - 4i)$

Solution

   a. $(6 + 4i) + (7 - 9i) = (6 + 7) + (4i - 9i) = 13 - 5i$

   b. $(1 - 2i) - (6 - 4i) = (1 - 6) + (-2i - (-4i)) = -5 + 2i$

You can extend the familiar principles of graphing on the real number plane to the *complex plane*. The **complex plane** is defined by a horizontal **real axis** and a vertical **imaginary axis**. Graph a complex number $a + bi$ on the complex plane as you graph the point (a, b) on the real number plane, associating a with the real axis and b with the imaginary axis.

EXAMPLE 2

Graph each complex number.

a. $4 - 3i$ **b.** -2 **c.** $2.5i$

Solution

a. Graph $4 - 3i$ by moving 4 units right on the real axis and 3 units down parallel to the imaginary axis.

b. Since $-2 = -2 + 0i$, graph -2 on the real axis.

c. Since $2.5i = 0 + 2.5i$, graph $2.5i$ on the imaginary axis.   ◄

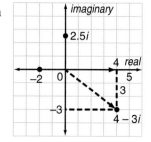

Recall that the absolute value of a real number is the distance of the number from zero on the real number line. Similarly, the absolute value of a complex number is the distance of the number from the origin of the complex plane.

EXAMPLE 3

Determine $|4 - 3i|$

Solution

By the Pythagorean theorem, the distance of the point $4 - 3i$ to the origin is $\sqrt{4^2 + 3^2} = 5$. Therefore, $|4 - 3i| = 5$. See graph above.   ◄

You can find the sum graphically.

EXAMPLE 4

Add graphically: $(-3 + i) + (2 + 4i)$

Solution

Plot $-3 + i$ and $2 + 4i$. Draw a line segment connecting each point to the origin. Then complete the parallelogram. The fourth vertex represents the sum $-1 + 5i$.

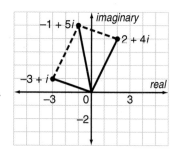

Check

$(-3 + i) + (2 + 4i) = (-3 + 2) + (1 + 4)i = -1 + 5i$ ◄

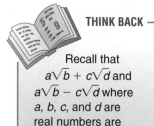

THINK BACK

Recall that $a\sqrt{b} + c\sqrt{d}$ and $a\sqrt{b} - c\sqrt{d}$ where a, b, c, and d are real numbers are *conjugates*.

You can multiply and divide complex numbers as you multiply and divide binomials. The complex numbers $a + bi$ and $a - bi$ are **conjugates**. To divide one complex number by another, use the conjugate c of the denominator to multiply by 1 in the form $\frac{c}{c}$.

EXAMPLE 5

Simplify.

a. $(5 + 3i)(-2 + 6i)$
b. $\dfrac{-2 + i}{3 - 5i}$

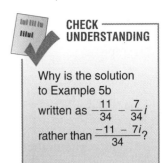
**CHECK
UNDERSTANDING**

Why is the solution
to Example 5b
written as $-\dfrac{11}{34} - \dfrac{7}{34}i$
rather than $\dfrac{-11 - 7i}{34}$?

Solution

a. $(5 + 3i)(-2 + 6i) = -10 + 30i - 6i + 18i^2$
$= -10 + 30i - 6i + 18(-1)$
$= (-10 - 18) + (30 - 6)i$
$= -28 + 24i$

b. $\dfrac{-2 + i}{3 - 5i} \cdot \dfrac{3 + 5i}{3 + 5i} = \dfrac{-6 - 10i + 3i + 5i^2}{9 + 15i - 15i - 25i^2}$

$= \dfrac{-11 - 7i}{9 + 25}$

$= -\dfrac{11}{34} - \dfrac{7}{34}i$ ◄

You are familiar with two- and three-dimensional objects such as
planes and spheres. **Fractal geometry** is the study of nonlinear
dimensions. Complex numbers lie at the heart of fractal geometry,
which has wide and surprising applications in fields as diverse as
meteorology, medicine, and literary analysis.

Fractal images like the one shown at the right are generated by
substituting an **initial value** into a complex function, plotting the
output, then using the output as the next value to substitute into the
function. This process of continual recycling of outputs is called
iteration, and each successive output is called an **iterate**.

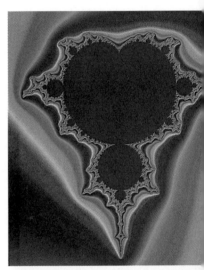

EXAMPLE 6

FRACTAL GEOMETRY Complex numbers are usually symbolized by
the variable z. Let $f(z) = 2iz$ represent a complex function. Beginning
with the initial value $z = 3 + 4i$, determine the first two iterates of
the function.

Solution
Begin by substituting the initial value $3 + 4i$ into the function
$f(z) = 2iz$.

First iterate

$f(3 + 4i) = 2i(3 + 4i) = 6i + 8i^2 = 6i - 8$

The first iterate in complex number form is $-8 + 6i$.

Second iterate

$f(-8 + 6i) = 2i(-8 + 6i) = -16i + 12i^2 = -16i - 12$

The second iterate is $-12 - 16i$.   ◄

Simplify.

1. $(3 - 2i) + (-4 - i)$ **2.** $(9 - 9i) - (12 - 3i)$ **3.** $(5 + i)(2 - 2i)$

4. $\dfrac{-1 + 4i}{2 + 6i}$ **5.** the product of $-2 + i$ and its conjugate

Graph each number on the complex plane.

6. $-3 + 2i$ **7.** $2 - 3i$ **8.** -4 **9.** i

10. Determine and simplify the absolute value of $6 - 4i$.

11. Add graphically: $(-2 - 4i) + (4 + i)$

12. FRACTAL GEOMETRY Let $f(z) = z + 5 + 2i$. Using $-3 + i$ as the initial value, determine the first four iterates of $f(z)$.

13. ELECTRICAL ENGINEERING In an electrical circuit, the voltage E in volts, the current I in amps, and the opposition to the flow of current, called impedance, Z in ohms, are related by the equation $E = I \cdot Z$. A circuit has a current of $(4 + 2i)$ amps and an impedance of $(-3 + 2i)$ ohms. Determine the voltage and determine $|Z|$, the magnitude of the impedance.

14. WRITING MATHEMATICS Explain how you could subtract two complex numbers graphically. Can you find more than one method?

PRACTICE

Simplify.

1. $(2 + 7i) + (6 + 4i)$ **2.** $(-5 + 5i) + (5 - 2i)$ **3.** $(8i) + (-1)$

4. $(14 - 7i) - (9 + 6i)$ **5.** $(-6 + i) - (13 - 2i)$ **6.** $(21.4 - 16.8i) - (32.9 + 11.4i)$

7. $(3 + 5i)(2 + 6i)$ **8.** $(10 + i)(4 - 3i)$ **9.** $(-2.5 + 12i)(6 - 14i)$

10. $\dfrac{2 + 2i}{4 + 3i}$ **11.** $\dfrac{7 - i}{7 + i}$ **12.** $\dfrac{6}{5 - 8i}$

13. What is the reciprocal of $5 - 2i$? Show that the product of $5 - 2i$ and its reciprocal is 1.

Graph each number on the complex plane.

14. $-6 + 2i$ **15.** $1 + i$ **16.** $4 - 5i$ **17.** $-3i$

Determine and simplify the absolute value of each number.

18. $8 + 6i$ **19.** $3 - 7i$ **20.** $-6 - 3i$ **21.** $20 + 21i$

Add graphically.

22. $(1 + 3i) + (3 + i)$ **23.** $(-4 - 2i) + (2 + 5i)$

24. $(4 - 3i) + (-3i)$ **25.** $(-5 + i) + (5 + i)$

26. **Fractal Geometry** Let $f(z) = iz$. Using $(1 + i)$ as the initial value, determine the first four iterates of $f(z)$. Predict the next four iterates of $f(z)$. Explain your reasoning.

27. Let $f(z) = z^2$. Using $(1 - i)$ as the initial value, find the first four iterates of $f(z)$.

28. **Electrical Engineering** The voltage E in an electrical circuit is $3 + 2i$ volts. The current I in amps is $1.6 + 0.2i$ amps. Use $E = I \cdot Z$ to find the impedance Z in ohms.

29. **Writing Mathematics** Explain why you cannot add $(9 + 6i)$ and $(-3 - 2i)$ using the graphical method. Find other pairs of complex numbers that cannot be added graphically.

EXTEND

Use the relationship between the sum and product of the solutions and the coefficients of a quadratic equation to determine the equation given these solutions.

30. $4 + 3i$ and $4 - 3i$ 31. $1 + 2i$ and $1 - 2i$ 32. $\dfrac{2 \pm i\sqrt{2}}{2}$ 33. $\dfrac{-5 \pm i\sqrt{7}}{4}$

34. The absolute value of $a + 16i$ is 34. Find a.

35. Show that $\sqrt{-5 - 12i} = 2 - 3i$. (*Hint*: What is $(2 - 3i)^2$?)

36. Simplify $\left(\dfrac{-1 + i\sqrt{3}}{2}\right)^3$.

THINK CRITICALLY

37. Determine the absolute value of the complex number $a + bi$.

38. Assume $a + bi$ is a nonzero complex number. Show that its reciprocal is
$$\dfrac{a}{a^2 + b^2} - \dfrac{b}{a^2 + b^2}i.$$

39. The binomial $a^2 + b^2$ cannot be factored over the real number field. However it can be factored over the complex number field. Find the factorization.

40. In Exercise 29, the two complex numbers could be added graphically by this method. Begin at $(9, 6i)$ and move 3 units to the left (-3) and down 2 units $(-2i)$. What is your answer? Compare to adding $9 + 6i$ and $-3 - 2i$ algebraically.

MIXED REVIEW

State whether the set is closed under the given operation.

41. $\{0, 1\}$, addition   42. $\{-1, 0, 1\}$, multiplication   43. {integers}, division

Determine the slope of the line.

44. $6x + 3y = 12$ 45. parallel to $x + y = 9$ 46. passing through $(1, 4)$ and $(0, 8)$

47. Find the probability that a point on an 8.5 in. $\times$ 11 in. sheet of paper is in a 3 in. $\times$ 5 in. rectangle drawn on the paper.

48. **Standardized Tests** What is the product of $3 - 4i$ and its conjugate?

   A. 25   **B.** $25i$ **C.** -7 **D.** $-7 - 24i$

Problem Solving File

Using Formulas: Annuities

People may save or invest money by depositing varying amounts at different times. If a person makes equal deposits or payments at regular intervals, he or she is contributing to an *annuity*. The payments may be made weekly, monthly, quarterly, yearly, or any other time period. For an *ordinary annuity*, payments are made at the end of each period. The sum of all payments plus all interest earned is called the amount of the annuity or the *future value* of the annuity. The future value F of an ordinary annuity is given by the formula

$$F = D\left(\frac{(1 + i)^n - 1}{i}\right)$$

where D is the payment (in dollars) at the end of each period,
n is the number of periods, and
i is the interest rate per period.

Problem

Roberto Ortega deposits $1000 at the end of each year in a mutual fund account that pays 6%, compounded annually. How much money will he have in the account at the end of 10 years?

Explore the Problem

1. What are the values of D, i, and n for this problem?

2. Write and evaluate the formula to determine F, the amount in Roberto's account at the end of 10 years.

3. Suppose that Roberto deposited $500 at the end of every 6 months (semiannually) in an account that paid 6% compounded semiannually. In this case, the interest rate per period is the rate divided by the number of periods per year. The number of periods n is the number of years multiplied by the number of periods per year. What values of D, i, and n should Roberto substitute in the formula for F to determine how much money he will have in 10 years with these new conditions?

4. How much would Roberto have in 10 years? Is this more or less than he had in Question 2? How much more or less?

5. WRITING MATHEMATICS Suppose you know that the future value of an annuity of $1 paid at the end of each year for 10 years is $14.49. Explain how you can use this information to determine the future value of an annuity of $750 paid at the end of each year for 10 years.

Investigate Further

Just as the term *annuity* is used to describe an account to which a person makes equal periodic payments, the term is also used to describe an account from which a person receives equal periodic payments or withdrawals. If you invest a lump sum of money today, so that at regular times you will receive a fixed amount of money, you have an annuity. The sum you must invest to get these payments is the *present value* of the annuity. The present value P of an ordinary annuity is given by the formula

$$P = D\left(\frac{1 - (1 + i)^{-n}}{i}\right)$$

where D is the amount of the payment from the annuity at the end of each period, n is the number of periods, and i is the interest paid per period.

> PROBLEM
> SOLVING PLAN
>
> - Understand
> - Plan
> - Solve
> - Examine

6. What sum would have to be invested at 6%, compounded annually, to provide an annuity of $1000 per year for 10 years?

   a. Identify the values of D, n, and i you will use.

   b. Write and evaluate the expression to determine P. Round to the nearest cent.

7. What is the present value of an annuity of $750 payable at the end of each 6-month period for 3 years if money earns 8%, compounded semiannually?

   a. Identify the values of D, n, and i you will use.

   b. Determine P to the nearest cent.

Apply The Strategy

8. INVESTING Joe Long deposits $1000 at the end of every 6 months into an annuity account earning 8% compounded semiannually. How much money will he have in the account at the end of 10 years?

9. RETIREMENT When Elaine Caruso was 35 she established a retirement account to which she contributed $1500 at the end of each year for 30 years. The interest rate on the account was 8% compounded annually.

   a. How much did Elaine actually contribute to the account?

   b. How much was the total account worth when she retired at 65 years of age?

10. **INVESTING**

   a. How much should Lee Soong deposit now in an annuity program paying 6% compounded monthly, in order to receive $200 per month for 5 years?

   b. How much money will he actually receive?

11. **COLLEGE COSTS** Parents of a college-age child want to create an annuity to cover part of the cost of college tuition. How much should they deposit now, at 8% compounded quarterly, for the annuity to pay $2000 every three months for 4 years?

12. **RETIREMENT** Mr. and Mrs. Taylor want to retire in 10 years with a fund of $100,000. They establish an annuity account at a bank whose annual interest rate is 12%, compounded quarterly. How much should the Taylors deposit at the end of each quarter in order to meet their goal? (When an annuity account is created to meet a future obligation or goal, the account is called a *sinking fund*.)

   a. Explain how this problem is different than the other future value problems you have solved so far.

   b. How can you use the formula for future value to solve this problem?

   c. Use the method you described in part b to determine the amount of the Taylor's periodic deposit.

13. **CAR REPLACEMENT** In 5 years, the Lopez family would like to have $20,000 to buy a new car. What fixed amount should they deposit every 3 months into an account paying 5% compounded quarterly to make their objective?

14. **WRITING MATHEMATICS** Explain the difference between solving a present value problem (such as Question 10) and solving a sinking fund problem (such as Question 13).

Review Problem Solving Strategies

CAREFUL PACKING

1. Terrence packed two identical glass spheres that are used as paperweights. He put 27 identical large spheres in one carton and 64 identical smaller spheres in the other carton. All the spheres are manufactured of the same glass. Terrence filled both cartons to the top. In each carton, each layer had the same number of spheres, and the outer spheres of each layer touched the sides. When Terrence put each packed carton on a scale to determine postage, what did he discover?

Here are some questions to help you think about the problem.

   a. How must the spheres be arranged in each carton?

   b. How does the diameter of a large sphere compare to the diameter of a small sphere?

   c. Since the spheres are made of the same material, you can use volume to represent weight. How does the volume (and therefore weight) of the large sphere compare to the smaller?

   d. How does the number of large spheres compare to the number of small ones? What can you conclude?

   e. Try the problem with other numbers that are perfect cubes. What general statement can you make?

2. Samantha and Sid Smith are sister and brother. Sid has the same number of brothers as he has sisters, but Samantha has twice as many brothers as sisters. How many children are in the Smith family?

   a. Could Sid have 1 brother and 1 sister? Explain.

   b. Could Sid have 5 brothers? Explain.

   c. How many Smith children are there? Explain how you know you found the only answer.

NUMBER PLEASE

3. Consider this equation: $[3(241 + x)]^2 = 594{,}y41$

   a. By what number must the expression on the right be divisible? Why?

   b. Use your reasoning from Question 1 to determine the number represented by x and the digit represented by y.

· · · CHAPTER REVIEW · · ·

VOCABULARY

Choose the word from the list that correctly completes each statement.

1. The letter i represents the ___?___ $\sqrt{-1}$.

2. In $\sqrt[5]{24}$, the number 24 is the ___?___ .

3. The ___?___ of $6 + 3i$ is $6 - 3i$.

4. For the equation $x^2 + 2x + 3$, the ___?___ is -8.

a. conjugate

b. discriminant

c. imaginary unit

d. radicand

Lessons 8.1 and 8.2   EXPONENTS AND RADICALS   pages 371–373, 374–380

- If n is an integer greater than 1 and $b^n = a$, then $b = \sqrt[n]{a}$. The **index** is n and a is the **radicand.**

- If n is odd, $\sqrt[n]{x^n} = x$. If n is even, $\sqrt[n]{x^n} = |x|$.

- $a^{\frac{m}{n}} = \left(\sqrt[n]{a}\right)^m = \sqrt[n]{a^m}$

Write each expression using an exponent or a radical.

5. $\sqrt[3]{k}$ 6. $\left(\sqrt[6]{9}\right)^5$ 7. $b^{\frac{1}{2}}$ 8. $(m + 4)^{\frac{2}{9}}$

Simplify.

9. $\sqrt[3]{x^3}$ 10. $\sqrt[5]{64x^{10}}$ 11. $125^{-\frac{1}{3}}$ 12. $\sqrt[6]{11^3}$

Lesson 8.3   GRAPHING RADICAL FUNCTIONS   pages 381–384

- The graph of $y = \sqrt{x} + c$ is the graph of $y = \sqrt{x}$ translated c units up or down.

- The graph of $y = \sqrt{x + c}$ is the graph of $y = \sqrt{x}$ translated c units left or right.

Sketch each graph.

13. $y = \sqrt{x}$ 14. $y = \sqrt{x} - 5$ 15. $y = \sqrt{x + 2}$

Lesson 8.4   EXPONENTIAL AND RADICAL EXPRESSIONS   pages 385–391

- For all real numbers a and b and integer $n > 1$, $\sqrt[n]{a} \cdot \sqrt[n]{b} = \sqrt[n]{ab}$ and $\dfrac{\sqrt[n]{a}}{\sqrt[n]{b}} = \sqrt[n]{\dfrac{a}{b}}, b \neq 0$

- To **rationalize a denominator** containing a radical, multiply the fraction by 1 in the form $\dfrac{r}{r}$, where r is an expression that will make the denominator a perfect nth power.

Simplify.

16. $\sqrt{6} + \sqrt{24}$ 17. $2\sqrt[3]{5} \cdot \sqrt[3]{25}$ 18. $\sqrt{\dfrac{1}{3}}$ 19. $\dfrac{\sqrt[3]{3}}{\sqrt[3]{24}}$ 20. $\dfrac{2}{2 + \sqrt{2}}$

- Solve a radical equation by raising both sides of the equation to the same power.

- Check the solutions in the original equation to eliminate **extraneous solutions**.

Solve.

21. $\sqrt{n + 6} = 8$ **22.** $\sqrt[3]{x} - 5 = -8$ **23.** $y^{\frac{3}{4}} = 8$ **24.** $\sqrt{5x - 1} + 3 = x$

- The **imaginary unit** i is defined to be $i = \sqrt{-1}$, where $i^2 = -1$.

- A **complex number** is a number of the form $a + bi$, where a and b are real numbers. The **real part** is a and the **imaginary part** is bi.

- Express square roots of negative numbers in terms of i before performing operations.

Simplify.

25. $\sqrt{-49}$ **26.** $\sqrt{-20}$ **27.** i^{21} **28.** $\sqrt{-3} + \sqrt{-75}$

29. $5i(2i)$ **30.** $12i - 10i$ **31.** $\sqrt{-2}\,\sqrt{-3}$ **32.** $\dfrac{15}{5i}$

- Add and subtract complex numbers by combining real parts and combining imaginary parts. Multiply and divide complex numbers as you multiply and divide binomials.

- The **complex plane** is defined by a horizontal **real axis** and a vertical **imaginary axis**. Graph $a + bi$ on the complex plane as you graph the point (a, b) on the real number plane.

Simplify.

33. $(4 + i) + (11 - 3i)$ **34.** $(2 - 5i) - (1 - 8i)$ **35.** $(-3 + i)(2 - 3i)$ **36.** $\dfrac{4 - 3i}{4 + 3i}$

37. Add $(-2 + 4i) + (3 + 2i)$ graphically.

- The *future value* of an *annuity* is given by the formula $F = D\left(\dfrac{(1 + i)^n - 1}{i}\right)$ where D is the payment, n is the number of periods and i is the interest rate per period.

- The *present value* of an annuity is given by the formula $P = D\left(\dfrac{1 - (1 + i)^{-n}}{i}\right)$ where D is the payment, n is the number of periods and i is the interest rate per period.

38. Darrell deposits $3,000 a year in a retirement account every year that pays 5% interest compounded annually. What will the value of the account be to the nearest dollar when Darrell retires in 45 years?

39. How much to the nearest dollar, do Ward and June have to invest at 6% annual interest in order to pay for Wally's college tuition of $10,000 over the next 4 years?

CHAPTER ASSESSMENT

CHAPTER TEST

Write using an exponent.

1. $\sqrt[4]{n^3}$

2. $(\sqrt{2p})^5$

Write using a radical symbol.

3. $k^{\frac{4}{7}}$

4. $(m + 5)^{\frac{1}{2}}$

Simplify.

5. $\sqrt{a^2}$

6. $32^{\frac{4}{5}}$

7. $-\sqrt[4]{81}$

8. $\sqrt[4]{32x^4}$

9. $\sqrt[8]{100^4}$

10. $\sqrt{5} + \sqrt{45}$

11. $\sqrt[3]{16y^6} - y^2\sqrt[3]{2}$

12. $\sqrt{6}\,(3\sqrt{6} + 2\sqrt{7})$

13. $\sqrt{18}\,\sqrt{8}$

14. $\dfrac{\sqrt{35}}{\sqrt{7}}$

15. $\sqrt{\dfrac{9}{16}}$

16. $\dfrac{3}{\sqrt[3]{y}}$

17. $\dfrac{11}{\sqrt{11}}$

18. $\dfrac{5x}{2 - \sqrt{5}}$

Sketch each graph.

19. $y = \sqrt{x - 4}$

20. $y = \sqrt{x} + 6$

21. **WRITING MATHEMATICS** Explain why $\sqrt{x^2} = |x|$.

Solve.

22. $2\sqrt{p} = 18$

23. $\sqrt{x - 3} = 5$

24. $\sqrt[3]{n} + 5 = 9$

25. $k^{\frac{2}{3}} - 1 = 3$

26. $\sqrt{x - 4} = \sqrt{x + 5} - 3$

27. $\sqrt{3x - 3} = x - 7$

Simplify.

28. $\sqrt{-25}$

29. i^{35}

30. $-6i(3i)$

31. $\sqrt{-5}\,\sqrt{-6}$

Simplify.

32. $\sqrt{-8n} + \sqrt{-18n} - i\sqrt{2n}$

33. $\dfrac{6}{i}$

34. $(8 + 11i) - (9 - 4i)$

35. $(2 + 5i)(-1 - 2i)$

36. $\dfrac{3 + i}{4 - 2i}$

37. Add $(5 - 3i) + (-2 - 2i)$ graphically.

38. **STANDARDIZED TESTS** The solutions of the equation $-x^2 + 12x - 36 = 0$ are:

A. real and unequal
B. real and equal
C. imaginary and equal
D. complex conjugates

39. The total reactance X_T of an electrical circuit is given by $X_T = X_L - X_C$, where X_L is the inductive reactance and X_C is the capacitive reactance. Find the capacitive reactance in a circuit with an inductive reactance of $15.5i$ ohms and a total reactance of $9.7i$ ohms.

40. Find and simplify the perimeter and area of a rectangle that measures $\sqrt{12}$ cm by $\sqrt{3}$ cm.

41. The volume V of a sphere with radius r is given by $V = \frac{4}{3}\pi r^3$. Find the radius of a baseball with volume $\dfrac{9\pi}{2}$ in³.

42. Mr. and Mrs. Jefferson start a college fund for their newborn daughter. They deposit $1000 on each birthday into an account that pays 5% interest, compounded annually. What will the value of the account be when the daughter goes to college after her 18th birthday.

PERFORMANCE ASSESSMENT

ROOT MODEL Use materials from your classroom or home to model each of the following equations:

a. $n = \sqrt{5}$

b. $n = \sqrt[3]{5}$

c. $n = 5^{\frac{2}{3}}$

THE LANGUAGE OF MATHEMATICS
Weather is a *function* of atmospheric conditions. In mathematics, the value of a variable expression is a *function* of the value of the variables. Is there any relationship between a *root* of a tree and a *root* of a real number? Choose three mathematical terms defined in this chapter that are the same or similar to real world terms. Describe the relationship of each mathematical term to its real world counterpart.

COMPLEX CIRCLE The figure shows a circle in the complex plane. The radius of the circle is 1 and the center is at the origin.

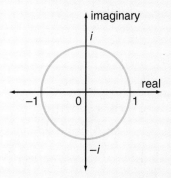

Sketch the circle on graph paper. Choose a point z that lies on the circle but not on the real or imaginary axis. Using z as your initial value, find and graph the first two iterates of the function $f(z) = z^2$. Describe your results.

FIND A FRACTAL Use books or magazines to find a real life example of a fractal. Make a copy of it. Then highlight portions of the figure that are small-scale replicas of the whole.

PROJECT ASSESSMENT

 Each group should plan a short presentation of the results of their experiments.

1. Explain each of the graphs you have made and the conclusions that can be drawn from the data. When all groups have presented, compare results and account for any differences.

2. Discuss the relation of the different experimental set-ups you used to conditions that might be present in a real rocket launching. For example, how did the string path modify the effects of gravity?

3. Design an experiment to explore a different relationship. For example, how could you find out if the launch angle affects the motion? Or, how does the shape of the balloon affect the motion? Remember that only one condition can vary in an experiment; all other factors must remain the same during the trials.

CUMULATIVE REVIEW

1. **STANDARDIZED TESTS** Which of the following are true for the complex number $2 - i$?

   I.   Its additive inverse is $-2 - i$.
   II. Its multplicative inverse is $\dfrac{2 + i}{5}$.
   III. Its conjugate is $2 + i$.

   A. I and II
   B. I and III
   D. II and III
   D. I, II, and III

2. Convert the expression $x^2 - 30x$ to a perfect square trinomial.

Factor.

3. $21x^2 - 29x - 10$

4. $x^3 + 5x^2 - 16x - 80$

Solve and check.

5. $\sqrt{y + 1} = 5$

6. $\sqrt{2x + 1} - 1 = -4$

7. $x^2 + 625 = 0$

8. $y^2 - 4y + 13 = 0$

9. **STANDARDIZED TESTS** The solution set of the inequality $x^2 + 4x + 3 < 0$ is

   A. $-3 < x < -1$
   B. $x < -3 \quad \text{or} \quad x > -1$
   C. $1 < x < 3$
   D. $x < 1 \quad \text{or} \quad x > 3$

10. Determine the equation of the graph.

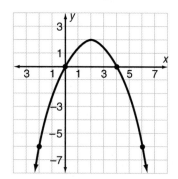

Evaluate each radical expression, if possible.

11. $64^{-\frac{1}{2}}$

12. $\sqrt[4]{-16}$

13. Use the formula $P = D \cdot \left(\dfrac{1 - (1 + i)^{-n}}{i} \right)$, where P is the present value of an ordinary annuity, D is the amount of the payment from the annuity at the end of each period, n is the number of periods, and i is the interest paid per period to calculate the sum that would have to be invested at 7%, compounded annually, to provide an annuity of $1500 per year for 15 years.

Graph each solution set in the coordinate plane.

14. $2y \geq 5x + 6$

15. $y < 3$

16. **WRITING MATHEMATICS** Describe how the graphs of $y = 3\sqrt{x}$, $y = \sqrt{x} + 3$, and $y = \sqrt{x + 3}$ are related to the graph of $y = \sqrt{x}$.

Solve each problem.

17. Jamie bought a 4-foot strip of wood to make a frame for an 8" × 10" photo she intends to enclose in a mat of uniform width before framing. What are her options for the width of the mat?

18. To provide wheelchair access to public buildings, a town paid $350 to widen each doorway and $540 to install each ramp. If the town spent $12,850 to widen doorways or install ramps at 34 locations, how many ramps were installed?

19. The 18th-century French scientist Jacques Charles was the first to notice that gases expand when heated and contract when cooled. Charles observed that a linear relation exists between volume and temperature. If a gas has a volume of 500 cc at 30°C and a volume of 600 cc at 90°C, write a linear equation to model the relation between volume and temperature. In theory, the volume of gas could contract until none remained. Estimate the coldest temperature possible to sustain a quantity of this gas.

STANDARDIZED TEST

QUANTITATIVE COMPARISON In each question compare the quantity in Column 1 with the quantity in Column 2. Select the letter of the correct answer from these choices.

A. The quantity in Column 1 is greater.
B. The quantity in Column 2 is greater.
C. The two quantities are equal.
D. The relationship cannot be determined by the information given.

Notes: In some questions, information which refers to one or both columns is centered over both columns. A symbol used in both columns has the same meaning in each column. All variables represent real numbers. Most figures are not drawn to scale.

Column I	Column II

1.
$$10x + 10y = 60$$

The mean of x and y 3

2. The solutions of $x^2 - 6x + 18 = 0$ are written in $a + bi$ form

a .... b

3.
$$x + y = z$$
$$z = 2y - 1$$

x .... y

4. x and y are even integers such that
$-5 < x < -3$ and $-7 < y < -5$

$$\frac{x + y}{x \cdot y} \qquad\qquad \frac{x \cdot y}{x + y}$$

5.
$$\sqrt{3x + 3} - \sqrt{6x + 7} + 1 = 0$$

The product of the solutions .... The sum of the solutions

6. $x^{\frac{3}{2}}y^{\frac{1}{2}}$ .... $\sqrt{x^3 y}$

Column I	Column II

7.
$$x^2 + 6x + 8 \leq 0$$

The lower bound of the solution set .... -6

8. $[x]$ denotes the greatest integer less than or equal to x.

$[5.2] + [-5.2]$ 0

9. The conjugate of $1 + \sqrt{3}$ .... The conjugate of $-1 - \sqrt{3}$

10.
$$3x^2 - 2x = -2$$

The sum of the solutions .... The product of the solutions

MACHINE-PART PRODUCTION, MAY 1			
			defective
	number not	number	as percent
	defective	defective	of total
company			
Acme	95	5	5%
Turbo	81	x	10%
Rheo	441	y	2%

11. x .... y

12. $\sqrt[3]{-d}$ .... $\sqrt[5]{-d}$

13. $\dfrac{1}{1 - i}$ is rewritten in $a + bi$ form

a .... b

14. $10x^2 + 11x - 6 = (2x + a)(5x + b)$

a .... b

15. product of the coordinates of the vertex of the graph of $y = 3|x - 3| - 3$ .... product of the coordinates of the vertex of the graph of $y = 3|x + 3| - 3$

9 Exponential and Logarithmic Functions

Take a Look AHEAD

Make notes about things that look new.

- Find the names of two types of logarithms.
- Identify an application of exponential growth. Then identify an application of exponential decay.
- Locate and copy the compound interest formula. What type of expression is involved?

Make notes about things that look familiar.

- What are the different parts of an exponential expression? What does each mean?
- What properties do you use to evaluate and simplify exponential expressions?
- When do you use a vertical line test? When do you use a horizontal line test?

DATA Activity

Counting Many Heads

The amount of data collected by the U.S. Bureau of Census during its decennial (10 year) population count is a statistician's dream or nightmare. Census takers attempt a complete count for every person, for each person's residence, and for other characteristics (gender, age, family relationships, and so on). Since 1940, the complete count has been supplemented by data obtained from representative samples of the population. In the 1990 census, one out of six households (about 17%) received the long form.

SKILL FOCUS

- Add, subtract, multiply, and divide real numbers.
- Determine functional relationships.
- Write an equation and use it to make predictions.
- Calculate percents.
- Construct pictographs, bar graphs, and circle graphs.

POPULATION

In this chapter, you will see how:

- **ONCOLOGISTS** use exponential growth curves to model the growth of tumors.
  (Lesson 9.2, page 435)
- **SEISMOLOGISTS** use a logarithmic scale to study the magnitude of earthquakes.
  (Lesson 9.3, page 442)
- **POPULATION ECOLOGISTS** use exponential and logarithmic functions to study the population of organisms.
  (Lesson 9.6, page 459)

U.S. Foreign-Born Population							
Percent of U.S. Population That Is Foreign-Born				Countries of Birth of Foreign-Born Population (1994)			
Year	Percent	Year	Percent	Country	Number (1000s)	Country	Number (1000s)
1910	14.7	1960	5.4	Mexico	6264	China	565
1920	13.2	1970	4.8	Philippines	4003	Dominican Republic	556
1930	11.6	1980	6.2	Cuba	805	Korea	533
1940	8.8	1990	7.9	El Salvador	718	Vietnam	496
1950	6.9	1994	8.7	Canada	679	India	494
				Germany	625		

Use the tables to answer the following questions.

1. Display the data about percentage of foreign-born population in a scatterplot.

2. Examine the scatterplot you made in Question 1 and suggest the type of function that might be a good fit for the data. Then use a graphing utility to determine an equation for the function. Explain what your variables represent. Round coefficients to three decimal places.

3. Compare the values predicted by your function to some of the actual values in the table. Then predict a percentage value for the year 2000.

4. To the nearest percent, what part of the foreign-born population for the countries shown is Mexican? What part is Chinese?

5. **WORKING TOGETHER** Use the data about countries of birth to make a pictograph, bar graph, and circle graph. Divide your group so that each member works on one type of graph. What decisions have to be made for each type of graph?

421

PROJECT

America: Land of How Many?

Reliable information about trends in the American population is essential for government and industry to plan our needs in areas such as housing, food supply, education, medical services, and transportation. Many factors—including immigration, age structure, and birth, death, and fertility rates—influence population trends. The National Census (taken once every ten years) as well as other population studies collect and analyze vast amounts of data. Different assumptions or interpretations of the data will lead to different models that, in turn, may yield widely varying predictions.

PROJECT GOAL

To create a mathematical model to predict the population of the United States in the year 2010.

Getting Started

Work in groups.

1. Discuss the information you need to research. For which years will you collect population figures? Can you find any population predictions for 2010 to use for comparison with your model's value? Why might you check several sources before selecting the numbers you will use?

2. Decide how you will record the population data, information about the methodology and assumptions of studies you use, and details about sources.

3. If a computer is available, discuss how you could use features such as graphics, spreadsheets, and curve-fitting in this project.

PROJECT *Connections*

Lesson 9.3, page 441:
Organize data collected; construct and analyze a linear population growth model.
Lesson 9.4, page 448:
Construct and analyze an exponential population growth model.
Lesson 9.5, page 453:
Investigate how a logistic growth curve models populations.
Chapter Assessment, page 467:
Prepare a group report on modeling population growth for class discussion.

Internet Connection
www.swpco.com/
swpco/algebra2.html

Algebra Workshop
Explore Exponential Functions

Think Back

1. Complete: The graph of a function $g(x)$ is the reflection of the graph of $f(x)$ over the x-axis if $g(x) = \blacksquare$.

Write a rule for the function that is the reflection of $f(x)$ over the x-axis.

2. $f(x) = 3x$

3. $f(x) = x^2$

4. $f(x) = |x|$

Write each expression with positive exponents.

5. 3^{-4}

6. $\dfrac{m^{-2}n^3}{p^{-5}}$

7. $4x^{-2}y$

Explore

An **exponential function** is a function that can be expressed in the form $f(x) = b^x$ where $b > 0$, and $b \neq 1$, and b is a constant.

8. Which of the following are exponential functions?

$$s(x) = 3^x \qquad t(x) = x^4 \qquad h(x) = \left(\frac{1}{2}\right)^x$$

9. Use a graphing utility to graph $f(x) = 2^x$ and $g(x) = 6^x$ on the same set of axes. Give the domain and range of each function.

10. Examine very large positive values of x on the graphs of f and g. As the value of x increases, what happens to the values of $f(x)$ and $g(x)$?

11. Examine very large negative values of x on the graphs of f and g. As the value of x decreases, what happens to the values of $f(x)$ and $g(x)$?

12. Discuss the similarities and differences between the graphs of f and g.

13. What point is common to both graphs? Why would this point be present on all graphs of the form $y = b^x$?

14. In Lesson 3.2, a function was called one-to-one if for any element in the domain of f,

$$f(x_1) = f(x_2) \quad \text{if and only if} \quad x_1 = x_2.$$

Use the horizontal line test to determine whether f and g are one-to-one functions.

15. Predict the shape of the graph of $h(x) = 15^x$ and its location on the coordinate plane in relation to the graphs of f and g.

The distance between the x-axis and the graph of $f(x) = b^x$ where $b > 1$ gets closer to 0 as x gets smaller. Therefore, the exponential function is *asymptotic* to the x-axis (the line $y = 0$) and the x-axis is a **horizontal asymptote** of the function $f(x) = b^x$.

Some exponential functions have graphs different from those just discussed.

16. Graph $j(x) = 2^{-x}$ and $k(x) = 6^{-x}$ on the same set of axes. Give the domain and range of each function.

17. Use the horizontal line test to determine whether j and k are one-to-one functions.

18. Examine very large positive values of x on the graphs of j and k. As the value of x increases, what happens to the values of $j(x)$ and $k(x)$?

19. Examine very large negative values of x on the graphs of j and k. As the value of x decreases, what happens to the values of $j(x)$ and $k(x)$?

20. Over what line can you reflect the graph of $f(x) = 2^x$ to obtain the graph of $j(x) = 2^{-x}$? Over what line can you reflect the graph of $g(x) = 6^x$ to obtain the graph of $k(x) = 6^{-x}$?

21. Express $j(x)$ and $k(x)$ with positive exponents.

22. What is the relationship between the graphs of $q(x) = b^x$ and $r(x) = \left(\frac{1}{b}\right)^x$?

23. Is the x-axis a horizontal asymptote of the function $f(x) = b^x$ when $0 < b < 1$? Explain or provide a counterexample.

You can also reflect the graphs of exponential functions over the x-axis.

24. Use your answer to Exercise 1 to write a function that is the reflection of $f(x) = 2^x$ over the x-axis. Do the same for $k(x) = 6^{-x}$ and for $p(x) = \left(\frac{1}{4}\right)^{-x}$.

Recall that a vertical translation moves a graph up or down.

25. Graph $f(x) = 3^x$, $g(x) = 3^x + 2$, and $h(x) = 3^x - 1$ on the same set of axes. Give the domain and range of each function.

26. Write an equation of the horizontal asymptote of each function.

27. Where is the graph of g in relation to the graph of f? Where is the graph of h in relation to the graph of f?

28. Graph several functions of the form $n(x) = b^x + c$ where c can be any real number. If a vertical translation of $m(x) = b^x$ is written as $n(x) = b^x + c$, write general statements about the location of the graph of n in relation to the graph of m when c is positive and when c is negative.

Recall that a horizontal translation moves a graph left or right.

29. Graph each of the following functions on the same set of axes.

$$f(x) = \left(\frac{1}{5}\right)^x \quad g(x) = \left(\frac{1}{5}\right)^{x+3} \quad h(x) = \left(\frac{1}{5}\right)^{x-2}$$

Give the domain and range of each function.

30. Write an equation of the horizontal asymptote of each function.

31. Where is the graph of g in relation to the graph of f? Where is the graph of h in relation to the graph of f?

32. Graph several functions of the form $n(x) = b^{x-d}$, where d can be any real number. If a horizontal translation of $m(x) = b^x$ is written as $n(x) = b^{x-d}$, write general statements about the location of the graph of n in relation to the graph of m when d is positive and when d is negative.

33. Work with a partner. Choose values of c and d and graph a function of the form $f(x) = 2^{x-d} + c$. Have your partner identify the values of c and d that you entered by studying the graph. Then switch roles with your partner.

Make Connections

34. Write a rule for a function g that is the reflection of f over the x-axis. Then write a rule for a function h that is the reflection of f over the y-axis.

a. $f(x) = 5^x$

b. $f(x) = 7^{-x}$

c. $f(x) = \left(\frac{1}{11}\right)^x$

d. $f(x) = \left(\frac{1}{3}\right)^{-x}$

35. State the direction(s) in which the graph of each function is translated from the graph of $f(x) = 2^x$.

    a. $g(x) = 2^{x-1} - 3$ **b.** $h(x) = 2^{x+2} + 4$ **c.** $j(x) = 2^{x-5} + 6$

36. Use the words "steeper" and "less steep" to describe how each graph differs from the graph of $f(x) = 5^x$.

    a. $g(x) = 8^x$ **b.** $h(x) = \left(\frac{1}{4}\right)^{-x}$ **c.** $j(x) = 3^x$

37. Match each function with its graph.

    a. $f(x) = \left(\frac{1}{2}\right)^{x+2}$ **b.** $g(x) = \left(\frac{1}{2}\right)^{x} + 2$ **c.** $h(x) = \left(\frac{1}{2}\right)^{x-2}$

I.
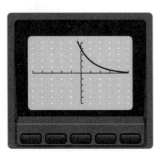

II.
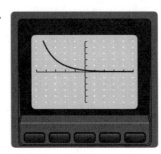

III.

38. Match each function with its graph.

    a. $f(x) = 3^{x+1} + 5$ **b.** $g(x) = 3^{x+5} - 1$ **c.** $h(x) = 3^{x-5} + 1$

I.
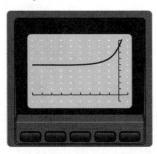

II.
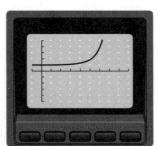

III.
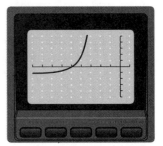

39. Match each function with its graph.

    a. $f(x) = 4^{-x} + 3$ **b.** $g(x) = \left(\frac{1}{4}\right)^{x} + 2$ **c.** $h(x) = 4^{-x} - 3$

I.

II.
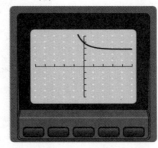

III.

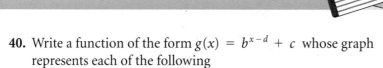

40. Write a function of the form $g(x) = b^{x-d} + c$ whose graph represents each of the following

    a. a translation of 1 unit down and 8 units left from the graph of $f(x) = b^x$

    b. a translation of 2 units up and 7 units right from the graph of $f(x) = b^x$

    c. a translation of 3 units down and 6 units right from the graph of $f(x) = b^x$

    d. a translation of 4 units up and 5 units left from the graph of $f(x) = b^x$

41. Determine the horizontal asymptote of each function.

    a. $f(x) = 5^{-x} + 1$ **b.** $g(x) = 6^{x+9} - 1$ **c.** $h(x) = \left(\frac{1}{7}\right)^x + \frac{1}{3}$

Summarize

42. WRITING MATHEMATICS Explain the relationships between the graphs of $y = 2^x$, $y = \left(\frac{1}{2}\right)^x$, and $y = 2^{-x}$.

43. WRITING MATHEMATICS Explain how the graph of $g(x) = b^{x-5} + 2$ compares to that of $f(x) = b^x$.

44. WRITING MATHEMATICS Explain how the graph of $h(x) = b^{-x}$ compares to that of $f(x) = b^x$.

45. WRITING MATHEMATICS Explain why the graph of $y = (-2)^x$ is undefined when $x = \frac{1}{2}$.

46. THINKING CRITICALLY Explain why the function $f(x) = b^x$ is not defined when $b < 0$.

47. THINKING CRITICALLY Compare the graphs of $f(x) = 9^x$ and $g(x) = 3^{2x}$. Provide an algebraic explanation for your result.

48. GOING FURTHER Describe some real world situations that can be modeled using exponential functions.

49. GOING FURTHER The natural exponential function is $f(x) = e^x$. Locate the e^x key on your graphing utility. Use $x = 1$ to find the approximate value of e. Name two exponential functions between whose graphs the graph of $f(x) = e^x$ will lie. Then graph $f(x) = e^x$ and the functions you named to see if you are correct.

9.2 Exponential Functions

SPOTLIGHT ON LEARNING

WHAT? In this lesson you will learn
- to evaluate exponential expressions.
- to solve exponential equations.

WHY? You can use exponential equations to model and solve problems that involve demographics, banking, marketing, depreciation, and radioactive decay.

Explore

Miko just began a job as a computer programmer at a salary of $25,000. She has a guaranteed salary increase of 5% each year for the next 8 years.

1. Write an equation in terms of her starting salary that will allow her to compute her salary after 1 year of employment.

2. Write an equation in terms of her starting salary to compute her salary after 3 years. After 5 years.

3. Find her salary after 5 years of employment, to the nearest cent.

4. If you know her salary after $n - 1$ years of employment where $1 < n < 8$, should you use addition, subtraction, multiplication, or division to determine her salary after n years of employment?

5. If her salary had been represented by a linear function, how would your answer to Question 4 change?

Build Understanding

As you learned in the previous lesson, an **exponential function** can be expressed in the form $f(x) = b^x$ where $b > 0$ and $b \neq 1$. Note that the base b is a constant and the exponent x is the variable.

In the previous lesson, you discovered the important properties of the exponential function $f(x) = b^x$.

PROPERTIES OF THE EXPONENTIAL FUNCTION $f(x) = b^x$

- **The domain is the set of all real numbers.**
- **The range is the set of positive real numbers.**
- **The y-intercept of the graph is 1.**
- **The function is one-to-one.**
- **The x-axis is an asymptote of the graph.**

The general appearance of the graphs of $g(x) = b^x$ where $b > 1$ and $h(x) = b^x$ where $0 < b < 1$ are shown below.

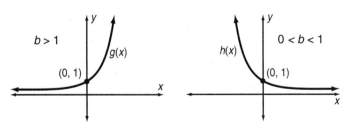

Since x can be any real number in an exponential function $f(x) = b^x$, x can be rational or irrational. Many of the points on the graph have irrational exponents. The properties of rational exponents in Chapter 8 also apply to irrational exponents. You can use a calculator or computer software to approximate the value of b^x when x is irrational.

EXAMPLE 1

Use a calculator to evaluate each expression to the nearest thousandth.

a. $5^{\sqrt{2}}$ **b.** $\sqrt{6}^{\sqrt{5}}$ **c.** 3^{π}

Solution

a. $5^{\sqrt{2}} = 9.739$ **b.** $\sqrt{6}^{\sqrt{5}} = 7.413$ **c.** $3^{\pi} = 31.544$ ◄

Since exponential functions are one-to-one, $b^m = b^n$ if and only if $m = n$. You can use this fact to determine an algebraic solution to an exponential equation. Begin by expressing both sides of the equation in terms of the same base.

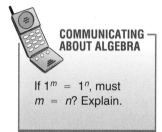
COMMUNICATING ABOUT ALGEBRA

If $1^m = 1^n$, must $m = n$? Explain.

EXAMPLE 2

Solve and check each equation.

a. $9 = 81^x$ **b.** $5^x = \dfrac{1}{125}$ **c.** $7^{6x-2} = 49^{2x-5}$

Solution

a.
$$9 = 81^x$$
$$3^2 = (3^4)^x$$
$$3^2 = 3^{4x}$$
$$2 = 4x$$
$$\frac{1}{2} = x$$

Check
$$9 = 81^x$$
$$9 \stackrel{?}{=} 81^{\frac{1}{2}}$$
$$9 = 9 \checkmark$$

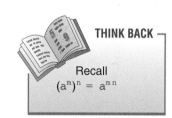
THINK BACK

Recall
$(a^m)^n = a^{mn}$

b.
$$5^x = \frac{1}{125}$$
$$5^x = \frac{1}{5^3}$$
$$5^x = 5^{-3}$$
$$x = -3$$

$$5^x = \frac{1}{125}$$
$$5^{-3} \stackrel{?}{=} \frac{1}{125}$$
$$\frac{1}{5^3} \stackrel{?}{=} \frac{1}{125}$$
$$\frac{1}{125} = \frac{1}{125} \checkmark$$

c.
$$7^{6x-2} = 49^{2x-5}$$
$$7^{6x-2} = (7^2)^{2x-5}$$
$$7^{6x-2} = 7^{4x-10}$$
$$6x - 2 = 4x - 10$$
$$2x = -8$$
$$x = -4$$

$$7^{6x-2} = 49^{2x-5}$$
$$7^{6(-4)-2} \stackrel{?}{=} 49^{2(-4)-5}$$
$$7^{-24-2} \stackrel{?}{=} 49^{-8-5}$$
$$7^{-26} \stackrel{?}{=} 49^{-13}$$
$$7^{-26} \stackrel{?}{=} (7^2)^{-13}$$
$$7^{-26} = 7^{-26} \checkmark$$
◄

You may not be able to express both sides of an exponential equation in terms of the same base. You can determine decimal approximations of the solutions to equations of this type using a graphing utility.

EXAMPLE 3

Use a graphing utility to solve $16 = 9^x$ to the nearest thousandth.

Solution
Graph $y_1 = 16$ and $y_2 = 9^x$ on the same axes. Set the range for the viewing window as follows: Xmin: -5, Xmax: 5, Xscl: 1, Ymin: -2, Ymax: 20, Yscl: 2. Use ZOOM and TRACE or the intersection feature to find the x-value of the point of intersection, which is 1.262. ◄

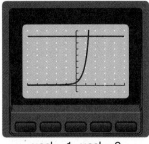

$x \text{scl} = 1 \quad y \text{scl} = 2$

The graph of an exponential equation of the form $y = ab^x$ where $b > 1$ and $a > 0$ is an **exponential growth curve**. It can be used to model situations in which a quantity is increasing at a constant rate, such as population growth, interest on a bank account, and economic growth. For example, consider a community with an initial population of P_i. If the growth rate of the community is represented by r where $0 < r < 1$, the population will grow to $P_i(1 + r)$ by the end of year 1, to $P_i(1 + r)^2$ by the end of year 2, and to $P_i(1 + r)^3$ by the end of year 3. At the end of t years, the population will be $P_i(1 + r)^t$.

EXAMPLE 4

DEMOGRAPHICS The 1990 Census gave the population of Dallas, Texas, as 1,006,831. It was determined that from 1980 to 1990, the population had increased at an average rate of 1.1% per year.

a. Construct a model to predict the population of Dallas n years after 1990 assuming that the population continues to increase at the same average rate. Use a graphing utility to graph the equation.

b. Use the model to predict the population of Dallas in the year 2000.

Solution
a. The population will have increased by 1.1% n times in n years.

$$P = 1,006,831(1 + 0.011)^n$$
$$P = 1,006,831(1.011)^n$$

b. The year 2000 is 10 years after 1990, so $n = 10$.

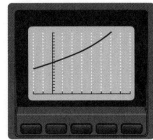

$x \text{scl} = 10 \quad y \text{scl} = 100,000$

$$P = 1,006,831(1.011)^{10} \approx 1,123,228.55$$

The predicted population is 1,123,229. ◄

The graph of an exponential equation of the form $y = ab^x$ where $0 < b < 1$ and $a > 0$ is an **exponential decay curve**. It can be used to model situations in which a quantity is decreasing at a constant rate, such as radioactive decay.

ALGEBRA: WHO, WHERE, WHEN

Professor Willard Libby of the University of California at Los Angeles received the Nobel Prize in Chemistry for devising the carbon-14 dating technique. This technique depends on the use of exponential functions.

EXAMPLE 5

NUCLEAR CHEMISTRY Radioactive elements "decay" and transform into other elements. The *half-life* of a radioactive element is the time that it takes for half of the initial quantity to decay. The half-life of francium-223 is 21 minutes.

a. If you have 80 mg of francium-223, write an equation that models the decay. Use a graphing utility to graph the equation.

b. How much will you have after 63 minutes?

Solution

a. Since 50% of the substance decays every 21 minutes, an equation for the decay is $y = 80(0.50)^t$ where t represents the number of half-lives that have elapsed.

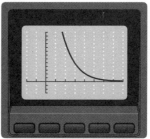

$x\,\text{scl} = 1 \quad y\,\text{scl} = 5$

b. Since 63 minutes represents $\frac{63}{21} = 3$ half-lives, $y = 80(0.50)^3$. You will have $80(0.50)^3$ mg = 10 mg of francium-223 after 63 minutes. ◄

TRY THESE

Use a calculator to evaluate each expression to the nearest thousandth.

1. 6^π **2.** $5^{\sqrt{5}}$ **3.** $13^{-\sqrt{3}}$ **4.** $7^{-\sqrt{2}}$

For the equation $y = a \cdot 3^x$, determine the value of a if the given point is on its graph.

5. $(1, 3)$ **6.** $(0, 6)$ **7.** $(2, 3)$ **8.** $(4, 9)$

9. **BANKING** If you invest an amount of money P in an account in which interest is compounded annually, the amount of money in the account after n years is $A = P(1 + r)^n$ where r is the yearly interest rate. Graph the equation and determine how much money you will have at the end of 5 years if you invest $1500 at 7% interest.

Solve each equation algebraically.

10. $7 = 49^x$ **11.** $2 = 16^x$

12. $4^x = \dfrac{1}{64}$ **13.** $3^x = \dfrac{1}{81}$

14. $5^{-x+2} = 625$ **15.** $2^{-2x} = 64$

16. $4^4 = 2^{x^2}$ **17.** $8^6 = 4^{x^2}$

18. MARKETING The demand curve for a new granola bar (in thousands of bars) can be represented by $D = 1300(0.94)^p$ where p represents the price per bar in cents. Graph the demand curve and find the demand for these granola bars if their price is 25 cents.

Use a graphing utility to solve each equation to the nearest thousandth. Check your answer with a calculator.

19. $10 = 3^x$ **20.** $15 = 8^x$ **21.** $5 = 8^x$ **22.** $6 = 11^x$

23. WRITING MATHEMATICS Discuss the drawbacks that may be involved with using a model such as the one described in Example 4 to predict population for more than a few years.

PRACTICE

Use a calculator to evaluate each expression to the nearest thousandth.

1. $8^{\sqrt{\pi}}$ **2.** $9^{\sqrt{8}}$ **3.** $100^{-\sqrt{2}}$ **4.** $25^{-\sqrt{3}}$

For the equation $y = a \cdot \left(\frac{1}{2}\right)^x$, find the value of a if the given point is on its graph.

5. $(2, 4)$ **6.** $(0, 9)$ **7.** $(-3, 4)$ **8.** $(-2, 32)$

9. WRITING MATHEMATICS Discuss how y changes each time x is increased by 1 in $y = ab^x$.

Solve each equation algebraically.

10. $8 = 4^x$ **11.** $128 = 4^x$ **12.** $\frac{1}{2} = 4^x$

13. $\frac{1}{4} = 16^x$ **14.** $6^{3x-5} = 36^{4x+10}$ **15.** $4^{3x-2} = 64^{3x+2}$

16. $9^8 = 3^{x^2}$ **17.** $\left(\frac{1}{2}\right)^6 = \left(\frac{1}{4}\right)^{x^2}$ **18.** $100^x = 0.01$

19. $1000^x = 0.1$ **20.** $25^{2x+1} = 125^{x+2}$ **21.** $9^x = 81^{2x+1}$

22. BIOLOGY The population of a particular type of bacteria doubles in number every 12 hours.

    a. If there are 8 bacteria initially, write an equation that models the number of bacteria present after t 12-hour periods.

    b. Use a graphing utility to graph the equation you found in part a.

    c. How many bacteria will be present after 5 days?

Use a graphing utility to solve each equation to the nearest thousandth. Check your answer with a calculator.

23. $8 = 5^x$ **24.** $18 = 3^x$

25. $2 = 7^x$ **26.** $4 = 13^x$

27. AUTOMOBILE DEPRECIATION A car that initially costs $18,000 depreciates at the rate of 15% each year.

    a. Write an equation that models this situation in the form $A = A_i(1 + r)^t$ where A_i is the initial cost of the car, r is the rate of increase (r is a negative number) in the value of the car, and t is the time in years.

    b. Use a graphing utility to graph the equation you found in part a.

    c. How much will the car be worth after 4 years?

28. WRITING MATHEMATICS Consider two functions $y = 2^x$ and $y = 2x$. Discuss the difference between exponential growth and linear growth.

EXTEND

The *exponential regression feature* of a graphing utility determines an exponential equation of the form $y = ab^x$ that models a given set of data. The graphing utility also provides the coefficient of correlation.

29. ECONOMICS The table below lists the gross domestic product of Switzerland from 1970 to 1991 in billions of dollars.

1970	1980	1985	1989	1990	1991
32	74	104	134	143	149

    a. Use exponential regression to find a curve that fits this data, letting 1970 be year 0, 1980 be year 10, and so on.

    b. How closely does this curve fit the data? Explain.

    c. Predict the gross domestic product of Switzerland in the year 2000.

30. MORTALITY RATES The table below lists the death rates from cancer per 100,000 females from 1970 to 1991.

1970	1980	1985	1990	1991
144.4	163.6	175.7	186.0	187.5

    a. Use exponential regression to find a curve that fits this data, letting 1970 be year 0, 1980 be year 10, and so on.

    b. How closely does this curve fit the data? Explain.

    c. Predict the death rate from cancer per 100,000 females in the year 2000.

    d. Will this model provide an accurate picture for the year 2025? Explain.

31. DEMOGRAPHICS The table below lists the population of Denver, Colorado, from 1950 to 1990.

1950	1960	1970	1980	1990
416,000	494,000	515,000	493,000	468,000

Do you think that it is appropriate to model the data with an exponential curve? Explain.

32. NUMERICAL METHODS

    a. Determine rational numbers h and k such that $h < \sqrt{3} < k$.

    b. Use the fact that if $h < x < k$, then $a^h < a^x < a^k$ to approximate $5^{\sqrt{3}}$.

    c. How can you improve your approximation of $5^{\sqrt{3}}$?

THINK CRITICALLY

33. Provide an example of an exponential equation in which both sides can be expressed in terms of the same base for which you would not obtain an exact answer if you used a graphing utility instead of algebraic techniques to solve it.

34. Describe the changes in the graph of $y = ab^x$ where $b > 1$ and $a > 1$ as a increases.

35. Describe the changes in the graph of $y = ab^x$ where $0 < b < 1$ and $a > 1$ as a increases.

36. What is the result if b equals 1 in an exponential equation of the form $y = ab^x$?

Solve each equation for x.

37. $(5c)^x = (25c^2)^{x+4}$

38. $(2c)^{x+3} = (8c^3)^{x-1}$

MIXED REVIEW

Determine the axis of symmetry of each graph.

39. $y = \frac{1}{2}x^2 - 2x + 6$

40. $y = 2x^2 + 32x + 125$

STANDARDIZED TESTS Choose the correct factorization for each polynomial.

41. $x^2 - 3x - 10$

    A. $(x + 5)(x - 2)$ **B.** $(x - 10)(x + 7)$

    C. $(x - 5)(x + 2)$ **D.** $(x - 10)(x + 1)$

42. $x^3 - 125$

    A. $(x - 5)(x^2 + 5x + 25)$ **B.** $(x^2 - 5)(x + 25)$

    C. $(x - 5)(x + 5)^2$ **D.** $(x + 5)(x^2 - 5x - 25)$

Write an equation for the inverse of each relation.

43. $y = 4x - 6$

44. $y = -x - 3$

Use a graphing utility to solve each equation to the nearest thousandth.

45. $11 = 6^x$

46. $17 = 7^x$

Perform each matrix operation.

47. $\begin{bmatrix} 5 & 7 & 1 & 2 \\ 3 & 6 & 4 & 8 \end{bmatrix} + \begin{bmatrix} 0 & 6 & -2 & 1 \\ 3 & 7 & 4 & 9 \end{bmatrix}$

48. $\begin{bmatrix} 0 & 1 & 3 & 2 \\ 8 & 6 & 2 & 7 \end{bmatrix} \begin{bmatrix} -1 & 6 \\ 2 & 1 \\ 5 & 8 \\ 3 & 7 \end{bmatrix}$

Oncologists are doctors who specialize in the study and treatment of tumors. They use blood work, physical examination, tissue examination, and radiographic imaging (x-rays, ultrasound scans, CT scans, and MRI scans) to identify the type, size, and extent of spread of a tumor.

Oncologists use the growth curve of the cancer cell population in a tumor to determine the timing and type of treatment for a malignant tumor. When a tumor is in the early stages of development, doctors use an exponential growth curve to model its growth. They refer to the "doubling time" that is characteristic of particular tumors.

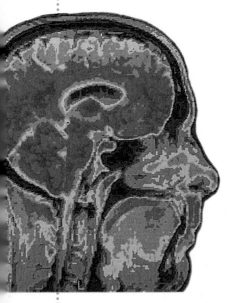

Decision Making

1. Write an exponential equation that models the cell population after x doubling times in a tumor that begins with one cell.

2. Use a graphing utility to graph this equation as x increases from 0 to 10 doubling times.

3. A tumor reaches the size of clinical detectability (approximately 1 cm^3) after it has undergone approximately 30 doublings. Use the model you developed in Question 1 to determine the cell population at this time.

In the later stages of tumor growth, an ever-increasing number of cells enter a nonproliferating (not multiplying) stage. For solid tumors, approximately 70% of the cells are nonproliferating at this time, and the population cannot be modeled with a standard exponential curve.

Write each answer in scientific notation.

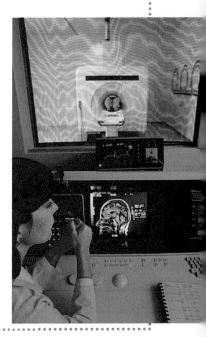

4. If the population of cancer cells in a solid tumor in the later stages of development is 1.1×10^{12}, how many cells are nonproliferating?

5. If the population of cancer cells in a solid tumor in the later stages of growth is 4.3×10^{13}, how many cells are proliferating?

9.3 Logarithmic Functions

Explore

1. Using a graphing utility, graph $y = 10^x$. Is this relation a function? Explain. Is $y = 10^x$ an exponential function?

2. Will the inverse of $y = 10^x$ be a function? Explain.

3. Determine an equation for the inverse of $y = 10^x$.

4. The inverse of $y = 10^x$ is a logarithmic function and is written $y = \log_{10} x$. Graph $\log_{10} x$ and $y = x$ on the same set of axes as $y = 10^x$. You may be able to use a draw inverse feature of your graphing utility to graph $y = \log_{10} x$.

Build Understanding

An exponential function of the form $y = b^x$ is a one-to-one function. Therefore its inverse, $x = b^y$, is also a function. The function $x = b^y$ is the **logarithmic function** written as $y = \log_b x$ and read as "y equals the logarithm to the base b of x." Note that a **logarithm** is defined to be an exponent. The general appearance of the graphs of $g(x) = \log_b x$ where $b > 0$ and $h(x) = \log_b x$ where $0 < b < 1$ are shown below with their inverse functions.

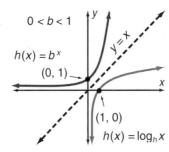 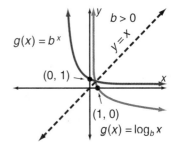

Logarithmic Function $f(x) = \log_b x$

- **The domain is the set of positive real numbers.**
- **The range is the set of all real numbers.**
- **The x-intercept of the graph is 1.**
- **The function is one-to-one.**
- **The y-axis is an asymptote of the graph.**

You will find it helpful to recognize the following special values:

$$\log_b 1 = 0 \quad \text{since} \quad b^0 = 1$$
$$\log_b b = 1 \quad \text{since} \quad b^1 = b$$
$$\log_b b^x = x \quad \text{since} \quad b^x = b^x$$

You can use the fact that $y = \log_b x$ is equivalent to $x = b^y$ to convert an equation expressed in exponential form to logarithmic form.

EXAMPLE 1

Express each equation in exponential form.

    a. $\log_5 125 = 3$ **b.** $\log_2 \dfrac{1}{2} = -1$ **c.** $\log_{10} 0.01 = -2$

Solution

    a. $5^3 = 125$ **b.** $2^{-1} = \dfrac{1}{2}$ **c.** $10^{-2} = 0.01$ ◄

You can also convert an equation expressed in logarithmic form to exponential form.

EXAMPLE 2

Express each equation in logarithmic form.

    a. $9^2 = 81$ **b.** $16^{\frac{1}{4}} = 2$ **c.** $196^{0.5} = 14$

Solution

    a. $\log_9 81 = 2$ **b.** $\log_{16} 2 = \dfrac{1}{4}$ **c.** $\log_{196} 14 = 0.5$ ◄

You can evaluate a logarithm by setting it equal to a variable and then rewriting the resulting equation in exponential form.

EXAMPLE 3

Evaluate each logarithm.

    a. $\log_4 64$ **b.** $\log_6 \dfrac{1}{36}$ **c.** $\log_2 \sqrt{2}$

Solution

    a. Let $x = \log_4 64$ **b.** Let $x = \log_6 \dfrac{1}{36}$ **c.** Let $x = \log_2 \sqrt{2}$

$$4^x = 64 \qquad\qquad 6^x = \dfrac{1}{36} \qquad\qquad 2^x = \sqrt{2}$$

$$4^x = 4^3 \qquad\qquad 6^x = 6^{-2} \qquad\qquad 2^x = 2^{\frac{1}{2}}$$

$$x = 3 \qquad\qquad\quad x = -2 \qquad\qquad\quad x = \dfrac{1}{2}$$

$$\log_4 64 = 3 \qquad \log_6 \dfrac{1}{36} = -2 \qquad \log_2 \sqrt{2} = \dfrac{1}{2} \;\;◄$$

CHECK UNDERSTANDING

You are given $\log x = 1.5441$. Use the 10^x key on your calculator to find the antilog of 1.5441.

Logarithms with base 10 are called **common logarithms**. They are often written without indicating their base. For example, an expression such as log 54 means $\log_{10} 54$. Most calculators have a log key that will allow you to evaluate an expression such as log 54. Use your calculator to find that $\log 54 = 1.7324$. You can also find x if you know $\log x$. The number x is called the **antilogarithm** of $\log x$.

ALGEBRA: WHO,
WHERE, WHEN

The unit for the
magnitude of sound
is the *bel*, named in
honor of Alexander
Graham Bell. The
small unit, the
decibel, is more
useful for most
problems.

Use the change-of-base formula to evaluate expressions such as $\log_3 8$, which cannot be expressed in terms of the same base.

Change-of-base formula $\quad \log_b x = \dfrac{\log_a x}{\log_a b}$

EXAMPLE 4

Evaluate $\log_3 8$ to the nearest ten-thousandth.

Solution

$$\log_3 8 = \frac{\log_{10} 8}{\log_{10} 3} = \frac{\log 8}{\log 3} = 1.8928 \qquad \blacktriangleleft$$

You can solve many logarithmic equations if you rewrite them in exponential form.

EXAMPLE 5

Solve each logarithmic equation.

    a. $\log_x 125 = 3$ **b.** $\log_2 x = 4$ **c.** $\log_4 64 = 3x + 6$

Solution

  a. $\log_x 125 = 3$ **b.** $\log_2 x = 4$ **c.** $\log_4 64 = 3x + 6$

$\qquad\qquad x^3 = 125 \qquad\qquad\quad 2^4 = x \qquad\qquad\qquad 4^{3x+6} = 64$

$\qquad\qquad\;\; x = 125^{\frac{1}{3}} \qquad\qquad 16 = x \qquad\qquad\qquad 4^{3x+6} = 4^3$

$\qquad\qquad\;\; x = 5 \qquad\qquad\qquad\qquad\qquad\qquad\qquad 3x + 6 = 3$

$\qquad\qquad\qquad\qquad\qquad\qquad\qquad\qquad\qquad\qquad\qquad\quad x = -1 \quad \blacktriangleleft$

Logarithms play a role in many real world situations.

EXAMPLE 6

ACOUSTICS The formula used to determine the magnitude of sound D in decibels (dB) is $D(I) = 10 \log \dfrac{I}{I_0}$, where I is the intensity of sound and I_0 is 10^{-16} W/cm^2, the intensity of the threshold of human hearing. Since a sound wave involves the transmission of energy or power, the intensity is measured in terms of the watt (W), a unit of power. Determine the magnitude of each sound to the nearest decibel.

    a. two-person conversation, $I = 10^{-10}$ W/cm^2

    b. jet at takeoff, $I = 10^{-4}$ W/cm^2

Solution

    a. $D(10^{-10}) = 10 \log \dfrac{10^{-10}}{10^{-16}} = 60$ dB

    b. $D(10^{-4}) = 10 \log \dfrac{10^{-4}}{10^{-16}} = 120$ dB $\qquad \blacktriangleleft$

Express in exponential form.

1. $\log_7 49 = 2$

2. $\log_{10} 0.1 = -1$

3. $\log_3 \frac{1}{27} = -3$

4. PHYSICS The electric current i in amperes in a particular circuit can be represented by $\log_2 i = -t$ where t is given in seconds. Express this equation in exponential form.

Express in logarithmic form.

5. $3^5 = 243$

6. $25^{\frac{1}{2}} = 5$

7. $10^{-3} = 0.001$

8. $8^{-\frac{1}{3}} = \frac{1}{2}$

9. WRITING MATHEMATICS Write your own definition of a logarithm.

Evaluate. Round answers to the nearest ten-thousandth.

10. $\log_6 216$

11. $\log_3 \frac{1}{81}$

12. $\log_{12} 12$

13. $\log_8 1$

14. $\log 62$

15. $\log 39$

16. $\log_5 10$

17. $\log_7 2$

Determine x to the nearest ten-thousandth.

18. $\log x = 3.2121$

19. $\log x = -1.4242$

20. $\log x = 15.5555$

21. ASTRONOMY The brightness of two stars can be compared using their magnitude difference d or brightness ratio r. The relationship between d and r can be approximated by $d = 2.5 \log r$. If the brightness ratio of two stars is 1000, find the magnitude difference.

Solve each logarithmic equation.

22. $\log_4 x = 4$

23. $\log_2 x = 8$

24. $\log_x 81 = 4$

25. $\log_x 512 = 9$

26. $\log_5 25 = 2x - 2$

27. $\log_6 \frac{1}{6} = \frac{x}{9}$

PRACTICE

Express in exponential form.

1. $\log_9 729 = 3$

2. $\log_{10} 0.0001 = -4$

3. $\log_4 \frac{1}{256} = -4$

4. $\log_8 1 = 0$

5. $\log_\pi \pi = 1$

6. $\log_5 5^x = x$

7. BIOLOGY An amoeba divides into two amoebas every hour. You can represent the number N of amoebas after t hours by the formula $t = \log_2 N$. How many hours will it take for a single amoeba to divide into 2048 amoebas?

Express in logarithmic form.

8. $7^3 = 343$

9. $625^{\frac{1}{4}} = 5$

10. $2^{-3} = \frac{1}{8}$

11. $32^{-\frac{1}{5}} = -\frac{1}{2}$

12. WRITING MATHEMATICS Compare and contrast the graphs of $y = 5^x$ and its inverse.

Evaluate. Round answers to the nearest ten-thousandth.

13. $\log_4 1024$

14. $\log_6 \dfrac{1}{216}$

15. $\log_9 1$

16. $\log_8 8$

17. $\log 119$

18. $\log 24$

19. $\log_6 6\sqrt{6}$

20. $\log_{\frac{1}{2}} 32$

21. $\log_8 10$

22. $\log_9 13$

23. $\log_7 5$

24. $\log_{12} 8$

Determine x to the nearest ten-thousandth.

25. $\log x = 2.3333$

26. $\log x = -2.7878$

27. $\log x = 11.8822$

28. ACOUSTICS Determine the magnitude of each sound to the nearest decibel.
  a. siren at 100 ft, $I = 10^{-6}$ W/cm²
  b. quiet radio, $I = 10^{-12}$ W/cm²
  c. busy street traffic, $I = 10^{-9}$ W/cm²
  d. leaves in a breeze, $I = 10^{-14}$ W/cm²

Solve each logarithmic equation.

29. $\log_5 x = 5$

30. $\log_9 x = 3$

31. $\log_x 256 = 4$

32. $\log_x 729 = 6$

33. $\log_x 6 = \dfrac{1}{2}$

34. $\log_7 49 = 3x - 3$

35. $\log_5 \dfrac{1}{5} = \dfrac{x}{8}$

36. $\log_8 x = \dfrac{4}{3}$

37. $\log_{16} x = \dfrac{3}{2}$

EXTEND

The graphs of logarithmic functions can be translated vertically and horizontally.

38. Graph $f(x) = \log_{10} x$, $g(x) = 2 + \log_{10} x$, and $h(x) = -1 + \log_{10} x$ on the same axes.
  a. Where is the graph of g in relation to the graph of f?
  b. Where is the graph of h in relation to the graph of f?

39. Graph $f(x) = \log_{10} x$, $j(x) = \log_{10} (x + 2)$, and $k(x) = \log_{10} (x - 3)$ on the same axes.
  a. Where is the graph of j in relation to the graph of f?
  b. Where is the graph of k in relation to the graph of f?

CHEMISTRY The pH of a solution indicates its acidity or alkalinity. It is calculated by the formula pH $= -\log$ [H+] where [H+] is the hydrogen ion concentration in moles per liter. A substance with a pH of less than 7 is acidic, a substance with a pH of 7 is neutral, and a substance with a pH of more than 7 is alkaline.

40. A soft drink has a hydrogen ion concentration of 10^{-3} moles per liter. Determine its pH to the nearest tenth.

41. An egg has a hydrogen ion concentration of 1.585×10^{-8} moles per liter. Determine its pH to the nearest tenth.

42. Pure water has a pH of 7.0. Find its hydrogen ion concentration.

THINK CRITICALLY

43. Use a calculator to evaluate log (log (log 20)).

44. Use your calculator to evaluate log (log (log 10)). Explain the result.

45. Why are negative numbers excluded from the domain of a logarithmic function?

46. If $x = \log_b a$ and $1 < a < b$, between what two integers does x lie? Explain.

47. Since $f(x) = \log_b x$ and $g(x) = b^x$ are inverse functions, their composites are the identity function. Show that $f(g(x)) = x$.

48. Most graphing calculators can graph $y = \log_{10} x$ using the log key. What function should you enter into your graphing utility to obtain the graph of $\log_4 x$?

MIXED REVIEW

Let $f(x) = 3x$ and $g(x) = x - 4$. Determine each function.

49. $(f + g)(x)$ **50.** $(f \cdot g)(x)$

Solve by completing the square.

51. $x^2 - 8x - 9 = 0$ **52.** $x^2 - 2x - 15 = 0$

STANDARDIZED TESTS Solve each logarithmic equation.

53. $\log_4 x = 3$ **54.** $\log_x 64 = 6$

  A. 81 **B.** 12 **C.** 64 **D.** 0.4771   **A.** 2 **B.** 384 **C.** 4 **D.** 10.67

PROJECT *Connection*   In this activity, you will construct a linear model of U.S. population growth.

1. Use the data you have collected to make a table showing population in millions for at least six years (starting from a base year such as 1990).

2. Let t represent the time elapsed in years since your base year. Make a scatterplot with t along the horizontal axis and the population along the vertical axis.

3. Use a graphing utility to determine the equation of the line of best fit for the data. Round values, including the coefficient of correlation, to three decimal places.

4. Compare the values predicted by your equation to actual values.

5. What does your linear model predict for the U.S. population in 2010? Compare this number to other estimates you may have found.

6. On what type of growth is your linear model based? Do you think this is a reasonable assumption for population growth? Explain.

Career
Seismologist

Seismologists are geophysicists who specialize in the study of the seismic waves (shock waves) produced by earthquakes or explosions. Earthquakes can cause buildings to fall, railroad tracks to buckle, water and gas pipelines to break, and bridges and dams to collapse. Natural disasters such as earthquakes can result in the loss of many lives and thereby influence the size of a population.

Seismologists study the magnitude of earthquakes using a logarithmic scale called the **Richter scale**, devised by Charles F. Richter, an American geologist. This scale is based on the equation $M(x) = \log \frac{x}{x_0}$, where x is the seismographic reading of the earthquake and x_0 is 1 micron or 0.001 mm (the seismographic reading of a zero-level earthquake).

The Richter Scale								
Magnitude	1	2	3	4	5	6	7	8
Intensity Level	10^1	10^2	10^3	10^4	10^5	10^6	10^7	10^8

Since the Richter scale is logarithmic, the energy released by an earthquake increases by powers of 10 in relation to the Richter numbers. For example, an earthquake of magnitude 8 is 10 times more powerful than an earthquake of magnitude 7 and 100 times more powerful than an earthquake of magnitude 6.

1. Determine the magnitude of an earthquake with a seismographic reading of 1 mm.

2. Is it true that an earthquake of magnitude 8 is twice as intense as an earthquake of magnitude 4? Explain.

3. Express the equation $M = \log \frac{x}{x_0}$ in exponential form.

4. An earthquake measuring 6.1 on the Richter scale occurred in Greece on June 15, 1995. Use your results from Question 3 to determine its seismographic reading.

Explore

1. Copy and complete the table below using a calculator. Round to the nearest hundred-thousandth. Then use the values in the table to answer Questions 2–7.

x	2	3	4	5	8	10	15
log x							

2. Determine each sum.

   a. $\log 2 + \log 5$ **b.** $\log 2 + \log 4$ **c.** $\log 3 + \log 5$

3. Compare the sums you found in Question 2 to the logarithms in the table. What pattern do you notice?

4. Determine each difference.

   a. $\log 8 - \log 4$ **b.** $\log 10 - \log 2$ **c.** $\log 15 - \log 5$

5. Compare the differences you found in Question 4 to the logarithms in the table. What pattern do you notice?

6. Determine each product.

   a. $2 \cdot \log 2$ **b.** $3 \cdot \log 2$

7. Compare the products you found in Question 6 to the logarithms in the table. What pattern do you notice?

Build Understanding

Since logarithms are exponents, the properties of logarithms follow directly from the properties of exponents.

> **PROPERTIES OF LOGARITHMS**
>
> If m, n, and b are positive real numbers, $b \neq 1$, and p is any real number, then
>
> $\log_b mn = \log_b m + \log_b n$ **Product Property**
>
> $\log_b \dfrac{m}{n} = \log_b m - \log_b n$ **Quotient Property**
>
> $\log_b m^p = p \log_b m$ **Power Property**

The expression on the right side of each of the properties of logarithms is the *expanded form* of the logarithm on the left side.

The proof of the product property is given below. You will be asked to prove the quotient and power properties in the exercises.

$$\text{Let } \log_b m = u \text{ and } \log_b n = v \qquad y = \log_b x \text{ if and only}$$
$$\text{Then } \quad b^u = m \text{ and } b^v = n \qquad \text{if } b^y = x$$
$$mn = b^u b^v \qquad\qquad\qquad \text{Multiply.}$$
$$mn = b^{u+v} \qquad\qquad\qquad \text{Product property of exponents}$$
$$\log_b mn = u + v \qquad\qquad\qquad y = \log_b x \text{ if and only if } b^y = x.$$
$$\log_b mn = \log_b m + \log_b n \qquad \text{Substitute.}$$

Note that $\log_b b^x = x$ and $b^{\log_b x} = x$ follow from the properties of logarithms.

EXAMPLE 1

Express in expanded form.

a. $\log_5 \dfrac{7}{x}$ **b.** $\log_3 x^5$ **c.** $\log_2 x^3 y^4$ **d.** $\log_6 \sqrt{\dfrac{x^2}{y^6}}$

Solution

a. $\log_5 \dfrac{7}{x} = \log_5 7 - \log_5 x$ \qquad Quotient property.

b. $\log_3 x^5 = 5 \log_3 x$ \qquad Power property.

c. $\log_2 x^3 y^4 = \log_2 x^3 + \log_2 y^4$ \qquad Product property.
$$= 3 \log_2 x + 4 \log_2 y \qquad \text{Power property.}$$

d. $\log_6 \sqrt{\dfrac{x^2}{y^6}} = \log_6 \left(\dfrac{x^2}{y^6}\right)^{\frac{1}{2}} = \log_6 \left(\dfrac{x}{y^3}\right)$
$$= \log_6 x - \log_6 y^3 \qquad \text{Quotient property.}$$
$$= \log_6 x - 3 \log_6 y \qquad \text{Power property.} \qquad \blacktriangleleft$$

You can also use the properties of logarithms to simplify or *condense* a logarithmic expression given in expanded form if the terms of the expression have the same base.

EXAMPLE 2

Express as a single logarithm.

a. $5 \log_4 x + 2 \log_4 y$ \qquad\qquad **b.** $\log_8 x - 3 \log_8 y + \log_8 6$

Solution

a. $5 \log_4 x + 2 \log_4 y = \log_4 x^5 + \log_4 y^2$ \qquad Power property
$$= \log_4 x^5 y^2 \qquad \text{Product property}$$

b. $\log_8 x - 3 \log_8 y + \log_8 6$
$$= \log_8 x - \log_8 y^3 + \log_8 6 \qquad \text{Power property}$$
$$= \log_8 6x - \log_8 y^3 \qquad \text{Product property}$$
$$= \log_8 \dfrac{6x}{y^3} \qquad \text{Quotient property} \qquad \blacktriangleleft$$

Recall that the logarithm function is a one-to-one function. Therefore, $\log_b m = \log_b n$ if and only if $m = n$. You can use this fact to solve logarithmic equations. Check to see whether your solution makes sense in the original equation.

PROBLEM SOLVING TIP

In Example 3, it is easier to solve a logarithmic equation if it is condensed first.

EXAMPLE 3

Solve each equation. Check your solution(s).

a. $\log_5 y = 6 \log_5 2 - \log_5 4$ **b.** $\log_2 x + \log_2 (x + 4) = 5$

Solution

a.
$\log_5 y = 6 \log_5 2 - \log_5 4$
$\log_5 y = \log_5 2^6 - \log_5 4$
$\log_5 y = \log_5 64 - \log_5 4$
$\log_5 y = \log_5 \dfrac{64}{4}$
$\log_5 y = \log_5 16$
$y = 16$

The solution is 16.

Check

$\log_5 y = 6 \log_5 2 - \log_5 4$
$\log_5 16 \overset{?}{=} \log_5 2^6 - \log_5 4$
$\log_5 16 \overset{?}{=} \log_5 64 - \log_5 4$
$\log_5 16 \overset{?}{=} \log_5 \dfrac{64}{4}$
$\log_5 16 = \log_5 16 \checkmark$

CHECK UNDERSTANDING

In Example 3b, why is $\log_2(-8)$ undefined?

b.
$\log_2 x + \log_2 (x + 4) = 5$
$\log_2 (x(x + 4)) = 5$
$\log_2 (x^2 + 4x) = 5$
$x^2 + 4x = 2^5$
$x^2 + 4x - 32 = 0$
$(x + 8)(x - 4) = 0$
$x = -8 \text{ or } x = 4$

The solution is 4.

When $x = -8$, $\log_2 x$ is undefined. Reject $x = -8$.

$\log_2 x + \log_2 (x + 4) = 5$
$\log_2 4 + \log_2 (4 + 4) \overset{?}{=} 5$
$\log_2 4 + \log_2 8 \overset{?}{=} 5$
$\log_2 (4 \cdot 8) \overset{?}{=} 5$
$\log_2 32 \overset{?}{=} 5$
$5 = 5 \checkmark$

COMMUNICATING ABOUT ALGEBRA

Discuss how you could use logarithms to evaluate $\dfrac{(278)(45)}{25}$. Carry out your plan then perform the computation as shown and compare the two results. If the results do not agree, give reasons why.

Recall from the AlgebraWorks in Lesson 9.3 the magnitude of an earthquake is determined using the Richter scale.

EXAMPLE 4

SEISMOLOGY Recall the formula for earthquake magnitude, $M = \log \dfrac{x}{x_0}$ where x is the seismographic reading of the earthquake and x_0 is 0.001 mm.

a. Express $M = \log \dfrac{x}{x_0}$ in expanded form.

b. Use your results from part a to determine the magnitude of an earthquake with a seismographic reading of 10 mm.

Solution

a. $M = \log x - \log x_0$

b. $M = \log 10 - \log 0.001 = 1 - (-3) = 4$

TRY THESE

Express in expanded form.

1. $\log_7 3x$

2. $\log_2 \dfrac{y}{5}$

3. $\log_8 x^7 y^3$

4. $\log_9 (3x)^3$

5. WRITING MATHEMATICS Explain how the properties of logarithms are related to the properties of exponents. Use examples.

Express as a single logarithm.

6. $3\log_3 y - 6\log_3 z$

7. $4\log_8 3 + 5\log_8 x$

8. $5\log_4 p + 2\log_4 q - 3\log_4 r$

9. $2\log_b 7 + 6\log_b d - 4\log_b e$

10. PHYSICS Isaac Newton found that any two particles exert a gravitational force on each other. The magnitude of the gravitational force of attraction between a particle of mass m_1 and one of mass m_2 is given by $F = k\dfrac{m_1 m_2}{d^2}$ where d is the distance between the particles and k is a constant. Express $\log F$ in expanded form.

Solve each equation.

11. $\log_6 t = 5\log_6 2 - \log_6 8$

12. $\log_7 v = 4\log_7 3 - \log_7 9$

13. $\log_3 (w^2 + 11) = 3$

14. $\log_2 (z^2 - 8) = 3$

15. $\log_4 (x - 12) = 3 - \log_4 x$

16. $\log_5 x = 1 - \log_5 (6x - 7)$

17. Prove the quotient property of logarithms, $\log_b \dfrac{m}{n} = \log_b m - \log_b n$.

ASTRONOMY In Lesson 9.3, you learned the formula for the magnitude difference d between two stars is $d = 2.5\log r$ where r represents the brightness ratio of the stars.

18. Express this formula in condensed form.

19. Use your results from part a to determine the magnitude difference between two stars if the brightness ratio of the stars is 10,000.

PRACTICE

Express in expanded form.

1. $\log_4 5y$

2. $\log_9 7z$

3. $\log_8 \dfrac{x}{6}$

4. $\log_6 \dfrac{t}{5}$

5. $\log_2 x^6 y^7$

6. $\log_5 p^2 q^3$

7. $\log_7 \sqrt{x^4 y^2}$

8. $\log_3 (x^6 y^9)^{\frac{1}{3}}$

9. ACOUSTICS Recall that the magnitude of a sound in decibels is $D(I) = 10\log \dfrac{I}{I_0}$ where I is the intensity of the sound and I_0 is 10^{-16} W/cm^2.

  a. Express this formula in expanded form.

  b. Determine the magnitude of busy street traffic, which has an intensity of 10^{-9} W/cm^2.

Express as a single logarithm.

10. $2 \log_5 y - 7 \log_5 z$

11. $3 \log_3 u + 8 \log_3 v$

12. $6 \log_7 e + 3 \log_7 8 - 4 \log_7 f$

13. $2 \log_d q + 6 \log_d r - 4 \log_d t$

14. $\frac{4}{3} \log_8 y$

15. $\frac{7}{2} \log_5 w$

16. CHEMISTRY Recall that the pH of a solution can be determined using the formula $pH = -\log[H+]$ where $[H+]$ is the hydrogen ion concentration in moles per liter.
   a. Express this formula in condensed form.
   b. Use the formula you found in part a to determine the pH of grapes, which have a hydrogen ion concentration of 10^{-4} moles per liter.

Solve each equation.

17. $\log_9 t = 4 \log_9 5 - \log_9 5$

18. $\log_8 v = 3 \log_8 4 - \log_8 32$

19. $\log_4 (w^2 + 15) = 3$

20. $\log_5 (z^2 + 225) = 4$

21. $\log_7 2x + \log_7 3x = \log_7 384$

22. $\log_3 7x + \log_3 8x = \log_3 224$

23. WRITING MATHEMATICS Explain why $\log \frac{x}{5}$ and $\frac{\log x}{\log 5}$ are not equivalent expressions. Use examples to justify your answer.

24. Prove the power property of logarithms, $\log_b m^p = p \log_b m$.

EXTEND

In Example 3b, you learned to solve equations of the form $\log_b u + \log_b v = k$ by writing them in exponential form. You can also solve equations of the form $\log_b u - \log_b v = k$ in the same manner. Solve each of the following equations.

25. $\log_4 (x + 8) - \log_4 (x - 1) = 2$

26. $\log_3 (2x + 5) - \log_3 (x - 3) = 1$

27. $\log_5 (x + 5) - \log_5 (x - 5) = 1$

28. $\log_2 (x + 4) - \log_2 (x - 3) = 3$

29. INVENTORY CONTROL The total cost of maintaining inventory on a specific food item is minimized when the size x of each order is $x = \sqrt{\frac{2cd}{e}}$ where c is the cost of placing an order, d is the monthly demand, and e is the monthly carrying cost. Express $\log x$ in expanded form.

30. DIFFUSION The average velocity v of a gas particle can be calculated using $v = k \sqrt{PV}$ where P is the pressure of the gas, V is its volume, and k is a constant. Express $\log v$ in expanded form.

31. PHYSICS The period of simple harmonic motion p can be calculated using $p = 2\pi \sqrt{\frac{m}{k}}$ where m is the mass and k is the proportionality constant between stress and strain. Express $\log p$ in expanded form.

THINK CRITICALLY

32. Prove that $\log_b b^x = x$.

33. Why is the property $b^{\log_b x} = x$ true?

34. Prove that $-\log_b x = \log_b \dfrac{1}{x}$.

35. Since $\log_b m^p = p \log_b m$, will the graphs of $y_1 = \log(x - 3)^2$ and $y_2 = 2 \log (x - 3)$ be equivalent? Use a graphing utility to verify your prediction.

36. If x is a positive real number, find $\log x \div \log \dfrac{1}{x}$.

MIXED REVIEW

37. STANDARDIZED TESTS Choose the letter of the relation that is a function.

    A. $\{(2, 1), (3, 1), (2, 5), (4, 0)\}$ **B.** $\{(6, 0), (9, 4), (8, 6), (6, 8)\}$

    C. $\{(3, 6), (4, 8), (5, 9), (6, 8)\}$ **D.** $\{(1, 1), (1, 2), (3, 1), (3, 2)\}$

Multiply.

38. $\begin{bmatrix} 3 & 0 & -2 \end{bmatrix} \begin{bmatrix} 2 & 9 \\ 3 & -6 \\ 1 & 0 \end{bmatrix}$

39. $\begin{bmatrix} 3 & 4 & 5 \\ 1 & -1 & -6 \\ 0 & 4 & 2 \end{bmatrix} \begin{bmatrix} 1 & 1 & -2 \\ 4 & -2 & 0 \\ 5 & -1 & 6 \end{bmatrix}$

Multiply.

40. $(5 + 3i)(4 - 2i)$ **41.** $(7 - 5i)(3 - 8i)$

Express in expanded form.

42. $\log_8 4t$ **43.** $\log_9 \dfrac{q}{2}$ **44.** $\log_3 4x^2 y^3$

PROJECT *Connection*   In this activity, you will construct an exponential model of U.S. population growth.

1. Use the same data as for the linear model. Determine the percent increase in the population from one year to the next and then calculate an average value for the percent increase. (Round to three decimal places.)

2. Use the average percent increase to express the population for any year as a function of t (years elapsed from base year) and the initial (base-year) population. Compare the values predicted by this model to the actual values. Also, make a prediction for 2010.

3. Use a graphing utility to fit an exponential model to your data. Compare this model to the one you constructed above. How well do predicted values match actual values? How well does the prediction for 2010 match other estimates you have found?

4. Compare the graphs of the linear and exponential models. Explain how the assumption about population growth in the exponential model is different than the linear model. Is this assumption more reasonable?

9.5 Natural Logarithms

Explore

1. Use a calculator to complete the table. Round answers to the nearest hundred-thousandth.

2. As n gets larger and larger, what number does the quantity $\left(1 + \frac{1}{n}\right)^n$ seem to approach?

3. Recall that when n is a whole number, $n! = n(n - 1)(n - 2)\ldots(2)(1)$. Evaluate
$$\frac{1}{0!} + \frac{1}{1!} + \frac{1}{2!} + \frac{1}{3!} + \frac{1}{4!} + \frac{1}{5!} + \frac{1}{6!} + \frac{1}{7!} + \frac{1}{8!}$$
to five decimal places.

4. Compare your results to Questions 2 and 3.

n	$\left(1 + \frac{1}{n}\right)^n$
1	2
10	
10^2	
10^3	
10^4	
10^6	
10^8	
10^{10}	

Build Understanding

In Explore, you discovered one of the most important numbers in the physical and mathematical world. The number e is defined to be the number that $\left(1 + \frac{1}{n}\right)^n$ approaches as n increases without bound. To ten decimal places, $e = 2.7182818285$. The number e is irrational because its decimal approximation neither repeats nor terminates.

The function $f(x) = e^x$ is the **natural exponential function** and is used extensively in science. The natural exponential function has an inverse, $f^{-1}(x) = \log_e x$ called the **natural logarithm** of x and written **ln x**.

Like all inverse functions, the graphs of $f(x) = e^x$ and $f^{-1}(x) = \ln x$ are reflections of each other over the line $y = x$.

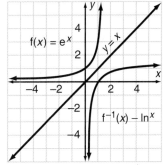

The properties of exponents apply to expressions with e.

EXAMPLE 1

Simplify each expression.

  a. $e^8 \cdot e^3$ **b.** $16e^6 \div 8e^2$ **c.** $(5e^4)^2$

Solution

  a. $e^8 \cdot e^3 = e^{11}$ **b.** $16e^6 \div 8e^2 = 2e^4$ **c.** $(5e^4)^2 = 25e^8$ ◄

PROBLEM SOLVING TIP

Use the factorial key (!) on your calculator. Remember that $0! = 1$.

PROBLEM SOLVING TIP

You may have to use an inverse or second function key to access the e^x key on your calculator.

THINK BACK

Recall that $\log_b b = 1$ and $\log_b 1 = 0$.

COMMUNICATING ABOUT ALGEBRA

Compare and contrast the numbers e and π.

The properties of logarithms also apply to natural logarithms.

EXAMPLE 2

Express in expanded form.

   a. $\ln 6x$ **b.** $\ln x^7$ **c.** $\ln x^5 y^3$ **d.** $\ln \sqrt{\dfrac{x^4}{y^8}}$

Solution

  a. $\ln 6x = \ln 6 + \ln x$ **b.** $\ln x^7 = 7 \ln x$

  c. $\ln x^5 y^3 = \ln x^5 + \ln y^3 = 5 \ln x + 3 \ln y$

  d. $\ln \sqrt{\dfrac{x^4}{y^8}} = \ln \left(\dfrac{x^4}{y^8}\right)^{\frac{1}{2}} = \ln \left(\dfrac{x^2}{y^4}\right) = \ln x^2 - \ln y^4 = 2 \ln x - 4 \ln y$ ◀

You can also use the properties of logarithms to simplify or condense natural logarithmic expressions.

EXAMPLE 3

Express as a single logarithm.

   a. $2 \ln r + 9 \ln q$ **b.** $\ln h - 4 \ln j + 3 \ln 2$

Solution

  a. $2 \ln r + 9 \ln q = \ln r^2 + \ln q^9 = \ln r^2 q^9$

  b. $\ln h - 4 \ln j + 3 \ln 2 = \ln h - \ln j^4 + \ln 2^3 = \ln \dfrac{8h}{j^4}$ ◀

There are many real world applications of natural logarithms.

EXAMPLE 4

SPACE SCIENCE The formula $\ln P = \ln 50 - \dfrac{t}{250}$ can be used to model the number of days left in the power output of a satellite where P is the power output in watts and t is the number of days left. Determine the number of full days left if the power output is 21.4 W.

  Solution

$$\ln 21.4 = \ln 50 - \frac{t}{250}$$

$$\ln 21.4 - \ln 50 = -\frac{t}{250}$$

$$\ln \frac{2.14}{50} = -\frac{t}{250} \qquad \text{Quotient property}$$

$$250\left(\ln \frac{2.14}{50}\right) = -t$$

$$212 = t$$

There are 212 full days of power left. ◀

Using a calculator, evaluate each expression to the nearest ten-thousandth.

1. e^5 **2.** $e^{-2.5}$ **3.** $\ln 12.8$ **4.** $4.2 \ln 8 + 6$

Simplify each expression.

5. $e^6 \cdot e^{-3}$ **6.** $e^8 \div e^5$ **7.** $(125e^{12})^{-\frac{1}{3}}$ **8.** $(3e^4)^3$

9. WRITING MATHEMATICS You use a calculator and find $\ln 650$ is approximately 3.4770. Could this answer be correct? Explain why or why not.

Express in expanded form.

10. $\ln \dfrac{a^9}{b^3}$ **11.** $\ln \sqrt{9u^8 v^{14}}$

12. PHARMACOLOGY The formula $C = 5e^{-0.4t}$ can be used to model the concentration C in milligrams of a drug in the bloodstream of a patient t hours after the drug has been administered. How much of the drug will be present in the patient's bloodstream after 2.5 h?

Express as a single logarithm.

13. $\ln 64 + 4 \ln \dfrac{1}{2} + \ln x$ **14.** $\ln 80 + 2 \ln \dfrac{1}{4} - \ln y$

15. MANUFACTURING The number of clocks N that a worker can assemble after t days of training can be modeled using the equation $N = 120 - 120e^{-0.09t}$. How many whole clocks can a worker assemble after 3 days of training?

Evaluate each expression to the nearest ten-thousandth.

16. $\log_9 14$ **17.** $\log_4 8$ **18.** $\log_7 4$ **19.** $\log_{10} 6$

20. WRITING MATHEMATICS Suppose you had a basic arithmetic calculator without an e^x key. What simpler function might you graph to approximate the graph of $y = e^x$?

Use a calculator to evaluate each expression to the nearest ten-thousandth.

1. $e^{\frac{1}{4}}$ **2.** $e^{4.1}$ **3.** $e^{-\frac{1}{2}}$ **4.** $\ln 5.4$ **5.** $3.2 \ln 7 + 2$ **6.** $\ln 6 + \ln 4.5$

7. WRITING MATHEMATICS Use a graphing utility to graph $y = 2^x$, $y = e^x$, and $y = 3^x$ on the same axes. Describe the relative position of each function.

8. FORENSICS The formula $y = (T_1 - T_0)e^{-0.031t} + T_0$ can be used to approximate a person's time of death where y is the body temperature in degrees F after t minutes, T_0 is the surrounding air temperature, and T_1 is the initial temperature of the body in degrees F. If a person has been dead for 90 min, what is the body temperature if the temperature of the surrounding air is 39°F and the initial body temperature was 98.6°F?

Simplify each expression.

9. $e^7 \cdot e^2$ **10.** $e^{12} \div e^8$ **11.** $(8e^{15})^{\frac{1}{3}}$ **12.** $(4e^5)^4$

Express in expanded form.

13. $\ln 8t$ **14.** $\ln u^4$ **15.** $\ln \dfrac{m^6}{n^5}$ **16.** $\ln \sqrt{4x^4 y^{10}}$

Express as a single natural logarithm.

17. $\ln 10 + \ln j + \ln k$ **18.** $\ln 40 + 2 \ln \dfrac{1}{2} + \ln w$ **19.** $\ln 56 + 3 \ln \dfrac{1}{2} - \ln z$

20. PHYSICS At a constant temperature, the atmospheric pressure can be modeled by $p = 101352e^{-0.000122h}$ where p is the pressure in Pascals (Pa) and h is the altitude in meters. Determine the pressure to the nearest hundred Pascals when the height is 2500 m.

Use natural logarithms. Evaluate each expression to the nearest ten-thousandth.

21. $\log_8 10$ **22.** $\log_3 11$ **23.** $\log_9 6$ **24.** $\log_{12} 2$

EXTEND

The logarithmic regression feature of a graphing calculator or computer software provides values for an equation of the form $y = a + b \ln x$ and the coefficient of correlation.

25. MUSIC A piano teacher found that her students' recall of a particular passage of music was related to the amount of time they spent practicing it. For example, a student who practiced for 40 minutes recalled 65% of the passage. She collected the following data.

Practice Time, minutes	10	20	30	40	50	60	70	80
Percent of Music Recalled	39	52	59	65	70	72	75	80

a. Use logarithmic regression to find a curve that fits the data.
b. How closely does the curve fit the data? Explain.
c. Predict the percent of the passage that the student knows after 120 min of practice.

26. HOUSING Jo has collected this information about rental prices of apartments in her town.

Monthly Rent, dollars	1000	760	450	1500	800	500	980
Number of Bedrooms	3	2	1	4	2	1	3

a. What type of curve or line best describes the data?
b. Write the equation for the curve or line.
c. How closely does the curve or line fit the data?

27. DEMOGRAPHICS The table shows the population of Chicago, IL from 1950 to 1990.

1950	1960	1970	1980	1990
3,621,000	3,550,000	3,369,000	3,005,000	2,784,000

a. What type of curve or line best describes the data?
b. Write the equation for the curve or line.
c. How closely does the curve or line fit the data?

28. **SPACE SCIENCE** A rocket is shot vertically into the air. The height that the rocket has attained after a given number of seconds is shown in the table.

Time, seconds	2	4	6	8	10
Height, feet	304	480	528	448	240

    a. What type of curve or line best describes the data?
    b. Write the equation for the curve or line.

THINK CRITICALLY

29. Explain why $\ln e = 1$.
30. Explain why $e^{\ln x} = x$.
31. Prove that $\ln e^x = x$.

32. It can be shown that $e^x = 1 + \dfrac{x}{1!} + \dfrac{x^2}{2!} + \dfrac{x^3}{3!} + \dfrac{x^4}{4!} + \ldots$. Use the first five terms of this sum to evaluate $e^{0.25}$. Compare with the value you obtain from a calculator.

33. Determine whether $\log 25$ or $\ln 25$ is larger without using a calculator. Explain.

MIXED REVIEW

34. **STANDARDIZED TESTS** Choose the equation of the line in slope-intercept form that has a slope of 5 and passes through the point $(-5, 10)$.

    A. $y = -5x + 10$ **B.** $y = 5x + 35$ **C.** $y = 10x - 35$ **D.** $y = \frac{1}{2}x + 35$

Use the quadratic formula to solve each equation.

35. $2x^2 + 6x - 5 = 0$
36. $3x^2 - 8x + 2 = 0$

Evaluate each expression to the nearest ten-thousandth.

37. $\log_5 26$
38. $\log_{12} 3$

PROJECT *Connection*   Exponential population growth does not continue indefinitely. Increased competition for available resources causes the growth rate to slow down. The population continues to grow, but at a smaller rate each year. When the population approaches a constant value, called the **carrying capacity** of the environment, there is almost no growth. A **logistic growth model** represents this situation.

$$l(t) = \frac{A}{1 + Be^{-kt}}$$

where $A, B, k > 0$ are constants determined by the initial population, the carrying capacity, and the rate at which the population approaches the carrying capacity.

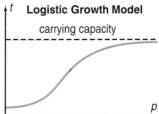

1. Determine the initial population by finding $l(0)$.

2. What happens to the term Be^{-kt} when t is very large?

3. When t is very large, $l(t)$ approaches the carrying capacity. Determine this value.

4. Discuss whether a logistics curve would better model the U.S. population. What are some advantages and some limitations of this model?

Explore

- In Lesson 9.2, you learned to solve an equation such as $3^x = 7$ by graphing $y_1 = 3^x$ and $y_2 = 7$ on the same set of axes and then finding the x-value of the point at which they intersect.

1. Use a graphing utility and solve $3^x = 7$ by graphing $y_1 = 3^x$ and $y_2 = 7$ and finding their intersection.

2. You will now explore another method of solving exponential and logarithmic functions. In this method, the logarithm of both sides of the equation is taken. Then the property that $\log_b b^x = x$ is applied. To apply this method to $3^x = 7$, what base logarithm would you use?

3. To maintain equality, what must you do to the right side of the equation if you take the logarithm of the left side of the equation?

4. If you take the base 3 logarithm of both sides of the equation $3^x = 7$, what is the resulting equation?

5. Evaluate the right side of the equation you wrote in Question 4. What formula will you apply?

6. Compare your results from Questions 1 and 5.

7. Which method of solving do you prefer? Why?

Build Understanding

- As you saw in Explore, you can solve an exponential equation of the form $b^x = y$ for x by taking the base b logarithm of each side of the equation and then evaluating $\log_b y$ using the change of base formula. You can also solve an equation of this form by taking the common logarithm or natural logarithm of both sides. Note how the method used in Example 1 differs from the second method you used in Explore.

EXAMPLE 1

Use common logarithms to solve $3^x = 7$ to the nearest ten-thousandth.

Solution

$$3^x = 7$$
$$\log 3^x = \log 7 \qquad \text{Take log of both sides.}$$
$$x \log 3 = \log 7 \qquad \text{Power property}$$
$$x = \frac{\log 7}{\log 3} \qquad \text{Divide both sides by log 3.}$$
$$x = 1.7712$$

◄

To solve an exponential equation of the form $e^x = y$, take the natural logarithm of both sides.

EXAMPLE 2

Solve $e^x = 50$ to the nearest ten-thousandth.

Solution

$$e^x = 50$$
$$\ln e^x = \ln 50 \qquad \text{Take ln of both sides.}$$
$$x = \ln 50 \qquad \ln e^x = x$$
$$x = 3.9120$$

To solve an equation of the form $ab^x + c = y$, isolate b^x on one side of the equation before taking the logarithm of both sides.

EXAMPLE 3

Solve each equation to the nearest ten-thousandth.

a. $3e^x + 5 = 26$ **b.** $10^{2x} + 30 = 1250$

Solution

a.
$$3e^x + 5 = 26$$
$$3e^x = 21$$
$$e^x = 7$$
$$\ln e^x = \ln 7$$
$$x = \ln 7$$
$$x = 1.9459$$

b.
$$10^{2x} + 30 = 1250$$
$$10^{2x} = 1220$$
$$\log 10^{2x} = \log 1220$$
$$2x = \log 1220$$
$$x = \frac{1}{2} \log 1220$$
$$x = 1.5432$$

To solve a logarithmic equation of the form $\log x = y$ or $\ln x = y$, write the equation in exponential form. For example,

$$\log x = 12 \quad \rightarrow \quad 10^{\log x} = 10^{12} \quad \rightarrow \quad x = 10^{12}$$
$$\ln x = 4 \quad \rightarrow \quad e^{\ln x} = e^{4} \quad \rightarrow \quad x = e^{4}$$

EXAMPLE 4

Solve each equation to the nearest ten-thousandth.

a. $6 + 4 \log x = 5$ **b.** $5 \ln 2x = 15$

Solution

a.
$$6 + 4 \log x = 5$$
$$4 \log x = -1$$
$$\log x = -\frac{1}{4}$$
$$10^{\log x} = 10^{-\frac{1}{4}}$$
$$x = 0.5623$$

b.
$$5 \ln 2x = 15$$
$$\ln 2x = 3$$
$$e^{\ln 2x} = e^{3}$$
$$2x = e^{3}$$
$$x = \frac{1}{2} e^{3}$$
$$x = 10.0428$$

There are many real world applications of logarithmic and exponential equations.

EXAMPLE 5

CONSUMERISM The value v of a brand new $28,000 car that is depreciating at a rate of 16% per year can be modeled by the equation $v = 28,000(1 - 0.16)^t$ where t is the time in years. After how many years will the car be worth $14,000?

Solution

$$v = 28,000(1 - 0.16)^t$$
$$14,000 = 28,000(0.84)^t$$
$$0.5 = (0.84)^t \qquad \text{Divide both sides by 28,000.}$$
$$\log 0.5 = \log (0.84)^t \qquad \text{Take the log of both sides.}$$
$$\log 0.5 = t \log 0.84 \qquad \text{Power property}$$
$$\frac{\log 0.5}{\log 0.84} = t \qquad \text{Divide both sides by } \log 0.84.$$
$$3.9755 = t$$

The value of the car will depreciate to $14,000 in about 4 years. ◄

TRY THESE

Solve and check each equation to the nearest ten-thousandth.

1. $5^x = 28$ 2. $6^x = 3$ 3. $e^x = 12$ 4. $e^x = 2$

5. WRITING MATHEMATICS Describe three methods that you can use to solve an equation of the form $b^x = y$. Provide an example and solve it using each of the three methods.

Solve and check each equation to the nearest ten-thousandth.

6. $2e^x + 8 = 40$ 7. $5e^x - 10 = 45$ 8. $7e^{2x} - 5 = 42$

9. $10^x - 17 = 50$ 10. $10^{5x} + 12 = 60$ 11. $10^{-3x} - 120 = 18$

12. FOOD SERVICE The number of appetizers a that an apprentice appetizer chef can prepare per day after t days of training can be modeled by the equation $a = 300 - 300e^{-0.3t}$. How many days of training will it take the chef to be able to prepare 175 appetizers?

Solve and check each equation to the nearest ten-thousandth.

13. $2 + 6 \log x = 10$ 14. $5 \log x - 8 = 4$

15. $3 \log 3x = 21$ 16. $7 + 8 \ln x = 2$

17. $9 - 3 \ln x = 6$ 18. $4 \ln 4x = 14$

Solve and check each equation to the nearest ten-thousandth.

1. $7^x = 35$
2. $8^{-x} = 3$
3. $4^{-3x} = 13$
4. $5^{\frac{1}{2}x} 12$

5. $e^{-x} = 31$
6. $e^x = 18$
7. $e^{\frac{1}{4}x} = 55$
8. $e^{-4x} = 5$

9. Which is the correct set of steps in solving $3 \log x = 7$? Explain your choice.

    a. $3 \log x = 7$

        $\log x = \dfrac{7}{3}$

            $x = \log \dfrac{7}{3}$

            $x = 0.3680$

    b. $3 \log x = 7$

        $\log x = \dfrac{7}{3}$

        $10^{\log x} = 10^{\frac{7}{3}}$

            $x = 215.4435$

Solve and check each equation to the nearest ten-thousandth.

10. $4e^x + 10 = 50$
11. $7e^x - 32 = 45$
12. $6e^{2x} - 6 = 66$

13. $10^x - 2 = 74$
14. $10^{\frac{1}{2}x} + 32 = 111$
15. $10^{-4x} - 88 = 4$

16. **PHYSICS** At a constant temperature, the atmospheric pressure can be modeled by the formula $p = 101,352e^{-0.000122h}$ where p is the pressure in Pascals and h is the altitude in meters. To the nearest meter, at what height will the pressure be 6000 Pascals?

Solve and check each equation to the nearest ten-thousandth.

17. $3 + 5 \log x = 18$
18. $6 \log x - 1 = 11$
19. $4 \log \frac{1}{2}x = 7$

20. $-9 + 2 \ln x = -5$
21. $10 - 7 \ln x = 5$
22. $5 \ln 5x = 23$

23. **DEMOGRAPHY** The population of Buffalo, New York, from 1950 through 1990 can be modeled by the equation $y = 600,744(0.9847)^x$ where $x = 0$ corresponds to 1950, $x = 10$ corresponds to 1960, $x = 20$ corresponds to 1970, and so on. If this model continues to hold through the year 2000, in what year will the population be 285,000?

24. **BIOLOGY** The number of bacteria in a specific culture can be modeled using the equation $B = 800(2)^{\frac{t}{40}}$ where B is the number of bacteria and t is the time in hours. Approximately how many hours will it take for 5000 bacteria to be present in the culture?

EXTEND

You can also solve equations of the form $b^x = c^y, b \neq c$, by taking the logarithm of both sides. For example, you can solve $3^{x-3} = 4^{x+2}$ using the steps at the right.

$$3^{x-3} = 4^{x+2}$$
$$\log 3^{x-3} = \log 4^{x+2}$$
$$(x - 3) \log 3 = (x + 2) \log 4$$
$$x \log 3 - 3 \log 3 = x \log 4 + 2 \log 4$$
$$x \log 3 - x \log 4 = 2 \log 4 + 3 \log 3$$
$$x(\log 3 - \log 4) = 2 \log 4 + 3 \log 3$$
$$x = \dfrac{2 \log 4 + 3 \log 3}{\log 3 - \log 4}$$
$$x = -21.0942$$

Solve and check each equation to the nearest ten-thousandth.

25. $6^{x-2} = 3^{x+1}$
26. $12^x = 5^{x+5}$

27. $9^{x-4} = 4^{4-x}$
28. $7^{2+x} = 5^{2-x}$

CARBON DATING Archaeologists approximate the age of certain specimens by measuring the amount of carbon-14, a radioactive substance, that remains inside them. The formula that is used for this process is $A = A_0 2^{\frac{-t}{5760}}$ where 5760 represents the half-life of carbon-14, A_0 is the original amount of carbon-14, and t is in years.

29. A specimen that originally contained 120 mg of carbon-14 now contains 100 mg of the isotope. Determine the age of the specimen to the nearest 100 years.

30. The longest bone in the human body is called the femur. A femur was found at a particular archaeological dig. The amount of carbon-14 that it contained had declined to $\frac{3}{4}$ of the initial amount. Determine the age of the bone to the nearest 100 years.

31. An animal bone that originally contained 180 g of carbon-14 was found to have 20 g of the isotope. Determine the age of the bone to the nearest 100 years.

THINK CRITICALLY

32. Debbie thinks that it will take longer for a bone that originally had 80 mg of carbon-14 to decay to 60 mg than it will take for a bone that originally had 40 mg of carbon-14 to decay to 30 mg. Steve thinks that the opposite is true. With whom do you agree? Explain.

33. Write a word problem that can be solved using the model $A = A_0 2^{\frac{-t}{3000}}$.

34. Write a word problem that can be solved using the model $y = 4500(1.02)^x$.

Solve each equation for x in terms of a.

35. $6^{3x-a} = 5^{a-2x}$

36. $4^{6x-a} = 3^{a-5x}$

MIXED REVIEW

Solve each equation using the quadratic formula.

37. $x^2 + 6x + 25 = 0$

38. $x^2 + 2x + 5 = 0$

39. Determine the probability of drawing a red face card from a standard deck of 52 playing cards. Round to the nearest thousandth.

Write a rule for each function. State the domain and range of the function.

40. $F = \{(2, 5), (-2, 1), (5, 8)\}$

41. $E = \{(0, 0), (1, 6), (-3, -18)\}$

STANDARDIZED TESTS Solve each equation to the nearest ten-thousandth.

42. $5^x = 44$

    A. 0.9445   **B.** 6332

    C. 0.4253   **D.** 2.3512

43. $3^{-5x} = 20$

    A. −4.0000   **B.** −0.5454

    C. −0.6021   **D.** 0.8239

HELP WANTED

Career
Population Ecologist

Population ecologists study populations of organisms. They define a population to be a group of individuals of the same species that interbreed and occupy a specific area at a specific time. Because natural populations have certain characteristics and properties, population ecologists can predict the way they will respond to competition, predators, and resource availability.

In 1967, Robert MacArthur of Princeton University and E.O. Wilson of Harvard University expressed the relationship between the number of species that live on an island and the area of the island. The formula $S = CA^z$ models this relationship, where S is the number of species of a given category of organism, A is the area of the island in square miles, C is a parameter that depends on the density of the species, and z is a parameter that is often in the range of 0.20–0.35.

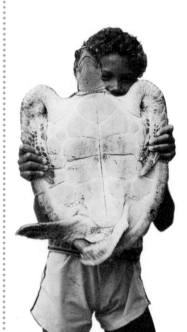

Decision Making

1. An island in the West Indies has nine species of herpetofauna (amphibians and reptiles) and an area of about 35 square miles. If the value of C for this island is 3.09, determine the value of z to the nearest thousandth.

2. Another island in the West Indies has an area of about 3980 square miles. If the values of C and z are the same as those of Question 1, how many species of herpetofauna does it have?

Population ecologists also study the relationship between percentage frequency (the chance of encountering one or more individuals of the population in a particular sample unit of area) and density. In a randomly dispersed population, the frequency can be modeled using the equation $F = 100(1 - e^{-D})$ where F is the percentage frequency and D is the density.

3. If the density of *Phlox pilosa*, an insect-pollinated plant, on a prairie in Illinois is 4.8 plants per square meter, determine the percentage frequency.

4. If a series of randomly selected samples is taken from a randomly dispersed population, the numbers of samples containing n individuals is $10,000 \dfrac{D^n}{n!} e^{-D}$. If the density of *Phlox pilosa* is 13 plants per square meter, how many randomly selected samples will contain 3 plants?

9.6   **Solve Exponential and Logarithmic Equations**   **459**

Compound Interest

If you invest a sum of money, called the **principal** P, for a time period t at an interest rate r per period, then you can determine the **simple interest** I you will receive by using the formula below.

$I = Prt$ Simple Interest

Many investments pay **compound interest**, where the interest for each period is added to the principal before interest is calculated for the next period. The more frequently interest is compounded, the larger the final amount will be.

Problem

You invest $10,000 for 3 years at 10% interest. How much will you have at the end of the three years if interest is compounded annually?

Explore the Problem

CHECK UNDERSTANDING

How much would the 3-year investment be worth at 10% simple interest?

1. The amount at the end of each year is equal to the principal plus the interest: $A = P + I$. Copy and complete the table below.

	Year 1	Year 2	Year 3
Principal	10,000		
Interest (at year end)	1,000		
Amount			

2. Look for a pattern in the last line of your table. By what number could you multiply the final amount from the previous year to determine the final amount for the current year? By what number could you multiply the original principal to determine the final amount after 3 years?

3. Suppose you kept the above investment for n years. Write an expression for the final amount.

If P dollars is invested at an interest rate r per year, compounded annually, the total compounded amount A at the end of n years is

$A = P(1 + r)^n$ Amount with Annual Compounding

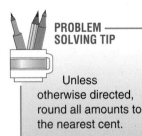

PROBLEM SOLVING TIP

Unless otherwise directed, round all amounts to the nearest cent.

4. If you invest $8000 for 5 years at 6% compounded annually, how much money will you have at the end of the 5 years? How much interest did your money earn?

Investigate Further

PROBLEM
SOLVING PLAN

• Understand
• Plan
• Solve
• Examine

● Some investments have interest compounded semianually, quarterly, monthly, or daily. For an annual interest rate r compounded m times per year, the interest rate i *per compounding period* is $i = \dfrac{r}{m}$. If t is the length of the investment in years, then the compound amount is

$$A = P\left(1 + \frac{r}{m}\right)^{mt} \quad \text{Amount with } m \text{ Compounding Periods}$$

5. Suppose you invest \$8000 for 5 years at 6% annual interest compounded quarterly. What is m? What is $\dfrac{r}{m}$? How much will the final amount be? Compare with your answer to Problem 4.

6. Complete the table below to explore what happens to the final amount as the number of compounding periods increases. It will help you identify a pattern to assume \$1 can be invested at 100% interest per year. Round answers to the nearest ten-thousandth.

Compounding Period	Number of Periods per Year	$P\left(1 + \frac{r}{m}\right)^{mt}$	Amount
Annually	1	$1\left(1 + \frac{1}{1}\right)^{1}$	
Monthly	12		
Daily	365		
Hourly			
Each minute			

7. What number do you recognize as the amount your \$1 investment approaches as the number of compounding periods increases?

The following formula tells you how much money you can expect from your investment even in the extreme case of *continuous* compounding. For P (dollars) invested for t years at an annual rate r compounded continuously, the total compounded amount A is

$$A = Pe^{rt} \quad \text{Amount with Continuous Compounding}$$

8. Determine the total compounded amount if \$10,000 is invested for 7 years at 8% annual interest, compounded continuously?

9. WRITING MATHEMATICS Two banks each advertise an annual interest rate of 5.8% on savings. Explain under what circumstances one bank may be offering a better deal than the other.

Apply the Strategy

10. Determine how much interest you would earn if you invest $9300 for 4 years at 6.6% annual interest compounded semiannually.

11. Determine the total compounded amount if $5800 is invested for 5 years at 9% annual interest compounded continuously.

12. How much more will your $20,000 investment be worth in 10 years if you receive 7% annual interest compounded continuously instead of quarterly?

13. For how many years would $11,000 have to be invested at 8% annual interest compounded continuously to grow to $19,257.39?

14. For how many years would $8300 have to be invested at 7.2% annual interest compounded semiannually to earn $7387.50 interest?

EFFECTIVE ANNUAL RATE Suppose $100 is invested for 1 year at 8% annual interest compounded quarterly. The compound amount will be $100(1.02)^4 = 108.24 and interest for the year will be $8.24. The annual rate of 8% is called the *nominal rate*. Interest compounded quarterly is equivalent to a simple interest rate of 8.24%. So the *effective annual rate* is 8.24%. For an annual interest rate of r with m compounding periods,

$$\text{Effective annual rate} = \left(1 + \frac{r}{m}\right)^m - 1$$

15. If you invest at a nominal rate of 5.4% compounded monthly, what is the effective annual rate? Round to the nearest hundredth.

16. The effective annual rate at a bank that compounds quarterly is 6.98%. What is the nominal rate? Round to the nearest tenth.

17. **WRITING MATHEMATICS** Explain how to determine the effective annual rate for a nominal rate that is compounded continuously. Provide an example.

DOUBLING TIME Suppose you have the opportunity to invest $1000 at 16%, compounded continuously.

18. Write an equation to model the growth of your investment.

19. Use a graphing utility to graph the equation you wrote. Determine how many years it would take for your investment to double. Explain your method.

REVIEW PROBLEM SOLVING STRATEGIES

AN INTERPLANETARY PUZZLE

1. It's 2096, and a typical business trip is from one planet to another. In fact, every day at noon, a shuttle flight leaves Earth for Venus and another shuttle leaves Venus for Earth. Shuttles in either direction travel the two-lane Interplanetary Zoomway at the same constant rate of speed and the trip lasts 9 days and 9 nights. How many Earth-to-Venus shuttles will the shuttle leaving Venus today pass during the journey to Earth?

a. How many Earth-to-Venus shuttles are already en route when today's shuttle leaves Venus?

b. How many Earth-to-Venus shuttles will take off before today's Venus-to-Earth shuttle arrives at it's destination?

c. How many meetings in all?

d. Describe when the meetings will take place.

e. How did you solve the problem? Can you solve it by drawing a picture?

SEE THE LIGHT

2. Nico has told his aunt he was bringing her three pairs of silver candleholders. When his aunt opened the box, she thought some items were missing. "No", said Nico, "that's exactly what I promised you." What is the minimum number of candleholders that Nico could have brought?

CHOCOLATE

OR

VANILLA

?

3. Felix, Sandra, and Evelyn usually stop for some frozen yogurt after school. One day, the store owner said, "I've noticed something interesting about the flavors you kids order each time you come in here. If Felix orders chocolate, Sandra orders vanilla. There will always be one order of chocolate from either Felix or Evelyn, but not two orders. And, Sandra and Evelyn never both choose vanilla." If one of the friends ordered chocolate yesterday and vanilla today, who was it? Explain the strategy you used to solve the problem.

CHAPTER REVIEW

Choose the word from the list that completes each statement.

1. A(n) __?__ can be expressed in the form $f(x) = b^x$, where $b > 0$ and $b \neq 1$.

2. The __?__ is the number that $\left(1 + \dfrac{1}{n}\right)^n$ approaches as n increases without bound.

3. The __?__ of x can be written as log x.

4. The __?__ is the inverse of the exponential function.

5. The __?__ of x can be written as ln x.

a. logarithmic function

b. natural logarithm

c. number e

d. common logarithm

e. exponential function

Lesson 9.1   EXPLORE EXPONENTIAL FUNCTIONS   pages 423–427

- You can translate the graph of an exponential function vertically and horizontally and reflect it over the x- and y-axes.

How can you transform the graph of $f(x) = 3^x$ to obtain the graph of each function?

6. $g(x) = 3^{x+1}$ 7. $h(x) = 3^x - 2$ 8. $j(x) = 3^{-x}$ 9. $k(x) = -3^x$

Lesson 9.2   EXPONENTIAL FUNCTIONS   pages 428–435

- To solve an exponential equation, express both sides of the equation in terms of the same base and solve the resulting equation. If both sides of the equation cannot be expressed in terms of the same base, use a graphing utility to obtain a decimal approximation.

Solve and check.

10. $11 = 121^x$ 11. $4^x = \dfrac{1}{64}$ 12. $32 = 4^x$ 13. $16^8 = 4^x$

14. $3^{8x+2} = 9^{2x-3}$ 15. $\left(\dfrac{1}{5}\right)^6 = \left(\dfrac{1}{25}\right)^x$ 16. $13 = 7^x$ 17. $8 = 82^x$

Lesson 9.3   LOGARITHMIC FUNCTIONS   pages 436–442

- Use the **change-of-base formula**, $\log_b x = \dfrac{\log_a x}{\log_a b}$ to evaluate logarithmic expressions.

- To solve a logarithmic equation, rewrite it in exponential form.

Evaluate to four decimal places.

18. $\log_9 31$ 19. $\log_8 15$ 20. $\log_{14} 7$ 21. $\log_6 5$

Solve each logarithmic equation.

22. $\log_x 729 = 3$ **23.** $\log_x 5 = \dfrac{1}{4}$ **24.** $\log_{16} x = \dfrac{5}{2}$ **25.** $\log_8 64 = 5x - 5$

Lesson 9.4   PROPERTIES OF LOGARITHMS   pages 443–448

- You can use the product, quotient, and power properties of logarithms to expand or condense logarithmic expressions.

Express in expanded form.   **Express as a single logarithm.**

26. $\log_6 \dfrac{j}{3}$ **27.** $\log_9 \sqrt{\dfrac{x^4}{y^8}}$ **28.** $3\log_4 x - \log_4 5$ **29.** $\dfrac{1}{2}\log a + \log b$

Solve and check.

30. $\log_6(x^2 - 9) = 3$ **31.** $\log_5 2x + \log_5 3x = \log_5 600$

Lesson 9.5   NATURAL LOGARITHMS   pages 449–453

- The **natural logarithm** of x is the base e logarithm of x. The properties of logarithms also apply to natural logarithms.

Express in expanded form.   **Express as a single logarithm.**

32. $\ln 3hk$ **33.** $\ln x^2 y^3$ **34.** $4\ln x - 2\ln y + \ln 9$

Lesson 9.6   SOLVE EXPONENTIAL AND LOGARITHMIC EQUATIONS   pages 454–459

- To solve an equation of the form $ab^x + c = y$, isolate b^x on one side of the equation and take the logarithm of both sides. To solve an equation of the form $\log x = y$ or $\ln x = y$, rewrite in exponential form.

Solve each equation to four decimal places.

35. $3e^x + 12 = 56$ **36.** $8e^{2x} + 10 = 72$ **37.** $10^x - 7 = 77$

38. $8 + 5\log x = 18$ **39.** $4\ln x - 1 = 11$ **40.** $3\ln 9x = 11$

Lesson 9.7   COMPOUND INTEREST   pages 460–463

- If you invest P dollars at an annual interest rate r compounded m times per year for t years, you will have a final amount A, where $A = P\left(1 + \dfrac{r}{m}\right)^{mt}$.

- If you invest P dollars at an annual interest rate r compounded continuously for t years, you will have a final amount A, where $A = Pe^{rt}$.

41. How much money will you have at the end of 5 years if you invest \$4500 at 8% annual interest compounded quarterly?

42. How much money will you have at the end of 4 years if you invest \$5200 at 7% annual interest compounded continuously?

CHAPTER ASSESSMENT

CHAPTER TEST

1. **STANDARDIZED TESTS** The graph of $f(x) = 6^x$ is shifted 3 units up and 4 units right. You can represent this translation by the graph of which of the following functions?

   A. $g(x) = 6^{x+4} + 3$ **B.** $h(x) = 6^{x-4} - 3$
   C. $j(x) = 6^{x+4} - 3$ **D.** $k(x) = 6^{x-4} + 3$

2. What is the relationship between the graphs of $f(x) = 8^x$ and $g(x) = -8^x$?

3. What is the relationship between the graphs of $h(x) = \frac{1}{2^x}$ and $j(x) = \frac{1}{2^{-x}}$.

Determine the horizontal asymptote of each function.

4. $f(x) = 3^x + 2$ 5. $g(x) = 7^{-x} - 3$

Solve by expressing each side of the equation in terms of the same base.

6. $3 = 27^x$ 7. $2^{-2x} = 256$

8. $9^6 = 3^{x^2}$ 9. $6^{2x} = 36^{3x-1}$

10. **WRITING MATHEMATICS** Write a paragraph that explains how you can use a graphing utility to solve an exponential equation in which both sides cannot be expressed in terms of the same base. Provide a numerical example.

Use a graphing utility to solve each equation to the nearest thousandth.

11. $9 = 7^x$ 12. $19 = 5^x$

13. $3 = 12^x$ 14. $7 = 13^x$

15. The population of a particular strain of bacteria doubles in number every 6 hours. If the initial number of bacteria is 12, how many bacteria will be present after 2 days?

16. **STANDARDIZED TESTS** Which of the following is (are) equal to $\log_4 6$?

   I. $\dfrac{\log_{10} 6}{\log_{10} 4}$ **II.** $\dfrac{\log 6}{\log 4}$ **III.** $\dfrac{\ln 6}{\ln 4}$

   A. I, II, and III   **B.** I and III
   C. II   **D.** I and II

Solve each logarithmic equation.

17. $\log_5 \sqrt{5} = x$ 18. $\log_x 343 = 3$

19. $\log_6 x = -3$ 20. $\log_5 625 = 2x - 1$

Evaluate.

21. $\log_7 16$ 22. $\log_9 2$

Express in expanded form.

23. $\ln ab^2c^3$ 24. $\log \sqrt{\dfrac{a^{10}}{b^{12}}}$

Express as a single logarithm.

25. $3 \log_8 x + 2 \log_8 y$ 26. $4 \ln q - \ln 3$

Solve each equation to four decimal places.

27. $5e^x - 11 = 77$ 28. $10^{2x} - 9 = 55$

29. $9 + 6 \log x = 25$ 30. $8 \ln x - 4 = 19$

31. Steven invests $5000 for 4 years at 6% interest. How much money will he have at the end of 4 years if interest is compounded monthly?

32. Sarah invests $3800 for 3 years at 10% interest. How much money will she have at the end of 3 years if interest is compounded continuously?

33. **WRITING MATHEMATICS** Mr. Ortega invested $8000 at 9% annual interest compounded continuously. He now has $13,728.05. Write and solve a problem using this information.

PERFORMANCE ASSESSMENT

CURVE YOUR CAR Select a particular car that is five years old and has been on the market for at least 5 years. Consult back copies of used car books or newspapers to determine how much money you would have paid for that car if you had purchased it new five years ago. Then find out how much it was worth four, three, two, and one year(s) ago as well as how much it is worth now. Graph your results. Connect the points with a smooth curve. What type of curve have you drawn? What is the average yearly rate of depreciation of your car?

USE FLOWCHARTS Construct a flowchart to show how to solve an equation of the form $ae^x + c = y$. Then use your flowchart to solve three different exponential equations.

EXPONENTIAL ORIGAMI Work with a partner. Graph an exponential equation on a set of coordinate axes and state its equation. Use tracing and paper folding techniques to show the graph of the equation that is its reflection over the x-axis. Ask your partner to state the equation for this reflection. Then repeat the procedure to show the graph of the equation that is its reflection over the y-axis. Ask your partner to state the equation for this reflection.

BANK RANKS Visit or call three banks in your area. Find their stated rates on savings accounts, 6-month certificates of deposit, and 1-year certificates of deposit. Calculate the effective rates on each of these accounts by bank and summarize your results in a table. Rank the banks from highest to lowest in terms of their effective rates in each category.

PROJECT ASSESSMENT

PROJECT *Connection*

Each group should write a report of its population study.

1. Summarize key research and cite sources. Explain the assumptions and methods used to construct different models and present the population predictions for 2010 (and further into the future). Discuss the advantages and limitations of each model.

2. Distribute copies of your report to other groups. Meet as a class to compare the work and conclusions of the groups and to discuss the project in general.

3. Make a list of extension activities, such as studying world population growth or population growth of third-world countries. Have students volunteer for activities that interest them.

4. Learn more about the field of population biology. Research and report on topics such as factors that affect population growth (including density-dependent and density-independent factors), survivorship curves, age, structure graphs, modern population trends, and predictions of scientists Paul and Anne Ehrlich.

CUMULATIVE REVIEW

1. Calculate the product matrix $B \cdot A$ if
   $$A = \begin{bmatrix} 1 & -2 \\ 6 & 5 \end{bmatrix} \text{ and } B = \begin{bmatrix} 3 & 7 \\ 8 & -4 \end{bmatrix}.$$

2. Use Cramer's rule to solve the following system of equations.
   $$\begin{cases} x = y + 1 \\ 2x + 3y - 12 = 0 \end{cases}$$

3. **WRITING MATHEMATICS** Describe the characteristic shape of $y = b^x$, discussing the domain of x, possible values for b, and the resulting range. Include sketches.

4. Solve the quadratic equation $x^2 + 4 = 6x$ by completing the square.

5. The equation $A_t = A_0 e^{rt}$ is used to model a situation at which an initial amount A_0 is increasing at a rate that is proportional to the amount present at any time, A_t. Find the length of time necessary for the value of an investment to double when invested at a yearly rate of 8% compounded continuously.

6. **STANDARDIZED TESTS** If $f(x) = \frac{x-1}{2}$ and $g(x) = 2x + 1$, then $f(g(3)) - g(f(3)) =$

   A. 0   **B.** 3
   C. 4   **D.** 5

7. Evaluate: $8^{\frac{2}{3}} + 8^0 + 8^{-1}$

Solve and check.

8. $125^x = 25$ 9. $\log_x 1331 = 3$

10. $x = e^{2 \ln 5}$ 11. $x = 10^{\frac{1}{3} \log 8}$

12. Mr. Chu wished to build a bin to hold 9 tons of coal. The bin is to be 6 feet deep and 4 feet longer than it is wide. Allowing 40 cubic feet for one ton of coal, what should the dimensions be?

13. Express $\log \frac{x^2}{y}$ in expanded form.

14. Eagle Electronics manufactures two-headed VCRs on which it makes a profit of $28 each and four-headed VCRs on which it makes a profit of $33 each. The production manager wants to produce at least 60 two-headed VCRs and 100 four-headed VCRs per day. If the factory can assemble no more than a total of 200 VCRs a day, what should the daily production be for a maximum profit?

15. **STANDARDIZED TESTS** The value of x that satisfies $2x^{-\frac{4}{3}} = 162$ is

   A. $-\frac{1}{9}$ **B.** $\frac{1}{9}$ **C.** $-\frac{1}{27}$ **D.** $\frac{1}{27}$

16. Find the solution set of $x^2 - 3x > 0$.

17. Write the equation of the line in slope-intercept form.

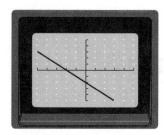

18. **STANDARDIZED TESTS** The graph below represents the solutions to which of the following inequalities?

   I. $3x - 18 \geq -6$ or $-4x \geq 16$
   II. $|x| \leq 4$
   III. $2x \geq 8$ or $5x + 2 \leq -18$

   A. I only   **B.** II only
   C. II and III   **D.** I and III

STANDARDIZED TEST

STANDARD FIVE-CHOICE Select the best choice for each question.

1. If the average of $3 + i$, $2 - 2i$, and x is $1 + i$, then x equals

   A. $-2 + 4i$ **B.** $-1 + 2i$ **C.** $5 - i$
   D. $3 + 3i$ **E.** $1 + i$

2. The graph of the solution of $x^2 + 2x \le 3$ is

   A.

   B.
   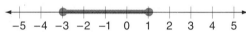

   C.

   D.
   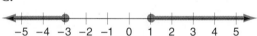

   E.

3. If $\log_x \frac{2}{z} = -1$, then x, expressed in terms of z, is

   A. $2z$ **B.** $\frac{2}{z}$ **C.** $-2z$
   D. $\frac{-2}{z}$ **E.** $\frac{z}{2}$

4. If $\left(x^{\frac{1}{2}}\right)^{\frac{1}{5}} = 3$, then $x =$

   A. 3^7 **B.** 3^{10} **C.** $3^{\frac{1}{10}}$
   D. $3^{\frac{10}{7}}$ **E.** $3^{\frac{7}{10}}$

5. If $2x + 3y = 3$ and $x + 2y = 1$, then $x + y$ is what percent of $3x + 5y$?

   A. 25 **B.** $33\frac{1}{3}$ **C.** 50
   D. 60 **E.** 75

6.

   If the graph above represents $y = x^2 + bx + c$, then the y-intercept of $y + 1 = x^2 + bx + c$ is

   A. 0 **B.** 1 **C.** 2
   D. 3 **E.** 4

7. Which of the following statements are true about the logarithmic function $y = \log_b x$?

   I. It is the reflection of $y = b^x$ over the line $y = x$.
   II. Its graph has the x-axis as an asymptote.
   III. Its y-intercept is 1.

   A. I, II, III **B.** I and II **C.** I and III
   D. II and III **E.** I only

8. If $\sqrt{\sqrt{\sqrt{y} + \sqrt{y} + \sqrt{y} + \sqrt{y}}} = \sqrt{2}$, then y equals

   A. 2 **B.** $\frac{1}{2}$ **C.** $\frac{1}{4}$
   D. $\frac{\sqrt{2}}{2}$ **E.** $2\sqrt{2}$

9. If $2 \log x = \log 2x$, then x is

   A. 0 or 2 **B.** 0 only **C.** 0, 2, or 4
   D. 2 only **E.** any real number

10.

   In the diagram above, if $PR = 4x + 5$, $RS = 2x + 3$, and Q is the midpoint of $\overline{PS}$, then QR equals

   A. $6x + 8$ **B.** $3x + 4$ **C.** $2x + 2$
   D. $x + 1$ **E.** $3x + 3$

10 Polynomial Functions

Take a Look
AHEAD

Make notes about things that look new.
- What special method for dividing polynomials will you learn about in this chapter?
- Copy the names of the different theorems you will study in this chapter. Try to write a one- or two-line explanation of the main focus of this chapter.
- What problem solving strategy will be covered?

Make notes about things that look familiar.
- Find and copy some examples of polynomials of degree greater than 2.
- Make a list of terms (for example, "solution") that you have used before, perhaps in your general study of functions in Chapter 3 or when you learned about quadratic functions in Chapter 6.

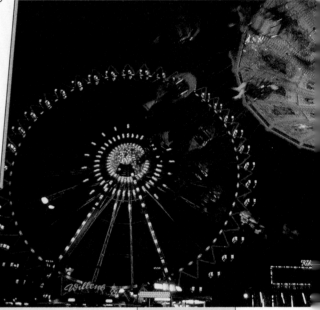

DATA Activity

Fun For All Ages

In the United States, amusement parks offering rides, games, and shows developed as a form of entertainment during the late 19th century. One of the earliest was located at the end of trolley routes to encourage weekend use of trolley car lines, hence the name *trolley park*. The first major amusement park was built in 1895 at Coney Island in New York City. Traditional amusement parks remained very popular until about 1950, when attendance began to decline. The trend was reversed with the opening of the first *theme park*, Disneyland, in 1955.

SKILL FOCUS

▶ Determine empirical probability.

▶ Use survey results to make predictions.

▶ Interpret statistical data.

▶ Make a histogram.

AMUSEMENT PARKS

In this chapter, you will see how:

- **DEMOGRAPHERS** use polynomials to model population data. (Lesson 10.3, page 490)

- **AMUSEMENT PARK RIDE ENGINEERS** use formulas to design exciting, but safe, rides. (Lesson 10.4, page 499)

Attendance at Amusement Parks*			
Age	**Percent**	**Education**	**Percent**
18 – 24	68	Grade school	24
25 – 34	68	Some high school	35
35 – 44	58	High school graduate	51
45 – 54	44	Some college	59
55 – 64	30	College graduate	58
65 – 74	29	Graduate school	54
75 – 96	14		

*Attendance at least once during past year

In the table, 100% is the total population in each category. For example, 24% for grade school represents the percentage of adults with a grade school education that attend amusement parks.

1. What is the probability that a randomly selected 50-year-old will have visited an amusement park during the past year?

2. In a group of 2000 randomly selected high school graduates, how many would you expect to have visited an amusement park?

3. Do you think age or education is the more significant factor affecting attendance? Explain.

4. A *histogram* is a bar graph in which the bars represent intervals. There are no spaces between the bars. Make a histogram for the data showing attendance by age group.

5. **WORKING TOGETHER** Survey students in your school to determine percent attendance at amusement parks. Isolate results for 13- to 17-year-olds and add a bar to the histogram constructed in Question 4.

PROJECT

Scream Machines

Your car seems to be speeding out of control as it races down one hill, climbs another, turns upside-down, and back-flips through a loop. You're not on an icy road, you're on a roller coaster, where a thrilling ride is expected. The first looping roller coaster was opened in California in 1976, and since then amusement parks have tried to out perform each other with scarier new rides.

PROJECT GOAL

To analyze physical models and data relating to roller coasters.

Getting Started

Work in groups.

1. As a group, list some topics relating to roller coasters that would involve mathematics. For example, you might research statistics about the top five U.S. roller coasters, investigate industry safety records, find out about construction costs, or compile attendance figures.

2. Try to obtain information directly from amusement parks. Consider arranging an interview with park officials.

3. The following two books list some famous roller coasters and the names and addresses of the parks where they are located.

> *The American Amusement Park Industry: A History of Technology and Thrills* by Judith A. Adams, Twayne Publishers, Boston, MA, 1991.
>
> *Scream Machines: Roller Coasters Past, Present and Future* by Herma Silverstein, Walker and Co., New York, NY, 1986.

PROJECT *Connections*

Lesson 10.1, page 478:
Construct a model of a roller coaster hill and collect roller coaster data.

Lesson 10.2, page 485:
Graph and analyze data collected from the roller coaster hill model.

Lesson 10.4, page 498:
Solve a system of equations to determine the coefficients of the cubic polynomial that describes the roller coaster hill.

Chapter Assessment, page 507:
Present results of data collection and modeling activities; speculate on the future of rides and the amusement park industry.

Internet Connection

www.swpco.com/
swpco/algebra2.html

10.1 Division of Polynomials

Explore

- Recall how to use Algeblocks to model a division problem like $6x^2 \div (-3x)$.

  1. Place the divisor $-3x$ in the horizontal axis of a Quadrant Mat.

  2. Place the dividend $6x^2$ in a quadrant with the divisor as one dimension.

  3. Make the other dimension of the rectangular area. This is the result, or quotient, $-2x$.

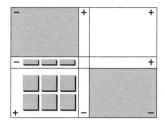

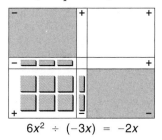

$$6x^2 \div (-3x) = -2x$$

  Use Algeblocks to model each division.

  4. $(2x^2 - 6x) \div 2x$ **5.** $(-3x^2 - 6x) \div (x + 2)$

  6. Division is the inverse operation of multiplication. Describe how you would check that $9 \div 2 = 4\frac{1}{2}$.

  7. Describe the steps in the division: $26{,}537 \div 38$

  8. Describe the steps that are repeated in division.

SPOTLIGHT ON LEARNING

WHAT? In this lesson you will learn
- to divide one polynomial by another.
- to factor higher degree polynomials.

WHY? Knowing how to divide polynomials and factor higher degree polynomials can help you solve problems in motion, geometry, and investments.

Build Understanding

- A **polynomial** in one variable x is an expression of the form

  $$a_n x^n + a_{n-1} x^{n-1} + \ldots + a_1 x + a_0$$

  where n is a nonnegative integer and the coefficients $a_0, \ldots, a_n$ are real numbers.

  The **degree** of a polynomial in one variable is the greatest exponent of the variable that appears in any term. Polynomials are generally written with exponents in descending order. The coefficient of the term with the greatest exponent is the **leading coefficient**. The term that is not multiplied by a variable is the **constant term**.

  Every polynomial expression can be written as a corresponding **polynomial function**.

  $$F(x) = a_n x^n + a_{n-1} x^{n-1} + \ldots + a_1 x + a_0$$

CHECK UNDERSTANDING

John told Carmen the polynomial $2x^3 + 8x^4 - x + 7$ has degree 3. Is he correct? Explain

The values of x for which $F(x)$ is 0 are the **zeros of the polynomial function**. The zeros of a polynomial function are the **solutions** of the corresponding polynomial equation when $F(x) = 0$.

In Chapter 6 you learned to find solutions of quadratic equations. To determine solutions of polynomial equations of higher degree, you need to know how to divide polynomials. The process, or *algorithm*, for dividing a polynomial by a polynomial is similar to the algorithm for long division in arithmetic.

$$\begin{array}{r} 26 \\ 32\overline{)853} \\ \underline{64} \\ 213 \\ \underline{192} \\ 21 \end{array}$$

Long Division Algorithm

Divide 85 by 32.
Multiply 32 by 2.
Subtract $85 - 64$.
Bring down the next digit. Repeat.
remainder → Stop when the remainder is 0 or less than the divisor.

So, $\dfrac{853}{32} = 26\dfrac{21}{32}$ or $26 + \dfrac{21}{32}$ or $853 = 26 \cdot 32 + 21$

In general: $\dfrac{\text{dividend}}{\text{divisor}} = \text{quotient} + \dfrac{\text{remainder}}{\text{divisor}}$

dividend $=$ quotient $\cdot$ divisor $+$ remainder

THINK BACK

The divisor is the quantity by which the dividend is being divided. The quotient is the result.

$$\text{divisor }\overline{)\text{dividend}}^{\text{quotient}}$$

EXAMPLE 1

Divide: $10x^3 + x^2 + 3x + 7$ by $5x - 2$

Solution

Apply the long division algorithm.

$$\begin{array}{r} 2x^2 \\ 5x - 2\overline{)10x^3 + 1x^2 + 3x + 7} \\ \underline{10x^3 - 4x^2} \\ 5x^2 + 3x \end{array}$$

Divide $10x^3$ by $5x$.
Multiply $(5x - 2)$ by $2x^2$.
Subtract; bring down the next term.

$$\begin{array}{r} 2x^2 + x \\ 5x - 2\overline{)10x^3 + 1x^2 + 3x + 7} \\ \underline{10x^3 - 4x^2} \\ 5x^2 + 3x \\ \underline{5x^2 - 2x} \\ 5x + 7 \end{array}$$

Divide $5x^2$ by $5x$.
Multiply $(5x - 2)$ by x.
Subtract; bring down the next term.

$$\begin{array}{r} 2x^2 + x + 1 \\ 5x - 2\overline{)10x^3 + 1x^2 + 3x + 7} \\ \underline{10x^3 - 4x^2} \\ 5x^2 + 3x \\ \underline{5x^2 - 2x} \\ 5x + 7 \\ \underline{5x - 2} \\ 9 \end{array}$$

Divide $5x$ by $5x$.
Multiply $(5x - 2)$ by 1.
Subtract. Stop, since the remainder is of lower degree than the divisor.

PROBLEM SOLVING TIP

In Example 1 when you subtract $10x^3 - 4x^2$ remember:
$10x^3 + 1x^2 - (10x^3 - 4x^2) = 10x^3 + 1x^2 - 10x^3 + 4x^2 = 5x^2$

$$\text{dividend} \stackrel{?}{=} \text{quotient} \times \text{divisor} + \text{remainder}$$
$$10x^3 + x^2 + 3x + 7 \stackrel{?}{=} (2x^2 + x + 1)(5x - 2) + 9$$
$$\stackrel{?}{=} (2x^2 + x + 1)(5x) + (2x^2 + x + 1)(-2) + 9$$
$$\stackrel{?}{=} 10x^3 + 5x^2 + 5x - 4x^2 - 2x - 2 + 9$$
$$= 10x^3 + x^2 + 3x + 7$$

So, $\dfrac{10x^3 + x^2 + 3x + 7}{5x - 2} = 2x^2 + x + 1 + \dfrac{9}{5x - 2}$ ◄

Before dividing, you may need to arrange the terms of the polynomials in descending or ascending order of the variable. Be sure that the terms of both polynomials are in the same order.

EXAMPLE 2

Divide: $6x - 2 + 7x^3 + 2x^4 - 3x^2$ by $1 + x^2 + 2x$

Solution

CHECK UNDERSTANDING

Check your solutions in Examples 2 and 3 by multiplying.

Arrange the terms of both polynomials in descending order.

$$
\begin{array}{r}
2x^2 + 3x - 11 \\
x^2 + 2x + 1\overline{)2x^4 + 7x^3 - 3x^2 + 6x - 2} \\
\underline{2x^4 + 4x^3 + 2x^2} \\
3x^3 - 5x^2 + 6x \\
\underline{3x^3 + 6x^2 + 3x} \\
-11x^2 + 3x - 2 \\
\underline{-11x^2 - 22x - 11} \\
25x + 9
\end{array}
$$

So, $\dfrac{2x^4 + 7x^3 - 3x^2 + 6x - 2}{x^2 + 2x + 1} = 2x^2 + 3x - 11 + \dfrac{25x + 9}{x^2 + 2x + 1}$ ◄

If the dividend or divisor has missing terms, insert these terms with coefficients of zero.

EXAMPLE 3

Divide: $x^3 + 8$ by $x - 2$

Solution

$$
\begin{array}{r}
x^2 + 2x + 4 \\
x - 2\overline{)x^3 - 0x^2 - 0x + 8} \\
\underline{x^3 - 2x^2} \\
2x^2 - 0x \\
\underline{2x^2 - 4x} \\
4x + 8 \\
\underline{4x - 8} \\
16
\end{array}
$$ In the dividend, insert $0x^2$ and $0x$.

So, $\dfrac{x^3 + 8}{x - 2} = x^2 + 2x + 4 + \dfrac{16}{x - 2}$ ◄

CHECK ——
UNDERSTANDING

In polynomial division, how does the degree of the quotient compare to the degree of the dividend? How does the degree of the divisor affect the degree of the quotient?

In a division, if the remainder is zero, then the divisor is a factor of the dividend. When the quotient is a quadratic polynomial, you may be able to factor it.

EXAMPLE 4

GEOMETRY A rectangular box has a width of $(x - 2)$ cm and a volume of $(x^3 - 3x^2 - 10x + 24)$ cm^3. Find the length and height in terms of x.

Solution

$$
\begin{array}{r}
x^2 - x - 12 \\
x - 2\overline{)x^3 - 3x^2 - 10x + 24} \\
\underline{x^3 - 2x^2} \\
-x^2 - 10x \\
\underline{-x^2 + 2x} \\
-12x + 24 \\
\underline{-12x + 24} \\
0
\end{array}
$$

Factor the quadratic polynomial: $x^2 - x - 12 = (x - 4)(x + 3)$

The length and height are $(x - 4)$ cm or $(x + 3)$ cm.   ◄

TRY THESE

1. If $\dfrac{10x^3 + x^2 + 3x + 7}{5x - 2} = 2x^2 + x + 1 + \dfrac{9}{5x - 2}$, identify the quotient, the remainder, the dividend, and the divisor.

2. **WRITING MATHEMATICS** To divide $3x - 1 + x^3 - 3x^2$ by $1 + x$, Ian wrote $1 + x\overline{)x^3 - 3x^2 + 3x - 1}$ and Jon wrote $x + 1\overline{)x^3 - 3x^2 + 3x - 1}$. With whom do you agree? Why?

Divide. Check by multiplying.

3. $(x^2 - 9x + 7) \div (x - 2)$

4. $(z^3 - 8z^2 - 6z + 9) \div (z - 2)$

5. $\dfrac{-5a^2 - 8a + 2a^3 + 2}{2a + 3}$

6. $(x^2 + 25) \div (x + 5)$

7. $\dfrac{x^3 + 2x^2 - 2x - 12}{x^2 + 4x + 6}$

8. $\dfrac{3x^4 - 23x^3 - 7x^2 + 42x - 10}{3x^2 + x - 5}$

9. Determine whether $(x - 2)$ is a factor of $x^3 - 2x^2 + 4x - 6$. Justify your answer.

10. **MOTION** Vic drove $(6y^3 + 11y^2 - 1)$ miles in $(3y + 1)$ hours. Find his average speed in terms of y.

11. **GEOMETRY** A rectangular box has a volume of $(x^3 + x^2 - 10x + 8)$ cm^3, a length of $(x - 2)$ cm and a width of $(x - 1)$ cm, find the height in terms of x.

1. **WRITING MATHEMATICS** After dividing $x^2 + x - 1$ by $x - 1$, Randi wrote the answer as $(x + 2)\left(\dfrac{1}{x - 1}\right)$ and Tanisha wrote $x + 2 + \dfrac{1}{x - 1}$. With whom do you agree? Why?

Divide. Check by multiplying.

2. $(4y^2 + 6y + 9) \div (2y - 5)$

3. $(3x^2 + 9x - 4) \div (3x + 3)$

4. $\dfrac{3z^3 + 14z^2 + 4z - 4}{4 + z}$

5. $\dfrac{4m^3 - 2m^2 + 6m - 3}{2m + 1}$

6. $\dfrac{7r + 3 + 13r^2 + 15r^3}{5r + 1}$

7. $\dfrac{x^3 + 16}{x + 2}$

8. $(64p^3 + 27) \div (4p + 3)$

9. $(y^4 - 1) \div (y - 1)$

10. $(y^4 - 6y^2 + 8) \div (y^2 - 4)$

11. $(4x^4 + 1) \div (2x^2 + 2x + 1)$

12. **INVESTMENTS** After one year, Ms. Chen found that the interest on the amount she had invested in the Growth Fund at her bank was $(x^3 - 2x^2 - 5x + 6)$ dollars. If the annual interest rate was $100(x + 2)\%$, find, in terms of x, the dollar amount that Ms. Chen had invested originally.

The first polynomial is a factor of the second. Find all the linear factors of the second polynomial. Check by multiplying.

13. $y + 3;\ y^3 + 4y^2 + y - 6$

14. $t - 2;\ t^3 + 3t^2 - 6t - 8$

15. $x + 1;\ x^3 - 21x - 20$

16. $z + 6;\ z^3 + 7z^2 - 36$

17. $a - 2;\ -a^3 + 6a^2 - 11a + 6$

18. $x + 2;\ 2x^3 + 5x^2 - x - 6$ ·

19. **GEOMETRY** The volume of a right circular cylinder is $\pi(x^3 - 3x^2 + 4)$ cm³. If the height of the cylinder is $(x + 1)$ cm, find the radius in terms of x.

The two binomials are factors of the quartic (fourth degree) polynomial. Find all the linear factors of the quartic polynomial. Check by multiplying.

20. $(x - 1), (x + 2);\ x^4 - 6x^3 + 3x^2 + 26x - 24$

21. $(x + 4), (x - 1);\ 2x^4 + 3x^3 - 19x^2 + 6x + 8$

22. $(x + 1)^2;\ -x^4 - 3x^3 + 3x^2 + 11x + 6$

Divide. Check your answer.

23. $\dfrac{\frac{3}{2}x^2 - \frac{5}{2}x - 1}{3x + 1}$

24. $\dfrac{3x^2 - x + \frac{3}{4}}{2x + 1}$

25. $\dfrac{10x^2 - 3xy + 9y^2}{2x + y}$

26. $\dfrac{2r^3 - 3r^2t - 8rt^2 - 3t^3}{r + t}$

27. $\dfrac{4a^4 - 4a^2b^2 - 13b^4}{2a^2 + 3b^2}$

28. $\dfrac{6x^3 + 4xy^2 - 13x^2y + 2y^3}{2x - 3y}$

THINK CRITICALLY

Determine whether the first polynomial is a factor of the second.

29. $x + 2$; $x^3 + 8$ **30.** $x + 2$; $x^4 + 16$ **31.** $x - 2$; $x^3 + 8$ **32.** $x - 2$; $x^4 + 16$

33. $x + 2$; $x^3 - 8$ **34.** $x + 2$; $x^4 - 16$ **35.** $x - 2$; $x^3 - 8$ **36.** $x - 2$; $x^4 - 16$

Use the results of Exercises 29–36 to determine which of the following statements are true. Assume n is an integer.

37. $(x - y)$ is a factor of $(x^n - y^n)$ if $n > 0$

38. $(x + y)$ is a factor of $(x^n - y^n)$ if $n > 0$

39. $(x + y)$ is a factor of $(x^n - y^n)$ if $n > 0$ and even

40. $(x + y)$ is a factor of $(x^n - y^n)$ if $n > 0$ and odd

41. $(x + y)$ is a factor of $(x^n + y^n)$ if $n > 0$ and even

42. $(x + y)$ is a factor of $(x^n + y^n)$ if $n > 0$ and odd

MIXED REVIEW

Solve each equation for the indicated variable.

43. $A = \dfrac{1}{2} h(a + b)$ for a

44. $mv = Ft + mv_0$ for m

Solve and check each system of equations.

45. $\begin{cases} 4x + 3y = 27 \\ 2x - y = 1 \end{cases}$

46. $\begin{cases} 7x = 5 - 2y \\ 3y + 2x = 16 \end{cases}$

PROJECT *Connection*

In this activity, your group will build a model of a roller coaster hill.

1. Assemble the following materials: 15 strands of uncooked spaghetti, masking tape, a plastic drinking straw, and a metric ruler.

2. Tape the straw to the edge of a table. The straw represents the top of the hill. Insert a strand of spaghetti into the straw, but do not tape it. As a group, decide how to tape additional spaghetti strands together so you can build a "hill" that stretches out as far as possible and does *not* touch the floor. Break strands into smaller pieces if you wish.

3. When you complete your model, take measurements A–W in centimeters as shown. Note that the distance A must be divided into 10 equal parts for measurements C–L. Save your data for the next Project Connection.

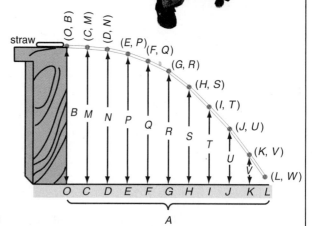

10.2 Synthetic Division and the Remainder and Factor Theorems

Explore

Polynomials in one variable are usually written in descending order of the exponents of the variable. Another useful way to write a polynomial is in *nested form*.

$$2x^3 - 5x^2 + x - 7$$
$$(2x^2 - 5x + 1)x - 7 \quad \text{Factor out an } x.$$
$$((2x - 5)x + 1)x - 7 \quad \text{Factor out } x \text{ again.}$$
$$(((2)x - 5)x + 1)x - 7 \quad \text{Factor out } x \text{ again.}$$

To write a polynomial in nested form, factor out the variable until the innermost nest contains only the leading coefficient.

Write each polynomial in nested form.

1. $5x^3 + 2x^2 - 3x + 6$ **2.** $3x^4 - 7x^3 + 2x^2 + 5x - 9$

The nested form is convenient for evaluating a polynomial.

3. If $F(x) = (((4)x - 3)x + 1)x - 2$, evaluate $F(3)$ mentally. Remember to follow the order of operations.

4. Using the nested form, determine $F(3)$ using a calculator. Remember to press ENTER or $=$ when a right-hand parenthesis appears in the nested form.

5. Write the expanded form of the nested polynomial in Question 3 in descending order. Then determine $F(3)$.

You can use the nested form to divide a polynomial by a divisor in the form $x - r$. Evaluate the nested form at r to find the quotient.

If $F(x) = 3x^3 - x^2 - 2x + 5$, divide $F(x)$ by $(x - 2)$.

$$F(x) = (((3)x - 1)x - 2)x + 5 \quad \text{Nested form}$$
$$F(2) = (((3)2 - 1)2 - 2)2 + 5 \quad \text{Substitute 2 for } x.$$
$$= ((5)2 - 2)2 + 5 \quad \text{Evaluate the first nest.}$$
$$= (8)2 + 5 \quad \text{Evaluate the next nest.}$$
$$= 21 \quad \text{Do the last evaluation.}$$

The blue numbers are the coefficients of the quotient and remainder.
$$(3x^3 - x^2 - 2x + 5) \div (x - 2) = 3x^2 + 5x + 8 + \frac{21}{x - 2}$$

6. Use the long division algorithm to check the above results.

Use the nested form to divide.

7. $3x^3 + 2x^2 - 4x - 8$ by $x - 3$

8. $5x^4 - 7x^3 - 2x^2 + 1$ by $x - 1$

Build Understanding

Another algebraic method of division that bypasses the repetition of the long division algorithm, and that is closely related to nested form, is called **synthetic division.** Here is how synthetic division works, using $3x^3 - x^2 - 2x + 5$ divided by $x - 2$.

The binomial divisor is of the form $x - r$. Write 2 at the left. Then write the coefficients of the terms of the dividend, with the terms in descending order.

$$\underline{2}\ \ \begin{array}{cccc} 3 & -1 & -2 & 5 \end{array}$$

Bring down the first coefficient. Multiply the divisor by this number. Write the product under the second coefficient and add.

$$\underline{2}\ \ \begin{array}{cccc} 3 & -1 & -2 & 5 \\ & 6 & & \\ \hline 3 & 5 & & \end{array}$$

Repeat multiplication and addition until there are no more coefficients.

$$\underline{2}\ \ \begin{array}{cccc} 3 & -1 & -2 & 5 \\ & 6 & 10 & 16 \\ \hline 3 & 5 & 8 & 21 \end{array}$$

The numbers in the bottom row are the coefficients of the quotient and remainder.

$$(3x^3 - x^2 - 2x + 5) \div (x - 2) = 3x^2 + 5x + 8 + \frac{21}{x - 2}$$

Synthetic division only works when the divisor is of the form $x - r$ and the synthetic divisor is r. The terms in the dividend should be in descending order. When you write the coefficients, use zeros for any missing terms.

CHECK UNDERSTANDING

Compare the synthetic division shown against the long division you did for Explore Question 6. Why is it clear which x's and exponents belong to which coefficients? Why can synthetic division use addition instead of subtraction?

EXAMPLE 1

Use synthetic division to divide.

a. $\dfrac{2x^3 + x^2 - 4x + 10}{x + 3}$

b. $\dfrac{x^4 - 3x^2 - 7}{x - 2}$

Solution

a. For the divisor $x + 3$, the synthetic divisor is -3 since $x + 3 = x - (-3)$.

$$\underline{-3}\ \ \begin{array}{cccc} 2 & 1 & -4 & 10 \\ & -6 & 15 & -33 \\ \hline 2 & -5 & 11 & -23 \end{array}$$

So, $\dfrac{2x^3 + x^2 - 4x + 10}{x + 3} = 2x^2 - 5x + 11 + \dfrac{-23}{x + 3}$

b. Use 0 for the missing coefficients.

$$\underline{2}\ \ \begin{array}{ccccc} 1 & 0 & -3 & 0 & -7 \\ & 2 & 4 & 2 & 4 \\ \hline 1 & 2 & 1 & 2 & -3 \end{array}$$

So, $\dfrac{x^4 - 3x^2 - 7}{x - 2} = x^3 + 2x^2 + x + 2 + \dfrac{-3}{x - 2}$ ◄

When the coefficient of the x-term of the divisor is other than 1, you can still use synthetic division. However, you must first write the divisor in a form equivalent to $x - r$.

EXAMPLE 2

NUMBER THEORY The product of three consecutive odd integers is represented by $8x^3 + 36x^2 + 46x + 15$. If the representation for one of the integers is $2x + 5$, find representations for the other two integers.

Solution

$$\frac{8x^3 + 36x^2 + 46x + 15}{2x + 5} \qquad \text{To find the other representations, divide.}$$

$$= \frac{1}{2}\left(\frac{8x^3 + 36x^2 + 46x + 15}{x + \frac{5}{2}}\right) \qquad \begin{array}{l}\text{Rewrite the division to get} \\ \text{the divisor in the form of} \\ x - r.\end{array}$$

$$\begin{array}{r|rrrr} -\dfrac{5}{2} & 8 & 36 & 46 & 15 \\ & & -20 & -40 & -15 \\ \hline & 8 & 16 & 6 & 0 \end{array} \qquad \begin{array}{l}\text{Use synthetic division with} \\ \text{the divisor } -\dfrac{5}{2}. \\ \text{The remainder is 0.}\end{array}$$

Substitute the result of the synthetic division.

$$\frac{8x^3 + 36x^2 + 46x + 15}{2x + 5} = \frac{1}{2}(8x^2 + 16x + 6)$$

$$= 4x^2 + 8x + 3$$

The quadratic quotient is factorable.

$$4x^2 + 8x + 3 = (2x + 3)(2x + 1)$$

The other two integers are represented by $2x + 3$ and $2x + 1$.
Check by multiplying $(2x + 5)(2x + 3)(2x + 1)$. ◀

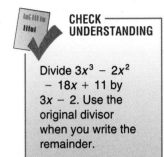

CHECK UNDERSTANDING

Divide $3x^3 - 2x^2 - 18x + 11$ by $3x - 2$. Use the original divisor when you write the remainder.

Synthetic division determines the quotient and the remainder. For example, on page 480 when the polynomial $F(x) = 3x^3 - x^2 - 2x + 5$ is divided by $x - 2$, the quotient is $3x^2 + 5x + 8$ and the remainder is 21.

Consider now the value of $F(x)$ at $x = 2$.

$$F(x) = 3x^3 - x^2 - 2x + 5$$
$$F(2) = 3(2)^3 - (2)^2 - 2(2) + 5 = 21$$

This connection between the remainder and $F(2)$ is expressed in the remainder theorem.

> **REMAINDER THEOREM**
>
> If a polynomial $F(x)$ is divided by $x - r$, then the remainder is equal to $F(r)$.

A consequence of the remainder theorem is that synthetic division can be used to evaluate a polynomial for a given value of the variable.

EXAMPLE 3

Use the remainder theorem to determine the value of $F(x) = 3x^3 - 4x^2 - 5x - 12$ at $x = 3$.

Solution

Use synthetic division to determine the value of the remainder. The synthetic divisor is 3.

$$
\begin{array}{r|rrrr}
3 & 3 & -4 & -5 & -12 \\
  &   & 9 & 15 & 30 \\
\hline
  & 3 & 5 & 10 & 18
\end{array}
$$
←Remainder.

So, by the remainder theorem, $F(3) = 18$.
Check this result by substituting $x = 3$ into $F(x)$.   ◀

A remainder of zero means that the divisor is a factor of the dividend. The following theorem is an important corollary of the remainder theorem.

FACTOR THEOREM

A polynomial $F(x)$ has a factor $x - r$ if and only if $F(r) = 0$.

COMMUNICATING ABOUT ALGEBRA

Explain how synthetic division could be used to solve part b of Example 4.

EXAMPLE 4

Use the factor theorem to determine whether the given binomial is a factor of $F(x)$.

  a. $x + 4$; $F(x) = x^3 + 6x^2 + 5x - 12$

  b. $x - i$; $F(x) = x^3 + 3x^2 + x - 3$

Solution

Since the binomials are of the form $x - r$, find $F(r)$ to determine the remainder. You can use synthetic division as shown in part a below or you can substitute into $F(x)$ as shown in part b below.

  a. $F(x) = x^3 + 6x^2 + 5x - 12$
  For a divisor of $x + 4$, the synthetic divisor is -4.

$$
\begin{array}{r|rrrr}
-4 & 1 & 6 & 5 & -12 \\
   &   & -4 & -8 & 12 \\
\hline
   & 1 & 2 & -3 & 0
\end{array}
$$
←Remainder

  Since the remainder is 0, $x + 4$ is a factor of $F(x)$.

  b. $F(x) = x^3 + 3x^2 + x - 3$
  $F(i) = i^3 + 3i^2 + i - 3$
  $ = -i - 3 + i - 3$
  $ = -6$

  Since $F(i)$ is not 0, $x - i$ is not a factor of $F(x)$.   ◀

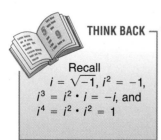

THINK BACK

Recall
$i = \sqrt{-1}$, $i^2 = -1$,
$i^3 = i^2 \cdot i = -i$, and
$i^4 = i^2 \cdot i^2 = 1$

1. **WRITING MATHEMATICS** Explain the advantages and disadvantages of synthetic division.

Use synthetic division to divide each polynomial by the given binomial. Check by multiplying.

2. $3x^3 + 2x^2 - 6x - 26$; $x - 2$

3. $2y^3 - y^2 + y - 3$; $y + 1$

4. $z^4 - 2z^3 + 5z + 13$; $z + 2$

5. $2a + a^3 - 2a^4 + 1$; $a - 1$

6. $2m^3 + 3m^2 + 4m + 5$; $m - \dfrac{1}{2}$

7. $6x^4 - 10x^3 - 4x^2 + 3x + 3$; $x + \dfrac{1}{3}$

8. $6t^3 - 13t^2 - 12t + 4$; $2t + 1$

9. $3c^4 + 7c^3 - 10c + 5$; $3c - 2$

10. **GEOMETRY** The volume of a rectangular prism is represented by $(2x^3 - 3x^2 - 11x + 6)$ cm³. If the representation for the length is $(x - 3)$ cm, find the representations for the other dimensions.

11. **MOTION** If Jorge drove $(6x^3 - 11x^2 - 46x - 24)$ mi at an average speed of $(2x + 3)$ mi/h, express the travel time in terms of x.

Use synthetic division to evaluate $F(x) = 6x^3 + 19x^2 + 8x - 5$ at the given value of x.

12. $x = 2$ 13. $x = -3$ 14. $x = \dfrac{1}{3}$ 15. $x = -\dfrac{5}{2}$

Determine whether the given binomial is a factor of each polynomial.

16. $x - 4$; $x^3 + 3x^2 - 18x + 38$

17. $y - 3$; $y^3 - 2y^2 + 2y - 15$

18. $z + 2$; $z^4 - 3z^2 - 4$

19. $3x - 2$; $3x^3 - 2x^2 - 18x + 12$

20. $x + i$; $2x^3 + 3x^2 + 2x + 3$

21. $x - \sqrt{3}$; $x^3 - \sqrt{3}x^2 - 2x + 2\sqrt{3}$

PRACTICE

Use synthetic division to divide each polynomial by the given binomial. Check by multiplying.

1. $x^3 + 3x^2 - 8x - 12$; $x - 3$

2. $a^3 + 6a^2 + 2a + 3$; $a + 1$

3. $2y^2 + 3y + y^3 - 1$; $y - 1$

4. $b^4 - 8b - 7b^2 + 2b^3 + 13$; $b - 2$

5. $3z^3 - 2z + 2$; $z + 2$

6. $2c^4 - 7c^2 - 10$; $c - 2$

7. $3w^3 + 5w^2 + 4w + 1$; $3w - 1$

8. $6v^4 - 3v^3 + 2v^2 + 3v - 3$; $2v - 1$

9. $4t^3 + 4t^2 + 5t - 3$; $2t + 1$

10. $4a^4 + 9a^2 + 21a + 9$; $2a + 3$

11. **NUMBER THEORY** The product of four numbers is represented by $6x^4 + 11x^3 - 3x^2 - 2x$. One of the numbers is represented by x and another by $2x - 1$. Find, in terms of x, the representations for the other two numbers.

12. GEOMETRY The volume of a right circular cylinder is represented by $\pi(3x^3 - 8x^2 - 4x + 16)$ cm^3. The radius is represented by $(x - 2)$ cm. Find, in terms of x, the representation for the height.

Evaluate each polynomial function $F(x)$ for the given value of x.

13. $F(x) = x^3 + 4x^2 - 3x + 7$ for $x = -3$ **14.** $F(x) = \frac{1}{2}x^4 - 3x^3 + 6x - 10$ for $x = -4$

Use synthetic division to evaluate the polynomial function
$F(x) = 6x^4 + 35x^3 + 13x^2 - 110x + 56$ **at each given value of x.**

15. $x = 2$ **16.** $x = -2$ **17.** $x = \frac{2}{3}$ **18.** $x = -\frac{7}{2}$

19. WRITING MATHEMATICS Explain why using synthetic division to evaluate a polynomial function at a particular value of the variable can be easier than substituting into the function. Give an example of a situation in which synthetic division is not the easier method for evaluation of a polynomial.

20. BUSINESS A manufacturer's total cost function in dollars for production of its new sailboat is $C(x) = 400 + 5x + x^3$. Use synthetic division to determine the total cost of producing 300 sailboats.

Determine whether the given binomial is a factor of each polynomial.

21. $x + 1;\ x^3 + 3x^2 + 3x + 1$

22. $x - 2;\ 2x^3 - x^2 - 4x + 5$

23. $x + 2;\ x^4 + 16$ **24.** $x + 3;\ x^5 + 243$

25. $3x + 1;\ 3x^3 - 2x^2 - 19x - 6$ **26.** $4x - 1;\ 4x^3 - x^2 - 4x - 1$

27. $x + \sqrt{5};\ x^3 + x^2 - 5x - 5$ **28.** $x - i;\ x^3 - 3x^3 + x - 3$

EXTEND

29. NUMBER THEORY The product of four numbers is represented by $6x^4 + 19x^3 - 23x^2 - 10x + 8$. The product of two of the numbers is represented by $6x^2 + x - 2$. Find the representation, in terms of x, for

    a. the product of the other two numbers   **b.** each of the four numbers

Using x as the variable, express each synthetic division in the form
$$\frac{\text{dividend}}{\text{divisor}} = \text{quotient} + \frac{\text{remainder}}{\text{divisor}}$$

30.

$$
\begin{array}{r|rrrrr}
-2 & 1 & -1 & -6 & 0 & 1 \\
   &   & -2 & 6 & 0 & 0 \\
\hline
   & 1 & -3 & 0 & 0 & 1 \\
\end{array}
$$

31.

$$
\begin{array}{r|rrrr}
\frac{1}{2} & 4 & 2 & 0 & -5 \\
            &   & 2 & 2 & 1 \\
\hline
            & 4 & 4 & 2 & -4 \\
\end{array}
$$

Determine the value of k so that the binomial is a factor of the polynomial.

32. $(x + 1)$; $2x^4 - 5x^2 + 2x + k$

33. $(x - 2)$; $2x^3 + 3x^2 - kx + 6$

34. $(x - 3)$; $kx^3 - 6x^2 + 24x - 36$

THINK CRITICALLY

35. If $F(x) = 2x^3 - 3x^2 + 3x - t$, determine the value of t so that $F(2) = 6$.

36. If $F(x) = 3x^3 - 2x^2 + mx + n$, determine values for m and n so that $F(1) = 2$ and $F(-2) = -37$.

37. Find the relation between a and b so that $x - 3$ is a factor of $2x^4 - 7x^3 + ax + b$.

38. WRITING MATHEMATICS Is $(x - 1)$ a factor of $1776x^{1776} - 1492x^{2001} - 284$? Explain your method.

39. Find the remainder when $F(x) = (x + 4)^2 + (x + 3)^3 + (x + 2)^4$ is divided by $(x + 1)$.

MIXED REVIEW

Evaluate each expression when $x = 2$.

40. $3x^2 \cdot 4x^3$ **41.** $(4x^2)^3$ **42.** $3x^0$ **43.** $8x^{-1}$

44. STANDARDIZED TESTS Which of the following are solutions to the equation $x^2 - 2x + 3 = 0$?

    A. $1 \pm \sqrt{2}$ **B.** $1 \pm i\sqrt{2}$ **C.** $-1 \pm \sqrt{2}$ **D.** $-1 \pm i\sqrt{2}$

Solve each equation. Check.

45. $2^{x+1} = 16$ **46.** $4^{3x} = 8^{x+1}$ **47.** $7^{x^2+4x} = 7^{-3}$

Write an equivalent expression in terms of x and y.

48. $\log \dfrac{x^2}{y}$ **49.** $\log \sqrt{mn}$ if $\log m = x$ and $\log n = y$

PROJECT *Connection*   In this activity, you will graph and analyze the data you collected from your roller coaster hill model.

1. Graph the following pairs on a sheet of centimeter graph paper. Connect the points to form a smooth curve.

    (O, B) (C, M) (D, N) (E, P) (F, Q) (G, R)
    (H, S) (I, T) (J, U) (K, V) (L, W)

2. Suppose your group wishes to use this curve in the design of a roller coaster. Decide on a scale that should be assigned to the graph. For example, you could let 1 centimeter represent 1 meter. Use your scale to determine the actual measurements that each ordered pair represents. What would B represent?

Graphs of Polynomial Functions

Think Back

Every polynomial expression can be written as a corresponding polynomial function and as a polynomial equation.

1. Consider the polynomial expression
   $x^3 - 4x^2 + 8x - 8$.
   a. Write a corresponding function.
   b. Write a corresponding equation.

Every polynomial function $y = F(x)$ has values of x for which the corresponding value of y is 0. These values of x are the **zeros** of the function and the **solutions** of the equation $F(x) = 0$.

2. **a.** Determine the solutions of the equation $x^2 + x - 6 = 0$.
   b. What are the zeros of the function $y = x^2 + x - 6$?

Explore/Working Together

Work with a partner and use a graphing utility. Recall that the real zeros of a function are the x-intercepts of its graph.

Graph each of the following functions. Find the zeros of the function. Describe the zeros as real or imaginary and equal or unequal.

3. $y = x^2 + 3x - 4$

4. $y = -x^2 + 10x - 21$

5. $y = (x - 5)^2$

6. $y = -x^2 - 6x - 9$

7. $y = x^2 - 2x + 5$

8. $y = -x^2 - 2x - 5$

As you know, the graph of the quadratic function $y = ax^2 + bx + c$ (where $a \neq 0$) is called a parabola. Its characteristic shape can be described as a "hill" or as a "valley."

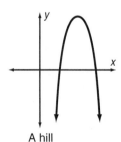

A hill
The graph rises, turns, and falls.

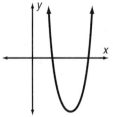

A valley
The graph falls, turns, and rises.

Study the graphs of Questions 3–8 to answer Questions 9–13.

9. Describe the nature of the leading coefficient a when the graph of $y = ax^2 + bx + c$ is
   a. a hill   **b.** a valley

10. Make a statement about the y-value
   a. at the top of a hill   **b.** at the bottom of a valley

11. If the two zeros of a quadratic function are real and unequal, how many x-intercepts does the graph have?

12. If the two zeros of a quadratic function are real and equal, describe fully how the graph relates to the x-axis.

13. If the two zeros of a quadratic function are imaginary, describe how the graph relates to the x-axis.

Graph and find the zeros of each of the following cubic functions.

14. $y = x^3 - x^2 - 16x + 16$ 15. $y = -x^3 + 3x^2 + 13x - 25$

16. In terms of "hill" and "valley," describe the characteristic shape of the graph of the cubic functions above.

17. Compare the characteristic shape of a cubic function whose leading coefficient a is positive to one whose leading coefficient is negative.

18. Graph: $y = x^3 + x^2 - 5x + 3$ What is the significance of the point of tangency? How many real zeros does this function have?

19. Graph: $y = x^3 + 4x^2 - 2x + 10$ How many real zeros does this function have? Explain.

20. Consider $y = (x - 5)^3$. Make a prediction about its zeros.

21. Graph $y = (x - 5)^3$. Relate the graph of this special case to other general observations you have made about cubic functions.

22. Predict how hills and valleys will combine to give the characteristic shape of a quartic (degree 4) polynomial function.

23. Verify your prediction by graphing the following functions.
   a. $y = x^4 - 3x^3 - 15x^2 + 19x + 30$
   b. $y = -x^4 + 6x^2 - 5$

24. State, by inspecting the given function, what the zeros are. Tell how you think the zeros of multiplicity 2 or more will show up on the graph. Graph the function. Describe the actual graph.
   a. $y = (x - 3)^2(x + 2)(x + 4)$
   b. $y = (x - 3)^3(x + 2)$
   c. $y = (x - 3)^4$

PROBLEM SOLVING TIP

If the zero of a function is not an integer, you can approximate the value of the zero from the graph using different features of a calculator.

COMMUNICATING ABOUT ALGEBRA

In what order should you describe a graph? Why?

25. Make a statement about how the number of turns in the graph of a polynomial function is related to the degree of the polynomial.

26. Consider the monomial functions $y = x^2$, $y = x^3$, $y = x^4$, and $y = x^5$. Make a statement about the zeros of these functions.

27. Graph $y_1 = x^2$ and $y_2 = x^4$ on one set of axes. Describe the similarities and differences in these graphs. Which points do the graphs have in common? Describe the behavior at the origin.

28. Repeat Question 27 using the graphs $y_1 = x^3$ and $y_2 = x^5$.

29. Predict the graphs of $y = x^6$ and $y = x^7$. Verify your predictions.

30. Make a statement about the shape of the graph of $y = x^n$, depending upon the nature of n.

> **THINK BACK**
>
> The degree of a polynomial in one variable is the greatest exponent of the variable.

31. Graph $y_1 = x^2$, $y_2 = (x - 5)^2$, and $y_3 = (x + 5)^2$ on one set of axes. Describe how the graphs of y_2 and y_3 are related to that of y_1.

32. Write a rule for translating the graph of $y = x^2$ in the horizontal direction by a distance of a units.

33. Predict how your rule would apply to the graph of $y = x^3$. Verify your prediction by graphing, on the same set of axes, $y_1 = x^3$, $y_2 = (x - 5)^3$, and $y_3 = (x + 5)^3$.

34. How do you think you could translate the graph of $y = x^2$ up 5 units? down 5 units? Verify your predictions.

Make Connections

35. If 0, 4, and -2 are zeros of a polynomial function, write factors of the function. Using the factored form, graph the function. On the same set of axes, graph $2 \cdot F(x)$. Make a statement about the zeros of $F(x)$ and the zeros of $2 \cdot F(x)$. Can this statement apply to other functions? Explain.

36. Write a polynomial function with these characteristics: odd degree, positive leading coefficient, zeros -2, 0, 3. Express your answer in expanded form. Graph the function. Make a sketch.

37. Using the function from Question 36, select an x-value less than -2, an x-value between -2 and 0, an x-value between 0 and 3, and an x-value greater than 3. Use synthetic division to determine the remainder at these x-values. Make a statement about these remainders.

38. Study the graph in Question 36. How does the pattern of the remainders you observed in Question 37 relate to the graph?

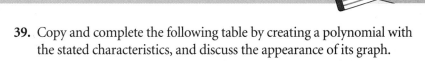

39. Copy and complete the following table by creating a polynomial with the stated characteristics, and discuss the appearance of its graph.

	Degree	Leading Coefficient	Function	Graph
	2 (even)	positive	$y = (x + 3)(x - 2)$	valley, graph points up at left and right
a.	2 (even)	negative		
	3 (odd)	positive	$y = x(x - 4)(x + 1)$	hill followed by valley, graph points down at left and up at right
b.	3 (odd)	negative		
c.	4 (even)	positive		
d.	4 (even)	negative		
e.	5 (odd)	positive		
f.	5 (odd)	negative		
g.	6 (even)	positive		
h.	6 (even)	negative		

Summarize

● **40.** WRITING MATHEMATICS Describe the relationship between a polynomial, a related function, a related equation, and the zeros.

41. GRAPHING Describe the effect that a change in sign of the leading coefficient of a monomial function, $y = ax^n$, has on the graph.

42. THINKING CRITICALLY Write three different polynomial functions that have -2, 3, and 5 as zeros.

43. CHEMISTRY The weight of a cube of ice is a function of the length of its side given by $W = 0.033s^3$ (W in pounds, s in inches).
 a. Graph the weight function.
 b. Use your graph to estimate the weight of a cubic foot of ice.

44. BUSINESS A banana exporting company has determined that its profit function is given by $P(x) = -\frac{1}{4}x^4 + \frac{1}{3}x^3 + 3x^2 - 8$ where $P(x)$ represents thousands of dollars and x represents tons of bananas.
 a. How must you restrict x for the problem to make sense?
 b. Graph the profit function. Estimate at what production level(s) the company's profit is $0.
 c. About how many tons of bananas should the company export to maximize its profit? What is that profit?

45. GOING FURTHER On the same set of axes, graph $y = x$ and $y = x^3$. Explain how to use these graphs to sketch the graph of $y = x - x^3$.

Career
Demographer

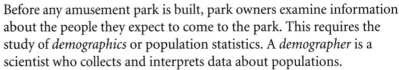

Before any amusement park is built, park owners examine information about the people they expect to come to the park. This requires the study of *demographics* or population statistics. A *demographer* is a scientist who collects and interprets data about populations.

The following table lists the actual and predicted percent of the U.S. population between ages 18 and 24 years.

Year	1960	1970	1980	1990	2000	2010	2020	2030
% of Population	8.9	12.1	13.3	10.4	9.4	9.6	8.5	8.4

Decision Making

A demographer might model the relationship by using an approximating polynomial. For this problem, let 1960 be Year 0. Then 1970 is Year 10, 1980 is Year 20, and so on.

1. List the values in the chart as ordered pairs (year, % of population).

2. Suppose the demographer uses the approximating polynomial
   $y = -0.0015x^2 + 0.068x + 10.39$.
   a. Graph the function. Tell how the demographer might use the graph to estimate the percent of the U.S. population that is 18 to 24 years of age in the year 1999.
   b. Estimate the percent to the nearest hundredth of a percent.

   c. Given the values in the table, is this a good estimate? Explain.

3. Suppose the demographer uses the cubic function
   $y = 0.00011x^3 - 0.013x^2 + 0.37x + 9.24$. Use your graphing utility to estimate, to the nearest hundredth of a percent, the percent of the U.S. population that will be
   a. 18 to 24 in the year 2012. Is this a good estimate? Explain.
   b. 18 and 24 in the year 2030. Is this a good estimate?

4. Of the following, which do you think might make the best model?
   a. the linear function $y = -0.04x + 11.475$
   b. the quadratic function $y = -0.0015x^2 + 0.068x + 10.39$
   c. the cubic function
      $y = 0.00011x^3 - 0.013x^2 + 0.37x + 9.24$
   d. the quartic function
      $y = -0.0000035x^4 + 0.0006x^3 - 0.034x^2 + 0.659x + 8.82$

10.4 Polynomial Equations

Explore/Working Together

● Work with a partner. Use a graphing utility.

1. Graph $y = x^3 - 7x^2 - x + 7$. Determine the zeros of the function. What is the relation between the zeros of the function and the solutions of the equation $x^3 - 7x^2 - x + 7 = 0$?

2. Repeat Question 1 for each of the following functions.
   a. $y = x^3 + 5x^2 - x - 5$ **b.** $y = x^3 + 5x^2 - 12x - 36$

3. Copy and complete the table.

Equation	Constant term	Factors of the constant term	Solutions
$x^3 - 7x^2 - x + 7 = 0$			
$x^3 + 5x^2 - x - 5 = 0$			
$x^3 + 5x^2 - 12x - 36 = 0$			

4. Make a conjecture about the relation between the solutions of a polynomial equation and the constant term of the equation.

SPOTLIGHT ON LEARNING

WHAT? In this lesson you will learn
• the fundamental theorem of algebra, the rational zero theorem, the location theorem, and other theorems about polynomial functions.

WHY? Solving polynomial equations of degree three or greater can help you solve problems in geometry, physics, engineering and business.

Build Understanding

● The results of Explore suggest that when the leading coefficient of a polynomial equation is 1, integral solutions of the equation are factors of the constant term. You can test the factors of the constant term by synthetic division. When the remainder is zero, you have found a solution and synthetic division shows you the factor that remains. Use this factor to find the remaining solutions.

EXAMPLE 1

Solve: $x^3 - 2x^2 - 5x + 6 = 0$

Solution

Possible integral solutions are the factors of 6: $\pm 1, \pm 2, \pm 3, \pm 6$.

Try $x = 1$.

$$
\begin{array}{r|rrrr}
1 & 1 & -2 & -5 & 6 \\
  &   & 1 & -1 & -6 \\
\hline
  & 1 & -1 & -6 & 0
\end{array}
$$

Since the remainder is 0, $x^2 - x - 6$ is a factor of the polynomial.

$$x^2 - x - 6 = (x - 3)(x + 2)$$

The three solutions of the equation are 1, 3 and -2. Check your answer by substituting $x = 1, 3, -2$ into the original equation.   ◄

PROBLEM SOLVING TIP

Some calculators will prepare a table of values for you. The table feature can then be used to display a table of x, y values. You can find the zeros in Example 2 by entering the function in your graphing utility and using the table feature to find the x-values where $y_1 = 0$.

What if not all the solutions are integers? Will there be solutions that are real numbers or complex numbers? What if there aren't any solutions? The German mathematician Gauss answered these questions by proving the following theorem.

FUNDAMENTAL THEOREM OF ALGEBRA

Every polynomial equation of degree 1 or more has at least one solution in the set of complex numbers.

The consequences of this theorem are:

- Every polynomial equation of degree n has exactly n solutions (where a solution of multiplicity k is counted as k solutions).
- The solutions of a polynomial equation are in the set of complex numbers (or any of its subsets).

The next result extends your observation that zeros of a function are factors of the constant term when the leading coefficient is 1.

THE RATIONAL ZERO THEOREM

If a polynomial function with integral coefficients has a rational zero $\frac{p}{q}$ (in simplest form), then p is a factor of the constant term and q is a factor of the leading coefficient.

EXAMPLE 2

Determine the rational zeros of $F(x) = 2x^3 - 7x^2 - 5x + 4$.

Solution

By the rational zero theorem, if $F(x)$ has rational zeros, then those zeros have numerators that are factors of 4 and denominators that are factors of 2. You can see all the possibilities in a tree diagram. When there are several possibilities to test by synthetic division, it is convenient to write the coefficients of the polynomial once and to perform the multiplication and addition steps mentally. You can also use a calculator to evaluate $F(x)$ or reduce the number of possible zeros to evaluate.

p	q	$\frac{p}{q}$
± 1	± 1	$\pm\frac{1}{1} = \pm 1$
	± 2	$\pm\frac{1}{2} = \pm\frac{1}{2}$
± 2	± 1	$\pm\frac{2}{1} = \pm 2$
	± 2	$\pm\frac{2}{2} = \pm 1$
± 4	± 1	$\pm\frac{4}{1} = \pm 4$
	± 2	$\pm\frac{4}{2} = \pm 2$

Test the possible zeros: $\pm\frac{1}{2}$, ± 1, ± 2, and ± 4.

x	2	−7	−5	4	← coefficients
−4	2	−15	55	−116	
−2	2	−11	17	−30	
−1	2	−9	4	0	−1 is a zero
$-\frac{1}{2}$	2	−8	−1	$4\frac{1}{2}$	
$\frac{1}{2}$	2	−6	−8	0	$\frac{1}{2}$ is a zero
1	2	−5	−10	−6	
2	2	−3	−11	−18	
4	2	1	−1	0	4 is a zero

↖ possible rational zeros

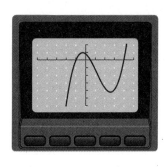

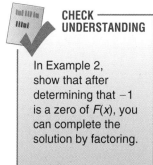

CHECK UNDERSTANDING

In Example 2, show that after determining that −1 is a zero of $F(x)$, you can complete the solution by factoring.

So, the rational zeros of $F(x)$ are −1, 4, and $\frac{1}{2}$. ◄

After you have located one rational solution of a cubic equation, synthetic division shows you the quadratic factor that remains. Solving the related quadratic equation, called the **depressed equation** (one degree less than the original), may require the quadratic formula. From the quadratic formula, you can see that, irrational and complex solutions occur in conjugate pairs.

CONJUGATE ZERO THEOREM

If a polynomial function with real coefficients has a complex zero $a + bi$, where a and b are real numbers, then the conjugate $a - bi$ is also a zero. Also, if $a + \sqrt{b}$ is a zero, then $a - \sqrt{b}$ is a zero.

EXAMPLE 3

Solve: $x^3 - 6x^2 + 6x + 8 = 0$

Solution

Since complex and irrational solutions occur in pairs, and the degree of this equation is odd, there must be at least one rational solution. Since the leading coefficient is 1, the rational solution must be a factor of 8: $\pm 1, \pm 2, \pm 4, \pm 8$. The rational solution is 4. So, the depressed equation is $x^2 - 2x - 2 = 0$.

$$\begin{array}{r|rrrr} 4 & 1 & -6 & 6 & 8 \\ & & 4 & -8 & -8 \\ \hline & 1 & -2 & -2 & 0 \end{array}$$

$$x = \frac{-b \pm \sqrt{b^2 - 4ac}}{2a} \quad \text{Quadratic formula.}$$

$$x = \frac{2 \pm \sqrt{4 - 4(1)(-2)}}{2(1)} \quad \text{Substitute } a = 1, b = -2, c = -2$$

$$x = 1 \pm \sqrt{3}$$

The solutions are $x = 4$, $x = 1 + \sqrt{3}$, and $x = 1 - \sqrt{3}$.
Check by substituting into the original equation. ◄

CHECK UNDERSTANDING

In Example 3, show how you use the sum and product of the solutions to check that $1 \pm \sqrt{3}$ are the solutions to the quadratic equation $x^2 - 2x - 2 = 0$.

The following theorem can be used to locate irrational solutions and nonintegral rational solutions of polynomial equations.

> **THE LOCATION THEOREM**
>
> If $F(x)$ is a polynomial function with terms in descending degrees of the variable, such that $F(a)$ and $F(b)$ have opposite signs, then there is at least one real zero between a and b.

Graphically, if the points $(a, F(a))$ and $(b, F(b))$ are on opposite sides of the x-axis, then there is at least one point between a and b where the graph crosses the axis. In the graph shown, there are three zeros.

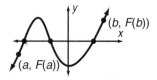

To limit the number of possible solutions, you can apply the following.

> **DESCARTES' RULE OF SIGNS**
>
> The number of positive real zeros of a polynomial $F(x)$ is equal to the number of sign changes of the coefficients of $F(x)$, or is less than this number by an even number. The number of negative real zeros is equal to the number of sign changes of the coefficients of $F(-x)$, or is less than this number by an even number.

PROBLEM SOLVING TIP

In applying Descartes' rule of signs, you ignore any missing terms, those with coefficients of 0.

COMMUNICATING ABOUT ALGEBRA

Using the results of Descartes' rule of signs in Example 4, give all the possibilities for the solutions.

EXAMPLE 4

Determine the possible number of positive and negative real zeros of $F(x) = x^4 + x^3 - 2x^2 + 3x - 4$. Then, locate, between successive integers, the real zeros for the function.

Solution

Examine the signs of the coefficients of $F(x)$.

There are 3 changes of sign.

By Descartes' rule, $F(x)$ can have 3 or 1 positive real zeros.

Substitute $-x$ to determine $F(-x)$.

$$F(-x) = (-x)^4 + (-x)^3 - 2(-x)^2 + 3(-x) - 4$$
$$= x^4 - x^3 - 2x^2 - 3x - 4$$

Examine the signs of the coefficients $F(-x)$:

There is 1 change of sign.

By Descartes' rule, $F(x)$ can have only 1 negative real zero.

Use synthetic division or a calculator to test values.

x	1	1	−2	3	−4	
−4	1	−3	10	−37	144	
−3	1	−2	4	−9	23	⎤ sign changes
−2	1	−1	0	3	−10	⎦
−1	1	0	−2	5	−9	
0	1	1	−2	3	−4	
1	1	2	0	3	−1	⎤ sign changes
2	1	3	4	11	18	⎦

$F(-3)$ is positive and $F(-2)$ is negative; $F(1)$ is negative and $F(2)$ is positive. So, by the location theorem, there must be at least one real zero for $F(x)$ between -3 and -2, and another real zero between 1 and 2. ◄

Descartes' rule reveals that a polynomial whose terms are all positive or all negative can have no real positive zeros. Using this fact, the next theorem helps you limit the possibilities you must try.

> ── **UPPER AND LOWER BOUND THEOREM** ──
>
> **For a positive number p, if $F(x)$ is divided by $(x − p)$ and the remainder and the coefficients of the quotient are all positive or all negative, then the zeros of $F(x)$ cannot be greater than p. The value p is an *upper bound* of $F(x)$. If q is an upper bound for $F(-x)$, then $-q$ is a *lower bound* for $F(x)$.**

PROBLEM SOLVING TIP

In Example 4, you could use a graphing utility to find upper and lower bounds.

EXAMPLE 5

Determine integral upper and lower bounds for the zeros of the polynomial function $F(x) = x^4 + x^3 + 70x^2 - 2x - 144$.

Solution

Use synthetic division to test positive integers for x. Since all the numbers are positive, 2 is an upper bound.

x	1	1	70	−2	−144	
1	1	2	72	70	−74	
2	1	3	76	150	156	← all positive

$$F(-x) = (-x^4) + (-x^3) + 70(-x^2) - 2(-x) - 144$$

As you can see from the synthetic division, 2 is also an upper bound for $F(-x)$.

x	1	−1	70	2	−144	
1	1	0	70	72	−72	
2	1	1	72	146	148	← all positive

Then, by the theorem, -2 is a lower bound for $F(x)$. So, the upper bound for the function is 2 and its lower bound is -2. It is unnecessary to try divisors outside this range. ◄

CHECK UNDERSTANDING

In Example 5, what are the only possible rational zeros of $F(x)$? Justify your answer.

Modeling insect populations can lead to an equation that is solved algebraically by combining results of this lesson.

EXAMPLE 6

ENTOMOLOGY An insect population P is represented by a part of the graph defined by $P(t) = t^3 + 5t^2 + 7t + 3$ where $t \geq 0$ in days. Is there a day on which the population will be three times what it is on day 3?

Solution

First, find $P(3)$. $P(3) = 3^3 + 5 \cdot 3^2 + 7 \cdot 3 + 3 = 96$

On day 3 the insect population is 96. To determine the day the population will be three times this number, you want to solve the equation.

$$3(96) = t^3 + 5t^2 + 7t + 3$$
$$0 = t^3 + 5t^2 + 7t - 285$$

The possible rational zeros are: $\pm 1, \pm 3, \pm 5, \pm 15, \pm 19, \pm 57, \pm 95, \pm 285$.

Descartes' rule indicates there is one positive zero. Try 5 as a possible zero.

$$F(x) = + + + -$$

So the insect population on day 5 is three times the population on day 3.

```
5| 1   5    7   -285
         5   50    285
   1  10   57      0
```

TRY THESE

1. Determine all the integral solutions of $x^3 - x^2 - 9x + 9 = 0$.

2. **WRITING MATHEMATICS** Explain which of these values, $\frac{3}{5}, \frac{5}{3}, -\frac{1}{3}, -5$ is *not* a possible solution of the polynomial equation $3x^3 + 17x^2 - 5x + 5 = 0$.

3. Determine all the rational solutions of $3x^3 - 2x^2 - 7x - 2 = 0$.

Solve and check each equation.

4. $x^3 - 3x^2 - 2x + 4 = 0$ 5. $x^3 - 2x^2 + 3x - 2 = 0$ 6. $x^4 + 2x^3 - 2x^2 + 8 = 0$

7. Locate, between successive integers, a real zero for $p(x) = 3x^4 + x^3 + 2x^2 + 2x - 6$.

Determine the possible number of positive and negative real solutions of each equation.

8. $x^3 - x^2 + 2x - 2 = 0$ 9. $x^5 - x^3 + 1 = 0$

10. Determine integral upper and lower bounds for the zeros of $p(x) = x^4 - 6x + 1$.

11. **GEOMETRY** After a slice 1 cm thick is cut off from one side of a cube, the volume that remains is 48 cm³. Determine the length of the original side of the cube.

12. **GEOMETRY** A box with an open top is constructed from a square piece of cardboard that is 5 cm on each side by cutting $s \times s$ squares from each corner and folding up the sides. If the volume of the open box is 8 cm³, determine the two possible values for s.

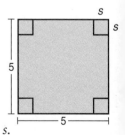

PRACTICE

Determine all the integer solutions of each equation.

1. $x^3 - 5x^2 - x + 5 = 0$

2. $x^4 - 3x^3 - 8x^2 + 12x + 16 = 0$

Determine all the rational solutions of each equation. Check.

3. $x^4 - 11x^2 - 18x - 8 = 0$

4. $5x^3 - 11x^2 - 13x + 3 = 0$

5. **WRITING MATHEMATICS** If $-3, 1 + \sqrt{2}$, and $2 - 3i$ are solutions of a polynomial equation, what is the lowest possible degree of the equation? Justify your answer.

Solve each equation. Check.

6. $x^3 + x^2 - 11x - 15 = 0$ 7. $x^3 + 4x^2 + 8x + 8 = 0$ 8. $x^3 - 3x^2 + 4x + 8 = 0$

Locate, between successive integers, any real zero for each polynomial.

9. $p(x) = 4x^3 + 4x^2 - 13x + 5$ 10. $p(x) = 4x^3 + 6x^2 - 23x - 15 = 0$

Determine the possible number of positive and negative real zeros of each polynomial.

11. $p(x) = x^3 - 4x^2 + x + 6$ 12. $p(x) = 2x^4 - 2x^3 + x^2 - 3$

Determine integral upper and lower bounds for the zeros of each polynomial.

13. $p(x) = x^3 - 2x^2 + 3x + 5$ 14. $p(x) = 2x^3 - 3x^2 - 3x + 3$

15. **BUSINESS** A manufacturer of packing cartons is interested in increasing the volume of some of his cartons. One of the cartons, a rectangular box, has dimensions 3 in. × 4 in. × 5 in. Can he increase each dimension by the same amount and produce a box whose volume would be double that of the original?

16. **MECHANICS** A 10-meter horizontal beam is built into a wall at one end and rests on a support at the other. Since weight is uniformly distributed along the length of the beam, it sags downward. A function that models the situation is: $y = -x^4 + 26x^3 - 160x^2$ where x is the distance (in meters) from the wall to a point on the beam and y is the amount of sag (in hundredths of a millimeter). What is an appropriate domain for x? Find the zeros. Interpret their meaning. Graph the function. Describe how the zeros show up on the graph. What is the maximum deflection of the beam? At what value of x?

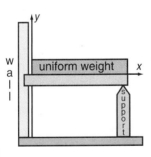

EXTEND

17. **GEOMETRY** If the volume of the box shown is 50 cm³, approximate the value of x to two decimal places.

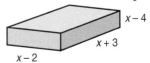

18. **GEOMETRY** If the volume of the cylinder shown is 100 cm³, approximate the value of x to two decimal places.

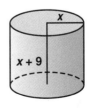

THINK CRITICALLY

19. If the solutions of a quadratic equation are r_1 and r_2, use the factor theorem to write the equation.

20. Expand the equation you wrote in Exercise 19 and simplify so you can describe the relationship between the solutions and the coefficients of a quadratic equation.

21. Use the relationship you described in Exercise 20 to write a quadratic equation with rational coefficients for each set of solutions.

    a. $-2, 3$ **b.** $1 \pm \sqrt{5}$ **c.** $1 \pm 2i$

22. If the three solutions of a cubic equation are r_1, r_2, r_3, use the factor theorem to write the equation.

23. Expand the equation you wrote in Exercise 22 and simplify so you can describe the relationship between the solutions and the coefficients of a cubic equation.

24. Use the relationship you described in Exercise 23 to write a cubic equation with rational coefficients for each set of solutions.

    a. $-2, 4, 7$ **b.** $3, 1 \pm \sqrt{3}$ **c.** $2, \pm i$

MIXED REVIEW

Determine each if $f(x) = x^2$ and $g(x) = x + 1$.

25. $(f \circ g)(x)$ **26.** $(f \circ g)(3)$ **27.** $(g \circ f)(x)$ **28.** $(g \circ f)(3)$

29. STANDARDIZED TESTS The solutions of the equation $x^2 + 3x + 1 = 0$ are

    A. real, rational, unequal   **B.** real, rational, equal   **C.** real and irrational   **D.** imaginary

---

PROJECT *Connection*

In this activity, you will find a third-degree polynomial of the form $F(x) = ax^3 + bx^2 + cx + d$ to describe the hill you constructed.

1. Use these four ordered pairs that you recorded earlier: (O, B), (E, P), (I, T), (L, W). Evaluate $F(x)$ for $x = O, E, I$, and L.

$$F(O) = aO^3 + bO^2 + cO + d = B$$
$$F(E) = aE^3 + bE^2 + cE + d = P$$
$$F(I) = aI^3 + bI^2 + cI + d = T$$
$$F(L) = aL^3 + bL^2 + cL + d = W$$

2. Since O equals zero, you can see that $d = B$. Also W equals zero. Verify that the remaining three equations can be written as shown.

$$aE^3 + bE^2 + cE = P - B$$
$$aI^3 + bI^2 + cI = T - B$$
$$aL^3 + bL^2 + cL = -B$$

3. Let X be the coefficients matrix, Z the constants matrix, and Y the variables matrix. Then $XY = Z$. Solve this system for Y by multiplying both sides of the matrix equation by the inverse of X on the left. Use the values you find for a, b, and c, along with d, to write the approximating polynomial. Use a graphing utility to graph your polynomial.

$$X = \begin{bmatrix} E^3 & E^2 & E \\ I^3 & I^2 & I \\ L^3 & L^2 & L \end{bmatrix} \quad Y = \begin{bmatrix} a \\ b \\ c \end{bmatrix} \quad Z = \begin{bmatrix} P - B \\ T - B \\ -B \end{bmatrix}$$

An **amusement park engineer** uses physical laws of motion in planning a safe ride. One such law is expressed as a polynomial equation in which distance traveled is related to speed, duration of the ride, and the acceleration of the vehicle.

A team of engineers is planning a water ride to consist of a coaster-type car that floats along a channel of rushing water. At several points during the ride, the car is pulled up an incline, poised at the top of a hill, and then sent gliding down the hill, resulting in a huge splash at the bottom.

The engineers want to study the effect that the length of the hill has on the velocity of the car when it hits the bottom. They want to keep the car's acceleration constant at 5 m/s^2. The flow of water is adjusted so that the car slides down the hill and reaches the bottom with a big splash.

To represent distance, the engineers use quadratic function $d = 0.5at^2$ in this investigation. They determine the velocity (speed) of the car using the formula, $v_f = v_i + at$ where v_f is the final velocity when the car reaches the bottom of the hill, v_i is the initial velocity, a is the acceleration of the car, and t is the time it takes to get from the top of the hill to the bottom.

Decision Making

1. If the acceleration of the car is 5 m/s^2, determine the missing variable in the chart below.

	Hill Distance (d)	Acceleration (a)	Time (t)
a.	40 m	5 m/sec^2	
b.	45 m	5 m/sec^2	
c.	50 m	5 m/sec^2	
d.		5 m/sec^2	6 sec
e.		5 m/sec^2	7 sec

2. What is the initial velocity if the car pauses at the top of the hill?

3. Use the values found in Question 1 to determine the final velocity v_f at the bottom of the hill in each case.

4. Why might an engineer be interested in the final velocity of the car?

10.5 Problem Solving File

The Method of Finite Differences

A set of data in two variables can sometimes be modeled by a polynomial function. The **method of finite differences** enables you to determine if a polynomial function is appropriate. This method can be used only when the x-values of the paired data increase in *constant* increments.

Problem

An engineer is studying the relationship between time and speed in the design of a new roller coaster. She measured the speed of the roller coaster over a six-second period with the following results. Can a polynomial function be used to model the data?

Time, seconds	1	2	3	4	5	6
Speed, m/s	3	5	8	12	17	23

Explore the Problem

1. Let time be represented by x and speed in m/s by y. Do the x-values of the paired data increase in constant increments?

2. To construct a *difference table*, begin by listing the y-values in order. The 1st order differences between the y-values are shown. Continue by writing the 2nd order and 3rd order differences.

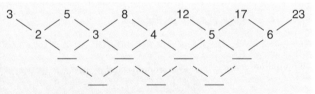

y – value	3		5		8		12		17		23
1st order diff.		2		3		4		5		6	
2nd order diff.											
3rd order diff.											

3. You interpret the results of a difference table by this rule.

   For any set of x-values that increases in constant increments, if the $(n + 1)$th order differences are equal to zero, then a polynomial of degree n can be found to relate the x-values and y-values of the data set.

   According to this rule, can a polynomial function be used to model this problem? If so, of what degree and what general form?

4. Construct a difference table for the following data set. If a polynomial function can be used to describe the relationship between the paired data, give the degree of the function.

   $(2, 1)$, $(3, 10)$, $(4, 33)$, $(5, 76)$, $(6, 145)$, $(7, 246)$

Investigate Further

- Once you have determined, from the results of a difference table, that a polynomial function can be used to model a set of paired data, you can write the function. To find a quadratic function that will model a particular sequence, where the x-values are consecutive integers, consider first the general difference in a general sequence.

PROBLEM SOLVING PLAN

- Understand
- Plan
- Solve
- Examine

x – value	1	2	3	4
$ax^2 + bx + c$	$a(1)^2 + b(1) + c$	$a(2)^2 + b(2) + c$	$a(3)^2 + b(3) + c$	$a(4)^2 + b(4) + c$
	$a + b + c$	$4a + 2b + c$	$9a + 3b + c$	$16a + 4b + c$
1st order diff.		$3a + b$	$5a + b$	$7a + b$
2nd order diff.			$2a$	$2a$

You can write a particular quadratic function if you:

- Use the 2nd order difference $2a$ to determine the value of a.
- Use the value of a and the 1st order difference $3a + b$ to find the value of b.
- Use the values of a and b, and the first y-value to find the value of c.

Apply this method to the data set for the speed of the roller coaster. Compare the 2nd order differences in Question 2 to the general 2nd order differences in the diagram above.

5. Use the 2nd order difference to determine the value of a.

6. Use the first 1st order difference and the value of a to determine b.

7. Use the first y-value and the values of a and b to determine c.

8. Use the values of a , b, and c to write the descriptive quadratic function. Check by substituting a value for x into the equation.

Apply the Strategy

- **Zoology** A laboratory assistant measured the growth of a pig fetus during the gestation period and recorded the data shown.

Time, weeks	3	6	9	12
Size, mm	41	74	107	140

9. If the elements of the data set are represented by (x, y), is the method of finite differences applicable to this data set? Explain.

10. Construct a difference table for the y-values of the data set.

11. Explain why a polynomial function can be used to model this data set. Tell the degree of the polynomial and the general form.

12. Explain how you would use the data to find the descriptive polynomial.

13. Determine the descriptive polynomial.

14. Use the descriptive polynomial to predict the week in which the fetus will first measure over 200 mm.

15. **WRITING MATHEMATICS** Prepare a set of guidelines showing the general form of the polynomial function that can be used to describe the relationship between paired data if
    a. the 1st order differences are 0   **b.** the 2nd order differences are 0
    c. the 3rd order differences are 0   **d.** the 4th order differences are 0

MOTION During the first 5 seconds after a model rocket was launched from a height of 5 ft above ground, a group of researchers obtained the following data.

Time, s	1	2	3	4	5
Height, ft	169	301	401	469	505

16. Construct a difference table.

17. Determine the descriptive polynomial.

18. Use the descriptive polynomial to find the height of the rocket after
    a. 7 seconds   **b.** 10 seconds

19. About when does the rocket hit the ground?

MANUFACTURING A company can produce 25,000 radios per week. Experience has shown that it is not profitable to work at maximum capacity. An efficiency expert has gathered the following data about profit and the number of radios manufactured weekly.

Number, thousands	1	2	3	4	5
Profit, hundreds of dollars	8	57	100	137	168

20. Construct a difference table and find the descriptive polynomial.

21. Use a graphing utility to graph the function.

22. Use the graph to determine the number of radios the plant should manufacture weekly to realize maximum profit.

23. What is the maximum profit?

REVIEW PROBLEM SOLVING STRATEGIES

SOLID FUN

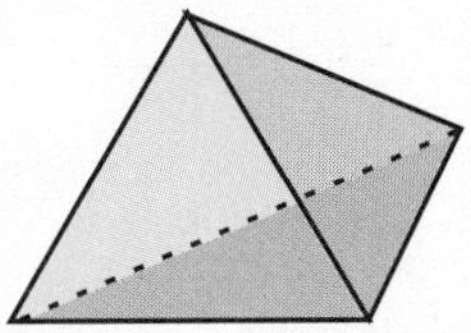

1. The simplest regular polyhedron is the regular *tetrahedron*—a solid with four equilateral triangles as faces. Three triangles must meet at each vertex of the tetrahedron. On thin cardboard, make two copies of the pattern shown below and cut them out. Fold each one along the solid lines and tape the edges so that you have two identical solid figures. The challenge is to put the two solids together to form a tetrahedron. Can you do it? What reasoning can you use to help you?

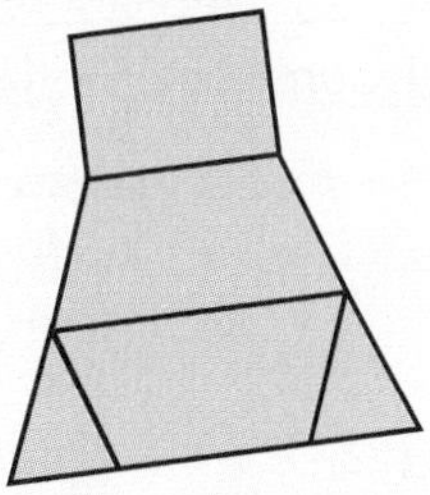

TICK-TOCK

2. The *big bang theory* states that about 20 billion years ago, the universe began forming out of a tremendous explosion. Suppose you wanted to represent a passage of this amount of time as a 24-hour day.

   a. To how many years of actual time does 1 hour correspond?

   b. To how many years of actual time does 1 minute correspond?

   c. Scientists estimate that Earth formed about 4.6 billion years ago. To what time of the 24-hour day does that correspond?

   d. The Cretaceous Period, during which dinosaurs became extinct, ended about 65 million years ago. To approximately what time of the 24-hour day does the end of this period correspond?

   e. Fossil discoveries in Ethiopia have revealed a human ancestor believed to be 4.4 million years old. To what approximate time of the 24-hour day does this correspond?

HOW CANOE DO IT?

3. A 200-lb man and his two sons, each weighing 100 lbs, want to cross a river. If they have only one canoe which can safely carry only 200 lbs, how can they cross the river?

· · · CHAPTER REVIEW · · ·

VOCABULARY

Choose the word from the list that completes each statement.

1. An expression of the form $a_n x^n + a_{n-1}x^{n-1} + \cdots + a_0$ where n is a nonnegative integer and $a_0, \ldots, a_n$ are real numbers is a __?__.

2. In the previous expression, a_n is the __?__.

3. The values of x for which the corresponding values of y are 0 are the __?__ of a function.

4. __?__ bypasses the repetition of long division by using only coefficients.

5. The __?__ states that every polynomial equation of degree 1 or more has a solution among the complex numbers.

a. leading coefficient

b. Fundamental Theorem of Algebra

c. synthetic division

d. polynomial

e. real zeros

Lesson 10.1 DIVISION OF POLYNOMIALS   pages 433–478

- You can use the Long Division Algorithm to divide a polynomial by another polynomial.

- Before dividing, arrange the terms of the dividend and the divisor in descending order. If the dividend or divisor has missing terms, insert these terms with coefficients of zero.

Divide. Check by multiplying.

6. $(6x^3 + x^2 + 7x + 10) \div (3x + 2)$ 7. $(36 + x^2) \div (x + 6)$

Lesson 10.2 SYNTHETIC DIVISION; REMAINDER AND FACTOR THEOREMS   pages 479–485

- Since the synthetic divisor is the opposite of the constant in the actual divisor, use addition instead of subtraction to carry out the process.

- When a function $F(x)$ is divided by a binomial $(x - r)$, the remainder $= F(r)$. The divisor $(x - r)$ is a factor of $F(x)$ if and only if $F(r) = 0$.

Use synthetic division to divide each polynomial by the given binomial.

8. $(3x^3 - 5x^2 - 7x - 20) \div (x - 4)$ 9. $(x^3 - 8) \div (x + 2)$

Determine whether the given binomial is a factor of the polynomial.

10. $x + 2; 4x^3 + 12x^2 + 10x - 15$ 11. $x - 3; 6x^3 - 17x^2 - 2x - 3$

Lesson 10.3 POLYNOMIAL GRAPHS   pages 486–490

- The real zeros of a function are the x-intercepts of its graph. A point of tangency indicates a zero of multiplicity 2 or more.

- Graphs of polynomial functions have characteristic shapes composed of "hills" and "valleys", which are related to the sign of the leading coefficient.

Use a graphing utility to display each graph. Determine the real zeros of each function.

12. $y = x^3 - 3x - 2$

13. $y = x^4 - x^3 - 6x^2$

Predict the shape of the graph of each function using the words "hill" and "valley". Then use a graphing utility to display the graph.

14. $y = x^2 - 3$

15. $y = -x^2 + 1$

16. $y = x^3 - 3x - 2$

17. $y = x^4 + 2x^3 - 8x^2$

Lesson 10.4 POLYNOMIAL EQUATIONS
pages 491–499

- If a polynomial function with integral coefficients has a rational zero $\frac{p}{q}$, then p is a factor of the constant term and q is a factor of the leading coefficient.

- Complex and irrational zeros of a polynomial with real coefficients occur in conjugate pairs.

- **Descartes' Rule of Signs** limits the number of positive and negative real zeros.

- For $p > 0$, if $F(x)$ is divided by $(x - p)$ and the remainder and the coefficients of the quotient are all positive or all negative, then p is an upper bound of the zeros of $F(x)$. If q is an upper bound for $F(-x)$, then $-q$ is a lower bound for $F(x)$.

Determine the rational zeros of each function.

18. $F(x) = x^3 - x^2 - 17x - 15$

19. $F(x) = 2x^3 + x^2 - 5x + 2$

Solve each equation.

20. $x^3 - x^2 + 2 = 0$

21. $x^3 - 8x - 8 = 0$

Determine the possible number of positive and negative real zeros.

22. $F(x) = 3x^5 + 2x^4 - 3x^2 - 3x + 1$

23. $F(x) = 2x^3 - 4x^2 + x - 7$

Find integral upper and lower bounds for the zeros.

24. $F(x) = 3x^3 - 5x^2 + 7x + 4$

25. $F(x) = x^3 - 4x^2 - 5x + 14$

Lesson 10.5 THE METHOD OF FINITE DIFFERENCES
pages 500–503

- You can use the **method of finite differences** to determine whether the relationship between paired data can be modeled by a polynomial equation.

Explain whether a difference table can be constructed. If so, tell the degree of the descriptive polynomial.

26.

x	1	2	3	4	5
y	2	9	22	41	66

27.

x	1	3	5	6	7
y	−1	2	9	20	35

Find the descriptive polynomial function for each data set.

28.

x	1	2	3	4	5
y	2	5	10	17	26

29.

x	1	2	3	4	5
y	0	−1	−4	−9	−16

Chapter Review **505**

CHAPTER TEST

Use the long division algorithm to divide.

1. $(10x^3 - 23x^2 + 5x + 5) \div (5x + 1)$

2. $(x^3 + y^3) \div (x + y)$

Use synthetic division to divide.

3. $(x^3 - 4x^2 + 9x - 12) \div (x - 2)$

4. $(4x^3 + 4x^2 - x - 7) \div (2x + 1)$

Determine whether the given binomial is a factor of the polynomial.

5. $x + 3; x^4 + 81$

6. $3m - 2; 9m^3 - 27m^2 - 4m + 12$

Evaluate $F(x)$ for the given value of x.

7. $F(x) = -x^4 + 2x^3 - 7x + 4$ for $x = -2$

8. $F(x) = -\frac{1}{3}x^3 - 3x^2 - 6x + 18$ for $x = 3$

Determine the zeros of the function.

9. $y = x^2 - 4x - 5$ 10. $y = x^2 + 2x + 1$

11. **STANDARDIZED TESTS** Describe the graph of $y = (x - 3)^2$ compared to the graph of $y = x^2$.

    A. 3 units left   **B.** 3 units up
    C. 3 units right   **D.** 3 units down

Use a graphing utility to display each graph. Determine the real zeros of the function.

12. $y = x^3 + x^2 - 16x - 16$

13. $y = x^3 - 9x^2 + 15x + 25$

14. $y = (x^2 + 1)(x - 3)^2$

Determine the rational zeros of each function.

15. $F(x) = x^3 - 4x^2 - 7x + 10$

16. $F(x) = 2x^4 - x^3 - 18x^2 + 9x$

Solve and check.

17. $x^3 - 7x^2 + 8x + 16 = 0$

18. $x^3 - x^2 - 7x + 15 = 0$

19. $x^3 + 3x^2 - 12x - 10 = 0$

20. **WRITING MATHEMATICS** Does the graph of a polynomial function of degree n with real coefficients always intersect the x-axis in n different points? Explain.

21. Locate a real zero between successive integers: $5x^3 - 13x^2 + 31x - 15 = 0$

Determine the possible kinds of solutions.

22. $2x^4 + x^3 - x^2 + x - 7 = 0$

23. $x^5 - 3x^4 + 5x^3 + x^2 - 4 = 0$

24. **STANDARDIZED TESTS** The upper and lower bounds for the zeros of $F(x) = -x^4 + 6x - 1$ are, respectively,

    A. $2, 0$ **B.** $0, 2$ **C.** $-2, 0$ **D.** $0, -2$

25. By what single value must all the dimensions of the box shown be increased so that the volume of the new box will be twice the volume of the original?

6"   10"

8"

Given the data set at the right.

x	1	2	3	4	5	6
y	−1	2	9	20	35	54

26. Construct a difference table and write a polynomial function that describes the data.

27. Use the descriptive polynomial to generate the next three pairs of data.

PERFORMANCE ASSESSMENT

USE ALGEBLOCKS Work with a partner to model division of a trinomial by a binomial. Follow the example below, which models
$$(6x^2 + 7x + 2) \div (2x + 1) = 3x + 2$$

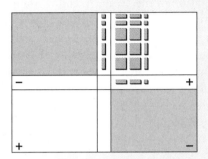

Have your partner give you a division problem to model. He/she can determine suitable trinomials by finding the product of two binomials. Then switch roles with your partner. Continue to switch roles until you have modeled four such divisions.

PROJECT A POLYNOMIAL Use an overhead projector. Draw the graph of a fifth degree polynomial with five distinct real zeros on a transparency. Superimpose your graph on a set of coordinate axes. Highlight the five zeros. Then translate your graph to show how a fifth degree polynomial can have only four real zeros. Repeat for three, two, and one real zero. Discuss the imaginary zeros each polynomial has.

USE FLOWCHARTS You can use the flowchart below to describe the synthetic division that corresponds to $(2x^2 + 3x + 4) \div (x + 1)$. The result yields the quotient and the remainder
$$2x + 1 + \frac{3}{x + 1}.$$

```
   ( Input 2 )          -1| 2   3   4
        |                       -2
        v                    _____
 [ Multiply by -1 ]           2
        |
        v                 -1| 2   3   4
   [ Add 3 ]                     -2  -1
        |                     _____
        v                      2   1
 [ Multiply by -1 ]
        |                  -1| 2   3   4
        v                        -2  -1
   [ Add 4 ]                  _____
        |                      2   1   3
        v
   ( Output
     2 1 3 )
```

Write a flowchart to show how to check this division. Include the corresponding algebraic display. Construct both types of flowcharts for a third, fourth, and fifth degree polynomial.

ASSUME ANOTHER'S IDENTITY Use a graphing utility to display clearly, on one set of axes, the graphs of any cubic polynomial and any quartic polynomial. Now form a new function that is the sum of your original two, and display its graph on the same set of axes. Describe your result and explain it.

PROJECT ASSESSMENT

PROJECT Connection   Plan a class meeting where each group will present project results and then there will be some general discussion.

1. Each group should decide on how the information researched or obtained from amusement parks can be organized and displayed. Explain to your classmates conclusions you have drawn from the data.

2. Each group should display its work on the spaghetti roller coaster model. After all groups have presented, compare polynomial models and attempt to account for differences in the results.

3. As a class, discuss roller coasters and the amusement park industry in general. Why do people enjoy rides that scare them? How do you think roller coaster rides will evolve in the future? What are your suggestions for theme parks?

• • • CUMULATIVE REVIEW • • •

1. Divide $(4 - 6i)$ by $(1 - i)$, then express the result in $a + bi$ form.

2. Determine the possible kinds of solutions of $x^5 - x^2 + 10 = 0$.

3. Farmer Perkins' rectangular wheat field is currently 200 m by 300 m. He plans to increase the area by increasing both dimensions by the same amount, but does not wish to devote more than 3000 m^2 to wheat. By how much can each dimension of the existing field be increased?

4. The directed graph below shows how managerial trainees can be transferred among four departments of a bank. Construct a matrix to represent the ways in which a trainee can be transferred between two departments without working in another department.

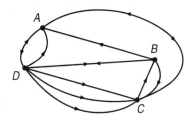

5. **STANDARDIZED TESTS** Which of the statements that follow are true about the data set shown in the table?

x	1	2	3	4	5
y	0	5	12	21	35

   I. Since the x-values increase in constant increments, a difference table can be constructed.
   II. Since the 3rd order differences are 0, the descriptive polynomial is cubic.
   III. The descriptive polynomial is $y = x^2 - 2x - 3$.

   A. I and II   **B.** I and III
   C. I only   **D.** II only

6. Having neared extinction, the prairie dog population of the Western United States began a comeback in the last quarter of the 20th century. In 1970 there were just 12 prairie dogs in Sheep Gulch and in 1980 there were 90. Assuming continued exponential growth, $P = P_0 e^{kt}$ where P_0 and k are constants, what is the expected prairie dog population for Sheep Gulch in the year 2010? Round your answer to the nearest ten.

7. Determine a system of inequalities whose intersection is the quadrilateral and its interior region shown below.

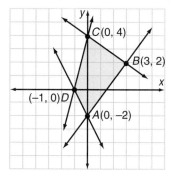

8. Use the long division algorithm to divide $(9x^2 + 2x^3 - 10) \div (x + 2)$.

9. **STANDARDIZED TESTS** The average of log 2 and log 18 is

   A. log 10   **B.** log 9
   C. log 36   **D.** log 6

10. Use synthetic division to evaluate the polynomial $F(x) = x^4 - 3x^2 - 10$ at $x = -\dfrac{3}{2}$.

11. Solve and check: $x^2 - 2x + 5 = 0$

12. **WRITING MATHEMATICS** Describe how the possible zeros of the quadratic function $y = ax^2 + bx + c$ relate to its graph.

Solve each equation for the indicated variable.

13. $A = \dfrac{1}{2}h(a + b)$ for a 14. $mv = Ft + mv_0$ for m

··· STANDARDIZED TEST ···

STUDENT PRODUCED ANSWERS Solve each question and on the answer grid write your answer at the top and fill in the ovals.

Notes: Mixed numbers such as $1\frac{1}{2}$ must be gridded as 1.5 or 3/2. Grid only one answer per question. If your answer is a decimal, enter the most accurate value the grid will accommodate.

1. The graph of the curve $xy = 12$ is symmetric about the origin. What is the x-value of the point on the curve that is symmetric to $(3, 4)$?

2. If $3 + (x - 2)i = 3\ 7i$, what is the value of x?

3. Use the descriptive polynomial for the data set shown in the table below to find the value of y that corresponds to $x = 12$.

x	1	2	3	4	5
y	1	12	27	46	69

4. What is the slope of a line that is perpendicular to the line $2y = 3x - 5$?

5. The half-life of radioactive carbon-14 is 5700 years. Use the formula $A(n) = A_0(0.5)^n$, where A_0 represents the original amount of carbon-14 in a substance and $A(n)$ represents the amount remaining after n 5700-year periods, to determine the percent of carbon-14 that would be in a substance after 3000 years.

6. Use the long division algorithm to divide $(x^2 + 2x - 30)$ by $(x + 7)$, and grid the remainder.

7. Ralph cuts a 1 mm-thick sheet of cardboard in half and stacks the pieces one on the other. Then he cuts the stack in half and stacks the pieces again. After how many cuts will the stack be more than 3 cm thick?

8. Solve for x: $\sqrt{2x + 3} - 5 = 1$.

9. Use synthetic division to divide $(x^3 + 4x^2 - 3x - 12)$ by $(x + 3)$ and grid the remainder.

10. Solve for x: $x^2 - 4x \le 21$. Grid the product of the smallest and largest values in the solution.

11. Turret Inc. manufactures two types of engines for mopeds, Model H and Model C. Production, shipping, sales, and labor considerations impose the following considerations per day. No more than 7 of Model H and no more than 11 of Model C can be manufactured. No more than a total of 12 engines can be manufactured. The number of Model H can be no more than twice the number of Model C. The plant must use more than 1000 person-hours of labor. (It takes 100 person-hours to make each Model C and 200 person-hours to make each Model H.) If Turret makes a profit of \$300 per Model C and \$200 per Model H, determine the greatest feasible daily profit.

12. Find the integral upper bound for the zeros of the polynomial function:
$F(x) = x^4 + 12x^2 - 10x - 60$

13. The average of three positive numbers x, y, and z is 15. If $y = 10$, what is the value of $x + z$?

14. The Mr. and Mrs. club of the local Y is going to hold its Spring Spree in the Y's Community Room, for which the rental fee is \$500. Additional charges per couple include \$40 for dinner and \$10 for souvenirs and miscellaneous expenses. What is the minimum number of couples that must attend in order to keep the cost per couple to no more than \$70?

15. **STANDARDIZED TESTS** A circular pool with a radius of 5 feet is in a circular garden with a radius of 20 feet. Choose the letter giving the probability that a leaf which lands within the boundaries of the garden will land in the pool.

    A. $\frac{1}{4}$ **B.** $\frac{\pi}{4}$ **C.** $\frac{1}{8}$ **D.** $\frac{1}{16}$

11 Rational Expressions and Equations

Take a Look
AHEAD

Make notes about things that look new.
- Describe the graphs in this chapter. Identify at least two ways that they are different than graphs of polynomial functions.
- Find and copy the definition of a rational function and its domain.
- Make a list of specialized vocabulary relating to photography.

Make notes about things that look familiar.
- What type of variation have you learned about before? What type of variation will you learn about in this chapter?
- What different operations can you perform with fractions? Will these operations be used with rational expressions?
- What is a useful method for approximating solutions to an equation?

DATA Activity

Importing and Exporting

International trade allows countries to specialize in producing the items they are best suited for with the resources they have. Imports and exports are significant factors in a country's *balance of payments*. If a country pays out more to other nations than it receives from them, that country will have a deficit in its balance of payments. Photographic equipment and optical goods is a major SITC (Standard International Trade Classification) commodity group.

SKILL FOCUS

- Add, subtract, multiply, and divide real numbers.
- Use number lines.
- Solve percent problems.
- Create visual displays of data.

Photography

In this chapter you will see how:

- **LENS DESIGNERS** can determine the amount of light that will enter a camera.
(Lesson 11.3, page 530)

- **PHOTOGRAPHERS** adjust several related variables to obtain optimal conditions for a shot.
(Lesson 11.5, page 542)

	Canada	Mexico	United Kingdom	Germany	France	Italy	Japan	South Korea	Other Countries
U.S. EXPORTS AND IMPORTS, 1994 **PHOTO APPARATUS, EQUIPMENT, AND OPTICAL GOODS** (in millions of dollars)									
Exports	868	340	401	278	194	73	620	76	1831
Imports	346	172	338	384	206	324	3655	1170	2618

Use the table to answer the following questions.

1. What is the total value of U.S. exports for this commodity group?

2. What is the value of imports? Does the U.S. have a favorable balance of payments for this commodity group?

3. What percent of the total exports went to Canada?

4. What percent of the total imports came from Japan?

5. Determine the difference between U.S. exports and imports for each country in the table. Create a graph that shows whether the U.S. has a positive or negative balance of payments with each country. Explain your graphs. For which country is the balance closest to 0?

6. **WORKING TOGETHER** Research and report on the top purchasers of U.S. exports and suppliers of U.S. imports for last year. Include at least two different types of graphs in your report. Include information on U.S. balance of payments since 1960.

511

Making Enlargements

You have probably noticed that the size of one frame on a roll of film is substantially smaller than the photograph you receive after the film is developed and printed. The size of the image on the film is changed using an enlarger that preserves the original proportions of the picture. Did you ever try to draw a logo of your favorite sports team or copy a picture? Did your drawing look as authentic as the original? Freehand copying often distorts the original proportions. You will explore several ways to enlarge drawings proportionally. The basis of any proportion is two rational expressions.

PROJECT GOAL

To create enlargements and use rational expressions to analyze properties of enlargements.

Getting Started

Divide the class into three large groups. There are three Project Connections and each group will work on one of them.

1. After each group selects an activity, meet as a class to make sure all the activities are covered.

2. Start gathering materials necessary for your group's activity.

3. Collect some simple designs or pictures to enlarge. Also, assemble sets of pictures that show the same image in different sizes. You might try advertisements for consumer products, such as cereals, or different materials from a company with a bold logo.

PROJECT *Connections*

Lesson 11.2, page 523:
Construct a pantograph and use it to copy figures.
Lesson 11.3, page 529:
Use an overhead projector to create enlargements and determine ratios between figure dimensions.
Lesson 11.5, page 541:
Use a photocopier to enlarge and reduce figures; check the accuracy of the photocopier.
Chapter Assessment, page 561:
Prepare a presentation summarizing your group's activities.

Internet Connection

www.swpco.com/
swpco/algebra2.html

Algebra Workshop
Explore Rational Functions

Think Back

1. What is a rational number?

2. Tell whether each of the following numbers is rational. If a number is not rational, tell why it is not.

   a. $\dfrac{4}{5}$ b. $-\dfrac{15}{7}$ c. 9

   d. $\sqrt{3}$ e. $\dfrac{0}{20}$ f. 0.666....

3. For the rational number $\dfrac{m}{n}$, what restriction must be placed on the value of n? Why?

4. Each expression represents a rational number. For each expression, tell the restriction(s) that must be placed on the value of x.

   a. $\dfrac{1}{x + 2}$ b. $\dfrac{11}{3x - 12}$ c. $\dfrac{5}{x^2 - 9}$

SPOTLIGHT
ON LEARNING

WHAT? In this lesson you will learn
• to use domain restrictions to find discontinuities in rational functions.
• to find asymptotes.
• to graph rational functions.

WHY? Rational functions can be used to model real world situations such as the cost of removing toxic chemicals from waste water.

Explore

A **rational function** is a function of the form $y = f(x) = \dfrac{p(x)}{q(x)}$, where $p(x)$ and $q(x)$ are polynomials in x, and $q(x) \neq 0$. The **domain** of a rational function is the set of all values of x for which the denominator $q(x)$ is not equal to zero.

5. Describe similarities between rational numbers and rational functions.

6. Tell whether each of the following functions is rational. If a function is not rational, tell why it is not.

   a. $f(x) = \dfrac{x}{x + 1}$ b. $f(x) = x^3 - 3x^2 + 11x - 14$

   c. $f(x) = \sqrt{\dfrac{x^2}{x - 2}}$ d. $f(x) = -6$

   e. $f(x) = \dfrac{3x + 5}{4x^2 - 5x + 1}$ f. $f(x) = \dfrac{2^x}{3}$

7. For the rational function $f(x) = \dfrac{x^2 - 25}{x - 5}$, what restriction(s) must be placed on the domain of x? Why?

8. Simplify the rational expression $\dfrac{x^2 - 25}{x - 5}$. Then describe what you would expect the graph of the expression to look like.

THINK BACK

A polynomial is an expression whose terms are of the form ax^k where k is a nonnegative integer and a is a real number.

$y = f(x) = \dfrac{p(x)}{q(x)}$

Algebra Workshop

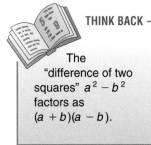

THINK BACK

The "difference of two squares" $a^2 - b^2$ factors as $(a + b)(a - b)$.

Graphs of rational functions may have discontinuities which means you can not draw the graph without lifting your pencil from the paper. The points on the graph where breaks occur are at values of the domain where the function is undefined.

9. Graph the rational function $f(x) = \dfrac{x^2 - 25}{x - 5}$. Describe any differences you discover between the graph of the function and the graph you expected to see. What might account for this? If you do not see any differences at first, use the ZOOM feature.

10. Let $f(x) = \dfrac{x + 3}{x - 2}$. You know that the domain is restricted to $x \neq 2$. Complete the following table to discover how the function behaves as x approaches 2 from the left.

x	0	1	1.5	1.8	1.9	1.99	1.999
$f(x)$							

11. Complete the table to discover how the function $f(x) = \dfrac{x + 3}{x - 2}$ behaves as x approaches 2 from the right.

x	4	3	2.5	2.2	2.1	2.01	2.001
$f(x)$							

12. Describe any trends you see in the range of the function $f(x) = \dfrac{x + 3}{x - 2}$ in the neighborhood of $x = 2$. How would you expect the graph of the function to appear in this region?

13. Evaluate $f(x) = \dfrac{x + 3}{x - 2}$ for several large values of x such as $x = 1000$. Repeat for several large negative values of x such as $x = -1000$. Based on your results, how will the graph look in regions far from the origin, both positive and negative?

14. Use your results from Exercises 12–13 to sketch the graph of $y = \dfrac{x + 3}{x - 2}$. Then graph the function. Was your sketch accurate? If not, give reasons why not.

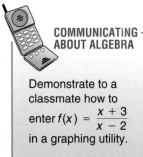

COMMUNICATING ABOUT ALGEBRA

Demonstrate to a classmate how to enter $f(x) = \dfrac{x + 3}{x - 2}$ in a graphing utility.

Discontinuous graphs of rational functions have sections called **branches**. The branches of the graph may get closer and closer to lines called **asymptotes**. An asymptote is not part of a graph but is useful for sketching a graph. A graph may have a vertical asymptote at a point of discontinuity. As $|x|$ increases, values of $f(x)$ may show that the function has a horizontal asymptote.

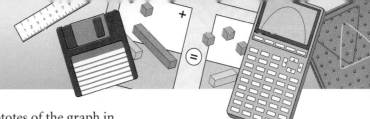

15. Name the horizontal and vertical asymptotes of the graph in Question 14.

Make Connections

● Where would you expect to find a discontinuity in the graph of each function?

16. $y = \dfrac{x^2 - 9}{x - 3}$

17. $y = x^2 - 5x + 6$

18. $y = \dfrac{x^2 + 7x + 10}{x + 2}$

19. $y = \dfrac{(x^2 - 1)(x + 3)}{x^2 + 3x - 4}$

For exercises 20–23 match the function with its graph.

20. $y = \dfrac{x^2 - 1}{x + 1}$

21. $y = x^3 - 2x^2 - x + 2$

22. $y = \dfrac{x - 1}{x - 2}$

23. $y = x - 1$

a.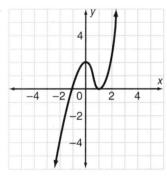

b.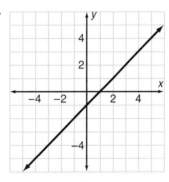

c.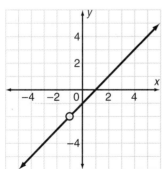

d.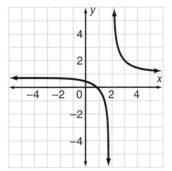

Determine the vertical asymptotes of the graphs of each of the following functions.

24. $y = \dfrac{x + 2}{x - 3}$

25. $y = \dfrac{2x - 1}{x + 5}$

26. $y = \dfrac{-x - 2}{x^2 + 2x - 3}$

Algebra Workshop

Summarize

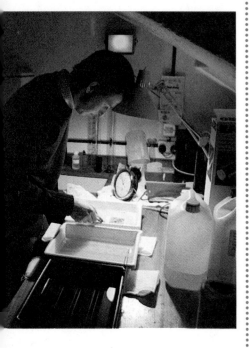

27. WRITING MATHEMATICS Create a rational function that is undefined when $x = 3$ and $x = 4$.

    a. Find the domain of the function. Explain your method.
    b. Find the asymptotes if there are any. Explain your method.
    c. Graph the function.

28. MODELING Star Finish charges $3.45 to develop a roll of film and $0.33 for each photo that it prints. Create a rational function giving the mean cost c per print when p prints are made.

29. MODELING Suppose the cost C in dollars of removing p percent of the toxic chemicals from the waste water of an industrial plant is

$$C(p) = \frac{10{,}000\,p}{101 - p}$$

Is this function continuous for all values of p for which the problem makes sense? Explain.

30. THINKING CRITICALLY Write a rational function which satisfies the requirement.

    a. The graph is discontinuous at $x = -6$.
    b. The graph has $x = 8$ as a vertical asymptote.
    c. The graph has $y = \dfrac{2}{3}$ as a vertical asymptote.

31. GOING FURTHER Let $f(x) = \dfrac{2x + 1}{x - 1}$.

    a. When $|x|$ is very large, which part of the expression $\dfrac{2x + 1}{x - 1}$ has little influence when you evaluate $f(x)$? Why?

    b. Write a simplified approximation of the expression, for very large values of $|x|$.

    c. Identify the horizontal asymptote of the graph. Explain.

    d. When $|x|$ is very large and x is positive, would you expect the graph to approach the horizontal asymptote from above or below? Explain.

    e. When $|x|$ is very large and x is negative, would you expect the graph to approach the horizontal asymptote from above or below? Explain.

    f. Graph $f(x)$. What other asymptote can you identify?

    g. Identify the horizontal asymptote of the graph of $f(x) = \dfrac{5x - 1}{2x + 1}$.

11.2 **Graph Rational Functions**

Explore

Photographers are concerned with the composition of their photos. They do not want the central feature of the photograph to appear too large or too small. A *subject* is located at a distance S in front of a camera lens. The *image* is focused at a distance I behind the lens. The *magnification* M of the subject is given by $M = \frac{I}{S}$. Let the image distance equal 1 in.

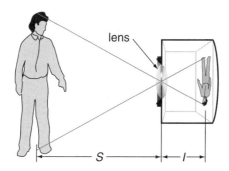

1. Make a table of values for M using a variety of values of S.

2. Graph M as a function of S for values of $S > 0$.

3. As S increases, how does your graph show what happens to M?

4. As S decreases, how does your graph show what happens to M?

5. The above formula does not give the magnification of a subject that touches the lens. Why?

6. The graph of $M = \frac{1}{S}$ approaches two lines, getting closer and closer to them as S increases and as S decreases, but never intersecting them. Identify the lines.

Build Understanding

The domain of a rational function is said to be *restricted* to values of x that produce nonzero denominators.

You can graph a rational function by plotting points, paying special attention to portions of the graph that are near values where the domain is restricted. A function is **continuous** if it passes through every point in the domain. This means its graph has no breaks or undefined range values. All polynomial functions are continuous.

CHECK
UNDERSTANDING

Why does the definition of the domain of a rational function specify that the denominator not be equal to zero?

PROBLEM
SOLVING TIP

A fraction cannot be simplified by dividing out a common factor unless both the numerator and denominator have that common factor.

THINK BACK

An asymptote is a line that a branch of a graph approaches.

A function is **discontinuous** when there is a break in the graph. If the discontinuity consists of only a single point, the function has **point discontinuity**. When using a graphing utility, you may need to use the ZOOM feature to see the break in the graph.

EXAMPLE 1

Graph: $f(x) = \dfrac{x^2 - 9}{x - 3}$

Solution

The domain of the function is restricted to values of x for which the denominator is not equal to zero.

$$x - 3 \neq 0 \text{ if } x \neq 3$$

By factoring the numerator, you can simplify the function.

$$f(x) = \frac{x^2 - 9}{x - 3}$$

$$= \frac{(x + 3)(x - 3)}{x - 3}$$

$$= x + 3$$

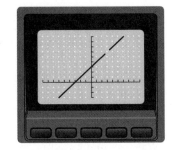

$f(x) = x + 3$ is a linear function. Its graph is a line with a slope of 1 and a y-intercept of 3. Since x cannot equal 3, there is a "break" in the line at $(3, 6)$.

Looking closely at what is happening near a domain restriction may aid you in graphing a rational function.

EXAMPLE 2

Graph: $f(x) = \dfrac{x + 1}{x - 1}$

Solution

The domain is restricted to values of x not equal to 1.

x	0	0.5	0.9	0.99	0.999	1.001	1.01	1.1	1.5	2
$f(x)$	-1	-3	-19	-199	-1999	2001	201	21	5	3

The table shows that as x approaches 1 from the left, $f(x)$ decreases rapidly moving closer and closer to the line $x = 1$. As x approaches 1 from the right, $f(x)$ increases rapidly moving closer and closer to the same line, $x = 1$. The vertical line $x = 1$ is an asymptote.

To complete the graph in the regions $x < 0$ and $x > 2$, consider how $f(x)$ behaves for very large values of $|x|$. If $x = 1000$, for example, $f(1000) = \frac{1001}{999} \approx 1.002$. If $x = -1000$, $f(-1000) \approx 0.998$.

As $|x|$ increases, $f(x)$ gets closer and closer to 1. You can see this by analyzing $\frac{x + 1}{x - 1}$. When $|x|$ becomes very large, the $+1$ in the numerator and the -1 in the denominator become insignificant compared with the value of x, and we can essentially disregard them. The fraction approaches $\frac{x}{x}$, or 1. Therefore, the horizontal line $f(x) = 1$ is also an asymptote of the graph.

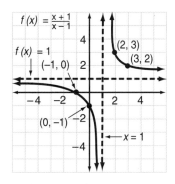

The following rules can help you find the asymptotes of rational functions. For any given rational function, first simplify by dividing out common factors.

HORIZONTAL AND VERTICAL ASYMPTOTES

Let $f(x) = \dfrac{p(x)}{q(x)}$ where $p(x)$ and $q(x)$ have no common factors.

1. The graph has a **vertical asymptote** at each real zero of $q(x)$ if p and q do not have the same real zeros.
2. The graph has a **horizontal asymptote**
   - at $f(x) = 0$ if the degree of $p(x)$ is less than the degree of $q(x)$.
   - at $f(x) = \dfrac{a}{b}$ if the degree of $p(x)$ equals the degree of $q(x)$, where a and b are the leading coefficients of $p(x)$ and $q(x)$, respectively.
3. The graph has **no horizontal asymptote** if the degree of $p(x)$ is greater than the degree of $q(x)$.

THINK BACK

Recall that a *zero* of a function $f(x)$ is a value of x for which $f(x) = 0$. The *degree* of a polynomial is the greatest power of the variable. The *leading coefficient* of a polynomial is the coefficient of the greatest power of the variable.

EXAMPLE 3

ECONOMICS On Wednesday, stock market prices rose so rapidly late in the day that an automatic computer shutdown went into effect, closing trading for the day. On Thursday, prices opened markedly down, climbed steadily till noon, then fell off again. On Friday, investor relief at the prospect of the coming weekend started prices sky high. They dropped quickly, leveling off at midday. A market analyst used the function $f(x) = \dfrac{3x^2}{x^2 - 4}$ to model prices for the three days. Graph the function.

Solution

The numerator and denominator of the function have no common factors. To determine vertical asymptotes, set the denominator equal to zero.

$$x^2 - 4 = 0$$
$$(x + 2)(x - 2) = 0$$
$$x = -2 \qquad \text{and } x = 2 \qquad \text{Vertical asymptotes.}$$

The degrees of the numerator and the denominator are equal. The leading coefficients of the numerator and denominator are 3 and 1, respectively, so the horizontal asymptote is $f(x) = \dfrac{3}{1} = 3$.

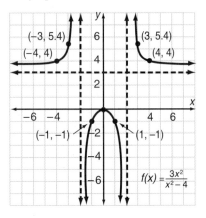

The three asymptotes divide the graph into six sections. Several other points are plotted to find out which of the six sections contain portions of the graph and to determine the shape of the graph.

TRY THESE

State the domain of each function.

1. $f(x) = \dfrac{1}{x + 1}$

2. $f(x) = \dfrac{x - 3}{4}$

3. $f(x) = \dfrac{x^2}{x^2 - 16}$

4. $f(x) = \dfrac{x - 2}{x^2 + x - 6}$

5. $f(x) = \dfrac{8x^2 - 4x + 11}{7x^3}$

6. $f(x) = \dfrac{3x + 5}{3x - 5}$

Identify the points of discontinuity, if any, for each graph.

7. $f(x) = \dfrac{2}{x - 4}$

8. $f(x) = \dfrac{x^2 - 16}{x + 4}$

9. $f(x) = \dfrac{(x^2 + 4x - 5)(x - 3)}{x^2 + 2x - 15}$

10. Complete the table for $f(x) = \dfrac{x - 1}{x - 2}$.

x	1	1.5	1.9	1.99	1.999	2.001	2.01	2.1	2.5	3
$f(x)$										

Determine the horizontal and vertical asymptotes of the graph of each function.

11. $f(x) = \dfrac{2}{x - 3}$

12. $f(x) = \dfrac{x}{x^2 - 5x + 6}$

13. $f(x) = \dfrac{3x - 1}{x + 2}$

Graph each rational function.

14. $f(x) = \dfrac{1}{x - 1}$

15. $f(x) = \dfrac{x^2 + 2x - 3}{x + 3}$

16. $f(x) = \dfrac{6x}{2x - 5}$

17. **Business** A publisher spent $10,000 to buy a book manuscript, $20,000 for editing, permissions, photographs, and other expenses prior to printing the book, and $2 to print each copy of the book. Five thousand copies were printed.

    a. Determine the total cost of the 5,000 copies.

    b. Determine the cost per book.

    c. Suppose that n copies had been printed. Write a rational function giving the cost per book C as a function of n.

    d. Describe the graph of the function for positive values of n.

18. **Writing Mathematics** Explain the meaning of *discontinuous*. Tell why the graphs of rational functions are often discontinuous but those of polynomial functions are not.

PRACTICE

State the domain of each function.

1. $f(x) = \dfrac{4}{x}$

2. $f(x) = \dfrac{x^2 - 16}{8}$

3. $f(x) = \dfrac{x + 1}{x - 5}$

4. $f(x) = \dfrac{2x + 3}{2x - 3}$

5. $f(x) = \dfrac{5x - 4}{x^2 - 100}$

6. $f(x) = \dfrac{12x^2 - 5x}{x^2 - 6x + 8}$

7. $f(x) = \dfrac{x^2 - 5x - 14}{x - 7}$

8. $f(x) = \dfrac{x^3 - 5x + 2}{6x^2 + x - 2}$

9. $f(x) = \dfrac{1}{x^4 - 13x^2 + 36}$

Identify the points of discontinuity, if any, for each graph.

10. $f(x) = \dfrac{x^2 - 1}{x + 1}$

11. $f(x) = \dfrac{x^2 - x - 12}{x - 2}$

12. $f(x) = \dfrac{(x^2 - 3x - 18)}{(x + 3)}$

13. Complete the table for $f(x) = \dfrac{x - 1}{x - 3}$.

x	2	2.5	2.9	2.99	2.999	3.001	3.01	3.1	3.5	4
$f(x)$										

Determine the horizontal and vertical asymptotes of the graph of each function.

14. $f(x) = \dfrac{5}{x - 5}$

15. $f(x) = \dfrac{1}{x^2 - 25}$

16. $f(x) = \dfrac{2x + 1}{x^2 - 7x + 10}$

17. $f(x) = \dfrac{6x + 5}{x - 3}$

18. $f(x) = \dfrac{x^2 + 6}{2x - 3}$

19. $f(x) = \dfrac{5}{(x - 4)(x^2 - 4)}$

20. **Writing Mathematics** Approximate the value of $\dfrac{2x + 3}{x - 1}$ for very large values of $|x|$. Explain how you made your approximation.

Graph each rational function.

21. $f(x) = \dfrac{1}{x + 1}$

22. $f(x) = \dfrac{4}{x^2 - 9}$

23. $f(x) = \dfrac{x^2 + 3x - 4}{x - 1}$

24. $f(x) = \dfrac{6x}{3x + 5}$

25. $f(x) = \dfrac{x + 3}{x^2 - 6x + 5}$

26. $f(x) = \dfrac{x - 3}{x - 4}$

27. $f(x) = \dfrac{-4}{x - 2}$

28. $f(x) = \dfrac{4x - 2}{2x + 1}$

29. $f(x) = \dfrac{1}{(x - 2)^2}$

30. PHOTOGRAPHY Recall from Explore the magnification M of a subject is $M = \dfrac{I}{S}$, where S is the distance of the subject in front of the camera lens and I is the distance of the image behind the lens.

   a. Sketch a graph showing magnification as a function of subject distance for a constant image distance of 4 cm.
   b. Use the graph to estimate the subject distance resulting in a magnification of 0.25.

EXTEND

31. A function $f(x)$ has $x = 5$ and $x = -2$ as vertical asymptotes and $f(x) = 2$ as a horizontal asymptote. Find a possible expression for $f(x)$.

32. TRANSPORTATION A truck driver drove at an average rate of v_1 mi/h for t_1 hours and then at an average rate of v_2 mi/h for t_2 hours.

   a. Find the total distance and the total time the truck driver drove.
   b. Write a rational expression for the average rate v for the entire trip.
   c. Graph v as a function of t_1 if $v_1 = 50$ mi/h, $v_2 = 60$ mi/h, and $t_2 = 4$ hours.

33. ECONOMICS On one recent day, the price p of one share of US Minerals stock could be modeled by the function $p(h) = \dfrac{9h^2 + h + 9}{h^2 + 1}$, where h represents the number of hours since the stock market opened.

   a. Find the price at opening and at closing 8 hours later to the nearest one-eighth dollar.
   b. Find the asymptote(s) of the function.
   c. Graph the function for the day.
   d. Estimate the stock's highest price for the day and the time it was attained.

THINK CRITICALLY

The function $f(x) = \dfrac{x^2 + 1}{x}$ has a *slant asymptote*, an asymptote that is neither horizontal nor vertical.

34. Use polynomial division to write $f(x)$ as a sum of two terms.

35. Approximate $f(x)$ when $|x|$ is very large. What is the slant asymptote for the function?

36. Approximate $f(x)$ when x is near zero. What is the vertical asymptote for the function?

37. Graph $f(x) = \dfrac{x^2 + 1}{x}$.

38. UTILITY BILLS The monthly water charge, in dollars, for residents of Pine Valley is given by

$$W(x) = \begin{cases} 25 & 0 \le x \le 30 \\ 25 + 0.6(x - 30) & x > 30 \end{cases}$$

where x represents hundreds of gallons of water.

a. What is $W(28)$? What is $W(32)$?

b. As x approaches 30 from the left, what does $W(x)$ approach?

c. As x approaches 30 from the right, what does $W(x)$ approach?

d. Graph $W(x)$. Is $W(x)$ continuous?

MIXED REVIEW

Let $A = \begin{bmatrix} 4 & -7 & 12 \\ 8 & 3 & -11 \end{bmatrix}$ and $B = \begin{bmatrix} -5 & 13 & 11 \\ -8 & -9 & 11 \end{bmatrix}$

Evaluate each expression.

39. $A + B$

40. $B - A$

41. $3B$

42. $2A - 2B$

State whether the given relation is a function.

43. $\{(1, 3), (2, 3), (2, 4)\}$

44. $\{(0, 0), (1, 0), (2, 0)\}$

45. $\{(-1, -1), (-2, -2)\}$

46. STANDARDIZED TESTS Choose the letter of the conjugate of $3 + 4i$.

A. $-4i$

B. $\dfrac{3 + 4i}{3 - 4i}$

C. $3 - 4i$

D. $\dfrac{3 - 4i}{3 + 4i}$

PROJECT *Connection*

You will construct a *pantograph* which can be used to enlarge or reduce drawings. You will need wood strips, a wood board, screws and nuts, nails, a screwdriver, hammer, and pencil.

1. Use the wood strips to create a parallelogram like ABCD in the figure. The dimensions should be exactly as shown. Join the strips so that they are secure but moveable. You must be able to change the corner angles by pulling at point A. Attach a pencil at P and nail the pantograph to the board at F.

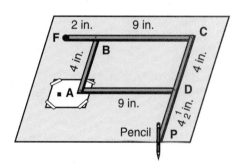

2. On a sheet of paper, draw a 2 in. square. Tape it to the board and trace it by going over it with point A. Measure the new square.

3. Determine the ratio of the side of the small square to the larger square.

4. Try tracing some other simple pictures or designs. What are some disadvantages of using the pantograph?

11.3 Inverse and Joint Variation

Explore

The distance from a camera lens to the focal point, the point where the light rays come together, is called the *focal length*, F. For a camera with lens diameter d, the *f-number*, f of an exposure is given by $f = \frac{F}{d}$. The f-number is also called the f-stop.

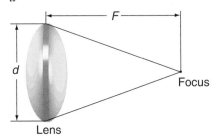

1. The table lists standard f-numbers. Determine the lens diameter that will give each f-number for a lens with a focal length of 200 mm. Give answers to the nearest tenth.

f-number	1.4	2	2.8	4	5.6	8	11	16	22	32
diameter (mm)										

2. Use your results to graph the function $f = \frac{200}{d}$.

3. Does the graph have asymptotes? How do you know?

4. Recall that when two quantities are related *directly*, they both increase together and decrease together. Are f and d related directly? Explain.

COMMUNICATING ABOUT ALGEBRA

When x and y vary directly, they are said to be *directly proportional* to each other. If (x_1, y_1) and (x_2, y_2) are two sets of values in a direct proportion, then $\frac{x_1}{y_1} = \frac{x_2}{y_2}$. Write a proportion relating $x_1, y_1, x_2,$ and y_2 when x and y are *inversely proportional*.

Build Understanding

A **direct variation** can be written in the form $y = kx$, where $k \neq 0$. For example, the direct variation $y = 6x$ might represent the amount of money y you earn in x hours at an hourly wage of \$6. In this relationship, the *constant of variation* is 6. Recall when y varies directly as x, the graph is a line with a slope of k.

The f-number function in Explore showed a different type of variation. An **inverse variation** is a relationship between x and y that can be written in the form $y = \frac{k}{x}$, or $xy = k$ where $k \neq 0$ and $x \neq 0$.

As with a direct variation, the number k is the **constant of variation** in an inverse variation.

EXAMPLE 1

OCEANOGRAPHY Boyle's Law states that the volume of a gas varies inversely with the pressure of the gas if the temperature remains constant. At sea level, where atmospheric pressure is 14.7 lb/in.^{2}, a balloon used for underwater rescues contains 50 ft^3 of air. Find the volume of air in the balloon when a scuba diver descends to a depth of 100 ft and fills the balloon, assuming that the temperature does not change. For each foot below sea level pressure increases by 0.44 lb/in.^{2}

Solution

Let $P_1 = 14.7$ lb/in.^{2} Pressure at sea level.
Let $V_1 = 50$ ft^3 Volume at sea level.

$k = P_1 V_1$ Inverse variation.
$k = 14.7(50)$ Substitute for P_1 and V_1.
$k = 735$ ft^3-lb/in.^{2} Solve for k.

Let $P_2 =$ pressure at a depth of 100 ft
      $= 14.7 + 100(0.44)$
      $= 58.7$ lb/in.^{2}

Let $V_2 =$ volume at a depth of 100 ft
    $P_2 V_2 = k$ Inverse variation.
$(58.7)V_2 = 735$ Substitute for P_2 and k.
    $V_2 \approx 12.5$ ft^3 Solve for V_2.

The volume of the balloon at a depth of 100 ft is approximately 12.5 ft^3. ◀

CHECK UNDERSTANDING

How can you determine that the graph of a direct variation is a line with a slope of k?

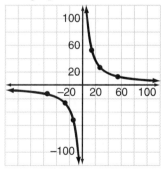

The graph shows the relationship between pressure and volume in Example 1. Negative volumes and pressures in Quadrant III are not relevant in an actual situation. Since $P = \dfrac{735}{V}$ is a rational function, the graph has a characteristic shape you will recall from Lesson 11.2. The asymptotes are $V = 0$ and $P = 0$. You can see that for two variables that vary inversely, as P and V do here, one variable increases as the other decreases.

Inverse square variation is widely used in real world situations. In this type of variation, one variable varies inversely as the *square* of the other variable. If y varies inversely as the square of x, the inverse square relationship can be written $y = \dfrac{k}{x^2}$, where $k \neq 0$ and $x \neq 0$.

ALGEBRA: WHO, WHERE, WHEN

Boyle's law is named for Irish chemist Robert Boyle who helped to establish the principles of the modern scientific method. His work on gases was supplemented by the French physicist Jacques Charles who showed that the product of the pressure and volume of a gas varies directly as the absolute temperature.

EXAMPLE 2

PHOTOGRAPHY *Illuminance* is the intensity of light that the eye perceives. As you move closer to or farther from a light source, the illuminance I varies inversely as the square of the distance d from the source. At 160 ft from a baseball pitcher, a sports photographer's light meter registers an illuminance of 20 foot-candles. Determine the meter reading at a distance of 80 ft from the pitcher.

Solution

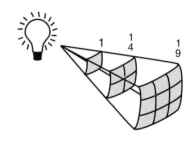

$$I_1 d_1^2 = k$$
$$20(160)^2 = k$$
$$k = 512{,}000$$
$$I_2 d_2 = k$$
$$I_2 d_2^2 = 512{,}000$$
$$I_2(80)^2 = 512{,}000$$
$$I_2 = 80$$

The meter shows an illuminance of 80 foot-candles. Because illuminance varies inversely as the *square* of the distance, moving to a point half as far from the pitcher *quadruples* (2^2) the intensity of light reaching the camera. ◄

In **joint variation**, one variable varies directly as the product of two *or more* other variables. If y varies jointly as x and z, the joint relationship can be written $y = kxz$. Variation problems can be solved by writing the relationships as ratios and solving a related proportion.

EXAMPLE 3

The simple interest I earned by an investment varies jointly as the rate r and the amount of time t. In 5 years, an investment earns \$1080 at a rate of 6%. How much will the investment earn in 8 years at 4%?

Solution

$$I = krt \qquad \text{Write the equation of variation.}$$
$$k = \frac{I}{rt} \qquad \text{Solve for } k.$$

Substitute given values in the expression $\frac{I}{rt}$ twice, using data from the problem.

5-year investment: $\dfrac{1080}{5(0.06)}$ 	 8-year investment: $\dfrac{I}{8(0.04)}$

Since both expressions are equal to k, write and solve a proportion.

$$\frac{1080}{5(0.06)} = \frac{I}{8(0.04)}$$
$$I = 1152$$

In 8 years the investment will have earned \$1152. ◄

CHECK UNDERSTANDING

Show that you get the same answer for Example 3 if you find the value of k, then solve the equation of variation.

TRY THESE

Tell whether the variables vary directly or inversely.

1. Your speed and the time it takes you to drive 500 mi.

2. The number of identical light bulbs purchased and the total cost of the purchase.

3. The number of people in an elevator and the amount of floor space per person.

4. Your monthly salary and the number of months it will take you to earn $50,000.

5. The radius of a circle and the circumference of the circle.

Tell whether x and y vary directly, inversely, or jointly. Then determine k.

6. $y = 6x$

7. $x = 0.5yz$

8. $y = \dfrac{1}{x}$

9. $\dfrac{y}{12} = x$

10. Suppose y varies inversely as x, and $y = 9$ when $x = 16$. Determine x when $y = 24$.

11. Suppose x varies jointly as y and z, and $x = 56.7$ when $y = 1.5$ and $z = 9$. Determine x when $y = 3$ and $z = 4.4$.

12. Suppose y varies inversely as the square of x, and $y = 4$ when $x = 3.2$. Determine y when $x = 4$.

13. **RECREATION** The force required to balance a person on a seesaw varies inversely with the person's distance from the balance point. A 150-lb person seated 6 ft from the balance point balances a 120-lb person on the other side. How far is the 120-lb person from the balance point?

14. **GEOMETRY** The volume of a rectangular prism varies jointly as the length and width of the prism. A rectangular prism with a volume of 16.2 cm³ has a length of 3.6 cm and a width of 2.5 cm. Find the volume of a rectangular prism with a length of 4 cm and a width of 1.5 cm.

15. **ACOUSTICS** The intensity of a radio signal varies inversely as the square of the distance from the transmitter. At a distance of 6 km from a transmitter, a station's signal has an intensity of 9 W/m². Find the intensity at a distance of 2 km from the transmitter.

16. **WRITING MATHEMATICS** Give a real world example of direct variation and another of inverse variation. Explain how you determined the type of variation represented by each example.

PRACTICE

Tell whether the variables vary directly or inversely.

1. The number of times you play an arcade game and the total cost of playing.

2. The number of apples you can buy for $5 and the cost of one apple.

3. The number of people working at the same rate and the length of time it takes to refinish the gym floor.

4. The number of pages you have read in a novel and the number of pages remaining.

5. The length of a side of a square and the perimeter of the square.

Tell whether x and y vary directly, inversely, or jointly. Then determine k.

6. $\dfrac{x}{yz} = 1.72$ 7. $\dfrac{y}{x} = 7.99$ 8. $4xy = 13$ 9. $x = \dfrac{20}{y}$

10. Suppose y varies inversely as x. If y is 12 when x is 0.4, determine y when x is 5.

11. Suppose a varies inversely as the square of b. If a is 100 when b is 6, determine a when b is 15.

12. Suppose w varies jointly as s and z. If w is 75 when s is 10 and z is 5, determine z when w is 115.5 and s is 7.

13. **HEALTH AND SAFETY** The length of time that safe air remains in an airtight room varies inversely with the number of people in the room. If the air will remain safe for 3 hours for 8 people stuck in an elevator, how long will it remain safe for 10 people in the same elevator?

14. **CONSTRUCTION** The length of rope L that can be wound onto a drum varies jointly with the diameter d, drum width w, and rim height r of the drum. A drum with diameter 30 in., width 36 in., and rim height 8 in. has a capacity of 9072 in. of $\frac{1}{2}$-in. rope. Find the $\frac{1}{2}$-in. rope capacity of a drum with diameter 24 in., width 20 in., and rim height 5 in.

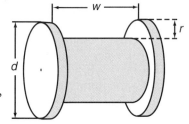

15. **WRITING MATHEMATICS** Use the table to describe the relationship between m and n. Explain your reasoning.

m	5	10	15	20
n	144	36	16	9

EXTEND

16. Suppose g varies jointly as h and l and inversely as m. If g is 40 when h is 10, l is 20, and m is 0.4, determine g when h is 5, l is 10, and m is 0.2.

17. **METEOROLOGY** The pressure of wind on a skyscraper varies jointly as the area of the building and the square of the wind's velocity. A 15-mph wind creates pressure of 1 lb/ft^2. Find the wind velocity that will create pressure of 5 lb/ft^2.

18. **PHYSICS** The time it takes a pendulum to swing back and forth once varies directly as the square root of its length. A 39-in. pendulum requires 1 second to make one complete swing. How long will one complete swing take if the length of the pendulum is increased by 3 in.?

19. **ENGINEERING** The load that a beam of constant depth can support varies directly as the width of the beam and inversely as the length. A beam of length 6 ft and width $1\frac{1}{2}$ in. can support a load of 5 tons. How great a load can be supported by a beam of the same material that is 4 ft long and 2 in. wide?

Sketch a graph representing each relationship.

20. y varies directly as x, and the constant of variation is negative.

21. y varies inversely as x, and the constant of variation is positive.

22. y varies directly as the square of x, and the constant of variation is positive.

THINK CRITICALLY

23. Suppose n varies inversely as the sum of the squares of s and t. If n is 3 when s and t are both 10, determine n when s and t are both 20.

24. **PHYSICS** The gravitational force between two objects varies directly as their masses and inversely as the square of the distance between them. If one mass is doubled, the second tripled, and the distance between them halved, how is the gravitational force affected?

PROJECT *Connection*   You will need an overhead projector, markers, blank transparencies, chalk or masking tape, and a ruler.

1. On a transparency, draw a right triangle with legs of 3 in. and 4 in. and hypotenuse 5 in.

2. Set the projector on a cart with wheels if possible. Project the triangle onto the chalkboard, and move the projector until the images of the sides measure 30 cm and 40 cm and the hypotenuse measures 50 cm.

3. Mark the spot on the floor where the projector produces the dimensions given in Question 2 with a line parallel to the front of the room. Name this Distance 1.

4. Use chalk or tape to mark off six parallel lines in 1 ft increments, measured from the Distance 1 line.

5. Move the projector back one foot at a time and measure the projected triangle in cm. Write a ratio that compares the 30 cm-40 cm-50 cm triangle's sides to the new triangle's sides. Do this for all six markers.

6. Compare the ratios you found in Question 5. Does the movement of one foot have the same effect each time? Predict the size of the triangle if you move 10 ft from Distance 1, then test your prediction.

Career
Lens Designer

The image on a negative is dependent on the amount of light that hits the film when the film is exposed. A camera's **aperture** controls the amount of light reaching the film. The standard camera aperture is a tribute to the lens designer's art. Simple in design and operation, it has the complex task of admitting exactly the right amount of light into a camera. By turning a ring, the photographer opens or closes the aperture to the chosen diameter. The size of the opening is indicated by the **f-number** of the lens.

f - number	1.4	2	2.8	4	5.6	8	11	16

Decision Making

1. You are about to photograph a friend on a cloudy day. You have chosen an aperture setting of f/4. Suddenly the sun comes out. Should you change the setting to f/2.8 or f/5.6? Explain.

2. Standard f-numbers (1.4, 2, 2.8, 4, 5.6, 8, 11, 16, 22, 32) are rounded approximations of a precisely defined sequence. To discover the rule that generates the sequence, you must see the terms before they are rounded: 1.414…, 2.000, 2.828…, 4.000, 5.656…, 8.000, 11.313…, 16.000, 22.627…, 32.000. The f-numbers normally stop at 32, but if they continued, what would the next one be? (Hint: Try squaring the terms of the sequence.)

3. Write a formula for the area of an aperture with diameter d.

4. The f-number and aperture diameter d are related by the formula $f = \dfrac{F}{d}$, where F is the focal length of the lens. Use your answer to Question 3 to write a formula for the area of an aperture with f-number f and focal length F.

5. Find the aperture areas of a lens with a 135 mm focal length at f-numbers f/2.8, f/4, f/5.6, and f/8. How is the area affected by each increase in f-number?

Explore

In each of the following, two methods for simplifying or operating on rational numbers are given. Choose the correct method and explain.

1. Simplify: $\dfrac{18}{24}$

a. $\dfrac{\overset{9}{\cancel{18}}}{\underset{12}{\cancel{24}}} = \dfrac{9}{12}$

b. $\dfrac{\overset{3}{\cancel{18}}}{\underset{4}{\cancel{24}}} = \dfrac{3}{4}$

2. Simplify: $\dfrac{3 + 9}{6}$

a. $\dfrac{3 + 9}{6} = \dfrac{3 + \overset{3}{\cancel{9}}}{\underset{2}{\cancel{6}}} = \dfrac{1 + 3}{2} = \dfrac{4}{2} = 2$

b. $\dfrac{3 + 9}{6} = \dfrac{3 + 9}{\underset{2}{\cancel{6}}} = \dfrac{1 + 9}{2} = \dfrac{10}{2} = 5$

3. Simplify: $\dfrac{4 + 5}{5 + 11}$

a. $\dfrac{4 + 5}{5 + 11} = \dfrac{9}{16}$

b. $\dfrac{4 + 5}{5 + 11} = \dfrac{4 + \overset{1}{\cancel{5}}}{\underset{1}{\cancel{5}} + 11} = \dfrac{5}{12}$

4. Multiply: $\dfrac{5}{8} \cdot \dfrac{7}{8}$

a. $\dfrac{5}{8} \cdot \dfrac{7}{8} = \dfrac{5 \cdot 7}{8} = \dfrac{35}{8}$

b. $\dfrac{5}{8} \cdot \dfrac{7}{8} = \dfrac{5 \cdot 7}{8 \cdot 8} = \dfrac{35}{64}$

SPOTLIGHT ON LEARNING

WHAT? In this lesson you will learn
- to simplify, multiply, and divide rational expressions.
- to find restrictions on the variable in rational expressions.

WHY? Multiplying or dividing rational expressions is often necessary to solve geometric problems such as determining the dimensions or volume of a rectangular prism.

Build Understanding

A **rational expression** is a quotient of two polynomials. Examples of rational expressions are

$$\dfrac{-13xy}{24x^3y^2} \qquad \dfrac{x + 3}{x^2 - 9} \qquad \dfrac{1}{4x^2 + 5x + 6}$$

For the same reason that the domain of a rational function $f(x) = \dfrac{p(x)}{q(x)}$ is restricted to $q(x) \neq 0$, a rational expression is not defined when its denominator equals zero.

A rational expression is **simplified** or in **simplest form** when its numerator and denominator contain no common factors. To simplify, use the same two-step process that you use with rational numbers. First factor the numerator and the denominator. Then use the identity property to remove any common factors.

EXAMPLE 1

Simplify $\dfrac{4x^3 - 2x^2 - 2x}{2x^2 - 2x}$. State any restrictions on the variable x.

Solution

First determine restrictions on x.

$$
\begin{aligned}
2x^2 - 2x &= 0 && \text{Set the denominator equal to zero.} \\
2x(x - 1) &= 0 && \text{Factor.} \\
2x = 0 \quad \text{or} \quad x - 1 &= 0 && \text{Zero product property.} \\
x = 0 \quad \text{or} \quad x &= 1 && \text{Solve for } x.
\end{aligned}
$$

So, $x \neq 1$ and $x \neq 0$.

Now Simplify.

$$
\begin{aligned}
\frac{4x^3 - 2x^2 - 2x}{2x^2 - 2x} &= \frac{2x(2x^2 - x - 1)}{2x(x - 1)} && \text{Factor out the common factor } 2x. \\[2mm]
&= \frac{2x(x - 1)(2x + 1)}{2x(x - 1)} && \text{Factor the trinomial.} \\[2mm]
&= \frac{\cancel{2x}}{\cancel{2x}} \cdot \frac{\cancel{x - 1}}{\cancel{x - 1}} \cdot (2x + 1) && \text{Divide common factors.} \\[2mm]
&= 2x + 1 && \text{Use the identity property.}
\end{aligned}
$$

So, $\dfrac{4x^3 - 2x^2 - 2x}{2x^2 - 2x} = 2x + 1$ when $x \neq 1$ and $x \neq 0$. ◄

In Example 1 the rational expression $\dfrac{4x^3 - 2x^2 - 2x}{2x^2 - 2x}$ is simplified to the binomial $2x + 1$. Many other rational expressions can be simplified to $2x + 1$ such as

$$
\frac{2x^2 + x}{x} \qquad \frac{4x^2 - 1}{2x - 1} \qquad \frac{2x^2 + 3x + 1}{x + 1}
$$

Expressions like these that have the same simplified form are called **equivalent expressions.**

To multiply two rational expressions, follow the rules for multiplying rational numbers.

EXAMPLE 2

Multiply: $\dfrac{5x^2 - 20}{3 - x} \cdot \dfrac{3x - 9}{10x + 20}$

Solution

$$
\begin{aligned}
&\frac{5x^2 - 20}{3 - x} \cdot \frac{3x - 9}{10x + 20} \\[2mm]
&= \frac{5(x + 2)(x - 2)}{3 - x} \cdot \frac{3(x - 3)}{5 \cdot 2(x + 2)} \qquad \text{Factor.}
\end{aligned}
$$

$$= \frac{5(x + 2)(x - 2)3(x - 3)}{(3 - x)(5 \cdot 2)(x + 2)}$$

Multiply numerators and multiply denominators.

$$= \frac{5(x + 2)(x - 2)3(x - 3)}{-(x - 3)(5 \cdot 2)(x + 2)}$$

$3 - x = -(x - 3)$

$$= \frac{3(x - 2)}{-2}$$

Simplify by dividing out common factors.

$$= -\frac{3(x - 2)}{2}$$

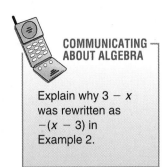

COMMUNICATING ABOUT ALGEBRA

Explain why $3 - x$ was rewritten as $-(x - 3)$ in Example 2.

To divide two rational expressions, follow the rules for dividing rational numbers.

EXAMPLE 3

BUSINESS The expression $\frac{5000x^2 - 5000x}{20x + 50}$ models actual monthly sales in thousands of dollars of Quicka multimedia computers during the first year of production (x = month number). The company's sales department had forecast monthly sales in thousands of dollars of $\frac{4000x^2 - 4000x}{8x + 20}$ multimedia computers for the first year. Find the ratio of actual sales to predicted sales during the first year.

Solution

$$\frac{5000x^2 - 5000x}{20x + 50} \div \frac{4000x^2 - 4000x}{8x + 20}$$

Write the ratio as a quotient.

$$= \frac{5000x^2 - 5000x}{20x + 50} \cdot \frac{8x + 20}{4000x^2 - 4000x}$$

Multiply by the reciprocal.

$$= \frac{5000(x)(x - 1)(4)(2x + 5)}{10(2x + 5)(4000)(x)(x - 1)}$$

Multiply numerators and denominators and factor.

$$= \frac{1}{2}$$

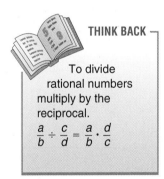

THINK BACK

To divide rational numbers multiply by the reciprocal.

$\frac{a}{b} \div \frac{c}{d} = \frac{a}{b} \cdot \frac{d}{c}$

Actual sales were only one-half of predicted sales.

TRY THESE
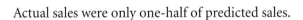

Simplify each rational expression, stating any restrictions on the variable.

1. $\frac{6x^5}{4x^2}$

2. $\frac{15(x + 3)}{5x^2(x + 3)}$

3. $\frac{x^2 - 3x - 4}{4x - 16}$

Perform the indicated operation. Assume no denominator is zero.

4. $\frac{5x^2}{3y^3} \cdot \frac{2xy^2}{x^3}$

5. $\frac{2x + 2}{3} \cdot \frac{9x}{4x + 4}$

6. $\frac{2x^5}{5y^2} \div \frac{4x^2}{15y^3}$

Perform the indicated operation. Assume no denominator is zero.

7. $\dfrac{2x - 10}{x + 10} \div \dfrac{x - 5}{3x + 3}$

8. $\dfrac{x^2 - 64}{x^2 - 9} \cdot \dfrac{x + 3}{x + 8}$

9. $\dfrac{2x^2 + x - 1}{2 - x} \div \dfrac{2x^2 - x}{x^2 - 5x + 6}$

10. **GEOMETRY** The width of a rectangle with an area of $2x^2 + 2xy$ is $4xy + 4y^2$. Determine the length of the rectangle.

11. **MANUFACTURING** Raoul worked for a hours and produced an average of $\dfrac{b}{a^2 - 1}$ door hinges per hour. If each hinge brings a profit of $\dfrac{a^2 - a}{b}$ cents, find the total profit on the sale of the hinges Raoul produced.

12. **WRITING MATHEMATICS** Give an example a rational expressions equivalent to $\dfrac{x^2 - 2x - 8}{x + 2}$. Then explain why they are equivalent.

PRACTICE

Simplify each rational expression, stating any restrictions on the variables.

1. $\dfrac{2abc}{6a^2c^2}$

2. $\dfrac{20xyz^3}{25x^3yz}$

3. $\dfrac{x^2 + 4x}{x + 4}$

4. $\dfrac{x^2 - 25}{x - 5}$

5. $\dfrac{20x^2 - 32x}{4x}$

6. $\dfrac{x^2 - 7x + 12}{x^2 - 5x + 6}$

7. $\dfrac{a^2 - b^2}{a^2 - ab}$

8. $\dfrac{x^2 - 9x + 20}{x^2 + x - 20}$

Perform the indicated operation. Assume no denominator is zero.

9. $\dfrac{2xy^2}{3z^2} \cdot \dfrac{6x^2z^2}{5y^3}$

10. $\dfrac{40mn}{18p^2} \div \dfrac{25n}{27p^3q}$

11. $\dfrac{x^2 + 5x - 6}{2x^2} \cdot \dfrac{6x}{x - 1}$

12. $\dfrac{7x + 7y}{y - x} \div \dfrac{21}{4x - 4y}$

13. $\dfrac{x^2 - 16}{2x + 6} \cdot \dfrac{x + 3}{x - 4}$

14. $\dfrac{x + 2}{x^3} \div \dfrac{6x + 12}{x}$

15. $xy \div \dfrac{x}{y + 1}$

16. $\dfrac{2x + 4y}{9} \cdot \dfrac{27}{6x + 8y}$

17. $\dfrac{2a^2 - 8}{a + 2} \div (2a - 4)$

18. $\dfrac{a^2 - b^2}{a^2 + 2ab + b^2} \cdot \dfrac{ab + b^2}{a^2 - ab}$

19. $\dfrac{x^2 + 8x + 16}{y^2 - 6y + 9} \div \dfrac{2x + 8}{3y - 9}$

20. $\dfrac{9x + 81}{x^2 - 9} \cdot \dfrac{2x^2 - 6x}{x^2 + 18x + 81}$

21. $\dfrac{x^3 - 27}{x^2 - 9} \cdot \dfrac{x^2 - 6x + 9}{x^2 + 3x + 9}$

22. $\dfrac{3x - 12}{20 - x - x^2} \div \dfrac{4x + 12}{x^2 + 7x + 10}$

23. $\dfrac{x^2 - 10x + 21}{x^2 - 12x + 27} \div \dfrac{x^2 - 11x + 30}{x^2 - 14x + 45}$

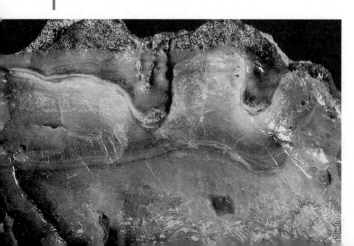

24. **GEOLOGY** A geologist collected $(n - 5)$ samples of a mineral each with a mean mass of $(n^2 - n - 6)$ g. The total volume of the samples was $(n^2 - 3n - 10)$ cm^3. Find the density of the sample in g/cm^3.

25. **WRITING MATHEMATICS** Describe similarities between rational numbers and rational expressions.

26. GEOMETRY A rectangular prism has a length of $\dfrac{3x^2 + 3x - 36}{x}$, a width of $\dfrac{x^2 + x - 6}{x + 4}$, and a height of $\dfrac{2x}{x + 3}$. Determine the volume of the prism.

27. TRANSPORTATION A plane flew from New York City to Chicago, a distance of m miles, at an average rate of 300 mi/h. Taking advantage of strong tail winds, the plane returned to New York at an average rate of 400 mi/h.

    a. Write expressions for the flying time for each leg of the trip.

    b. Write an expression for the total flying time.

    c. Write and simplify a rational expression for the average rate of the entire journey.

    d. What is unusual about your result?

28. CONSTRUCTION A rectangular window in a rectangular door is half as tall as it is wide. The door is four times as tall as the window is wide. The width of the door is two-thirds the height. Find the ratio of the area of the window to the area of the door.

EXTEND

Simplify each rational expression.

29. $\dfrac{x^{10m}y^{8n}}{x^{5m}y^{4n}}$

30. $\dfrac{x^{2a} + 10x^a + 24}{x^{2a} - x^a - 20}$

31. $\dfrac{x^{4m} - 9x^{2m} + 14}{x^{4m} - 8x^{2m} + 12}$

In Exercises 32–34, determine $\dfrac{f(x + h) - f(x)}{h}$ for each function.

32. $f(x) = 5x$

33. $f(x) = x^2$

34. $f(x) = x^2 - 4x$

THINK CRITICALLY

In Exercises 35–36, determine P.

35. $\dfrac{x^2 - 2x - 15}{x^2 - 3x - 18} \cdot \dfrac{P}{x^2 - 3x - 10} = 1$

36. $\dfrac{x^2 - 16}{2x^2} \div \dfrac{2x^2 - 11x + 12}{P} = \dfrac{x + 4}{2x - 1}$

MIXED REVIEW

Solve each system. Check your answer.

37. $\begin{cases} y = 2x \\ 3x + 2y = 21 \end{cases}$

38. $\begin{cases} x + 2y = 7 \\ 3x - 2y = -11 \end{cases}$

39. $\begin{cases} 4x - 3y = -7 \\ 3x + 2y = 16 \end{cases}$

40. $\begin{cases} 4(1 - x) = 8x + 7y \\ 6x + y + 8 = 0 \end{cases}$

Solve each equation. Check your answer.

41. $x^4 - 13x^2 + 36 = 0$

42. $x - 7\sqrt{x} - 8 = 0$

43. $x^{\frac{4}{3}} - 29x^{\frac{2}{3}} + 100 = 0$

44. STANDARDIZED TESTS Choose the letter of the correct quotient: $\dfrac{3x - 3}{3} \div \dfrac{x^2 - 1}{x + 1}$

    A. $(x - 1)^2$ **B.** 1 **C.** $3x^2 - 3x$ **D.** $x + 1$

Adding and Subtracting Rational Expressions

Explore

• The figure depicts the relationship between distances affecting how a photograph turns out. The focal length is F. The distances from the lens to the subject being photographed and to the image that forms on the film inside the camera are labeled S and I, respectively, in the figure. The variables are related by the formula $\frac{1}{S} + \frac{1}{I} = \frac{1}{F}$.

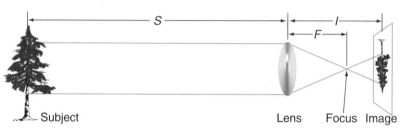

Subject   Lens   Focus   Image

1. For a given lens, the focal length is a constant. Suppose that a photographer moves farther away from a subject. Will the image distance increase or decrease? Explain.

2. Choose values for S and I and find the sum $\frac{1}{S} + \frac{1}{I}$. Repeat for several other pairs of values. Then decide which of the following is the algebraic sum of $\frac{1}{S} + \frac{1}{I}$. Give reasons for your answer.

    a. $\frac{2}{SI}$ **b.** $\frac{2}{S + I}$ **c.** $\frac{S + I}{SI}$

3. Describe how to add $\frac{1}{3}$ and $\frac{1}{2}$. Then describe a similar method you use to add the rational expressions $\frac{1}{S}$ and $\frac{1}{I}$.

Build Understanding

• To add and subtract rational expressions, follow the same procedure you use to add and subtract rational numbers. When the rational expressions have common denominators, add or subtract the numerators and write the sum or difference over the denominator. After adding or subtracting, check to see whether you can simplify further. If the denominators are different, rewrite the rational expressions as equivalent expressions with the LCD.

EXAMPLE 1

Subtract: $\dfrac{5c}{6a^2b^3} - \dfrac{c}{9a^3b}$

Solution

$\dfrac{5c}{6a^2b^3} - \dfrac{c}{9a^3b}$

$= \dfrac{5c}{2 \cdot 3 \cdot a^2 \cdot b^3} - \dfrac{c}{3^2 \cdot a^3 \cdot b}$ Write the prime factorization of each denominator.

$= \dfrac{5c(3a)}{6a^2b^3(3a)} - \dfrac{c(2b^2)}{9a^3b(2b^2)}$ Write equivalent fractions with the LCD $18a^3b^3$.

$= \dfrac{15ac - 2b^2c}{18a^3b^3}$ Subtract the numerators.

CHECK UNDERSTANDING

In the LCD in Example 2, why are both a and b cubed?

Although the above expression is in simplified form after subtracting the numerators, further simplification is sometimes possible after you rewrite a sum or difference using a common denominator.

EXAMPLE 2

Add: $\dfrac{20 - 16x}{3x^2 - 12x - 15} + \dfrac{2x}{3x - 15}$

Solution

$\dfrac{20 - 16x}{3x^2 - 12x - 15} + \dfrac{2x}{3x - 15}$

$= \dfrac{20 - 16x}{3(x - 5)(x + 1)} + \dfrac{2x}{3(x - 5)}$ Write the prime factorizations of each denominator.

$= \dfrac{20 - 16x}{3(x - 5)(x + 1)} + \dfrac{2x(x + 1)}{3(x - 5)(x + 1)}$ Write equivalent fractions with the LCD.

$= \dfrac{20 - 16x + 2x^2 + 2x}{3(x - 5)(x + 1)}$ Add the numerators.

$= \dfrac{2x^2 - 14x + 20}{3(x - 5)(x + 1)}$ Simplify.

$= \dfrac{2(x - 2)(x - 5)}{3(x - 5)(x + 1)}$ Factor.

$= \dfrac{2(x - 2)}{3(x + 1)}$ Simplify.

THINK BACK

Recall that the prime factorization of a polynomial changes a sum or difference of terms to a product of terms.

You may need to add or subtract rational expressions when you use real world formulas. One common type of problem involves the formula relating rate, time, and distance.

EXAMPLE 3

TRANSPORTATION Data collected by traffic engineers in one city indicate that during morning commute hours on the 12-mile Descartes Freeway, the difference between 50 mi/h and the speed of traffic traveling into the city equals, on average, the difference between the speed of traffic traveling out of the city and 50 mi/h.

a. On average, how much longer do incoming cars take to travel the freeway during the morning rush hour than outgoing cars?

b. The average incoming speed on the Descartes Freeway is 44 mi/h. How much longer does it take to drive the freeway coming into the city than it does leaving the city?

Solution

a. Let x = the difference between 50 mi/h and the traffic speed.

Then $50 - x$ = speed of incoming traffic.

$50 + x$ = speed of outgoing traffic.

$\dfrac{12}{50 - x}$ = incoming time to drive the freeway $\qquad t = \dfrac{d}{r}$

$\dfrac{12}{50 + x}$ = outgoing time to drive the freeway

$$\boxed{\begin{array}{c}\text{incoming}\\\text{speed}\end{array}} - \boxed{\begin{array}{c}\text{outgoing}\\\text{speed}\end{array}}$$

$$\frac{12}{50 - x} - \frac{12}{50 + x}$$

$$= \frac{12(50 + x)}{(50 - x)(50 + x)} - \frac{12(50 - x)}{(50 + x)(50 - x)}$$

$$= \frac{600 + 12x - 600 + 12x}{(50 + x)(50 - x)}$$

$$= \frac{24x}{(50 + x)(50 - x)}$$

b. Use the value of x in the difference found in Part a as 50 mi/h minus the average incoming speed of 44 mi/h.

$$\frac{24x}{(50 + x)(50 - x)}$$

$$= \frac{24(6)}{(50 + 6)(50 - 6)} \qquad x = 50 - 44 = 6$$

$$= \frac{144}{2464}$$

$$\approx 0.058$$

The time difference is about 0.058 hours, or about 3.5 minutes. ◄

PROBLEM SOLVING TIP

To confirm that the fractions $\dfrac{12}{50 - x}$ and $\dfrac{12}{50 + x}$ are written in the correct order in Example 3, think:

$50 - x$ is *less than* $50 + x$. Therefore, when 12 is divided by $50 - x$, the quotient will be *greater than* the quotient when 12 is divided by $50 + x$. That means that $\dfrac{12}{50 - x} - \dfrac{12}{50 + x}$ will be positive for $x > 0$.

Determine the least common denominator for each set of expressions.

1. $\dfrac{1}{xy^2}, \dfrac{2}{xy^3}, \dfrac{y}{x^2}$

2. $\dfrac{2}{9abc}, \dfrac{cd}{12a}, \dfrac{-5}{4c^3d}$

3. $\dfrac{3}{4x - 4y}, \dfrac{x}{x^2 - y^2}$

Add or subtract. Simplify if possible.

4. $\dfrac{3x}{y} + \dfrac{4x}{y}$

5. $\dfrac{-3}{m} + \dfrac{2 - k}{m}$

6. $\dfrac{x}{x - y} - \dfrac{y}{x - y}$

7. $\dfrac{3y - 1}{y - 2} + \dfrac{y + 3}{2 - y}$

8. $\dfrac{3x + 2}{3x - 6} - \dfrac{x + 2}{x^2 - 4}$

9. $\dfrac{n}{n^2 - n - 6} - \dfrac{n + 2}{n^2 + 5n + 6}$

10. **PHOTOGRAPHY** A camera's focal length F is related to the subject distance S and image distance I by the formula $\dfrac{1}{S} + \dfrac{1}{I} = \dfrac{1}{F}$.

   a. Express the sum on the left as a single fraction.

   b. Use your answer to write F as a function of S and I.

11. **WRITING MATHEMATICS** When you rewrite a fraction using the LCD, do you change the value of the fraction? Explain.

12. **BICYCLING** Kim bicycles back and forth between her home and Sunrise High School. Going to school, she rides mostly downhill and averages 15 mi/h. Returning home, the same route is mostly uphill, so she averages only 10 mi/h.

   a. Let $x =$ distance from home to school. Write expressions for the time it takes Kim to ride to school and the time it takes to return home.

   b. If Kim lives 12 miles from school, how much longer is the ride home from school than to school?

PRACTICE

Determine the least common denominator for each set of expressions.

1. $\dfrac{11}{8x^3}, \dfrac{5}{12x^2y}$

2. $\dfrac{7}{12a^4b}, \dfrac{8}{27a^3c^2}$

3. $\dfrac{h}{m + n}, \dfrac{3}{m}$

4. $\dfrac{x}{2x - 4y}, \dfrac{y}{10x - 20y}$

5. $\dfrac{x + 3}{x^2 + 9x + 20}, \dfrac{1}{3x + 12}$

6. $\dfrac{1}{(x + 1)(x - 2)^4}, \dfrac{1}{x(x - 2)^3}$

7. **WRITING MATHEMATICS** To subtract $\dfrac{2x + 1}{x - 3}$ from $\dfrac{5x}{x - 3}$, Vincent wrote these steps:

$$\dfrac{5x}{x - 3} - \dfrac{2x + 1}{x - 3} = \dfrac{5x - 2x + 1}{x - 3}$$

$$= \dfrac{3x + 1}{x - 3}$$

Vincent made an error. Find the error and write the correct answer.

Add or subtract. Simplify if possible.

8. $\dfrac{x}{4} + \dfrac{2x}{5}$

9. $\dfrac{3}{p+1} - \dfrac{5}{p+1}$

10. $\dfrac{4}{x^2} + \dfrac{3}{xy}$

11. $\dfrac{2x}{2-x} + \dfrac{x}{2}$

12. $\dfrac{2}{3x} + \dfrac{3}{4x} - \dfrac{4}{5x}$

13. $\dfrac{2}{x-1} - \dfrac{3}{x-2}$

14. $\dfrac{2}{3a-3b} + \dfrac{3}{2a+2b}$

15. $1 + \dfrac{2}{m}$

16. $12 - \dfrac{4x-3y}{x-y}$

17. $\dfrac{y-1}{3y+15} - \dfrac{y+3}{5y+25}$

18. $\dfrac{x}{x^2-2x+1} - \dfrac{1}{x-1}$

19. $\dfrac{x+3}{x^2-4x} + \dfrac{x-2}{x-4}$

20. $\dfrac{6ab}{a^2-b^2} - \dfrac{a+b}{ab}$

21. $\dfrac{x+3}{x^2-6x+9} - \dfrac{8x-24}{9-x^2}$

22. $\dfrac{x^2+2}{x^2-x-2} + \dfrac{1}{x+1} - \dfrac{x}{x-2}$

23. $\dfrac{1}{x+y} + \dfrac{1}{x-y} - \dfrac{1}{x^2-y^2}$

24. **AERONAUTICS** Flying into a headwind, an aircraft's net velocity equals its velocity in still air minus the velocity of the wind. Flying with the wind, a plane's net velocity equals its velocity in still air plus the velocity of the wind. A jet liner's velocity in still air equals 500 mi/h.

   a. For a distance of d miles across Utah, the jet stream flows west-to-east at a speed of 100 mph. Write expressions for the length of time it takes the plane to fly west-to-east and fly east-to-west.

   b. Find the length of time it takes the plane to cross the distance in both directions.

   c. Suppose $d = 360$ miles. How long does it take the plane to cross in both directions?

25. **ENVIRONMENTAL SCIENCE** Mining engineers have placed three pumps in a tailings pond to pump out a spill of hazardous waste. Working alone, pump A could empty the pond in 16 h, pump B in 12 h, and pump C in h h. A breakdown in the electrical system causes the pumps to shut down after $h - 8$ h of operation.

   a. Write expressions for the fraction of the total job each pump has completed when the electricity shuts off.

   b. Write an expression for the fraction of the job completed by all three pumps.

   c. The electrical breakdown occurred after 2 h. How much of the job had been completed?

26. **WRITING MATHEMATICS** Find a rational expression for the sum of a number n and its reciprocal. Explain how you found the expression.

EXTEND

Perform the indicated operations.

27. $\left(\dfrac{m^2+4m-5}{2m^2+m-3} \cdot \dfrac{m-1}{m+5}\right) - \dfrac{2}{m+2}$

28. $\left(4 + \dfrac{1}{n+4}\right)\left(\dfrac{n+4}{n-3}\right)$

29. The number h is the *harmonic mean* of the numbers a and b if the reciprocal of h equals the average of the reciprocals of a and b. Find the harmonic mean of each pair.

a. 5 and 20   **b.** 2 and 6   **c.** 5 and 15

d. Write a rational expression for the harmonic mean of a and b.

THINK CRITICALLY

30. Write the solutions of the quadratic equation $ax^2 + bx + c = 0$. Find the sum of the solutions of the above equation. Find the difference of the solutions of the equation.

In Exercises 31-32, determine A.

31. $\dfrac{x}{x^2 - 1} + \dfrac{A}{x^2 - 1} = \dfrac{1}{x - 1}$

32. $\dfrac{1}{x + 2} + \dfrac{A}{x - 3} = \dfrac{4x + 3}{(x + 2)(x - 3)}$

MIXED REVIEW

Simplify each expression.

33. $12 - 9 \div 3$ **34.** $7 + 8 \cdot 2$ **35.** $64 \div 8 \div 2$ **36.** $5 \cdot 2^3$

37. Sᴛᴀɴᴅᴀʀᴅɪᴢᴇᴅ Tᴇsᴛs The axis of symmetry of the graph of $y = x^2 + 6x - 9$ is

    A. $x = 3$ **B.** $x = -3$ **C.** $x = \dfrac{1}{3}$ **D.** $z = 0$

Add or subtract. Simplify if possible.

38. $\dfrac{x}{x - 3} - \dfrac{3}{x - 3}$ **39.** $\dfrac{2}{x^2 - 1} + \dfrac{2x}{x + 1}$ **40.** $\dfrac{-5x + 5}{x^2 - x - 6} + \dfrac{3}{x + 2} - \dfrac{x - 1}{x - 3}$

PROJECT *Connection*   You will need access to a photocopier that can reduce and enlarge an original.

1. Draw a right triangle with legs 90 mm and 120 mm, and hypotenuse 150mm.

2. Use a photocopier to enlarge your drawing. Record the percent setting you used.

3. Predict the measurements of the enlarged triangle, then measure it. Determine the percent error of the enlargement from what it should have been.

4. Next, try to reduce the enlargement to the original size. Determine the percent setting you should use and explain how you found it.

5. Place your original and the reduction made in Question 4 on top of each other. Hold them up to light and see how well they match.

6. Compute the area of the original triangle and the enlargement in Question 2. By what percent did the area increase?

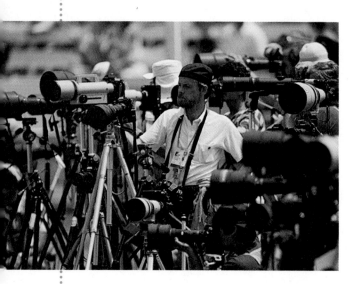

Career
Photographer

For the professional photographer achieving first-class photographs demands a thorough understanding of the complexities of photography. A good photographer must be part artist, part engineer, and part mathematician. The artist composes the photograph, the engineer operates highly technical equipment, and the mathematician strikes a balance among an array of conflicting variables.

Decision Making

1. You have decided that there is too much light illuminating a subject you would like to photograph. Describe how you can change each of the following variables to reduce the amount of light striking the film.

   a. the camera's distance from the subject

   b. the size of the *aperture*, the opening in the lens through which light passes on its way to the film

   c. the *shutter speed*, the length of time the aperture is open and the film is exposed.

2. The photographer must strike a balance among several variables because a change in one variable usually affects the other variables. Choose one of the variables listed in Question 1 or another variable common to photography. How does a change in the variable affect another variable with which the photographer must deal?

3. The figure shows the shutter speed selections x, where $\frac{1}{x}$ is the shutter speed. Express each marking in fractions of a second.

4. A sports photographer wants to create an unblurred "stop action" shot of a high jumper in mid-jump. Which of the settings on the selection dial should the photographer choose? Explain.

Explore

- Let S be the distance from a camera lens to the subject, and let I be the distance from the lens to the image on the film inside the camera. Recall the focal length F of the camera is given by

$$F = \frac{SI}{S + I}$$

1. If a fraction has one or more fractions in the numerator or denominator, it is called a **complex fraction**. Let $S = \frac{3}{4}$ ft and let $I = \frac{1}{12}$ ft. Then $\dfrac{\frac{3}{4} \cdot \frac{1}{12}}{\frac{3}{4} + \frac{1}{12}} = \dfrac{\frac{1}{16}}{\frac{3}{4} + \frac{1}{12}}$. Simplify this complex fraction by adding the rational numbers in the denominator. Then divide the numerator by the denominator.

2. Another method to simplify the above complex fraction is to multiply the numerator and the denominator by the LCD of $\frac{1}{16}, \frac{3}{4}$, and $\frac{1}{12}$. What is the LCD? Simplify using this method.

3. Use both of the above methods to simplify these complex fractions.

   a. $\dfrac{\frac{1}{2} + \frac{1}{3}}{\frac{1}{4} + \frac{1}{6}}$

   b. $\dfrac{\frac{2}{5} + \frac{3}{4}}{\frac{3}{10} + \frac{7}{20}}$

4. Which of the two methods you used to simplify complex fractions was more efficient? Why?

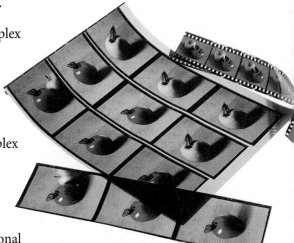

Build Understanding

- A complex rational expression contains one or more rational expressions in its numerator or denominator. Examples include

$$\frac{\frac{1}{x}}{\frac{5}{y}} \qquad \frac{n}{n + \frac{2}{3}} \qquad \frac{\frac{x}{x^2 - 1}}{\frac{x - 1}{x + 1}}$$

As you discovered in Explore, one method to simplify a complex rational expression is to simplify the numerator and denominator separately, then multiply the numerator by the reciprocal of the denominator.

Usually, however, simplification is streamlined by multiplying the numerator and denominator of the complex fraction by the LCD of all the rational expressions in both the numerator and the denominator. This reduces the complex fraction to a simple fraction in one step, and makes any additional simplification easier.

EXAMPLE 1

Simplify: $\dfrac{2 - \dfrac{1}{x}}{1 + \dfrac{1}{x^2}}$

Solution

The LCD of $\dfrac{1}{x}$ and $\dfrac{1}{x^2}$ is x^2.

$$\frac{x^2}{x^2} \cdot \frac{\left(2 - \dfrac{1}{x}\right)}{\left(1 + \dfrac{1}{x^2}\right)} \qquad \text{Multiply by the LCD.}$$

$$= \frac{2x^2 - x}{x^2 + 1} \qquad \text{Simplify.} \qquad \blacktriangleleft$$

When the denominators in a complex rational expression are polynomials, factor to find the LCD of all the fractions in the expression.

EXAMPLE 2

MANUFACTURING The Northridge division of Consolidated Hardware can produce $n^2 + 5n + 6$ door knobs for $n^2 + 2n + 1$ cents. The South Valley division can produce $n^2 + 7n + 12$ knobs for $n^2 - 1$ cents. Find the ratio of the cost-per-knob ratios of the two divisions.

Solution

Northridge cost/knob: $\dfrac{n^2 + 2n + 1}{n^2 + 5n + 6}$ South Valley cost/knob: $\dfrac{n^2 - 1}{n^2 + 7n + 12}$

$$\frac{\dfrac{\text{Northridge}}{\text{cost/knob}}}{\dfrac{\text{South Valley}}{\text{cost/knob}}} = \frac{\dfrac{n^2 + 2n + 1}{n^2 + 5n + 6}}{\dfrac{n^2 - 1}{n^2 + 7n + 12}}$$

$$= \frac{\dfrac{(n + 1)^2}{(n + 2)(n + 3)}}{\dfrac{(n + 1)(n - 1)}{(n + 3)(n + 4)}}$$

$$= \frac{(n+2)(n+3)(n+4)}{(n+2)(n+3)(n+4)} \cdot \frac{\dfrac{(n+1)^2}{(n+2)(n+3)}}{\dfrac{(n+1)(n-1)}{(n+3)(n+4)}}$$ Multiply by the LCD.

$$= \frac{(n+4)(n+1)^2}{(n+2)(n+1)(n-1)}$$ Multiply and simplify.

$$= \frac{(n+4)(n+1)}{(n+2)(n-1)}$$ ◄

COMMUNICATING ABOUT ALGEBRA

For any $n > 1$, is the value of the ratio you found in Example 2 less than or greater than 1? Explain.

TRY THESE

Simplify.

1. $\dfrac{\dfrac{2}{3}}{\dfrac{8}{15}}$

2. $\dfrac{1 + \dfrac{5}{6}}{2 - \dfrac{1}{4}}$

3. $\dfrac{3 + \dfrac{1}{y}}{2 - \dfrac{2}{y}}$

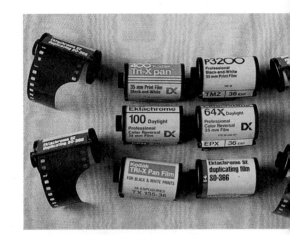

4. $\dfrac{\dfrac{2}{x}}{\dfrac{8}{x^2}}$

5. $\dfrac{\dfrac{1}{a}}{1 - \dfrac{1}{a}}$

6. $\dfrac{5 - \dfrac{2}{m-1}}{3 - \dfrac{1}{m-1}}$

7. $\dfrac{1}{\dfrac{1}{a} + \dfrac{1}{b}}$

8. $\dfrac{\dfrac{2x-2}{x^2+2x}}{\dfrac{4x-4}{3x+6}}$

9. $\dfrac{\dfrac{x^2-1}{3x+3}}{\dfrac{x-1}{x+1}}$

10. **PHOTOGRAPHY** A photographer is comparison shopping for two types of film. At Deluxe Camera she can buy x rolls of Type 1 film for $200 and y rolls of Type 2 film for $300. At Cameraworks, the same quantities cost $240 for Type 1 and $320 for Type 2.

    a. Write an expression for the total cost of one roll of each type at Deluxe Camera. Write another expression for the total cost of one roll of each type at Cameraworks.

    b. Write and simplify a complex rational expression comparing the total cost of one roll of each type of film at Deluxe Camera with the total cost at Cameraworks.

11. **WRITING MATHEMATICS** Describe two methods you could use to simplify $\dfrac{\dfrac{1}{x} - 1}{\dfrac{2}{x} + 1}$ and

explain which method you think would be easiest. Simplify the expression each way and confirm or change your decision.

PRACTICE

Simplify.

1. $\dfrac{\dfrac{3}{4}}{\dfrac{3}{16}}$

2. $\dfrac{1 + \dfrac{5}{6}}{1 - \dfrac{3}{8}}$

3. $\dfrac{\dfrac{5}{4} - \dfrac{7}{8}}{\dfrac{1}{16} + \dfrac{1}{2}}$

Simplify.

4. $\dfrac{\dfrac{6}{a}}{\dfrac{7}{a}}$

5. $\dfrac{\dfrac{5}{x} + 5}{\dfrac{3}{2x} + 2}$

6. $\dfrac{\dfrac{9}{c - d}}{\dfrac{4}{c - d}}$

7. $\dfrac{2 - \dfrac{1}{x + y}}{3 - \dfrac{2}{x + y}}$

8. $\dfrac{\dfrac{x^2 - y^2}{xy}}{\dfrac{x - y}{y}}$

9. $\dfrac{\dfrac{1}{xy} - \dfrac{1}{y^2}}{\dfrac{1}{x^2 y} - \dfrac{1}{xy^2}}$

10. $\dfrac{\dfrac{1}{y + 2} - 3}{2 + \dfrac{2}{y + 2}}$

11. $\dfrac{\dfrac{1}{x - 4} + 5}{\dfrac{2}{x - 4} + 3}$

12. $\dfrac{\dfrac{2}{n - 3} + \dfrac{3}{n - 2}}{\dfrac{2}{n - 2} + \dfrac{3}{n - 3}}$

13. $\dfrac{\dfrac{1}{x + 5} + \dfrac{1}{x - 3}}{\dfrac{2x^2 - 3x - 5}{x^2 + 2x - 15}}$

14. $\dfrac{x + \dfrac{xy}{y - x}}{\dfrac{y^2}{x^2 - y^2} + 1}$

15. $\dfrac{\dfrac{n^2 - n - 6}{n^2 - 5n - 14}}{\dfrac{n^2 + 6n + 5}{n^2 - 6n - 7}}$

16. $\dfrac{\dfrac{t + 6}{t} - \dfrac{1}{t + 2}}{\dfrac{t^2 + 4t + 3}{t^2 + t}}$

17. $\dfrac{\dfrac{5x}{x^2 - 6x + 8}}{\dfrac{2}{x - 4} + \dfrac{3}{x - 2}}$

18. $\dfrac{\dfrac{1}{y + 3} + \dfrac{2y}{(y + 3)^3}}{\dfrac{5}{(y + 3)^2} + \dfrac{4}{y + 3}}$

	Earnings, $	Wages, $/h
Week 1	100	$n^2 - 4$
Week 2	40	$n + 2$
Week 3	50	$n + 2$
Week 4	25	$n - 2$

19. **BUSINESS** The table above lists Jeremy's earnings and average hourly wages for his summer lawn mowing service.

   a. Write expressions for the number of hours he worked each week.

   b. Write and simplify a complex rational expression giving the ratio of the total number of hours he worked the first two weeks to the total number he worked the last two weeks.

20. WRITING MATHEMATICS Explain how you would find the reciprocal of $\frac{1}{x} + 1$.

Simplify.

21. If $a = \frac{b + 1}{b - 1}$ and $b = \frac{1}{1 - c}$, express a in terms of c.

22. $\dfrac{\dfrac{1}{a} - \dfrac{a - x}{a^2 + x^2}}{\dfrac{1}{x} - \dfrac{x - a}{x^2 + a^2}} + \dfrac{\dfrac{1}{a} - \dfrac{a + x}{a^2 + x^2}}{\dfrac{1}{x} - \dfrac{x + a}{x^2 + a^2}}$

23. $\dfrac{\dfrac{x}{2}\left(\dfrac{1}{x - y}\right)\left(\dfrac{1}{x + y}\right)\left(\dfrac{x^2 - y^2}{x^2y + xy^2}\right)}{\dfrac{1}{x + y}}$

24. ELECTRONICS Resistors are common elements of electrical circuits. If three resistors with resistances R_1, R_2, and R_3 are connected in parallel, their total combined resistance is given by $R_T = \dfrac{1}{\dfrac{1}{R_1} + \dfrac{1}{R_2} + \dfrac{1}{R_3}}$

   a. Simplify the expression for R_T.

   b. If $R_1 = 1$ ohm, $R_2 = 2$ ohms, and $R_3 = 6$ ohms, determine their combined resistance in a parallel circuit.

THINK CRITICALLY

The sequence of complex fractions 1, $1 + \dfrac{1}{2}$, $1 + \dfrac{1}{2 + \dfrac{1}{2 + \dfrac{1}{2}}}$, can be represented by the **continued fraction** at the right.

$1 + \dfrac{1}{2 + \dfrac{1}{2 + \dfrac{1}{2 + \ldots}}}$

25. Use a calculator to approximate the first six terms in the sequence.

26. Describe patterns you see in the terms of the sequence.

27. Square your approximation of the sixth term in the sequence. Find the value of the continued fraction.

MIXED REVIEW

Find the inverse of each matrix, if it exists.

28. $\begin{bmatrix} 8 & 5 \\ 5 & 3 \end{bmatrix}$ **29.** $\begin{bmatrix} 4 & 6 \\ 6 & 9 \end{bmatrix}$ **30.** $\begin{bmatrix} 2 & 5 \\ -1 & 3 \end{bmatrix}$ **31.** $\begin{bmatrix} 0 & -1 \\ 1 & 0 \end{bmatrix}$

Solve each inequality.

32. $x^2 + 2x \geq 24$ **33.** $x^2 + 4x + 3 \geq 0$ **34.** $x^2 > 25$ **35.** $x^2 \geq 8x - 16$

36. STANDARDIZED TESTS Simplify $\dfrac{\dfrac{2x + 4}{x^2 - 9}}{\dfrac{x + 2}{x + 3}}$. Choose the letter of the correct answer.

   A. $\dfrac{x}{x - 3}$ **B.** $\dfrac{2(x + 2)^2}{(x - 3)(x + 3)^2}$ **C.** $\dfrac{2}{x - 3}$ **D.** 1

Solve Rational Equations

Explore

1. Recall that the solutions to an equation can be approximated and sometimes determined exactly by drawing a graph. Solve the equation $2x + 1 = x + 4$ graphically. Explain your method.

2. You can extend graphical methods of solving equations which contain one or more rational expressions. Explain how you would solve the equation $\dfrac{x}{x + 2} = 3$ graphically.

3. Use the graphical method you described to solve the equation in Question 2.

4. For the functions $y_1 = \dfrac{x^2}{x - 4}$ and $y_2 = \dfrac{16}{x - 4}$, what restrictions must be placed on the domain? Explain.

5. Solve $\dfrac{x^2}{x - 4} = \dfrac{16}{x - 4}$ graphically. What happens to the graphs near the domain restrictions?

Build Understanding

A **rational equation** is an equation that contains one or more rational expressions. To solve a rational equation algebraically, first note restrictions on the domain of the variable. Then simplify the equation by multiplying both sides of the equation by the LCD of all the rational expressions. Solve the resulting equation. Finally, check solutions against the domain restrictions you have noted and by substituting the solutions into the original equation.

EXAMPLE 1

Solve: $\dfrac{5}{x} + \dfrac{3}{2x} = \dfrac{13}{4}$

Solution

Algebraic Method

Note the domain restriction: $x \neq 0$

$$\frac{5}{x} + \frac{3}{2x} = \frac{13}{4} \qquad \text{The LCD is } 4x.$$

$$4x\left(\frac{5}{x} + \frac{3}{2x}\right) = 4x\left(\frac{13}{4}\right) \quad \text{Multiply each term by the LCD.}$$

$$20 + 6 = 13x$$

$$x = 2$$

Since $x = 2$ does not violate the domain restriction, it is the solution.

$$\frac{5}{x} + \frac{3}{2x} = \frac{13}{4}$$

$$\frac{5}{2} + \frac{3}{4} = \frac{13}{4}$$

$$\frac{10}{4} + \frac{3}{4} = \frac{13}{4}$$

$$\frac{13}{4} = \frac{13}{4} \checkmark$$

Solution

Graphical Method

$$\frac{5}{x} + \frac{3}{2x} = \frac{13}{4}$$

Write and graph a function representing each side of the equation. The solutions, if any, are the x-values of the points where the graphs intersect.

$$y_1 = \frac{5}{x} + \frac{3}{2x}$$

$$y_2 = \frac{13}{4}$$

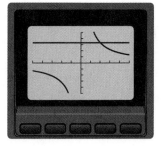

The solution is $x = 2$. ◀

The following example illustrates the importance of noting restrictions on the variable before you start to solve an equation.

EXAMPLE 2

Solve: $\dfrac{1}{x + 5} = \dfrac{2}{x - 3} + \dfrac{2x + 2}{(x + 5)(x - 3)}$

Solution

Note the domain restriction: $x \neq -5, 3$

$$\frac{1}{x + 5} = \frac{2}{x - 3} + \frac{2x + 2}{(x + 5)(x - 3)} \qquad \text{The LCD is } (x + 5)(x - 3).$$

$$(x + 5)(x - 3)\left(\frac{1}{x + 5}\right) = (x + 5)(x - 3)\left(\frac{2}{x - 3} + \frac{2x + 2}{(x + 5)(x - 3)}\right)$$

$$x - 3 = 2(x + 5) + (2x + 2)$$

$$x - 3 = 2x + 10 + 2x + 2$$

$$3x = -15$$

$$x = -5$$

Because of the domain restriction, the solution $x = -5$ is extraneous. The equation has no solution. ◀

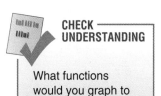

CHECK UNDERSTANDING

What functions would you graph to solve the equation

$$3 + \frac{1}{x - 1} = \frac{x + 2}{x^2 - 2x + 1}?$$

THINK BACK

An *extraneous* solution can appear while you are solving an equation. This is a value that satisfies the simplified equations you are working with, but does not satisfy the original equation.

PROBLEM SOLVING TIP

In Example 3, you can use dimensional analysis to check that the units make sense. Does time equal

$$\frac{\text{earnings}}{\text{hourly wage}} \quad ?$$

The units for earnings are $. The units for hourly wage are $/h.

$$\frac{\text{earnings}}{\text{hourly wage}} =$$

$$\frac{\$}{\$/h} = \frac{\$}{1} \cdot \frac{h}{\$} = h.$$

Since hours are the units for time, the units check.

Real world problems sometimes lead to rational equations.

EXAMPLE 3

ECONOMICS Monica is deciding between two jobs. One offers convenient hours and easy traveling. The other pays $2.50 more per hour and she would earn $1000 in 10 hours less than it takes her to earn $900 at the first job. Find the hourly wage for both jobs.

Solution

Let x = higher hourly wage ($/h)
So, $x - 2.5$ = lower hourly wage ($/h).

time to earn $1000 at higher wage	=	time to earn $900 at lower wage	−	10h

Domain restrictions: $x \neq 0$ and $x \neq 2.5$

$$\frac{1000}{x} = \frac{900}{x - 2.5} - 10 \qquad \text{Time} = \frac{\text{earnings}}{\text{hourly wage}}.$$

$$x(x - 2.5)\left(\frac{1000}{x}\right) = x(x - 2.5)\left(\frac{900}{x - 2.5} - 10\right) \qquad \text{Multiply by the LCD.}$$

$$1000(x - 2.5) = 900x - 10(x)(x - 2.5)$$

$$1000x - 2500 = 900x - 10x^2 + 25x$$

$$10x^2 + 75x - 2500 = 0 \qquad \text{Collect like terms}$$

$$5(2x^2 + 15x - 500) = 0 \qquad \text{Factor.}$$

$$5(x + 20)(2x + 25) = 0 \qquad \text{Factor.}$$

$$x = -20 \text{ or } x = 12.5 \qquad \text{Solve.}$$

Neither value of x is excluded from the domain, although the negative value is not relevant in a real world sense. Therefore, the wages are $12.50 per hour and $10.00 per hour. ◀

When a rational equation is difficult to solve algebraically, try a graphical solution.

EXAMPLE 4

BIOLOGY A wildlife biologist used the function

$$y = \frac{x^2 - 105x - 200}{x^3 - 12x^2 + 11x - 98}$$

to model the growth of an Alaskan caribou herd from 1985 to 1995, where x is the year number and y is the size of the herd, in thousands. When did the population first surpass 4000? Let 1985 = 0.

Solution

There is no straightforward method for solving a cubic equation, as there is for a quadratic. Therefore, the equation $\dfrac{x^2 - 105x - 200}{x^3 - 12x^2 + 11x - 98} = 4$ can be difficult to solve algebraically. However, the equation can be graphed using a graphing utility.

Using TRACE, the graph shows that the population curve first rises above 4 at $x \approx 8.5$. This represents the year $1985 + 8.5 = 1993.5$.

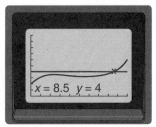

x = 8.5   y = 4

TRY THESE

Solve and check. If an equation has no solution, write no solution.

1. $\dfrac{x}{3} + \dfrac{x}{8} = 11$

2. $\dfrac{x + 2}{5} + \dfrac{x - 1}{5} = \dfrac{11}{5}$

3. $\dfrac{m + 4}{6} = \dfrac{m - 3}{5}$

4. $3 - \dfrac{x - 1}{4} = \dfrac{x}{9}$

5. $\dfrac{5}{x} + x = -6$

6. $\dfrac{18}{2x} - 1 = 2$

7. $\dfrac{1 - x}{1 + x} = 4$

8. $2 + \dfrac{3}{y - 1} = \dfrac{5}{2y - 2}$

9. $\dfrac{x - 2}{x - 4} = \dfrac{2}{x - 4}$

10. **MANUFACTURING** Two workers can put together nine microwave ovens in an hour. The faster worker can assemble an oven three minutes faster than the slower worker.

   a. Let x = the number of minutes it takes the faster worker to assemble an oven. Write a rational equation that states that the number of ovens the faster worker assembles in an hour plus the number the slower worker assembles in an hour equals nine.

   b. Solve the equation to find the faster worker's assembly time for one oven.

11. **BUSINESS** The function $y = \dfrac{2x^3 - x^2 + 5}{x^2 + 2}$ models daily sales of Roller Zone skates, in thousands of pairs, during the first six months following the introduction of the skates (x = month number).

   a. Write an equation to find when sales first surpassed 5000 pairs per day.

   b. Approximate the solution by graphing.

12. **WRITING MATHEMATICS** Examine these rational equations. Would you solve them algebraically or graphically? Explain.

   a. $\dfrac{x^3 + x + 4}{2x^2 + 6x + 1} = 7$

   b. $\dfrac{4}{2 - x} = 16$

   c. $\dfrac{x}{x^2 + 4x + 3} = 2$

PRACTICE

Solve and check.

1. $\dfrac{y}{2} + \dfrac{y}{4} = 12$

2. $3x + \dfrac{3}{4} = \dfrac{5}{6}$

3. $\dfrac{1}{x} + \dfrac{2}{x} = 1$

4. $2x + 3 = \dfrac{x-1}{4}$

5. $\dfrac{3}{n+2} + \dfrac{15}{n+2} = 2$

6. $\dfrac{c+1}{c-3} = \dfrac{4}{c-3}$

7. $\dfrac{2}{x} = \dfrac{3}{x-4}$

8. $\dfrac{5}{2x+5} = \dfrac{4}{x+5}$

9. $\dfrac{3x+12}{3x+23} = \dfrac{3}{4}$

10. $\dfrac{1}{x+3} = \dfrac{2}{x} - \dfrac{3}{4x}$

11. $\dfrac{1}{x-5} + \dfrac{1}{x+5} = \dfrac{6}{x^2-25}$

12. $\dfrac{5}{x} + \dfrac{7}{x^2} = 2$

13. $\dfrac{6}{x-1} - \dfrac{4}{x-2} = \dfrac{2}{x+1}$

14. $\dfrac{6x^2+5x-11}{3x+2} = 2x - 5$

15. $\dfrac{x}{x^2-1} = \dfrac{1}{2x-2} - \dfrac{2}{x+1}$

16. **RECREATION** The velocity of the Rio Grande River in Albuquerque, New Mexico, is 3 mi/h. Moving with the current, a canoe can travel 10 mi in the same amount of time it would take to go 4 miles moving against the current.

   a. Let x be the canoe's rate in still water. Assume that the rate of the current is added to the still water velocity downstream and subtracted from it upstream. Write an equation expressing the conditions stated in the problem.

   b. Solve the equation to find the canoe's rate in still water.

17. **SPORTS** Rounded to the nearest mile per hour, the speed of the fastest Indianapolis 500 winner to date, Arie Luyendyk in 1990, was 2.48 times as fast as the slowest winner, Ray Harroun in 1911. Luyendyk finished the 500-mi race in 4 hours less time than Harroun. Find the average velocities of both drivers.

18. **RECREATION** A swimming pool can be filled by a large inlet pipe in 10 h, by a small inlet pipe in 15 h, and by a garden hose in 24 h.

   a. How long will it take the large and small pipes, working together, to fill the pool?
   b. How long will it take both pipes and the hose to fill the pool?

19. **NUMBER THEORY** The sum of a number and its reciprocal is 4. Find the number.

20. **ENERGY** The function $f(x) = \dfrac{x^3 + x^2 - 5x + 8}{x^2 + x + 1}$ approximates daily coal production during month x, in tons, at one mine in West Virginia, during 1994 (January 1 = 0).

   a. When did production equal 4 tons per day?
   b. When does it appear that a new vein of coal was discovered?

21. **WRITING MATHEMATICS** Solve: $\dfrac{1}{x-3} + \dfrac{1}{x+3} = \dfrac{2x}{x^2-9}$. Explain your results.

EXTEND

Solve each formula for the given variable.

22. $\frac{r_1}{L - d} = \frac{r_2}{d}$ for d (distance d to a ground in a wire of length L with resistors r_1 and r_2)

23. $I = \frac{E}{R_L + r}$ for r (current I for electromotive force E, internal resistance r, and external resistance R_L)

24. $\left(P + \frac{a}{V_2}\right)(V_1 - b) = RT$ for P (relationship of pressure P, volume V, and temperature T under real gas conditions; a, b are constants depending on gas, R is universal gas constant)

25. GEOMETRY   Recall that corresponding sides of *similar figures* are in proportion. Triangles *ABC* and *DEF* are similar. Determine the length of side *AB* and side *EF*.

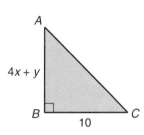 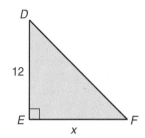

THINK CRITICALLY

26. HYDROLOGY   A municipal water tank can be emptied through its single outlet in 48 minutes. With the outlet closed, the tank can be filled in 36 minutes.

    a. If the outlet is left open by mistake, how long will it take to fill the tank?

    b. Change one number in the problem so that the tank never fills.

Solve for x.

27. $\frac{m - x}{m - n} = 2 + \frac{p - x}{n - p}$

28. $\frac{x - a}{x - b} = \frac{b - x}{a - x}$

MIXED REVIEW

Solve each inequality.

29. $|x| < 2$ **30.** $|4x + 2| \geq 6$ **31.** $|2x - 7| > 10$ **32.** $|3x + 5| \geq 0$

Simplify.

33. i^2 **34.** i^3 **35.** $4i^{13}$ **36.** $\frac{2}{i}$

37. STANDARDIZED TESTS   Solve $64 = 32^x$. Choose the letter of the correct answer.

    A. 1.2   **B.** 2   **C.** $\frac{5}{6}$ **D.** 1.5

Solve and check each equation.

38. $\frac{c}{c + 2} = \frac{7}{9}$ **39.** $x + \frac{3}{x} = 4$ **40.** $\frac{5}{x} + \frac{3}{x} = 1$ **41.** $2 - \frac{1}{a} = \frac{1}{a^2}$

Problem Solving File

Area Under a Curve

One way to determine the total production, such as the number of tons of wheat from a farm, over a period of time is to graph the rate of production as a function of time and find the area under the curve for the given interval of time. For example, suppose the farm produced at the rate of 8 tons per day. Then you can represent the total production for 10 days, which is $10 \cdot 8 = 80$ tons, by the area under the line $y = 8$ between $x = 0$ and $x = 10$.

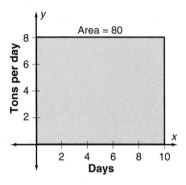

If the rate of production is not constant, then the production function will not be linear. In order to determine the total production over time, you need a general method for finding areas under curves.

Problem

If $y = x^2, 0 \le x \le 10$ represents the number of tons of ore that a mine produces, where x is the number of weeks since the digging began, then the area under the curve and above the x-axis from $x = 0$ to $x = 2$ represents the total ore for the first 2 weeks. How many tons of ore were mined the first two weeks?

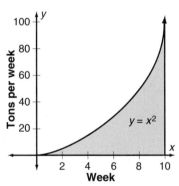

Explore the Problem

To develop a method for finding area under a curve, explore a simpler problem first. Suppose you want to find the area under the line $y = x$ and above the x-axis from $x = 0$ to $x = 1$. One way to approximate this area is to use areas of rectangles whose bases are on the x-axis and whose heights are the vertical distances from their bases to the curve.

1. If the interval $0 \leq x \leq 1$ is divided into n equal subintervals that are used as bases for rectangles, what is the width of each rectangle?

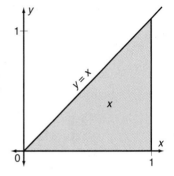

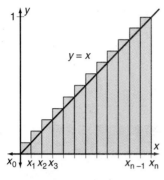

2. Note that $x_1 = \dfrac{1}{n}$, $x_2 = \dfrac{2}{n}$, $x_3 = \dfrac{3}{n}$, $\ldots$, $x_n = \dfrac{n}{n}$. If the functional value at the right endpoint of each subinterval is used as the height of the rectangle, what is the height of the first rectangle? the second rectangle? the nth rectangle?

3. What is the area of the first rectangle? The second rectangle? The third rectangle? The nth rectangle?

4. How will the sum of the areas of all the rectangles compare to the actual area under the curve? Explain.

PROBLEM SOLVING PLAN

- Understand
- Plan
- Solve
- Examine

From what you discovered above, you can see that the area A under the curve is approximated by the sum

$$A \approx \frac{1}{n^2} + \frac{2}{n^2} + \frac{3}{n^2} + \cdots + \frac{n}{n^2} \qquad \text{or}$$

$$A \approx \frac{1 + 2 + 3 + \cdots + n}{n^2} \qquad \text{Use the common denominator } n^2.$$

The numerator is the sum of the numbers 1 to n which is equal to $\frac{n}{2}(n + 1)$. So,

$$A \approx \frac{n + 1}{2n} \qquad\qquad A \approx \frac{\frac{n}{2}(n + 1)}{n^2} = \frac{n + 1}{2n}$$

5. Determine the approximate value of the area if the number of rectangles n equals 5, 10, and 100. List what you observe about the values.

6. You can divide the numerator and denominator of $\dfrac{n + 1}{2n}$ to get $\dfrac{1 + \dfrac{1}{n}}{2}$. As the number of rectangles n gets very large, what value does this expression approach? How do you know this is the correct value for the area under the curve $y = x$ for $x = 0$ to $x = 1$?

7. **WRITING MATHEMATICS** Show the approximating rectangles if you use the functional value at the left endpoint of each of the n subintervals. How would the approximation compare to the actual area. As n gets very large, what value would the sum of the rectangular areas approach?

PROBLEM SOLVING TIP

To find the sum of the first 10 numbers, you can form 5 pairs of 11

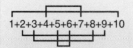

1+2+3+4+5+6+7+8+9+10

so the sum is $5 \cdot 11 = 55$. In general, the number of pairs is $\frac{n}{2}$ and the pair sum is $n + 1$, so the sum of the n numbers is $\frac{n}{2}(n + 1)$.

Investigate Further

- Now you can return to the original problem. To approximate the area under $y = x^2$ from $x = 0$ to $x = 2$, use 4 equal subintervals. Refer to the figure.

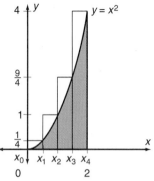

  8. What is the width of each subinterval? What are the values of x_0, x_1, x_2, x_3, and x_4?

  9. What is the value of the function $y = x^2$ at each right endpoint?

  10. What is the sum of the areas of the four approximating rectangles?

  11. How does this approximation compare to the actual area? How could you improve your approximation?

  12. Refer to the figure that shows the approximating rectangles obtained if you use the functional value at the left endpoint of each subinterval. Why are there only 3 rectangles? What is the sum of the rectangular areas?

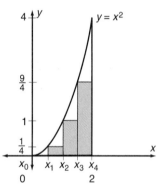

  13. Use your results from Questions 10 and 12 to write an inequality for the actual area AA.

Apply the Strategy

- 14. **a.** Approximate the area under $y = x^2$ from $x = 0$ to $x = 2$ using 8 equal subintervals. Evaluate the function at right endpoints.

     b. By using a general sum formula and letting n get very large, the actual area $y = x^2$ from $x = 0$ to $x = 2$ is found to be $\frac{8}{3}$. Write an inequality relating the actual area to the values you determined for it in Part a and Question 10.

  15. Approximate the area under $y = 4x - x^2$ from $x = 0$ to $x = 2$ using 4 equal subintervals and

     a. evaluate the function at right end points and at left endpoints.

     b. write an inequality relating the actual area to the two values you found.

Review Problem Solving Strategies

STEP BY STEP

1. Mrs. Ortega was telling her husband how she had chased after their three-year old daughter, Rosa, that morning. "Rosa had taken twenty-seven steps when I started to go after her. You know, I noticed that she takes eight steps to my five." "Well then," said Mr. Ortega, "if that is so, how did you ever catch her?" Oh, I have a longer stride than little Rosa. In fact, two of my steps are equal to five of hers." How many steps did Mrs. Ortega take to catch Rosa?

LAKE WISE, I'M SURE

2. The motorboat Explorer leaves the eastern shore of a lake at exactly the same time as another motorboat, Voyager, leaves the western shore. Each boat travels across the lake at a constant speed, and the boats meet for the first time 500 yd from the eastern shore. Each boat continues until it reaches the opposite shore from which it started, immediately turns and heads back. This time the boats meet 300 yd from the western shore. How wide is the lake? Draw a diagram and answer the following questions to help you solve the problem.

a. When the boats meet for the first time, what does the combined distance they have traveled represent?

b. When the boats meet for the second time, what does the combined distance they have traveled represent? How does the elapsed time at the second meeting compare to the time of the first meeting?

c. What total distance in yards has Explorer traveled at the second meeting? How does this distance compare to the width of the lake? How wide is the lake?

d. What is the ratio of the distance traveled by each boat at the first meeting? What is the ratio of their speeds?

COUNTING CUBES

3. You have probably solved problems where you had to determine the different sizes of squares or triangles in a figure in order to find a total. Now try some similar problems with cubes. Assume all cubes are made up of unit cubes.

a. What is the total number of cubes (all sizes) in a cube with side 3 units?

b. What is the total number of cubes (all sizes) in a cube with side 4 units?

c. Explain how you can determine the total number of cubes in a cube with side 10 units? Use a calculator to find the number.

d. If the total number of cubes is 44,100, what is the size of the largest cube?

••• CHAPTER REVIEW •••

VOCABULARY

Match the letter of the word in the right column with the description at the left.

1. a limitation on the values of the variable in a rational function

2. a line to which a graph moves arbitrarily close for very large or very small values of the variable

3. a rational expression containing one or more rational expressions in its numerator and its denominator

4. the set of values for which the denominator of a rational function is nonzero

5. the separation between two branches of a graph

a. asymptote

b. discontinuity

c. domain

d. restriction

e. complex

Lessons 11.1 and 11.2 EXPLORE AND GRAPH RATIONAL FUNCTIONS   pages 513–523

- If the **rational function** $f(x) = \dfrac{p(x)}{q(x)}$, and if $p(x)$ and $q(x)$ have no common factors, then the graph of $f(x)$ has a vertical asymptote at each real zero of $q(x)$.

- The graph of $f(x)$ has a horizontal asymptote at $y = 0$ if the degree of $p(x)$ is less than the degree of $q(x)$, and at $y = \dfrac{a}{b}$ if the degree of $p(x)$ equals the degree of $q(x)$, where a and b are the leading coefficients of $p(x)$ and $q(x)$, respectively. It has no horizontal asymptote if the degree of $p(x)$ is greater than the degree of $q(x)$.

State the domain of each function.

6. $f(x) = \dfrac{1}{x - 4}$

7. $f(x) = \dfrac{x + 5}{x^2 - 25}$

8. $f(x) = x^2 + 2x - 5$

9. Identify the point(s) of discontinuity, if there are any, for the graph of $f(x) = \dfrac{x^2 - 64}{x - 8}$.

10. Determine the horizontal and vertical asymptotes of the graph of $f(x) = \dfrac{4x + 1}{x - 1}$.

Lesson 11.3 INVERSE AND JOINT VARIATION   pages 524–530

- If y varies **inversely** as x, then $y = \dfrac{k}{x}$ or $xy = k$, where k is the **constant of variation** and $k \neq 0$ and $x \neq 0$.

- If y varies **jointly** as x and z, then $y = kxz$.

11. If y varies inversely as x, and $y = 4$ when $x = 9$, determine y when $x = 12$.

12. If y varies jointly as x and z and $y = 100$ when $x = 2.5$ and $z = 10$, determine y when $x = 14$ and $z = 7$.

- A rational expression is **simplified** or in **simplest form** when its numerator and denominator contain no common factors (other than ± 1).

- Multiply and divide rational expressions as you multiply and divide rational numbers.

Perform the indicated operation and simplify.

13. $\dfrac{5x^2y}{3z^2} \cdot \dfrac{6z^2}{10xy^2}$ 14. $\dfrac{a^2 - b^2}{a^2 + 2ab + b^2} \cdot \dfrac{5a + 5b}{2a}$ 15. $\dfrac{8e}{f} \div \dfrac{16e^2}{15f^2}$

- When rational expressions have like denominators, add or subtract the expressions by adding or subtracting the numerators and writing the sum or difference over the denominator.

- If the denominators are different, rewrite the rational expressions as equivalent expressions with a least common denominator. Then add or subtract as above.

Perform the indicated operation and simplify.

16. $\dfrac{n + 3}{5} + \dfrac{9n + 2}{5}$ 17. $\dfrac{x + 3}{x - 3} - \dfrac{x - 3}{x + 3}$ 18. $\dfrac{2}{y^2 - 7y} + \dfrac{3}{y - 7} - \dfrac{3}{y}$

- To simplify a complex rational expression, multiply the numerator and denominator by the LCD of all the rational expressions in both the numerator and the denominator.

Simplify.

19. $\dfrac{1 + \dfrac{7}{10}}{2 - \dfrac{3}{5}}$ 20. $\dfrac{5}{\dfrac{1}{x} - \dfrac{1}{y}}$ 21. $\dfrac{\dfrac{3x - 3}{x + 1}}{\dfrac{x^2 - 1}{x^2 + 2x + 1}}$

- To solve a rational equation algebraically, multiply both sides of the equation by the LCD of all the rational expressions. Solve the resulting equation. Check the solutions in the original equation to eliminate extraneous solutions.

Solve.

22. $\dfrac{n}{3} + \dfrac{n}{2} = 10$ 23. $4 - \dfrac{x + 1}{3} = \dfrac{x}{8}$ 24. $\dfrac{x + 1}{x - 2} = \dfrac{3}{x - 2}$

- One way to find the total production over a period of time is to find the area under the curve.

25. Write an inequality to approximate the total production of oil if $y = -x^2 + 3x$, $0 \le x \le 2$ represents the barrels of oil produced in x weeks. Use 4 equal subintervals and right and left-hand function values.

CHAPTER ASSESSMENT

CHAPTER TEST

State the domain of each function.

1. $\dfrac{2}{x}$

2. $\dfrac{x+2}{x+3}$

3. $\dfrac{x^2-25}{x^2-5x+4}$

4. $x^3+2x^2-5x+12$

5. Name the point(s) of discontinuity, if there are any, for the graph of $f(x)=\dfrac{x^2-4x-12}{x-6}$.

Name the asymptotes of the graph of each function.

6. $f(x)=\dfrac{2}{x+1}$

7. $f(x)=\dfrac{3x-4}{x^2+x-20}$

8. $f(x)=\dfrac{8x-11}{x+6}$

9. $f(x)=\dfrac{x^2+x+1}{3x-6}$

10. If y varies inversely as x and $y=5$ when $x=20$. Determine x when $y=200$.

11. If x varies jointly as y and z and $x=128$ when $y=3.2$ and $z=5$, determine x when $y=24$ and $z=\dfrac{1}{3}$.

12. If y varies inversely as the square of x and $y=0.25$ when $x=12$, determine y when $x=3$.

13. The intensity of a TV signal varies inversely as the square of the distance from the transmitter. At a distance of 24 miles from a transmitter, a station's signal has an intensity of 5 units/ft². Find the intensity at a distance of 6 miles from the transmitter.

14. **WRITING MATHEMATICS** Explain how to add two rational expressions.

15. Solve graphically: $\dfrac{x+3}{x}=2$.

Perform the indicated operation and simplify.

16. $\dfrac{m}{4}\cdot\dfrac{6m}{4}$

17. $\dfrac{10}{x+2}\div\dfrac{5}{2x+4}$

18. $\dfrac{x^2+2x+1}{x-2}\cdot\dfrac{x^2-4}{x^2+3x+2}$

19. $\dfrac{2k^2-2k}{k^2-25}\div\dfrac{k^3}{2k^2-9k-5}$

20. $\dfrac{y}{3}-\dfrac{6-2y}{3}$

21. $\dfrac{5}{x}+\dfrac{2}{3x}+\dfrac{1}{3}$

22. $\dfrac{3}{x-2}+\dfrac{8}{x+1}$

23. $\dfrac{1}{x+y}+\dfrac{1}{x-y}-\dfrac{1}{x^2-y^2}$

24. One day a salesperson drove 300 miles in x^2-4 hours. The following day she drove 325 miles in $x+2$ hours. Write and simplify a ratio comparing her average rate the first day with her average rate the second day.

Simplify.

25. $\dfrac{\dfrac{1}{x}+\dfrac{1}{y}}{\dfrac{2}{x}-\dfrac{2}{y}}$

26. $\dfrac{\dfrac{x}{x+1}+\dfrac{1}{x}}{\dfrac{x}{x^2+x}}$

Solve.

27. $\dfrac{12}{x}=\dfrac{x+2}{4}$

28. $\dfrac{6}{x-4}+\dfrac{4}{x-2}=\dfrac{9}{x-4}$

29. **STANDARDIZED TESTS** Which values are solutions of the equation $\dfrac{4}{x-2}=\dfrac{x^2}{x-2}$?

   A. 2 only   **B.** -2 only

   C. 2 and -2 **D.** no solution

PERFORMANCE ASSESSMENT

SPORTS STORY Shown below is a graph of a real world situation involving sports.

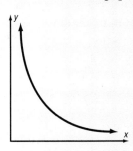

Explain as much as you can about what the graph represents. Discuss the meaning and value of the variables, type of variation, and asymptotes.

EASY SIMPLIFICATION? Each of the fractions shown below can be simplified by a curious method of "digit cancellation."

$$\frac{1\cancel{6}}{\cancel{6}4} = \frac{1}{4} \qquad \frac{1\cancel{9}}{\cancel{9}5} = \frac{1}{5}$$

Does this method always work? Use what you know about rational expressions to explain your answer. Give examples to illustrate your explanation.

RATIONAL THOUGHTS As you have seen in this chapter, rational numbers and rational expressions have much in common. Work with a partner. Find as many similarities between rational numbers and rational expressions as you can. Describe each similarity and give examples to illustrate your description.

THE VARIATION OF CURRENT EVENTS Do research to find data that illustrates variation in a current event. Then do each of the following:

a. Identify the type of variation.

b. Use your data to draw a graph.

c. Use the graph to make a prediction about some future value of the data.

PROJECT ASSESSMENT

 Plan a day when each group can make a presentation.

1. Your presentation should include a description of your activity. Exhibit materials and explain what was learned about making enlargements. In addition, your group should prepare a handout summarizing your work and discoveries. Include questions for other students to answer.

2. After presentations are completed, have a class discussion about how coordinates can be used to accurately enlarge or reduce a figure without using proportions.

3. Explore mathematical relationships such as enlargement and areas using some of the designs collected in Getting Started.

··· CUMULATIVE REVIEW ···

1. Is the set of imaginary numbers closed under addition? If not, give a counterexample.

2. **WRITING MATHEMATICS** Define *rational function* and its *domain,* and explain how to find the domain of a rational function.

3. Find the inverse of matrix A. $A = \begin{bmatrix} 6 & 5 \\ 5 & 4 \end{bmatrix}$

4. Solve for x: $3x^2 - 8x - 3 \geq 0$

5. The wind's force on a sail varies jointly as the area of the sail and the square of the wind's velocity. If a 6-mile-per-hour wind exerts 3 pounds of force on 2 square yards of sail, how much force will a 20-mile-per-hour wind exert on a 5-square-yard sail?

6. Find the value of k for which the solutions of $x^2 + 8x + k = 0$ are equal.

Write and solve an equation for each problem.

7. Some boys chipped in equally to buy an $80 used couch for their clubhouse. Just before they made the purchase, another boy joined them. This reduced by $4 the amount that each boy in the original group had expected to pay. How many boys were in the original group?

8. The dimensions of a rectangular box are 5 in., 6 in., and 7 in. If you want to double the volume of the box, by what constant amount c should the dimensions of each side be increased? Round your answer to the nearest hundredth of an inch.

9. **STANDARDIZED TESTS** The expression $\dfrac{3}{3x - 1} - \dfrac{1}{x}$ is equivalent to

   A. $\dfrac{6x - 1}{x(3x - 1)}$ B. $\dfrac{-1}{x(3x - 1)}$

   C. 1 D. $\dfrac{1}{x(3x - 1)}$

10. Graph the rational function $y = \dfrac{x}{x - 1}$.

11. The value, v, of a brand new $24,000 car that is depreciating at a rate of 15% per year can be modeled by the equation $v = 24{,}000(1 - 0.15)^t$, where t is the time in years. After how many years will the car be worth $10,000? Round your answer to the nearest year.

12. The graph shown is a frequency distribution for the grades of all the students in a math class. If a student is selected at random from the class, what is the probability that the student received a B or better?

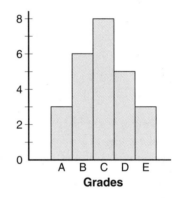

13. The average of 1 and x equals the average of 1, x, and x^2. Find x.

14. If $3x + 2y = 7$ and $kx - y = 5$, express x in terms of k.

15. Simplify: $\dfrac{1 - \dfrac{1}{x}}{1 - \dfrac{1}{x^2}}$

16. **STANDARDIZED TESTS** If $x = \log_2 5$, then 4^x is

   A. 10 B. 7

   C. $\sqrt{10}$ D. 25

17. Determine as completely as possible the number and nature of the solution of $x^4 - x^2 - 2 = 0$.

18. Multiply: $\dfrac{x^3 - 64}{x^2 - 16} \cdot \dfrac{x^2 + 8x + 16}{x^2 + 4x + 16}$

• • • STANDARDIZED TEST • • •

Select the best choice for each question.

1. The conjugate of $2\sqrt{3} - 3\sqrt{5}$ is

   A. $3\sqrt{5} - 2\sqrt{3}$ **B.** $3\sqrt{2} - 5\sqrt{3}$

   C. $\dfrac{1}{2\sqrt{3} + 3\sqrt{5}}$ **D.** $2\sqrt{3} + 3\sqrt{5}$

   E. $3\sqrt{2} + 5\sqrt{3}$

2. The number of rational solutions of the function $x^3 - 8x - 3 = 0$ is

   A. 0   **B.** 1   **C.** 2   **D.** 3   **E.** 0 or 1

3. The equation $x^3 + 2x^2 + x + k = 0$ has a real solution between 0 and 1 if

   A. $k < -4$ **B.** $k > 4$ **C.** $k = 0$
   D. $0 < k < 4$ **E.** $-4 < k < 0$

4. Which of the statements that follow are true for the function shown at the right? $y = \dfrac{x}{x + 1}$

   I. The function is a rational function.
   II. The line x = −1 is a vertical asymptote.
   III. The domain is the set of real numbers greater than −1.

   A. I and II   **B.** I and III   **C.** II and III
   D. II only   **E.** I, II, and III

5. The solution to the system $\begin{cases} ax + by = 11 \\ bx + ay = 10 \end{cases}$
   is $x = 2, y = 1$. Find the values of a and b?

   A. $a = 4$ **B.** $a = 5$ **C.** $a = 6$
      $b = 3$ $b = 1$ $b = -1$
   D. $a = 3$ **E.** $a = 1$
      $b = 4$ $b = -1$

6. $\dfrac{1 - x^{-2}}{1 + x^{-1}}$ is equivalent to

   A. $\dfrac{1}{1 - x}$ **B.** $-\dfrac{1}{1 - x}$ **C.** $\dfrac{x - 1}{x}$

   D. $\dfrac{1 - x}{x}$ **E.** $-\dfrac{x - 1}{x}$

7. The one real zero of the polynomial function $F(x) = 3x^3 - 7x^2 - 7x - 10$ lies between which two consecutive integers?

   A. −1 and 0   **B.** 0 and 1   **C.** 1 and 2
   D. 2 and 3   **E.** 3 and 4

8. If $\log x^3 - \log y = 4$ and $\log x + \log y = 2$ then $\log \dfrac{x}{y}$ equals

   A. $\dfrac{1}{2}$ **B.** 1   **C.** −1   **D.** 2   **E.** −2

9. If y varies inversely as x^2 and x is doubled, then y is

   A. doubled   **B.** halved   **C.** quadrupled
   D. unchanged   **E.** divided by 4

10. Which of the inequalities that follow is represented by the graph shown at the right?

    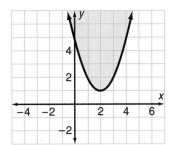

    A. $y > x^2 - 5x + 4$
    B. $y < x^2 - 5x + 4$
    C. $y < x^2 - 4x + 5$
    D. $y \geq x^2 - 4x + 5$
    E. $y > x^2 - 4x + 5$

11. If $xyz \neq 0$, then $\dfrac{9x^2yz^6}{27xy^2z^3}$ equals

    A. $3xyz^3$ **B.** $\dfrac{3xz^2}{y}$ **C.** $\dfrac{xz^2}{3y}$

    D. $\dfrac{xz^3}{3y}$ **E.** $\dfrac{x^2z^3}{3y^2}$

12. The solution set of
    $\dfrac{2x}{x + 5} + \dfrac{1}{x - 5} = \dfrac{10}{x^2 - 25}$ is

    A. $\{-2, 5\}$ **B.** $\left\{-\dfrac{1}{2}, 5\right\}$ **C.** 0

    D. $\{-2\}$ **E.** $\left\{-\dfrac{1}{2}\right\}$

12 Conic Sections

Take a Look AHEAD

Make notes about things that look new.
- Create a list of new vocabulary terms that you will learn in this chapter.
- How do you think solving quadratic systems might be similar to solving linear systems? What differences might there be?

Make notes about things that look familiar.
- Identify the relationship that is often helpful in determining an unknown length in a right triangle.
- What type of equation will you be concerned with in most of the lessons in this chapter?
- Which conic sections(s) have you studied before?

DATA Activity

Hello, Hello!

One of the most significant changes in communications is the use of cellular telephone service. The basic *cell site*, which may be a city or county, can vary in size depending on geography or service demand. Each is equipped with a low-powered radio transmitter/receiver so the service company can limit the radio frequencies assigned to one cell. At the switching office, a computer monitors movement of a cellular telephone and transfers or *hands off* the call to the new cell and another frequency.

In this chapter you will see how:

- **ACOUSTICAL ENGINEERS** use properties of an ellipsoid to design a concert hall. (Lesson 12.4, page 586)

- **TELECOMMUNICATIONS ENGINEERS** use the focus of a paraboloid to produce a strong signal from incoming waves. (Lesson 12.6, page 597)

- **PHYSICIANS** use sound waves to break up a patient's kidney stones. (Lesson 12.8, page 609)

Cellular Telephone Systems								
	1987	**1988**	**1989**	**1990**	**1991**	**1992**	**1993**	**1994**
Systems	312	517	584	751	1252	1506	1529	1581
Customers, thousands	1231	2069	3509	5283	7557	11,033	16,009	24,134
Cell sites	2305	3209	4169	5616	7847	10,307	12,805	17,920
Revenue, $ in millions	1151	1959	3340	4548	5708	7822	10,891	14,229
Average monthly bill, $	96.83	98.02	89.30	80.90	72.74	68.68	61.48	56.21
Average call length, min	2.33	2.26	2.48	2.20	2.38	2.58	2.41	2.24

Use the table to answer the following questions.

1. What was the average revenue for cellular systems in 1987? What was the average revenue in 1994? Round answers to the nearest hundred thousand dollars.

2. Estimate the increase in cell sites from 1991 to 1992.

3. What is the range for the average call lengths? What is the median?

4. Let 87 represent the year 1987, 88 represent 1988, and so on. How will you represent 2002?

5. Use a graphing utility to determine the equation of the line of best fit for the number of customers as a function of the year. Give the coefficient of correlation. Round coefficient to the nearest hundredth.

6. Use your equation to predict the number of customers in the year 2000.

7. **WORKING TOGETHER** Construct a graph to display the data about the average monthly bill for cellular service. Give reasons for your choice of graph. Discuss the trend shown by the data and whether you think this trend will continue.

Conics on Camera

The cornerstone of education, in or out of the school setting, is communication. Whether you learn from another person, book, museum display, television show, CD-ROM, billboard, audio tape, or other means, communication is taking place. For decades, communication and learning in schools revolved around a teacher's lecture. The technological revolution has brought other means of communicating information into the classroom setting and changed the face of education. If a picture is worth a thousand words, imagine the power of an educational video!

PROJECT GOAL

To create a presentation featuring manipulative activities about the conic sections.

Getting Started

The whole class will work on this production as a group.

1. Choose a chairperson to lead the discussions and keep track of assignments. Work will involve planning and script writing, design and graphics, narration or acting, props and set arrangement, and filming.

2. Check on the availability of a video camera, either from school or private sources. The project can also be done using a regular camera, slide projector, and tape player.

3. Research additional information about the history or applications of conic sections to make your video more interesting. Decide on a unifying theme for your presentation.

PROJECT Connections

Lesson 12.2, page 574:
Create and film demonstrations about conics in general and circles in particular.
Lesson 12.3, page 580:
Create and film demonstrations about ellipses and hyperbolas.
Lesson 12.6, page 596:
Create and film a demonstration about parabolas.
Chapter Assessment, page 617:
Plan a video festival that will include screenings of student productions, discussions, and related activities.

Internet Connection
www.swpco.com/
swpco/algebra2.html

Explore

Microwave relay towers are spaced equally in a straight line on a level section of land. On a grid map of the area, Tower 37 is located at (14, 23) and Tower 39 is located at (6, 17). A repair crew is searching for Tower 38, which has been toppled in a windstorm.

1. Draw a coordinate plane and plot towers 37 and 39. Using another sheet of graph paper as a ruler, measure the distance between the towers.

2. Sketch a right triangle joining Tower 37, Tower 39, and the point (14, 17). Determine the distance between the towers. Describe your method.

3. Tower 38 is located halfway between Towers 37 and 39. Use your graph paper ruler to plot the position of Tower 38. Estimate the coordinates of the tower.

4. Confirm that Tower 38 is midway between towers 37 and 39. Describe your method.

5. Study the coordinates of the three towers. Describe any patterns that you see.

SPOTLIGHT ON LEARNING

WHAT? In this lesson you will learn
- to use the Pythagorean theorem to determine the distance between two points in the coordinate plane.
- to find the midpoint of a line segment.

WHY? You can use the distance formula to solve problems about mileage, communication, and surveying.

Build Understanding

Suppose that (x_1, y_1) and (x_2, y_2) are two points in the plane separated by a distance d. Together with (x_2, y_1), the points form a right triangle with legs measuring $|x_2 - x_1|$ and $|y_2 - y_1|$. You can use the Pythagorean theorem to find d.

$$d^2 = |x_2 - x_1|^2 + |y_2 - y_1|^2$$
$$d^2 = (x_2 - x_1)^2 + (y_2 - y_1)^2$$
$$d = \sqrt{(x_2 - x_1)^2 + (y_2 - y_1)^2}$$

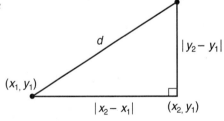

The above equation is the **distance formula**. You can use it to find the distance between any two points in a plane.

> **DISTANCE FORMULA**
>
> The distance between two points with coordinates (x_1, y_1) and (x_2, y_2) is $d = \sqrt{(x_2 - x_1)^2 + (y_2 - y_1)^2}$.

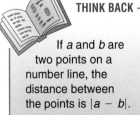

THINK BACK

If a and b are two points on a number line, the distance between the points is $|a - b|$.

EXAMPLE 1

GEOMETRY The points $A(-6, 1)$, $B(0, 4)$, $C(2, 0)$, and $D(-4, -3)$ form the vertices of a rectangle. Find the lengths of the sides and the diagonals of the rectangle.

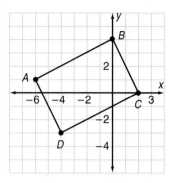

Solution

The opposite sides of a rectangle are the same length, so you can find the lengths of any pair of adjacent sides.

$$d = \sqrt{(x_2 - x_1) + (y_2 - y_1)}$$

$$\overline{AB} = \sqrt{(0 - (-6))^2 + (4 - 1)^2} \qquad \text{Use } (-6, 1) \text{ and } (0, 4).$$

$$= \sqrt{45}$$

$$= 3\sqrt{5} \qquad \text{Simplify radical.}$$

$$d = \sqrt{(x_2 - x_1) + (y_2 - y_1)}$$

$$\overline{BC} = \sqrt{(0 - 2)^2 + (4 - 0)^2} \qquad \text{Use } (0, 4) \text{ and } (2, 0).$$

$$= \sqrt{20}$$

$$= 2\sqrt{5} \qquad \text{Simplify radical.}$$

The diagonals are congruent, so you can find the length of either one.

$$d = \sqrt{(x_2 - x_1) + (y_2 - y_1)}$$

$$\overline{AC} = \sqrt{(-6 - 2)^2 + (1 - 0)^2} \qquad \text{Use } (-6, 1) \text{ and } (0, 2).$$

$$= \sqrt{65}$$

You can use a calculator to approximate side AB as 6.71, side BC as 4.47, and the diagonals as 8.06.   ◄

The **midpoint** of a line segment is the point equidistant from the endpoints of the segment. You can determine the coordinates of the midpoint of a segment by averaging the coordinates of the endpoints.

MIDPOINT FORMULA

The midpoint of the line segment with endpoints (x_1, y_1) and (x_2, y_2) is the point $\left(\dfrac{x_1 + x_2}{2}, \dfrac{y_1 + y_2}{2}\right)$.

EXAMPLE 2

GEOMETRY The points $A(-5, 7)$, $B(-9, 1)$, $C(5, 1)$, and $D(9, 7)$ are the vertices of a parallelogram. The diagonals of a parallelogram intersect at their midpoints. Find the coordinates of the intersection of the diagonals of the parallelogram.

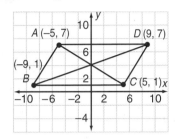

Solution

Use the midpoint formula.

$$\text{midpoint of } \overline{AC} = \left(\frac{-5 + 5}{2}, \frac{7 + 1}{2}\right) = (0, 4) \qquad \left(\frac{x_1 + x_2}{2}, \frac{y_1 + y_2}{2}\right)$$

You can find the midpoint of $\overline{BD}$ to check your answer. ◄

When a problem involves the distance formula, you may need to solve a radical equation.

EXAMPLE 3

COMMUNICATIONS On a coordinate grid of Kansas, radio station KALG is located at $K(39, 21)$. The station has a range of 75 mi. Find the eastern most point on the Kansas-Oklahoma border where the station can be heard.

Solution

Designate the border point 75 mi southeast of KALG as $B(x, 0)$.

$$
\begin{aligned}
75 &= \sqrt{(39 - x)^2 + (21 - 0)^2} &&\text{Use distance formula.} \\
&= \sqrt{1521 - 78x + x^2 + 441} \\
5625 &= 1521 - 78x + x^2 + 441 &&\text{Square both sides.} \\
0 &= x^2 - 78x - 3663 &&\text{Collect like terms.} \\
0 &= (x - 111)(x + 33) &&\text{Factor.} \\
x - 111 &= 0 \quad \text{or} \quad x + 33 = 0 &&\text{Zero products property.} \\
x &= 111 \quad \text{or} \quad\quad\; x = -33
\end{aligned}
$$

The eastern most point where the station can be heard is $(111, 0)$. The point $(-33, 0)$ lies west of Kansas, on the southern border of Colorado. ◄

PROBLEM SOLVING TIP

If you cannot factor the trinomial $x^2 - 78x - 3663$ in Example 3, use the quadratic formula to solve the equation. Another method is to use a graphing utility to plot $y = x^2 - 78x - 3663$. Find where the curve intersects the x-axis.

Determine the distance between the two points. Then determine the midpoint of the segment with the two points as endpoints.

1. $(2, 7)$ and $(14, 12)$ 2. $(-9, 15)$ and $(11, -6)$

3. $(11, 3)$ and $(9, -3)$ 4. $(0, 5)$ and $(-5, 0)$

5. Which point is farther from the origin, $(0, 8)$ or $(-4, 7)$?

6. Determine the coordinates of the point(s) on the y-axis that are 17 units from the point $(15, 6)$.

7. **GEOGRAPHY** On a coordinate grid of Yosemite National Park, Half Dome is located at $(20, 18)$ and Tuolumne Meadows is located at $(32, 28)$. Find the distance between the two points to the nearest tenth of a mile.

8. **GEOMETRY** Find the perimeter of the triangle with vertices $A(-1, 1)$, $B(7, 1)$, and $C(7, 7)$.

9. **WRITING MATHEMATICS** When finding the distance between two points $(2, 3)$ and $(7, 1)$, explain why it makes no difference which point you use as (x_1, y_1) and (x_2, y_2).

PRACTICE

Determine the distance between the two points. Then determine the midpoint of the segment with the two points as endpoints.

1. $(9, 5)$ and $(3, -3)$ 2. $(-8, 12)$ and $(2, -12)$ 3. $(6, 2)$ and $(-1, 3)$

4. $(-11, -4)$ and $(-3, 0)$ 5. $(15.2, 7)$ and $(-11.8, 7)$ 6. $(-6, 2)$ and $(7, 6)$

7. $(2.3, -5.1)$ and $(2.3, 6.5)$ 8. $\left(\frac{1}{2}, 2\right)$ and $\left(1, -\frac{1}{2}\right)$ 9. $\left(2\sqrt{6}, 3\sqrt{5}\right)$ and $\left(4\sqrt{6}, 7\sqrt{5}\right)$

10. **WRITING MATHEMATICS** Paul found the distance between $(92, 5)$ and $(92, 1)$ using the distance formula. Janet found the distance using mental math. Explain how Janet found the distance and write a general method for this type of problem.

11. The point $(3, 5)$ is the midpoint of a segment that has $(7, 11)$ as one endpoint. Find the other endpoint.

12. Find the value(s) of k so that the point $(17, -6)$ is 29 units from the point $(-3, k)$.

13. Find the value(s) of n so that the point $(-11, 9)$ is 15 units from the point $(n, 0)$.

14. **GEOMETRY** The points $(5, 6)$, $(-2, 10)$, and $(2, 4)$ are joined to form a triangle. Give the lengths of the sides. Then state whether the triangle is a right triangle.

15. **GEOMETRY** Show the triangle with vertices $(1, 3)$, $(8, 5)$, and $(3, -4)$ is isosceles.

16. **SURVEYING** Surveyors determine the corners of a quadrilateral lot are located at $(0, 0)$, $(28, 45)$, $(36, 60)$, and $(47, 0)$ on a coordinate grid. Find the perimeter of the lot.

GEOGRAPHY The intersection of the equator and the prime meridian is the origin. The coordinates, in degrees, of Lagos, Nigeria and Pointe-Noire, Congo, are approximately (3°, 6°) and (12°, −5°), respectively.

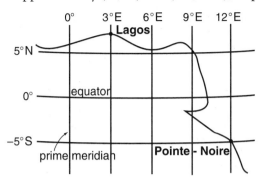

17. Find the distance between the cities. Round to the nearest degree.

18. One degree equals about 70 miles at the equator. About how far apart are the cities?

EXTEND

19. **TRANSPORTATION** A truck driver drove south on Interstate 79 in Pennsylvania, averaging 55 mi/h for 0.6 h, then east on Interstate 80, averaging 62 mi/h for 1.4 h. Approximately how far was the truck driver from the starting point?

20. **GEOMETRY** The vertices of a triangle are $A(0, 0)$, $B(4, 6)$, and $C(6, 2)$. Show that the segment joining the midpoints of $\overline{AB}$ and $\overline{BC}$ is parallel to $\overline{AC}$ and half the length of $\overline{AC}$.

The position of a point in space can be specified by the ordered triple (x, y, z). The distance between two points (x_1, y_1, z_1) and (x_2, y_2, z_2) is given by $d = \sqrt{(x_2 - x_1)^2 + (y_2 - y_1)^2 + (z_2 - z_1)^2}$. Use this formula to determine the distance between each pair of points.

21. $A(-3, 5, -13)$ and $B(-6, 17, -9)$

22. $C(-2, -5, 8)$ and $D(10, 15, -1)$

THINK CRITICALLY

23. Point P is 10 units from point M and 10 units from point N. Is point P necessarily the midpoint of line segment $\overline{MN}$? If your answer is no, provide a counter example.

24. **GEOMETRY** Right triangle $\triangle ABC$ has coordinates $C(0, 0)$, $A(a, 0)$, and $B(0, b)$. Show that the midpoint of the hypotenuse is equidistant from the vertices.

MIXED REVIEW

Solve for x.

25. $\log_2 x = 5$

26. $\log_x (125) = 3$

27. $\log_{49} x = \dfrac{3}{2}$

28. **STANDARDIZED TESTS** The endpoints of the diameter of a circle are $(-12, 7)$ and $(-2, -17)$. Choose the letter giving the coordinates of the center.

A. $(-7, 5)$ B. $(-10, 24)$ C. $(-14, -10)$ D. $(-7, -5)$

Algebra Workshop
Explore Conic Sections

Think Back

Geometric definitions or descriptions often involve distance.

1. Draw two parallel lines. Mark several points that are equidistant from the two lines. Use your drawing to name and describe the set of all points that are equidistant from two parallel lines.

2. Describe the set of all points on a line segment that are equidistant from the endpoints of the segment.

3. Describe the set of all points equidistant from the endpoints of a line segment but not necessarily on the segment.

Explore

The four most important **conic sections** are the *circle*, the *ellipse*, the *parabola*, and the *hyperbola*.

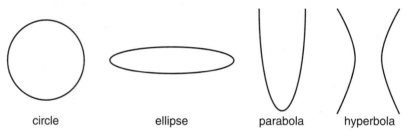

circle   ellipse   parabola   hyperbola

4. Use a compass to draw a large circle on heavy paper. Cut out the circle. Cut a pie-shaped wedge from the edge of the circle to the center. Tape the sides of the wedge together to create a cone.

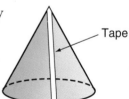

Tape

Use a sheet of paper to represent a plane. Imagine the plane slicing through the cone at various angles. Describe how to cut the cone with the plane to produce a cross-section shaped like

5. a circle   6. an ellipse   7. a parabola

8. Two cones placed point to point form a *double-napped cone*. Describe how you could cut a double-napped cone with a plane so that the cross-section forms a hyperbola.

9. A flashlight emits light in the shape of a cone. Describe how you could create a circle, an ellipse, and a parabola by shining a flashlight against a flat surface. Explain why this works.

Make Connections

10. Place a point at the approximate center of a blank sheet of paper. Label the point C.

11. Sketch and describe the set of all points on the paper that are 5 cm from C.

12. Describe the set of points in a plane equidistant from a point.

Use double concentric circle graph paper.

13. Mark point P on your graph. Find the sum of the distances $PF_1 + PF_2$. Explain.

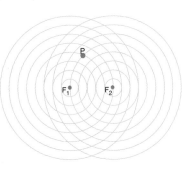

14. Determine and mark other points with the property that the sum of their distances from F_1 and F_2 is the same as the sum you obtained above. How does double concentric circle graph paper help to locate such points?

15. Look at the points you have marked and identify a pattern. Use the pattern to draw the graph of the set of all points the sum of whose distances from F_1 and F_2 is the sum you obtained in Question 14. Your graph should include points that do not lie on concentric circles as well as points that do.

16. Describe the figure you drew and any symmetry you observe.

17. Use a different double concentric circle graph to mark point P on your graph. Find $|PF_1 - PF_2|$. Explain.

18. Find other points with the property that the absolute value of the difference of their distances from F_1 and F_2 is the same as the difference you obtained above. How does double concentric circle graph paper help to find such points?

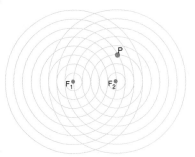

19. Mark enough points so that you can see a pattern in their location. Then graph the set of all points so that the value of the difference of distances from F_1 and F_2 is the difference you obtained in Question 17.

ALGEBRA: WHO, WHERE, WHEN

Apollonius, another third century B.C. Greek showed the curves could be produced by slicing a cone in different directions in his book named *The Conics*. Edmund Halley, the English astronomer, translated Apollonius' book. The comet named for Halley travels along an elliptical path, and Halley used this conic section to predict the time of its return.

Algebra Workshop

20. Describe the figure you drew and any symmetry you observe.

Use concentric circle graph paper with lines.

21. Mark point P and line L as shown.

22. Find a point that is 3 units from both P and L. Find several other points that are equidistant from P and L. How does concentric circle graph paper with lines help to find such points?

23. Mark enough points to identify a pattern. Then graph the set of all points that are equidistant from P and L.

24. Describe the figure you have drawn. Identify any symmetry you observe.

Summarize

25. **MODELING** The conic sections are so named because each can be "sliced" from a cone. Give a real world example that models each of the conic sections you studied in this workshop.

26. **WRITING MATHEMATICS** Explain how each conic section can be defined as the intersection of a cone and a plane and as a set of points that satisfy a distance condition.

27. **GOING FURTHER** The circle, the ellipse, the hyperbola, and the parabola are the major conic sections. There are also three **degenerate conic sections,** the point, the line, and two intersecting lines. Explain how each degenerate conic section can be formed by the intersection of a plane and a double-napped cone.

PROJECT *Connection*   The following activities demonstrate ideas about the conic sections in general and circles in particular. Write a script that a narrator, who may speak off camera, can use while an actor performs the activity. Create posters or any other explanatory material that will enhance your presentation. Rehearse, then film your work.

1. Mold clay inside of a funnel to create clay cones that can be easily cut with a plastic knife. Show how each conic section can be cut from a cone.

2. Use a string, marker, push pin, and piece of cardboard or wood to demonstrate how a circle can be drawn. Emphasize how the properties of the circle relate to the set-up and steps of the manipulative activity.

12.3 Circles

Explore

1. Draw a circle with center at $(0, 0)$ and radius 10 units.

2. Complete the table below by estimating the missing x- or y-coordinates for points on the circle.

x	y	$x^2 + y^2$	$\sqrt{x^2 + y^2}$
6	■	■	■
6	■	■	■
■	0	■	■
■	0	■	■
−3.5	■	■	■
−3.5	■	■	■

3. Explain the patterns you see in the table.

4. Predict the x-coordinates of points on the circle with a y-coordinate of 5.4. Check by referring to the graph.

5. A circle of radius 37 is drawn with its center at $(0, 0)$. Find the y-coordinates of points on the circle with x-coordinates of -35.

SPOTLIGHT ON LEARNING

WHAT? In this lesson you will learn
- to use the Pythagorean theorem to write the standard equation of a circle.
- to apply the standard equation of a circle.

WHY? The equation of a circle can help you solve problems in sports, navigation, and communications.

Build Understanding

A **circle** is the set of all points in a plane that are equidistant from a given point, called the **center** of the circle. The distance between the center and any point on the circle is the **radius.** You can use the center of the circle, the radius, and the distance formula to write the equation of a circle.

EXAMPLE 1

Write the equation of a circle with center at $(2, -1)$ and radius of 3.

Solution

Let (x, y) represent any point on the circle. Since the radius is 3, the distance between (x, y) and the center $(2, -1)$ is 3.

$$\sqrt{(x - 2)^2 + (y - (-1))^2} = 3 \quad \text{Use the distance formula.}$$
$$\sqrt{(x - 2)^2 + (y + 1)^2} = 3 \quad \text{Simplify.}$$
$$(x - 2)^2 + (y + 1)^2 = 9 \quad \text{Square both sides.}$$

The equation of the circle is $(x - 2)^2 + (y + 1)^2 = 9$. ◄

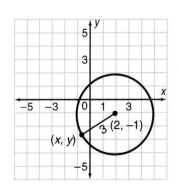

Look at the equation of the circle in Example 1. Notice that the equation tells you both the center and the radius of the circle.

−2 is the opposite of the *x*-coordinate of the center. ↓   +1 is the opposite of the *y*-coordinate of the center. ↓   9 is the square of the radius. ↓

$$(x - 2)^2 \qquad + (y + 1)^2 \qquad = 9$$

> **STANDARD FORM OF THE EQUATION OF A CIRCLE**
>
> **The standard form of the equation of a circle with radius *r* and center (*h*, *k*) is**
>
> $$(x - h)^2 + (y - k) = r^2$$

If the center of a circle is at (0, 0), then $h = 0$ and $k = 0$, and the equation of the circle is

$$(x - 0)^2 + (y - 0)^2 = r^2$$
$$x^2 + y^2 = r^2$$

The graph of the equation $x^2 + y^2 = 16$ is a circle with center at (0, 0) and radius 4. If you translate every point of this circle to the right 3 units and up 2 units, then the translated circle has center at (3, 2) and this circle's equation is $(x - 3)^2 + (y - 2)^2 = 16$.

Use the standard form to write the equation of a circle when you know the center and the radius.

EXAMPLE 2

INDUSTRIAL DESIGN Find the equation of the knob on this audio designer's sketch for the front panel of a radio.

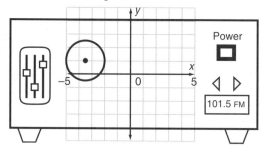

Solution

The circle has its center at (−3.5, 1) and a radius of 1.5.

$$(x - (-3.5))^2 + (y - 1)^2 = (1.5)^2 \qquad \text{Use the standard form.}$$
$$(x + 3.5)^2 + (y - 1)^2 = 2.25 \qquad \text{Simplify.}$$

The equation of the circle is $(x + 3.5)^2 + (y - 1)^2 = 2.25$. ◀

If the equation of a circle is given to you in expanded form, you may have to complete the square in order to find the center and the radius.

EXAMPLE 3

Find the center and radius of the circle $x^2 + y^2 - 8x + 4y = 29$. Then graph the circle.

Solution

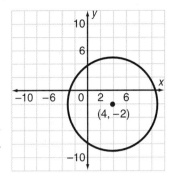

$$x^2 + y^2 - 8x + 4y = 29$$
$$(x^2 - 8x) + (y^2 + 4y) = 29 \quad \text{Group } x \text{ and } y \text{ terms.}$$
$$(x^2 - 8x + 16) + (y^2 + 4y + 4) = 29 + 20 \quad \text{Complete the square.}$$
$$(x - 4)^2 + (y + 2)^2 = 49 \quad \text{Factor and simplify.}$$

The circle has its center at $(4, -2)$. Since $7^2 = 49$, the radius is 7. ◄

As you can see from the graph of a circle, there are two values of y for each value of x. Therefore, the equation of a circle does not represent a function. Since most graphing utilities graph only functions, you must separate the equation of a circle into two equations, each of them a function, and then graph both equations.

EXAMPLE 4

NAVIGATION A bush pilot was forced to land at the point $(-7, 12)$ on a coordinate map of a section of the Alaska wilderness. After hiking in a straight line for 5 h at an average speed of 2.2 mi/h, the pilot was spotted by a search plane. Using a graphing utility, show the pilot's possible positions at the time of rescue.

Solution

After 5 h, the pilot was somewhere on a circle with center $(-7, 12)$ and radius equal to his speed times the time traveled. So, $r = 11$ mi. The equation of the circle is

$$(x + 7)^2 + (y - 12)^2 = 121$$

Solve for y as a function of x.

$$(y - 12)^2 = 121 - (x + 7)^2$$
$$y - 12 = \pm\sqrt{121 - (x + 7)^2}$$
$$y = 12 \pm \sqrt{121 - (x + 7)^2}$$

To show the possible rescue points, graph both equations

$$y = 12 + \sqrt{121 - (x + 7)^2}$$
$$y = 12 - \sqrt{121 - (x + 7)^2}$$

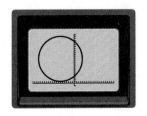

◄

Write the equation of the circle with the given center and radius.

1. center, $(0, 0)$; radius, 5

2. center, $(-2, 3)$; radius, 9

3. center, $(3.7, -6.5)$; radius, π

4. center, $(15, 0)$; radius, $\sqrt{3}$

Determine the center and radius of each circle. Then graph the circle.

5. $x^2 + y^2 = 100$

6. $x^2 + (y + 3)^2 = 4$

7. $(x + 12)^2 + (y - 5)^2 = 20$

8. $x^2 + y^2 + 8x - 2y + 15 = 0$

9. COMMUNICATIONS Many communications satellites circle the earth in *synchronous* orbits that exactly match the rotation period of the earth. Synchronous satellites orbit at an altitude of 22,300 mi above the earth's surface. Assuming that the earth's center is at the origin and that satellite orbits are circular, write an equation of a synchronous satellite orbit. (The radius of the earth is about 3,960 mi.)

Write an equation for each graph in a form you could use to enter into a graphing utility.

10.

11.

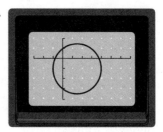

12. WRITING MATHEMATICS Explain the similarities and differences between the graphs of $x^2 + y^2 = 36$, $(x - 2)^2 + (y - 2)^2 = 36$, and $(x - 2)^2 + (y - 2)^2 = 64$.

PRACTICE

Determine the equation of the circle with the given center and radius.

1. center, $(0, 0)$; radius, 14

2. center, $(6, 7)$; radius, 1

3. center, $(-9, -3)$; radius, 7

4. center, $(0, -4)$; radius, 2.8

5. center, $(4.3, -7.7)$, radius, 3π

6. center, $(\frac{2}{3}, \frac{1}{2})$, radius, $\frac{2}{5}$

Determine the center and radius of each circle. Then graph the circle.

7. $x^2 + y^2 = 9$

8. $x^2 + (y - 3)^2 = 4$

9. $(x + 1)^2 + y^2 = 25$

10. $(x - 2)^2 + (y + 3)^2 = 35$

11. $x^2 + y^2 + 6x - 4y - 15 = 0$

12. $x^2 + y^2 - 6x + 6y + 2 = 0$

13. $x^2 + y^2 + 14y = -48$

14. $2x^2 + 2y^2 - 20x + 32y = -130$

NAVIGATION The illustration depicts the positions of two planes on a radar screen in an airport control tower located at the origin. Units are in miles. The pilots of both planes have been told to continue circling and to maintain airspeeds of 240 mi/h. The planes are at the same altitude.

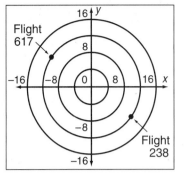

15. Give the equation of the circle on which the two planes are flying.

16. The planes' positions are the endpoints of a diameter of the circle. How far apart are they on a line? How far apart are they along their flight routes?

17. How many hours will it take Flight 238 to make one complete circuit of the tower?

18. The x-coordinate of Flight 617's current position is -9. What is the y-coordinate?

19. **SPORTS** Concentric circles are circles with the same center but different radii. Write a set of equations to model the design of a dart board with concentric circles of diameter 5.2 cm, 15.6 cm, 26.0 cm, 36.4 cm, and 46.8 cm.

20. **WRITING MATHEMATICS** Write a word problem that requires finding the coordinates of the center of a circle.

┌ EXTEND

Determine the equation of each circle described.

21. A circle with its center at the origin that passes through the point $(5, -12)$.

22. A circle with its center at $(-3, 4)$ that has an area of 36π.

23. A circle with center $\left(\frac{1}{21}, -\frac{3}{2}\right)$ and a diameter of 10.

24. The circle is the translation of $x^2 + y^2 = 15$ right 3 units and down 7 units.

25. A circle has its center at $(-2, 5)$ and a radius of $\sqrt{40}$.

   a. Write the equation of the circle.

   b. Find the x- and y-intercepts of the circle.

26. A circle has a diameter with endpoints $(-3, 4)$ and $(1, 2)$.

   a. Determine the center of the circle. Explain your method.

   b. Determine the radius of the circle. Explain your method.

   c. What is the equation of the circle?

27. Write the equation of a circle that has a diameter with endpoints $(-2, 7)$ and $(4, -3)$.

THINK CRITICALLY

28. A circle tangent to a line intersects that line in exactly one point. Determine the equation of a circle with center at $(-4, 2)$ that is tangent to the x-axis.

29. The equation of a circle is $(x - h)^2 + (y - k)^2 = 16$. Explain how you can determine the quadrant in which the center of the circle lies by examining h and k.

30. A *semicircle* (half-circle) with radius a has its center at the origin. Point B (b, c) is on the circle. Prove that $\angle ABC$ is a right angle.

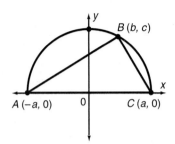

MIXED REVIEW

Give the slope and y-intercept of each line.

31. $y = 3x - 5$

32. $4x - 2y + 5 = 0$

33. $4(x - 2y) = 4y - 2$

Determine the inverse of $f(x)$.

34. $f(x) = 5x$

35. $f(x) = x + 4$

36. $f(x) = \dfrac{1}{x + 1}$

37. STANDARDIZED TESTS Choose the letter giving the radius of the circle $x^2 + y^2 = 10x + 24$.

A. 5 **B.** 7 **C.** $2\sqrt{6}$ **D.** $\sqrt{5}$

PROJECT *Connection* The following activities demonstrate ideas about ellipses and hyperbolas.

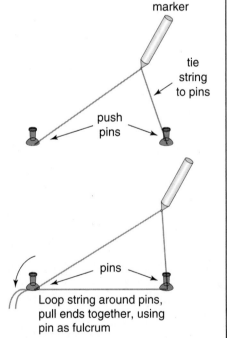

1. Refer to the diagram to construct a manipulative using push pins, a piece of wood or cardboard, and a marker for drawing an ellipse. Drag the marker around the pins, keeping the string taut. Be sure your narration explains why this manipulative demonstrates the properties of an ellipse.

2. Using similar materials as above, follow the diagram to construct a manipulative for drawing a hyperbola. You must pull on the two ends of string together. As you do this, the marker will be dragged towards the transverse axis. Once the marker reaches the transverse axis, continue the hyperbola by letting up on the string as you drag the marker away from the transverse axis. Be sure your narration explains why this manipulative demonstrates the properties of a hyperbola.

12.4 Ellipses

Explore

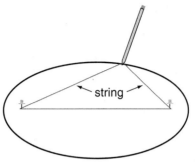

The diagram shows a method of drawing an ellipse. A loop of string is around two thumb tacks. A pencil held tight against the string is moved around the tacks, creating the ellipse.

1. Explain why the method works.

2. How far apart are the tacks? How is the length of the loop related to the distance between the tacks?

3. What part of the ellipse is represented by the tacks?

SPOTLIGHT ON LEARNING

WHAT? In this lesson you will learn
• to write and apply the standard form of the equation of an ellipse.

WHY? Ellipses are fundamental to an understanding of planetary motion, and the reflective properties of ellipses are widely applied in architectural designs.

Build Understanding

An ellipse can be described as a "flattened" circle. Planets and satellites travel in elliptical orbits. When a bat hits a baseball, the circular ball is momentarily compressed into an ellipse with a longer axis that is as much as three times the length of its shorter axis.

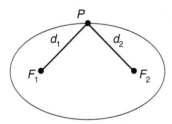

$$d_1 + d_2 = \text{constant}$$

An **ellipse** is the set of all points in a plane the sum of whose distances from two fixed points F_1 and F_2, the **foci**, is constant.

An ellipse has two axes of symmetry. The longer axis is the **major axis** $\overline{RS}$. The shorter axis is the **minor axis** $\overline{AB}$. Each axis intersects the ellipse at two **vertices**. The axes intersect each other at the **center** of the ellipse. The foci lie on the major axis, equidistant from the center.

ALGEBRA: WHO, WHERE, WHEN

Before the 16th century, planets were believed to be "perfect" heavenly bodies that moved in perfect circles. The German astronomer Johannes Kepler proved in 1609 that the planets move in elliptical orbits with the sun at one focus.

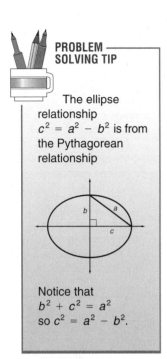
The **standard form of the equation of an ellipse** with center $(0, 0)$, foci at $(-c, 0)$ and $(c, 0)$, and major and minor axes of lengths $2a$ and $2b$, where $a > b$, is

$$\frac{x^2}{a^2} + \frac{y^2}{b^2} = 1, \text{ where } c^2 = a^2 - b^2.$$

This ellipse has a horizontal major axis on the x-axis.

The **standard form of the equation of an ellipse** with center $(0, 0)$, foci at $(0, -c)$ and $(0, c)$, and major and minor axes of lengths $2a$ and $2b$, where $a > b$, is

$$\frac{x^2}{b^2} + \frac{y^2}{a^2} = 1, \text{ where } c^2 = a^2 - b^2$$

This ellipse has a vertical major axis on the y-axis.

EXAMPLE 1

COMMUNICATIONS A communications satellite is launched into an elliptical orbit with Earth at one focus. The major axis is 32,400 mi long. The minor axis is 24,800 mi long. Write and graph the equation of the orbit, showing the position of Earth. Assume the foci lie on the x-axis and the origin is the center of the ellipse.

Solution

length of major axis: $2a = 32,400$
$a = 16,200$

length of minor axis: $2b = 24,800$
$b = 12,400$

$$\frac{x^2}{a^2} + \frac{y^2}{b^2} = 1 \qquad \text{Write the standard form.}$$

$$\frac{x^2}{(16,200)^2} + \frac{y^2}{(12,400)^2} = 1 \qquad \text{Substitute.}$$

$$\frac{x^2}{262,440,000} + \frac{y^2}{153,760,000} = 1 \qquad \text{Simplify.}$$

Use $c^2 = a^2 - b^2$ to determine the foci.
$$c^2 = (16,200)^2 - (12,400)^2$$
$$c = \sqrt{(16,200)^2 - (12,400)^2}$$
$$c \approx 10,425$$

The equation of the orbit is

$$\frac{x^2}{262,440,000} + \frac{y^2}{153,760,000} = 1.$$

Note that Earth appears at the focus $(10,425, 0)$.

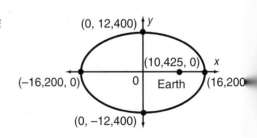

If you are given the graph of an ellipse, you can use it to find its equation.

EXAMPLE 2

Write the equation of the ellipse shown.

Solution

The center of the ellipse is $(0, 0)$. The vertical major axis is $a = 12$ and $c = 8$.

$$8^2 = 12^2 - b^2 \qquad c^2 = a^2 - b^2$$
$$80 = b^2$$

$$\frac{x^2}{b^2} + \frac{y^2}{a^2} = 1 \qquad \text{Write the standard form.}$$

$$\frac{x^2}{80} + \frac{y^2}{12^2} = 1 \qquad \text{Simplify.}$$

$$\frac{x^2}{80} + \frac{y^2}{144} = 1 \qquad \text{Write the equation of the ellipse.} \quad \blacktriangleleft$$

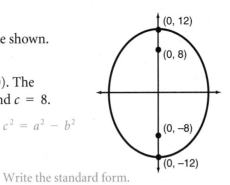

Recall that the terms $(x - h)^2$ and $(y - k)^2$ appear in the standard form of the equation of a circle with its center at (h, k). The same terms appear in the standard form of the equation.

> **THE STANDARD FORM OF THE EQUATION OF AN ELLIPSE**
>
> **The standard form of the equation of an ellipse with center at (h, k), major axis 2a, and minor axis 2b is**
> $$\frac{(x - h)^2}{a^2} + \frac{(y - k)^2}{b^2} = 1 \text{ or } \frac{(x - h)^2}{b^2} + \frac{(y - k)^2}{a^2} = 1$$

EXAMPLE 3

Determine the foci, vertices, center, and lengths of the major and minor axes of the ellipse $4x^2 - 8x + 9y^2 + 36y = -4$. Then graph.

Solution

$$4x^2 - 8x + 9y^2 + 36y = -4$$
$$4(x^2 - 2x) + 9(y^2 + 4y) = -4 \qquad \text{Factor.}$$
$$4(x^2 - 2x + 1) + 9(y^2 + 4y + 4) = -4 + 4 + 36 \qquad \text{Complete the squares.}$$
$$4(x - 1)^2 + 9(y + 2)^2 = 36 \qquad \text{Factor.}$$
$$\frac{(x - 1)^2}{9} + \frac{(y + 2)^2}{4} = 1 \qquad \text{Divide both sides by 36.}$$
$$\frac{(x - 1)^2}{3^2} + \frac{(y + 2)^2}{2^2} = 1$$

The center is at $(1, -2)$. The major axis vertices are at $(4, -2)$ and $(-2, -2)$. The minor axis vertices are at $(1, 0)$ and $(1, -4)$. The major axis is $2a = 6$. The minor axis is $2b = 4$. The foci are $\sqrt{5}$ units left and right of the center, at $(1 + \sqrt{5}, -2)$ and $(1 - \sqrt{5}, -2)$.

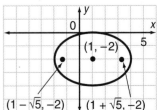

COMMUNICATING ABOUT ALGEBRA

Discuss how you would graph the ellipse $\dfrac{x^2}{9} + \dfrac{y^2}{4} = 1$ with a graphing utility.

CHECK UNDERSTANDING

In Example 2, why is $\dfrac{x^2}{b^2} + \dfrac{y^2}{a^2} = 1$ used as the standard form of the equation?

PROBLEM SOLVING TIP

Remember to multiply the term being added to complete the square by the coefficient of the trinomial and add the correct number to the opposite side.

Determine each value for the ellipse shown at the right.

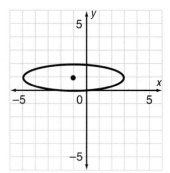

1. The length of the major axis
2. The length of the minor axis

3. The coordinates of the vertices
4. The coordinates of the center

5. The coordinates of the foci
6. The equation of the ellipse

Find the foci, vertices, center, and lengths of the major and minor axes. Then graph.

7. $\dfrac{x^2}{100} + \dfrac{y^2}{36} = 1$

8. $\dfrac{x^2}{9} + \dfrac{y^2}{25} = 1$

9. $x^2 + 4y^2 = 36$

10. $\dfrac{(x + 2)^2}{169} + \dfrac{(y - 3)^2}{25} = 1$

11. Write the equation of the ellipse $9x^2 + 54x + 16y^2 - 32y = 47$ in standard form. Then give the center, vertices, and foci of the ellipse.

12. TRAVEL Island Tours operates sightseeing cruises among several Hawaiian islands. Due to fuel limitations, the tour ship can travel a maximum of 30 mi during one cruise between two islands 16 mi apart.

   a. Describe the shape of the area the ship can cover during the tour.
   b. Write the equation of the perimeter of the area you have described.

13. WRITING MATHEMATICS Imagine an ellipse. Suppose you move the foci toward the center. What happens to the shape of the ellipse?

PRACTICE

Write the equation of each ellipse.

1.
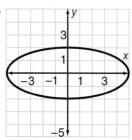

2.
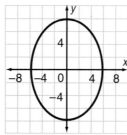

3.
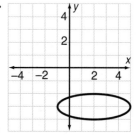

Find the foci, vertices, center, and lengths of the major and minor axes. Then graph.

4. $\dfrac{x^2}{25} + \dfrac{y^2}{16} = 1$

5. $\dfrac{x^2}{36} + \dfrac{y^2}{20} = 1$

6. $2x^2 + y^2 = 8$

7. $169x^2 + 25y^2 = 4225$

8. $\dfrac{(x + 5)^2}{9} + \dfrac{(y - 2)^2}{3} = 1$

9. $9x^2 + 72x + 25y^2 - 150y = -144$

10. **ARCHITECTURE** A whisper spoken at one focus of the elliptical Statuary Hall in the United States Capitol Building can clearly be heard at the other focus. The major and minor axes of the hall are 96 feet and 46 feet, respectively.

    a. Find the equation of the ellipse.   **b.** Find the coordinates of the foci.

11. **WRITING MATHEMATICS** You are given the equation of an ellipse in standard form. Explain how to find the coordinates of the vertices.

Write the equation of each ellipse described below.

12. The foci are at $(6, 0)$ and $(-6, 0)$. The length of the major axis is 20 and the center is $(0, 0)$.

13. The endpoints of the major axis are at $(-6, -6)$ and $(4, -6)$. The endpoints of the minor axis are at $(-1, -8)$ and $(-1, -4)$.

EXTEND

The **eccentricity**, e, of an ellipse is a measure of its "flatness". Use $e = \dfrac{c}{a}$ to find the eccentricity of each ellipse. Round to the nearest hundredth.

14. $\dfrac{x^2}{25} + \dfrac{y^2}{16} = 1$ 15. $\dfrac{x^2}{169} + \dfrac{y^2}{144} = 1$ 16. $\dfrac{x^2}{289} + \dfrac{y^2}{64} = 1$ 17. a circle

18. Find the greatest and least possible eccentricities for an ellipse. Explain your reasoning.

19. **ASTRONOMY** Planets move in elliptical orbits, the most eccentric being that of Pluto. Pluto's distance from the sun, one of the foci of the planet's orbit, ranges from 2.75 billion mi at its closest, to 4.55 billion mi at its farthest. Find the eccentricity of Pluto's orbit. (Hint: Sketch the orbit. Where on the ellipse is Pluto closest to the sun? Where is it farthest? What is the sum of these distances?)

20. Write the equation of Pluto's orbit.

THINK CRITICALLY

To show that a circle is a special type of ellipse, rewrite the equation in elliptical form.

21. Show that the equation of the circle $x^2 + y^2 = 25$ can be written in the form of an equation of an ellipse.

MIXED REVIEW

Use synthetic division to find each quotient.

22. $(x^2 - x - 12) \div (x - 4)$ 23. $(x^3 - 7x^2 - 13x + 10) \div (x + 2)$

24. $(3x^4 - 25x^2 - 18) \div (x - 3)$ 25. $(x^3 - 1) \div (x - 1)$

26. **STANDARDIZED TESTS** Choose the letter giving the lengths of the major and minor axes of the ellipse $9x^2 + 54x + y^2 - 10y + 97 = 0$.

    A. 9 and 1   **B.** 5 and -3 **C.** 6 and 2   **D.** 25 and 9

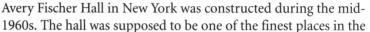

Career
Acoustical Engineer

Avery Fischer Hall in New York was constructed during the mid-1960s. The hall was supposed to be one of the finest places in the world for listening to music. Instead, some members of the opening night audience could barely hear, while others were nearly blasted out. After analyzing their mistakes, acoustical engineers changed the shape of the hall and the materials used in its construction. Avery Fischer Hall eventually became the sound showcase it was intended to be. The difficulties demonstrated the complexity of designing a concert hall in which every member of the audience will hear the same thing.

Decision Making

1. Why might the acoustics in Avery Fischer Hall have been different on opening night than they had been during earlier tests?

2. Give an example of how the shape, building material, or some other factor has affected the acoustical properties of a room.

3. The shape and size of a room greatly affect its acoustical properties. Suppose that a sound is made at one focus of an **ellipsoid**, a three-dimensional figure obtained by rotating an ellipse around its major axis. No matter where the sound strikes the ellipsoid, it will be reflected to the other focus. Describe the acoustical properties of a room with a semi-ellipsoid for a roof.

4. Speaker A is 6 ft from the center of a room, represented by the y-axis. Speaker B, is twice as loud as speaker A and 12 ft from the center. Answer the following questions to find the points in the room where volume V from both speakers is the same.

   a. Find expressions d_A and d_B, the distance of the point $P(x, y)$ from each speaker.

   b. The volume of sound varies inversely as the square of the distance from a sound source. Therefore, $V_P = \dfrac{kV_A}{d_A{}^2}$ and $V_P = \dfrac{kV_B}{d_B{}^2}$ for some constant k. Using the fact that $V_B = 2V_A$, write an expression relating d_A and d_B.

   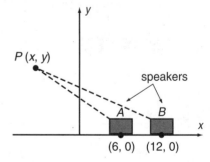

   c. Simplify the expression and interpret your results.

Explore

1. Graph $y_1 = 2\sqrt{x^2 - 4}$ and $y_2 = -2\sqrt{x^2 - 4}$. Which conic sections do these graphs represent?

2. Graph $y_3 = 2x$ and $y_4 = -2x$ on the same graph as y_1 and y_2. Do these graphs intersect the graphs in Question 1?

3. For $x = 3$, calculate $y_1 = 2\sqrt{x^2 - 4}$ and $y_3 = 2x$, and examine the graphs of these functions at $x = 3$. Calculate the difference between $y_1(3)$ and $y_3(3)$. Repeat for $x = 5, 10, 100,$ and 1000. What do you notice about y_1 and y_3?

Build Understanding

Hyperbolas have much in common with ellipses. The definitions and standard equations of these two conic sections are similar.

A **hyperbola** is the set of all points in the plane where the difference of the distance between two fixed points F_1 and F_2, is constant. The two fixed points are called foci.

A hyperbola has two **branches**. The **transverse axis** joins the two vertices of a hyperbola. At right angles to the transverse axis and intersecting it at the **center** is the **conjugate axis**. As the hyperbola moves away from the foci, the two branches approach but never reach the asymptotes, which intersect the center and corners of the transverse axis–conjugate axis rectangle. The asymptotes are not a part of the hyperbola, but they are used as an aid in graphing.

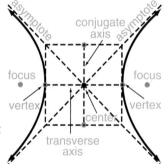

─── THE STANDARD FORM OF THE EQUATION OF A HYPERBOLA ───

The standard form of the equation of a hyperbola with center at (h, k), foci at $(h + c, k)$ and $(h - c, k)$, and a horizontal transverse axis parallel to the x-axis is

$$\frac{(x - h)^2}{a^2} - \frac{(y - k)^2}{b^2} = 1, \text{ where } c^2 = a^2 + b^2$$

The standard form of the equation of a hyperbola with center at (h, k), foci at $(h, k + c)$ and $(h, k - c)$, and a vertical transverse axis parallel to the y-axis is

$$\frac{(y - k)^2}{a^2} - \frac{(x - h)^2}{b^2} = 1, \text{ where } c^2 = a^2 + b^2$$

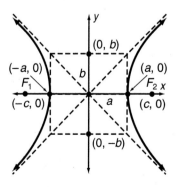

If the hyperbola has its center at $(0, 0)$, replace $(x - h)^2$ with x^2 and $(y - k)^2$ with y^2.

EXAMPLE 1

Determine the equation of the hyperbola.

Solution

$$c^2 = a^2 + b^2$$
$$13^2 = 5^2 + b^2 \qquad a = 5, c = 13$$
$$144 = b^2 \qquad \text{Substitute.}$$
$$12 = b \qquad \text{Solve for } b.$$

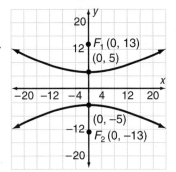

The transverse axis is vertical, so the

equation is $\dfrac{y^2}{25} - \dfrac{x^2}{144} = 1$. ◀

The center is not part of the hyperbola, but it is very useful for graphing. Draw asymptotes with slopes of $\pm\dfrac{b}{a}$ for a horizontal hyperbola or $\pm\dfrac{a}{b}$ for a vertical hyperbola through the center of the hyperbola. Then use the asymptotes and the vertices to graph.

EXAMPLE 2

Graph $4x^2 - 16x - 9y^2 - 54y = 101$ showing the foci.

Solution

$$4x^2 - 16x - 9y^2 - 54y = 101 \qquad \text{Convert to standard form.}$$
$$4(x^2 - 4x) - 9(y^2 + 6y) = 101 \qquad \text{Group terms and factor.}$$
$$4(x^2 - 4x + 4) - 9(y^2 + 6y + 9) = 36 \qquad \text{Complete the squares.}$$
$$4(x - 2)^2 - 9(y + 3)^2 = 36 \qquad \text{Group terms and factor.}$$
$$\frac{(x - 2)^2}{9} - \frac{(y + 3)^2}{4} = 1 \qquad \text{Divide by 36.}$$
$$\frac{(x - 2)^2}{3^2} - \frac{(y + 3)^2}{2^2} = 1$$

The equation shows that the center of the hyperbola is at $(2, -3)$, $a = 3$, $b = 2$, and the transverse axis is parallel to the x-axis.

Since $c^2 = a^2 + b^2$, $c = \sqrt{3^2 + 2^2}$. So $c = \sqrt{13}$. The foci are at $\left(2 - \sqrt{13}, -3\right)$ and $\left(2 + \sqrt{13}, -3\right)$. To aid in graphing the hyperbola, sketch asymptotes with slopes of $\pm\dfrac{2}{3}$ and $-\dfrac{2}{3}$ through the center of the hyperbola. ◀

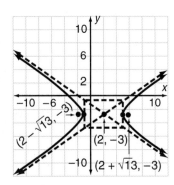

A **rectangular** hyperbola is a special type of hyperbola with its center at the origin, the coordinate axes as asymptotes, and $xy = k$ for some constant k as its equation. If $k > 0$, the branches lie in Quadrants I and III. If $k < 0$, the branches lie in Quadrants II and IV.

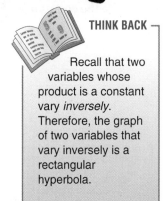

EXAMPLE 3

TRANSPORTATION A truck driver must make a delivery 360 mi away by the end of the day. Graph the truck driver's possible average rate of speed compared to travel time.

Solution

Let $x =$ speed in miles per hour and let $y =$ time in hours. Since speed $\cdot$ time $=$ distance, $xy = 360$.

The graph is a rectangular hyperbola. Since $k = 360$, the branches are in Quadrants I and III. The points in Quadrant III are not relevant to the problem. Points in Quadrant I represent an average rate of speed and a length of time the driver could travel. The real world solutions speeds between 0 and the speed limit. ◄

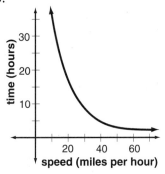

THINK BACK

Recall that two variables whose product is a constant vary *inversely*. Therefore, the graph of two variables that vary inversely is a rectangular hyperbola.

TRY THESE

Write the equation of each hyperbola.

1.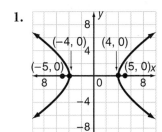

2.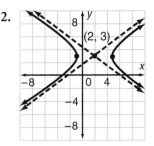

3.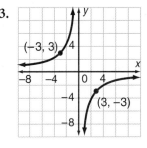

4. A hyperbola has its center at $(0, 0)$, a transverse axis of length 12, and foci at $\left(0, 2\sqrt{10}\right)$ and $\left(0, -2\sqrt{10}\right)$. Find the equation of the hyperbola.

Graph the hyperbola and show the location of the vertices.

5. $\dfrac{x^2}{225} - \dfrac{y^2}{64} = 1$

6. $\dfrac{(y - 3)^2}{9} - \dfrac{(x + 1)^2}{1} = 1$

7. $xy = 4$

8. Determine the center, vertices, foci, and slopes of the asymptotes of the hyperbola $y^2 + 8y - 2x^2 + 4x = -6$.

9. Write the equations you would enter in a graphing utility in order to graph the hyperbola $8x^2 - 15(y - 3)^2 = 120$. Then graph the hyperbola with a graphing utility.

10. **ASTRONOMY** A comet travels along one branch of a hyperbola with the Sun at its focus. Using a graph with a scale of 1 unit = 5 million mi, the vertex of the comet's path is $(4, 0)$ and the focus is $(6, 0)$. Find the equation of the comet's path.

11. **WRITING MATHEMATICS** You are given the slopes of the asymptotes of a horizontal hyperbola with its center at the origin. Explain how you could find the vertices and foci.

PRACTICE

Write the equation of each hyperbola.

1.
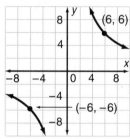

2.
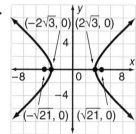

3.
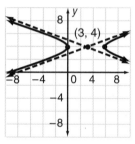

4.
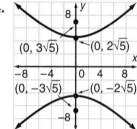

5.
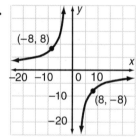

6.
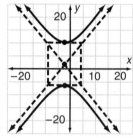

Write the equation of each hyperbola described below.

7. center at $(0, 0)$, vertical transverse axis of length 36, conjugate axis of length 20.

8. center at $(5, -3)$, foci at $(13, -3)$ and $(-3, -3)$, transverse axis of length 10

Graph the hyperbola and show the location of the vertices.

9. $\dfrac{x^2}{9} - \dfrac{y^2}{4} = 1$

10. $25y^2 - 16x^2 = 400$

11. $\dfrac{(y - 1)^2}{6} - \dfrac{(x - 4)^2}{9} = 1$

12. $\dfrac{(x + 2)^2}{10} - \dfrac{y^2}{4} = 1$

13. $xy = 4$

14. $0.01xy = -1$

Determine the center, vertices, foci, and slopes of the asymptotes of the hyperbola.

15. $25x^2 - 9y^2 - 100x - 72y = 269$

16. $y^2 - 3x^2 + 6y + 6x = 18$

17. Write the equations you would enter in a graphing utility in order to graph the hyperbola $\dfrac{(y - 3)^2}{16} - \dfrac{(x + 5)^2}{20} = 1$. Then graph the hyperbola with a graphing utility.

18. **OCEANOGRAPHY** Boyle's Law states that at a constant temperature, the volume of a gas varies inversely with the pressure. At sea level, the atmospheric pressure is 14.7 lb/in². A SCUBA diver is at a depth of 100 ft where the pressure is 58.7 lb/in.². If the diver exhales a bubble of 10 cm³, what is the volume of the bubble just before it reaches the surface of the water? Round to the nearest tenth.

19. **WRITING MATHEMATICS** Explain why rectangular hyperbola branches are in Quadrants I and III when $k > 0$ and in Quadrants II and IV when $k < 0$.

EXTEND

Determine whether the graph of the equation will be a circle, ellipse, or hyperbola.

20. $3x^2 + 3y^2 = 12$ 21. $6x^2 - 6y^2 = 36$ 22. $9x^2 + 16y^2 = 144$

23. **COMMUNICATIONS** Rangers in fire towers 8800 ft apart are speaking by telephone. One ranger hears thunder 5 s after hearing it over the phone. Sound travels at 1,100 ft/s. Find the equation of all possible locations of the lightning strike that caused the thunder.

24. **GEOMETRY** Show that the two asymptotes of the hyperbola $x^2 - y^2 = 9$ are perpendicular to each other.

25. **GEOMETRY** Find values of a and b for which the asymptotes of $\dfrac{x^2}{a^2} - \dfrac{y^2}{b^2} = 1$ are perpendicular.

THINK CRITICALLY

26. If you are given only the equation of the two asymptotes of a hyperbola, can you determine if the hyperbola has a vertical or horizontal orientation? Explain.

27. A hyperbola with a horizontal transverse axis passes through $(7, -2)$. The equations of its asymptotes are $3x - 4y = 17$ and $3x + 4y = 1$. Determine the equation of the hyperbola.

28. **WRITING MATHEMATICS** Suppose you graph the equation $\dfrac{x^2}{a^2} - \dfrac{y^2}{b^2} = 1$, where $a > b$. Then you interchange the values of a and b and graph the new equation. Explain the similarities and differences between the two graphs. Also explain how you reached your conclusions.

MIXED REVIEW

A spinner has eight equal sections numbered from 1 to 8. Find the probability of spinning the given number(s).

29. 3   30. an odd number

31. 6 or 7   32. a number less than 4

33. **STANDARDIZED TESTS** $f(x) = 2x - 3$ and $g(x) = 3x + 1$. Choose the letter giving $f(g(x))$.

   A. $6x - 9$ **B.** $6x - 1$ **C.** $6x + 2$ **D.** $6x - 8$

12.6 Parabolas

Explore

1. Using a graphing utility, graph the equation $y = \dfrac{x^2}{4p}$ where $p = 1$. Which conic section does this graph represent?

2. Let $p = -1$ and graph $y = \dfrac{x^2}{4p}$. How is this graph related to the graph in Question 1?

3. To examine equations of the form $y^2 = 4px$, you must graph $y_1 = \sqrt{4px}$ and $y_2 = -\sqrt{4px}$ on the same coordinate axes. Let $p = 1$ and $p = -1$. How are the graphs related?

Build Understanding

In your study of quadratic functions in Chapter 6, you learned that the graph of any function of the form $y = ax^2 + bx + c$ is a *parabola*. The reflective properties of parabolas are employed in flashlights, car headlights, radio antennas, solar energy collectors and satellite dishes. Like the other conic sections, the parabola can be defined geometrically.

A **parabola** is the set of all points in a plane that are equidistant from a line called the **directrix** and a point not on the line called the **focus**.

ALGEBRA: WHO, WHERE, WHEN

Every incoming ray of light is directed to a single point in front of a parabolic mirror, the point where the greatest light and heat intensity is achieved. Astronomers chose the term *focus* for this point, after the Latin word meaning "fireplace."

The **axis of symmetry** of the parabola is the line through the focus, perpendicular to the directrix. The **vertex** of the parabola is the point on the axis of symmetry midway between the focus and the directrix. A parabola of this form can have a horizontal axis of symmetry or a vertical axis of symmetry.

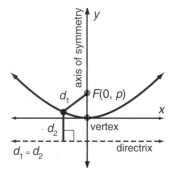

The **standard form of the equation of a parabola** with focus $(0, p)$, directrix $y = -p$, and vertex $(0, 0)$ is $x^2 = 4py$. When $p > 0$, the parabola opens upward. When $p < 0$, the parabola opens downward.

The **standard form of the equation of a parabola** with focus $(p, 0)$, directrix $x = -p$, and vertex $(0, 0)$ is $y^2 = 4px$. When $p > 0$, the parabola opens to the right. When $p < 0$, the parabola opens to the left.

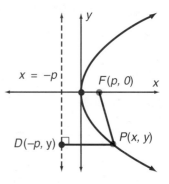

EXAMPLE 1

Determine the focus, and directrix of the parabola $y^2 = -8x$. Then graph the parabola.

Solution

The equation has the form $y^2 = 4px$, so the vertex is $(0, 0)$. The y-term is squared, so the axis of symmetry is horizontal. The coefficient of x is -8, so $4p = -8$ and $p = -2$.

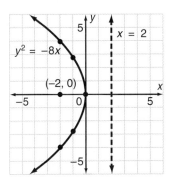

The focus is $(-2, 0)$ and the directrix is $x = 2$. Since $p < 0$, the parabola opens to the left. The shape of the graph—how wide or narrow it is—can be gauged by plotting a few points. ◀

THE STANDARD FORM OF THE EQUATION OF A PARABOLA

The standard form of the equation of a parabola with a vertical axis of symmetry and vertex at (h, k) is

$$(x - h)^2 = 4p(y - k)$$

The focus is at $(h, k + p)$. The directrix is $y = k - p$.

The standard form of the equation of a parabola with a horizontal axis of symmetry and vertex at (h, k) is

$$(y - k)^2 = 4p(x - h)$$

The focus is at $(h + p, k)$. The directrix is $x = h - p$.

EXAMPLE 2

CIVIL ENGINEERING A valley in the road is modeled by the parabola $y = \frac{1}{100}x^2 - 2x + 130$. If the flat section of the road is the x-axis, how far below the level section of the road is the valley at its lowest point?

Solution

The parabola's lowest point is its vertex. To find the vertex, complete the square on the equation of the parabola. Then write it in standard form.

$$y = \frac{1}{100}x^2 - 2x + 130$$
$$100y = x^2 - 200x + 13{,}000 \qquad \text{Multiply by 100.}$$
$$100y - 13{,}000 = x^2 - 200x$$
$$100y - 3000 = x^2 - 200x + 10{,}000 \qquad \text{Complete the square.}$$
$$100(y - 30) = (x - 100)^2 \qquad \text{Write in standard form.}$$

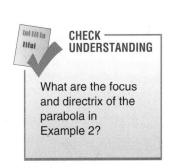

CHECK UNDERSTANDING

What are the focus and directrix of the parabola in Example 2?

The vertex of the parabola is at $(100, 30)$, so the lowest point of the valley is 30 ft below the level section of the road. ◀

If you know the focus or directrix of a parabola with its vertex at the origin, you can determine the equation of the parabola.

EXAMPLE 3

The point $(0, 3)$ is the focus of a parabola that has its vertex at the origin. Determine the equation of the parabola.

Solution

The focus is $(0, 3)$, so $p = 3$. The focus is on the y-axis, so the equation has the form $x^2 = 4py$.

So, the equation of the parabola is $x^2 = 12y$. ◄

TRY THESE

Express each equation in the form $(x - h)^2 = 4p(y - k)$.

1. $x^2 + 20y = 0$
2. $x^2 = y$
3. $x^2 - 12x - 8y = -60$
4. $x^2 + 4x + 2y + 10 = 0$

Determine the vertex, focus, and directrix of the parabola. Then graph the parabola.

5. $x^2 = 12y$
6. $-20x = y^2$
7. $0.25y^2 = 2x$
8. $24y = -3x^2$

Write the equation of the parabola.

9. vertex $(0, 0)$, focus $(0, -5)$
10. vertex $(0, 0)$, focus $(7, 0)$
11. vertex $(0, 0)$, directrix $y = -9$
12. vertex $(4, -5)$, focus $(1, -5)$

13. Determine the vertex, focus, and directrix of the parabola $8y = x^2 + 4x + 20$.

14. **COMMUNICATIONS** A parabolic TV satellite dish directs all incoming signals to a receiver located 3 feet above the vertex. Find the equation of the dish, assuming that its open end is pointed straight up and that its vertex is at the origin.

15. **WRITING MATHEMATICS** A parabola has its vertex at the origin. Explain how to use the equation of the parabola to tell if the parabola opens up, down, right, or left.

PRACTICE

Determine the vertex, focus, and directrix of the parabola. Then graph the parabola.

1. $y^2 = 24x$
2. $x^2 = 32y$
3. $0.5x^2 = 10y$
4. $3y^2 = -48x$
5. $4y = -x^2$
6. $-6x = 0.75y^2$
7. $y^2 - 2x - 8 = 0$
8. $(x - 2)^2 = 4(y - 3)$
9. $y^2 - 4y - 12x = 8$
10. $x^2 + 8x + 16 - 8y = 0$

Write the equation of the parabola.

11. vertex $(0, 0)$, focus $(3, 0)$

12. vertex $(0, 0)$, focus $(0, -10)$

13. vertex $(0, 0)$, directrix $x = 6$

14. vertex $(0, 0)$, directrix $y = -5$

15. vertex $(1, -1)$, focus $(1, 7)$

16. focus $(-3, 6)$, directrix $x = -5$

17. focus $(8, 0)$, directrix $y = 4$

18. directrix $y = -3$, focus $(-2, 3)$

19. WRITING MATHEMATICS The graph of the equation $y = ax^2 + bx + c$ is a parabola. Explain how you can find the vertex, focus, and directrix of the parabola with the equation $y = ax^2 + bx + c$.

Determine the vertex, focus, and directrix of the parabola.

20. $x^2 + 6x + 4y + 5 = 0$

21. $y^2 - 6y - 4x + 17 = 0$

22. $2y^2 + 4y - 2x = -2$

23. $y = x^2 + 4x + 1$

24. ENERGY Parabolic mirrors can be used to power steam turbines to generate electricity. The sun's rays are focused on pipes containing oil. When the oil reaches a high enough temperature, it is used to boil water. The equation of the parabola, which includes the cross-section of the mirror illustrated, is $y = 0.032x^2$. (Units are in feet.) Determine the distance above the mirror of the oil pipe.

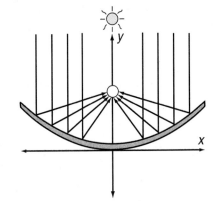

EXTEND

Determine the equation of the axis of symmetry of the parabola.

25. $(x - h)^2 = 4p(y - k)$

26. $(y - k)^2 = 4p(x - h)$

Use your results from Exercises 25-26 to determine the axis of symmetry of the parabola.

27. $y = (x - 1)^2 + 3$

28. $x = (y + 5)^2 - 2$

29. $y = x^2 - 8x$

30. $x = -y^2 - 4y - 1$

31. ARCHITECTURE A parabolic arch with a vertical axis of symmetry has a width of 40 ft at the base and a height of 15 ft at the center.

    a. Determine the equation of the arch. Assume that the vertex is at $(0, 0)$.

    b. At what height above the base is the width of the arch 16 ft?

32. Use a graphing utility to graph $x^2 = 4py$ for several values of $p > 0$. Then describe how the absolute value of p affects the shape of the parabola. Check to see whether you obtain similar results for the parabola $y^2 = 4px$.

33. For what value of b will the parabola with equation $by = 2(x - 3)^2 - 4$ pass through the point $(9, 20)$?

34. Two parabolas have $(2, -2)$ as their vertex and pass through the point $(6, -6)$.

    a. On the same coordinate axes, graph the two parabolas.
    b. Determine the equations of the parabolas.

35. The *latus rectum* of a parabola is the line segment through the focus perpendicular to the axis of symmetry with endpoints on the parabola. In the figure, $\overline{MN}$ is the latus rectum of the parabola $(x - h)^2 = 4p(y - k)$. Determine the length $\overline{MN}$.

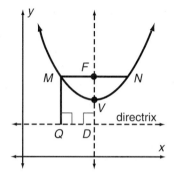

MIXED REVIEW

Find each product. $A = \begin{bmatrix} 1 & -1 \\ 3 & 0 \end{bmatrix}$ $B = \begin{bmatrix} 1 & 0 \\ 0 & 1 \end{bmatrix}$ $C = \begin{bmatrix} 2 & 3 \\ -1 & 4 \end{bmatrix}$

36. AC **37.** CA **38.** BC **39.** CB

Find the sum and product of the solutions of each equation.

40. $x^2 + 9x + 20 = 0$ **41.** $x^2 - 5x - 13 = 0$ **42.** $3x^2 + 9x + 2 = 0$ **43.** $-5x^2 + x + 3 = 0$

Solve each equation.

44. $\sqrt{x} = 7$ **45.** $\sqrt{y} = -4$ **46.** $\sqrt{2x + 3} = 5$ **47.** $n - 5 = \sqrt{n + 7}$

48. STANDARDIZED TESTS Choose the letter giving the vertex of the parabola $x^2 - 6x - 4y + 29 = 0$.

    A. $(3, 5)$ **B.** $(3, -5)$ **C.** $(3, 6)$ **D.** $(3, -6)$

PROJECT *Connection*   The following activity demonstrates ideas about parabolas.

1. Refer to the diagram to construct a manipulative using a push pin, wood, string, cardboard, and a marker for drawing a parabola. Make sure the string is exactly the length between the focus and directrix.

2. Secure the string to the right-angled cardboard and place the marker at this point. Slowly drag the marker down the side of the cardboard. As you do this, the marker will drag the string down the side of the cardboard.

3. When you reach the focus, trace out half a parabola, and the marker should be halfway down the side of the cardboard. Reflect your drawing to complete the parabola. Explain why this demonstrates the properties of parabolas.

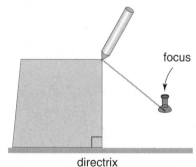

Career
Telecommunications Engineer

Telstar I, the first commercially financed telecommunications satellite, was put into Earth's orbit on July 10, 1962. Telstar allowed viewers in the United States to see live television pictures broadcast from Europe, a feat which had never before been possible.

Today hundreds of satellites circle the globe. Anyone with a few hundred dollars can purchase a receiving dish not much larger than a catcher's mitt. In designing more and more sophisticated receiving dishes, telecommunications engineers have taken advantage of some of the special properties of the parabola and the hyperbola.

Decision Making

1. A satellite dish is a **paraboloid**, a three-dimensional surface formed by rotating a parabola about its axis of symmetry. Ideally, incoming signals from a satellite that are parallel to the axis of symmetry are reflected off the parabolic surface to the focus. Draw a diagram illustrating this.

2. If the dish were a perfect parabola, all of the incoming parallel waves, each quite weak on its own, would be concentrated at the focus, producing a strong signal. In practice, parabolic dishes usually are not perfect. Instead, they are distorted due to their own weight, wind and temperature fluctuations, and so on. Draw a diagram showing how incoming waves parallel to the axis of symmetry fail to concentrate at the focus of an imperfect parabolic surface.

3. The solution to the distortion problem utilizes a small **hyperboloid**, a three-dimensional surface formed by rotating a branch of a hyperbola about its axis of symmetry. Any incoming wave directed toward the focus of one branch of a hyperbolic surface will be reflected toward the focus of the other branch. Draw a diagram showing this reflective property.

4. Now put your diagram for a satellite dish together. Use a large parabolic surface for a main dish. Fix a small hyperbolic surface so that it and the parabola have a common focus. At the hyperbola's other focus, place the "receiving cone," the electronic device that gathers the incoming rays and translates them into sounds and pictures. Draw a diagram of your dish showing the path of an incoming wave.

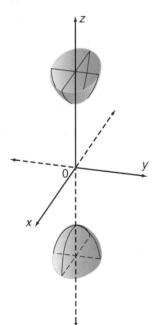

Explore

1. Write the equations of three different circles in the form $Ax^2 + Bxy + Cy^2 + Dx + Ey + F = 0$ where $B = 0$. What generalization can you make about the values of A and C?

2. Write the equations of three different ellipses in the form $Ax^2 + Bxy + Cy^2 + Dx + Ey + F = 0$ where $B = 0$. What generalizations can you make about the values and signs of A and C.

3. Write the equations of three different hyperbolas in the form $Ax^2 + Bxy + Cy^2 + Dx + Ey + F = 0$ where $B = 0$. What generalization can you make about the signs of A and C?

4. Write the equations of three different parabolas in the form $Ax^2 + Bxy + Cy^2 + Dx + Ey + F = 0$ where $B = 0$. What generalization can you make about the values of A and C?

5. If you are given the equation of a conic section in the form $Ax^2 + Bxy + Cy^2 + Dx + Ey + F = 0$ where $B = 0$. Without using standard form, how can you determine which conic section is represented.

Build Understanding

Recall, conic sections are formed by intersecting a plane with a cone and can be written in a standard form that specifies the center.

Circle $(x - h)^2 + (y - k)^2 = r^2$

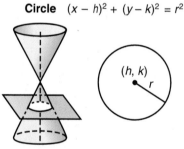

Ellipse $\dfrac{(x - h)^2}{a^2} + \dfrac{(y - k)^2}{b^2} = 1$

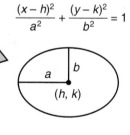

Hyperbola $\dfrac{(x - h)^2}{a^2} - \dfrac{(y - k)^2}{b^2} = 1$

slopes: $\pm \dfrac{b}{a}$

(h, k)

Parabola $(x - h)^2 = 4p(y - k)$

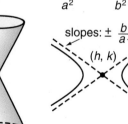

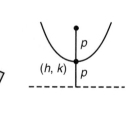

Each standard conic section equation is a special case of the **general form of the second-degree equation** $Ax^2 + Bxy + Cy^2 + Dx + Ey + F = 0$. If A, B, and C all equal zero, the equation will not be second-degree. But if at least one of the coefficients A, B, or C is nonzero, you can use the values of the coefficients to determine which conic section the equation represents.

If at least one of A, B, or C is nonzero, then the graph of $Ax^2 + Bxy + Cy^2 + Dx + Ey + F = 0$ is

- a hyperbola if $B^2 - 4AC > 0$
- a parabola if $B^2 - 4AC = 0$
- an ellipse if $B^2 - 4AC < 0$; if $A = C$, the ellipse is a circle.

Recall that the expression $b^2 - 4ac$ is called the *discriminant* of a quadratic equation. For that reason, the above theorem is sometimes called the *discriminant theorem* of conic sections.

CHECK UNDERSTANDING

If $A = B = C = 0$ in the general form of the second-degree equation, but D, E, or both are nonzero, what graph will the equation represent?

EXAMPLE 1

Each equation represents a conic section. Identify each conic section.

a. $8x^2 + 12x + 8xy = -2y^2 - 13y + 20$

b. $3x^2 + 5xy + 3y^2 + 14x - 9y - 24 = 0$

c. $5xy - 45 = 0$

Solution

a. Write equation in the general form.

$$8x^2 + 12x + 8xy = -2y^2 - 13y + 20$$
$$8x^2 + 8xy + 2y^2 + 12x + 13y - 20 = 0$$

$B^2 - 4AC$
$8^2 - 4(8)(2) = 0$ $A = 8$, $B = 8$, $C = 2$

The equation represents a parabola.

b. $3x^2 + 5xy + 3y^2 + 14x - 9y - 24 = 0$

$B^2 - 4AC$
$5^2 - 4(3)(3) < 0$ $A = 3$, $B = 5$, $C = 3$

The equation represents an ellipse.

c. $5xy - 45 = 0$

$B^2 - 4AC$
$5^2 - 4(0)(0) > 0$ $A = 0$, $B = 5$, $C = 0$

The equation represents a hyperbola.

COMMUNICATING ABOUT ALGEBRA

Are the discriminants of the general quadratic equation and the general second-degree equation the same? Explain your answer to another student.

◄

By writing the equation from Example 1 Part c in the form $xy = 9$, you recognize it as the equation of a rectangular hyperbola with branches in the first and third Quadrants. Except for rectangular hyperbolas, none of the equations of conic sections that you studied earlier in this chapter contained xy-terms.

With the exception of rectangular hyperbolas, all of the conic sections you graphed had axes of symmetry parallel to the x-axis or y-axis. In general, second-degree equations containing xy-terms represent ellipses, hyperbolas, and parabolas with their axes of symmetry rotated around the intersection of their axes of symmetry.

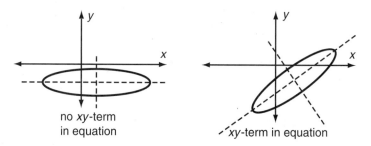

There are three **degenerate conic sections;** the point, the line, and two intersecting lines.

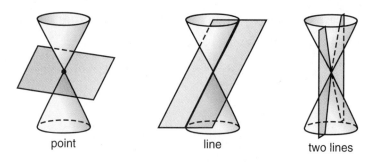

Each degenerate conic section is an extreme case of one of the primary conic sections. A point is formed when the axes of an ellipse is zero. A line is formed when the width of a parabola is zero. Two intersecting lines are formed when the vertices of a hyperbola meet and the branches coincide with the asymptotes.

EXAMPLE 2

TECHNOLOGY Graph $x^2 - y^2 = 0$ and identify the conic section represented.

Solution
$A = 1$, $B = 0$, and $C = -1$, so $B^2 - 4AC > 0$. The graph is a hyperbola. The graph is a degenerate hyperbola, or two straight lines. ◄

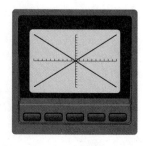

Write the equation in the general form of a second-degree equation. Give the values of
A, B, C, D, E, and F.

1. $5x^2 - 3x + 2 = -4y^2 - 6xy + 2y + 1$

2. $6x + 5y = 12 - x + 5y + x^2$

3. $\dfrac{(x - 1)^2}{9} + \dfrac{(y + 2)^2}{4} = 1$

4. $y = -3(x^2 + 2xy - 5) + 2(x^2 - xy + 1)$

Identify the conic section represented by the equation.

5. $25x^2 + 16y^2 = 400$

6. $3x^2 - 6x + 3 = y - 2$

7. $x^2 - y^2 = 8 - 4y$

8. $x^2 + 2x + y^2 - 6y = 15$

9. $6x^2 + 4xy + 6y^2 + 10x - 2y + 5 = 0$

10. $x^2 - 5xy - 2y^2 - 5x + 11 = 0$

11. $-2x^2 - 3xy + 3y^2 + 4x - y - 19 = 0$

12. $-3x^2 + 6xy - 3y^2 + x - 8y + 5 = 0$

13. Graph $x - y + 3 = 0$ and identify the conic section represented.

14. **WRITING MATHEMATICS** You are given a second-degree equation. Explain how you can
tell which conic section is represented by the equation.

PRACTICE

Identify the conic section represented by the equation.

1. $2x^2 - 4x - 3y + 3 = 0$

2. $x^2 + 8y^2 = 8$

3. $4y^2 - 8y + 3 = 9x^2$

4. $x^2 - 6x + y^2 + 10y = 66$

5. $-2x^2 + 8xy - 8y^2 + 5x - 3y + 11 = 0$

6. $x^2 + xy - 3y^2 + 5 = 0$

7. $25x^2 - 10xy + 12x - 11y + 16 = 0$

8. $9x^2 + 2xy + y^2 - 7y = 0$

9. $12xy + 5y = 0$

10. $x^2 - xy + y^2 + 3x = 0$

11. $\dfrac{x^2}{225} - \dfrac{y^2}{64} = 1$

12. $(x + 5)^2 + (y - 3)^2 = 40$

13. $15x^2 + 15y^2 - 5x + 4y + 1 = 0$

14. $y = 3x^2 - 11x + 4$

15. $3x(x + 1) - 2y(y - 2x) = 2(y - 1)$

16. $5y(y - x - 1) + 9 = -2x(x + 4)$

17. $(x - 2)^2 = 4(y - 3)$

18. $\dfrac{(x - 2)^2}{10} + \dfrac{(y + 5)^2}{18} = 1$

Graph the equation and identify the conic section represented.

19. $x + y = 5$ 20. $(x - 1)^2 + (y - 2)^2 = 0$ 21. $x^2 - y^2 + 6x + 4y + 5 = 0$

22. A parabola was graphed on a chalkboard. A passing student accidentally rubbed off the
coefficient of xy in the equation of the parabola, $-5x^2 + \boxed{?}\, xy - 2y^2 + 5x - 3y - 8 = 0$.
Find the missing coefficient.

23. **WRITING MATHEMATICS** Write specific equations of a hyperbola, a parabola, and an
ellipse in the general form of a second-degree equation. Explain how you found the
coefficients.

EXTEND

The graph of each of the following equations is a degenerate conic section. Without drawing the graph, identify the degenerate conic section represented. Explain.

24. $x + y = 2$

25. $x^2 + y^2 = 0$

26. $4x^2 - y^2 = 0$

27. The graph of the equation $4x^2 + Bxy + 9y^2 - 3x - 2y + 5 = 0$ is a parabola. Find B.

28. The graph of the equation $3x^2 + Bxy + 2y^2 + 2x - y - 6 = 0$ is a hyperbola. What restrictions can you place on the value of B?

29. Can the graph of the equation $-3x^2 + Bxy + 5y^2 + x - 6y + 1 = 0$ be an ellipse? Explain your reasoning.

THINK CRITICALLY

30. To use a graphing utility to graph the equation $x^2 + 4xy + 4y^2 - 2x + y - 9 = 0$, you must first solve the equation for y.

    a. Write the equation $My^2 + Ny + P = 0$, where M, N, and P are functions of x as the quadratic equation in y.

    b. Solve the equation for y. Write the two equations you would input into a graphing utility to graph the equation. Graph the equation.

31. Find the y-intercept of the graph of the general second-degree equation $Ax^2 + Bxy + Cy^2 + Dx + Ey + F = 0$.

32. Use your answer to find the y-intercept(s) of the graph of the equation $7x^2 - 2xy + 2y^2 + 9x + 10y + 8 = 0$.

MIXED REVIEW

State whether the given set is closed under the given operation.

33. $\{0, 1\}$, addition

34. $\{$integers$\}$, multiplication

35. $\{-1, 0, 1\}$, division

Write each expression using rational exponents in simplest form.

36. $\sqrt{5}$

37. $\sqrt[6]{x^3}$

38. $\sqrt[3]{2n^5}$

39. $\sqrt[5]{h^3 k^{15}}$

Solve for x.

40. $\log_2 16 = x$

41. $9^x = 27$

42. $\log_{10} x = -2$

43. $\log_x 32 = \dfrac{5}{3}$

44. STANDARDIZED TESTS Choose the letter of the conic section represented by the equation $x^2 + 2xy + y^2 - 3x + y = 25$.

    A. circle   **B.** ellipse   **C.** hyperbola   **D.** parabola

12.8 Graph and Solve Quadratic Systems

Explore

- You know that when two lines intersect in
  - no points, the lines are parallel
  - one point, the lines are different and not parallel
  - an infinite number of points, the lines are the same

  1. In how many points can a line and a nondegenerate conic section intersect? Make sketches of possible intersections involving a line and a primary conic section. For each answer that you find, show at least one arrangement of figures that produces that answer.

  2. Repeat Question 1 but this time look for intersection possibilities involving two primary conic sections. Again, show at least one example for each possible number of intersection points.

SPOTLIGHT ON LEARNING

WHAT? In this lesson you will learn
- to solve systems of second-degree equations.

WHY? You can use quadratic systems for determining the position of an ocean vessel or pinpointing the epicenter of an earthquake.

Build Understanding

- Many of the methods you developed for solving systems of linear equations can also be used to solve second-degree or quadratic systems. A **quadratic system** is a set of equations that involves at least one quadratic equation. Quadratic systems can be solved using graphing, substitution, and linear combinations.

EXAMPLE 1

Solve the quadratic system by graphing.

$$\begin{cases} x^2 + y^2 = 25 \\ y = 3x - 5 \end{cases}$$

Solution

The circle $x^2 + y^2 = 25$ and the line $y = 3x - 5$ appear to intersect at the points $(3, 4)$ and $(0, -5)$. The following check confirms that those are the intersection points.

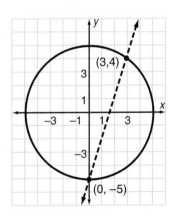

Check

$$\begin{array}{ll} & (3, 4) & (0, -5) \\ x^2 + y^2 = 5: & 3^2 + 4^2 \overset{?}{=} 25 & 0^2 + (-5)^2 \overset{?}{=} 25 \\ & 25 = 25 \checkmark & 25 = 25 \checkmark \\ y = 3x - 5: & 4 \overset{?}{=} 3(3) - 5 & -5 \overset{?}{=} 3(0) - 5 \\ & 4 = 4 \checkmark & -5 = -5 \checkmark \end{array}$$

CHECK UNDERSTANDING

In Example 1, why is it necessary to check to be certain that $(3, 4)$ and $(0, -5)$ are the solutions?

If you can make a sketch of the graphs of the equations in a quadratic system, you can estimate the number of solutions and their coordinates. Often, however, graphing does not allow you to find exact solutions. When that happens, you will have to use an algebraic method to solve the system. If you can easily solve one equation for one of the variables, substitution may be the best method. Substitution is usually appropriate for solving a quadratic-linear system since the linear equation may be easily solved for x or y.

EXAMPLE 2

Solve the system.

$$\begin{cases} x - 2y = -1 \\ xy = 6 \end{cases}$$

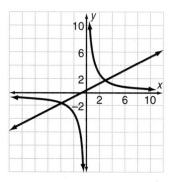

Solution

The equations represent a line and a rectangular hyperbola. The maximum number of intersections is two, so there are at most two real solutions. Either equation can easily be solved for either variable. So, use the substitution method. Since solving the second equation for x or y will produce fractions, there may be an advantage to solving the first equation for x or y.

$$
\begin{array}{ll}
x - 2y = -1 & \text{First equation.} \\
x = 2y - 1 & \text{Solve for } x. \\
xy = 6 & \text{Second equation.} \\
(2y - 1)y = 6 & \text{Substitute } x \text{ from the first equation.} \\
2y^2 - y = 6 & \text{Distributive property.} \\
2y^2 - y - 6 = 0 & \text{Simplify.} \\
(2y + 3)(y - 2) = 0 & \text{Factor.} \\
2y + 3 = 0 \quad \text{or} \quad y - 2 = 0 & \\
2y = -3 \quad \text{or} \quad y = 2 & \\
y = -\dfrac{3}{2} \quad \text{or} \quad y = 2 &
\end{array}
$$

Each value of y can now be substituted in either original equation to determine the corresponding value of x.

$$
\begin{array}{ll}
y = -\dfrac{3}{2}: \qquad xy = 6 & \qquad y = 2: \qquad xy = 6 \\[2mm]
\qquad\quad x\left(-\dfrac{3}{2}\right) = 6 & \qquad\qquad x(2) = 6 \\[2mm]
\qquad\qquad\quad x = -4 & \qquad\qquad\quad x = 3
\end{array}
$$

The solutions are $\left(-4, -\dfrac{3}{2}\right)$ and $(3, 2)$. You should check both solutions in the original equations. You can graph the system with a graphing utility and note the two intersection points. ◄

PROBLEM SOLVING TIP

Remember, you can use the discriminant of an equation in standard form to identify the conic section if you are given the equation in a conic section and not a degenerate or undefined.

Since both equations in a quadratic-quadratic system are quadratics, it is usually difficult to solve either equation for one of the variables. Therefore, the method of linear combinations may be a better choice for solving a quadratic-quadratic system.

EXAMPLE 3

Solve the system using linear combination.

$$\begin{cases} (x - 5)^2 + y^2 = 64 \\ \dfrac{x^2}{9} - \dfrac{y^2}{36} = 1 \end{cases}$$

Solution

The equations represent a circle and a hyperbola. There may be from zero to four real solutions.

Begin by writing both equations in standard form.

$$(x - 5)^2 + y^2 = 64$$
$$x^2 - 10x + 25 + y^2 = 64$$
$$x^2 + y^2 - 10x - 39 = 0$$

$$\frac{x^2}{9} - \frac{y^2}{36} = 1$$
$$4x^2 - y^2 = 36$$
$$4x^2 - y^2 - 36 = 0$$

$$\begin{array}{l} x^2 + y^2 - 10x - 39 = 0 \\ \underline{4x^2 - y^2 \qquad - 36 = 0} \\ 5x^2 \qquad - 10x - 75 = 0 \\ \qquad x^2 - 2x - 15 = 0 \\ \qquad (x - 5)(x + 3) = 0 \\ \qquad x = 5 \text{ or } x = -3 \end{array}$$

Add the equations.
Factor out common factor.
Factor.

To determine corresponding values of y, substitute each value of x in either of the original equations.

Use $x = 5$
$$(x - 5)^2 + y^2 = 64$$
$$(5 - 5)^2 + y^2 = 64$$
$$y^2 = 64$$
$$y = \pm 8$$

Use $x = -3$
$$(x - 5)^2 + y^2 = 64$$
$$(-3 - 5)^2 + y^2 = 64$$
$$64 + y^2 = 64$$
$$y = 0$$

There are three solutions, $(5, 8)$, $(5, -8)$, and $(-3, 0)$. Check each solution in the original equations. Verify your results by graphing. ◄

The LORAN navigation system is a useful application of quadratic systems. Radio pulses broadcast simultaneously from two widely separated locations arrive at a ship at different times. An on-board computer measures the difference in arrival times and places the ship somewhere on a hyperbola. The computer repeats the process with a second pair of signals, finding a second hyperbola. The ship's position is the point where the hyperbolas intersect.

EXAMPLE 4

NAVIGATION A cargo ship is sailing in the Indian Ocean somewhere off the coast of Somalia. LORAN signals indicate that the ship's position is at the intersection of the hyperbolas $xy = 200$ and $x^2 - 25y^2 = 1600$. Determine the location of the ship.

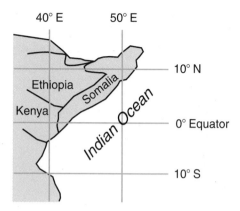

Solution

To determine the points of intersection of the hyperbolas, solve the first equation for x and substitute the resulting expression in the second equation.

$$xy = 200 \qquad \text{First equation.}$$
$$x = \frac{200}{y}$$
$$x^2 - 25y^2 = 1,600 \qquad \text{Second equation.}$$
$$\left(\frac{200}{y}\right)^2 - 25y^2 = 1,600 \qquad \text{Substitute.}$$
$$40,000 - 25y^4 = 1,600y^2 \qquad \text{Simplify.}$$
$$25y^4 + 1,600y^2 - 40,000 = 0 \qquad \text{Rewrite.}$$
$$y^4 + 64y^2 - 1,600 = 0 \qquad \text{Divide by 25.}$$

The equation is in quadratic form. Solve for y^2 using the quadratic formula.

$$y^2 = \frac{-64 \pm \sqrt{64^2 - 4(1)(-1,600)}}{2(1)}$$
$$y^2 \approx 19.225 \text{ or } y^2 \approx -83.225$$

The first value gives $y \approx \pm 4.38$. Substitution gives $x \approx \pm 45.66$. The second value has no real solutions. The hyperbolas intersect at approximately $(45.66, 4.38)$ and $(-45.66, -4.38)$. The map shows that only the first solution is in the Indian Ocean. The ship is at $(45.66, 4.38)$, or 45.66 E longitude, 4.38 N latitude. ◀

TRY THESE

Identify the graphs and solve the system.

1. $\begin{cases} y = 2x - 1 \\ 4x^2 + y^2 = 25 \end{cases}$

2. $\begin{cases} xy = 15 \\ x^2 + y^2 = 34 \end{cases}$

3. $\begin{cases} x^2 + 4y^2 = 16 \\ 9x^2 - 16y^2 = -64 \end{cases}$

4. $\begin{cases} x + y = 5 \\ x^2 + y^2 = 17 \end{cases}$

5. $\begin{cases} y^2 - x^2 = 9 \\ 2x - y = 3 \end{cases}$

6. $\begin{cases} 5x^2 + y^2 = 30 \\ y^2 - 9x^2 = 16 \end{cases}$

7. GEOMETRY A rectangular garden has a perimeter of 42 yards. A string attached between posts at opposite corners of the garden has a length of 15 yards. Find the area of the garden.

8. WRITING MATHEMATICS Describe how you can decide whether you should solve a quadratic system graphically or algebraically.

PRACTICE

Identify the graphs and solve the system.

1. $\begin{cases} y = x + 2 \\ x^2 + y^2 = 100 \end{cases}$

2. $\begin{cases} y = 5x \\ y = x^2 \end{cases}$

3. $\begin{cases} xy = 54 \\ 3x = 2y \end{cases}$

4. $\begin{cases} x^2 + 3y = 18 \\ 2x^2 - 5y = 3 \end{cases}$

5. $\begin{cases} 4x^2 - 3y - 10 = 0 \\ 2x - 3y = 10 \end{cases}$

6. $\begin{cases} 3x^2 + 5y^2 = 95 \\ 4x^2 - 3y^2 = 88 \end{cases}$

7. $\begin{cases} x^2 - xy + y^2 = 43 \\ x + y = 8 \end{cases}$

8. $\begin{cases} y = -3x + 1 \\ x^2 - 2xy - 5 = 0 \end{cases}$

9. $\begin{cases} 5x + 2y = 4 \\ x^2 + 4y^2 = 16 \end{cases}$

10. $\begin{cases} 3x + 5y = 44 \\ \dfrac{(x - 3)^2}{25} + \dfrac{(y - 4)^2}{9} = 1 \end{cases}$

11. $\begin{cases} 2xy + x + 36 = 0 \\ xy - 3y + 5 = 0 \end{cases}$

12. $\begin{cases} x + y = 11xy \\ 24xy = 1 \end{cases}$

13. GEOMETRY A right triangle with a hypotenuse of 41 cm has an area of 180 cm². Find the lengths of the legs of the triangle.

14. SEISMOLOGY Seismological stations are located at $A(5, 22)$, $B(5, -8)$, and $C(40, 10)$. (Units are in miles.) The epicenter of an earthquake is located 40 miles from A, 50 miles from B, and 13 miles from C. Find the coordinates of the epicenter.

15. NAVIGATION The navigator of a freighter nearing the Brazilian coast receives LORAN signals indicating that the ship is located at the intersection of the hyperbolas $xy = 320$ and $64y^2 - 5x^2 = 1,280$. Find the ship's position.

16. WRITING MATHEMATICS Can a quadratic system have an infinite number of solutions? Explain your answer.

EXTEND

Solve the system.

17. $\begin{cases} y = 2^x \\ y = 2^{2x} - 12 \end{cases}$

18. $\begin{cases} y = 4^x \\ y = 4^{2x+1} - 3 \end{cases}$

19. Explain how you can solve the system without graphing or performing any calculations.

$$\begin{cases} (x - 5)^2 + y^2 = 16 \\ (x + 5)^2 + y^2 = 16 \end{cases}$$

20. NUMBER THEORY Find two numbers such that the sum of their squares is 1145 and the difference of their squares is 423.

21. NUMBER THEORY Find two numbers such that their sum is 24, and three times their product is 357.

22. NAVIGATION Radar at $(0, 0)$ in an air traffic control tower detects all planes within a 50-mile radius of the airport.

    a. A plane flies on the course $x + 2y = 110$. Will it be picked up on the radar screen? If so, when?

    b. The plane's speed is 240 mi/h. How long will it appear on radar?

23. WRITING MATHEMATICS Write a word problem that requires determining the point of intersection of a line and a parabola.

THINK CRITICALLY

24. The equation of a hyperbola with its center at the origin is $\dfrac{x^2}{a^2} - \dfrac{y^2}{b^2} = 1$. The equations of its asymptotes are $y = \pm\dfrac{b}{a}x$.

    a. Explain how you could solve a second-degree system to prove that a hyperbola with center at the origin does not intersect its asymptotes.

    b. Show that such a hyperbola does not intersect its asymptotes.

25. GEOMETRY A rectangle has an area of A and a perimeter of P. Find formulas giving the length l and the width w of the rectangle in terms of A and P.

MIXED REVIEW

Determine the mean, median, and mode of each set of data.

26. $\{14, 23, 11, 9, 22, 23, 10\}$

27. $\{427, 481, 491, 450\}$

28. $\{-5.6, -6.8, -7.1, 3.4, 0.7\}$

29. $\left\{\dfrac{3}{4}, \dfrac{1}{2}, \dfrac{5}{12}, \dfrac{2}{3}\right\}$

Determine the equation of each line.

30. slope $= 3$, y-intercept $= -2$

31. slope $= 2$, contains $(-3, 4)$

32. parallel to $x - 2y = 5$, contains $(4, 1)$

33. perpendicular to $y = -\dfrac{1}{3}x - 5$, contains $(1, 1)$

34. STANDARDIZED TESTS A variable x varies directly as y and inversely as z. When $y = 4$ and $z = -1$, $x = 8$. Choose the letter that gives the value of x when $y = -2$ and $z = 4$.

    A. 1   **B.** -4 **C.** 2   **D.** 4

Solve each system.

35. $\begin{cases} xy = 1 \\ y = -x^2 \end{cases}$

36. $\begin{cases} y = x^2 \\ 3x = y + 2 \end{cases}$

37. $\begin{cases} x^2 + y^2 = 25 \\ 25x^2 + 16y^2 = 400 \end{cases}$

Healthy kidneys act to maintain water and electrolyte balance in the body and to filter wastes from the blood. Occasionally, improper functioning of the kidneys causes small hard masses called kidney stones to form inside the organs. Kidney stones can be very painful and until recently had to be removed by surgery. Then in 1980, Dr. Christian Chaussy, a urologist at the University of Munich, developed a new way to remove kidney stones. Dr. Chausy's method is safer and less expensive than traditional surgery, and allows patients to get back on their feet in one-fourth the time. This method utilizes sound waves, which break the kidney stones into particles small enough to be excreted in the patient's urine. A semi-ellipsoidal device, called a *lithotripter*, generates and focuses the sound waves.

Decision Making

1. A lithotripter is about the size and shape of a pointed half of a football. Sketch an ellipse with a major axis about 50% longer than its minor axis. Erase half the ellipse on one side of its minor axis. Estimate the position of the focus of the remaining half of the ellipse and mark it. Draw a second sketch showing the semi-ellipsoid that would result if you rotated the semi-ellipse around its major axis. This semi-ellipsoid should resemble a lithotripter.

2. Sound waves are emitted by a lithotripter at its focus. Because of the reflective properties of an ellipse, all the waves are reflected off the sides of the device and concentrated at the second focus. Observing the kidney stone with an x-ray machine, the physician moves the patient until the stone is positioned precisely at the second focus. Then the stone is bombarded with sound waves until it breaks up.

Draw a sketch of a lithotripter in operation.

3. The distance from the focus to the vertex of a lithotripter is 2.5 cm. The minor axis measures 6 cm.

   a. Determine the equation of the ellipse from which the semi-ellipsoid is created.

   b. Determine the distance from the emitter to the kidney stone.

Problem Solving File

Use Conics in Astronomy

The Laws of Elliptical Orbits, known as Kepler's laws, apply to any
object orbiting another object.

First Law   The orbit of each planet is an ellipse with the sun at
one focus.

Second Law   The *radius vector* is the line segment from the Sun to
an orbiting planet and sweeps over equal areas in
equal times.

Third Law   The cube of a planet's semi-major axis is equal to the
square of the planet's orbital period, where the semi-
major axis and the orbital period are measured
relative to those of Earth. Orbital period is the time to
complete one orbit. Earth's orbital period is one year.

Problem   At *perihelion*—a planet's closest approach to the Sun—
Mercury is 28.6 million miles from the Sun. At *aphelion*—a planet's
greatest distance—Mercury is 43.4 million miles from the Sun.

a. Write the equation of Mercury's orbit.

b. At what point in its orbit is Mercury's orbital velocity greatest?

c. How long does it take Mercury to orbit the Sun?

Explore the Problem

1. The **semi-major axis** of an ellipse is the length *a*, one-half the
major axis. Determine the lengths of the major and semi-major
axes of Mercury's orbit.

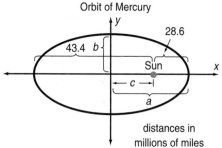

Orbit of Mercury

distances in
millions of miles

2. Determine *c*, the Sun's distance from the center of the ellipse.

3. Use the relationship $a^2 = b^2 + c^2$ to determine *b*, the length of the
semi-minor axis.

4. Write the equation of Mercury's orbit.

5. The figure shows three positions of Mercury during one of the planet's orbits. Copy the figure. Use Kepler's Second Law to estimate the location of M_4, the position to which Mercury will move from M_3 in the amount of time it took the planet to move from M_1 to M_2. Explain.

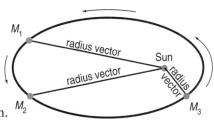

6. At what point is Mercury's orbital velocity greatest? Explain.

7. To use Kepler's Third Law, you must express the length of Mercury's semi-major axis relative to that of Earth as a_{Earth} = 93 million miles = **1 astronomical unit (A.U.)** Give Mercury's semi-major axis in A.U.s. Round to three decimal places.

8. Use Kepler's Third Law.

   a. Find Mercury's orbital period in Earth years. Round to three decimal places.

   b. To the nearest whole day, express Mercury's orbital period.

9. The formula $C = 3\pi(a + b) - \pi\sqrt{(a + 3b)(3a + b)}$ can be used to approximate the circumference C of an ellipse with semi-major and semi-minor axes of lengths a and b respectively.

   a. Determine the circumference of Mercury's orbit to the nearest hundred thousand miles.

   b. Determine Mercury's mean orbital velocity in miles per second. Round to one decimal place.

10. WRITING MATHEMATICS A comet has an elliptical orbit that takes the comet from a maximum of 1 billion miles from the Sun at aphelion to a mere 10 million miles at perihelion. Could the comet collide with Mercury? Explain.

Investigate Further

The path of a comet may be an ellipse, a hyperbola, or a parabola. All comets travel along paths that have the Sun at one focus. One survey of comets showed that about 40% have elliptical orbits, about 12% have hyberbolic paths, and about 48% have parabolic paths.

11. Some comets such as Halley's Comet are *periodic*, meaning they return again and again at predictable times. The remainder of comets are *nonperiodic*, visiting the Sun once and then never returning. About a dozen comets are observed each year. About how many of them would you expect to be periodic? Explain.

The equations of the paths of three comets are given below. (Units are in millions of miles.) For each equation, tell whether the path is elliptical, hyberbolic, or parabolic. Then determine the comet's perihelion distance.

12. $x^2 = 32y + 256$

13. $16y^2 - 9x^2 + 800y + 4600 = 0$

14. $9x^2 + 25y^2 = 562,500$

Apply the Strategy

ASTRONOMY Among the many objects that have elliptical orbits around the Sun are planets, comets, asteroids, and meteors. Determine the missing information for each object in the following table.

	Name	Type	Location of Focus	Semi-major axis (millions of miles)	Orbital period (years)
15.	Mars	Planet	■	141.8	■
16.	Halley	Comet	■	■	76.05
17.	Crommelin	Comet	■	■	27.68
18.	Tuttle	Comet	■	530.1	■
19.	Ceres	Asteroid	■	257.0	■
20.	Taurids	Meteors	■	■	3.26

The terms aphelion and perihelion refer to the distance from the Sun. When referring to objects orbiting the earth you must use the terms *apogee* in place of aphelion and *perigee* in place of perihelion.

21. Sputnik I, the first artificial earth satellite, was launched on October 4, 1957. It orbited Earth in 96 minutes at altitudes ranging from 146 miles to 560 miles. Determine Sputnik I's apogee and perigee distances from Earth's center. Earth's radius is 3,963 miles.

22. Determine the length of Sputnik I's semi-major axis.

23. Use your data on Sputnik I and the above adaptation of Kepler's Third Law to find the moon's orbital period in minutes. Then express the period in days.

Review Problem Solving Strategies

LOCO - MOTION

1. A railroad engineer needs to switch the portions of the freight car and passenger car shown below, and return the locomotive to its original position. The cars and the locomotive can hook up in any combination and can move in any direction, and the locomotive can push or pull either car or both cars together. The engineer can use a side track with a tunnel on it; the cars, but not the locomotive, can pass through the tunnel. Also, each car is longer than the tunnel and can be accessible to the locomotive from either end. How can the engineer switch the positions of the freight and passenger cars?

TUNNEL

PASSENGER

LOCOMOTIVE

FREIGHT CAR

Here are some suggestions to help you solve the problem.

a. Make a model by drawing the tracks and using paper slips, counters, or coins to represent the cars and locomotive.

b. Try solving a simpler problem first. For example, suppose there was only the freight car and the locomotive, and the engineer wanted to switch their positions. Then think about how your solution to this problem can help you with the original problem.

RACING AROUND

2. A 1500 - meter race is being run on a circular track. The total number of runners on the track is 1/3 the number of runners in front of Lucas, plus 3/4 the number of runners behind Lucas. How many runners are there? Explain your method.

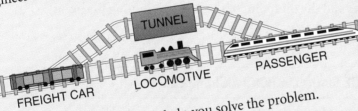

PASSING PASSENGER PUZZLE

3. The Yankee Clipper, a train traveling 60 miles per hour meets and passes the Empire Special, another train traveling 60 miles per hour in the opposite direction. A passenger in the Yankee Clipper sees the Empire Special pass in 6 seconds. How long is the Empire Special?

··· CHAPTER REVIEW ···

VOCABULARY

Match the letter of the word in the right column with the description at the left.

1. the set of points the sum of whose distances from two fixed points is constant
2. the set of all points where the absolute value of the difference of the distances from two fixed points is constant
3. the set of points that are equidistant from a given point
4. the set of points equidistant from a line and a point not on the line

a. circle

b. parabola

c. ellipse

d. hyperbola

Lesson 12.1   THE DISTANCE AND MIDPOINT FORMULAS   pages 567–571

- The distance between the points (x_1, y_1) and (x_2, y_2) is $\sqrt{(x_2 - x_1)^2 + (y_2 - y_1)^2}$.
- The midpoint of the segment with endpoints (x_1, y_1) and (x_2, y_2) is $\left(\dfrac{x_1 + x_2}{2}, \dfrac{y_1 + y_2}{2}\right)$.

Find the distance between the points and the midpoint of the segment joining the points.

5. $(5, 2)$ and $(9, -1)$ 　　　　6. $(-12, 7)$ and $(-4, -8)$ 　　　　7. $(3, 3)$ and $(-3, -1)$

Lessons 12.2 and 12.3   EXPLORE CONIC SECTIONS AND CIRCLES   pages 572–580

- The **standard form of the equation of a circle** with radius r and center (h, k) is $(x - h)^2 + (y - k)^2 = r^2$.
- If the equation of a conic section is given in expanded form, complete the square to rewrite the equation in standard form.

8. Determine the standard form of the equation of the circle with center $(3, -7)$ and radius $\sqrt{5}$.

9. Find the center and radius of the circle $x^2 - 2x + y^2 + 6y = 39$.

Lesson 12.4   ELLIPSES   pages 581–586

- The standard form of the equation of a horizontal ellipse with center (h, k) is $\dfrac{(x - h)^2}{a^2} + \dfrac{(y - k)^2}{b^2} = 1$. For a vertical ellipse, the standard form is $\dfrac{(x - h)^2}{b^2} + \dfrac{(y - k)^2}{a^2} = 1$. In both cases, $a > b$.
- The foci are located c units left and right of the center (horizontal ellipse), where $c^2 = a^2 - b^2$, or c units above and below the center (vertical ellipse), where $c^2 = b^2 - a^2$.

10. Find the foci, vertices, and center of the ellipse $9x^2 - 90x + 16y^2 - 96y + 225 = 0$. Then graph the ellipse.

- The standard form of the equation of a horizontal hyperbola with center (h, k) is $\frac{(x - h)^2}{a^2} - \frac{(y - k)^2}{b^2} = 1$. For a vertical hyperbola, the equation is $\frac{(y - k)^2}{a^2} - \frac{(x - h)^2}{b^2} = 1$.

- The foci are located c units left and right of the center (horizontal hyperbola), where $c^2 = a^2 + b^2$, or c units above and below the center (vertical hyperbola), where $c^2 = b^2 + a^2$.

- The asymptotes of a hyperbola pass through the center of the hyperbola and have slopes $\pm \frac{b}{a}$ if the hyperbola is horizontal, or $\pm \frac{a}{b}$ if it is vertical.

11. Find the center, vertices, foci, and slopes of the asymptotes of the hyperbola $36y^2 - 360y - 64x^2 - 256x = 1660$.

12. Write the equation of the hyperbola with center $(0, 0)$, x-intercepts ± 5 and asymptotes $y = \pm \frac{2}{5}x$ in standard form.

- The standard form of the equation of a parabola with a vertical axis of symmetry and vertex (h, k) is $(x - h)^2 = 4p(y - k)$. The focus is $(h, k + p)$. The directrix is $y = k - p$.

- The standard form of the equation of a parabola with a horizontal axis of symmetry and vertex (h, k) is $(y - k)^2 = 4p(x - h)$. The focus is $(h + p, k)$. The directrix is $x = h - p$.

Find the vertex, focus, and directrix of each parabola.

13. $y^2 = 48x$

14. $12y = x^2 + 10x + 61$

- If A, B, or C is nonzero, the graph of $Ax^2 + Bxy + Cy^2 + Dx + Ey + F = 0$ is a hyperbola if $B^2 - 4AC > 0$; a parabola if $B^2 - 4AC = 0$; or an ellipse if $B^2 - 4AC < 0$. If the graph is an ellipse and $A = C$, the ellipse is a circle.

Identify the conic section represented by the equation.

15. $3x^2 - 5xy + 2y^2 + 3x - 5y + 13 = 0$

16. $5x - 3y + 7xy - 9 = 2(x^2 + y^2) + 3xy$

- A **quadratic system** is a set of equations that involves at least one quadratic equation. Quadratic systems can be solved using graphing, substitution, or elimination.

Solve the system.

17. $\begin{cases} x^2 + y^2 = 25 \\ x + y = 1 \end{cases}$

18. $\begin{cases} 2x - y - 1 = 0 \\ 4x^2 + y^2 = 25 \end{cases}$

19. $\begin{cases} x^2 + 3y = 18 \\ 2x^2 - 5y = 3 \end{cases}$

20. Venus orbits the sun at a mean distance of 0.72 astronomical units (AU) in 0.62 years. How long does it take Mars to orbit the sun at a mean distance of 1.52 AU?

CHAPTER ASSESSMENT

CHAPTER TEST

Use the points (2, 4) and (−10, 9).

1. Find the distance between the points.

2. Find the midpoint of the segment joining the points.

3. Find the lengths and the point of intersection of the diagonals of the rectangle formed by joining the points (2, 5), (6, 9), (0, 7), and (4, 11).

4. Find the center and radius of the circle $x^2 + y^2 - 8y - 84 = 0$.

5. Find the equation of the circle with radius 6.2 and center at $(-4, 0)$.

6. Find the foci, vertices, center, and lengths of the major and minor axes of the ellipse $9x^2 + 72x + 25y^2 - 50y = 56$. Then graph the ellipse.

7. Find the equation of the ellipse with vertices at (3, 4), (7, 5), (11, 4), and (7, 3).

8. A communications satellite is in an elliptical orbit around the earth. The major axis is 28,000 miles long. The minor axis is 20,000 miles long. Write the equation of the orbit.

9. Find the center, vertices, foci, and slopes of the asymptotes of the hyperbola $9x^2 - 54x - 16y^2 - 32y = 79$.

10. Find the equation of the horizontal hyperbola with center at (2, 3) and asymptotes with slopes of $\pm\dfrac{2}{5}$.

11. Let y represent the length of time it will take Kyle to type a 2,000-word term paper at a rate of x words per minute. Draw a graph showing the relationship between x and y.

Find the vertex, focus, and directrix of each parabola.

12. $x^2 = 16y$

13. $y^2 = x$

14. $8y = x^2 - 8x$

15. Find the equation of the parabola that has its vertex at (0, 0) and directrix $y = -7$.

16. **STANDARDIZED TESTS** The conic section represented by the equation $5x^2 + 3xy + 5y^2 - 4x - 4y + 1 = 0$ is a:

   I. hyperbola   **II.** parabola
  III. ellipse   **IV.** circle

   A. I   **B.** II   **C.** III   **D.** III and IV

Solve the system.

17. $x^2 + y^2 = 106$
     $x + y = 14$

18. $4x^2 - xy = 0$
     $2x - 3y = 10$

19. The area of a rectangular garden is 1200 yd². By increasing the width by $1\frac{1}{2}$ yd and decreasing the length by 3 yd, the area can be increased by 60 yd². Find the dimensions of the garden.

20. **WRITING MATHEMATICS** Explain why the circle, the ellipse, the hyperbola, and the parabola are called "conic sections." Refer to each of the four classes of curves in your answer.

21. Jupiter orbits the sun at a mean distance of 5.2 astronomical units (AU) in 11.87 years. How long does it take Mercury to orbit the sun at a mean distance of 0.39 AU? (Hint: Use the formula $\dfrac{a_1{}^3}{a_2{}^3} = \dfrac{P_1{}^2}{P_2{}^2}$.)

PERFORMANCE ASSESSMENT

MAKE A PRESENTATION Work with a partner. Use a flashlight, a paper model, or some other model of a cone. Use your model to demonstrate each of the following conic sections to your class:

a. circle   **b.** ellipse
c. parabola   **d.** hyperbola
e. degenerate conic sections

GRAPH YOUR FAVORITE CONIC On graph paper draw an ellipse, a hyperbola, or a parabola.

a. Identify the features of each conic section.
b. Choose three points on the graph and give their coordinates.
c. Show that the points satisfy the geometrical definition of the conic section.

CONIC SECTIONS IN YOUR LIFE Find an example of a conic section in everyday life. Take appropriate measurements and then sketch a graph of the section. The graph may be drawn full-size or to scale. Use your graph to find the equation of the conic section. Explain how you found the equation.

MAP YOUR CHOICE The map below is marked with a coordinate system that can be used to locate any point on the map.

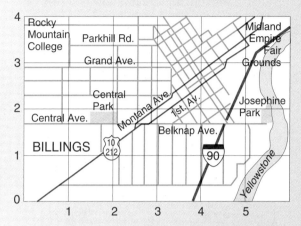

Use a state map, city map, bus route map, or another map of your choice. Draw a simple coordinate system on the map. Identify two points and their coordinates. Then use the distance and midpoint formulas to find the distance between points and the midpoint of the segment connecting them. Check your answers by measuring.

PROJECT *Connection*   Plan a video festival. Events may include watching and discussing your own production, inviting other classes or groups to a viewing, and screening and comparing related videos from professional filmmakers. Here are some discussion topics and activities. Be sure to have all manipulatives available for use.

1. Explain the benefits of watching the video as opposed to just being told how to draw the conic sections.

2. Challenge volunteers to show how the push pins in the ellipse manipulative can be moved to create a flatter ellipse or an ellipse that is more circular.

3. Challenge volunteers to use the hyperbola manipulative to explain why a hyperbola can never pass through its focus.

4. Use clay cones like the ones in the first Project Connection to demonstrate and explain how each of the degenerate conics can be cut.

· · · CUMULATIVE REVIEW · · ·

1. Which of the functions that follow is shown on the graph below?

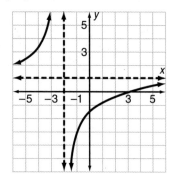

    A. $y = \dfrac{x - 3}{x - 2}$ **B.** $y = \dfrac{x - 3}{x - 1}$

    C. $y = \dfrac{x - 3}{x + 2}$ **D.** $y = \dfrac{x - 3}{x + 1}$

2. STANDARDIZED TESTS Which of the statements that follow are true about an ellipse?

    I. An ellipse is the set of all points in a plane the sum whose distances from two given points is constant.

    II. An ellipse has horizontal line and vertical line symmetries.

    III. An ellipse has point symmetry.

    A. I, II and III   **B.** I and II

    C. I and III   **D.** II and III

3. WRITING MATHEMATICS Explain the steps you would use to solve the equation $5^x = 17$. Carry out the steps to find the value of x to the nearest tenth.

4. STANDARDIZED TESTS The value of $\sqrt{-4} \cdot \sqrt{25}$ is

    A. 10   **B.** -10 **C.** ± 10 **D.** $10i$

5. Find the center and radius of the circle $3x^2 + 3y^2 + 4x - 10y - 7 = 0$.

Solve each equation or system of equations, and check.

6. $x^4 - 1 = 0$

7. $x^3 + 6x^2 - 5x - 2 = 0$

8. $2x^2 - 2x + 5 = 0$

9. $x^3 + x^2 - 16x + 20 = 0$

10. $9^{2x} = 3^{3x + 1}$

11. $\dfrac{x}{x - 1} + \dfrac{2}{x^2 - 1} = \dfrac{8}{x + 1}$

12. $2y = x + 6$
    $x + y = 6$

13. $x^2 + y^2 = 26$
    $x - y = 6$

Write and solve an equation or system of equations for each problem.

14. Stacey must accept one of two job offers. She prefers job A. However, job B pays $3.00 more per hour and would enable her to earn $1200.00 in 20 hours less than it would take her to earn $1080.00 at job A. Determine the hourly rate for each job.

15. Jessie wants to use 150 feet of fencing to build a rectangular dog run whose area is 1000 square feet. Find the dimensions to the nearest tenth of a foot.

16. Ms. Moreno invested a total of $20,000 in three stocks, one paying a 4% annual dividend, the second 5%, and the third 6%. The annual dividend from the 4% stock was equal to the annual dividend from the 6% stock. If the total annual dividend from the three stocks was $980, how much did she invest at each rate?

Tell if x and y vary directly, inversely, or jointly.

17. $y = \dfrac{5}{x}$ 18. $xy = -50$

19. $x = \dfrac{y}{4}$ 20. $x = -10yz$

STANDARDIZED TEST

QUANTITATIVE COMPARISON In each question compare the quantity in Column 1 with the quantity in Column 2. Select the letter of the correct answer from these choices:

A. The quantity in Column 1 is greater.
B. The quantity in Column 2 is greater.
C. The two quantities are equal.
D. The relationship cannot be determined by the information given.

Notes: In some questions, information which refers to one or both columns is centered over both columns. A symbol used in both columns has the same meaning in each column. All variables represent real numbers. Most figures are not drawn to scale.

	Column I		Column II

1. Some ordered pairs that satisfy this relation when x varies inversely as y^2 are shown in the chart.

x	$\frac{1}{4}$	x_2	$\frac{1}{16}$
y	± 4	$\pm\frac{\sqrt{2}}{2}$	y_3

Column I: x_2 Column II: y_3

2. The major axis of $9x^2 + 16y^2 = 144$ —— The transverse axis of $9y^2 - 16x^2 = 144$

3. $x^2 - 2x - 2 = 0$

The sum of the roots —— The product of the roots

4. $F(x) = x^4 + x^3 + 20x^2 - 2x - 40$

The integral upper bound for the zeros of $F(x)$ —— The integral upper bound for the zeros of $F(-x)$

5. $i = \sqrt{-1}$

i^{18} —— i^{12}

Column I **Column II**

6. The area of the region in the first quadrant that is the solution set of the system.
$$\begin{cases} x^2 + y^2 < 36 \\ x^2 + y^2 > 16 \end{cases}$$
—— The area of the region in the first quadrant that is the solution set of the system.
$$\begin{cases} x^2 + y^2 > 36 \\ x < 6 \\ y < 6 \end{cases}$$

7.

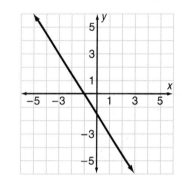

The value of the slope —— The value of the y-intercept

8. Triangle ABC, with $A(-4, 6)$, $B(3, 1)$, and $C(4, 7)$

AC —— AB

9. $$2x + y = 7$$ $$x + 2y = 2$$

$x + y$ —— $x - y$

10. The parabola with equation $(x + 4)^2 = 8(y + 3)$

The y-value of the vertex —— The y-value of the focus

11. The value of x in $2^x = 32$ —— The value of x in $\log_x 625 = 4$

12. The circle with equation $x^2 + y^2 - 8x + 2y - 19 = 0$

The x-coordinate of the center —— The y-coordinate of the center

13 Sequences and Series

Take a Look AHEAD

Make notes about things that look new.
- In which lesson will you learn the difference between a recursive formula and an explicit formula for a sequence?
- Locate and copy a symbol that you use for the first time in this chapter.

Make notes about things that look familiar.
- What are some of the relationships you look for when you try to determine the pattern for a given sequence of numbers?
- Use the word converge in an ordinary English sentence. Define or give some synonyms for this word. How do you think the mathematical meaning of this term might be similar to ordinary usage?

DATA Activity

If the Ring Fits, Wear It

Jewelry designers may work creating one-of-a-kind pieces or they may be part of a large operation that mass produces for a national market. Especially in the latter case, it is important that rings be manufactured in standard sizes. In the 1920s, the National Bureau of Standards established the system shown in the table on the next page.

SKILL FOCUS
- Add, subtract, multiply, and divide real numbers.
- Determine circumference.
- Identify and use a pattern.
- Write a linear equation.

DESIGN

In this chapter, you will see how:

- **QUILT DESIGNERS** use designs that can be modeled by arithmetic and geometric sequences.
  (Lesson 13.2, page 632)

- **COMPUTER GRAPHIC ARTISTS** incorporate arithmetic and geometric series, as well as other algebraic methods, in their work.
  (Lesson 13.4, page 645)

- **ARCHITECTS** incorporate geometry into architectural designs.
  (Lesson 13.5, page 652)

RING SIZES	
	Inside measurement in inches
Size	**Diameter**
$2\frac{1}{2}$	0.538
3	0.554
$3\frac{1}{2}$	0.570
4	0.586
$4\frac{1}{2}$	0.602
5	0.618
$5\frac{1}{2}$	0.634
6	0.650
$6\frac{1}{2}$	0.666
7	0.682
$7\frac{1}{2}$	0.698
8	0.714
$8\frac{1}{2}$	0.730
9	0.746
$9\frac{1}{2}$	0.762

1. What is the inside circumference of a size $6\frac{1}{2}$ ring? Round to the nearest thousandth of an inch.

2. Identify the pattern in the table and use it to write the diameters for sizes 10 to $13\frac{1}{2}$.

3. If size is graphed as a function of diameter, what would the slope of the resulting line be?

4. Write an equation that could be used to determine ring size given the diameter.

5. **WORKING TOGETHER** Use a piece of string to measure the ring finger diameter of several people. Determine the ring size for each person. What percent of the people had a measurement that fell about midway between two standard sizes? Discuss if you think this system of sizing is adequate.

Exploring Fractals

"Clouds are not spheres, mountains are not cones, coastlines are not circles, and bark is not smooth, nor does lightning travel in a straight line . . . patterns of nature are so irregular . . . that, compared with . . . standard geometry—Nature exhibits . . . an altogether different level of complexity." So begins Benoit Mandelbrot's extraordinary book, *The Fractal Geometry of Nature* (1982, W.H. Freeman), which changed the way mathematicians model the real world.

One way to define a *fractal* is as an object whose characteristics are not lost as it is enlarged.

PROJECT GOAL

To explore the algebra of some simple fractals.

PROJECT *Connections*

Lesson 13.2, page 631:
Explore the sequences that result from a trisection and construction procedure.
Lesson 13.4, page 644:
Construct a self-similar "snowflake" figure and examine sequences.
Lesson 13.6, page 657:
Create a manipulative fractal model and examine sequences.
Chapter Assessment, page 665:
Present research; study and create box fractals.

Getting Started
Work in groups of three or four.

1. As part of this project, each group member will do some research on fractals and share the information. The research topics are:
   - Benoit Mandebrot and his work
   - H. von Koch and the "Koch snowflake"
   - Fractals in nature
   - Geometric versus random fractals
   - The Cesaro Curve
   - The Peano curve

2. Fractal images appear regularly in magazines and other print materials. Collect as many examples as you can find for display and discussion.

Internet Connection

www.swpco.com/
swpco/algebra2.html

Algebra Workshop
Exploring Patterns and Sequences

Think Back

1. If $f(x) = 2x + 2$, find $f(0), f(1), f(2), f(3), f(4), f(5)$.

2. Create a table of values to show x and $f(x)$ from Question 1.

3. If $g(x) = 2^{x-1}$, find $g(1), g(2), g(3), g(4), g(5)$.

4. Create a table of values to show x and $g(x)$ from Question 3.

5. Explain how you found the function values in Questions 1 and 3.

6. How can you use the table to find other values for each function?

SPOTLIGHT ON LEARNING

WHAT? In this lesson you will learn
• to recognize patterns in sequences.
• to use patterns to find additional terms in a sequence.

WHY? Recognizing patterns in sequences can help you write formulas about motion, design, and depreciation.

Explore

Work with a partner.

7. Examine the following set of figures.

What pattern or patterns can you identify regarding the number of line segments in each figure?

8. If the pattern shown in Question 7 were to continue, how many line segments would the next figure in the set contain? Explain how you found your answer.

9. Examine the following set of figures.

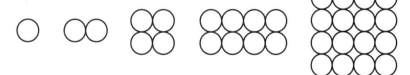

What pattern or patterns can you identify regarding the number of circles in each figure?

10. If the pattern shown in Question 9 were to continue, how many circles would the next figure in the set contain? Explain.

11. Suppose Question 9 contained ten figures in all. How many circles would be in the eighth figure? in the tenth figure?

12. How are the function $f(x)$ in Question 1 and the number of line segments in Question 7 related?

13. How are the function $g(x)$ in Question 3 and the number of circles in Question 9 related?

Make Connections

A **sequence** is an ordered set of numbers that are related mathematically. Each value in a sequence is called a **term**. The first term in a sequence is symbolized by a_1, the second term by a_2, and so on to the nth term, a_n. The nth term is also called the **general term**. The subscript n represents a positive integer. For the first term of a sequence, n is always 1, unless otherwise stated.

Use the following sequences to answer Questions 14–20.

Sequence X: 1, 2, 3, 4, 5 . . .
Sequence Y: 2, 9, 16, 23, 30 . . .
Sequence Z: 1, 3, 9, 27, 81 . . .

14. What is a_1 in sequence X? What is a_2 in sequence X? If $n = 5$, what is a_n in sequence X?

15. In sequence X, what is the relationship between each term and the term that follows it? If a_n is the term that follows a_{n-1} in this sequence, write a formula to show how to obtain a_n from a_{n-1}.

16. Using the formula you wrote in Question 15, find the sixth term of sequence X.

17. In sequence Y, what is the relationship between each term and the term that follows it? Write a formula to show how to obtain a_n from a_{n-1}.

18. Using the formula you wrote in Question 17, find the sixth and seventh terms of sequence Y.

19. In sequence Z, what is the relationship between each term and the term that follows it? Write a formula to show how to obtain a_n from a_{n-1}.

20. Using the formula you wrote in Question 19, find the sixth and eighth term of sequence Z.

ALGEBRA: WHO, WHERE, WHEN

One of the most famous recursively defined sequences is the Fibonacci sequence, named after its discoverer, Leonardo Fibonacci in 1201. The Fibonacci sequence describes many patterns in nature and can be used to derive mathematical relations. In fact, the sequence is so interesting and useful, there is a special journal, called the *Fibonacci Quarterly*, that publishes articles about the sequence.

Summarize

21. Look back at the pattern shown in Question 7. Write a formula that shows how to obtain a_n from a_{n-1} for that pattern. Then use the formula to find the first ten terms of the sequence.

22. Look back at the pattern shown in Question 9. Write a formula that shows how to obtain a_n from a_{n-1} for that pattern. Then use the formula to find the first ten terms of the sequence.

23. **LEAP YEARS** Every 4 years, an additional day, February 29, is on the calendar.

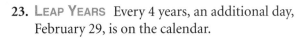

   a. Recall 1980 was a leap year. List the sequence of leap years from 1980 to 1996.

   b. Write a formula to show how to obtain a particular leap year, if you know the leap year that precedes it.

   c. If you continued the leap year sequence, would the year 2007 be a term in the sequence? Explain.

24. In the sequence 2, 7, 17, 37, 77, . . . what is the relationship between each term and the term that follows it? Look for a pattern that involves both addition and multiplication. Write a formula to show how to obtain a_n from a_{n-1}.

25. Write the next three terms in the sequence in Question 24.

26. **WRITING MATHEMATICS** Write a description of the pattern or patterns you can identify in the following sequence: 1, 4, 9, 16, 25, . . . Then write the next three terms in the sequence.

27. Find the first five terms of a sequence if $a_1 = 5$ and $a_n = 2a_{n-1} - 3$.

28. Write the first five terms of a sequence if $a_1 = 5$ and $a_n = \frac{1}{2}\left(a_{n-1}\right)$

29. **WRITING MATHEMATICS** Create three different sequences in which the terms are related in different ways. Describe each sequence.

30. **THINKING CRITICALLY** Suppose you were told that the first three terms in a sequence were 1, 2, and 4. Have you been given enough information to determine the next term? If so, find the next term in the sequence. If not, explain why not.

31. **GOING FURTHER** The sequence 1, 1, 2, 3, 5, 8, 13, 21, . . . is known as the Fibonacci sequence. How is each term of this sequence generated? What are the next six terms of this sequence?

13.2 Sequences

Explore

As part of a school project, Risa Belau is studying how bacteria reproduce. Each single-celled bacterium splits in half to form two new bacteria in the next generation, as shown.

1. Assuming the pattern shown in the diagram continues, copy and complete the table to show the number of bacteria in each generation for six generations.

Generation Number	0	1	2	3	4	5	6
Number of Bacteria	1	2					

2. Look for a pattern that shows the relationship between the number of bacteria in one generation and the number of bacteria in the previous generation. Describe the pattern.

3. Let a_n represent the number of bacteria in a new generation and a_{n-1} represent the number of bacteria in the previous generation. Write a formula that allows you to obtain a_n from a_{n-1}.

4. Look for a pattern that shows the relationship between the number of bacteria in a generation and the generation number to which the bacteria belong. Describe the pattern.

5. Let a_n represent the number of bacteria in a generation, and n represent the generation number. Write a formula that allows you to obtain a_n if you know n.

6. Which of the two formulas in Questions 3 and 5 would you use to determine the number of bacteria in the seventh generation? Why? Which of the two would you use to determine the number of bacteria in the fifteenth generation? Why? What are the number of bacteria in the seventh and fifteenth generations?

7. Assuming that the pattern continues, use a calculator to determine approximately how many generations it takes for a single bacterium to produce one million descendants.

8. Under optimal conditions, a population of bacteria can double every 30 minutes. How long will it take for a single bacterium to produce one million descendents?

Build Understanding

If a sequence does not have a last term, it is called an **infinite sequence**. Three dots called an **ellipsis** are used to show that a sequence goes on indefinitely. If a sequence stops at a particular term, it is a **finite sequence**.

In Explore, you saw that there can be more than one type of formula for obtaining the terms in a sequence. The first formula you found gave a rule for a term a_n using the preceding term a_{n-1}. This type of formula is called a **recursive formula**. You can find the terms in a recursively defined sequence if you are given the first term a_1 and the recursive formula for the sequence.

EXAMPLE 1

Write the first five terms of the sequence if

$$a_1 = 5 \quad \text{and} \quad a_n = 2a_{n-1} - 3$$

Solution

$a_1 = 5$ Use $2a_{n-1} - 3$ to find next four terms.
$a_2 = 2(5) - 3 = 7$ $a_3 = 2(7) - 3 = 11$
$a_4 = 2(11) - 3 = 19$ $a_5 = 2(19) - 3 = 35$

The first five terms of this sequence are 5, 7, 11, 19, 35.

In Question 5 of Explore, you found a rule for finding terms in a sequence by expressing the general term a_n as a function of the position n of the term within the sequence. This type of formula is called an **explicit,** or **closed, formula**.

EXAMPLE 2

Write the first four terms of a sequence whose nth term is $a_n = \dfrac{1}{2^n}$.

Solution

To write the first four terms of the sequence, replace n in the formula with 1, 2, 3, and 4, respectively.

$$a_1 = \frac{1}{2^1} = \frac{1}{2} \qquad a_2 = \frac{1}{2^2} = \frac{1}{4}$$

$$a_3 = \frac{1}{2^3} = \frac{1}{8} \qquad a_4 = \frac{1}{2^4} = \frac{1}{16}$$

The first four terms of the sequence are $\dfrac{1}{2}, \dfrac{1}{4}, \dfrac{1}{8},$ and $\dfrac{1}{16}$.

PROBLEM SOLVING TIP

To see a pattern more easily, make a function table of n and a_n.

CHECK UNDERSTANDING

Explain why you need to know at least one term in a sequence in order to use a recursive formula to find other terms in the sequence.

CHECK
UNDERSTANDING

Is the formula
$a_n = -4n^3(3n + 1)$
recursive or explicit?
Explain your answer.

If you are given the first few terms of a sequence, you can look for a pattern that helps you determine a formula for the general term.

EXAMPLE 3

Write a recursive formula for the sequence:

$$4, 20, 100, 500, \ldots$$

Solution

Identify the relationship between two consecutive terms. Then write the recursive formula for the general term.

$$a_n = 5a_{n-1} \quad \text{Multiply previous term by 5.} \qquad \blacktriangleleft$$

Use a similar approach to find the explicit formula for a sequence.

PROBLEM
SOLVING TIP

   Recognizing the pattern in a sequence may involve some trial and error. Don't be afraid to guess at the formula for the general term—an incorrect guess may lead to the correct formula.

EXAMPLE 4

Write an explicit formula for the sequence:

$$3, 12, 27, 48, \ldots$$

Solution

Look for a relationship between each term and its position n in the sequence. Each of the first four terms is 3 times a perfect square, and each perfect square is the square of n for that term.

$$a_1 = 3(1) = 3(1)^2 \qquad a_2 = 3(4) = 3(2)^2$$
$$a_3 = 3(9) = 3(3)^2 \qquad a_4 = 3(16) = 3(4)^2$$

If you continue this pattern, then a_n will equal 3 times n^2. The explicit formula for the general term is $a_n = 3n^2$ $\qquad \blacktriangleleft$

TRY THESE

Classify each formula as recursive or explicit.

1. $a_n = 2n - 1$
2. $a_n = \left(1 + \dfrac{1}{n}\right)^2$
3. $a_n = 3a_{n-1} - 1$

4. $a_n = 1 + \dfrac{1}{a_{n-1}}$
5. $a_n = \sqrt{n + 1} - \sqrt{n}$
6. $a_n = (-1)^n \sqrt{a_{n-1}}$

Write the first five terms of each sequence.

7. $a_n = n - 1$
8. $a_n = \dfrac{n}{n + 3}$
9. $a_n = (-2)^n$

10. $a_1 = 7; a_n = a_{n-1} - 4$
11. $a_1 = 2; a_n = 7 - 2a_{n-1}$
12. $a_1 = 4; a_n = 1 + \dfrac{1}{a_{n-1}}$

13. **HOURLY WAGES** A student is paid $5.65 per hour for working in a clothing manufacturer's warehouse. Each month, the student receives a $0.25 hourly raise.

   a. Write a sequence that lists the hourly salary of the student over a 6-month period.

   b. Write a recursive formula for this sequence.

Use the given formula to determine the indicated term in each sequence.

14. $a_n = 5n - 6$; determine a_8 **15.** $a_n = 5n^3$; determine a_{11} **16.** $a_n = \dfrac{3n + 7}{2n - 5}$; determine a_{14}

17. WRITING MATHEMATICS Describe the difference between an infinite and finite sequence, and give an example of each.

Write a formula for the general term in each sequence.

18. 4, 8, 12, 16, 20, . . .   **19.** $\dfrac{2}{3}, \dfrac{3}{4}, \dfrac{4}{5}, \dfrac{5}{6}, \dfrac{6}{7}, \ldots$ **20.** 1, 4, 16, 64, 256, 1024, . . .

PRACTICE

Write the first five terms of each sequence.

1. $a_n = 4n + 5$ **2.** $a_n = \dfrac{n + 3}{n}$ **3.** $a_n = (-1)^n$

4. $a_n = 9$; $a_n = 2 + 10a_{n-1}$ **5.** $a_1 = 2$; $a_n = (a_{n-1})^2$ **6.** $a_1 = 4$; $a_n = n + \dfrac{a_{n-1}}{2}$

Use the given formula to determine the indicated term in each sequence.

7. $a_n = 9n + 2$; determine a_8

8. $a_n = (3n - 4)(2n + 5)$; determine a_7

9. $a_n = 3n^2(4n - 5)$; determine a_5

10. UNIFORM MOTION A model car traveling at a constant speed moves $3n$ ft in n seconds. Write a sequence for the distance the car travels in each of the first 5 seconds. If it continues traveling at the same rate, how far will it move in 10 seconds?

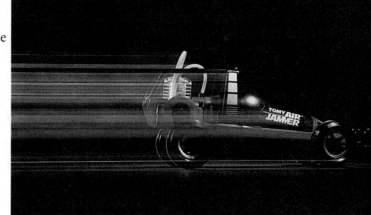

Write an explicit formula for each sequence.

11. 1, 3, 5, 7, 9, . . .

12. $-1, -4, -7, -10, -13, \ldots$

13. $\dfrac{1}{3}, \dfrac{1}{9}, \dfrac{1}{27}, \dfrac{1}{81}, \ldots$

Write a recursive formula for each sequence.

14. 1, -7, 49, -343, 2401, . . .   **15.** 1, 3, 6, 10, 15, . . .   **16.** 4, 1, $\dfrac{1}{4}, \dfrac{1}{16}, \dfrac{1}{64}, \ldots$

17. WRITING MATHEMATICS You are given the sequence $\{3, 9, 27, 81, \ldots\}$. Can you write both a recursive formula and an explicit formula for this sequence? Explain.

18. COMPUTER DEPRECIATION Carlos buys a new computer that costs $4500. Each year, the value of the computer is 10% less than its value the year before.

    a. If Carlos owns it for 5 years, what is the value of the computer at the start of each year?

    b. Write a recursive formula for this sequence.

EXTEND

19. WRITING MATHEMATICS Write the first five terms of the sequence $a_n = n^2 - n + 41$. What do you observe about these terms? Find a_{41} for the sequence. What does this tell you about the pattern you observed in the first five terms? Explain how you might have predicted this for $n^2 - n + 41$ when $n = 41$?

Write an explicit formula for each sequence.

20. $0, \sqrt{x^2}, 2\sqrt{x^3}, 3\sqrt{x^4}, 4\sqrt{x^5}$

21. $x, \dfrac{x^2}{2}, \dfrac{x^3}{4}, \dfrac{x^4}{8}, \dfrac{x^5}{16}$

22. $-x, x^3, -x^5, x^7$

23. PLANETARY DISTANCES Bode's sequence can be used to approximate distances of planets from the Sun. For example, the third term in the sequence corresponds to the distance of Earth from the Sun. These distances are measured in astronomical units with 1 AU = 93,000,000 mi. Bode's sequence is defined by

$$a_1 = 0.4 \qquad a_n = 0.1[3(2^{n-2}) + 4]$$

Find the first four terms of the sequence. Then convert the astronomical units into miles.

24. VACUUM PUMPS A vacuum is created in a container by a pump that removes $\dfrac{1}{4}$ of the air remaining in the container with each stroke.

    a. Write a sequence of terms representing the fractional part of the original amount of air *being removed* on each of the first five strokes of the pump.

    b. Write an explicit formula for the sequence in Part a.

THINK CRITICALLY

25. WRITING MATHEMATICS Consider the recursively defined sequence $a_n = \sqrt{a_{n-1}}$ where $a_1 = 5$. Approximate the successive terms to the nearest thousandth.

    a. List the first ten terms and describe what happens as n increases.

    b. Substitute any positive real number for a_1. Describe what happens as n increases.

    c. Explain the results in Parts a and b. (*Hint*: Use the laws of exponents.)

26. Suppose the general term of a sequence is $a_n = mn + b$, for some constants m and b. What kind of function does this suggest?

27. Suppose the general term of a sequence is $a_n = k^n$, for some constant k. What kind of function does this suggest?

28. FIBONACCI SEQUENCE Recall that each term of the Fibonacci sequence is the sum of the previous two terms; $a_1 = a_2 = 1$ and $a_n = a_{n-1} + a_{n-2}$.

   a. Write the first seventeen terms of the Fibonacci sequence.

   b. Find the sum of the first five terms of the sequence. Compare it to the seventh term.

   c. Find the sum of the first eight terms of the sequence. Compare it to the tenth term.

   d. Describe the pattern you observed in Parts a and b.

   e. Without adding, predict the sum of the first fifteen items. Add to check.

MIXED REVIEW

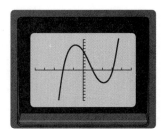

29. The graph of $y = x^3 - 2x^2 - 5x + 6$ is shown with Xscl $= 1$ and Yscl $= 1$. What are the solutions to the equation $x^3 - 2x^2 - 5x + 6 = 0$?

Determine the sum, difference, and product of each pair of binomials.

30. $2n - 5, 3n + 2$ **31.** $a - 6b, 2c + d$

32. STANDARDIZED TESTS Which gives an explicit formula for the sequence 6, 12, 24, 48, 96, . . . ?

   A. $a_n = a_{n-1} + 6$ **B.** $a_n = 6n$ **C.** $a_n = 6(2^{n-1})$ **D.** $a_n = 6 + 6n$

PROJECT *Connection*   You will need a centimeter ruler and a compass.

  1. On each of three sheets of paper, draw a line segment that is 27 cm long. Use the first segment to construct an equilateral triangle, the second to construct a square, and the third to construct a regular pentagon. What is the perimeter of each figure?

  2. Use your ruler to divide the base of each figure into three congruent parts. Construct an equilateral triangle, square, and pentagon on the middle thirds. What is the perimeter of each smaller figure?

  3. Trisect each of the thirds and construct concentric figures as shown. Repeat this procedure one more time. Copy and complete the following table.

Length of Segment	Number of Segments	Perimeter of		
		Triangle	Square	Pentagon
27 cm	1	■	■	■
■	3	■	■	■
■	■	■	■	■
■	■	■	■	■

  4. Examine the sequence in each column. Develop a formula for the *n*th terms of each sequence.

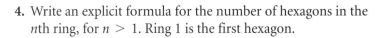

Career
Quilt Designer

Quilt-making has been practiced for thousands of years. A few traditional types such as the assortment of colors and shapes called a "crazy quilt" are randomly designed. The majority, however, are carefully composed arrangements of simple geometrical shapes and bright colors. Quilt designs can generate numerical sequences.

Decision Making

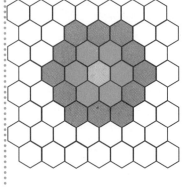

1. Choose a hexagon in the design. How many other hexagons touch it?

2. Move one "ring" further out to include all the hexagon used in Question 1. How many hexagons are in this "ring" of the design?

3. Continue outward. Write the sequence that represents the number of hexagons in each "ring", beginning with the first hexagon.

4. Write an explicit formula for the number of hexagons in the nth ring, for $n > 1$. Ring 1 is the first hexagon.

5. Write a recursive formula for the number of hexagons in ring $n + 1$, if there are a_n hexagons in ring n.

6. **a.** Count the number of 1×1 squares in the pattern; the number of 2×2 squares; the number of 3×3 squares. Write your answers as a three-term sequence.

   b. Sketch a 4×4 grid. Count the number of 1×1, 2×2, 3×3, and 4×4 squares in the grid and write your answers as a sequence of four terms.

   c. Suppose you followed the above steps for a 7×7 grid. What sequence of 7 terms would you write? Why?

7. The diagram shown illustrates the traditional "windmill" quilt design. Find and describe a way to generate a sequence of numbers using the pattern. Give either an explicit formula or a recursive formula for finding the terms of the sequence.

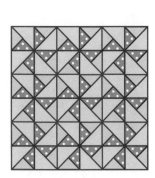

Explore

A photographer offers new assistants two salary plans. Plan 1 has a starting salary of $17,000 per year. Each year the salary increases $3715. Plan 2 also has a starting salary of $17,000 per year, but the salary increases each year to $\frac{6}{5}$ of the previous year's salary.

1. Write a recursive formula that will allow you to create a sequence of salaries for each salary plan.

2. For each salary plan, use the formulas from Question 1 to create a table that shows an assistant's salary each year for 5 years. The first year is year 0. Round each to the nearest dollar.

3. Plot the salary for each year for Salary Plan 1. Show the year on the *x*-axis and the salary on the *y*-axis. Graph Salary Plan 2 on a separate graph, but use the same scales.

4. If the points on the graph for Salary Plan 1 were connected, what kind of equation would the graph represent?

5. If the points on the graph for Salary Plan 2 were connected, what kind of equation would the graph represent?

6. Which salary plan would give an employee the best salary over a period of 3 years? over a period of 8 years?

Build Understanding

When each term in a sequence is obtained by *adding* a fixed number to the previous term, the sequence is an **arithmetic sequence.** The constant added to obtain the next term is called the **common difference** and is represented by the variable *d*. You can determine the common difference of an arithmetic sequence by subtracting any two consecutive terms.

$$d = a_{n+1} - a_n$$

If you know the first term a_1 of an arithmetic sequence, and the common difference *d* you can find the *n*th, or general, term of the sequence using this formula.

$$a_n = a_1 + (n - 1)d$$

EXAMPLE 1

Write the general term of the arithmetic sequence in which the first term is −4 and the common difference is 3. Then write the first five terms of the sequence. Graph with n on the horizontal axis and a_n on the vertical axis.

Solution

$$a_n = a_1 + (n - 1)d \qquad \text{Formula for general term.}$$
$$= -4 + (n - 1)3 \qquad \text{Substitute } -4 \text{ for } a_1 \text{ and } 3 \text{ for } d.$$
$$= -7 + 3n \qquad \text{Simplify.}$$

Now use $a_n = -7 + 3n$ to write the first five terms of the sequence.

$$a_1 = -4$$
$$a_2 = -7 + 3(2) = -1$$
$$a_3 = -7 + 3(3) = 2$$
$$a_4 = -7 + 3(4) = 5$$
$$a_5 = -7 + 3(5) = 8$$

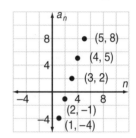

The first five terms of the sequence are −4, −1, 2, 5, 8.

You can also use the formula for the general term of an arithmetic sequence to obtain other terms in the sequence if you know any two nonconsecutive terms in the sequence. The terms between any two such nonconsecutive terms of an arithmetic sequence are called **arithmetic means**.

EXAMPLE 2

Insert three arithmetic means between 22 and 50.

Solution

Let $a_1 = 22$ and $a_5 = 50$. Then the sequence can be represented as follows.

$$22, a_2, a_3, a_4, 50$$

First use the formula for the general term to determine the common difference d.

$$a_n = a_1 + (n - 1)d$$
$$a_5 = a_1 + (5 - 1)d \qquad \text{Substitute } n = 5$$
$$50 = 22 + (5 - 1)d \qquad \text{Substitute } a_1 = 22 \text{ and } a_5 = 50$$
$$7 = d \qquad \text{Solve for } d.$$

Now you can use the common difference to determine the missing terms.

$$a_2 = a_1 + d = 22 + 7 = 29$$
$$a_3 = a_2 + d = 29 + 7 = 36$$
$$a_4 = a_3 + d = 36 + 7 = 43$$

The three arithmetic means are 29, 36, and 43.

A sequence in which each term after the first is a constant multiple of the preceding term is a **geometric sequence**. The constant multiplier is called the **common ratio** and is represented by r. To find the common ratio, divide any term by the preceding term.

$$r = \frac{a_{n+1}}{a_n}$$

The formula for the general term of a geometric sequence is

$$a_n = a_1 r^{n-1}$$

where a_1 is the first term and r is the common ratio.

ALGEBRA: WHO, WHERE, WHEN

The English clergyman Thomas R. Malthus (1766–1834) predicted famine because he found that human populations grow in a geometric sequence whereas food production increases in an arithmetic sequence.

EXAMPLE 3

MOTION A ping-pong ball is dropped from a height of 16 ft. After the first bounce, the ball rebounds to a height of 4 ft. After the second bounce, the ball rebounds to a height of 1 ft. After the third bounce, the ball rebounds to a height of $\frac{1}{4}$ ft.

Determine the height that the ball rebounds after the sixth bounce. Graph the result with the number of the bounce on the horizontal axis and the height on the vertical axis.

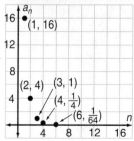

Solution

The first four terms of the sequence are

$$16, \quad 4, \quad 1, \quad \frac{1}{4}$$

first second third
bounce bounce bounce

To find the common ratio r, divide any term by its predecessor.

$$r = \frac{a_{n+1}}{a_n} = \frac{4}{16} = \frac{1}{4}$$

Now use the general formula for the sequence.

$$a_n = a_1 r^{n-1} \qquad \text{General formula.}$$

$$a_n = 16\left(\frac{1}{4}\right)^{n-1} \qquad \text{Substitute 16 for } a_1 \text{ and } \frac{1}{4} \text{ for } r.$$

$$a_6 = 16\left(\frac{1}{4}\right)^{6-1} = 16\left(\frac{1}{4}\right)^5 = \frac{1}{64} \qquad \text{Substitute 6 for } n.$$

The ball rebounds to a height of $\frac{1}{64}$ ft after the sixth bounce. ◄

Geometric means are the terms between any two nonconsecutive terms of a geometric sequence. You can use the formula for the general term to insert geometric means into a sequence.

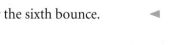

EXAMPLE 4

Insert three geometric means between 3 and 48.

Solution

Let $a_1 = 3$ and $a_5 = 48$. The sequence can be represented as

$$3, a_2, a_3, a_4, 48$$

Determine the common ratio, r.

$$a_n = a_1 r^{n-1} \qquad \text{General formula.}$$
$$48 = 3r^{(5-1)} \qquad \text{Substitute } a_1 = 3, a_5 = 48, \text{ and } n = 5.$$
$$16 = r^4 \qquad \text{Solve for } r.$$
$$\pm\sqrt[4]{16} = r$$
$$\pm 2 = r$$

Because you obtained both a positive and negative fourth root for r, you should substitute both $r = 2$ and $r = -2$ in $a_1 r^{n-1}$.

when $r = 2$ when $r = -2$

$a_2 = 3(2^1) = 6$ $a_2 = 3(-2^1) = -6$
$a_3 = 3(2^2) = 12$ $a_3 = 3(-2^2) = 12$
$a_4 = 3(2^3) = 24$ $a_4 = 3(-2^3) = -24$

The three geometric means are 6, 12, and 24 or -6, 12, and -24. ◀

TRY THESE

Determine if each sequence is arithmetic or geometric. For each arithmetic sequence, find the common difference d. For each geometric sequence, find the common ratio r.

1. 6, 11, 16, 21, 26, . . .

2. 20, 30, 45, 67.5, 101.25, . . .

3. 2, 10, 50, 250, 1250, . . .

4. 100, 250, 400, 550, 700, . . .

5. $2, 2\frac{1}{2}, 3, 3\frac{1}{2}, 4, \ldots$

6. $1, \frac{1}{2}, \frac{1}{4}, \frac{1}{8}, \frac{1}{16}, \ldots$

7. STADIUM SEATING A section of seats in a stadium has 20 seats in the first row, 22 in the second row, 24 in the third, and so on for 25 rows. How many seats are there in the last row?

Use the given information to write the formula for the nth term of each sequence. Then determine the indicated term.

8. $a_1 = 5, d = -3; a_{12}$

9. $a_5 = 2, d = 4; a_{17}$

10. $a_1 = 4, a_7 = 76; a_{30}$

11. $a_4 = -8, r = -\frac{1}{2}; a_{10}$

12. WRITING MATHEMATICS If you wanted to determine whether a sequence is geometric, is it sufficient to find the ratio between only one pair of successive terms? Explain and give examples.

13. Insert two arithmetic and two geometric means between 4 and 500.

Find the common difference, _d_, for each arithmetic sequence. Then write the next term.

1. $2, 5, 8, 11, \ldots$ **2.** $9, 5, 1, -3, \ldots$ **3.** $3.42, 5.57, 7.72, 9.87, \ldots$

Find the common ratio, _r_, for each geometric sequence. Then write the next term.

4. $4, 16, 64, 256, \ldots$ **5.** $9, -3, 1, -\frac{1}{3}, \ldots$ **6.** $5200, 3900, 2925, 2193.75, \ldots$

7. WRITING MATHEMATICS Suppose you are told that the first two terms of a sequence are 7 and 14. Write the first five terms of the sequence if the sequence is arithmetic. Write the first five terms of the sequence if the sequence is geometric. Then graph each sequence. What do you notice about the resulting graphs?

Use the given information to determine the indicated value for each arithmetic sequence.

8. $2, 6, 10, 14, \ldots ; a_{10}$ **9.** $a_1 = 6, d = -2; a_{12}$

10. $a_1 = 11, d = 1.4, n = 4; a_n$ **11.** $a_6 = 17, a_{12} = 29; a_{30}$

12. $a_2 = \frac{5}{6}, a_6 = \frac{13}{6}; a_{10}$ **13.** $a_1 = 7.5, a_9 = -8.5; a_6$

Use the given information to find the indicated value for each geometric sequence.

14. $4, -6, 9, -13.5 \ldots$, determine a_5 **15.** $a_1 = -6, r = \frac{1}{2}$ determine a_4

16. $a_1 = 81, r = \frac{1}{2}, n = 10$ determine a_n **17.** $a_5 = 48, a_8 = -384$ determine a_{10}

18. $a_3 = \frac{1}{9}, r = \frac{1}{3}$ determine a_1 **19.** $a_1 = 300, a_5 = 0.03$ determine a_n

20. PAPER MANUFACTURE A piece of paper is 0.01 in. thick. The paper is folded repeatedly, and its thickness doubles each time it is folded. If the paper were folded 15 times, how thick would be the result? (_Hint_: Let a_0 represent the original thickness of the paper.)

21. MOTION A small object that is dropped from the top of a tall building falls 16 ft during the first second, 48 ft during the second second, 80 ft during the third second, 112 ft during the fourth second, and so on. Find the formula that will let you calculate the distance the object falls during the _n_th second. Then use the formula to determine the distance the object falls during the tenth second.

22. Insert five arithmetic means between 2 and 10.

23. Insert three geometric means between 2 and 512.

EXTEND

24. In the arithmetic sequence that begins $-3, 4, 11, \ldots$, which of the following is a term?

     a. 276   **b.** 277   **c.** 278   **d.** 279

25. In the geometric sequence that begins $2, 6, 18, \ldots$, which of the following is a term?

     a. 13,120   **b.** 39,368   **c.** 118,098   **d.** 354,293

26. **WRITING MATHEMATICS** Describe how you can find the number of terms in an arithmetic sequence if you are given the first term, the last term, and the common difference. Then write an example to show the steps you described.

27. **BUILDING BLOCKS** A child using building blocks places 35 blocks in the first row, 31 in the second row, 27 in the third row, and so on. If the child continues this pattern, how many blocks will the last row contain? How many rows can the child build in all?

28. Determine a_1 and d for an arithmetic sequence where the second term is $4p - 3q$ and the fourth term is $10p + q$.

29. **VACUUM PUMPS** One particular vacuum pump removes one-half the air in a container with each stroke. After 11 strokes, what percentage of the original amount of air is left in the container?

THINK CRITICALLY

30. **WRITING MATHEMATICS** A single arithmetic mean between any two given numbers in a sequence is called the **arithmetic mean** of the numbers. Explain why the arithmetic mean of two real numbers, a and b is $\dfrac{a + b}{2}$.

31. **WRITING MATHEMATICS** A single geometric mean between any two given numbers in a sequence is called the **geometric mean** of the numbers. Explain why the geometric mean of two real numbers, a and b is $\sqrt{ab}$ or $-\sqrt{ab}$.

32. **ELECTRICITY USAGE** Fairview's consumption of electricity has increased by 6%/y. If the community is using 1.1 billion kilowatt-hours of electricity now, how much will it use 5 y from now? How many years will it take for the consumption of electricity to double?

33. **INVESTMENTS** An advertisement for a mutual fund claimed that investments made 5 y ago have doubled in value. If the fund increased at a constant rate annually, what was the rate? Round to the nearest tenth of a percent.

MIXED REVIEW

State whether the graph of the equation is symmetric about the x-axis, the y-axis, the origin, or none of these.

34. $y = x^2$ **35.** $x^4 = y^2 + 5$ **36.** $y = 3x - 2$ **37.** $y = x^3$

Rewrite each expression using a rational exponent.

38. $\sqrt{19}$ **39.** $\sqrt[4]{c^3}$ **40.** $\left(\sqrt[6]{11}\right)^5$ **41.** $\sqrt[3]{(n - 1)^7}$

42. **STANDARDIZED TESTS** Choose the letter that gives the 18th term of the sequence $-30, -24, -18, -12, \ldots.$

     A. 72   **B.** 78   **C.** -132 **D.** $-30\left(\dfrac{4}{5}\right)^{17}$

Explore

Work with a partner. You will need graph paper and scissors.

1. Examine the sum $1 + 2 + 3$. Cut out two copies of the shape shown in figure A. Notice that the first row of the shape is the first addend, the second row is the second addend, and the third row is the third addend.

A

2. Join the two cut-out shapes as shown in figure B. Notice that a rectangle is formed. Record the dimensions of the rectangle and the area of the rectangle. What is the relationship between the sum $1 + 2 + 3$ and the area of the rectangle?

B

3. Repeat Questions 1 and 2 for $1 + 2 + 3 + 4$. Record the dimensions of the rectangle and the area of the rectangle. What is the relationship between the sum $1 + 2 + 3 + 4$ and the area of the rectangle?

4. Repeat Questions 1 and 2 for $1 + 2 + 3 + 4 + 5 + 6$. What is the relationship between the sum and the area of the rectangle?

5. Examine the sum $1 + 2 + 3 + 4 + 5 + 6 + \cdots + n$. Suppose that you were to make a cut-out of this sum. How many rows of squares would you cut down? How many columns?

6. What would be the dimensions of the rectangle formed when you and your partner joined your cut-outs together?

7. Write and express, in terms of n, for the area of the rectangle formed in Question 6. What is the relationship between this area and the sum $1 + 2 + 3 + 4 + 5 + 6 + \cdots + n$?

8. For the sequence in Question 5, $a_1 = 1$ and $a_n = n$. Show that the sum of the terms of the sequence equals $\frac{1}{2}$ the number of terms times the sum of the first and last terms.

SPOTLIGHT ON LEARNING

WHAT? In this lesson you will learn

- to determine the sum and specific terms of an arithmetic series.
- to determine the sum and specific terms of a finite geometric series.
- to use sigma notation to express the sum of a series.

WHY? Sums of arithmetic and geometric sequences can help you solve problems in employments, seating capacity, and law.

Build Understanding

The sum of the terms of a sequence, such as the sums you found in Explore, is called a **series**. For example, if 2, 4, 6, and 8 are the terms of a sequence, then $2 + 4 + 6 + 8$ is the series. If the sequence is infinite, the corresponding series is an **infinite series**. If the sequence is finite, then the corresponding series is a **finite series**, or a **partial sum**, because it is the sum of the first n terms. Partial sums are symbolized by S_n.

The sum of the terms of an arithmetic sequence is called an **arithmetic series**. The sum of the first n terms of an arithmetic series can be determined using the formula

$$S_n = \frac{n}{2}(a_1 + a_n)$$

When the last term is not known, you can rewrite the formula for the partial sum as an explicit formula in terms of a_1, n, and d, by substituting the expression $a_1 + (n - 1)d$ for a_n into the formula.

$$S_n = \frac{n}{2}[2a_1 + (n - 1)d]$$

EXAMPLE 1

Use a formula for S_n to find the partial sum of each arithmetic series.

a. Find S_5 of the arithmetic sequence $-2, -4, -6, -8, -10$.

b. Find S_6 of the arithmetic sequence with $a_1 = 2$ and $d = 4$.

Solution

a. For this series, $a_1 = -2$, $n = 5$, so $a_n = a_5 = -10$.

$$S_n = \frac{n}{2}(a_1 + a_n)$$

$$S_5 = \frac{5}{2}[-2 + (-10)]$$

$$= \frac{5}{2}(-12) = -30$$

The sum of the first five terms of the sequence is -30.

b. In this case, $n = 6$, $a_1 = 2$, and $d = 4$. You do not know a_n, so use the formula

$$S_n = \frac{n}{2}[2a_1 + (n - 1)d]$$

$$S_6 = \frac{6}{2}[2(2) + (6 - 1)4]$$

$$S_6 = 72$$

The sum of the first six terms of the sequence is 72. ◄

The sum of the terms of a geometric sequence is a **geometric series**. The partial sum S_n of a geometric series is the sum of the first n terms of the series. Therefore,

$$S_n = a_1 + a_1 r + a_1 r^2 + \cdots + a_1 r^{n-1}$$

By multiplying S_n by r and subtracting the product from S_n, you find a formula for the partial sum of a geometric series.

$$S_n = \frac{a_1(1 - r^n)}{(1 - r)}, r \neq 1$$

EXAMPLE 2

Determine the sum of the first six terms of the geometric series with a first term of -2 and a common ratio of 3.

Solution

$$S_n = \frac{a_1(1 - r^n)}{(1 - r)} \quad \text{Substitute } n = 6, a_1 = -2, \text{ and } r = 3.$$

$$S_6 = \frac{-2(1 - 3^6)}{(1 - 3)}$$

$$= \frac{-2(1 - 729)}{(1 - 3)}$$

$$= -728$$

The sum of the first six terms of the sequence is -728.

When you know the first and last terms of a geometric series, as well as the common ratio, you can use the formula

$$S_n = \frac{a_1 - ra_n}{1 - r} \quad r \neq 1$$

A series may be written using **sigma** or **summation notation**. This notation uses the uppercase Greek letter sigma, Σ. For example, if the general term of an arithmetic series is $a_n = 3n + 1$, the series can be expressed in sigma notation as

$$\sum_{n=1}^{6} (3n + 1)$$

The variable n is called the **index of summation**, 1 is the lower limit, and 6 is the upper limit.

EXAMPLE 3

Write each sum using summation notation. Then evaluate.

a. $2 + 5 + 8 + 11$ **b.** $12 + 24 + 48 + 96 + 192$

Solution

a. The series is an arithmetic series. First, determine the general term a_n. In this case, $a_n = 3n - 1$. The series starts with $n = 1$ and ends with $n = 4$, so the sum can be written as

$$\sum_{n=1}^{4} (3n - 1)$$

To determine the sum, calculate a_1 and a_4. Then use the formula for the arithmetic series.

$$S_n = \frac{n}{2}(a_1 + a_n)$$

$$S_4 = \frac{4}{2}(2 + 11) = 26 \quad a_1 = 2, a_4 = 11$$

COMMUNICATING ABOUT ALGEBRA

What happens if $r = 1$ in the formula for a partial sum of a geometric series? What kind of series would result if $r = 1$?

CHECK UNDERSTANDING

Show that $\frac{a_1(1 - r^n)}{1 - r}$ is equivalent to $\frac{a_1 - ra_n}{1 - r}$.

CHECK UNDERSTANDING

Identify the index of summation, the lower and upper limits of summation, and the general term in the following sigma notation:

$$\sum_{k=1}^{4} \left(2^k + 5\right)$$

Then write the series in *expanded form* by replacing the index successively with the integers starting with the lower limit and ending with the upper limit.

b. The series is a geometric series. Notice that the numbers in this series are equal to 3 times 2^n. Therefore, the general term can be written as $3(2^n)$. In this case, the series starts with $n = 2$ and ends with $n = 6$, and the sum can be written as

$$\sum_{n=2}^{6} 3(2^n)$$

To determine this sum, use the formula for a geometric series.

$$S_n = \frac{a_1(1 - r^n)}{(1 - r)}$$

$$S_5 = 12\frac{(1 - 2^5)}{(1 - 2)} = 372 \qquad \text{Substitute } n = 5, \\ a_1 = 12, \text{ and } r = 2.$$

◀

TRY THESE

Determine the indicated sum for each arithmetic series.

1. $2 + 10 + 18 + 26 + \cdots$; find S_{10}

2. $5 + 8 + 11 + 14 + \cdots$; find S_{20}

3. $-7 - 4 - 1 + 2 + 5 + \cdots$; find S_{26}

4. $a_1 = 6, a_6 = 21$; find S_6

5. $a_1 = 2, d = 5, n = 20$; find S_n

6. $a_1 = 1, a_{100} = 100$; find S_{100}

7. Writing Mathematics Explain how you would decide which of the two formulas for the sum of an arithmetic series to use. Use examples to explain your answer.

Determine the indicated sum for each geometric series.

8. $3 + 15 + 75 + 375 + \cdots$; find S_7

9. $2 - 1 + \frac{1}{2} - \frac{1}{4}\cdots$; find S_{10}

10. $1 + 0.3 + 0.09 + 0.027 + \cdots$; find S_5

11. $a_1 = 1, r = 2$; find S_{20}

12. $a_1 = \frac{1}{64}, r = -2, n = 14$; find S_n

13. $a_1 = -2, r = -\frac{1}{2}, a_5 = -\frac{1}{8}$; find S_5

14. Writing Mathematics The formula for the nth partial sum of a geometric series can be written $S_n = \dfrac{a_1(r^n - 1)}{(r - 1)}$ $(r \neq 1)$. Explain how you could derive this formula. (*Hint*: Begin with $S_n = a_1 + a_1 r + a_1 r^2 + \cdots + a_1 r^{n-1}$.)

15. Lumber A stack of logs has 24 logs in the bottom layer, 23 in the second layer, 22 in the third layer, and so on. The top layer contains 10 logs. Describe the stack as a series and find the total number of logs in the pile.

16. Salary Suppose someone offered you a job for the month of June under the following conditions: You will be paid 1¢ for the first day, 2¢ for the second day, 4¢ for the third day, and so on, doubling your previous day's salary each day. How much would you earn after working all 30 days in June?

Write each sum using summation notation. Then evaluate each sum.

17. $1 + 3 + 5 + 7 + 9 + 11 + 13$

18. $-2 - 5 - 8 - 11 - \cdots - 56$

19. $2 + 4 + 8 + 16 + 32 + 64$

20. $1 + 4 + 16 + 64 + \cdots + 65536$

PRACTICE

Determine the indicated sum to the nearest tenth for each arithmetic series.

1. $1 + 3 + 5 + \cdots$; find S_{13}

2. $2 + 4 + 6 + \cdots$; find S_{16}

3. $2 - 1 - 4 - \cdots$; find S_8

4. $-3 + 1 + 5 + \cdots$; find S_{20}

5. $15 + 16.5 + 18 + \cdots$; find S_{10}

6. $2\frac{2}{3} + 3\frac{1}{2} + 4\frac{1}{3} + \cdots$; find S_{27}

7. $a_1 = 3, a_{13} = 51$; find S_{13}

8. $a_1 = 0.8, d = -0.2$; find S_9

9. $a_1 = 7, a_{15} = 63$; find S_{15}

10. $9 + 3 + 1 + \cdots$; find S_5

11. $1 + \frac{1}{2} + \frac{1}{4} + \cdots$; find S_9

12. $16 - 8 + 4 - \cdots$; find S_7

13. $2 + 2\sqrt{2} + 4 + \cdots$; find S_{10}

14. $4 + 0.4 + 0.04 + \cdots$; find S_8

15. $4 + 5 + 6.25 + \cdots$; find S_{12}

16. $a_1 = 1, r = -\frac{2}{3}$; find S_6

17. $a_1 = 3, a_3 = 12$; find S_8

18. $a_1 = 2, r = -3$; find S_7

19. $a_1 = 36, a_5 = 2.25$; find S_5

20. $a_7 = 128, r = 2$; find S_7

21. EMPLOYMENT Arturo takes a job as a computer programmer at a starting salary of $32,000/y, with a guaranteed 5% annual raise. Assuming no other increases, how much will his total income be for his first 6 y?

22. SEATING CAPACITY The first row in a semicircular arena contains 72 seats. Moving back, each row has two more seats than the row in front of it. How many seats are there in the 50-row arena?

Write each sum using sigma notation. Then evaluate each sum.

23. $1 + 5 + 9 + 13 + 17$

24. $20 + 18 + 16 + \cdots + 2$

25. $5 + 8 + 11 + \cdots + 65$

26. $1 + 3 + 9 + \cdots + 2187$

27. $10 + 20 + 40 + \cdots + 2560$

28. $16 - 8 + 4 - \cdots -0.5$

29. COMMERCE At a used-book sale, the Carnegie Library charged $4.00 for the first hardcover book purchased by a buyer, $3.75 for the second, $3.50 for the third, and so on. Megan spent $27 on books at the sale. How many books did she buy?

30. LAW A contractor failed to meet his deadline for completing a project for the city. A judge ordered the contractor to pay a fine of $2,000 for day 1, $2500 for day 2, $3000 for day 3, and so on, until the project was completed. The contractor ultimately paid a fine of $92,000. How late was the project finished?

31. WRITING MATHEMATICS Explain the difference between a sequence and a series.

EXTEND

32. Determine the sum of the first 10 terms of the sequence
$4x - y,\ x + 3y,\ -2x + 7y,\ \cdots$

33. Determine the sum of the first n terms of the sequence $1, 3, 5, \cdots$

34. Determine the sum of the first 12 terms of the sequence $2, 2a^2, 2a^4, \cdots$

35. Determine the first term of an arithmetic sequence in which the common difference is 5 and the sum of the first 348 terms is 3534.

36. Explain when you would use each of the two formulas for the sum of a geometric series.

THINK CRITICALLY

37. The sum of three consecutive terms in an arithmetic sequence is 30, and their product is 360. What are the three terms?

38. The sum of three consecutive terms in geometric sequence is 52. The third term is 2.25 times the sum of the other two. What are the terms?

39. The three terms of an arithmetic sequence have a sum of 42. If the first is diminished by 4, the second by 2, and the third is increased by 2, the resulting numbers form a geometric sequence. What is the arithmetic sequence?

MIXED REVIEW

Divide.

40. $(x^2 - 2x - 48) \div (x + 6)$

41. $(x^3 - 4x^2 + 3x - 6) \div (x - 2)$

42. $(x^4 - x^2 - 42) \div (x^2 - 7)$

43. $(x^4 + x^3 + x^2 + 2x + 1) \div (x + 1)$

44. STANDARDIZED TESTS Which is the correct sum for $\sum_{n=1}^{6} 2^n$?

A. 42   **B.** 12   **C.** 126   **D.** 128

PROJECT *Connection* A fractal is *self-similar* because the parts of the figure are similar to the entire figure. You will now create a fractal. You will need a large sheet of paper or posterboard, a ruler, and a compass.

1. Construct a large equilateral triangle. Determine the side length and perimeter.

2. Trisect each of the sides of the triangle. Construct a new equilateral triangle on the middle segment of each side as shown. Determine the new side length, the number of sides, and the new perimeter.

3. Repeat Question 2. You may want to erase some lines to make a cleaner figure.

4. Examine the sequence of perimeters. Predict the perimeter of the new figure if you were to repeat Question 2 one more time. Explain.

5. Discuss why the figure formed each time is self-similar. Does the sequence of perimeters approach a limit?

Career
Computer Graphic Artist

Artists have always used mathematics. Cave paintings of France and Spain are the oldest paintings known. They date back 15,000 years, yet they show a solid understanding of proportion. Renaissance artists developed perspective, and artists of recent decades created highly mathematical op-art and photo-realism techniques.

Mathematics and art are thoroughly integrated in the field of computer graphics. The computer graphic artist uses a monitor for a canvas and software for paint and brushes. Computer art ranges from simple logos to the special effects in motion pictures.

Decision Making

As you move toward the center of Spiral A, each segment decreases in length by 3 units. As you move inward in Spiral B, each segment is 0.9 times the length of the previous segment. The longest segment in each spiral measures 180 units.

1. Add the lengths of the line segments of each spiral. Which series is arithmetic? Which is geometric?

2. Approximate the length of the 30th segment in each spiral.

3. Sum the lengths of the first 30 segments in each spiral.

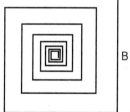

A computer graphic artist wants to create a replica of the spiral track of a CD. Answer the following questions to determine the length of the track.

4. Find the area of the recorded (shaded) portion of the CD to the nearest mm².

5. Imagine that the spiral is a long and narrow rectangle wound onto the CD. The width of the rectangle is the width of the CD track, about 0.0015 mm. The area is the area of the recorded portion of the CD. Determine the length of the track that the artist will have to replicate to the nearest 1000 mm.

6. Write the length of an entire CD track to the nearest tenth of a mile. (1 in. = 25.4 mm).

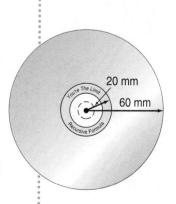

20 mm

60 mm

Explore

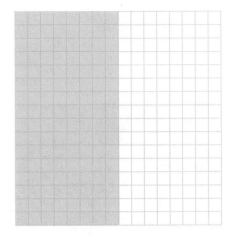

1. Shade half of a 16×16 grid as shown. How many squares did you shade? What fractional part of the grid did you shade?

2. Shade half of the unshaded region. How many squares did you shade this time? What fractional part of the original grid did you shade?

3. Repeat Question 2 four more times.

4. Write one series expressing the total number of squares you shaded. Then write another series expressing the fraction of the original square you shaded. Are the series arithmetic or geometric? Explain.

5. Determine the sum of each of the above series.

6. Suppose you repeated this process of shading indefinitely. Would the entire 16×16 grid ever be shaded? Explain.

Build Understanding

Suppose the first generation of a virus has 10 members. Each new generation is double the size of the last. The number of members of the population after n generations (assuming none die) is given by the geometric series

$$S_n = 10 + 20 + 40 + \cdots + 10(2^{n-1})$$

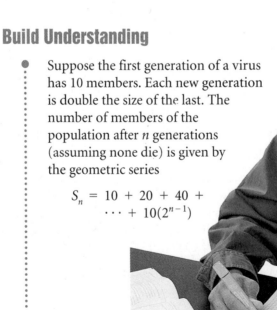

The graph at the right shows the partial sums for 10 generations. Notice that the partial sums increase without *limit*. This means that no matter how great a number A is named (10 billion, for example), it is always possible to find a value of n for which the size of generation n exceeds A.

$$S_{30} \approx 10{,}737{,}000{,}000 > 10 \text{ billion}$$

In this case, the infinite geometric series $10 + 20 + 40 + \cdots$ is said to **diverge**, or increase without limit.

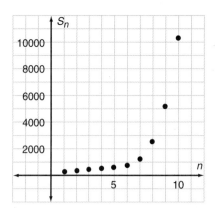

Suppose that each virus begins with a diameter of 1 unit. Following a mutation, each virus grows in such a way that its diameter after n hours is given by the geometric series

$$D_n = 1 + \frac{1}{2} + \frac{1}{4} + \frac{1}{8} + \cdots + \frac{1}{2n}$$

The graph at the right shows that, although the diameter increases each hour, the amount of increase diminishes continually. The result is that there appears to be a limit to the diameter of a virus. The broken line suggests that the diameter can get close to 2 units but can never exceed 2 units. In this case, the infinite geometric series $1 + \frac{1}{2} + \frac{1}{4} + \frac{1}{8} + \cdots$ is said to **converge** to the sum 2, or approach 2 as a limit.

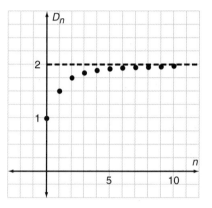

You have seen that the infinite series $10 + 20 + 40 + \cdots$, which has a common ratio of 2, diverges, but that the series $1 + \frac{1}{2} + \frac{1}{4} + \cdots$, which has a common ratio of $\frac{1}{2}$, converges.

> **THE CONVERGENCE RULE**
>
> An infinite geometric series with common ratio r converges to a sum if $|r| < 1$. Otherwise, the series diverges and has no sum.

EXAMPLE 1

Determine whether the infinite geometric series converge or diverge.

a. $486 - 162 + 54 - 18 + \cdots$ **b.** $24 + 36 + 54 + 81 + \cdots$

Solution

a. To determine r, divide any term by the preceding term.

$$r = \frac{-162}{486} = -\frac{1}{3} \qquad |r| = \left|-\frac{1}{3}\right| = \frac{1}{3} < 1$$

Since $|r| < 1$, the series converges. ◄

ALGEBRA: WHO, WHERE, WHEN

During the seventeenth century, both the Scotsman James Gregory and the German Gottfried Leibnitz proved that π can be written as the infinite series $4\left(1 - \frac{1}{3} + \frac{1}{5} - \frac{1}{7} \cdots\right)$. In 1761, German physicist Johann Lambert showed that π is irrational and can only be expressed as an infinite series.

b. $r = \dfrac{36}{24} = 1.5$ $|1.5| = 1.5 > 1$

Since $|r| > 1$, the series diverges.

If an infinite geometric series converges, you can use the formula $S_n = \dfrac{a_1(1 - r^n)}{1 - r}$ to determine the sum. Since $|r| < 1$, r^n approaches 0 as a limit for large values of n. So, $a_1(1 - r^n)$ will approach $a_1(1 - 0)$, or a_1.

> The sum S of a convergent infinite geometric series, where $|r| < 1$, is given by
> $$S = \frac{a_1}{1 - r}$$

An infinite series has an infinite number of terms, so there is no upper limit for n. In sigma notation, the symbol ∞ is used to represent *infinity*.

EXAMPLE 2

Determine the sum of the series if it exists.

a. $9 - 6 + 4 - \dfrac{8}{3} + \cdots$ **b.** $\displaystyle\sum_{n=1}^{\infty} 3(2^n)$

Solution

a. $r = \dfrac{-6}{9} = -\dfrac{2}{3}$

$|r| < 1$, so the sum exists.

$$S = \frac{a_1}{1 - r}$$
$$= \frac{9}{1 - \left(-\dfrac{2}{3}\right)}$$
$$= \frac{27}{5} \qquad 9 \times \frac{3}{5}$$

The sum of the series is $\dfrac{27}{5}$.

b. $\displaystyle\sum_{n=1}^{\infty} 3(2^n) = 6 + 12 + 24 + 48 + \cdots$

The common ratio is 2. So, $|r| \geq 1$ and the series diverges.

You can use the formula for the sum of an infinite series to write a repeating decimal as a rational number.

EXAMPLE 3

Write $0.\overline{15}$ as a rational number.

Solution

$$0.151515\ldots = 0.15 + 0.0015 + 0.000015 + \cdots$$

The decimal is an infinite geometric series with $a_1 = 0.15$ and $r = 0.01$.

$$S = \frac{0.15}{1 - 0.01} = \frac{0.15}{0.99} = \frac{5}{33} \qquad S = \frac{a_1}{1 - r}$$

THINK BACK

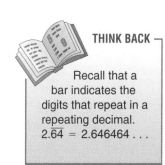

Recall that a bar indicates the digits that repeat in a repeating decimal.
$2.\overline{64} = 2.646464\ldots$

An infinite series can be used to model real world phenomena involving motion that changes by a constant ratio. Examples include the slowing of a pendulum, the acceleration of a rocket, and the "decay" in the successive high points of a bouncing ball.

EXAMPLE 4

AEROSPACE A satellite in orbit 400 km above the surface of Mars ejects a payload towards the planet's surface. The payload is programmed to descend 80 km the first 10 min, 60 km the second 10 min, 45 km the third 10 min, and so on. Using this program, will the payload reach the surface of Mars?

Solution

distance traveled $= 80 + 60 + 45 + \cdots$

$$r = \frac{60}{80} = \frac{3}{4} \qquad \text{The series converges.}$$

$$S = \frac{80}{1 - \left(\frac{3}{4}\right)} = 320 \qquad S = \frac{a_1}{1 - r}$$

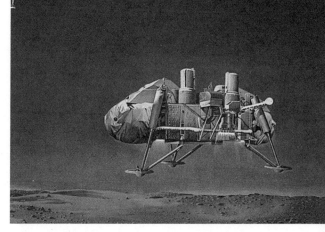

The payload will descend a total of 320 km, leaving it $400 - 320 = 80$ km above the surface of Mars. The payload will not land using this program.

TRY THESE

Determine whether the infinite geometric series converges or diverges.

1. $18 + 6 + 2 + \cdots$ **2.** $-125 + 25 - 5 + \cdots$ **3.** $40 + 60 + 90 + \cdots$

Determine the sum of the series if it exists.

4. $9 + 6 + 4 + \cdots$ **5.** $8 - 4 + 2 - \cdots$ **6.** $1 + \frac{1}{4} + \frac{1}{16} + \cdots$

7. $10 - 20 + 40 - \cdots$ **8.** $\sum_{n=1}^{\infty} 6(0.4)^{n-1}$ **9.** $10 - 1 + 0.1 - \cdots$

Write the decimal as a rational number.

10. $0.636363\cdots$ **11.** $3.\overline{2}$ **12.** $0.0\overline{9}$ **13.** $1.\overline{285}$

14. PHYSICS On each swing of a pendulum, the tip travels 0.8 times as far as it traveled on the preceding swing. If the tip traveled 40 cm on the first swing, how far will it travel before coming to rest?

15. WRITING MATHEMATICS Explain how you can determine whether an infinite geometric series converges. If it does, what is the sum?

PRACTICE

Determine whether the infinite geometric series converges or diverges.

1. $1 - \dfrac{1}{2} + \dfrac{1}{4} - \cdots$ **2.** $1 + 2 + 4 + \cdots$ **3.** $1 + 0.1 + 0.01 + \cdots$

4. $25 + 0.25 + 0.0025 + \cdots$ **5.** $-\dfrac{5}{2} - \dfrac{5}{3} - \dfrac{10}{9} - \cdots$ **6.** $2 - 4 + 8 - \cdots$

Determine the sum of the series if it exists.

7. $4 + 2 + 1 + \cdots$ **8.** $1 - \dfrac{1}{3} + \dfrac{1}{9} - \cdots$ **9.** $a_1 = 15, r = \dfrac{5}{3}$

10. $12 - 4 + \dfrac{4}{3} - \cdots$ **11.** $2 + \dfrac{1}{2} + \dfrac{1}{8} + \cdots$ **12.** $10 - 2 + 0.4 - \cdots$

13. $\dfrac{1}{2} - \dfrac{2}{3} + \dfrac{8}{9} - \cdots$ **14.** $-5 - 3 - 1.8 - \cdots$ **15.** $108 - 81 + 60.75 - \cdots$

16. $6 + 6 + 6 + \cdots$ **17.** $27 - 9 + 3 - \cdots$ **18.** $2 - 2 + 2 - \cdots$

19. $\displaystyle\sum_{n=1}^{\infty}\left(\dfrac{1}{2}\right)^{n-1}$ **20.** $\displaystyle\sum_{n=1}^{\infty} 3(0.9)^{n-1}$ **21.** $\displaystyle\sum_{n=1}^{\infty} 100\left(\dfrac{2}{5}\right)^{n-1}$ **22.** $\displaystyle\sum_{n=1}^{\infty} \dfrac{1}{2}\left(\dfrac{7}{8}\right)^{n-1}$

Write the decimal as a rational number.

23. $0.939393\ldots$ **24.** $6.\overline{5}$ **25.** $0.\overline{75}$ **26.** $0.\overline{123}$

27. $0.\overline{657}$ **28.** $0.\overline{702}$ **29.** $0.2\overline{8}$ **30.** $0.2\overline{740}$

31. ADVERTISING Research conducted by an advertising agency suggest that, of every 50 people who learn about a new product, three will tell someone else about the product. Suppose a television commercial about a new brand of jeans reaches 3 million people. If the research is correct, how many people will learn about the jeans through the ad and word-of-mouth?

32. MECHANICAL ENGINEERING The power is shut off to a flywheel making 80 revolutions per second. As the wheel slows, it makes three-quarters as many revolutions each second as it did the preceding second. How many revolutions will it make before coming to rest?

33. PHYSICS On each bounce, a rubber ball dropped from a height of 60 ft rebounds to three-fifths of its previous high point. Find the total vertical distance traveled by the ball before it comes to rest.

34. WRITING MATHEMATICS Compare and contrast finite and infinite geometric series.

EXTEND

Determine the sum of each series.

35. $\dfrac{1}{6} + \dfrac{1}{6^2} + \dfrac{1}{6^3} + \cdots$

36. $\sqrt{2} + 1 + \dfrac{\sqrt{2}}{2} + \cdots$

37. $\dfrac{\sqrt{3}}{\sqrt{3} + 1} + \dfrac{\sqrt{3}}{\sqrt{3} + 3} + \cdots$

38. Write $0.12353535\ldots$ as a rational number.

THINK CRITICALLY

Write each repeating decimal as a rational number.

39. $0.\overline{41}$

40. $0.\overline{67}$

41. $0.\overline{89}$

42. $0.\overline{04}$

43. Use the above results to write a rule for expressing a two-digit repeating decimal as a rational number.

44. Write $0.\overline{781}$ as a rational number. Then write a rule for expressing a three-digit repeating decimal as a rational number.

45. Find the sum: $\displaystyle\sum_{n=1}^{\infty} a_1 r^{n-1}$

46. Find a value of n so that $S_n = \dfrac{1}{2} + \dfrac{1}{4} + \dfrac{1}{8} + \cdots \dfrac{1}{2n} + \cdots$ is within one-millionth of 1.

MIXED REVIEW

Determine what must be added to each expression to produce a perfect square trinomial.

47. $x^2 + 8x$

48. $v^2 - 18v$

49. $n^2 + \dfrac{1}{2}n$

50. $9y^2 - 12y$

Simplify, rationalizing the denominator if necessary.

51. $\dfrac{2}{\sqrt{3}}$

52. $\sqrt{\dfrac{1}{2}}$

53. $\dfrac{3}{2 + \sqrt{7}}$

54. $\sqrt[3]{\dfrac{16}{9}}$

55. STANDARDIZED TESTS Which value gives the sum of the infinite geometric series
$9 - 6 + 4 - \dfrac{8}{3} + \cdots?$

   A. 27   **B.** $\dfrac{27}{5}$ **C.** $\dfrac{27}{4}$ **D.** does not exist

Solve for x. Check your answer.

56. $x^2 - 7x + 12 = 0$ **57.** $6x^2 - x - 1 = 0$ **58.** $18x^2 + 3x = 10$ **59.** $3x^2 - 5x - 3 = 0$

From the earliest human habitation to the most modern skyscraper, every building begins in the mind of its designer as geometry. In the words of the celebrated French architect Le Courbusier "Geometry is the foundation." The following problems illustrate how the geometry of conic sections has been incorporated into architectural design through the ages.

Decision Making

1. The Roman Colosseum is an ellipse 620 ft long and 510 ft wide. The elliptical arena floor measures 290 ft by 180 ft. Write the equations of the Colosseum and of the arena floor.

2. Cheyenne, Blackfoot, Sioux, and other Plains Indians lived in conical homes called tipis. The floor of a typical tipi measured about 3 m across.

   a. What is the shape of a tipi floor? Give a mathematical reason why this should be so.

   b. Write the equation of the floor of a tipi with the origin at the center.

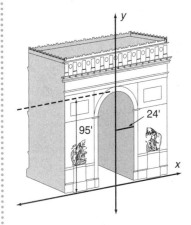

3. The Arc de Triomphe in Paris, which celebrates the victories of Napoleon, was designed by Jean Chalgrin. The 95 ft high central archway is topped by a semicircular arch 24 ft in radius. With the origin in the position shown, what equation would you use to graph the arch on a graphing utility?

4. The Gateway Arch in St. Louis is approximately a parabola that can be closely modeled by the equation $y = -\frac{1}{160}x^2 + 4x$. Write the equation in standard form.

5. By moving a plane figure through space, a three-dimensional figure is created. For example, a sphere is created by rotating a circle about one of its diameters. Nuclear power plant cooling towers are often constructed in the shape of a hyperboloid, a space figure obtained by rotating a hyperbola about its conjugate axis. Draw a sketch of a hyperboloid.

The Binomial Theorem

Explore

- The number pattern shown at the right is called **Pascal's triangle**.

  1. Describe patterns you see in the triangle.

  2. Give the numbers in the next row of the triangle. How did you find them?

  3. Determine the sum of the numbers in each of the first six rows of the triangle. Describe the pattern and predict the sum of the numbers in row 10 of the triangle.

```
        1
      1   1
    1   2   1
  1   3   3   1
1   4   6   4   1
1  5  10  10  5  1
```

SPOTLIGHT ON LEARNING

WHAT? In this lesson you will learn
- to use patterns to find powers of binomials.
- to determine powers of binomials using the binomial theorem.
- to determine specific terms of powers of binomials.

WHY? The binomial theorem can help you solve problems in geometry and numerical analysis.

Build Understanding

- Expansions of simple binomials are connected to Pascal's triangle.

$$(a + b)^0 = 1$$
$$(a + b)^1 = a + b$$
$$(a + b)^2 = a^2 + 2ab + b^2$$
$$(a + b)^3 = a^3 + 3a^2b + 3ab^2 + b^3$$
$$(a + b)^4 = a^4 + 4a^3b + 6a^2b^2 + 4ab^3 + b^4$$
$$(a + b)^5 = a^5 + 5a^4b + 10a^3b^2 + 10a^2b^3 + 5ab^4 + b^5$$

Notice these patterns in the binomial expansion of $(a + b)^n$ where n is a positive integer.

- There are $n + 1$ terms.

- The coefficients are the numbers in row $(n + 1)$ of Pascal's triangle.

- The first term is a^n and the last term is b^n.

- In succeeding terms, exponents of a decrease by 1 and exponents of b increase by 1.

- The sum of the exponents in each term is n.

ALGEBRA: WHO, WHERE, WHEN

Pascal's triangle is named after Blaise Pascal, a 17th century French mathematician and philosopher.

EXAMPLE 1

Expand $(a + b)^7$.

Solution

Row 8 of Pascal's triangle is 1, 7, 21, 35, 35, 21, 7, 1.

$$(a + b)^7 = a^7 + 7a^6b + 21a^5b^2 + 35a^4b^3 + 35a^3b^4 + 21a^2b^5 + 7ab^6 + b^7$$

As a check, note that there are $n + 1$, 8, terms in the expansion and that the sum of the exponents in each term is 7. ◄

CHECK UNDERSTANDING

Determine each of the following.
1. 2!
2. 3!
3. 6!
4. Does 2!3! equal 6! ?
5. Does $\frac{6!}{2!}$ equal 3! ?

You can use Pascal's triangle to find the coefficients of a binomial expansion. For higher powers of a binomial, however, a more practical method for determining coefficients is to use factorial notation. The expression $n(n - 1)(n - 2) \ldots (2)(1)$ means the product of all positive integers from n to 1 and is called **n factorial**, written as **n!**

In the above expansion of $(a + b)^7$, note that the coefficient of the third term, 21, can be written in terms of the factorials of 7 (the power of the binomial), 5 (the exponent of a), and 2 (the exponent of b).

$$\frac{7!}{5!2!} = \frac{7 \cdot 6 \cdot 5 \cdot 4 \cdot 3 \cdot 2 \cdot 1}{(5 \cdot 4 \cdot 3 \cdot 2 \cdot 1)(2 \cdot 1)} = \frac{5040}{120 \cdot 2} = 21$$

In general, the coefficient of any term in the expansion of $(a + b)^n$ is

$$\frac{n!}{(\text{exponent of } a)!(\text{exponent of } b)!}$$

EXAMPLE 2

Write the fifth term of $(a + b)^{10}$.

Solution

The exponent of a in the first term is 10 and decreases by 1 each term. Therefore, the exponent of a in the fifth term is 6. The exponent of b is $10 - 6 = 4$. So, the coefficient is

$$\frac{10 \cdot 9 \cdot 8 \cdot 7 \cdot \overbrace{6 \cdot 5 \cdot 4 \cdot 3 \cdot 2 \cdot 1}^{6!}}{\underbrace{(6 \cdot 5 \cdot 4 \cdot 3 \cdot 2 \cdot 1)}_{6!}\underbrace{(4 \cdot 3 \cdot 2 \cdot 1)}_{4!}} = \frac{10 \cdot 9 \cdot 8 \cdot 7}{4 \cdot 3 \cdot 2 \cdot 1} = 210$$

The fifth term is $210\, a^6 b^4$. ◀

The patterns you have seen are reflected in the **binomial theorem**.

CHECK UNDERSTANDING

In Example 2, once you know that the exponent of a in the seventh term is 6, why can you conclude that the exponent of b is $10 - 6 = 4$?

THE BINOMIAL THEOREM

Let n be a positive integer. Then

$$(a + b)^n = a^n + \frac{n}{1} a^{n-1}b^1 + \frac{n(n - 1)}{1 \cdot 2}a^{n-2}b^2 +$$
$$\frac{n(n - 1)(n - 2)}{1 \cdot 2 \cdot 3} a^{n-3}b^3 + \cdots + b^n$$

The binomial theorem can also be written using these notations:

Factorial notation

$$(a + b)^n = a^n + \frac{n!}{(n - 1)!1!} a^{n-1}b^n + \frac{n!}{(n - 2)!2!} a^{n-2}b^2 + \cdots + b^n$$

Sigma notation

$$(a + b)^n = \sum_{r=0}^{n} \frac{n!}{(n - r)!r!} a^{n-r}b^r$$

EXAMPLE 3

Expand $(3m - n)^4$.

Solution

$$(3m - n)^4 = [3m + (-n)]^4 \qquad \text{Rewrite in form } (a + b)^n$$

$$(3m - n)^4 = (3m)^4 + 4(3m)^3(-n) + \frac{4 \cdot 3}{1 \cdot 2}(3m)^2(-n)^2 + \frac{4 \cdot 3 \cdot 2}{1 \cdot 2 \cdot 3}(3m)(-n)^3 + (-n)^4$$

$$= 81m^4 - 108m^3n + 18m^2n^2 - 12mn^3 + n^4 \qquad \blacktriangleleft$$

The binomial theorem in sigma notation is useful for determining a given term of an expansion.

EXAMPLE 4

Determine the sixth term of the expansion of $(x - 2)^{11}$.

Solution

$$(x - 2)^{11} = \sum_{r=0}^{11} \frac{11!}{(11 - r)!r!} x^{11-r}(-2)^r$$

Because r begins at zero, $r = 5$ for the sixth term.

$$\text{sixth term} = \frac{11!}{6!5!} x^{11-5}(-2)^5 = 462x^6(-32) = -14{,}784x^6 \qquad \blacktriangleleft$$

PROBLEM SOLVING TIP

Most calculators have a feature to calculate factorials automatically.

TRY THESE

Answer these questions about the expansion of $(a + b)^{17}$.

1. How many terms are there?   **2.** What is the first term?

3. What is the exponent of b in the term containing a^{11}?

4. What are the exponents of a and b in the eighth term.

5. Find the coefficient of the term containing m^3 in the expansion of $(m + n)^9$.

Expand each binomial.

6. $(a + b)^4$ **7.** $(d - 2)^6$ **8.** $(3p + q)^5$

Write the indicated term of the expanded polynomial.

9. 4th, $(x + y)^5$ **10.** 5th, $(m + n)^6$

11. 5th, $(2c - d)^9$ **12.** 10th, $(2a + b)^{12}$

13. GEOMETRY The edges of a cube are each x in. in length. If each edge is increased by 4 in., write the binomial expansion that represents the volume of the new cube.

14. WRITING MATHEMATICS Explain how to use Pascal's triangle to determine the coefficients in the expansion of $(x + y)^{10}$.

PRACTICE

Expand each binomial.

1. $(a + b)^6$

2. $(2a + x)^5$

3. $(c - 5d)^4$

4. $(1 - x)^8$

5. $(2 - 3b)^4$

6. $(a - b^2)^7$

7. $(b^2 - ac)^3$

8. $(3a^3 - 2b^3)^4$

9. $(30 + 1)^3$

10. $(100 - 2)^4$

11. $\left(a - \dfrac{3}{b}\right)^4$

12. $\left(n + \dfrac{1}{2}\right)^5$

Write the indicated term of the expanded polynomial.

13. 4th, $(a + b)^7$

14. 5th, $(a - b)^8$

15. 7th, $(3 - x)^9$

16. 4th, $(x - 3y)^6$

17. 6th, $(2a + b)^7$

18. 2nd, $(3a^2 - 2b)^4$

19. 12th, $(n - 2)^{14}$

20. 6th, $(x^2 - 2y)^5$

21. 3rd, $(3a - b)^6$

22. 4th, $(3x - 2y)^5$

23. 5th, $(2m^2 - 3n^2)^6$

24. 4th, $\left(\dfrac{a}{b} + \dfrac{b}{a}\right)^6$

25. **NUMERICAL ANALYSIS** Use the first four terms of the binomial expansion of $(1 + 0.02)^{10}$ to approximate $(1.02)^{10}$. Evaluate $(1.02)^{10}$ using a calculator. What is the sum of the remaining terms of the binomial expansion?

26. **NUMERICAL ANALYSIS** If you wanted to approximate $(0.97)^{12}$ using the first four terms of a binomial expansion, write the binomial you would use.

27. **WRITING MATHEMATICS** One term in the binomial expansion of $(a + b)$ is a^4b^7. Determine the next term, including the coefficient, explaining each step.

EXTEND

Simplify.

28. $\dfrac{20!}{18!}$

29. $\dfrac{45!}{42!}$

30. $\dfrac{25!\,32!}{24!\,33!}$

31. $36 \cdot 35!$

32. $\dfrac{n!}{(n - 1)!}$

33. $(p + 5)!(p + 6)$

34. $\dfrac{(x + 1)!}{(x - 1)!}$

35. $\dfrac{(x - y)!}{(x - y - 1)!}$

36. **WRITING MATHEMATICS** Without using a calculator prove that $(1.015)^{30} \geq 1.45$.

37. Write the term in the expansion of $(c + 2d)^9$ that contains d^5.

38. Write the term that does not contain x in the expansion of $\left(x - \dfrac{1}{x}\right)^8$.

39. Evaluate $(1 - i)^5$, where $i = \sqrt{-1}$.

THINK CRITICALLY

40. **WRITING MATHEMATICS** Does $(n - 4)!$ equal $(n - 4)(n - 5)(n - 6)!$ for $n \geq 6$? Explain your answer and support it with an example.

41. What are the first terms of $(1 + b)^{\frac{1}{2}}$?

42. You can use your answer from Exercise 38 and the binomial theorem to approximate $\sqrt{6}$. Write $\sqrt{6}$ as follows. Round to the nearest hundredth.

$$\sqrt{6} = \sqrt{4 + 2} = \sqrt{4\left(1 + \dfrac{1}{2}\right)} = 2\left(1 + \dfrac{1}{2}\right)^{\frac{1}{2}}$$

Let $f(x) = 2x + 1$ and $g(x) = x^2 + x$. Determine each of the following.

43. $(f + g)(x)$ **44.** $(f \cdot g)(x)$ **45.** $f(g(x))$ **46.** $g(f(x))$

Let $y = 24$ when $x = 6$. Determine each constant of variation.

47. y varies directly as x **48.** y varies inversely as x

49. y varies directly as the square of x **50.** y varies inversely as the square of x

51. Determine the center of the ellipse $9x^2 + 4y^2 + 54x - 8y + 49 = 0$.

52. STANDARDIZED TESTS Which is the seventh term of the expansion of $(x + y)^9$?

     A. $36x^2y^7$ **B.** $84x^3y^6$ **C.** $126x^4y^5$ **D.** $36x^7y^2$

Solve each system of linear equations.

53. $\begin{cases} x + y - z = 0 \\ -x + y + z = 4 \\ x - y + z = 2 \end{cases}$ **54.** $\begin{cases} 2x + 2y + z = 13 \\ x - 3y + 2z = -8 \\ 5x - 2y + 3z = 5 \end{cases}$ **55.** $\begin{cases} x - y + 3z = 9 \\ 2x + 2y + 5z = -7 \\ 4x - 2y - 7z = 5 \end{cases}$

PROJECT *Connection*

You will need a large quantity of congruent square tiles or pieces of paper.

1. Arrange three squares as shown.

2. Replace each of the three squares with an exact duplicate of the first figure.

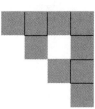

3. Continue the replacement two more times.

4. Predict how many squares will be used if you continue this procedure one more time.

5. Have some members of your group create a design using four squares and the other members create a design using five squares. Repeat Questions 2–4 for each design.

6. Explain how you can zoom in on a part of the fractal and reveal the original picture.

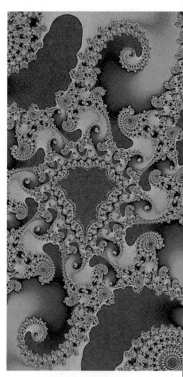

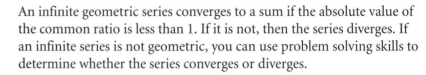
Convergence and Divergence

An infinite geometric series converges to a sum if the absolute value of the common ratio is less than 1. If it is not, then the series diverges. If an infinite series is not geometric, you can use problem solving skills to determine whether the series converges or diverges.

Problem

ENVIRONMENTAL SCIENCE During its first year of operation, Synergy Computer Corporation released 1 g of toxic mercury into the soil surrounding the plant. Concerned about safety, the city council passed a law requiring Synergy to cut mercury contamination to $\frac{1}{2}$ g during year 2, $\frac{1}{3}$ g during year 3, and so on. The council's intent was to keep total contamination under 4 g, regardless of how long Synergy continued to build computers. Did the council succeed?

Explore the Problem

1. Write a series, including the general term, that expresses total mercury contamination after n years.

2. Show that the series is neither arithmetic nor geometric.

3. Copy and complete the following table, giving the first ten partial sums of the series. Write sums to the nearest hundredth.

n	1	2	3	4	5	6	7	8	9	10
$\sum\limits_{n=1}^{n} \frac{1}{n}$										

4. What conclusions, if any, can you reach from studying the table?

5. Graph the data in the table and the line $y = 4$. What conclusions, if any, can you reach from studying the graph?

6. Sometimes you can tell whether a series is convergent or divergent by looking at the sums of *groups* of terms within the series. Explain the grouping scheme that has been followed below.

$$(1) + \left(\frac{1}{2}\right) + \left(\frac{1}{3} + \frac{1}{4}\right) + \left(\frac{1}{5} + \frac{1}{6} + \frac{1}{7} + \frac{1}{8}\right) + \left(\frac{1}{9} + \cdots \frac{1}{16}\right) + \cdots$$

7. **a.** Explain how you know without adding that $\frac{1}{3} + \frac{1}{4} > \frac{1}{2}$.

   b. How do you know without adding that $\frac{1}{5} + \frac{1}{6} + \frac{1}{7} + \frac{1}{8} > \frac{1}{2}$?

8. Why is $(1) + \left(\frac{1}{2}\right) + \left(\frac{1}{3} + \frac{1}{4}\right) + \left(\frac{1}{5} + \frac{1}{6} + \frac{1}{7} + \frac{1}{8}\right) +$

   $\left(\frac{1}{9} + \frac{1}{16}\right) + \cdots > \frac{1}{2} + \frac{1}{2} + \frac{1}{2} + \frac{1}{2} + \cdots?$

9. How do you know that $\frac{1}{2} + \frac{1}{2} + \frac{1}{2} + \frac{1}{2} + \cdots$ diverges?

10. Use the results of Questions 6–9 to show total contamination by Synergy Computer Corporation will exceed 4 g.

11. Interpret the graph at the right of the first 40 partial sums of the infinite series $1 + \frac{1}{2} + \frac{1}{3} + \cdots + \frac{1}{n} + \cdots$.

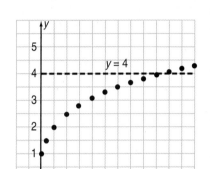

PROBLEM SOLVING PLAN

- Understand
- Plan
- Solve
- Examine

Investigating Further

If an infinite series is not geometric, the ratio of succeeding terms is not constant. You may be able to use the succeeding-term ratio to determine whether a nongeometric series converges. To do so, write the ratio as an algebraic expression. Then think about what happens to the expression when n gets very large.

- If for a large n the expression is less than 1, then the series converges.
- If for a large n the expression is greater than 1, the series diverges.

This method is called the **ratio test.** Answer Questions 12–15 to show that the infinite series $\frac{1}{2} + \frac{2}{2^2} + \frac{3}{2^3} + \cdots + \frac{n}{2^n} + \cdots$ converges.

12. What term follows $\frac{n}{2^n}$?

13. Determine the ratio of the succeeding terms.

14. Divide each term in your answer to Question 13 by n to obtain a complex fraction.

15. Suppose that n is a very large number. Explain why the answer to Question 14 must equal a number between 0 and 1.

Determine whether the series converges or diverges.

16. $\frac{2}{1} + \frac{3}{2} + \frac{4}{3} + \cdots + \frac{n + 1}{n} + \cdots$

17. $\frac{1}{10} + \frac{2}{10^2} + \frac{3}{10^3} + \cdots + \frac{n}{10^n} + \cdots$

18. $\frac{1}{2} + \frac{1}{4} + \frac{1}{6} + \cdots + \frac{1}{2n} + \cdots$

19. $\frac{2}{1} + \frac{2^2}{2^2} + \frac{2^3}{3^2} + \cdots + \frac{2^n}{n^2} + \cdots$

20. $\frac{1}{1!} + \frac{1}{2!} + \frac{1}{3!} + \cdots + \frac{1}{n!} + \cdots$

CHECK UNDERSTANDING

Give a number that each expression is approximately equal to when n is very large.

a. $\frac{1}{n}$

b. $\frac{n}{n + 1}$

c. $\frac{4 + n}{7 - n}$

d. $\dfrac{3 + \frac{1}{n}}{4 + \frac{2}{n}}$

21. WRITING MATHEMATICS Explain how you would decide whether each of the following types of infinite series converges or diverges.

   a. arithmetic

   b. geometric

   c. neither arithmetic nor geometric

Apply the Strategy

22. MARINE BIOLOGY During one year, $\frac{1}{3}$ of the rocks on a reef previously unaffected by barnacles became covered by shellfish. The following year, another $\frac{2}{9}$ of the reef was covered, and the year after that another $\frac{3}{27}$ of the reef. Assume that additional fractions of the reef are covered annually according to the already established pattern, and that portions of the reef already covered remain so. Will the reef ever be completely covered? Explain.

23. THINK CRITICALLY An infinite geometric series with common ratio r converges. Does the series comprised of the reciprocals of the terms of the series also converge? Explain.

ALGEBRA: WHO, WHERE, WHEN

The *curie* is named after Polish-born French physicist Marie Curie (1867–1934) who twice won the Nobel Prize.

24. ENVIRONMENTAL SCIENCE The annual radioactive discharge at a nuclear power plant during year n of operation is $\frac{1}{2} + \frac{2}{3} + \frac{3}{4} + \cdots + \frac{n}{n+1}$ curies. The Nuclear Regulatory Commission (NRC) has ordered that total cumulative radioactive discharge may not exceed 20 curies. Assuming that the plant remains in operation, can it meet the NRC standards? Explain.

25. NUMBER THEORY The infinite series $1 + \frac{1}{1!} + \frac{1}{2!} + \frac{1}{3!} + \cdots + \frac{1}{n!} + \cdots$ is used to calculate exponential growth and decay, although in your study of growth and decay in Chapter 9 you used the series in a different form. To discover the more familiar form, answer the following questions.

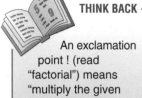

THINK BACK

An exclamation point ! (read "factorial") means "multiply the given natural number by every natural number less than itself." $4! = 4 \cdot 3 \cdot 2 \cdot 1$

   a. Make a table of values giving the first 8 partial sums of the series. Round sums to the nearest one-thousandth.

   b. Graph the data in the table.

   c. What sum does the series appear to be converging on?

   d. What letter is used to symbolize this important mathematical constant?

REVIEW PROBLEM SOLVING STRATEGIES

What's the Difference?

1. a. The integers 1 through 15 can be arranged in five arithmetic sequences of three integers so that the common differences are 7, 5, 4, 2, and 1. For example, the first triplet is 1, 8, 15 with $d = 7$.

Arrange the remaining twelve integers so that the differences for the triplets are 5, 4, 2, and 1. (You might want to write the integers on pieces of paper so you can move them around.)

b. Keep the same integers for the first triplet as above. Find four entirely new triplets using the same twelve integers also with differences 5, 4, 2, and 1.

c. On your own, experiment with other arrangements of the integers from 1 through 15. What types of sequences can you create?

WHO WENT WHERE WHEN?

2. One of four friends, Marsha, Ned, Olivia, or Paul, went to the box office to buy concert tickets. If Marsha left school before Paul, then Marsha went. If Marsha left before Ned, then Ned went. If Ned left before Olivia, then Olivia went. If Olivia left before Paul, then Paul went. Marsha left school before Olivia. Ned left before Paul. Who bought the tickets? Explain your reasoning.

MATCH WITS

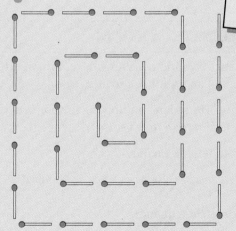

3. The spiral figure at right is made up of thirty five matches.

Show how to move exactly four matches so that you create three squares.

CHAPTER REVIEW

VOCABULARY

Match the letter of the word in the right column with the description at the left.

1. a series without a limit
2. a sequence for which the value of each term is found from the previous term
3. a rule for finding each term of a sequence using n, the number of the term
4. a series with a sum
5. an unlimited number of terms

a. convergent

b. divergent

c. explicit

d. infinite

e. recursive

Lessons 13.1 and 13.2   PATTERNS AND SEQUENCES   pages 623–625, 626–632

- A **sequence** is an ordered set of numbers that are related mathematically.

- A rule for finding each term of a sequence from the previous term is a **recursive formula**. A rule for finding a term based on its position n within the sequence is an **explicit formula**.

Classify each formula as recursive or explicit.

Write the first five terms of each sequence.

6. $a_n = n^2 + 2$ 7. $a_n = 4a_{n-1} - 1$

8. $a_n = n^2 - 1$ 9. $a_1 = 3, a_n = a_{n-1} - 1$

Write an explicit formula for each sequence.

10. $1, 2, 4, 8, 16, \ldots$ 11. $-1, -3, -5, -7, -9, \ldots$

Write a recursive formula for each sequence.

12. $4, 8, 12, 16, 20, \ldots$ 13. $1, 2, 3, 5, 8, \ldots$

Lesson 13.3   ARITHMETIC AND GEOMETRIC SEQUENCES   pages 633–638

- Each term in an **arithmetic sequence** is found by adding the **common difference** to the previous term. The general term a_n of an arithmetic sequence is $a_n = a_1 + (n - 1)d$.

- Each term in a **geometric sequence** is found by multiplying the previous term by a constant called the **common ratio**. The general term a_n of a geometric sequence is $a_n = a_1 r^{n-1}$.

Find the indicated value for each arithmetic sequence.

14. $a_{15}: 3, 7, 11, 15, \ldots$ 15. $a_{20}: a_1 = 70, d = -3$ 16. $d: a_1 = 5, a_{18} = 56$

Find the indicated value for each geometric sequence.

17. $a_6: a_1 = -64, r = -\dfrac{1}{2}$ 18. $a_8: \dfrac{1}{27}, \dfrac{1}{9}, \dfrac{1}{3}, 1, \ldots$ 19. $n: 1, 2, 4, 8, \ldots, 1024$

- A **series** is the indicated sum of the terms of a sequence. A **partial sum**, S_n, of an arithmetic or geometric series is the sum of the first n terms of the series.

- The partial sum of an arithmetic series is given by $S_n = \frac{n}{2}[2a_1 + (n - 1)d]$.

- The partial sum of a geometric series is given by $S_n = \frac{a_1(1 - r^n)}{1 - r}$, $r \neq 1$.

Identify each series as arithmetic or geometric. Then find the indicated sum.

20. S_{30}: $1 + 6 + 11 + 16 + \ldots$

21. S_{17}: $a_1 = 102$, $d = -2$

22. S_{200}: $a_1 = 1$, $a_{200} = 200$

23. S_{10}: $32 - 16 + 8 - 4 + \ldots$

24. S_7: $a_1 = 1$, $r = 3$

25. S_6: $1280 + 320 + 80 + \ldots$

Write each sum using sigma notation.

26. $4 + 7 + 10 + 13 + 16 + 19 + 22$

27. $1 + 3 + 7 + 15 + 31 + 63$

- An **infinite geometric series** with common ratio r will **converge** to a **limit** or sum if $|r| < 1$. Otherwise, the series will **diverge**.

- The sum S of an infinite geometric series, where $|r| < 1$, is given by $S = \frac{a_1}{1 - r}$.

Determine whether the infinite geometric series converges or diverges.

28. $200 + 100 + 50 + \ldots$

29. $0.25 + 0.5 + 1 + \ldots$

30. $\frac{81}{16} - \frac{27}{8} + \frac{4}{9} + \ldots$

Find the sum of the infinite series.

31. $64 + 16 + 4 + 1 + \ldots$

32. $12 - 6 + 3 - 1.5 + \ldots$

33. $\sum_{n=1}^{\infty} 5(0.3)^n$

34. Write $0.454545 \ldots$ as a rational number.

- The **Binomial Theorem** states that if n is a positive integer, then

$$(a + b)^n = a^n + na^{n-1}b^1 + \frac{n(n - 1)}{1 \cdot 2}a^{n-2}b^2 + \frac{n(n - 1)(n - 2)}{1 \cdot 2 \cdot 3}a^{n-3}b^3 + \ldots + b^n$$

- The coefficient of any term in the binomial expansion of $(a + b)^n$ is given by

$$\frac{n!}{(\text{exponent of } a)!(\text{exponent of } b)!}$$

35. Expand $(a + b)^4$.

Find the indicated term.

36. 3rd, $(m + n)^7$

37. 6th, $(a - b)^8$

38. 5th, $(2x^2 - y^3)^5$

CHAPTER ASSESSMENT

CHAPTER TEST

Classify each formula as recursive or explicit.

1. $a_n = 1 + \dfrac{1}{n-1}$

2. $a_n = 1 + \sqrt{n-1}$

Write the first five terms of each sequence.

3. $a_n = 20 - n$

4. $a_1 = 2, a_n = (4 - a_{n-1})^2$

5. Write an explicit formula for the sequence
   $3, 8, 13, 18, 23, \ldots$

6. Write a recursive formula for the sequence
   $1, 3, 7, 15, 31, \ldots$

Find the indicated value for each arithmetic sequence.

7. $a_{30}: 5, 9, 13, 17, \ldots$

8. $a_{21}: a_1 = -98, d = 7$

Find the indicated value for each geometric sequence.

9. $a_7: a_1 = 243, r = -\dfrac{1}{3}$

10. $a_7: 52{,}000, 5200, 520, 52, \ldots$

11. Insert 3 arithmetic means between 12 and 22.

12. Insert 2 geometric means between 4 and 500.

Find the indicated sum for each arithmetic series.

13. $S_{40}: 2 + 8 + 14 + 20 + \ldots$

14. $S_{25}: a_1 = 209, d = -7$

Find the indicated sum for each geometric series.

15. $S_8: 1200 + 120 + 12 + 1.2 + \ldots$

16. $S_8: a_1 = 1, r = 2$

17. Use sigma notation to write the sum
    $12 + 18 + 15 + 21 + 24$

18. A triangular toothpaste display has 1 carton in the top row, 3 in the second row, 5 in the third row, and so on. There are 41 cartons in the bottom row.

    a. How many rows are there?

    b. How many cartons are in the display?

19. **WRITING MATHEMATICS** Write a paragraph that distinguishes between the following: arithmetic sequence, geometric sequence, arithmetic series, and geometric series.

Determine whether the infinite series converges or diverges.

20. $-40 + 20 - 10 + 5 - \ldots$

21. $0.03 + 0.3 + 3 + 30 + \ldots$

Find the sum of the infinite series.

22. $1 - \dfrac{1}{3} + \dfrac{1}{9} - \dfrac{1}{27} + \ldots$

23. $\displaystyle\sum_{n=1}^{\infty} 8(0.2)^n$

24. Write $0.2222\ldots$ as a rational number.

25. On each bounce, a rubber ball dropped from a height of 100 inches rebounds to 56% of its previous high point. Find the distance traveled by the ball before it comes to rest.

26. Expand $(m - n)^5$.

27. **STANDARDIZED TESTS** The third term in the binomial expansion of $(3x - y)^6$ is

    A. $1215x^4y^2$ B. $-1458x^4y^2$

    C. $15x^4y^2$ D. $81x^4y^2$

PERFORMANCE ASSESSMENT

PARTNERS IN SEQUENCE Write the first five terms of an arithmetic sequence. Ask your partner to write a recursive formula and an explicit formula for your sequence. Then find S_5, S_{10}, and S_{20} for your sequence. Change roles with your partner and repeat the steps for a geometric sequence.

INFINITE SERIES

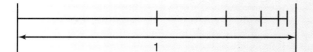

a. Describe the infinite series suggested by the figure.

b. Describe a real-world situation that can be modeled by the series.

c. In your own words, explain why the sum will always be less than 1.

GEOMETRIC PATTERNS As you learned in this chapter, a simple pattern such as

```
o    oo    ooo    oooo
      o     o      o
            o      o
                   o
```

can be described algebraically. Select a pattern found in nature or in a substance produced by humans from a book, magazine, or newspaper. Describe the pattern algebraically, giving as much information about it as you can.

MODEL A SEQUENCE Use pencils, thumb tacks, books, or some other objects of your choice to model a sequence. Describe the sequence and name its important features. Write a formula for the general term of the sequence, and use the formula to predict a term that is not in your model.

PROJECT ASSESSMENT

PROJECT *Connection* Get ready for a class discussion of fractals by decorating the room with the designs you created during the project and the graphics you collected from print materials. Add captions to explain the special properties of the figures.

1. Form new student groups according to the topic researched. Share and discuss all the information, and then work together to prepare a comprehensive presentation that will last about 10 minutes. You may wish to use a panel format, in which each member gets a chance to speak.

2. With your original group, study and discuss the development of the *box fractal* below. Predict the number of squares that would be in the next stage. Then create a box fractal of your own by removing the smaller squares according to a pattern. Develop a rule for determining the number of squares in your fractal design at the nth stage.

CUMULATIVE REVIEW

1. The figures are the first three stages of a pattern.

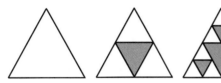

   How many unshaded triangles are in the figure at Stage 10?

2. Bob Simmons bought a 1996 sports utility vehicle for $19,995. If the car depreciates 13%/y, write an equation to give its value n after 1996. What is the car's approximate value in the year 2001? Round to the nearest dollar.

3. Find the value of x in the equation below, where i is the imaginary unit.
$$\frac{\dfrac{i}{2i^2}}{3i^3} = 3x$$

4. Find the sum of the series shown at the right, if it exists.
$$\sum_{n=1}^{\infty} \left(\frac{1}{4}\right)^n$$

5. Determine if $(2x - 1)$ is a factor of $2x^3 + 7x^2 - 14x - 5$.

6. Determine the domain of the rational function.
$$y = \frac{x - 1}{x^2 - x - 6}$$

7. Write a recursive formula for the sequence $1, 4, 19, 364, \ldots$.

8. **STANDARDIZED TESTS** If the area of the rectangle shown is 1 square unit, then the width is

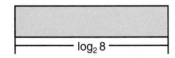

   A. $\log_2 3$ **B.** $\log_2 \frac{1}{3}$ **C.** $\log_8 2$ **D.** $\log_8 \frac{1}{2}$

9. Determine the equation of an ellipse with center at the origin, one focus at $(4, 0)$, and minor axis equal to 8.

10. Find the first term of an arithmetic sequence if the common difference is -2 and the 80th term is -56.

11. From her summer jobs after high school graduation and during college, Zoe Gold saved money for law school, which she planned to attend in the academic year after college.

Summer After	Amount Saved
h.s. graduation	$1100
college freshman	2000
college sophomore	2200
college junior	2700
college senior	3000

   She put these savings in a money fund that paid 6% annually at the end of each summer. If no other monies were added to or withdrawn from this account, write a polynomial to describe the amount of money Zoe had in this account when she was ready for law school. What was the total?

12. The safe load of a horizontal beam varies jointly as the width and the square of the depth, and inversely as its length. A beam 15 ft long, 3 in. wide, and 6 in. deep can take a load of 1800 lbs. What is the safe load for a beam 10 ft long, 4 in. wide, and 2 in. deep?

13. $\triangle OAB$ is located in the plane as shown.

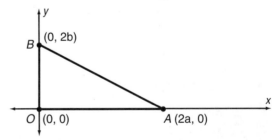

   If M is the midpoint of $\overline{AB}$, find the lengths of $\overline{OM}$, $\overline{MB}$, and $\overline{MA}$, and draw a conclusion about the general situation.

STANDARDIZED TEST

STANDARD FIVE-CHOICE   Select the best choice for each question.

1. If $\log_2 x$ is 25% of $\log_2 y$, then y expressed in terms of x is

   A. $\dfrac{x}{4}$ **B.** $4x$ **C.** x^4 **D.** $\dfrac{4}{x}$ **E.** $\sqrt[4]{x}$

2. The series $7 + 14 + 28 + 56 + 112 + 224 + 448 + 896 + 1792$ can be written as

   A. $\displaystyle\sum_{n=1}^{9} 7n$ **B.** $\displaystyle\sum_{n=1}^{8} 7 \cdot 2^n$ **C.** $7\displaystyle\sum_{n=1}^{9} 2^{n-1}$

   D. $2\displaystyle\sum_{n=1}^{9} 7^{n-1}$ **E.** $7\displaystyle\sum_{n=1}^{9} \dfrac{n(n+1)}{2}$

3. If $f(x) = x + 1$, then $f(f(f(x)))$ equals

   A. $x^3 + 3$ **B.** $x^3 + 1$ **C.** $3x + 1$
   D. $x + 3$ **E.** $3x + 3$

4. One leg of a right triangle is 6 ft longer than the other. If the area of the triangle is 10 ft^2, then the length of the hypotenuse to the nearest tenth of a foot is

   A. 7.5 ft   **B.** 8.7 ft   **C.** 8.4 ft
   D. 7.4 ft   **E.** 16.8 ft

5. If $\dfrac{1}{x} + \dfrac{1}{y} = 1$, then $\dfrac{1}{x+y}$ equals

   A. xy **B.** $\dfrac{x}{y}$ **C.** $\dfrac{1}{xy}$ **D.** $\dfrac{y}{x}$ **E.** $\dfrac{1}{x} + \dfrac{1}{y}$

6. If $P(1) = 0$ for a polynomial equation $P(x) = 0$, then which of the following must be true?

   A. x is a factor of $P(x)$.
   B. $x + 1$ is a factor of $P(x)$.
   C. $x - 1$ is a factor of $P(x)$.
   D. -1 is a zero of $P(x) = 0$.
   E. 0 is a zero of $P(x) = 0$.

7. The third term of the binomial expansion of $(a + 4b)^5$ is

   A. $160a^3b^2$ **B.** $640a^2b^3$ **C.** $20a^4b$
   D. $1024a^3b^3$ **E.** $1024a^2b^3$

8. For all values of n for which the expressions are defined, $\left(1 + \dfrac{1}{n}\right) \div \left(\dfrac{n+1}{n^2}\right)$ is equivalent to

   A. $\dfrac{(n+1^2)}{n^3}$ **B.** $\dfrac{2n}{n+1}$ **C.** $\dfrac{n^2}{n+1}$

   D. n **E.** 1

9. The equation of the circle that is tangent to each vertex of the hyperbola $\dfrac{x^2}{9} - \dfrac{y^2}{4} = 1$ is

   A. $x^2 + y^2 = \dfrac{1}{4}$ **B.** $x^2 + y^2 = 4$

   C. $x^2 + y^2 = -1$ **D.** $x^2 + y^2 = 9$

   E. $x^2 + y^2 = \dfrac{1}{9}$

10. $\sqrt{-3} \cdot \sqrt{-5}$ equals

    A. $\sqrt{15}$ **B.** $-\sqrt{15}$ **C.** $\sqrt{15}i$
    D. $i\sqrt{15}$ **E.** $\pm\sqrt{15}$

11. A sequence is defined as $a_1 = 10$ and

    $$a_n = \begin{cases} \dfrac{a_{n-1}}{2} & \text{if } a_{n-1} \text{ is even} \\ 3(a_{n-1}) + 1 & \text{if } a_{n-1} \text{ is odd} \end{cases}$$

    The tenth term of the sequence is

    A. $\dfrac{697}{8}$ **B.** 1   **C.** 2   **D.** 4   **E.** 16

Use the graph to answer Exercises 12–14.

12. What is the x-intercept?

    A. 1.5   **B.** -1.5
    C. -2 **D.** -0.5

13. What is the x-intercept?

    A. 3   **B.** -3
    C. -1.5 **D.** 0

14. What is the slope?

    A. -2 **B.** $\dfrac{1}{2}$
    C. 2   **D.** 6

14 Probability and Statistics

Take a Look
AHEAD

Make notes about things that look new.
- The Latin word *permutare* means "to change thoroughly." How do you think the term *permutation* relates to a group of objects?
- Examine the formulas for permutations, combinations, and probability. Copy the symbols used.

Make notes about things that look familiar.
- How would you define the term *probability* in your own words? Is the mathematical meaning similar to ordinary usage?
- What is a tree diagram and when can it help you?
- Identify a few common devices or materials that are often used in discussions about probability. Why are these items useful?

DATA *Activity*

Fire! Fire!
Whatever the cause — faulty electrical wiring, lightning, or carelessness — fire is a potential disaster that can result in injuries or loss of life and extensive property damage. It is no surprise that insurance companies have many customers seeking coverage against any financial hardship created by fire. To assess the risk they incur by offering fire insurance, companies analyze data relating to frequency and effects of fires.

SKILL FOCUS
- Add, subtract, multiply, and divide real numbers.
- Solve percent problems.
- Convert units of time.
- Determine frequency of occurrence.
- Apply place value for large numbers.

668

Insuring
AMERICA

In this chapter, you will learn how

- **LOCKSMITHS** use permutations in the construction and adjustment of house locks.
  (Lesson 14.1, page 678)

- **INSURANCE AGENTS** use probability to determine the risk an insurance company is taking by issuing a policy.
  (Lesson 14.3, page 694)

- **TRAVEL AGENTS** analyze flight and trip cancellation insurance for clients.
  (Lesson 14.6, page 716)

Fires in the United States Selected Data, 1994	
Number of Fires	**2,054,500**
• Building fires	614,000
• Vehicle fires	422,000
Civilian Fire Deaths	**4275**
• Fire deaths in home	3425
Civilian Injuries	**27,250**
Property Damages	**$ 8.151 billion**
• Building damage	$ 6.867 billion
• Vehicle damage	$ 156 million

1. The total number of fires for 1994 represents a decrease of about 5.2% from the 1993 figure. Determine the number of fires in 1993.

2. Approximately what percent of civilian fire deaths occurred in the home?

3. How often did someone die in a fire? How often was someone injured? Give your answers in terms of minutes. Use 1 year = 365 days.

4. What was the value of property damage in categories other than buildings or vehicles? Express your answer in billions.

5. About 21.1% of all building damage was from fires suspected of being deliberately set. What was the cost of property loss due to these suspicious fires?

WORKING TOGETHER Work in groups to complete at least one of the following activities.

6. Contact your fire department to obtain data about fires in your city or town. Summarize the data in a table. Include some conclusions you can determine from the data.

7. Prepare a poster with tips for preventing fires in the home. Obtain information from a fire department representative or from reference books.

Taking a Home Inventory

The purpose of insurance is to protect against financial loss "in case something happens." Different types of insurance are designed to cover different emergencies; for example, there is health insurance, fire insurance, and unemployment insurance to name just a few. Insurers, including private companies or government agencies, charge an annual fee called a *premium* in return for the coverage they provide. In this project, you will learn about the costs and arrangements involved in insuring items of value.

PROJECT GOAL

To collect data and information about insuring valuable property.

Getting Started

For the actual project, you will work in groups of four or five students.

1. As a class, create a list of as many different types of insurance as you can. Discuss the events for which each type of insurance offers protection. Are some events more likely than others? How do you thinks this affects the cost of the insurance? What other factors influence this cost?

2. The insurance industry offers many career opportunities. As part of the project, each group should research possible jobs within this field. Find out what is involved in performing each job, the education and/or special training required, and the approximate salary range.

PROJECT *Connections*

Lesson 14.1, page 677:
Compile a detailed list of items owned by a hypothetical family.

Lesson 14.4, page 702:
Assign a replacement value to items listed and prepare a scrapbook of documentation.

Lesson 14.7, page 723:
Interview an insurance agent and obtain an estimate of insurance cost.

Chapter Assessment, page 731:
Listen critically as each group makes a presentation; evaluate crime and accident statistics provided by a police officer.

Internet Connection

www.swpco.com/
swpco/algebra2.html

14.1 Permutations

Explore

1. Alicia, Bob, and Carol are entered in a race. If there are no ties, write all the possible orders in which they can finish the race.

2. If Paul also entered the race and there are no ties, write all possible orders in which they can finish the race.

3. How many possibilities were there for finishing orders when there were three runners?   four runners?

4. How is the number of possible results for four runners related to the result for three runners?

5. How do you think you can use the number of possible results for four runners to determine the number of possible results for five runners? What is the total number of possible ways five runners can finish?

SPOTLIGHT ON LEARNING

WHAT? In this lesson you will learn
- the fundamental counting principle.
- to determine the number of linear permutations of a set.
- to determine the number of circular permutations of a set.

WHY? Permutations can help you solve problems about music, retailing, and nutrition.

Build Understanding

A tree diagram can help you *list the elements* in a sample space. If you want to know the number of elements in a sample space, you can use the following rule.

> **THE FUNDAMENTAL COUNTING PRINCIPLE**
>
> If one event can occur in m different ways and for each of these ways a second event can occur in n different ways, then together the events can occur in $m \cdot n$ different ways.

The fundamental counting principle can be extended to include more than two events.

EXAMPLE 1

MUSIC In how many different ways can a musical director select a trio consisting of a violinist, a pianist and a drummer from 5 violinists, 3 pianists, and 6 drummers?

Solution

Apply the fundamental counting principle.

$$\underset{\text{violinists}}{5} \cdot \underset{\text{pianists}}{3} \cdot \underset{\text{drummers}}{6} = 90 \text{ possible trios}$$

There are 90 possible ways in which a trio can be formed. ◀

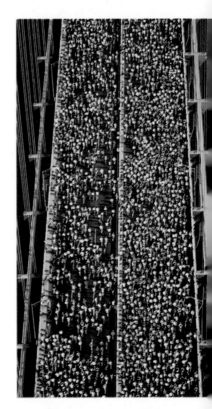

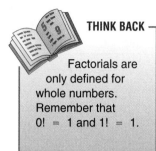
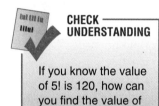
In Example 1, the order in which the instrumentalists are chosen for the trio is not important. In Explore, however, the order in which the runners finish is important. An arrangement of the elements of a set in specific order is called a **permutation**.

The fundamental counting principle also applies to permutations.

EXAMPLE 2

In how many different ways can you arrange the letters of the word GLOBE?

Solution

There are 5 letter positions to be filled. Any of the 5 letters can be used in the first letter position. After a letter has been placed, there are only 4 letters available for the second letter position and so on. Apply the fundamental counting principle.

$$\underset{\text{1st letter}}{5} \cdot \underset{\text{2nd letter}}{4} \cdot \underset{\text{3rd letter}}{3} \cdot \underset{\text{4th letter}}{2} \cdot \underset{\text{5th letter}}{1} = 120$$

There are 120 different ways to arrange the 5 letters. ◄

In solving permutations, the result is often a sequence of factors, each being one less than the preceding factor. Recall from Lesson 13.6 that this special sequence of factors is called **n factorial**.

$$n! = n(n - 1)(n - 2) \ldots (2)(1)$$

PERMUTATIONS OF n ELEMENTS

The number of permutations of n different elements of a set, symbolized by $_nP_n$, is given by the formula

$$_nP_n = n!$$

Applying this formula to Example 2, you can immediately write the number of different ways in which the 5 letters of the word can be arranged as 5!.

Sometimes you will want to determine the number of permutations when not all the elements of a set are used.

PERMUTATIONS OF A PART OF A SET

The number of permutations of n different elements taken r at a time, symbolized $_nP_r$, is given by the formula

$$_nP_r = \frac{n!}{(n - r)!} \text{ where } 0 \leq r \leq n$$

EXAMPLE 3

RETAILING A store owner has received a shipment of lamps in 7 new styles. In how many different ways can the owner display 3 of the new styles on a shelf?

Solution

$$_7P_3 = \frac{7!}{(7-3)!}$$

$$= \frac{7 \cdot 6 \cdot 5 \cdot 4 \cdot 3 \cdot 2 \cdot 1}{4 \cdot 3 \cdot 2 \cdot 1} = 210$$

There are 210 different ways to display 3 of the 7 lamps. ◄

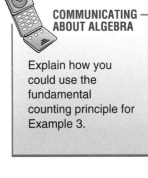

COMMUNICATING ABOUT ALGEBRA

Explain how you could use the fundamental counting principle for Example 3.

In the next example, the elements of a set are used more than once.

EXAMPLE 4

Using the digits 1, 2, . . . , 9, how many 4-digit numbers can you form if the digits can be used more than once?

Solution

Because digits can be repeated, 9 digits are available for each place.

$$\underset{\text{thousands place}}{9} \cdot \underset{\text{hundreds place}}{9} \cdot \underset{\text{tens place}}{9} \cdot \underset{\text{ones place}}{9} = 9^4, \text{ or } 6561$$

There are 6561 possible 4-digit numbers. ◄

PROBLEM SOLVING TIP

Check your graphing utility or scientific calculator for the factorial feature.

The result of Example 4 can be generalized.

> ### PERMUTATIONS WITH REUSE
>
> **With reuse allowed, the number of permutations of n different elements taken r at a time is**
>
> $$n^r$$

In the examples so far, the elements to be arranged were all different. When this is not the case, the permutation formula is adjusted to account for the repetitions.

PROBLEM SOLVING TIP

Check your graphing utility for the $_nP_r$ feature. Try computing $_7P_3$ and see if you get the correct answer, 210.

> ### PERMUTATIONS WITH REPETITIONS
>
> **The number of different permutations P of n elements taken n at a time, with r like elements of one kind, s like elements of another kind, and so on, is given by the formula**
>
> $$P = \frac{n!}{r!s!\ldots}$$

EXAMPLE 5

Determine the number of permutations of the letters of the word CONCENTRIC.

Solution

$$P = \frac{n!}{r!s!\ldots}$$

$$= \frac{10!}{3!2!} \quad \begin{array}{l} \leftarrow \text{There are 10 letters in all.} \\ \leftarrow \text{There are 3 C's and 2 N's.} \end{array}$$

$$= 302{,}400$$

There are 302,400 different ways to arrange the letters of the word. ◄

An arrangement may be in a line (**linear permutation**) or in a circle (**circular permutation**). For example, the number of ways in which 4 people can be seated in a row is 4! or 24. If the 4 people are seated at a round table, the number of different arrangements is less. The four circular arrangements shown are really all the same because the position of each person relative to the others has not changed.

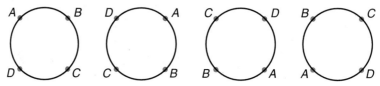

For each of the 4! permutations, there are 4 that are alike. So, the total number of circular permutations is

$$\frac{{}_4P_4}{4} = \frac{4!}{4} = \frac{4 \cdot 3 \cdot 2 \cdot 1}{4} = 3! \text{ or } 6 \quad \text{Note that 3! is } (4 - 1)!$$

CIRCULAR PERMUTATIONS

The number of different circular permutations P of n distinct elements is

$$P = (n - 1)!$$

TRY THESE

1. **FASHION** If an outfit consists of a pair of pants, a shirt, and a jacket, in how many ways can a designer show her coordinated line that has 3 pairs of pants, 4 shirts, and 2 jackets?

Evaluate each expression.

2. $\frac{6!}{1!}$

3. $\frac{9!}{3!}$

4. $\frac{12!}{2!\,4!}$

5. $\frac{n!}{(n-1)!}$

6. **WRITING MATHEMATICS** Which of the following is equivalent to $n!$? Explain your choice.

    a. $n(n-1)$

    b. $n(n-1)!$

    c. $\frac{n!}{(n+1)!}$

Evaluate each expression.

7. $_6P_6$ **8.** $_6P_1$ **9.** $_{10}P_3$ **10.** $_nP_{n-1}$

11. ATHLETICS The coach of the Yale Rowing Club has to determine positions for his 8-person scull team. How many different positions are possible?

12. ACADEMICS How many different ways can 7 classes be scheduled in a 7-period day?

13. NUMBER THEORY How many different 2-digit numbers can you form from the digits 3, 4, 5, 6 if a digit can appear just once in a number?

14. WORD PLAY How many different 3-letter arrangements can you make from the letters of the word OLYMPICS if a letter is used only once in each arrangement?

15. GOVERNMENT How many 5-digit zip codes can be formed using odd digits when

    a. digits cannot be reused?   **b.** digits can be reused?

Find the number of different arrangements you can make using all the letters of each word.

16. INTERNET   **17.** CONSTITUTION

18. MULTICULTURAL CELEBRATIONS A school fair will have a snack bar with items from Brazil, Ethiopia, Greece, Haiti, Italy, Korea, Mexico, and Thailand. In how many ways can the flags of these nations be displayed in a circular arrangement around the snack bar?

PRACTICE

1. GEOGRAPHY The towns Waterdown and Millgrove are connected by 4 roads. Millgrove and Hopewell are connected by 3 roads. In how many different ways can a tourist travel from Waterdown to Hopewell via Millgrove?

2. NUTRITION The lunch menu at a school cafeteria lists 3 soups, 6 sandwiches, and 4 fruits. In how many different ways may a student choose a lunch that consists of a soup, a sandwich, and a fruit?

3. SPORTS A golf pro has 5 pairs of brown golf gloves. To rotate usage, in how many different ways can she select a right glove and a left glove?

4. SCHOOL MANAGEMENT A school has 12 doors. During a fire drill, in how many different ways could a class leave by one door and reenter by another?

5. DEMOGRAPHY Rockville has 5 elementary schools. In how many different ways can 3 sisters attend these schools so that no 2 of the sisters are in the same school?

6. TOURISM A town has 8 hotels. In how many different ways can 4 tour groups stay at a hotel so that no 2 of the groups are in the same hotel?

Evaluate each expression.

7. $\dfrac{9!}{3!\,2!}$ **8.** $\dfrac{11!}{2!\,5!}$ **9.** $\dfrac{(8-2)!}{6!}$ **10.** $\dfrac{(r+2)!}{r!}$

11. WRITING MATHEMATICS Mac says that $(6 - 2)!$ is equivalent to $6! - 2!$. Do you agree?

Evaluate each expression.

12. $_8P_8$ **13.** $_8P_1$ **14.** $_{12}P_4$ **15.** $_{n+1}P_{n-1}$

Find the number of permutations of each seating arrangement.

16. 5 people in 5 chairs in a row

17. 5 people in 5 chairs at a round table

18. 4 of 6 people on a bench that seats 4

19. 4 people to take any of 6 chairs in a row

Find the number of different arrangements of the letters if letters cannot be reused.

20. DOT   **21.** CONVEX   **22.** POWWOW   **23.** SLEEVELESS

Determine the number of ways in which each number arrangement can be accomplished if digits cannot be reused.

24. 4-digit numbers from the digits of 1492

25. 2-digit numbers from the digits of 345

26. 3-digit numbers from the digits 4, 7, and 9

27. 3-digit numbers from {odd digits}

28. EMPLOYMENT Of 10 candidates, 6 have morning appointments and 4 have interviews in the afternoon. In how many different ways can the interviewer arrange a schedule?

29. MARITIME In how many different ways can a ship's signal sender arrange 5 red flags, 3 white flags, and 6 blue flags, using all of the flags each time?

30. AMUSEMENT In how many different ways can a carpenter arrange 11 wooden horses on a merry-go-round?

EXTEND

Using all the letters in the word SQUARE without reuse, tell how many different 6-letter arrangements are possible if

31. the first letter must be S

32. vowels and consonants alternate, beginning with a vowel

Use all the letters in the word DECAGON without reuse, and tell how many different 6-letter arrangements are possible that

33. begin with D and end with N

34. begin with D, end with N, and have a vowel in the middle place

35. begin with D, end with N, and have a consonant in the middle place

Use the digits 1, 2, 3, 4, 5, 6 without reuse to form 3-digit numbers, and tell how many have

36. a value greater than 300   **37.** a value less than 300   **38.** a value between 200 and 400

39. digits that are alternately odd and even, beginning with an odd digit

DAYCARE The girls Amy, Anaise, and Doris and the boys Al, Abe, Ti Hua, and Roy are in a group. Find the number of different ways the teacher can arrange the children in a line so

40. a girl is at the head of the line

41. Roy is at the head of the line

42. a child whose name begins with the letter A is at the head of the line

43. A child whose name begins with the letter A is at the head and at the rear of the line

44. the boys and the girls alternate, beginning with a boy

45. **AUTOMOTIVE** A V-6 engine has its cylinders numbered 1, 3, 5 on the right and 2, 4, 6 on the left. A good firing order is back and forth from left to right and vice versa. How many good firing orders are possible?

46. **PARTY PLANNING** The seating for a dinner party is 4 men and 4 women at a round table for 8. How many different arrangements are possible if men and women alternate?

THINK CRITICALLY

47. Solve for x: $_xP_2 = 20$

48. Solve for n: $_nP_r = k(_{n-1}P_{r-1})$

49. Consider the equation $_xP_2 = 4$. Is it possible to find a value for x? Explain.

50. If the number of permutations of n things taken 3 at a time is $\frac{2}{5}$ of the number of permutations of $(n-1)$ things taken 4 at a time, find n.

51. Show that: $_nP_r - {_nP_{r-1}} = (n-r) \cdot {_nP_{r-1}}$

MIXED REVIEW

Use the spinner shown to determine the theoretical probability that you will spin each of the following.

52. an even number

53. a 3 or a 4

54. **STANDARDIZED TESTS** The solution of $|2x - 1| < 7$ is

A. $x > -3$
B. $x < 4$
C. $-3 < x < 4$
D. $x > -3$ or $x < 4$

PROJECT *Connection* In this activity, you are going to assemble a list of valuables for a hypothetical family. This will be used to obtain an estimate of how much the premium would be to protect these items under home insurance.

1. Decide on the members of your "family" and then list their items of value according to category. Try to include everything of value—clothes, furniture, toys, appliances, jewelry, and so on.

2. Divide the categories so that group members can work at adding important details for items. Use newspaper and magazine ads, or visit stores, to enhance your descriptions of items, obtain model numbers and so on. Also, read the next Project Connection, since you may be able to complete some of that assignment at the same time.

Career
Locksmith

Think of all the locks you and your family use—for your house, car, bicycle, gym locker, and so on. *Locksmiths* install and repair locks. Permutations play an important role in the construction and adjustment of the typical house-entry lock. Examine your house key. It most likely has five "low spots". When you insert the key into the *lock cylinder*, the key pushes the pins out of the way, against a spring. When the key is all the way in the lock, these pins rest in the low spots.

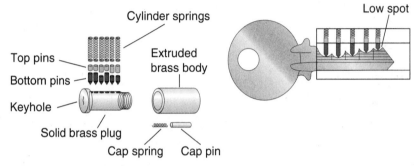

Each pin rests in a position above the lock cylinder. Each pin is actually composed of two parts —a *bottom pin* and a *top pin*. When the pins extend from the body of the lock to the lock cylinder, the cylinder cannot be turned and the lock cannot be opened. When the correct key is inserted, the bottom pins are pushed up so that the bottom and top pins meet where the cylinder meets the body of the lock. The line up of "breaks" between the top and bottom pins allow the lock to be turned and opened.

Decision Making

Different-sized pins allow the locksmith to "key" the lock, that is, to change the pins to match a new or old key. There are nine basic pin sizes, #1 to #9, with different heights.

1. If a lock has five pins, how many different keys (*pin permutations*) can be made?

2. How many more pin permutations does a 6-pin lock have over a 3-pin lock?

3. The bottom pins are tapered so when you slide the key in, it doesn't jam against the pins. When the key is fully inserted, the tapered parts of the pins sit in the low spots of the key. Explain why it would be impractical to have a key set with all #5 pins.

4. Explain why a pin permutation of #1 – #4 – #3 – #2 – #5 would make a cylinder harder to "pick" than a pin arrangement of #1 – #2 – #3 – #4 – #5. Picking a lock is using an object other than the key to open it.

14.2 Combinations

Explore/Working Together

• Work in a group of four, and record your answers.

1. From your group of 4, select 3 officers, a President, a Vice President, and a Treasurer. List all the different possibilities.

2. From your group of 4, select a committee of 3. List all possibilities.

3. **a.** Were your lists the same for Questions 1 and 2? Explain.
   b. Can you find a relationship between the two lists? Explain.

4. Suppose your group had 5 members.

   a. In how many different ways could you select the 3 officers?
   b. Use the relationship you wrote in Question 3b to predict the number of different ways you could select 3 members from this group of 5.
   c. Verify by making lists. Were you correct in your prediction?

Build Understanding

• As you saw in Explore, there are more ways to select elements from a larger set and arrange them in a special order than there are ways to simply select the elements from the set.

How many first and second place finishes are possible from 5 runners A, B, C, D, and E? In Lesson 14.1, you solved this problem by finding the number of *permutations* of 5 elements taken 2 at a time.

$$_5P_2 = \frac{5!}{(5-2)!} = \frac{5!}{3!} = \frac{5 \cdot 4 \cdot 3 \cdot 2 \cdot 1}{3 \cdot 2 \cdot 1} = 20 \text{ ways}$$

How many pairs of runners can be chosen from 5 runners A, B, C, D, and E to be sent on to the City Finals?

In this case, the order of the 2 to be chosen is not important.

Possible Outcomes
AB AC AD AE BC BD BE CD CE DE
BA CA DA EA CB DB EB DC EC ED

In the list of 20 possible outcomes, note that the second row is a repeat of the first row. To determine the number of *combinations* of 2 runners, divide by 2, written as 2!.

$$\text{combinations} = \frac{_5P_2}{2!} = \frac{\dfrac{5!}{(5-2)!}}{2!} = \frac{\dfrac{5!}{3!}}{2!} = \frac{5!}{3! \, 2!} = \frac{5 \cdot 4 \cdot 3 \cdot 2 \cdot 1}{3 \cdot 2 \cdot 1 \cdot 2 \cdot 1} = 10 \text{ ways}$$

Selections in which order does not matter are called **combinations**.

COMMUNICATING ABOUT ALGEBRA

What is the value of $_nC_0$ and $_nC_n$? Explain.

COMBINATIONS

The number of combinations of n different elements taken r at a time, symbolized by $_nC_r$ or $\binom{n}{r}$, is given by the formula

$$_nC_r = \binom{n}{r} = \frac{n!}{r!(n-r)!} \qquad \text{where } 0 \le r \le n$$

EXAMPLE 1

BOOK CLUBS A new book-club member can select 3 free books from a list of 10 books. If Jana just joined the club, in how many different ways can she select her free books?

Solution

Since order does not matter, determine the number of combinations.

$$_nC_r = \frac{n!}{r!(n-r)!}$$

$$_{10}C_3 = \frac{10!}{3!(10-3)!} \qquad \text{Substitute } n = 10, r = 3.$$

$$= \frac{10 \cdot 9 \cdot 8 \cdot 7!}{3! \cdot 7!} \qquad \text{Rewrite 10! in terms of 7!.}$$

$$= \frac{10 \cdot 9 \cdot 8}{3 \cdot 2 \cdot 1} \qquad \text{Divide numerator and denominator by 7!.}$$

$$= 120$$

Jana may choose her free books in 120 different ways. ◄

Sometimes you may need to consider more than one event.

CHECK UNDERSTANDING

In Example 2, how many ways can a coach choose a 5 person team if there are no gender restrictions on the team?

EXAMPLE 2

SPORTS In how many ways can a coach choose a gymnastics team of 5 members from a group consisting of 8 males and 7 females if the team must consist of 3 males and 2 females?

Solution

First event: Choose the male members of the team.

There are $_8C_3$ or $\binom{8}{3}$ possibilities.

Second event: Choose the female members of the team.

There are $_7C_2$ or $\binom{7}{2}$ possibilities.

Possible teams $= \binom{8}{3} \cdot \binom{7}{2}$ Fundamental counting principle

$$= 56 \cdot 21 = 1176$$

There are 1176 ways to form the required team. ◄

Carefully analyze the conditions of a problem.

EXAMPLE 3

PLAYING CARDS From a standard deck of 52 cards, how many 5-card hands are possible if

a. there are no restrictions?
b. the queen of hearts is in the hand?
c. the hand contains exactly 3 spades?

Solution

a. $\binom{52}{5} = 2{,}598{,}960$

There are 2,598,960 possible unrestricted 5-card hands.

b. If the queen of hearts must be in the hand, the problem is equivalent to selecting 4 cards from 51 available cards.

$\binom{51}{4} = 249{,}900$

There are 249,900 possibilities for a hand with the queen of hearts.

c. The number of ways to choose the 3 spades is $\binom{13}{3}$.

Removing the 13 spades, there are $52 - 13$ or 39 cards remaining, from which you want to choose 2 to complete the hand.

$$\binom{13}{3} \cdot \binom{39}{2} = 286 \cdot 741 = 211{,}926 \qquad \text{Fundamental counting principle}$$

There are 211,926 possible 5-card hands with exactly 3 spades. ◄

To satisfy the condition of *at least* or *at most*, you will have to consider different cases and add the combinations for each case.

EXAMPLE 4

ACCESSORIES From a group of 7 necklaces, find the number of different ways to select

a. at least 4 necklaces   **b.** at most 4 necklaces

Solution

a. To choose *at least* 4 necklaces, you can have 4, 5, 6, or 7 necklaces. Find the possible combinations for each case and find the sum.

$$_7C_4 + {_7C_5} + {_7C_6} + {_7C_7} = 35 + 21 + 7 + 1 = 64$$

There are 64 ways to choose at least 4 necklaces from the set.

b. To choose *at most* 4 necklaces, you can have 0, 1, 2, 3, or 4 necklaces.

$$_7C_0 + {_7C_1} + {_7C_2} + {_7C_3} + {_7C_4} = 1 + 7 + 21 + 35 + 35 = 99$$

There are 99 ways to choose at most 4 necklaces from the set. ◄

PROBLEM SOLVING TIP

Sometimes, as in Example 3, when a combinations problem has a condition, you can break the problem into two events. Satisfy the condition as the first of the two events. Note the effect that the first event has on the second event. Apply the fundamental counting principle.

PROBLEM SOLVING TIP

When working with an *at most* condition, you may or may not have to consider the 0 case. In Example 4b, the 0 case is appropriate. If you were choosing a committee of at most 4 people, the 0 case is not appropriate since a committee must have at least 1 person to exist.

Evaluate each expression.

1. $_7C_3$ **2.** $\binom{6}{5}$ **3.** $_{20}C_{20}$ **4.** $\binom{20}{1}$ **5.** $_{20}C_0$

6. WRITING MATHEMATICS Maria asked for the value of $_nC_n$. Kaya said the value is n, Sara said the value is $n!$, and Gus said the value is 1. With whom do you agree? Explain.

7. GEOMETRY There are 7 different points on a circle. How many straight lines can be drawn through pairs of these points?

8. RETAILING A specialty shop received a shipment containing 4 different paisley ties, 8 different striped ties, and 5 different solid ties. In how many ways can the window dresser choose 4 of these ties for a window display?

POLITICS In how many ways can an affirmative action committee of 4 people be chosen from a group of 5 men and 4 women if

9. There are no restrictions?

10. Marie-Jean Buyle, one of the group, is always on the committee?

11. Sanford Crosley, one of the group, is never on the committee?

PLAYING CARDS In the game of *bridge*, a hand has 13 cards. Determine the number of bridge hands possible from a standard deck of 52 cards under each of the following conditions. You may answer in notation form without evaluating.

12. There are no restrictions.   **13.** The hand contains exactly two aces.

14. The hand contains exactly 5 diamonds.   **15.** The hand does not contain any hearts.

16. The hand contains both the ace of spades and the ace of diamonds.

SPORTS During a game, Coach Lisa Marie is evaluating the nine members of her baseball team who are now playing on the field for possible selection as "most valuable player." Determine the number of selections possible under each of the following conditions.

17. There are at least 3 candidates for MVP.   **18.** There are at most 5 candidates for MVP.

Evaluate each expression.

1. $_8C_6$ **2.** $_{10}C_5$ **3.** $\binom{12}{4}$ **4.** $_{100}C_{100}$ **5.** $\binom{1000}{1}$

6. WRITING MATHEMATICS Evaluate $_7C_2$ and $_7C_5$, and give a reason for the result. Generalize the result.

7. PRIZES A radio station chose 22 postcards from the many listeners who entered their monthly Concert Contest. In addition to their prize, 4 of those chosen will also receive backstage passes. In how many different ways can the pass winners be selected?

8. **GEOMETRY** There are 5 points in a plane, no 3 of which are collinear. How many straight lines are determined?

9. **SAFETY** In how many ways can a 6-person neighborhood nightwatch be selected from 12 men and 8 women who are on the safety patrol?

10. **TESTING** On a math exam with 10 questions and 2 bonus questions, students are required to answer 5 questions. In how many ways can Joel select his 5 questions?

JOURNALISM *The Seattle Sun* is doing an article on the set design of a new musical. The writer has limited time and can only interview 4 of the 12 people who designed and built the set. In how many ways can the 4 interviews be done if the writer

11. has no further restrictions?

12. wants the head designer to be included?

13. does not want to include the 2 assistant stage hands?

14. wants the head designer to be included and does not want to include the 2 assistant stage hands?

MUSIC A school jazz club consists of 5 males and 7 females. Six members need to be chosen to represent the school at the Chicago Jazz Festival. In how many ways can the music director make the selection if the director

15. has no further restrictions?

16. wants to include Elicia, the female section leader?

17. does not want to include Seymour, the least experienced member of the club?

18. wants to include Elicia and does not want to include Seymour?

19. wants to send 3 males and 3 females?

20. wants to send 3 males and 3 females, and wants one of the females to be Elicia?

21. wants to send 3 males and 3 females, wants to include Elicia, and does not want to include Seymour?

PLAYING CARDS Determine the number of 13-card bridge hands possible from a 52-card deck under each of the following conditions. You may answer in notation form.

22. The hand contains no spades.

23. The hand contains exactly two kings.

24. The hand contains only red cards.

25. The hand contains exactly 4 diamonds and 3 clubs.

ENVIRONMENT Ten new members of a conservation club wish to join the members of the Animal Protection Committee. Determine the number of ways the chairperson can select the committee members if the number of new members chosen is

26. at least 7

27. at most 4

28. 6 or more

29. 5 or less

EXTEND

PLAYING CARDS Determine the number of 13-card bridge hands from a 52-card deck under each of the following conditions.

30. The hand contains only 2 suits, with 5 cards in one suit and the remaining 8 cards in the other suit.

31. The hand contains two 5-card suits and the remaining 3 cards from another suit.

32. Solve for n: $_nC_2 = 45$

33. Consider the equation $_nC_2 = 4$. Is it possible to find a value for n? Explain.

For some problems, you may use an equation involving combinations to model the situation.

34. **SOCIALIZING** After shaking hands all around at a diplomatic reception, 91 handshakes had taken place. How many people were in the group?

35. **TELEPHONE OPERATIONS** In a small beach complex, each cottage has one telephone. If 435 telephone communications are possible within the complex, how many cottages are in the complex?

PASCAL'S TRIANGLE Recall that the coefficients of the binomial expansion can be displayed in a triangular pattern.

36. Verify that you can also write this pattern using combinations notation.

$$
\begin{array}{cccccccccc}
(a + b)^0 & & & & & 1 & & & & \\
(a + b)^1 & & & & 1 & & 1 & & & \\
(a + b)^2 & & & 1 & & 2 & & 1 & & \\
(a + b)^3 & & 1 & & 3 & & 3 & & 1 & \\
(a + b)^4 & 1 & & 4 & & 6 & & 4 & & 1
\end{array}
$$

37. Continue Pascal's triangle to write the numerical coefficients in the expansion of $(a + b)^8$. (Keep the full display you have written for reference.)

38. **a.** Use a calculator to evaluate $\binom{8}{3}$ and $\binom{8}{5}$.

b. Use Pascal's triangle to explain the result.

c. Generalize the result.

39. **a.** Use a calculator to evaluate $\binom{8}{3}$, $\binom{7}{3}$, and $\binom{7}{2}$.

b. From the values you have seen in part a, write a relationship between $\binom{8}{3}$ and $\binom{7}{3}$ and $\binom{7}{2}$.

c. Use Pascal's Triangle to explain the relationship.

d. Generalize the result.

THINK CRITICALLY

40. **MONEY** What is the greatest amount of change you can have and not be able to give change for a dollar?

Simplify each expression.

41. $_nC_{n-1}$

42. $_nC_{n-2}$

43. $_{n+3}C_{n+2}$

44. Explain why $\binom{n}{r} = \binom{n}{n-r}$ using the example $\binom{5}{3} = \binom{5}{2}$.

45. If $_nC_r = 6$ and $_nP_r = 12$, determine n and r.

Write an algebraic proof for each statement.

46. $_nC_r + {_nC_{r-1}} = {_{n+1}C_r}$

47. $_nC_r = {_{n-1}C_r} + {_{n-1}C_{r-1}}$

MIXED REVIEW

Determine the theoretical probability of each event.

48. You roll an even number in one roll of a number cube.

49. You pick a red card in one draw from a standard deck of 52 cards.

50. **STANDARDIZED TESTS** For which set of data do the mean, median, and mode all have the same value?

    A. 2, 6, 6, 10   **B.** 2, 2, 4, 10, 12   **C.** 2, 2, 2, 4, 10   **D.** 2, 2, 6, 10, 20

51. **STANDARDIZED TESTS** The set of integers under the operations of addition and multiplication is not a field because

    A. addition is not associative
    B. addition is not commutative
    C. multiplication is not distributive over addition
    D. except for 1, the nonzero integers do not have inverses under multiplication

Tell if the given relation for the set shown satisfies the transitive property. If not, give a counterexample.

52. "is greater than" for {integers}

53. "sits next to" for {5 people seated in a row}

54. "is a factor of" for {natural numbers}

Write the given expression as a single matrix in simplest form.

55. $-2\begin{bmatrix} -4 & 7 \\ 3 & 5 \\ 0 & 1 \end{bmatrix}$

56. $3\begin{bmatrix} 4 & -2 \\ -2 & 1 \end{bmatrix} + \begin{bmatrix} 7 & 9 \\ 3 & -5 \end{bmatrix}$

Explore

● Consider the letters A, B, C, D, E, and the numbers 1, 2, 3, 4, 5.

1. a. List the ways you can choose one letter and one number?
    b. Write a second list of all the items whose letter is a vowel.
    c. How many items are on each list?
    d. Suppose each item in the sample space were written on a separate piece of paper. If you draw one of the papers at random, what is the probability you get a letter that is a vowel?

2. Refer to Question 1 to determine the probability of getting a paper with a letter that is a consonant.

3. a. How are the events in Questions 1 and 2 related?
    b. How are the probabilities related?

4. List the items in Question 1 that have the letter A.

Consider the A-items from Question 4 as a new sample space.

5. a. What is the probability of choosing an *A*-paper with an even number less than 4? Explain.
    b. What is the probability of choosing an *A*-paper with a number that is even or is less than 4? Explain.

Build Understanding

● The **probability of an event** $P(E)$ is the *ratio* of the number of outcomes in the event, $n(E)$, to the number of outcomes in the sample space, $n(S)$.

$$P(E) = \frac{n(E)}{n(S)} \quad \begin{array}{l} \leftarrow \text{number of outcomes in event} \\ \leftarrow \text{number of outcomes in sample space} \end{array}$$

Probability is a number from 0 through 1 inclusive that expresses the likelihood that an event can occur. In a finite sample space,

$$P \text{ (impossibility)} = 0 \quad P \text{ (certainty)} = 1$$

The events E and *not E* are **complementary events**. Complementary events have probabilities whose sum is 1.

$$P(E) + P(\text{not } E) = 1 \text{ or } P(\text{not } E) = 1 - P(E)$$

The likelihood that an event will occur is sometimes expressed as **odds**. Odds is a ratio obtained from the probability the event will occur to the probability the event will not occur.

EXAMPLE 1

ALGEBRA: WHO, WHERE, WHEN

In 1812, the French mathematician Pierre LaPlace published *Theorie Analytique des Probabilities*, in which the probability of an event was first meaningfully defined.

SCOUTING Six children from a Cub Scout den are to be arranged in a line to form a color guard that carries the American flag to the pack meeting. Each of the 6 hopes to be the one to carry the flag.

a. What is the probability that Andy, one of the 6, will carry the flag?

b. What is the probability that Andy will not carry the flag?

c. What are the odds that Andy will not carry the flag?

Solution

a.
$$P(E) = \frac{n(E)}{n(S)}$$

$$P(\text{Andy with flag}) = \frac{{}_5P_5}{{}_6P_6}$$

← If Andy has the flag, 5 children remain to be arranged in line.
← the number of ways to arrange 6 children in a line

$$= \frac{5!}{6!} = \frac{5!}{6 \cdot 5!} = \frac{1}{6}$$

The probability that Andy will carry the flag is $\frac{1}{6}$.

b. "Andy will carry the flag" and "Andy will not carry the flag" are complementary events, since there are no other outcomes.

$$P(\text{Andy will not}) = 1 - P(\text{Andy will})$$

$$= 1 - \frac{1}{6} = \frac{5}{6}$$

The probability that Andy will not carry the flag is $\frac{5}{6}$.

c. Odds (will not) $= \dfrac{P(\text{will not})}{P(\text{will})}$

$$= \frac{\frac{5}{6}}{\frac{1}{6}} = \frac{5}{1}$$

The odds that Andy will not carry the flag are 5:1. ◄

THINK BACK

In Lesson 14.1, you learned that an arrangement in which order is important is called a permutation. The number of permutations of n different elements is ${}_nP_n$ or $n!$.

Suppose that from 9 students, Max, Neil, Oprah, Pam, Raul, Rita, Tom, Will, and Zoe, the teacher will choose a student at random to work a problem at the chalkboard.

Event A: Calculate the probability that the student chosen is a girl.

$$P(\text{girl}) = \frac{4}{9}$$ ← (Oprah, Pam, Rita, Zoe)

Event B: Calculate the probability that the student chosen has a name beginning with R.

$$P(\text{R-name}) = \frac{2}{9}$$ ← (Raul, Rita)

What is the probability that the student chosen is a girl *and* has an R-name?

$$P(\text{girl } and \text{ R}) = \frac{1}{9} \leftarrow \text{(Rita)}$$

What is the probability that the student chosen is a girl *or* has an R-name?

$$P(\text{girl } or \text{ R}) = \frac{5}{9} \leftarrow \text{(Oprah, Pam, Zoe, Rita, Raul)}$$

A **Venn diagram** may be used to model the problem. The sample space is represented by a rectangular region and the events are represented by circular regions.

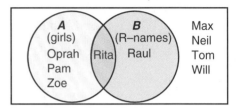

The **intersection** of sets A and B, symbolized $A \cap B$, contains the elements that belong to *both sets.*

$$P(A \text{ and } B) = \frac{n(A \cap B)}{n(S)}$$

The **union** of the sets A and B, symbolized $A \cup B$, contains the elements that belong to *either set or both sets.*

$$P(A \text{ or } B) = \frac{n(A \cup B)}{n(S)}$$

In the model, sets A and B have 5 outcomes in their union: the girls Oprah, Pam, Zoe, Rita, and the students with R-names, Rita and Raul. Since we want to count Rita only once, the outcomes in the intersection of A and B must be subtracted.

Probability with multiple events is called a **compound probability**.

COMPOUND PROBABILITY

For two events A and B in a sample space
$$P(A \text{ or } B) = P(A) + P(B) - P(A \cap B)$$

EXAMPLE 2

HANDICRAFTS In a class of 20 crafters, there are 6 woodworkers and 4 ceramicists. The instructor must select participants for an exhibition.

a. If 3 crafters are to be chosen from the class at random, what is the probability that 2 woodworkers and 1 ceramicist will be chosen?

Sanji, one of the woodworkers, is also a ceramicist, and Lou and Anna, two of the ceramicists, are also woodworkers. If one crafter is chosen from the class at random, what is the probability that the crafter will be,

b. a woodworker and a ceramicist?
c. a woodworker or a ceramicist?

Solution

a. Determine the number of outcomes in the event.

$$\underset{\text{wood}}{_6C_2} \cdot \underset{\text{ceramic}}{_4C_1} \qquad \text{Fundamental counting principle}$$

Determine the total number in the sample space.

$$_{20}C_3$$

Form a probability ratio.

$$P(2 \text{ wood}, 1 \text{ ceramic}) = \frac{_6C_2 \cdot {_4C_1}}{_{20}C_3}$$

$$= \frac{15 \cdot 4}{1140} = \frac{1}{19}$$

The probability of choosing 2 woodworkers and 1 ceramicist is $\frac{1}{19}$.

Draw a Venn diagram. The two sets, woodworkers and ceramicists, have 3 people in common Sanji, Lou, and Anna.

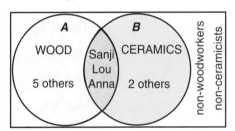

b. The people who are woodworkers *and* ceramicists are represented by the number in the *intersection*.

$$P(A \text{ and } B) = \frac{n(A \cap B)}{n(S)} = \frac{3}{20}$$

The probability of choosing a woodworker and a ceramicist is $\frac{3}{20}$.

c. The people who are woodworkers *or* ceramicists are represented by the number in the *union*.

$$P(A \text{ or } B) = P(A) + P(B) - P(A \cap B)$$

$$= \frac{6}{20} + \frac{4}{20} - \frac{3}{20} = \frac{7}{20}$$

The probability of choosing a woodworker or a ceramicist is $\frac{7}{20}$. ◄

THINK BACK

The empty set (the set with no members) is symbolized Ø.

Events that cannot occur together are called **mutually exclusive**.

In this case, the circular regions of a Venn diagram do not overlap. Such sets are called **disjoint**. Their intersection is empty and $P(A \cap B) = 0$.

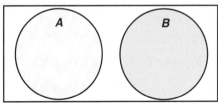

Disjoint Sets
$A \cap B = \emptyset$

┌─ **MUTUALLY EXCLUSIVE EVENTS** ─────────────┐

For two events A and B that are mutually exclusive
$$P(A \text{ or } B) = P(A) + P(B)$$

└──┘

EXAMPLE 3

In one spin on the board shown, what is the probability of getting an even number or the number 5?

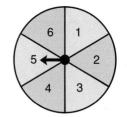

Solution

Let event A = getting an even number and event B = getting the number 5.

Events A and B are mutually exclusive. A and B are disjoint sets.

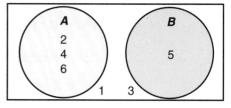

$A = \{2, 4, 6\}$ and $n(A) = 3$
$B = \{5\}$ and $n(B) = 1$

$$P(A \text{ or } B) = P(A) + P(B)$$
$$P(\text{even or } 5) = P(\text{even}) + P(5)$$
$$= \frac{n(A)}{n(S)} + \frac{n(B)}{n(S)}$$
$$= \frac{3}{6} + \frac{1}{6} = \frac{2}{3}$$

In one spin, the probability of getting an even number or the number 5 is $\frac{2}{3}$. ◀

In some situations, you may know $P(A \text{ or } B)$ and want to determine the probability of one of the simple events.

EXAMPLE 4

QUALITY CONTROL At Cora's Cookie Company, inspectors reject cookies that are either overbaked or have too few chocolate chips. It is estimated that the probability of a cookie being overbaked is 0.03, the probability that the cookie is overbaked or has too few chips 0.06, but the probability of both defects is 0.01.

What is the probability that a cookie has too few chips?

Solution

The events are not mutually exclusive.

Let event A = the cookie is overbaked. $P(A) = 0.03$
Let event B = the cookie has too few chips $P(A \text{ or } B) = 0.06$
                                                   $P(A \text{ and } B) = 0.01$

$$P(A \text{ or } B) = P(A) + P(B) - P(A \text{ and } B)$$
$$0.06 = 0.03 + P(B) - 0.01$$
$$0.04 = P(B)$$

The probability that a cookie has too few chips is 0.04. ◀

1. **WRITING MATHEMATICS** Explain why or why not the pairs of events are complementary.
   a. Tossing a coin: flipping *heads*, flipping *tails*
   b. Rolling a numbered cube: rolling a 2, rolling a 6
   c. Drawing one card from a deck of cards: drawing a king, not drawing a king.

In Exercises 2–6, one marble is to be selected at random from a jar that contains 3 red marbles, 4 white marbles, and 5 blue marbles.

Determine the probability that

2. the marble is not red   3. the marble is not black   4. the marble is red or white.

Determine the odds that

5. the marble is white   6. the marble is not blue

In Exercises 7–13, the spinner is spun once. Determine the probability that the number is

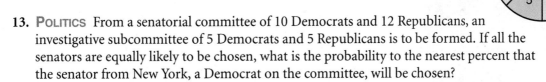

7. even and greater than 5   8. odd and not less than 5

9. less than 2 and even   10. less than 5 or even

11. odd or is 4   12. less than 3 or is 7

13. **POLITICS** From a senatorial committee of 10 Democrats and 12 Republicans, an investigative subcommittee of 5 Democrats and 5 Republicans is to be formed. If all the senators are equally likely to be chosen, what is the probability to the nearest percent that the senator from New York, a Democrat on the committee, will be chosen?

14. **COMMUTING** Lisa estimates that the probability of meeting only her friend Kim on the bus is $\frac{1}{2}$, the probability of meeting either Kim or Jamal is $\frac{3}{4}$, but the probability of meeting them both is $\frac{1}{3}$. If Lisa boards the bus on a particular day, what is the probability that she will meet only Jamal?

PRACTICE

1. **WRITING MATHEMATICS** Considering the integers 1-10 and the random selection of one integer, Louise said "getting an odd number" and "getting an even number" are complementary. Jim said they are not complementary. Arnie agreed the events are complementary and also mutually exclusive. With whom do you agree? Explain.

In Exercises 2–11, one card is to be selected at random from a standard deck of 52 cards.

Determine the probability that the card is

2. not a 7   3. a king or a queen   4. not a jack or an ace   5. a red 10   6. black or a 9

Determine the odds that the card is

7. red   8. a 3 or a queen   9. a black 8   10. not a 6 or a jack or a queen   11. not red or a 7

In Exercises 12–16, the name of a month is drawn at random from a box with the names written one to a slip of paper. Determine the probability that the name

12. ends in Y and begins with J   **13.** ends in Y or R   **14.** contains R and at least 2 vowels

15. begins with a vowel or ends with R   **16.** has fewer than 10 letters or begins with K

In Exercises 17–20, the letters of the word POINT are arranged at random. Determine the probability that the arrangement

17. begins with P   **18.** begins and ends in a vowel

19. begins with P and ends in a vowel   **20.** begins with P or ends in a vowel

COLLEGE ADMINISTRATION A five-person committee is to be formed at random from liberal arts professors and 9 science professors. Find the probability that the committee has

21. all liberal arts professors   **22.** three liberal arts and two science professors

23. all science professors   **24.** four science and one liberal arts professors

SCHOOL CAFETERIA Twenty-six students from Ms. Wong's homeroom had lunch in the school cafeteria. Nineteen of the students had a hamburger, seventeen had a slice of pizza, and thirteen had both. If Jason was among the students, determine the probability that

25. he did not have a hamburger

26. he did not have either a hamburger or pizza

METEOROLOGY Weather records for a certain city show that on average, during the month of June, it rains on 12 days and the daily high temperature is over 65°F on 15 days. However, on only 8 days in June, there is rain and daily high temperatures over 65°F. Based on this information, what is the probability that

27. it will rain on a day in June?

28. the daily high will be over 65°F?

29. it will rain or the daily high will be over 65°F?

30. MEDICINE A certain drug causes water retention or hair loss in 35% of patients. Twenty-five percent experience only water retention, and 5% experience both water retention and hair loss. What is the probability that a prospective patient might experience hair loss only?

EXTEND

The table at the right shows the gender and grade distribution of all the students in Mr. Li's 4th-period Algebra 2 class. Find the probability that a student chosen at random is

31. female   **32.** a senior

33. male and a junior   **34.** female or a senior

35. male or a junior

	Junior	Senior
Male	3	8
Female	9	5

FOOD DISTRIBUTION A fortune cookie producer has four different categories of fortunes: romance, wealth, longevity, and career. One fortune is placed in each cookie. There are three times as many fortunes about romance as there are about wealth, and twice as many wealth fortunes as each of longevity and career. If a cookie is chosen at random from a bag of 30, find the probability that it contains a fortune about

36. romance

37. either longevity or career

38. HOSPITALITY A hotel laundry bag contains only washcloths and towels. There are 5 more towels than washcloths. If the probability of randomly selecting a towel from the bag is $\frac{2}{3}$ how many towels are in the bag?

MERCHANDISING A stock clerk is sorting a carton of sweaters that has gotten mixed up. The carton contains a dozen cotton sweaters and a dozen wool sweaters. Of the cotton sweaters, 4 are red, 4 are green, and 4 are blue. Of the wool sweaters, 4 are red, 4 are green, and 4 are blue. If the clerk reaches in and picks a sweater at random, find the probability that it is

39. a wool sweater or a green sweater

40. a cotton sweater or a sweater that is not red

THINK CRITICALLY

41. The table shows the gender and grade distribution of the students in the Computer Club. The probability the student is female or in Grade 10 is $\frac{3}{5}$. Find x.

Grade Level	9	10	11	12
Male	3	5	5	6
Female	4	5	x	4

42. Use the value you found for x in Question 41. If a student is chosen at random, find the probability that the student is male or is not in Grade 12.

43. LINGUISTICS A language program has 25 students taking French, 23 students taking Spanish, and 28 taking Italian. Of these, 4 students are taking all three languages, 5 are taking only Spanish and Italian, 8 are taking only French, and 11 are taking only Italian. If Tómas is in the program, what is the probability that he is studying all three languages?

MANUFACTURING There are 5 types of cartons in a packing plant, A, B, C, D, and E. Of the 360 cartons packed in one hour, the number of A-cartons exceeds twice the number of B-cartons by 80. The number of B-cartons was 40 less than the number of C-cartons. The number of C-cartons was equal to the sum of the D-cartons and the E-cartons, which were equal in number. If an inspector chose a packed carton at random, find the probability it was

44. an A
45. not a C
46. either B or D
47. neither C nor A

MIXED REVIEW

Use Cramer's rule to solve each system of equations.

48. $\begin{cases} 2m - 5n = 2 \\ 3m - 7n = 1 \end{cases}$

49. $\begin{cases} 6a - 2b + c = 16 \\ 2a - b + 5c = 2 \\ 2a - 3c = 8 \end{cases}$

50. STANDARDIZED TESTS Which of the following statements involving exponents is true?

A. $2^{-1} + 3^{-1} = 5^{-1}$
B. $5^0 - 3^0 = 2^0$
C. $\left(\frac{1}{2}\right)^{-1} + \left(\frac{1}{3}\right)^{-1} = \left(\frac{1}{5}\right)^{-1}$
D. $\frac{1}{2^0} = \frac{2}{1^0}$

Career
Insurance Agent

An *insurance agent* uses statistics related to the science of risk-taking. Among an insurance agent's tools is a *mortality table* like the one at the left. It lists the likelihood that a person will die within a year. The death rate per thousand is converted to a decimal and is the probability of death within one year.

TABLE OF MORTALITY

Age	Number Living	Deaths Each Year	Deaths Per 1,000
67	6,355,865	241,777	38.04
68	6,114,088	254,835	41.68
69	5,859,253	267,241	45.61
70	5,592,012	278,426	49.79
71	5,313,586	287,731	54.15
72	5,025,855	294,766	58.65
73	4,731,089	299,289	63.26
74	4,431,800	301,894	68.12
75	4,129,906	303,011	73.37
76	3,826,895	303,014	79.18
77	3,523,881	301,997	85.70
78	3,221,884	299,829	93.06
79	2,922,055	295,683	101.19

Decision Making

An agent for an insurance company found that, over the past year, 28 of 23,456 thirty-year-old male policyholders had died.

1. What percent of these policyholders had died? Round to the nearest ten-thousandth of a percent.

2. Convert this percent to a death rate per 1000 policyholders.

3. If the company takes on no new policyholders in this category, how many 31-year-old male policyholders does the company have?

4. If the mortality rate per thousand is 1.33 for the 31-year-old males, how many life insurance claims can the company expect to pay for this group of policyholders? What degree of accuracy makes sense for this answer?

5. If each of the policies held by the 31-year-old males who died was $100,000, how much must the company pay on these claims?

6. If the company wanted to just "break even" in this category, what should they charge their 31-year-old male policyholders as an annual premium? Round to the nearest cent.

7. Based on the table what is the insurance company's estimate of the probability that a randomly selected 60-year-old female will die within one year?

Age Nearest Birthday Graduated Mortality Rates Per 1,000		
Attained Age	Male Deaths	Female Deaths
60	11.89	7.37
61	13.17	8.00
62	14.57	8.67
63	16.07	9.38
64	17.71	10.15

Independent and Dependent Events

14.4

Explore

1. Use two number cubes of different colors, if available. If not, remember that the two cubes are different so that, for example

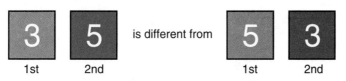

3 5   is different from   5 3

1st   2nd   1st   2nd

   a. Construct the sample space to show all the possible outcomes when rolling two cubes. Save for future reference.
   b. How many possible outcomes are there?

Use the sample space you constructed to answer Questions 2–5.

2. a. How many outcomes show a 4 on the first cube?
   b. What is the probability of getting a 4 on the first cube?

3. a. How many outcomes show a 4 on the second cube?
   b. What is the probability of getting a 4 on the second cube?

4. a. How many outcomes show a 4 on both cubes?
   b. What is the probability of getting a 4 on both cubes?

5. How are the probabilities in Questions 2 and 3 related to the probability in Question 4?

Build Understanding

When the first event has no effect on the probability of occurence of the second event, the two are called **independent events**.

Rolling two number cubes are independent events. As you saw in Explore, the probability of getting a 4 on both cubes is $\frac{1}{36}$. This probability is equivalent to taking the *product* of the probabilities of getting a 4 on each of the individual cubes.

$$\frac{1}{6} \quad \cdot \quad \frac{1}{6} \quad = \quad \frac{1}{36}$$

probability of   probability of   probability of
4 on 1st cube   4 on 2nd cube   4 on both cubes

> **INDEPENDENT EVENTS**
>
> *A* and *B* are *independent events* if and only if
>
> $$P(A \text{ and } B) = P(A) \cdot P(B)$$

When the first of two events does have an effect on the second, the two are called **dependent events**.

> **DEPENDENT EVENTS**
>
> If A and B are *dependent events*, then the probability of both occurring is
>
> $$P(A \text{ and } B) = P(A) \cdot P(B \text{ after } A)$$

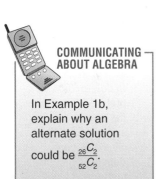

COMMUNICATING ABOUT ALGEBRA

In Example 1b, explain why an alternate solution could be $\frac{{}_{26}C_2}{{}_{52}C_2}$.

EXAMPLE 1

a. Julia draws a single card at random from a standard deck of 52 cards. She identifies the card, returns it to the deck, and then draws a second card. What is the probability that both cards are red?

b. Suppose, after drawing the first card, Julia does *not* return the card to the deck, and then draws a second card. What is the probability that both cards are red?

Solution

a. Since the first card is replaced before the second card is drawn, the drawings of the two cards are independent events.

$$P(\text{red } and \text{ red}) = P(\text{red}) \cdot P(\text{red})$$

$$= \frac{{}_{26}C_1}{{}_{52}C_1} \cdot \frac{{}_{26}C_1}{{}_{52}C_1}$$ For the second card, there is still a complete deck, 52 cards with 26 red.

$$= \frac{26}{52} \cdot \frac{26}{52} = \frac{1}{4}$$

With replacement, the probability of getting two red cards is $\frac{1}{4}$.

b. Since the first card is *not* replaced before the second card is drawn, the drawings of the two cards are dependent events.

$$P(\text{red } and \text{ red}) = P(\text{red}) \cdot P(\text{red after red})$$

$$= \frac{{}_{26}C_1}{{}_{52}C_1} \cdot \frac{{}_{25}C_1}{{}_{51}C_1}$$ For the second card, only 51 cards remain, with 25 red cards.

$$= \frac{26}{52} \cdot \frac{25}{51} = \frac{25}{102}$$

Without replacement, the probability of getting two cards is $\frac{25}{102}$. ◄

A tree diagram is helpful in reviewing the items of a sample space.

EXAMPLE 2

Sarah has 5 books in her locker: math, physics, French, history, and English. Without looking, Sarah pulls out one book. Then, without replacing the first book, she pulls out a second book. What is the probability Sarah pulled first the math book and then the French book?

Solution

Make a tree diagram to find the sample space.

1st book	M	P	F	H	E
2nd book	P F H E	M F H E	M P H E	M P F E	M P F H

Possible Outcomes:

There is only one desired outcome out of 20 possibilities.

MP	MF	MH	ME
PM	PF	PH	PE
FM	FP	FH	FE
HM	HP	HF	HE
EM	EP	EF	EH

By formula, since the events are dependent

$$P(\text{math } and \text{ then French}) = P(\text{math}) \cdot P(\text{French } after \text{ math})$$
$$= \frac{1}{5} \cdot \frac{1}{4} \qquad \text{After math, there are 4 books remaining.}$$
$$= \frac{1}{20}$$

The probability that Sarah pulled first her math book and then her French book is $\frac{1}{20}$. ◄

$P(B \text{ after } A)$ is also referred to as the **conditional probability** that B will occur given that A has occurred, denoted $P(B|A)$.

$$P(B|A) = \frac{P(A \text{ and } B)}{P(A)}$$

In problems, knowing that event A has occurred may change the sample space you use to determine the probability that B will occur.

EXAMPLE 3

A coin is tossed and a number cube is rolled. Determine the probability that the number cube will show a 5, given that heads shows.

Solution

Method 1   Let   event F = roll a 5   and   event H = heads
$$P(H) = \frac{1}{2} \qquad P(H \text{ and } F) = \frac{1}{2} \cdot \frac{1}{6} = \frac{1}{12}$$

$$P(F|H) = \frac{P(H \text{ and } F)}{P(H)} = \frac{\frac{1}{12}}{\frac{1}{2}} = \frac{1}{6}$$

Method 2   You know that heads came up, so the sample space is

(H, 1) (H, 2) (H, 3) (H, 4) (H, 5) (H, 6)

Only one outcome is favorable, (H, 5), so $P(F|H) = \frac{1}{6}$. ◄

Conditional probability is sometimes referred to as *Bayes' probability*, named for English clergyman and mathematician Thomas Bayes (1702–1761) who developed the rules in his pioneering work, *Essay Towards Solving a Problem in the Doctrine of Chances.*

Other important applications also make use of conditional probability and may ask you to consider events in a "backwards" order.

EXAMPLE 4

SURVEYS A survey asked 100 males and 100 females if they work the daily newspaper crossword puzzle. 30 males and 60 females said they work the puzzle. Based on this survey, if a person is selected at random

a. what is the probability that the person works the puzzle?
b. what is the probability that the person is female, given that he or she works the puzzle?

Solution

a. Out of 200 people surveyed, $30 + 60 = 90$ worked the puzzle. So,

$$P(\text{works puzzle}) = \frac{90}{200} = \frac{9}{20}$$

b. Let event F = person is female
event W = person works puzzle.

Since $P(F) = \frac{1}{2}$ and 60 out of 100 females work the puzzle,

$$P(F \text{ and } W) = \frac{1}{2} \cdot \frac{60}{100} = \frac{3}{10}$$

$$P(F \mid W) = \frac{P(F \text{ and } W)}{P(W)}$$

$$= \frac{\frac{3}{10}}{\frac{9}{20}} = \frac{2}{3}$$

TRY THESE

1. **WRITING MATHEMATICS** Without looking, Pia selects two marbles from a jar containing a set of colored marbles. Explain when these events are *independent* and when they are *dependent*.

Tracy selects 2 cards at random one at a time from a standard deck. Determine each probability **a.** when the first card is replaced and **b.** when the first card is *not* replaced.

2. both cards are black

3. the first card is red and the second card is black

4. both cards are red queens

Chamique picks 2 marbles at random one at a time from a jar with 3 orange marbles numbered 1, 2, and 3, and 2 blue marbles numbered 1 and 2. Determine each probability **a.** when the first marble is replaced and **b.** when the first marble is *not* replaced.

5. both marbles are orange   **6.** neither marble shows a 3   **7.** one of the marbles is green

GARDENING   Ti Hua bought 3 packets of vegetable seeds: beans, carrots, and radishes; and 2 packets of flower seeds: marigolds and petunias. Finding that his garden was too small, he decided to give away one packet of vegetable seeds and one packet of flower seeds, each selected at random.

8. List all possible pairs of packets in the sample space that Ti Hua might select to give away.

9. Determine the probability that he did not give away the packet of bean seeds.

10. Determine the probability that he gave away the packet of petunia seeds.

In Exercises 11–12, a coin is tossed three times. The given event A has happened. Determine $P(B|A)$.

11. Event A = heads on the first two tosses   **12.** Event A = at least one head
Event B = heads on the third toss   Event B = at least two heads

13. FAMILY PORTRAIT   Mother, father, and son line up at random for a family photo. Find the probability that the father is in the middle, given that the son is on one end.

In Exercises 14–18, two number cubes are rolled once. Find the probability of getting:

14. a 5 on both cubes   **15.** two even numbers   **16.** 4 or lower on both cubes

17. an even number on the first and on odd number on the second

18. an odd number on the first and a number less than 3 on the second

PRACTICE

1. WRITING MATHEMATICS   Explain the difference between events that are *mutually exclusive* and those that are *independent*.

In Exercises 2–4, two letters are chosen at random and without replacement from the word HEXAGON. Determine the probability that:

2. both letters are vowels   **3.** both letters are consonants

4. the first letter is a vowel and the second is a consonant

In Exercises 5–7, five-letter arrangements are made using the letters of the word PRISM, with no reuse of letters. Two arrangements are chosen at random and without replacement. Determine the probability that

5. both arrangements begin with P   **6.** both arrangements begin with a consonant

7. the first arrangement begins with P and the second arrangement begins with M

In Exercises 8–12, three cans of cat food are selected at random and without replacement from a case that contains a mixture of flavors: 6 liver, 6 tuna, 7 beef, and 5 veal. Determine the probability that

8. all three are beef

9. all three are liver

10. there are two veal and one beef

11. there are one tuna and two liver

12. there are one beef, one tuna, one veal

TERM PAPERS Professor Kellogg wrote topics for a term project on slips of paper, 8 in the humanities and 5 in the sciences. Chris chose 2 slips at random. Determine the probability that

13. both slips had humanities topics

14. one slip had a humanities topic and the other had a topic in the sciences

SEMAPHORE A ship's signal is made by arranging 2 red and 3 white flags on a vertical pole. The signal officer chose 2 flags at random. Determine the probability that

15. both flags were red

16. both flags were of the same color

17. the first flag was red and the second was white

In Exercises 18–19, two coins are tossed once. The given event A has happened. Find $P(B|A)$.

18. Event A = one coin has heads
    Event B = one coin has tails

19. Event A = there are no heads
    Event B = there are no tails

20. HEALTH A survey found the probability that a person who does not floss regularly has healthy gums is 0.2 and the probability that a person who does floss has healthy gums is 0.6. The probability that a person chosen at random flosses is 0.3. What is the probability a person chosen at random flosses if it is known the person has healthy gums?

DOMINOES The tiles used to play the game *Dominoes* have two numbers on each tile. The numbers are all possible combinations of the integers 1–6 and "blank," including doubles of these numbers. Dominoes are drawn at random.

21. List the elements in the sample space for the tiles. How many tiles are there in all?

22. If a player draws a hand of 5 tiles, what is the probability that the tiles are all doubles?

23. If a player draws 1 domino, what is the probability that it will contain a blank or a 6?

24. If a player draws 2 dominoes, what is the probability that they will both contain a 3?

25. If a player draws 1 domino, what is the probability that the sum of the values on the domino is at most 3?

EXTEND

26. **MEDICAL TESTS** A study found that a home monitoring kit correctly detects high blood pressure in 85% of people with the condition, but fails to detect the condition in 15% of the cases. The study also found that the monitor registers an incorrect high reading in 10% of the people with a normal blood pressure level. Finally, the study estimated that 18% of the people in the region actually had high blood pressure. What is the probability that a person chosen at random actually has high blood pressure given that the home monitor indicated the condition? Round your answer to the nearest hundredth.

CANDY WRAP A dish contains 10 candies. Three candies are covered with gold foil and 7 with silver foil. If 2 candies are chosen at random, determine the probability that

27. both candies will be covered with the same color foil

28. the two candies will be covered with different color foil

29. If 5 candies are chosen at random from the dish, determine the probability that at least 3 will be covered with silver foil.

ASSESSMENT Abbie, Bobbie, and Carrie independently answer a question on a test. The probability that Abbie answers correctly is 0.8, the probability that Bobbie answers correctly is 0.6, and the probability that Carrie answers correctly is 0.3. Determine the probability that

30. all three answer correctly

31. at least two of them answer correctly

32. at least one answers correctly

THINK CRITICALLY

TETRAHEDRON TOSS When a 4-sided die is tossed, the number that is face down is the outcome. The sample space for tossing a 4-sided die twice is shown.

$(1, 1)$ $(1, 2)$ $(1, 3)$ $(1, 4)$
$(2, 1)$ $(2, 2)$ $(2, 3)$ $(2, 4)$
$(3, 1)$ $(3, 2)$ $(3, 3)$ $(3, 4)$
$(4, 1)$ $(4, 2)$ $(4, 3)$ $(4, 4)$

33. If event $A = $ sum exceeds 5 and event $B = $ faces match, determine $P(B \mid A)$.

34. If event $A = $ the first toss is 4 and event $B = $ the second toss is 2, determine $P(B \mid A)$.

35. **RED AND BLACK** Three bags contain red and black disks. Bag I has 4 red and 4 black. Bag II has 2 red and 6 black. Bag III has 7 red and 1 black. Event A has happened. Determine $P(B \mid A)$.

    Event $A = $ a disk has been drawn from one of the bags at random and found to be black.
    Event $B = $ a black disk came from Bag I.

36. **ARCHERY** On the target shown, the smallest of the concentric circles has a radius of 2 units. The radius of each of the other circles is 2 units greater than that of the next smaller circle. Points are allocated for the placement of a dart as shown. If two darts hit the board at random, what is the probability the score will be exactly 25 points? Answer to the nearest percent.

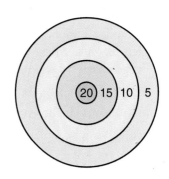

Solve each quadratic equation by completing the square. Check your answer.

37. $x^2 - 6x - 7 = 0$

38. $x^2 + 6x - 4 = 0$

39. **STANDARDIZED TESTS** Which of the following could represent the graph of $y = ax^2 + bx + c$ when $a < 0$?

A.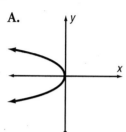

B.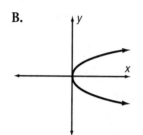

C.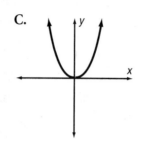

D.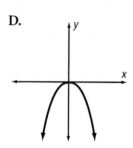

Solve each quadratic inequality.

40. $x^2 + 9x + 9 > 0$

41. $x^2 + 4x < 0$

Evaluate each expression.

42. $2p^0 + p^{\frac{2}{3}}$ for $p = 27$

43. $x^{-1}\left(3x^{\frac{1}{3}} + x^0\right)$ for $x = 8$

Factor each expression completely.

44. $2x^2 - 50$

45. $ax^2 + 7ax + 12a$

46. $g^2 - 4g + 12h - 3gh$

47. **STANDARDIZED TESTS** Which of the following polynomial equations has the solutions 0, -1, and 2?

A. $x^3 + x^2 - 2x = 0$

B. $x^3 - x^2 - 2x = 0$

C. $x^3 - x^2 + 2x = 0$

D. $x^3 + x^2 + 2x = 0$

PROJECT *Connection* In this activity, you will systematically gather documentation of the value of the items you listed earlier.

1. Use copies of owner's manuals, newspaper or magazine ads, or sales receipts to establish replacement cost for the items on your list of valuables. (You may adjust items in your list to match those for which you can find prices.)

2. Compile a scrapbook of all items listed. Put each item on a separate page with its picture, replacement cost, model number, and other important details. If this was a real home inventory, where should it be kept? Why?

3. Indicate which items "leave the house" and which do not. For example, a camera or bicycle could be stolen at a park, but a sofa never leaves the house.

4. Compute the total replacement cost for the items in each category. Then determine the grand total of all categories.

The Normal Distribution

Explore/Working Together

SPOTLIGHT ON LEARNING

WHAT? In this lesson you will learn
- the characteristics of a normal curve.
- to use a normal distribution to determine probability.

WHY? Normal distribution helps you solve problems about environmentalism, consumerism, and sports.

● Work with a partner.

1. Find the height of each student in your class to the nearest inch. Record the data in a frequency table like the one shown.

Height	4'9"	4'10"	4'11"	5'0"	5'1"	5'2"
Number of Students						

2. Plot the data from your frequency table. Connect the points using a smooth curve, and describe the shape.

3. Write another frequency table with the heights recorded in intervals of 3 inches, as shown.

Height	$4'9" \leq h < 5'0"$	$5'0" \leq h < 5'3"$
Number of Students		

4. Using the grouped data from your frequency table for Question 3, draw a vertical bar graph in which the bars are placed next to each other to show that as one interval ends, the next interval begins. Comment on the relative heights of the bars.

5. If a student were chosen at random from your class, in what interval would the height of that student be most likely to occur.

Build Understanding

● Mindy McDonald recorded the weights of 50 trout chosen at random from her fish farm. She grouped the data in 8 oz intervals, and graphed the results as a **histogram**, a vertical bar

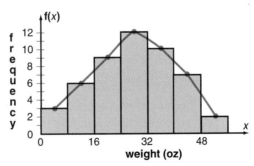

graph with intervals of equal width. She connected the midpoints of the bars by a smooth curve.

The result is a *bell-shaped curve,* called the **normal curve** or **normal distribution**. Refer to the data of Explore to note that the heights of a group of people exhibit a normal distribution.

You can use a histogram to calculate a probability using the areas of the bars of the histogram. Since the width of each bar is the same, the area of each bar is proportional to the corresponding frequency.

EXAMPLE 1

FISHING If a fish is chosen at random from Mindy McDonald's sample, what is the probability that it weighs

a. between 2 and 2.5 pounds?   **b.** less than 1.5 pounds?

Solution

Use the histogram on page 703. Let f represent the weight of a fish.

a. $P(2 < f < 2.5) = \dfrac{\text{Area of Bar 5}}{\text{Total Area}}$

$= \dfrac{(0.5)(10)}{25} = \dfrac{5}{25}$ or $\dfrac{1}{5}$ or 0.20

The probability a fish weighs between 2 and 2.5 pounds is $\dfrac{1}{5}$ or 20%.

b. $P(f < 1.5) = \dfrac{\text{Area of (Bar 1 + Bar 2 + Bar 3)}}{\text{Total Area}}$

$= \dfrac{(0.5)(3) + (0.5)(6) + (0.5)(9)}{25} = \dfrac{9}{25}$ or 0.36

The probability a fish weighs less than 1.5 pounds is $\dfrac{9}{25}$ or 36%. ◄

For a normal curve, the highest point is directly over the **mean** $\overline{x}$. The curve is symmetric with respect to a vertical line through the mean, and the curve approaches the x-axis asymptotically as $|x|$ increases.

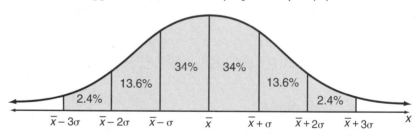

Normal curves vary in height and width. The height of a normal curve indicates the frequency of occurrence. The width of a normal curve indicates how spread out the data are. This dispersion is called the **standard deviation**, denoted σ (sigma).

In a normal distribution, about 68% of the data fall within one standard deviation of the mean, and 95% within 2 standard deviations; 99.8% of the data fall within 3 standard deviations. The remaining 0.2% represents rare data, such as a person who is 8 feet tall.

Using the percentages of area under a normal curve, you can calculate **percentile values**, those values that indicate what percent of the total frequency scored *at* or *below* that measure.

The 50th percentile is at the mean. Since 34% of the scores correspond to 1 standard deviation, you can find the locations of the 84th percentile (50 + 34) and the 16th percentile (50 − 34) and so on.

PROBLEM SOLVING TIP

EXAMPLE 2

On a certain standardized test where scores are normally distributed, the mean score was 50 and the standard deviation 3.
  a. What percent of those taking the test scored between 44 and 47?
  b. If Judy's score was 59, how did she compare to the rest of the group?

Check your graphing calculator for one-variable statistics. If you enter a set of data, it may calculate the mean and standard deviation, and other statistical measures as well.

Solution

Sketch a normal curve with the given mean and standard deviation.

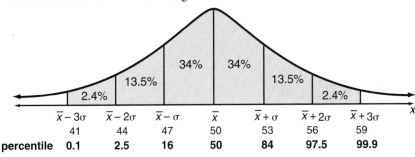

	$\bar{x}-3\sigma$	$\bar{x}-2\sigma$	$\bar{x}-\sigma$	$\bar{x}$	$\bar{x}+\sigma$	$\bar{x}+2\sigma$	$\bar{x}+3\sigma$
	41	44	47	50	53	56	59
percentile	0.1	2.5	16	50	84	97.5	99.9

  a. The interval between 44, which is $\bar{x} - 2\sigma$, and 47, which is $\bar{x} - \sigma$, contains the scores of about 13.5% of the group.
  b. Judy's score of 59 puts her at about the 99th percentile. ◄

A normal distribution with a mean of 0 and a standard deviation of 1 is called a **standard normal distribution**. The area between this curve and the x-axis is 1. The mean is 0 and $\sigma = 1$.

Standard Normal Curve

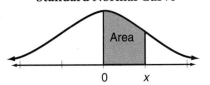

Area Under Standard Normal Curve

x	Area	x	Area
0.0	0.0000	1.6	0.4452
0.2	0.0793	1.8	0.4641
0.4	0.1554	2.0	0.4772
0.6	0.2257	2.2	0.4861
0.8	0.2881	2.4	0.4918
1.0	0.3413	2.6	0.4953
1.2	0.3849	2.8	0.4974
1.4	0.4192	3.0	0.4987

The table shown gives the areas under the standard normal curve for values of x from 0 (the mean) through 3.0 (3σ).

You can use a standard normal curve to approximate probability.

EXAMPLE 3

Use the table to determine each value for a standard normal curve.

  a. $P(0 < x < 1.4)$ **b.** $P(-1.4 < x < 0)$ **c.** $P(x > 1.4)$

Solution
  a. $P(0 < x < 1.4) = 0.4192$ Use the table.

b. Use the symmetry of the curve.

$$P(-1.4 < x < 0) = P(0 < x < 1.4) = 0.4192$$

c. The total area under the curve is 1.0. So the area to the right of $x = 0$ is 0.5000. The area between $x = 0$ and $x = 1.4$ is 0.4192.

$$P(x > 1.4) = 0.5000 - 0.4192 = 0.0808 \quad \blacktriangleleft$$

You can use the standard normal curve to determine probabilities relating to normal distributions that are *not* standard. The probability that a value x falls between $\bar{x}$ and σ in a nonstandard normal distribution is the same as the probability that a value x falls between 0 and 1 in a standard normal distribution.

$$\overset{non\ standard}{P(\bar{x} < x < \sigma)} = \overset{standard}{P(0 < x < 1)}$$

EXAMPLE 4

Mustafa was chosen at random from all students who took an exam with normally distributed scores, with a mean of 100 and a standard deviation of 10. What is the probability that his score is between 90 and 120?

Solution
For this curve, 90 is 1σ below $\bar{x} = 100$.
120 is 2σ above $\bar{x}$.

Use the table for the standard normal curve on page 705.

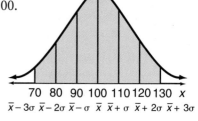

$$P(90 < x < 100) = 0.3413$$

$$P(100 < x < 120) = 0.4772$$

$$P(90 < x < 120) = 0.3413 + 0.4772 = 0.8185.$$

The probability that Mustafa's score is between 90 and 120 is 0.8185 or about 82%. $\quad \blacktriangleleft$

TRY THESE

ENVIRONMENTALISM An environmental control group conducted a pollution study near a steel mill over a 5-week period. The 35 pollution counts recorded were

5.4	10.2	15.2	20.3	25.7	31.0	6.1
10.6	16.4	20.4	26.1	31.4	18.4	18.3
10.4	17.1	20.6	26.4	31.9	32.6	27.3
20.7	20.9	17.4	11.1	11.4	17.6	22.1
27.9	22.4	17.9	13.0	14.5	18.0	23.8

1. Group the data in a frequency table using the intervals 0–4.9, 5–9.9, . . . , 30–34.9.

2. Display the grouped data in a histogram. Let the horizontal axis represent the pollution count. Let the vertical axis represent the frequency.

Use the histogram from Exercise 2 to answer the following questions. A day is chosen at random from the 35 days for which the pollution counts were recorded.

3. In which interval is it most likely to fall?

4. What is the probability that its pollution count is between 5 and 15?

5. What is the probability that its pollution count is greater than 20?

6. WRITING MATHEMATICS Stan said that either graph *A* or *B* could represent a normal curve, while Sally said only graph *B* could. With whom do you agree? Explain.

A 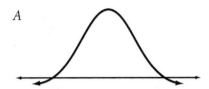   *B*

CONSUMERISM The mean volume for a milk carton at the Modern Mart is 0.95 l and the standard deviation is 0.05 l. There are 50 milk cartons on the shelves.

7. Sketch a normal curve to represent the distribution of the volume of these 50 milk cartons.

8. About what percent of the milk cartons contain between 0.90 and 1.00 l?

9. About how many milk cartons would contain less than 0.90 l?

10. If Ms. Samson chooses one milk carton at random from a shelf, determine the probability that her carton contains

   a. between 0.825 l and 0.875 l   **b.** more than 1.075 l

ASSESSMENT In 1995, the national mean for math scores on the SAT was 482 and the standard deviation was 127.

11. Sketch a normal curve to represent the distribution of math scores.

12. About what percent of those taking the SAT had math scores between 482 and 736?

13. If Nina's math score on the SAT was 609, in what percentile does she fall?

14. Joel's math score on the SAT was 595. The college he is interested in attending requires a minimum math score in the 67th percentile. Is Joel eligible? Explain.

Use the table of Area Under Standard Normal Curve on page 705 to answer the following questions about math scores on the 1995 SAT in which the mean was 482 and the standard deviation was 127. Sergé, Liam, and Elsie are three students chosen at random from those who took the 1995 SAT. Find the probability that

15. Sergé's math score was between 482 and 609.

16. Liam's math score was between 355 and 482.

17. Elsie's math score was higher than 609.

PRACTICE

SPORTS A group of 40 runners competed in a one-kilometer sprint. Their finish times in minutes were recorded as

3.2	3.2	3.6	3.6	4.1	4.1	4.1	4.2	3.5	4.0
5.6	5.6	4.5	4.5	2.9	3.3	3.4	4.9	3.9	4.2
4.8	4.6	4.6	5.3	5.2	5.0	3.7	4.3	4.1	4.2
4.3	4.3	5.1	5.8	4.4	3.8	3.9	3.8	4.7	3.4

1. Group the data in a frequency table using the intervals 2.5–2.9, 3.0–3.4, . . . , 5.5–5.9.

2. Display the grouped data in a histogram. Let the horizontal axis represent the finish time. Let the vertical axis represent the frequency.

Use the histogram from Exercise 2. A runner is chosen at random from the group.

3. In which interval is the runner's time most likely to fall?

4. What is the probability that the runner's time is between 3.0 and 4.5 min?

5. What is the probability that the runner's time is less than 4 min?

RESEARCH AND DEVELOPMENT Engineers found life spans of 500 samples of a light bulb had a normal distribution with a mean of 1580 h and a standard deviation of 250 h.

6. Sketch a normal curve to represent the distribution of the life of these bulbs.

7. About what percent of the light bulbs would have a life span between 1080 and 1580 hours?

8. About how many bulbs would have a life span greater than 1830 hours?

9. If one bulb is chosen at random, determine the probability that its life span is

   a. about 1330 h   b. at least 1830 h   c. no more than 1955 h

STANDARDIZED TESTS Tests that are called *standardized* have a normal distribution of scores.

10. For a standardized test that has a mean of 63 and a standard deviation of 5, which of the following scores can be expected to occur most frequently?
    A. 45   B. 55   C. 65   D. 74

11. For a standardized test that has a mean of 75 and a standard deviation of 5.8, which of the following scores can be expected to occur less than 5% of the time?
    A. 90   B. 80   C. 70   D. 65

12. For a standardized test that has a mean of 20 and a standard deviation of 2.6, which of the following scores can be expected to occur fewer that 5 times out of 100?
    A. 25.1   B. 22.6   C. 18.7   D. 14.6

13. On a standardized test with a standard deviation of 2, a score of 26 will occur fewer than 5 times out of 100. Which of the following scores could be the mean for this test?
    A. 20   B. 23   C. 24   D. 25

14. On a standardized test, $\bar{x} + 2\sigma = 80$ and $\bar{x} - 2\sigma = 40$. Which of the following is the value of σ?
    A. 30   B. 20   C. 10   D. 1

15. WRITING MATHEMATICS Seymour's results on a standardized test showed that he was in the 70th percentile of his class. He told his parents that he got 70% on his test. Did Seymour give his parents correct information? Explain.

ASSESSMENT In 1995, the national mean for verbal scores on the SAT was 428 and the standard deviation was 113.

16. Sketch a normal curve to represent the distribution of verbal scores.

17. About what percent of those taking the SAT had verbal scores between 315 and 541?

18. If Juan's verbal score on the SAT was 315, in what percentile does he fall?

19. Cesar's verbal score on the SAT was 715. The college he is interested in attending requires a minimum verbal score in the 90th percentile. Is Cesar eligible? Explain.

Use the table of Area Under Standard Normal Curve on page 705 to answer the following questions about verbal scores on the 1995 SAT. Annie, Tòmas, Sue, and Ted are four students chosen at random from those who took the 1995 SAT. Find the probability that

20. Annie's verbal score is between 428 and 541.   **21.** Sue's verbal score is between 428 and 631.

22. Tòmas's verbal score is between 541 and 654.   **23.** Ted's verbal score is lower than 315.

EXTEND

DeMoivre, in the 1730's, discovered that the equation of the normal curve is: $y = \dfrac{1}{\sigma\sqrt{2\pi}} e^{-\frac{1}{2}\left(\frac{x - \bar{x}}{\sigma}\right)^2}$

Sir Francis Galton recorded the heights of 8585 men in Great Britain, and demonstrated a normal distribution with a mean of 67 in. and a standard deviation of 2.5 in.

24. Substitute Galton's values for $\bar{x}$ and σ in the equation for the normal curve and simplify.

25. Use a graphing utility to graph Galton's distribution. Describe your graph.

26. Find the percentage of men in Galton's study who were between 66 and 68 inches tall.

THINK CRITICALLY

A number that shows the spread of data is called a **measure of dispersion**. Three measures of dispersion are shown below.

mean absolute deviation	variance	standard deviation
$\dfrac{\sum\limits_{i=1}^{n} \lvert x_i - \bar{x} \rvert}{n}$	$\dfrac{\sum\limits_{i=1}^{n} (x_i - \bar{x})^2}{n}$	$\sqrt{\dfrac{\sum\limits_{i=1}^{n} (x_1 - \bar{x})^2}{n}}$

For the data set 15, 32, 57, 44, calculate the

27. mean   **28.** mean absolute deviation   **29.** variance   **30.** standard deviation

MIXED REVIEW

31. STANDARDIZED TESTS If the discriminant is 37, the solutions of the quadratic equation are.

A. rational and equal   **B.** rational and unequal   **C.** irrational and unequal   **D.** not real

The Binomial Distribution

Explore/Working Together

● Work with a partner.

1. Toss 6 coins 50 times and record the frequencies of the number of *heads* that occur in a table such as the one shown.

Number of Heads	0	1	2	3	4	5	6
Frequency							

2. Display the data in a histogram. Let the horizontal axis represent the number of heads and the vertical axis represent the frequency.

3. What number of heads had the greatest frequency?

4. Discuss any symmetries you observe about your histogram.

5. Repeat Question 1 using 100 trials, and compare the results.

6. Predict the nature of the results as the number of trials increases.

THINK BACK

A histogram is a graph that shows a frequency distribution.

Build Understanding

● A **binomial experiment** has the following characteristics:

- There are exactly two possible outcomes for any trial.
- There are a fixed number of trials.
- The trials are independent.
- The probability of success on each trial is the same.

EXAMPLE 1

Maria spins 4 times on the spinner shown and keeps track of whether the outcome is an odd number or an even number. Determine if this is a binomial experiment.

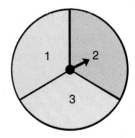

Solution

Test the characteristics.

- There are exactly two possible outcomes for any trial: odd or even.
- There are a fixed number of trials: 4.
- The trials are independent: the first spin does not affect the second spin, and so on.
- The probability for each trial is $P(\text{odd}) = \frac{2}{3}$ and $P(\text{even}) = \frac{1}{3}$.

Yes, this is a binomial experiment. ◀

Usually, in a binomial experiment, you are interested in determining the probability of a particular occurrence.

EXAMPLE 2

In Maria's binomial experiment, determine the probability of getting exactly 3 odd numbers in the 4 spins.

Solution

There are $_4C_3$, or 4, possible ways of getting exactly 3 odd numbers in 4 spins, as shown at the right.

For each of the ways: $P(3 \text{ odd}, 1 \text{ even}) = \left(\frac{2}{3}\right)^3\left(\frac{1}{3}\right)^1$

$$\begin{array}{cccc} O & O & O & E \\ O & O & E & O \\ O & E & O & O \\ E & O & O & O \end{array}$$

Taking into account the 4 possibilities:

$$P(3 \text{ odd}, 1 \text{ even}) = 4\left(\frac{2}{3}\right)^3\left(\frac{1}{3}\right)^1 = 4 \cdot \frac{8}{27} \cdot \frac{1}{3} = \frac{32}{81}$$

So, the probability that Maria gets 3 odd numbers in 4 spins is $\frac{32}{81}$. ◀

Since Maria's experiment is *binomial* in nature, for the 4 spins, you can use the terms of the binomial expansion of $(O + E)^4$ to determine the probabilities of each combination of odd and even.

$$(O + E)^4 = O^4 + 4O^3E + 6O^2E^2 + 4OE^3 + E^4$$

Term	Meaning	Coefficient of Term	
O^4	1 way to get 4 odd	$_4C_4 = 1$	Note that the
$4O^3E$	4 ways to get 3 odd and 1 even	$_4C_3 = 4$	values of these
$6O^2E^2$	6 ways to get 2 odd and 2 even	$_4C_2 = 6$	coefficients are
$4OE^3$	4 ways to get 1 odd and 3 even	$_4C_1 = 4$	the 4th row of
E^4	1 way to get 4 even	$_4C_0 = 1$	Pascal's triangle.

To determine the probability of getting exactly 3 odd numbers, substitute $\frac{2}{3}$ for O and $\frac{1}{3}$ for E in the term $4O^3E$.

$P(\text{exactly 3 odd}) = 4O^3E$

$= 4\left(\frac{2}{3}\right)^3\left(\frac{1}{3}\right)$

$= \frac{32}{81}$

Number of Odd Outcomes	0	1	2	3	4
Probability	$\frac{16}{81}$	$\frac{32}{81}$	$\frac{24}{81}$	$\frac{8}{81}$	$\frac{1}{81}$

By substituting into the other terms of the expansion, you evaluate the probabilities for all possible odd outcomes.

A graph of these results shows that a binomial distribution approximates the bell-shaped curve of a normal distribution. The greater the number of trials, the better the approximation.

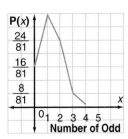

ALGEBRA: WHO, WHERE, WHEN

Attributed to the work of the Swiss mathematician Jacques (James) Bernoulli (1654–1705), the Binomial experiment is also called the Bernoulli experiment.
No family in history has produced as many celebrated mathematicians as did the Bernoulli family. A dozen members of the family achieved distinction in mathematics and physics. Many common studies of probability are binomial experiments.

CHECK UNDERSTANDING

Write out the rows of Pascal's triangle that correspond to the coefficients of the binomials $(a + b)^0$ through $(a + b)^5$.

To find the probability of a particular occurrence in a binomial experiment, find the corresponding term of the binomial expansion.

> **FORMULA FOR BINOMIAL PROBABILITIES**
>
> **The probability of exactly r successes in a sample of n repeated, independent trials is $_nC_r \cdot P^r \cdot (1 - P)^{n-r}$ where P is the probability of success on each trial.**

EXAMPLE 3

QUALITY CONTROL On the assembly line at Marcos Manufacturing, the probability that an item is defective is $\frac{1}{20}$. If an inspector checks 5 items at random, what is the probability of finding

a. exactly 2 defects? **b.** at least 2 defects?

Solution

There are only two possible outcomes: an item is defective or not defective. There are a fixed number of trials: 5. The trials are independent: the manufacture of one item does not affect the manufacture of another. The probability of each trial is the same: $P\,(\text{defective}) = \frac{1}{20}$. So, this is a binomial experiment.

a. $P\,(r \text{ successes}) = {}_nC_r\, P^r\, (1 - P)^{n-r}$

$P\,(\text{exactly 2 defective}) = {}_5C_2 \left(\frac{1}{20}\right)^2 \left(\frac{19}{20}\right)^3 \approx 0.0214343$

The probability that exactly 2 items are defective is about 2.14%.

b. The probability of *at least* 2 defective items is the sum of the probabilities of 2, 3, 4, or 5 defective items.

$P\,(\text{at least 2 defective}) = P(2) + P(3) + P(4) + P(5)$

$= {}_5C_2 \left(\frac{1}{20}\right)^2 \left(\frac{19}{20}\right)^3 + {}_5C_3 \left(\frac{1}{20}\right)^3 \left(\frac{19}{20}\right)^2 + {}_5C_4 \left(\frac{1}{20}\right)^4 \left(\frac{19}{20}\right) + {}_5C_5 \left(\frac{1}{20}\right)^5$

$\approx 0.0214343 + 0.0011281 + 0.0000296 + 0.0000003$

≈ 0.0225923

The probability that at least 2 items are defective is 2.26%. ◀

TRY THESE

Determine if each experiment is a binomial experiment. Explain how each characteristic of a binomial experiment is satisfied or not.

1. tossing a coin 8 times

2. guessing a correct answer from a 10-item multiple choice test where each item has 5 choices and 1 choice is correct

3. selecting 6 balls from a box that contains exactly 10 red balls and 13 white balls

Describe a binomial experiment using each device.

4. rolling a number cube

5. drawing a card from a standard deck

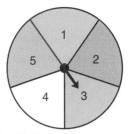

Martin spins 3 times on the board shown and keeps track of whether the outcome is an odd number or an even number.

6. Determine the probability that Martin got an odd number exactly 2 times in the 3 spins.

7. List the probabilities for all possible numbers of odd outcomes in Martin's 3 spins.

8. Graph the probability curve from the list in Exercise 7.

9. **WRITING MATHEMATICS** Describe the probability curve you drew in Exercise 8.

COIN TOSS Sue tosses a coin 8 times. Calculate the probability to the nearest tenth of a percent that she gets

10. *heads* exactly 3 times   11. *tails* at least 6 times   12. *heads* no more than 4 times

SPORTS The probability that baseball player Keenan Jones gets a hit is $\frac{3}{10}$. For his next 9 times at bat, calculate the probability that he gets

13. exactly 6 hits   14. no more than 3 hits   15. at least 7 hits

PRACTICE

Determine if each experiment is a binomial experiment. Explain how each characteristic of a binomial experiment is satisfied or not.

1. Joe, whose free-throw average in basketball is $\frac{3}{8}$, has 7 free throws in this game.

2. Sally draws 10 cards from a standard deck and keeps track of whether the outcome is red or black.

3. Zach selects 8 marbles from a jar that contains exactly 15 red marbles and 13 white marbles, with replacement of the selection each time.

Describe a binomial experiment using a spinner that is

4. equally divided into 7 parts numbered 1–7

5. equally divided into 3 colors: red, white, blue

6. Bert tosses 3 number cubes. Determine the probability he gets the number 4 on exactly 2 of the cubes.

7. Samantha tosses 4 number cubes. Determine the probability that she gets the number 6 on exactly 2 of the cubes.

8. Jason tosses 8 coins. List the probabilities of all possible ways in which heads could appear.

9. Graph the probability curve from the list in Exercise 8.

The "probability demonstrator" shown is also known as a Galton Board.

10. **WRITING MATHEMATICS** Explain how you would determine the probability that the marble at the top of the board will land in one of the four compartments at the bottom. What is the probability for each compartment?

11. Write the probabilities for a Galton Board of 8 compartments.

SPORTS The probability Kennedy High will beat Adams High each time they play soccer is $\frac{2}{3}$. Find the probability of each of the following wins for Kennedy.

12. exactly 1 win in 3 games   13. at least 1 win in 3 games   14. no wins in 5 games

15. exactly 3 wins in 5 games   16. no more than 3 wins in 5 games

GENETICS The probability that a child born to the Cho family will inherit a certain trait is 0.27. If there are 4 children, find the probability the number of children who inherit the trait is

17. exactly 1   18. exactly 3   19. no more than 1   20. zero

SPORTS Mindy's basketball record for the season shows that she scored an average of 38% for all her foul shots. Find the probability that in her next 10 foul shots, she would make

21. exactly half of them   22. at least half of them

AIR TRAVEL An airline books 16 passengers on a commuter flight that has a capacity for only 12 passengers. The probability a booked passenger will not show for the flight is $\frac{1}{5}$.

23. Find the probability that exactly 12 booked passengers will show for the flight.

24. Find the probability that the airline has "overbooked," causing a seat shortage.

EXTEND

The study of *linked chains* of events was initiated in 1905 by the Russian, Andrei Andreyevich Markov. If two self-aligning instruments begin in alignment, then the probability that they will remain aligned after one use is 0.9. If two self-aligning instruments begin unaligned, then the probability that they will become aligned after use is 0.8. What is the probability that the two instruments are aligned after two uses if they begin in alignment?

You can use a transition **matrix** to find the probabilities. This matrix has three characteristics: it is square, all entries are between 0 and 1, the sum of the entries in any row is 1. To calculate the probability after two uses, multiply the transition matrix by itself.

$$
\begin{array}{c}
\text{next stage} \\
\begin{array}{cc} A & U \end{array} \\
\text{beginning state} \begin{array}{c} A \\ U \end{array}
\begin{bmatrix} 0.9 & 0.1 \\ 0.8 & 0.2 \end{bmatrix}
\end{array}
\qquad
\begin{bmatrix} 0.9 & 0.1 \\ 0.8 & 0.2 \end{bmatrix}
\begin{bmatrix} 0.9 & 0.1 \\ 0.8 & 0.2 \end{bmatrix}
=
\begin{array}{c}
\begin{array}{cc} A & U \end{array} \\
\begin{array}{c} A \\ U \end{array}
\begin{bmatrix} 0.89 & 0.11 \\ 0.88 & 0.12 \end{bmatrix}
\end{array}
$$

Read the answer to the required probability from *AA*, 0.89. Thus, if the instruments begin in alignment, the probability that they will be in alignment after two uses is 0.89 or 89%.

ALIGNMENT Using the conditions of the self-aligning instruments on the previous page, find the probability that after 3 uses, the instruments are

25. aligned if they begin aligned

26. aligned if they begin unaligned

27. unaligned if they begin aligned

28. unaligned if they begin unaligned

POLITICS In Centre City, if the current mayor is a Democrat, then the probability is $\frac{5}{6}$ that the next mayor is a Democrat. If the current mayor is a Republican, then the probability is $\frac{2}{3}$ that the next mayor is a Republican.

29. If the current mayor is a Democrat, find the probability of a Republican mayor after two elections.

30. If the current mayor is a Republican, find the probability of a Democratic mayor after three elections.

THINK CRITICALLY

TRAFFIC CONTROL Three widely-separated traffic lights on a road operate independently of each other. The probability that a motorist will be stopped by any one of them is 40%.

31. Calculate the probabilities that a motorist will not be stopped and the probabilities that a motorist will be stopped at exactly one, exactly two, and all three lights.

32. How could you know if your answers to Exercise 31 are reasonable?

33. Which is more probable, being stopped at more than one light or at one or fewer lights? Explain.

BASEBALL The World Series is a contest between the two "best" baseball teams, one from the American League, and one from the National League, and is won by the team that wins 4 of a possible 7 games.

34. If the two teams are evenly matched, what is the probability that the series will end in exactly 5 games?

35. If the probability that the National League will win is $\frac{3}{5}$ for every game, find the probability that the National League team will win the World Series in exactly 5 games. What is the most probable length of the World Series: 4, 5, 6, or 7 games?

MIXED REVIEW

Write each irrational number in simplest radical form.

36. $2\sqrt{108}$

37. $\sqrt{72a^6b^9}$

38. $5\sqrt[3]{16}$

39. **STANDARDIZED TESTS** The expression i^{47}, where $i = \sqrt{-1}$, is equivalent to

    A. -1 **B.** 1 **C.** $-i$ **D.** i

Solve each logarithmic equation for x.

40. $\log_5 x = 2$

41. $\log_9 x = \frac{3}{2}$

42. $\log_x 81 = 4$

In addition to helping you plan a trip, a travel agent offers various insurance policies. Some policies cover accidents that may occur in connection with the trip, and other policies reimburse you for prepayment if you need to cancel or interrupt a trip.

Decision Making

Annual Travel Accident Insurance

Plan A covers you for injuries when riding as a passenger on

- regularly scheduled flights.
- specified military carriers.
- other public carriers, such as a train, bus, taxi, ship, or airport limousine.

Plan B covers you for injuries described under Plan A. *In addition*, this plan covers you for injuries received on vacation

- while driving or riding in automobiles and mobile homes.
- as a pedestrian.

1. What is the cost of $200,000 worth of Plan A insurance?

2. What is the cost of Plan B for $200,000 coverage?

3. Why is $200,000 worth of coverage of Plan B so much more than $200,000 worth of Plan A?

4. Visit a travel agent and ask for the rates on baggage insurance, trip cancellation, and flight insurance. Compare the frequency of missing baggage to the other occurrences.

	Benefit Amount	Annual Premium
Plan A		
☐	$ 300,000.00	$ 75.00
☐	$ 200,000.00	$ 50.00
☐	$ 100,000.00	$ 25.00
Plan B		
☐	$ 300,000.00	$ 225.00
☐	$ 200,000.00	$ 150.00
☐	$ 100,000.00	$ 75.00

14.7 Use Simulations

Explore/Working Together

• Work with a partner.

Perform each experiment 1–4 for 50 trials and record your results in a frequency table.

1. Toss a folded index card.

2. Flip a thumbtack.

3. Toss a cylindrical paper cup.

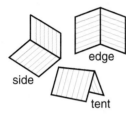

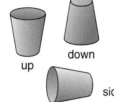

4. Select a card from a standard deck of 52 cards and record whether the result is red or black. Replace the card before each selection.

5. For each of the previous experiments, draw a conclusion about the likelihood of various outcomes.

6. What is the theoretical probability that one card drawn from a standard deck will be red?

7. How did the results of your experiment 4 compare to the theoretical probability?

8. Repeat experiment 4 another 50 times and add these results to your prior results. Are you closer to the theoretical probability?

9. In 100 draws of a card, would you expect to get red exactly 50 times? Explain.

Build Understanding

• The *theoretical probability* of getting heads during a coin toss is $\frac{1}{2}$. You can also calculate an *experimental probability*. The experimental and theoretical probabilities may not be equal, but these values approach each other as the number of experimental trials increases. So, you may not get exactly 5 heads in 10 tosses, but in 10,000 tosses you should get close to 5000 heads.

In many real world problems, you must depend on experimental probability. By using a variety of devices, you **simulate** the real situation, carry out the trials, collect data, and calculate probabilities. It is essential that a simulation possess the same characteristics as the problem.

ALGEBRA: WHO, WHERE, WHEN

The French scientist, Georges-Louis Leclerc, Comte de Buffon, (1707-1788), devised a simulation using probability to calculate the value of π. You will repeat his simulation in Practice.

EXAMPLE 1

Describe a model that could be used to simulate each problem.

a. The Frye family is planning to have 4 children. Find the probability that there will be an equal gender split, 2 boys and 2 girls.

b. Yolanda has a batting average of 0.250. In her next 10 times at bat, what is the probability that she gets exactly 3 hits?

Solution

a. To model gender, use a device that has 2 equally likely outcomes. Your simulation may be tossing a coin, in which getting *heads* represents a child is a girl and getting *tails* represents a child is a boy. Since there are 4 children, flip 4 coins at a time.

b. To model a batting average of 0.250 $\left(\text{that is, } \frac{250}{1000} \text{ or } \frac{1}{4}\right)$, use a device that has outcomes in the ratio of $\frac{1}{4}$. Your simulation could be drawing a particular suit from a standard deck. Since the ratio of the number of hearts, 13, to the number of cards in a full deck, 52, is $\frac{1}{4}$, let drawing a heart represent getting a hit and drawing a card of any other suit represent getting no hit. Since the player has 10 turns at bat, draw 10 cards, with replacement of cards each time. ◄

When designing a simulation, you may use familiar devices in a nonroutine fashion.

EXAMPLE 2

Two-thirds of the seniors at Lakota High have part-time jobs. If three seniors are polled, what is the probability that at least one of them has a part-time job?

a. Design an experiment to simulate the situation.
b. Perform 20 trials of the experiment, and calculate the probability.

Solution

a. To model a ratio of $\frac{2}{3}$, use a number cube so that

　　4 of the outcomes, 1, 2, 3, 4, represent having a job
　　2 of the outcomes, 5, 6, represent not having a job
　　Thus, $P(\text{job}) = \frac{4}{6}$ or $\frac{2}{3}$.

　　Since there are 3 seniors polled, roll three cubes at a time.

b. Sam rolled 3 number cubes 20 times. Here are his results.

　　*251　*213　*264　556　*343　*422　*152　*125　555　*562
　　*361　*661　*242　*354　*246　*265　*445　*546　*221　*345

　　*indicates trials containing at least one of the numbers 1, 2, 3, 4.

Since 18 of the 20 trials resulted in at least one employed student,

$$P \text{ (at least 1 employed)} = \frac{18}{20} \text{ or } \frac{9}{10} \text{ or } 90\%$$

The simulation shows that if 3 seniors are polled at Lakota High, the probability at least one of them has a part-time job is 90%. ◄

Another useful device in a simulation is **random numbers**. These are numbers that have no detectable pattern associated with them. Computers or calculators can generate random numbers, and tables are available. Here is a portion of a table of random numbers.

24130 48360 22527 97265 76393 64216 79309 30624 36168
61637 57039 97581 83716 65606 12197 79210 69071 10084
77565 34094 29939 69526 36927 37889 74103 65611 29875
89482 37071 19973 36710 48081 78772 33135 10851 27655

To use a random number table, begin by randomly choosing an entry. Then choose a second random digit by selecting the digit next to it — right, left, up, down, or diagonally.

EXAMPLE 3

a. Repeat the problem of Example 2 using random numbers.
b. Calculate the theoretical probability and compare the results of the two simulations to it.

Solution

a. Here is one way to establish the ratio $\frac{2}{3}$. Consider the digits 1–9.

Let the digits 1–6 represent those seniors with jobs.
Let the digits 7–9 represent those seniors without jobs.

Jody selected these triplets from a table of random numbers.

679 549 372 651 852 236 662 195 643 315
836 596 877 919 991 353 264 631 128 795

Only one number, 877, does not contain the digits 1–6.

$$P \text{ (at least 1 employed)} = \frac{19}{20} \text{ or } 95\%$$

This simulation shows that if 3 seniors are polled at Lakota High, the probability at least one of them has a part-time job is 95%.

b. Since this is a binomial experiment

$$P \text{ (at least 1 employed)} = {}_3C_1\left(\frac{2}{3}\right)\left(\frac{1}{3}\right)^2 + {}_3C_2\left(\frac{2}{3}\right)^2\left(\frac{1}{3}\right) + {}_3C_3\left(\frac{2}{3}\right)^3$$
$$\approx 0.2222 + 0.4444 + 0.2963 \text{ or } 96\%$$

The theoretical probability is 96%, and the results using numbered cubes was 90% and using random numbers was 95%. ◄

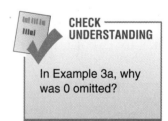

CHECK UNDERSTANDING

In Example 3a, why was 0 omitted?

THINK BACK

In example 3b, in order to calculate at least 1, the probability of 1, 2, and 3 are added.

Using the device listed, describe a model that could simulate each problem situation.

1. **PHARMACEUTICALS** A medicine has a 3 in 5 chance of curing the condition for which it is prescribed. If two patients are chosen at random to use the medicine, what is the probability it will cure their condition? *Use a number cube.*

2. **HIGHER EDUCATION** At Bramson High, 75% of graduating seniors are college bound. What is the probability four seniors polled at random are college bound? *Use a standard deck of playing cards.*

3. **PETS** A litter has 6 solid-colored kittens who are either black or white. What is the probability that there is an equal color split, 3 black and 3 white? *Use coins.*

4. **SPORTS** The likelihood that a baseball player will get a hit is $\frac{1}{3}$. In his next 5 times at bat, what is the probability that he will get exactly 2 hits? *Use a standard deck of playing cards.*

5. **HEALTH** Adults living in Park Village have a 2 in 7 chance of catching a cold each day during the month of December. If Ms. Simpson, a resident, is chosen at random, what is the probability that she will catch a cold in December? *Use random numbers.*

METEOROLOGY The probability of rain in New City during the first week in April is 60% each day. For her trip to New City during this time period, Mary Jo plans for 3 days of rain.

6. Explain how you could use number cubes to simulate the conditions of the problem.

7. Perform 20 trials using number cubes to find the probability of exactly 3 days of rain.

8. Explain how you could use random numbers to simulate the conditions of the problem.

9. Using random numbers, perform 20 trials to find the probability of exactly 3 days of rain.

10. Calculate the theoretical probability and compare the results of your simulations to it.

PRACTICE

1. **WRITING MATHEMATICS** To simulate a problem situation with two outcomes, Magee said you could only use a device with two outcomes, such as a coin. Maury said you could also use a device that has more than two outcomes, such as a number cube. With whom do you agree? Explain.

In Exercises 2–4, describe a model for the problem. Perform 25 trials of the simulation, and find the indicated probability. Find the theoretical probability and compare to the simulation.

2. **SAFETY** A glass has a 3 in 4 chance of breaking if dropped onto a wooden floor. If 4 glasses are dropped, calculate the probability that just one of them breaks.

3. **EDUCATION** At a certain university, a freshman has a 4 in 5 chance of returning for the sophomore year. If 5 freshmen are polled, find the probability exactly 4 will be returning.

4. **PHARMACEUTICALS** An ointment is found to work 5 times out of 6 in curing a certain skin condition. If 20 people require treatment, find the probability exactly 14 of them are cured.

In Exercises 5–7, use random numbers to model the problem and calculate the indicated probability. Calculate the theoretical probability and compare the results of your simulation to it.

5. EDUCATION At a certain college, only 40% of the freshmen will pursue, as sophomores, the major they declared at college entry. If 20 sophomores are polled at random, calculate the probability that exactly 10 of them will be pursuing their declared majors.

6. QUALITY CONTROL Of appliances that are used as floor models in Thrifty Mart, there is a 1 in 10 chance that the appliance is defective. If 10 floor models are tested at random, calculate the probability that exactly 2 of them are defective.

7. SPORTS Tom, a football quarterback, has a 60% pass completion rate. If he throws 30 passes, calculate the probability that he will complete 20 of them.

TECHNOLOGY Jon ran a computer program to simulate the rolling of two number cubes. He noted the following frequencies for the sums indicated.

Sum of Number Cubes	2	3	4	5	6	7	8	9	10	11	12
Frequency	906	1824	2664	3630	4524	5420	4518	3644	2660	1798	888

8. Describe some patterns you observe.

9. Use the sample space you itemized in Explore of Lesson 14.4 to calculate the theoretical probabilities related to the observations you made from the table above.

MERCHANDISING Each box of HITECH Cereal contains a single card featuring an athlete. There are 36 different cards in the set, randomly distributed in equal quantities.

10. Describe a simulation to estimate the probability that a person will have at least 34 of the cards after purchasing 100 boxes of the cereal.

11. Perform 20 trials of your simulation to estimate the probability.

ASSESSMENT Consider a true-false test of 10 items.

12. Describe a simulation to estimate the probability of guessing exactly 7 answers correctly.

13. Explain how to find the theoretical probability of guessing exactly 7 answers correctly.

AIR TRAVEL An airline books 96 passengers on a flight that can hold only 85 passengers.

14. If there is only 1 chance in 6 that a booked passenger will not show, describe a simulation to estimate the probability that more than 85 of the 96 passengers will show.

15. In notation form, show how to calculate the theoretical probability that the airline has "overbooked," causing a seat shortage.

16. Repeat Leclerc's method to calculate π. On a sheet of paper, draw a number of parallel lines d units apart. Drop a needle or pin of length l where $l < d$ onto the paper. The probability P that the needle touches a line is related to π by $P = 1 - \dfrac{2d}{\pi l}$. Perform the experiment 10 times and solve for π. Perform the experiment 50 and 100 times and solve for π. Does your experimental value approach the true value of π?

MONTE CARLO You can use Monte Carlo methods to approximate the area under certain curves.

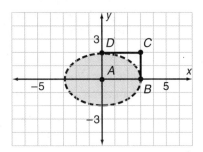

To approximate the area of the region enclosed by the ellipse $\dfrac{x^2}{9} + \dfrac{y^2}{4} = 1$, choose a point at random in the region enclosed by the rectangle $ABCD$. The probability that this point also lies in the first quadrant region enclosed by the ellipse is

$$\dfrac{\text{Area of quarter of ellipse}}{\text{Area of rectangle } ABCD}$$

You can find this probability experimentally by generating the coordinates of R random points in the region enclosed by the rectangle $ABCD$ and counting the number of points N that also lie in the interior of the ellipse. The probability that a point lies in the region enclosed by the ellipse is $\dfrac{N}{R}$. The area of $ABCD$ is 6. Therefore

$$\dfrac{\text{Area of quarter of ellipse}}{\text{Area of rectangle } ABCD} = \dfrac{N}{R}$$

$$\text{Area of quarter of ellipse} = 6\left(\dfrac{N}{R}\right)$$

To get area of the whole ellipse, multiply by 4.

$$\text{Area of ellipse} = 24\left(\dfrac{N}{R}\right)$$

You can use the BASIC program at the right to randomly generate the coordinates of R points in rectangle $ABCD$. Test the coordinates in the inequality that defines the area enclosed by the ellipse, and count the number of points N that satisfy the inequality.

```
Program: MONTECARLO
Disp "HOW MANY POINTS?"
Input R
0→N
For (P, 1, R)
3 * rand → X
2 * rand → Y
If X²/9 + Y²/4 < 1
Then
N + 1 → N
End
End
24 * N/R → A
Disp "AREA IS", A
```

Run the program several times and find the average of the areas generated. Sample results for $R = 100$ (that is, the program tested 100 points chosen at random) are 18.96, 18.24, 18.96, 17.52, 20.16, with an average of 18.77.

The theoretical area can be formed by the given formula where $a = x$-intercept and $b = y$-intercept.

$$A_{\text{Ellipse}} = \pi ab = \pi(3)(2) = 6\pi \approx 18.85 \text{ square units}$$

For the area of each of the regions indicated in Exercises 17 and 18, describe a simulation to calculate the area. Carry out your simulation 5 times and take the average of the areas obtained. Find a theoretical approximation of the area using the indicated formula.

17. the area enclosed by the circle $x^2 + y^2 = 9$
   Formula: Area of circle $= \pi r^2$

18. the area enclosed by the parabola $y = -x^2 + 4$ and the x-axis
   Formula: $A = \dfrac{h}{3}(y_0 + 4y_1 + y_2)$ where 3 points on the parabola are $P(-h, y_0)$, $Q(0, y_1)$, and $R(h, y_2)$

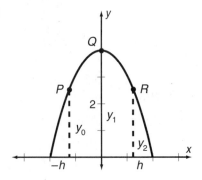

THINK CRITICALLY

COMPUTER PROGRAMMING The computer language *TURING* is named after a British mathematician, Alan Turing, who was a key figure in solving the "Enigma Code" used by the Germans during World War II.

The program at the right, written in TURING, rolls a number cube 20 times, and keeps track of the frequency of the results. After examining the program in detail to determine what each command does, tell the changes necessary to do the following.

19. Roll the number cube 100 times.

20. Print only the number of odd-numbered rolls.

21. Simulate the tossing of a coin.

22. Simulate drawing a card from one suit of a standard deck of cards.

```
% the following are counters to count the frequency of each roll
var count1, count2, count3, count4, count5, count6: int : = 0
var cube : int
randomize
for i : 1 . . 20
        randint (cube, 1, 6)
        if cube = 1 then
          count1 : = count1 + 1% increments the counter of '1' by 1
        elsif cube = 2 then
          count2 : = count2 + 1
        elsif cube = 3 then
          count3 : = count3 + 1
        elsif cube = 4 then
          count4 : = count4 + 1
        elsif cube = 5 then
          count5 : = count5 + 1
        elsif cube = 6 then
          count6 : = count6 + 1
end for
% output the results of the rolls
        put " # of 1's is ," count1
        put " # of 2's is ," count2
        put " # of 3's is ," count3
        put " # of 4's is ," count4
        put " # of 5's is ," count5
        put " # of 6's is ," count6
```

MIXED REVIEW

Simplify and combine.

23. $5\sqrt{27} - \sqrt{108} - \sqrt{75}$

24. $\sqrt[3]{192} - \sqrt[3]{0.375}$

25. STANDARDIZED TESTS The value of $\sqrt{-4} \cdot \sqrt{-25}$ is

    A. 10   **B.** -10 **C.** ±10 **D.** $10i$

PROJECT *Connection* For this activity, you will need the name and address of a local insurance agent. Perhaps your family uses such an agent.

1. Contact the insurance agent and set up an office appointment.

2. Take your scrapbook to the office visit. Ask the insurance agent to give you an estimate of the annual premium to insure your items under a home insurance policy.

3. Find out the details of the policy, including coverage limits, deductibles, and payment schedule. Also what constitutes proof of ownership if something is lost in a fire or theft.

4. Prepare a report on the information you learn from the agent. Make a list of key terms and definitions. Indicate any mathematical relationships you observe, such as whether the ratio of premium to amount of coverage is constant or not.

Problem Solving File

Polls and Sampling

SPOTLIGHT ON LEARNING

WHAT? In this lesson you will learn

- about selecting representative samples.
- how the mean and standard deviation are affected by sampling.

WHY? Polls and sampling can help you solve problems about campus services and agriculture.

Decision makers often want to know the opinions of the public. Since it is not practical to ask every person, leaders may rely on information-gathering surveys. These **polls** are conducted by interviewing **sample groups** of people who are representative of a general population. Sample groups must be selected carefully to avoid **biased** results that are not representative of a population.

Problem

A group of parents who propose offering after-school SAT preparation want to determine the willingness of people in the neighborhood to contribute time and money to the project. How can they select a sample of 20 families to determine if there is sufficient interest?

Explore The Problem

1. You would expect some people to be more interested in the proposal than others. What is the influential factor in this case?

2. When the data is influenced by some factor, it may be appropriate to have the sample reflect the makeup of the population with regard to the influential factor. Such a sample is called a **stratified sample**. If 30% of the families in the neighborhood have high-school age children (SAT candidates), describe an appropriate breakdown for the stratified sampling of 20 families.

3. Using the numbers appropriate to a stratified sample, describe a procedure for selecting the 20 families for the sample.

4. It may be appropriate to limit the sample to a particular segment of the population. This type of sample is called a **clustered sample**. Discuss using a clustered sample for the SAT proposal.

Investigate Further

After a method of sampling has been selected and data has been collected, it is important to organize, display, and analyze the data appropriately and communicate the results.

5. Suppose, in the neighborhood under discussion, there are 500 families with high-school age children and the 20 families of the sample are to be randomly selected from this population. Discuss the likelihood that the children of these families will want SAT preparation.

If samples of the same size are taken from a population, the set of means of the samples is itself a set of data.

6. Here are the math averages of 4 high school students selected at random from the families of the neighborhood: 75, 85, 80, 97.

   a. Write all possible ways of selecting 3 of these 4 values, and determine the mean for each of these samples of size 3.

   b. Find the mean of the sample means. Compare it to the mean of the population of 4 values. Make a general observation.

   c. Make a conjecture about the relative sizes of the standard deviation for this set of sample means and the standard deviation for the population of 4. Explain your reasoning.

   d. Use a calculator to determine the standard deviation for the set of the sample means and the standard deviation for the population of 4. Do the results verify your conjecture?

> PROBLEM SOLVING PLAN
>
> - Understand
> - Plan
> - Solve
> - Examine

> PROBLEM SOLVING TIP
>
> There are two different calculations for standard deviation.
>
> σ_x assumes the data to be population.
>
> S_x is based on the data as a sample.

Apply the Strategy

Should a sample or the entire population be studied? Explain.

7. At Manor High, the senior class president wants to know how many seniors are coming to the prom.

8. A pastry chef wants to know if the cookie dough is sweet enough.

WRITING MATHEMATICS Explain why the sample obtained may not be representative of the population.

9. A neighborhood newsletter wants to predict whether Eduardo Ruiz will be elected to City Council. A staff member asks the first 5 adults who walk by if they intend to vote for Mr. Ruiz.

10. An employment agency is interested in learning what percentage of women holding full-time jobs would pursue training for advanced skills. They conduct random telephone interviews of women using a telephone directory each morning for a week.

11. Every 20th appliance off the assembly line is inspected and the inspectors find nothing wrong. The dealers of these appliances, however, report that many appliances do have defects.

Describe a method for obtaining a random sample of 25.

12. There are 300 households in a neighborhood.

13. You have a large carton filled with different-colored candies of the same size. You want to estimate the percentage of red candies.

Describe how a device could be used to choose random samples from the given population.

14. a random sample of 6 people from a class of 36

15. a random sample of 4 weeks of the year

CAMPUS SERVICES A college is considering increasing its library facilities. The library committee considers choosing a random sample of 250 students from the population of 5000 students for suggestions for change. The student population is distributed as follows.

	Male	Female
Undergraduate	1600	1400
Graduate	1000	500
PhD	300	200

16. Suggest a way to improve a random sample of the entire population. Explain your choice.

17. To implement the method you suggested, how many of each type of student should be in the sample?

AGRICULTURE To draw some conclusions about the weights of red potatoes grown on his farm this season, Adam Gomez weighed 30 potatoes selected at random with the following results.

Weight (oz)	4.4	4.3	4.2	4.1	4.0	3.9	3.8	3.7	3.6	3.5
Frequency	1	0	1	3	6	8	5	4	1	1

18. Calculate the mean weight of the sample.

19. Comment on how representative you think this mean weight is with respect to all the red potatoes that Adam Gomez grew this season.

20. Use a calculator to determine the standard deviation. Explain the values the calculator displays for standard deviation.

21. Given the data set {12, 15, 16, 17}.

    a. Write all possible selections for a sample of size 2, and determine the mean for each sample.

    b. For the set of sample means determined in part a, calculate the mean and standard deviation.

    c. What is the mean and standard deviation of the data set?

    d. How do the statistics in parts b and c compare?

REVIEW PROBLEM SOLVING STRATEGIES

FAST CAR OR BURNT TOAST ?

1. A game show contestant is given 10 green disks, 10 red disks, and 2 boxes. The prize that is awarded depends on the color of the disk that the blindfolded contestant picks from one of the boxes. The prize for a green disk is a new sports car, while the prize for a red disk is a toaster oven. What arrangement of the disks in the boxes will give the contestant the best chance for winning the car? Note that there are no restrictions on how many disks must be in each box, but all disks must be in one box or the other.

   a. Suppose the contestant places all 10 red disks in one box and all 10 green disks in the other. What is the probability of winning the car?

   b. Suppose the contestant places 5 red and 5 green disks in each box. What is the probability of winning the car?

   c. Suppose the contestant places 5 red in one box and 5 red and 10 green in the other box. What is the probability of winning the car?

   d. Determine the best arrangement. What is the probability of winning the car with this arrangement?

SUGAR, SUGAR

2. Eileen has a balance scale and only two weights: one is 1 ounce and the other is 4 ounces. What is the minimum number of weighings necessary if Eileen wants to divide 180 ounces of sugar into two batches of 40 and 140 ounces? Explain the procedure Eileen should use.

GIFT BASKETS

3. A boutique owner has some leftover items:
   4 boxes of scented soap, each worth $11
   9 tubes of hand lotion, each worth $13
   5 bottles of perfume, each worth $15
   How can the owner arrange the items to make three gift baskets as nearly equal in value as possible? At least one of each item must be in each basket.

CHAPTER REVIEW

VOCABULARY

Choose the word from the list that completes each statement.

1. _____ involves two or more events.

2. _____ have probabilities whose sum is one.

3. Events that cannot occur together are _____.

4. A _____ has exactly two possible outcomes.

5. The width of a normal curve depends on the size of the _____.

a. complementary events

b. mutually exclusive

c. standard deviation

d. compound probability

e. binomial experiment

Lessons 14.1 and 14.2   PERMUTATIONS AND COMBINATIONS   pages 671–678, 679–685

- The **permutations** of n elements taken r at a time is $_nP_r = \dfrac{n!}{(n-r)!}$, where $0 \le r \le n$.

  The **combinations** of n elements taken r at a time is $_nC_r = \dfrac{n!}{r!(n-r)!}$, where $0 \le r \le n$.

6. In how many different ways can you arrange the letters of the word SQUARE?

7. In how many ways can you select a committee of 2 from a group of 6 people?

Lesson 14.3   COMPOUND PROBABILITY   pages 686–694

- For two events A and B in a sample space $P(A \text{ and } B) = \dfrac{n(A \cap B)}{n(S)}$.

- For two events A and B in a sample space $P(A \text{ or } B) = \dfrac{n(A \cup B)}{n(S)} = P(A) + P(B) - (A \cap B)$.

8. If 3 numbers are chosen from {positive integers ≤ 10}, what is the probability they are odd or greater than 6?

Lesson 14.4   INDEPENDENT AND DEPENDENT EVENTS   pages 695–702

- When one event does not effect the second, they are **independent events**. When one event does effect the second, they are **dependent events**.

- The **conditional probability** that B will occur if A has occurred is $P(B|A) = \dfrac{P(A \text{ and } B)}{P(A)}$.

If two cards are selected from a standard deck, find the P(both black) if the first card

9. is replaced

10. is not replaced

11. The first of two bags contains 4 red marbles and 4 black marbles. The second bag contains 2 red marbles and 6 black marbles. A black marble is drawn at random from one of the bags. Find the probability that the marble came from the first bag.

- A normal distribution with a mean of 0 and a **standard deviation** of 1 is called a **standard normal distribution**. The area between the curve and the x-axis is 1.

A standardized test had a mean score of 100 and a standard deviation of 8.

12. Approximately what percent of the scores were between 92 and 108?

13. Determine the probability that a score chosen at random will be between 116 and 124?

- The probability of r successes in a sample of n trials is $_nC_r \cdot P^r \cdot (1 - P)^{n-r}$ where p is the probability of success on each trial.

Zoe rolls a numbered cube 4 times. Find the probability that she gets 5

14. exactly 2 times   15. at least 2 times

16. The probability a single tornado detection device will detect a tornado is 0.9. If three devices are running simultaneously, what is the probability all three will fail to detect a tornado?

- You can calculate the **experimental probability** for real life problems by **simulating** the real situation using a variety of devices.

Describe a model that can be used to simulate the situation.

17. Ms. Roth's doctor prescribes a medicine that has a 75% cure rate. What is the probability that it will cure Ms. Roth?

Describe and perform a simulation to model the situation.

18. The probability that a student at Adams County High School is a freshman is 0.25. If 40 students are chosen at random, what is the probability that 10 of them are freshmen?

- You can use **stratified sampling** to establish a sample group that is representative of various factions within a population.
- The standard deviation for a sample is affected by the size of the sample and is different from the standard deviation for the general population.

Tell how a stratified sample of 50 people should be selected.

19. The target population of 1000 people divides into three percentages, 60%, 30%, and 10%, with respect to factions relevant to the issue.

Use the data {7, 10, 13, 18} to write all possible selections for a sample of size 2 and find the mean for each sample.

20. For the set of sample means, use a calculator to find the mean and standard deviation.

CHAPTER ASSESSMENT

CHAPTER TEST

Tell how many different outcomes are possible.

1. Six-letter sets containing 2 vowels ad 6 consonants are chosen at random from the English alphabet.

2. Six-letter sets are chosen from the word SYZYGY.

3. Seven people are seated in chairs arranged
   a. in a line   b. in a circle

4. Seven men and 7 women are seated in 14 chairs in a row so that women and men alternate.

5. **STANDARDIZED TESTS** If the probability that an event will occur is $\frac{m}{n}$, then the probability that the event will not occur is

   A. $-\frac{m}{n}$ **B.** $-\frac{n}{m}$ **C.** $\frac{1-m}{n}$ **D.** $\frac{n-m}{n}$

6. **WRITING MATHEMATICS** Write a paragraph to explain the difference among events that are independent, that are dependent, and that are mutually exclusive. Give an example of each type using a standard deck of cards.

In Questions 7–10, a jar contains exactly 10 red marbles, 8 white marbles, and 12 blue marbles. A marble is removed at random from the jar. The choice is identified, the marble is returned to the jar, and a second marble is drawn. Find the probability

7. both marbles are red

8. neither marble is red

9. the first marble is red and the second is blue

10. If the first marble is not returned before the second draw, what is the probability that both are red?

11. A number cube is rolled three times. Find the probability a 4 is rolled third, given the first and second rolls were 6 and 5, respectively.

12. If two boys and two girls line up at random and a girl is on an end, determine the probability that the girls are separated.

13. **STANDARDIZED TESTS** In a family of 6 children, the probability of exactly one male child is

    A. $\frac{58}{64}$ **B.** $\frac{32}{64}$ **C.** $\frac{7}{64}$ **D.** $\frac{6}{64}$

Find each probability.

14. If Joanne guesses the answers to 5 true/false questions, what is the probability that she will get exactly one wrong answer?

15. A letter is selected at random from the word MATH and replaced before another letter is chosen. What is the probability of picking exactly 2 A's in 3 attempts?

16. In Storm City, the probability of rain on any given day is $\frac{2}{3}$. What is the probability of at most one day of rain during the next 3 days?

The spinner shows circle O divided into 4 regions with $\overline{ROS}$ a diameter. In 3 spins, find the probability that the spinner will stop

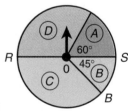

17. in Region A exactly twice

18. in Region D at least twice

19. If the mean score of 30 and the standard deviation is 3.1, which of the following scores could be expected to occur most frequently?

    A. 24.1   **B.** 26.7   **C.** 33.0   **D.** 36.2

PERFORMANCE ASSESSMENT

PRIZE SIMULATION A television game show requires the two contestants who do not win the grand prize run through a maze to determine their parting gifts. Today's maze is pictured at the right. At point A, the probabilities of choosing each of the three paths are equal. At points B, C, and D, the probabilities of choosing each of the two paths are equal. Determine the theoretical probabilities that contestants receive each prize. Explain how you can simulate this situation using a die and a coin. Perform the simulation 12 times, recording the prize each time. Compare with the theoretical probabilities.

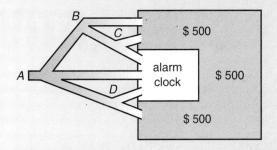

DO YOU SHARE YOUR BIRTHDAY? Work in groups of three. Determine the smallest number of people for which the probability at least two have the same birthday is greater than 50%. Explain how you can use the complement to solve the problem. Have the first group member calculate the probability two people will have different birthdays, the second group member calculate the probability three people will have different birthdays, and so on until one member finds the number of people for which the probability they have different birthdays is under 50%.

POLL THE PUPILS Select an issue that is relevant to the students in your high school. Discuss whether or not the students in your mathematics class are a representative sample of the high school population. Should you use a stratified sample? a clustered sample? Design a method to obtain a random sample that is representative of the high school population. Collect and analyze the data and then present your conclusions to your class.

PERMUTATION OR COMBINATION? Explain how to determine if a situation describes a permutation or a combination. Provide a problem in which a permutation is involved and another in which a combination is involved. Demonstrate how to solve both problems.

A NORMAL DEMONSTRATION Use an overhead projector to demonstrate each of the following:
a. three normal curves that have the same standard deviation and different means.
b. three normal curves that have the same mean and different standard deviations.
c. the standard normal curve.

PROJECT ASSESSMENT

Each group will make an oral/visual presentation of the information gathered about insurance.

1. As each group presents, other students should write notes and questions. Be prepared to compare coverage offered by different insurers. Think of reasons to explain differences.

2. Discuss how probability enters into both the individual's decision about purchasing insurance and the insurance company's premium schedule. When might a company refuse coverage?

3. Work with your teacher to arrange a class talk by a police officer or have a group of students interview an officer at the police station. Prepare a list of questions to help you determine the frequency of different types of thefts, accidents, fires, and so on.

CUMULATIVE REVIEW

1. In how many ways can 7 children be arranged on the 7 horses of the merry-go-round at Farrington Fair?

2. Determine the integral lower bound for the zeros of the function $f(x) = x^4 - 2x^3 + x^2 - 2$.

3. **STANDARDIZED TESTS** Which of the following statements are true about the function $y = 3^{-x}$?

   I. The graph is asymptotic to the x-axis.
   II. It is the inverse of $y = \log_3 x$.
   III. Its graph is the reflection over the y-axis of the graph of $y = 3^x$.

   A. I and II   **B.** I and III
   C. II and III   **D.** I, II, and III

4. **WRITING MATHEMATICS** Explain how you can use number cubes to simulate the following problem. The chance of rain in Kings Township over the next 5 days is 40%. What is the probability that it will rain on exactly 2 of the days?

5. Write an equation of the parabola whose focus is $(4, 1)$ and whose directrix is $y = -3$.

6. Determine the descriptive polynomial for the data set shown in the table below and find the value of y when x is 20.

x	−3	−2	−1	0	1	2	3
y	50	30	16	8	6	10	20

7. Ms. Korman administered a standardized test to the 25 students in her class. The mean score was 78 and the standard deviation was 5. About how many of her students received a score between 73 and 83?

8. Solve the system $\begin{cases} y = -x^2 \\ \dfrac{y+1}{x+1} = \dfrac{1}{2} \end{cases}$

9. Bella planned a trip from Walla Walla, WA to Pensacola, FL, a distance of about 3240 miles. She found out that travel by train would take 66 hours more than travel by plane. If the plane travels 12 times the rate of the train on average, what is the average rate of the train?

10. Rosa has 11 coins in her pocket: 4 quarters, 5 nickels, and 2 pennies. If she withdraws 2 coins at random, what is the probability that they are both of the same kind?

11. Evaluate: $\displaystyle\sum_{n=1}^{5} [(n+1)^3 - n^3]$

12. **STANDARDIZED TESTS** For all values of x for which the expressions are defined, $\dfrac{x^2}{x-3} + \dfrac{9}{3-x}$ is equivalent to

   A. -1 **B.** $x - 3$
   C. $\dfrac{1}{3-x}$ **D.** $x + 3$

13. The weights of four astronaut trainees are 145, 149, 168, and 154. Write all possible ways of selecting 3 of these 4 values. Determine the mean for each of these samples of size 3, and determine the mean of the sample means.

14. Write the 5th term in the sequence whose recursive formula is shown below.
   $$\begin{cases} a_1 = 6 \\ a_n = \dfrac{2}{3}a_{n-1} \end{cases}$$

15. **WRITING MATHEMATICS** Suppose that you know the coordinates of the vertices of a quadrilateral. Explain two different ways to determine if the quadrilateral is a parallelogram.

Find the remainder of the indicated division.

16. $(x^2 - x - 15) \div (x - 3)$

17. $(x^3 - 1) \div (x + 1)$

STANDARDIZED TEST

STUDENT PRODUCED ANSWERS Solve each question and on the answer grid write your answer at the top and fill in the ovals.

Notes: Mixed numbers such as $1\frac{1}{2}$ must be gridded as 1.5 or 3/2. Grid only one answer per question. If your answer is a decimal, enter the most accurate value the grid will accommodate.

1. Evaluate $i^3 \cdot i^7$ where i is the imaginary unit.

2. How many values are excluded from the domain of the rational function shown below?
$$y = \frac{\frac{x-1}{x+2}}{x}$$

Evaluate each expression.

3. $\dfrac{8!}{2! \cdot 5!}$ 4. $_8P_3 + {}_4P_0$ 5. $_5C_2 + {}_{10}C_3$

6. Find the sum of the sum and product of the solutions of the equation $3x^2 - 4x - 4 = 0$.

7. Find the length of the major axis of the ellipse $16x^2 + y^2 = 64$.

8. How many points are in the solution of the system $\begin{cases} (x-5)^2 + y^2 = 25 \\ y - x = 5 \end{cases}$?

9. Solve: $\dfrac{2x}{x+5} + \dfrac{1}{x-5} = \dfrac{10}{x^2 - 25}$

10. A silver dollar, a half-dollar, a quarter, a dime, and a nickel are in a box. Sophie draws one coin at random from the box and does not replace it before drawing a second coin. What is the probability that the value of the two coins is greater than one dollar?

11. Determine the coefficient of xy^3 in the expansion of $(3x + y)^4$.

12. Radium decays at a rate of 0.04% per year. Find the amount of radium that will be present after 50 years in a sample of material that initially contains 54 grams of radium.

13. Evaluate the expression $A^0 + A^{-\frac{1}{2}}$ when $A = 16$.

14. Solve for x: $1.95^x = 54$

15. The midpoints of the sides of a 3-4-5 right triangle are joined to form a smaller triangle, and then the midpoints of the smaller triangle are joined to form an even smaller triangle. If this process is continued without end, what is the sum of the perimeters of all the triangles formed?

16. Pat has a biased coin for which the probability of getting heads is 1:3. If Pat tosses the coin 5 times, what is the probability that he will get exactly 3 heads?

17. Determine the vertices of the hyperbola $9x^2 - 4y^2 - 144 = 0$.

18. What is the value of the largest rational solution of the equation $2x^5 - 11x^4 + 14x^3 - 2x^2 + 12x + 9 = 0$?

19. Simplify: $\dfrac{3 + 3i}{2 - 4i}$

20. Sheila Kaye wants to buy leather suits and suede suits for her chain of boutiques. Her cost per leather suit is $80 and her cost per suede suit is $160. She can buy between 20 and 90 leather suits, inclusive, and up to 100 suede suits. The manufacturer can supply anywhere between 51 and 120 suits, inclusive, but requires that the number of suede suits ordered be at least half the number of leather suits. What is Sheila's minimum feasible investment?

21. The altitude to the hypotenuse of a right triangle divides the hypotenuse into two segments, one of which is 4 cm longer than the other. If the length of the altitude is 6 cm, find the length of the hypotenuse to the nearest tenth of a centimeter.

15 Trigonometric Functions and Graphs

Take a Look AHEAD

Make notes about things that look new.

- The degree is a unit of angle measure. What is the name of the other unit of angle measure you will use in this chapter?

- You have worked with angles having measures from 0° to 360°. Copy several examples of angles that do not fall into this category. Explain what is different about each angle.

Make notes about things that look familiar.

- What is the Pythagorean theorem and what type of problems does it help you solve?

- Look at the graphs in Lesson 15.7. Which ways of transforming the graph of a function do you recognize? Give brief descriptions.

- Copy and identify some Greek letters used in this chapter.

DATA Activity

When Your Ship Comes In

Shipping is a multibillion-dollar industry that depends on the safe and timely arrival of cargo-laden vessels at their port destinations. Guiding a large ship through a busy port area requires navigational skill, experience, and expert knowledge of the waterway. Each ship enters and leaves port with a local harbor pilot aboard who knows every obstacle that could threaten the ship.

The table on the next page shows how busy some major U.S. ports are.

SKILL FOCUS

- Read and interpret a table.
- Add, subtract, multiply, and divide real numbers.
- Determine range of a data set.
- Use estimation.
- Solve percent problems.
- Use physical models.

In this chapter you will see how:

- **PILOTS** determine the bearings that describe the heading and course of aircraft. (Lesson 15.3, page 753)

- **NAVIGATORS** use trigonometry to calculate distances between locations on the earth. (Lesson 15.5, page 764)

Navigation

TOP 20 U.S. PORTS			
Rank	**Total Tonnage**	**Rank**	**Total Tonnage**
1 South Louisiana, LA	193,796,104	11 Norfolk Harbor, VA	45,543,792
2 Houston,TX	141,476,979	12 Lake Charles, LA	45,436,380
3 New York, NY and NJ	116,735,760	13 Tampa, FL	44,992,777
4 Valdez, AK	85,722,337	14 Pittsburgh, PA	44,490,094
5 Baton Rouge, LA	85,078,863	15 Mobile, AL	43,959.704
6 New Orleans, LA	67,037,285	16 Los Angeles, CA	43,622,807
7 Corpus Christi, TX	59,649,751	17 Philadelphia, PA	42,707,684
8 Long Beach, CA	54,320,932	18 Port Arthur, TX	38,326,902
9 Texas City, TX	53,652,781	19 Duluth-Superior, MN and WI	37,679,398
10 Plaquemine, LA	53,110,120	20 Baltimore, MD	37,170,223

Use the table to answer the following questions.

1. Find the range of tonnages for the top 20 ports.

2. Estimate the total tonnage handled by the ports of New York, Norfolk Harbor, Pittsburgh, Philadelphia, and Baltimore. Explain.

3. What percent of the top 20 ports are either in Louisiana or Texas?

4. If the list included the top 25 ports, 20% of them would be in Texas. How many Texas ports would be on the extended list?

5. Of the total tonnage handled at the port of Long Beach, 24,202,970 tons were domestic goods. To the nearest tenth, determine the percent of domestic goods.

6. **WORKING TOGETHER** Select information from the table and create a display to communicate the information. For example, you might use proportionally-sized cartons to show how total tonnage compares at three or four ports.

735

Using Navigation Instruments

In the Northern Hemisphere, all the stars appear to move in circles around a stationary star called *Polaris*. Early navigators used this star and an instrument called a *quadrant* to navigate the Atlantic and Pacific oceans.

PROJECT GOAL

To build and use two early navigational instruments to map a locale.

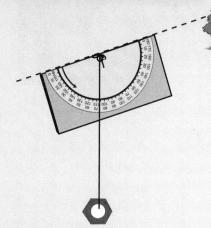

Getting Started

Work in groups of four students.

1. To build a quadrant you will need a photocopy of a protractor, a cardboard rectangle slightly larger than the protractor, a 50-cm length of string, a heavy washer, and glue.

2. Glue the photocopy of the protractor onto the cardboard as shown. Tie the washer to one end of the string. Pierce a hole in the cardboard at the center of the protractor and pass the free end of the string through the hole. Tie a knot in the string to prevent it from slipping from the hole.

3. Hold the quadrant at eye level with the string hanging down. Sight a reference point, such as the top of a flagpole, along the edge of the protractor and read the angle as shown.

4. Move around and use your quadrant to find other positions that have the same angle measurement to the reference point. What is the relationship of all points with this property?

PROJECT *Connections*

Lesson 15.1, page 742:
Use the quadrant to determine latitude.
Lesson 15.3, page 752:
Use the quadrant and right triangle trigonometry to determine height and distance.
Lesson 15.7, page 777:
Construct and use a sextant and law of sines to determine distance.
Chapter Assessment, page 785:
Summarize the project and use skills to map a locale using quadrant and sextant.

✉ **Internet Connection**

**www.swpco.com/
swpco/algebra2.html**

Explore

1. Find the measure of each angle on the drawing of Earth.

   a. ∠EON

   b. ∠EOW

   c. ∠EOS (> 180°)

2. Find the fractional part of the circumference of the circle intercepted by each of the above angles.

3. Give the circumference of the circle in terms of π if the radius is r.

4. Use the answers to Questions 2 and 3 to determine the length of the arc intercepted by each of the three angles.

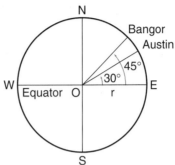

5. The radius of Earth is about 3960 mi. Determine the length in miles of each of the three arcs, in terms of π.

6. Describe how you can determine the length of an arc of a circle if you know the radius and the fraction of the circle represented by the arc.

7. Determine the distances to the nearest mile along the circle of longitude from Bangor, ME, and from Austin, TX, to the equator.

THINK BACK

Lines of longitude are perpendicular to the equator and pass through the North and South Poles.

Build Understanding

Until now, most of the angles you have encountered have been parts of geometric figures and have been measured in degrees. In **trigonometry** it is often convenient to position angles on the coordinate plane and to use a different system of measurement.

The angle illustrated is in **standard position** with its vertex at the origin, its **initial side** on the positive x-axis, and its **terminal side** in the plane. Because the angle is formed by the rotation of the terminal side about the vertex, it is called an **angle of rotation**. The **measure of an angle** is determined by the amount of rotation from the initial side to the terminal side.

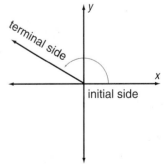

The notation m∠A means the measure of angle *A*, but is often shortened to ∠A. A Greek letter, such as θ (theta), is used to represent an angle and also to represent its measure. You should be able to tell from the context whether the angle or its measure is being discussed.

EXAMPLE 1

Find the quadrant in which the terminal side of each angle appears.

a. 333°

b. $\frac{1}{6}$ of a complete rotation

Solution

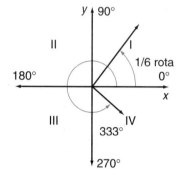

a. 333° is between 270° and 360° (or 0°), so the terminal side of a 333° angle is in Quadrant IV.

b. $\frac{1}{6} \cdot 360° = 60°$. The terminal side is in Quadrant I. ◀

It is sometimes useful to measure angles in radians instead of degrees. In a circle of radius *r*, one **radian** is the measure of a central angle θ whose sides intercept an **arc** the same length as the radius.

PROPERTIES OF EQUALITY

The radian measure of a central angle θ is the ratio where *s* is the length of the arc intercepted by the angle and *r* is the radius.

$$\theta = \frac{s}{r}$$

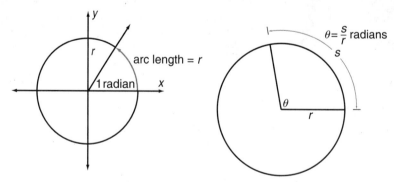

Since the circumference of a circle, the length of its complete arc, is $2\pi r$, there are $\frac{2\pi r}{r}$ or 2π radians in one 360° rotation of an angle.

$$2\pi \text{ radians} = 360° \qquad \pi \text{ radians} = 180°$$

$$\text{So, 1 radian} = \left(\frac{180}{\pi}\right)° \text{ and } 1° = \left(\frac{\pi}{180}\right) \text{ radians.}$$

EXAMPLE 2

Convert to radian measure: **a.** 45°   **b.** 210°

Convert to degree measure: **c.** $\dfrac{3\pi}{2}$ radian   **d.** 1 radian

Solution

a. 45 degrees $\cdot \left(\dfrac{\pi}{180}\right) \dfrac{\text{radians}}{\text{degree}} = \dfrac{45\pi}{180} = \dfrac{\pi}{4}$ radians

The answer can be written $\dfrac{\pi}{4}$ since, by convention, radian measure is implied when no unit of measure is specified.

b. $210\left(\dfrac{\pi}{180}\right) = \dfrac{7\pi}{6}$

c. $\dfrac{3\pi}{2}$ radians $\cdot \left(\dfrac{180}{\pi}\right) \dfrac{\text{degrees}}{\text{radian}} = \dfrac{3\pi \cdot 180}{2\pi} = 270°$

d. $1 \cdot \left(\dfrac{180}{\pi}\right) = \dfrac{180}{\pi} \approx 57.3°$

This shows that a 1 radian angle measures slightly less than 60°.   ◄

Multiply by r to get the expression $s = \theta r$ for **arc length** s on a circle.

EXAMPLE 3

NAVIGATION A freighter F is located at 40°N in the Atlantic Ocean. Determine the freighter's distance measured along the freighter's circle of longitude from

a. the equator E
b. the North Pole N
c. the South Pole S

Solution

a. $s = \theta r$

$s = 40 \cdot \dfrac{\pi}{180} \cdot 3960$ $\theta = 40, r = 3960$ mi

To the nearest tenth, $s = 2764.6$ mi.

b. $m\angle NOF = 90 - 40 = 50$ $\angle NOF$ is the complement of $\angle EOF$.

$s = 50 \cdot \dfrac{\pi}{180} \cdot 3960$

To the nearest tenth, $s = 3455.8$ mi.

c. $m\angle SOF = 180 - 50 = 130$ $\angle SOF$ is the supplement of $\angle NOF$.

$s = 140 \cdot \dfrac{\pi}{180} \cdot 3960 = 2860\pi$

To the nearest tenth, $s = 8985.0$ mi.   ◄

PROBLEM SOLVING TIP

Learn the radian measure of common angles.

$30° = \dfrac{\pi}{6}$, $45° = \dfrac{\pi}{4}$,

$60° = \dfrac{\pi}{3}$, $90° = \dfrac{\pi}{2}$,

$180° = \pi$,
$360° = 2\pi$. Since
$210° = 180° + 30°$,
you can solve
Example 2b by

adding $\pi + \dfrac{\pi}{6} = \dfrac{7\pi}{6}$.

THINK BACK

The sum of the measures of two *complementary* angles is 90° (or $\dfrac{\pi}{2}$ radians). The sum of the measures of two *supplementary* angles is 180° (or π radians).

1. Give the range of degree measures of an angle θ with its terminal side in Quadrant III.

Find the quadrant in which the terminal side of the angle appears.

2. 265°   **3.** $\frac{\pi}{5}$ **4.** $\frac{3}{8}$ of a complete rotation

Convert to radian measure.   **Convert to degree measure.**

5. 30°   **6.** 315°   **7.** 200°   **8.** $\frac{\pi}{3}$ **9.** $\frac{5\pi}{4}$ **10.** 0.7π

11. RECREATION A defective section of a bicycle tire is intercepted by spokes that create a central angle of $\frac{5\pi}{12}$. The radius of the tire is 13 in. Find the length of the defective section.

12. PHYSICS A pendulum swings through an angle of 30°, describing an arc 6.0 m long. Determine the length of the pendulum to the nearest tenth of a meter.

13. WRITING MATHEMATICS Do you think it is more convenient to measure angles in degrees or radians for applications involving arc length? Explain.

PRACTICE

1. Give the range of degree measures of an angle θ with its terminal side in quadrant IV.

Find the quadrant in which the terminal side of the angle appears.

2. 100°   **3.** $\frac{6\pi}{5}$ **4.** $\frac{5}{6}$ of a complete rotation

Convert to radian measure.

5. 90°   **6.** 150°   **7.** 330°   **8.** 60°

9. 225°   **10.** 72°   **11.** 340°   **12.** 6°

13. 54°   **14.** 255°   **15.** 81°   **16.** 2°

Convert to degree measure.

17. 2π **18.** $\frac{2\pi}{3}$ **19.** $\frac{3\pi}{2}$ **20.** $\frac{3\pi}{5}$

21. $\frac{\pi}{4}$ **22.** $\frac{7\pi}{6}$ **23.** $\frac{11\pi}{8}$ **24.** $\frac{5\pi}{9}$

25. $\frac{16\pi}{9}$ **26.** $\frac{\pi}{90}$ **27.** 0.95π **28.** 1.66π

Find the complement of the angle.

29. 48°   **30.** $\frac{\pi}{8}$ **31.** $\frac{4\pi}{9}$

Find the supplement of the angle.

32. 35.72°

33. $\dfrac{5\pi}{12}$

34. 0.87π

35. TIME Through how many radians does the minute hand of a clock turn from 2:30 P.M. until 2:55 P.M.? The minute hand is 6 in. long. How far does the tip travel from 2:30 until 2:55?

36. GEOMETRY The measures of the angles of a triangle are in the ratio 3:4:5. Write the measures in radians.

37. ASTRONOMY Assume that the Earth travels around the sun in a circular orbit. The radius of the orbit is 9.3×10^7 mi and the time needed to make one revolution is 365 days.

    a. Determine the angle, in radians, that the Earth moves through in 1 day.

    b. Determine the distance that the Earth travels in 1 day.

38. NAVIGATION A passenger liner at latitude 62.5° N is due south of a region of icebergs at latitude 62.7° N.

    a. How far is the liner from the iceberg region?

    b. Should the liner strike an iceberg, the nearest help is a coast guard vessel 20 mi south of the liner's present position. Determine the latitude of the coast guard ship.

39. WRITING MATHEMATICS Laura drew an angle measuring 3°. Carrie drew an angle measuring 3 radians. Why are they different?

EXTEND

For greater precision, degrees can be subdivided into minutes (′) and seconds (″).

    $1° = 60′$ $1′ = 60″$

Convert to decimal degrees.

40. 35° 15′

41. 80° 24′

42. 114° 18′ 48″

43. Convert 48° 46′ 12″ to decimal degrees and to radians.

44. Convert 1.49π to decimal degrees and to degrees, minutes, and seconds.

45. TRAVEL The tires of a truck each measure 30 in. in diameter. How many revolutions does each wheel make while the truck travels 1 mi?

46. ASTRONOMY The diameter of the moon is about 2160 mi. The angle intercepted by radii drawn from the center of the earth to the edges of the moon is 0.00872. How far is the moon from the center of the earth?

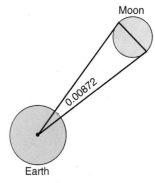

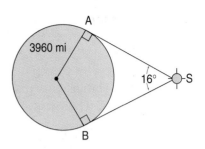

47. An angle has a measure of d degrees and r radians. Write an equation expressing the relationship between d and r.

48. NAVIGATION A marine navigation satellite can signal to any ship on arc AB, where $\overline{SA}$ and $\overline{SB}$ are tangents from the satellite to Earth. The effective signal angle is 16°. Determine arc length AB.

MIXED REVIEW

Determine the mean, median, and mode of each set of data.

49. 28, 19, 33, 40, 19

50. 1.6, 1.8, −1.9, −0.5, 0.2

51. $4\frac{1}{2}, 3\frac{3}{4}, 6\frac{1}{4}, 2\frac{1}{2}$

Find $f^{-1}(x)$.

52. $f(x) = 5x$

53. $f(x) = x - 6$

54. $f(x) = \frac{-x}{3}$

55. $f(x) = \frac{3x + 2}{4}$

The points $(-3, 4)$ and $(9, -1)$ are graphed on a coordinate plane.

56. Determine the distance between the points.

57. Determine the midpoint of the segment connecting the points.

58. STANDARDIZED TESTS Choose the letter giving the radian measure of a 240° angle.

    A. 240π **B.** $\frac{4\pi}{3}$ **C.** $\frac{2\pi}{3}$ **D.** $\frac{\pi}{120}$

PROJECT *Connection*   Use your quadrant to determine latitude.

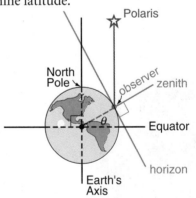

1. Look at the figure showing how the degree of latitude θ is related to the observed angle that *Polaris* makes with the horizon at that latitude. Assume the line from the observer to Polaris is parallel to Earth's axis.

2. Note below how to use the quadrant.

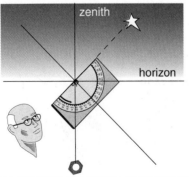

3. How is latitude related to the angle measurement on the quadrant?

4. Use a reference book to find out how to locate *Polaris*. Use a night sighting of *Polaris* to determine the latitude of your sighting. Verify your latitude observation by using an atlas.

Think Back

● The diagram shows an angle of rotation indicated by a curved arrow. Identify the following parts of the angle.

  1. the vertex

  2. the initial side

  3. the terminal side

  4. Identify the quadrant in which the terminal side is located.

The measure of θ is 30°. Determine the measure of the angle of rotation in the given units.

  5. degrees

  6. radians

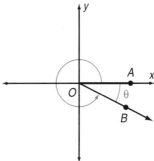

Explore

● The angles of rotation you have worked with so far have had measures ranging from 0° to 360° or 0 to 2π but other measures are possible.

7. The terminal side may move more than once in a *counterclockwise* direction through the four quadrants, adding 360° to its measure with each complete rotation. The resulting measure will be positive and greater than 360°. Determine the measure of the angle of rotation $\angle AOB$.

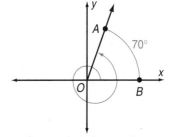

Counterclockwise rotation: angle measure is positive

8. The terminal side may move in a *clockwise* direction through part of a rotation or more than one rotation. The resulting measure will be negative. Determine m$\angle DOC$.

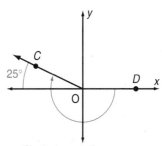

Clockwise rotation: angle measure is negative

Algebra Workshop

CHECK UNDERSTANDING

The terminal side of an angle makes two complete counterclockwise rotations plus 5°. What is the measure of the angle?

Determine the quadrant in which the terminal side of each angle lies.

9. 600°   **10.** $\frac{11\pi}{4}$ **11.** 400°   **12.** $-35°$

13. $-\frac{4\pi}{3}$ **14.** $-\frac{13\pi}{6}$ **15.** 4.2π **16.** $-1200°$

Angles of rotation in standard position with coinciding terminal sides are **coterminal**. You can use $\theta + n \cdot 360°$ where n is an integer and θ is measured in degrees, to find angles that are coterminal with the angle θ. Draw each angle to determine if the angles are coterminal.

17. 40°, 400°   **18.** $-90°$, 270°   **19.** $\frac{3\pi}{2}, \frac{7\pi}{2}$

20. 250°, 950°   **21.** $-1.6\pi, -7.6\pi$ **22.** $\frac{7\pi}{6}, -210°$

23. Use the drawing at the right to determine y and z if $x = 60°$.

For each angle, find a coterminal angle measuring between 0° and 360°.

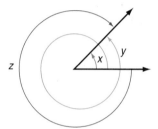

24. $-234°$ **25.** 638°

26. 1101°   **27.** $-977°$

28. 2351°   **29.** $-3601°$

PROBLEM SOLVING TIP

Learn the measures of angles coterminal with common angles: for example, 90° is coterminal with 450° and 810°, 180° is coterminal with 540° and 900°, and so on. Learn radian measures as well: π is coterminal with 3π and 5π, and $\frac{\pi}{2}$ is coterminal with $\frac{5\pi}{2}$ and $\frac{9\pi}{2}$.

Make Connections

30. Assume m $\angle A = 30°$. Write a sequence of numbers representing the measures of the five angles coterminal with $\angle A$ that have the least positive measures greater than 30°.

31. Describe the pattern in Question 30.

32. Assume m $\angle B = \frac{\pi}{2}$. Write a sequence of numbers representing the measures of the five angles coterminal with $\angle B$ that have the least positive measures greater than $\frac{\pi}{2}$. Express measures in radians.

33. Describe the pattern in Question 32.

34. Outline a method you could use to determine the measures of all the angles, positive and negative, that are coterminal with a given angle. Explain how you could determine both degree measures and radian measures of the coterminal angles.

35. Determine the measure of an angle coterminal with a 50° angle and with a measure between 100,000° and 100,500°.

36. Determine the measure of an angle coterminal with a $\frac{\pi}{4}$ angle and with a measure between 500π and 501π.

Summarize

37. WRITING MATHEMATICS Describe coterminal angles. Give examples to clarify your description.

38. THINKING CRITICALLY Determine an angle with measure between 0 and -2π that is coterminal with the given angle.

    a. $\frac{3\pi}{8}$ **b.** $\frac{17}{16}\pi$ **c.** n where $0 < n < 2\pi$

39. THINKING CRITICALLY Suppose the sum of the measures of two angles in standard position is 360°.

    a. Sketch two examples of the angle pairs.

    b. Describe the relationship between the terminal sides.

40. ASTRONOMY Earth completes an orbit around the Sun in the direction shown in 1 year. Saturn completes an orbit around the Sun in the same direction in 29.5 Earth years. Suppose that the three bodies are aligned as shown on January 1, 1996.

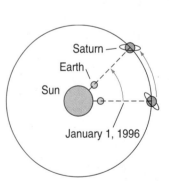

    a. When will Earth next be at an angle of rotation coterminal with its January 1, 1996 position? (Disregard leap years.)

    b. Explain why the Sun, Earth, and Saturn will not be aligned on the answer to Part a.

    c. Let t represent the number year after January 1, 1996. When $t = 1$, will the planets be realigned? Will they be realigned on the time interval $1 < t < 2$? Explain.

    d. Determine the measure of the angle of rotation through which Earth will move in t years and the angle through which Saturn will move in t years.

    e. Write an expression that represents the angle of rotation of the Earth for $1 < t < 2$.

    f. Use your results from Parts c, d, and e to write an equation whose solution represents the first time at which Earth and Saturn will realign. Solve your equation for t and provide the date of realignment.

The Trigonometric Functions

Explore

1. In equilateral $\triangle ABC$, $\overline{BD}$ bisects $\angle ABC$ to create $\triangle ABD$. Find the lengths of the sides of $\triangle ABD$.

2. Repeat Question 1 several times beginning with an equilateral triangle with sides of different lengths from those used before. Study the patterns in your results. Then describe the relationship between the lengths of the sides in any 30°-60°-90° triangle.

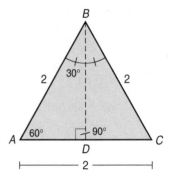

3. In $\triangle ABC$, $AB = 1$. Find the lengths of the sides of $\triangle ABC$.

4. Repeat Question 3 several times for $AB \neq 1$. Study the patterns in your results. Then describe the relationships among the lengths of the sides of any 45°-45°-90° triangle.

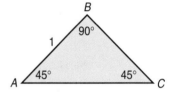

Build Understanding

In the diagram, θ is the measure of an angle in standard position. Let $P(x, y)$ be a point (not the origin) on the terminal side of the angle. Draw a perpendicular from P to the x-axis creating a **reference triangle** with hypotenuse $r = \sqrt{x^2 + y^2}$. Then x, y, and r can be used to define the six ratios called the **trigonometric functions**.

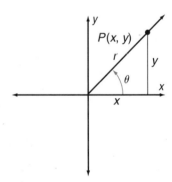

TRIGONOMETRIC FUNCTIONS

sine θ (sin θ) $= \dfrac{y}{r}$ cosecant θ (csc θ) $= \dfrac{r}{y}$

cosine θ (cos θ) $= \dfrac{x}{r}$ secant θ (sec θ) $= \dfrac{r}{x}$

tangent θ (tan θ) $= \dfrac{y}{x}$ cotangent θ (cot θ) $= \dfrac{x}{y}$

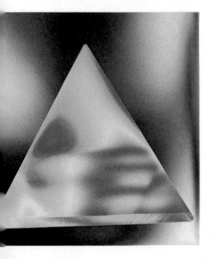

Since $\sin \theta \cdot \csc \theta = \dfrac{y}{r} \cdot \dfrac{r}{y} = 1$, the sine and cosecant are called **reciprocal functions.**

$$\sin \theta = \frac{1}{\csc \theta}$$

Similarly, the cosine and secant, and the tangent and cotangent are reciprocal functions.

$$\cos \theta = \frac{1}{\sec \theta} \qquad \tan \theta = \frac{1}{\cot \theta}$$

THINK BACK

Two numbers are *reciprocals* if their product is 1.

EXAMPLE 1

Determine the sine, cosine, and tangent for the angle θ with the given point on its terminal side.

a. $(3, 4)$ **b.** $(0, -2)$

Solution

a. $r = \sqrt{3^2 + 4^2} = 5$ $x = 3, y = 4$

$$\sin \theta = \frac{y}{r} = \frac{4}{5} \qquad \cos \theta = \frac{x}{r} = \frac{3}{5} \qquad \tan \theta = \frac{y}{x} = \frac{4}{3}$$

For $(3, 4)$, $\sin \theta = \dfrac{4}{5}$, $\cos \theta = \dfrac{3}{5}$, and $\tan \theta = \dfrac{4}{3}$.

b. $r = \sqrt{0^2 + (-2)^2} = 2$ $x = 0, y = -2$

$$\sin \theta = \frac{y}{r} = \frac{-2}{2} \qquad \cos \theta = \frac{x}{r} = \frac{0}{2} \qquad \tan \theta = \frac{y}{x} = \frac{-2}{0}$$

For $(0, -2)$, $\sin \theta = 1$, $\cos \theta = 0$, and $\tan \theta$ is undefined. ◄

Notice that if a different point on the terminal side of the angle is chosen, the values of the functions do not change. Suppose in Example 1 Part a the point $(6, 8)$ is chosen. Then $x = 6, y = 8$, and $r = 10$.

$$\sin \theta = \frac{8}{10} = \frac{4}{5} \qquad \cos \theta = \frac{6}{10} = \frac{3}{5} \qquad \tan \theta = \frac{8}{6} = \frac{4}{3}$$

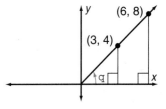

For a given angle there is exactly one value for each trigonometric ratio. So, each ordered pair $(\theta, \sin \theta)$, $(\theta, \cos \theta)$, $(\theta, \tan \theta)$, $(\theta, \csc \theta)$, $(\theta, \sec \theta)$, and $(\theta, \cot \theta)$ is unique. Therefore, the trigonometric ratios are functions that depend only on the measure of the angle.

An angle whose terminal side falls on an axis such as Example 1 Part b is a **quadrantal angle**. Some of the trigonometric functions are defined for all angles and some are undefined at certain angles.

THINK BACK

A *function* is a relation in which each element of the domain is paired with exactly one element of the range.

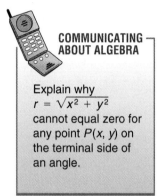

COMMUNICATING ABOUT ALGEBRA

Explain why
$r = \sqrt{x^2 + y^2}$
cannot equal zero for
any point $P(x, y)$ on
the terminal side of
an angle.

EXAMPLE 2

Examine the definitions of the six trigonometric functions and
determine the angles for which they are undefined.

Solution

Since division by zero is undefined,
the tangent and secant functions are
undefined when $x = 0$. This will
occur at angles of 90° and 270°. For
the same reason, cotangent and
cosecant will be undefined when
$y = 0$. This occurs at angles of 0°
and 180°. The sine and cosine
functions are defined everywhere
since $r = \sqrt{x^2 + y^2}$ cannot
equal zero.

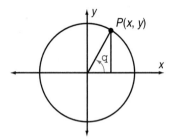

If you are given a point on the terminal side of an angle, you can
determine the trigonometric functions of the angle from the
definitions of the functions. If you are given the measure of the angle,
you may need to draw the reference triangle in order to evaluate the
functions. The two right triangles, the 30°−60°−90° triangle and the
45°−45°−90° triangle are useful for evaluating functions.

The measurements in the figure represent the proportions relating the
sides of any 30°−60°−90° triangle and any 45°−45°−90° triangle. Use
them to evaluate the trigonometric functions of any angles that are
multiples (positive or negative) of 30° or 45°.

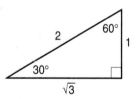

When you know the lengths of the sides of a reference triangle, you can
use these alternative definitions.

CHECK UNDERSTANDING

Compare the two
definitions of the
trigonometric
functions. Then
identify the side that
is "adjacent" to θ and
the side "opposite" θ
below.

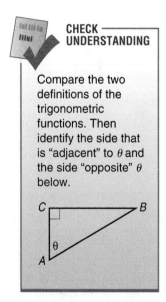

┌─ **TRIGONOMETRIC RATIOS** ─────────────

$\sin \theta = \dfrac{\text{opposite side}}{\text{hypotenuse}}$ $\csc \theta = \dfrac{\text{hypotenuse}}{\text{opposite side}}$

$\cos \theta = \dfrac{\text{adjacent side}}{\text{hypotenuse}}$ $\sec \theta = \dfrac{\text{hypotenuse}}{\text{adjacent side}}$

$\tan \theta = \dfrac{\text{opposite side}}{\text{adjacent side}}$ $\cot \theta = \dfrac{\text{adjacent side}}{\text{opposite side}}$

EXAMPLE 3

Evaluate the function.

a. $\sin \dfrac{5\pi}{4}$ **b.** $\sec(-930°)$ **c.** $\csc 90°$ **d.** $\tan 3.672$

Solution

a. The terminal side of an angle measuring $\dfrac{5\pi}{4}$ (225°) lies in Quadrant III. The **reference angle** is a 45° angle in a 45°−45°−90° triangle. The triangle proportions show that $(-1, -1)$ is on the terminal side of the angle. You can use the point to find the lengths of the sides of the triangle. To be consistent with the values of x and y, the "lengths" of the legs are negative in this example. The hypotenuse is always positive.

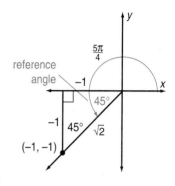

$$\sin \dfrac{5\pi}{4} = \dfrac{-1}{\sqrt{2}} \qquad \dfrac{\text{opposite side}}{\text{hypotenuse}}$$

$$= -\dfrac{\sqrt{2}}{2} \qquad \text{rationalize denominator}$$

b. An angle of −930° is coterminal with an angle of 150° with terminal side in Quadrant II. The reference angle is the 30° angle in a 30°−60°−90° reference triangle.

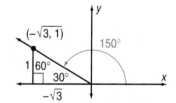

$$\sec(-930°) = \dfrac{2}{-\sqrt{3}} \qquad \dfrac{\text{hypotenuse}}{\text{adjacent side}}$$

$$= -\dfrac{2\sqrt{3}}{3} \qquad \text{rationalize denominator}$$

c. The y-axis is the terminal side of a 90° angle. No reference triangle can be drawn, but any point on the positive y-axis can be chosen to obtain values for x and y. Use $(0, 1)$ so $x = 0, y = 1, r = 1$

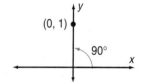

$$\csc 90° = \dfrac{1}{1} = 1 \qquad \dfrac{r}{y}$$

d. You can use a calculator to evaluate trigonometric functions. Be sure the calculator is in the correct mode, degrees or radians,

$$\tan 3.672 \approx 0.5856$$

When trigonometric values are given as ratios as in Example 3 parts a, b, and c, the value is called the **exact value**. In part d, the value is an approximate value. Trigonometry is used to model many real world applications.

EXAMPLE 4

GEOGRAPHY All circles of longitude have the same radius, but circles of latitude grow smaller as their distance from the equator increases. The formula $r = R \cos \theta$ gives the radius of the circle of latitude at latitude θ, where R is the equatorial radius of the earth. Determine the radius of the circle of latitude that passes through Minneapolis, Minnesota (45° N, 93° W). Use $R = 3960$ mi.

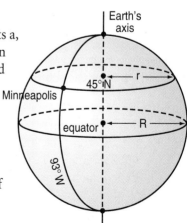

Solution

$\cos 45° \approx 0.7071$

$r = R \cos \theta \approx 3960(0.7071) \approx 2800$.

The radius is approximately 2800 mi. ◄

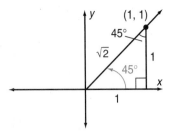

A **unit circle** is a circle centered at the origin whose radius is 1 unit long. The unit circle is useful in evaluating trigonometric functions.

 THINK BACK

Recall the side opposite the 30° angle is one-half the hypotenuse.

EXAMPLE 5

Find $\sin 210°$.

Solution

The triangle formed by the negative x-axis, the radius of the unit circle and the perpendicular from (x, y) to the x-axis forms a $30°-60°-90°$ triangle with a hypotenuse of 1.

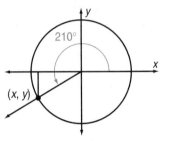

$$\sin \theta = \frac{-\frac{1}{2}}{1} = -0.5$$

So, $\sin 210° = -0.5$. ◄

TRY THESE

1. Determine the values of the six trigonometric functions for the angle of rotation with the point $(12, -5)$ on its terminal side.

Evaluate without using a calculator. Give answers in simplest form.

2. $\tan 30°$ 3. $\csc(-45°)$ 4. $\sec \frac{4\pi}{3}$ 5. $\sin(-1080°)$

Evaluate using a calculator. Round answers to the nearest ten-thousandth.

6. $\cos 73°$ 7. $\sin 1.55$ 8. $\cot 371.66°$ 9. $\csc \frac{7\pi}{9}$

Tell whether the value is positive or negative. Do not evaluate.

10. $\cot 211°$ 11. $\sec(-301°)$ 12. $\cot 2.6\pi$ 13. $\sin\left(-\frac{\pi}{9}\right)$

14. **NAVIGATION** The length of a nautical mile L at the equator is 6087 ft. Since the earth is slightly flattened, L changes, depending on the latitude. At latitude θ, $L = 6087 - 30 \cos 2\theta$. Without using a calculator, determine L in Rio de Janeiro, Brazil (latitude 22.5° S).

15. **WRITING MATHEMATICS** Describe how to find the sine of a 330° angle.

PRACTICE

Determine the values of the six trigonometric functions for the angle with the given point on its terminal side.

1. $(15, -8)$
2. $(-4, 3)$
3. $(1, 2)$
4. $(4, 0)$

Evaluate without using a calculator. Give answers in simplest form.

5. $\sin 225°$
6. $\cos 150°$
7. $\cot \dfrac{7\pi}{6}$
8. $\sin (-30°)$

9. $\tan (-60°)$
10. $\sec 405°$
11. $\csc (-480°)$
12. $\sin \dfrac{13\pi}{6}$

13. $\tan 300°$
14. $\sin 180°$
15. $\cos 0°$
16. $\sec (-630°)$

Evaluate using a calculator. Round answers to the nearest ten-thousandth.

17. $\cos 9°$
18. $\csc 275.4°$
19. $\tan -19.31$
20. $\sec \dfrac{11\pi}{14}$

21. OPTICS Light rays are bent as they travel from one medium, such as air, to another, such as water. The *index of refraction, n,* of a medium is given by $n = \dfrac{\sin i}{\sin r}$, where i is the incoming *angle of incidence* of the light ray and r is the outgoing *angle of refraction.* Measurements taken using a piece of crown glass gave an angle of incidence of 42.7° and an angle of refraction of 27.1°. Determine the index of refraction of crown glass.

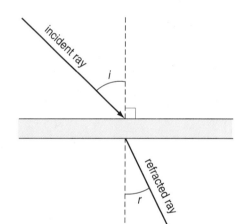

22. PHYSICS A projectile launched at a velocity of v ft/sec and an angle θ will reach a maximum height h, in feet, given by $h = \dfrac{(v \sin \theta)^2}{64}$. Determine the maximum height of a baseball that flies off a bat at an angle of 0.18π at a velocity of 150 ft/sec.

23. WRITING MATHEMATICS Explain why the cosine of any angle in quadrant III is negative.

EXTEND

24. The angle θ is in Quadrant I and $\sin \theta = \dfrac{2}{5}$. Determine $\cos \theta$ and $\tan \theta$.

25. The angle θ is in Quadrant IV and $\cot \theta = -\dfrac{5}{12}$. Find $\sec \theta$ and $\csc \theta$.

Let $0 \le \theta \le 360°$. Determine all values of θ for the given value.

26. $\tan \theta = \sqrt{3}$
27. $\sec \theta = -\sqrt{2}$
28. $\sin \theta = -1$

29. Use the definitions of $\sin \theta$ and $\cos \theta$ to show that $\dfrac{\sin \theta}{\cos \theta} = \tan \theta$.

THINK CRITICALLY

30. What is the maximum value of sin θ? Explain your reasoning.

31. If tan θ is positive, state the sign of the given function.

a. cot θ **b.** cos θ

32. GEOGRAPHY Circles of latitude are parallel to the equator. The
latitude of a circle of latitude is the angle θ that a segment
from the circle to the center of the earth makes with the plane
of the equator. Let d represent the distance from the center of
the plane of the equator to the center of the plane of circle of
latitude θ. Let R represent the radius of the earth. Show that
$r = R\cos \theta$, where r is the radius of the circle of latitude.

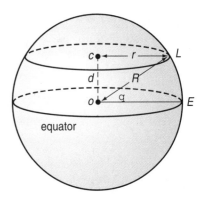

MIXED REVIEW

33. A 6-ft diameter bull's eye is painted on a 15-by-24 ft rectangular baseball scoreboard.
What is the probability that a home run that hits the scoreboard will land in the bull's
eye? Give the answer in terms of π.

Determine the slope of each line.

34. $y = -4x + 7$ **35.** $x = -3$ **36.** $y = 8$ **37.** $5x - 10y - 4 = 0$

38. the line containing $(0, -6)$ and $(4, 4)$ **39.** a line parallel to $x + y = 9$

40. STANDARDIZED TESTS The angle θ is in Quadrant II and sin $\theta = \frac{2}{5}$. Choose the letter
that gives cos θ.

A. $-\dfrac{\sqrt{21}}{5}$ **B.** $\dfrac{3}{5}$ **C.** $\dfrac{5}{2}$ **D.** $-\dfrac{2\sqrt{21}}{21}$

PROJECT *Connection*   Your quadrant can help you determine heights and distances.

1. Use a tape measure to carefully measure a convenient distance (such as 20.0 ft or
10.0 m) from the base of the reference point you previously chose.

2. The angle of elevation is the angle between the horizontal and the line of sight. At this
distance determine the angle of elevation of the reference point. Remember, the angle of
elevation is the complement of the angle read from the quadrant.

3. Use the angle of elevation and distance in Step 1
to determine the height of the reference point.
Repeat this step for several other distances.

4. Using the average height of the reference point
and elevation angles read from the quadrant,
determine the distance from the base of the
reference point of four other locations. Use the
tape measure to verify these distances.

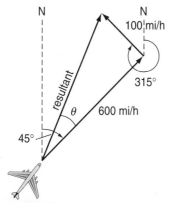

The *heading* of an airplane is the direction in which it is pointed. The *course* of the plane is its path over the ground. The course may be different from the heading because of the effect of wind. The ground speed of an airplane is the speed of the plane relative to the Earth. Both heading and course are given as *bearings* measured clockwise from true north. For example, a plane heading southwest has a bearing of 225°.

A **vector** is a quantity having both size and direction. Vectors can be used to determine the course and speed of a plane affected by wind and are represented by arrows. The plane vector in the figure shows that the plane is heading on a bearing of 45° at 600 mi/h. The wind vector shows a bearing of 315° and a speed of 100 mi/h. Note that the wind vector is drawn tail-to-head with the plane vector. The *resultant* vector represents the course of the plane.

Decision Making

1. Why is the triangle a right triangle?

2. Use the Pythagorean theorem to determine the resultant ground speed of the plane.

3. Determine the ratio of the side opposite θ to the side adjacent to θ in the diagram. Find the angle whose tangent is this ratio.

4. Suppose the bearing of the wind were 135° rather than 315°. How would the resultant vector be affected?

5. Estimate the bearing of each of the planes at the left.

6. A plane is heading due east at a speed of 500 mi/h. The wind has a speed of 90 mi/h and a bearing of 0°. Determine the plane's ground speed and course. Use your calculator.

7. The plane in Question 6 continued on its heading at 500 mi/h. The wind speed increased, still on a 0° bearing, setting the plane on a 72.3° course. Determine the speed of the wind and the ground speed of the plane.

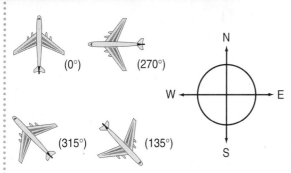

Problem Solving File

Right Triangle Trigonometry

When you solve real world problems involving trigonometry, draw a diagram and label all the known information. Write the equation of the angle(s) or side(s) of a right triangle that you need to determine using one of the trigonometric functions. You will find the sine, cosine, and tangent functions most helpful to use. Check the solutions to be certain they are reasonable.

To define the right triangle in some diagrams, an angle *subtended* at a point by a line segment is the angle created by lines drawn from the point to the endpoints of a given segment.

Problem

NAVIGATION A ship approaches the left end of New York's 0.807-mi long Verrazano-Narrows Bridge on a perpendicular. The navigator measures the angle subtended by the bridge as 10.3°. How far is the ship from the bridge?

Explore the Problem

1. Let *d* represent the distance from the ship to the bridge. In the figure, the angle and sides that you know or want to know are $\angle S$, $\overline{LR}$ and $\overline{LS}$. Use the word *opposite*, *adjacent*, or *hypotenuse* to describe the relationship of $\overline{LR}$ to $\angle S$; of $\overline{LS}$ to $\angle S$.

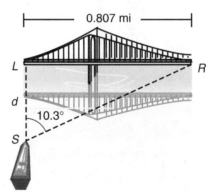

2. Consider the definitions of the sine, cosine, and tangent ratios. Which ratio uses the sides that you named above?

3. Write an equation involving 10.3°, 0.807 mi, *d*, and the trigonometric ratio you have named.

4. Solve the equation. To the nearest tenth of a mile, how far is the ship from the bridge?

Investigate Further

● For each right triangle, use the sine, cosine, or tangent function to write an equation using θ, 5, and x. Do not solve the equation.

5.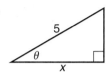

6.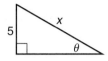

7.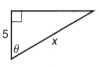

8.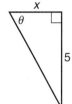

Sometimes information about a right triangle that is not given in a problem can be determined.

9. In $\triangle ABC$, $\angle C$ is a right angle and m$\angle A = 27.9°$.

   a. How are $\angle A$ and $\angle B$ related to one another?

   b. If you know the measure of one acute angle in a right triangle, how can you determine the measure of the other acute angle?

   c. What is m$\angle B$?

10. In $\triangle ABC$, $\angle C$ is a right angle, $\overline{AB} = 6.5$, and $\overline{AC} = 5.6$.

   a. If you know the lengths of two sides of a right triangle, how can you determine the length of the third side?

   b. Find $\overline{BC}$.

If you know the value of a trigonometric function of an angle, you can use the appropriate inverse function to determine the angle. Use the figure at the right for Questions 11 and 12.

11. Write sin A as a decimal.

12. Determine $\angle A$ and $\angle B$ to the nearest tenth of a degree.

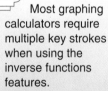

If you know the length of the hypotenuse and either the cosine or sine of an acute angle of a right triangle, use the definitions of the functions and multiply each side of the equation by r to get an equation you can use to find the lengths of the legs.

$$x = r \cos A \quad \text{and} \quad y = r \sin A$$

13. **a.** Write an equation to determine the height of the kite in the figure at the right.

   b. Determine the height of the kite to the nearest tenth of a foot.

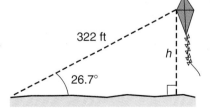

PROBLEM SOLVING PLAN

- Understand
- Plan
- Solve
- Examine

PROBLEM SOLVING TIP

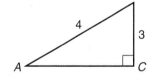

Most graphing calculators require multiple key strokes when using the inverse functions features.

The label sin^{-1} means the inverse function of sine, not $\frac{1}{\sin}$.

PROBLEM SOLVING TIP

Before evaluating sin or sin^{-1} of an angle, check if you are in the proper mode, degree or radian.

Apply the Strategy

Round answers to the nearest tenth unless instructed otherwise.

14. **SURVEYING** A surveyor is at a perpendicular distance of 162.25 ft from the left end of a parking lot. He measures the angle subtended by the lot as 47.3°. Determine the width of the lot.

15. **NAVIGATION** Angles of elevation and depression are measured from the horizontal. From a plane, the **angle of depression** of an airport control tower is 21.9°. The **angle of elevation** of the plane from the airport is ∠*PAG*. The straight-line distance from the plane to the tower is 8.6 mi. Find the height of the plane.

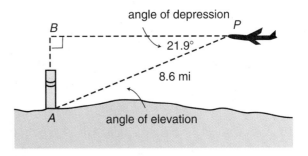

16. **MEASUREMENT** A 230-ft redwood tree casts a 398.4-ft shadow. Determine the angle of elevation of the sun.

17. **WILDLIFE MANAGEMENT** From point *A*, the angle of elevation of an eagle's nest on a cliffside is 60°. From point *B*, the angle of elevation is 30°. *B* is 300 ft from *A* on a perpendicular with the line joining *A* and *C*, the point beneath the nest. Determine *h*, the height of the nest. (Hint: Let $n = \overline{AC}$. Write two equations involving *h* and *n*.)

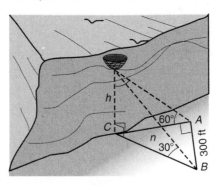

18. **WRITING MATHEMATICS** To *solve a right triangle*, you must determine the measures of all angles and sides. What is the least amount of information you need to solve a right triangle?

REVIEW PROBLEM SOLVING STRATEGIES

POINT TO POINT

1. Every dot on the grid represents a street corner in Squareville. Samantha is standing at the point S indicated by the star. Al, Bob, Charlie, and Dan are at points A, B, C and D respectively. Each is meeting Samantha at point S. How many blocks must each travel in order to get to the point where Samantha is waiting?

Here are some questions to help guide your thinking.

a. Draw line segments AB, CD, BC, and AD. Are triangles CBS and DAS similar? Why?

b. What is the ratio of the sides of triangle DAS to the corresponding sides of triangle CBS? How do you know?

c. Without measuring, find the distance in blocks to Samantha from Al, Bob, Charlie, and Dan.

1 block {

A CRAZY CUBE

2. Use each of the numbers 1, 2, 3, 4, 5, 6, 7, 8, 9, 10, 11 12, and 13 exactly once. One number will not be used. Fill the circles around the cube such that:

• the sum around the four edges that determines the face is 28, and

• the sum along the three edges that meet at any vertex is 21.

(Hint: you may want to write the numbers on paper circles so you can move them around.)
Compare your solution with others. What different strategies were used to solve the problem?

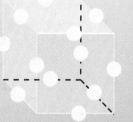

DOUBLE OR NOTHING

3. On Monday morning, Paul stopped at an ATM machine and withdrew enough money to double the amount he had when he left home. During the day, he spent a total of $36 and received no other cash. On Tuesday morning, Paul again stopped at the ATM and withdrew enough money to triple the amount he had remaining from the day before. During the day, he spent $36 and received no other cash. On Wednesday, he stopped at the ATM again, quadrupled the money that remained from the day before, and spent $36 during the day. On Wednesday evening, Paul discovered he didn't have any money on him at all! How much money did he have when he left home Monday morning?

Explore

- Descriptions of five figures are given below. Tell whether you could construct a unique triangle based on the description. If not, explain why not. Experimenting with a protractor and straws or other manipulatives may help you decide.

  1. a triangle with sides of length 3, 9, and 4

  2. a triangle with two sides each of length 5 and an included angle measuring 60°

  3. a triangle with two angles each measuring 45° and an included side of length 4

  4. a triangle with an angle of 30°, a side of length 5 adjacent to the 30° angle, and a side of length 3 opposite the 30° angle

  5. a triangle with angles measuring 50°, 60°, and 70°

Build Understanding

- The usefulness of trigonometry would be limited if it applied only to right triangles. Fortunately, trigonometry can also be applied to **oblique triangles**, triangles that do not contain right angles.

  $\triangle ABC$ is an oblique triangle with height h drawn to side AC. Let K represent the area of $\triangle ABC$. Then $K = \frac{1}{2} bh$.

  Note that $\sin A = \frac{h}{c}$, which can be rewritten as $h = c \sin A$. Substitute for h to obtain

  $$K = \frac{1}{2} bc \sin A$$

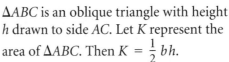

> **TRIANGLE AREA USING SINES**
>
> **The area of any triangle equals one-half the product of any two sides times the sine of the included angle.**

You can also derive area formulas by drawing heights to the other two sides of $\triangle ABC$ above.

$$K = \frac{1}{2} ac \sin B \quad \text{and} \quad K = \frac{1}{2} ab \sin C$$

SPOTLIGHT ON LEARNING

WHAT? In this lesson you will learn
- to determine the unknown sides and angles of oblique triangles using the laws of sines and cosines.
- to determine the area of a triangle using Heron's formula.

WHY? You can use the law of sines, the law of cosines, and Heron's formula to solve problems that involve surveying, advertising, navigation, forest management, and manufacturing.

Dividing the expressions for area by $\frac{1}{2}abc$ gives the laws of sines.

THE LAW OF SINES

Let $\triangle ABC$ be a triangle with sides measuring a, b, and c opposite the vertices A, B, and C, respectively. Then

$$\frac{\sin A}{a} = \frac{\sin B}{b} = \frac{\sin C}{c}$$

Solving a triangle means to determine the measures of all the unknown sides and angles.

EXAMPLE 1

In $\triangle ABC$, $\angle A = 40°$, $\angle B = 60°$, and $b = 100$.

  a. Solve the triangle.

  b. Determine the area of the triangle.

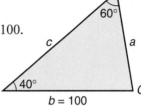

Solution

a.
$$\frac{\sin A}{a} = \frac{\sin B}{b} \qquad\qquad \frac{\sin C}{c} = \frac{\sin B}{b}$$

$$\frac{\sin 40°}{a} = \frac{\sin 60°}{100} \qquad\qquad \frac{\sin 80°}{c} = \frac{\sin\ 60°}{100}$$

$$a = \frac{100\sin 40°}{\sin 60°} \approx 74.2 \qquad c = \frac{100\sin 80°}{\sin 60°} \approx 113.7$$

$$\angle C = 180° - 40° - 60° = 80°$$

In $\triangle ABC$, $a \approx 74.2$, $c \approx 113.7$, and $\angle C = 80°$.

b. area $= \frac{1}{2}(74.2)(100)\sin 80° \qquad \frac{1}{2}ab\sin C$

$$\approx 3653.6$$

The area is approximately 3653.6 square units. ◀

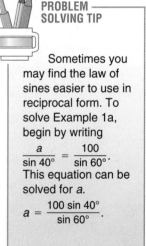

PROBLEM SOLVING TIP

Sometimes you may find the law of sines easier to use in reciprocal form. To solve Example 1a, begin by writing

$$\frac{a}{\sin 40°} = \frac{100}{\sin 60°}.$$

This equation can be solved for a.

$$a = \frac{100\sin 40°}{\sin 60°}.$$

The law of sines can be used to determine unknown parts of a triangle when you know two angles and a side of the triangle, as in Example 1, or when you know two sides and the angle opposite one of them. The latter, however, is known as the *ambiguous case* because it does not always determine a unique triangle.

Given $a = 6$, $c = 10$, and $\angle A = 30°$, two triangles can be constructed, as the figure shows. $\triangle ABC$ and $\triangle ABC'$ both contain sides of length 6 and 10 and a 30° angle not included between the sides.

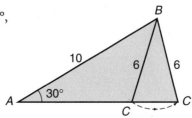

You should draw a quick sketch before using the law of sines. This will allow you to estimate how many triangles satisfy the given information, one or two or perhaps none. When you are given three sides of an oblique triangle, or two sides and the included angle, you cannot apply the law of sines. The following leads to a method you can use.

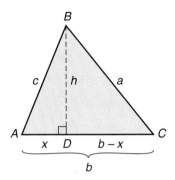

In $\triangle ABC$, let $x = \overline{AD}$. Then $\overline{DC} = b - x$.

By the Pythagorean theorem

$$\begin{aligned}
a^2 &= (b - x)^2 + h^2 \\
&= b^2 - 2bx + x^2 + h^2 && \text{Multiply } (b - x)^2 \\
&= b^2 - 2bx + c^2 && \text{Substitute } c^2 \text{ for } x^2 + h^2 \\
&= b^2 - 2b(c \cos A) + c^2 && \text{Substitute } c \cos A \text{ for } x \\
&= b^2 + c^2 - 2bc \cos A
\end{aligned}$$

Similar expressions can be derived for b^2 and c^2 as well.

THE LAW OF COSINES

Let $\triangle ABC$ be a triangle with sides measuring a, b, and c opposite the vertices A, B, and C, respectively. Then

$$a^2 = b^2 + c^2 - 2bc \cos A$$
$$b^2 = a^2 + c^2 - 2ac \cos B$$
$$c^2 = a^2 + b^2 - 2ab \cos C$$

The laws of cosines can be used to find the measure of unknown angles in a triangle.

EXAMPLE 2

In $\triangle ABC$, $a = 6$, $b = 7$, and $c = 8$. Determine the measures of the angles of the triangle.

Solution

By the law of cosines,

$$\begin{aligned}
a^2 &= b^2 + c^2 - 2bc \cos A \\
6^2 &= 7^2 + 8^2 - 2(7)(8) \cos A \\
36 &= 49 + 64 - 112 \cos A \\
-77 &= -112 \cos A \\
\frac{77}{112} &= \cos A && \cos A = 0.6875
\end{aligned}$$

Use a calculator to determine the angle to the nearest tenth of a degree.
$\angle A \approx 46.6°$.

To determine the measure of another angle, you can use either law.

$$\sin B = \frac{7 \sin 46.6°}{6} \approx 0.8477 \qquad \frac{\sin B}{b} = \frac{\sin A}{a}$$

$$\angle B \approx 58.0° \qquad\qquad\qquad \text{Calculate angle.}$$

Now you can use either law to determine $\angle C$. It is easier, however, to subtract the measures of the known angles from 180°

$$\angle C \approx 180° - 46.6° - 58.0° \approx 75.4°$$

Therefore, $\angle A \approx 46.6°$, $\angle B \approx 58.0°$, $\angle C \approx 75.4°$. ◄

CHECK UNDERSTANDING

How could you use the law of cosines to determine $\angle B$ in Example 2?

The law of cosines is used to prove Heron's formula for the area of a triangle, where s is the *semiperimeter*, one half the perimeter of the triangle.

┌─── **TRIANGLE AREA USING HERON'S FORMULA** ───────┐

The area A of a triangle with sides measuring a, b, and c is given by

$$A = \sqrt{s(s - a)(s - b)(s - c)}, \text{ where}$$

$$s = \frac{1}{2}(a + b + c).$$

└──┘

The laws of sines and Heron's formula can be used to solve real world problems.

ALGEBRA: WHO, WHERE, WHEN

While Heron of Alexandria made a few contributions to mathematical theory, his most significant insight concerned the path of a light ray. He deduced that light always travels between two points in the shortest possible distance, even when the path is interrupted by mirrors.

EXAMPLE 3

SURVEYING One of the principal tasks of surveyors is determining areas of plots of land. Having laid out baseline $AB = 114.6$ m, a surveyor measured the angle of point C from A and B at 49.7° and 56.1°, respectively. Determine $\overline{AC}$, $\overline{BC}$, and the area of $\triangle ABC$.

Solution

$$\frac{AC}{\sin 56.1°} = \frac{114.6}{\sin 74.2°} \qquad\qquad \frac{BC}{\sin 49.7°} = \frac{114.6}{\sin 74.2°}$$

$$AC \approx 98.9 \qquad\qquad\qquad\qquad BC \approx 90.8$$

$$\angle C = 180° - 49.7° - 56.1° = 74.2°$$

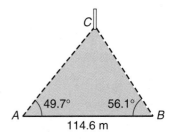

To use Heron's formula, first calculate the semiperimeter s.

$$s = \frac{1}{2}(114.6 + 98.9 + 90.8) = 152.2$$

$$\text{area} = \sqrt{s(s - a)(s - b)(s - c)}$$

$$= \sqrt{152.2(37.6)(53.3)(61.4)}$$

$$\approx 4327.6$$

The area of the triangle is approximately 4327.6 m². Check your answer using one of the sine formulas for the area of a triangle. ◄

CHECK UNDERSTANDING

Find the area of a right triangle with sides 5, 12, and 13 using $\frac{1}{2} bh$ and Heron's formula.

Solve △ABC. Round answers to the nearest tenth unless otherwise indicated.

1. $a = 12$, $\angle A = 36°$, $\angle B = 48°$ **2.** $a = 6.6$, $b = 5.4$, $c = 5.6$

3. In △ABC, $a = 3$ cm, $b = 4$ cm, and $\angle C = 115°$. Determine the area of the triangle.

4. ADVERTISING Triangle Communications uses a scalene triangle for a logo. The company estimates the cost of building the logo at \$8.50 per ft^2. If the logo will be 9 ft by 10 ft by 13 ft, what is the total cost?

5. MEASUREMENT From a point 60 yd from one end of a pond and 40 yd from the other, the pond subtends an angle of 60°. How long is the pond?

6. Use a calculator to verify the law of sines for a 45°−45°−90° triangle with legs of length 2.

7. WRITING MATHEMATICS Describe how you would decide whether to use $\frac{1}{2}bh$, an area formula with a sine, or Heron's formula to determine the area of a triangle.

PRACTICE

Solve △ABC. Round answers to the nearest tenth unless otherwise indicated.

1. $b = 8$, $\angle B = 125.2°$, $\angle C = 30°$ **2.** $\angle A = 60°$, $b = 12$, $c = 10$

3. $a = 14$, $b = 12.5$, $c = 20$ **4.** $\angle A = 30.7°$, $\angle B = 31.3°$, $c = 80$

5. $a = 8$, $c = 9$, $\angle B = 74.3°$ **6.** $a = 12$, $\angle A = 30°$, $\angle B = 120°$

7. $\angle A = 30°$, $a = 6$, $c = 8$ **8.** $a = 8$, $b = 15$, $c = 17$

Determine the area of △ABC.

9. $b = 12$, $c = 15$, $\angle A = 42°$ **10.** $a = 19.6$, $c = 55.5$, $\angle B = 104.0°$

11. NAVIGATION A coast guard cutter receives radar reflections from two lighthouses, 62.7 mi and 115.9 mi from the cutter, respectively. The angle from the cutter to the two lighthouses measures 125.3°. How far apart are the lighthouses?

12. FOREST MANAGEMENT Forest rangers at A and B, 6 miles apart, spot a fire at F. $\angle FAB = \frac{2\pi}{15}$ and $\angle FBA = \frac{4\pi}{9}$. How far is each ranger from the fire?

13. MANUFACTURING A company manufactures triangular shelf braces that measure 11.5 cm by 9.6 cm by 4.3 cm and cost \$0.0023/cm^2. Determine the cost of the material for producing 1000 braces.

14. MEASUREMENT The base of a cliff and the base of a lighthouse are both at sea level. An observer at the top of the cliff measures angles of depression of the top and the bottom of the lighthouse at 14.9° and 18.3° respectively. The lighthouse is 50 ft tall. How tall is the cliff?

15. **NAVIGATION** Airport B is 72 mi west of airport A. A plane 51 mi from A is picked up on radar at B, 24.8° north of due east. How far is the plane from B?

16. **WRITING MATHEMATICS** Summarize how you would decide whether first to apply the law of sines or the law of cosines to solve a triangle, and describe precautions you should take when planning your solution.

EXTEND

GEOMETRY Complete the table to discover what the five triangles listed have in common. Think about the best area formula to use in each case.

		Side 1	Side 2	Side 3	Area	Perimeter
17.	Triangle 1	6	8	10		
18.	Triangle 2	5	12	13		
19.	Triangle 3	6	25	29		
20.	Triangle 4	7	15	20		
21.	Triangle 5	9	10	17		

22. Describe the property shared by the five triangles in Exercises 17–21.

THINK CRITICALLY

23. By the triangle inequality, a triangle cannot have sides measuring 1, 2, and 10. Suppose you were to attempt to solve the triangle using the law of cosines. What would alert you to the fact that the triangle is impossible?

Determine the number of solutions for each triangle.

24. $\angle A = 60°, a = 10, c = 6\sqrt{3}$ 25. $\angle C = 60°, b = 12, c = 5\sqrt{3}$

MIXED REVIEW

26. Monthly payments on a 4-year $6000 loan are $151.57. Determine the cost of the loan.

Find the sum and product of the solutions to the equation.

27. $x^2 - 5x + 8 = 0$ 28. $5x^2 - 9x - 3 = 0$ 29. $-2x + 6 = 3x - 4x^2$

30. Write the equation for the parabola $y = x^2 - 6x + 8$ in standard form. Identify the vertex and any x-intercepts and y-intercepts.

31. **STANDARDIZED TESTS** In $\triangle ABC$, $b = 3$, $c = 2$, and $\angle A = 30°$. Choose the letter giving the length of a.
    A. $\sqrt{3}$ B. $\sqrt{7}$ C. $\sqrt{13 - 6\sqrt{3}}$ D. $\sqrt{13}$

Career
Navigator

Trigonometry owes its development as much to explorers and sea captains as to mathematicians. When long sea voyages began, navigation by plane trigonometry failed, necessitating the development of spherical trigonometry. The difference between the two was a simple matter of distance. Over short distances the curvature of the Earth can reasonably be approximated by straight lines.

Decision Making

1. The angle formed by segments from the Earth's center to the points of Columbus' departure from Spain and landing in the Bahamas in 1492 measures about 58°.

   a. Determine the arc distance and the straight-line distance between P and W.

   b. At his 57 mi/day average, how much quicker would Columbus have reached Watling's Island if the Earth had been flat than the 70.3 days it took him on a spherical Earth?

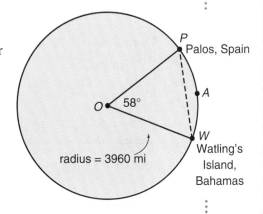

2. A navigator is likely to use plane trigonometry over short distances where the error will be small. The distance from S, St. Augustine, Florida, to D, Daytona Beach, Florida is 50 mi. After leaving Daytona and sailing 35 mi at an angle of 23° to a point B, how far is the sailboat from St. Augustine? (*Hint*: Draw a diagram.)

3. When the angle d at the Earth's center between two points A and B on the globe is not known, the angle can be determined using the formula

   $$\cos d = (\sin a \sin b) + (\cos a \cos b \cos L)$$

   where a is the latitude of A, b is the latitude of B, and L is the difference in longitudes between the points.

   a. Determine the central angle between San Francisco, California (38° N, 122° W) and Honolulu, Hawaii (21° N, 158° W).

   b. Determine the distance between the cities in miles.

15.6 Algebra Workshop

Graphs of the Trigonometric Functions

Think Back

1. Find the sine and cosine of each angle.

   a. $0°$ **b.** $90°$ **c.** π **d.** $\dfrac{3\pi}{2}$

2. Recall the definitions of the sine and cosine functions. How can you tell that each is a function?

3. Describe the domain and range of the sine and cosine functions. Explain your reasoning.

4. List three angles of rotation that are coterminal with $30°$. Give the sine and cosine of each angle.

5. Is it true that for each member of the range of the sine and cosine functions, there is one and only one member of the domain? Explain.

Explore/Working Together

Work with a partner. Use a graphing utility in degree mode.

6. Graph the function $y = \sin x$ on the intervals $0° \le x \le 720°$ and $-2 \le y \le 2$ with Xscl $= 90$ and Yscl $= 0.5$.

   a. What patterns do you observe?

   b. What are the maximum and minimum y-values?

   c. What is the horizontal width, in degrees, of the pattern that appears to repeat itself? Why should the pattern repeat?

   d. What are the x- and y- intercepts.

   e. Change the horizontal interval to $-720° \le x \le 720°$. Does what you see confirm your answers to Parts a–d? If not, how would you change the answer(s)?

7. Compare and contrast the graph of $y = \cos x$ with the graph of $y = \sin x$.

8. On the same coordinate axes, graph $y = A \sin x$ for several values of A, both positive and negative. Change your viewing window, if needed. What effect does A appear to have on the graph?

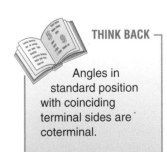

THINK BACK

Angles in standard position with coinciding terminal sides are coterminal.

ALGEBRA: WHO, WHERE, WHEN

The study of *spherical triangles*—triangles drawn on spheres, especially the sphere of the earth—led to the development of trigonometry, which means "triangle measurement." The most prominent founders of trigonometry, the Greek astronomers Hipparchus (ca 140 BC) and Ptolemy (ca 150 AD), applied their study to the sphere of the heavens in an effort to describe the positions and movements of the planets and the stars.

9. On the same coordinate axes, graph $y = A \cos x$ for several values of A both positive and negative. Does the value of A appear to have the same effect on the graph of the cosine function as it has on the graph of the sine function? If not, describe the differences.

You can also graph in terms of radians. Change your graphing utility to radian mode and for Questions 10-12, graph the given set of functions on the same coordinate axes. Use the intervals $0 \leq x \leq 2\pi$ and $-2 \leq y \leq 2$. Clear the screen after each question.

10. Graph $y_1 = \sin x$, $y_2 = \sin 2x$, and $y_3 = \sin 0.5x$. What effect does the coefficient of x appear to have on the graph? How can you predict the number of complete sine curves that appear in a 2π interval?

11. Graph $y_1 = \sin x$, $y_2 = \sin x + 1$, and $y_3 = \sin x - 1$. What effect do the added constants appear to have on the graph of $y = \sin x$?

12. Graph $y_1 = \sin x$, $y_2 = \sin\left(x - \dfrac{\pi}{4}\right)$, and $y_3 = \sin\left(x + \dfrac{\pi}{4}\right)$. What effect do the added constants appear to have on the graph of $y = \sin x$?

13. Explore the cosine function by graphing variations similar to the ones you used for the sine function in Questions 8 and 10-12. Compare your results with those you obtained for the sine functions.

14. Graph $y = \tan x$ on the intervals $0 \leq x \leq 2\pi$ and $-10 \leq y \leq 10$. Compare and contrast the graph with those of the sine and cosine functions.

15. Describe what happens to the graph of $y = \tan x$ at $\dfrac{\pi}{2}$ and $\dfrac{3\pi}{2}$. Why do you think this happens?

Graph each of the following functions on the same coordinate axes with the graph of $y = \tan x$. Describe how the way in which each equation has been altered affects the graph of $y = \tan x$.

16. $y = 2 \tan x$ 17. $y = \tan 2x$

18. $y = \tan\left(x - \dfrac{\pi}{4}\right)$ 19. $y = \tan x + 2$

20. The cotangent, cosecant, and secant functions do not appear on the keys of your graphing utility. Explain how you can graph each of these functions.

21. Where would you expect to find asymptotes for the graphs of the cotangent, cosecant, and secant functions? Give reasons.

Graph each of the following pairs of functions on the same coordinate axes. Use the intervals $0 \le x \le 360°$ and $-5 \le y \le 5$. For each pair, describe any patterns or relationships that you observe.

22. $y_1 = \sin x \qquad y_2 = \csc x$

23. $y_1 = \cos x \qquad y_2 = \sec x$

24. $y_1 = \tan x \qquad y_2 = \cot x$

THINK BACK

An *asymptote* is a line that a curve approaches.

Make Connections

● Describe how you would change the graphs of $y = \sin x$ and $y = \cos x$ to produce each of the following graphs.

25. $y = N \sin x$
$\quad\ y = N \cos x$

26. $y = \sin Nx$
$\quad\ y = \cos Nx$

27. $y = \sin (x + N°)$
$\quad\ y = \cos (x + N°)$

28. $y = \sin x + N$
$\quad\ y = \cos x + N$

Match each graph with its equation. The viewing windows have Xscl = 90 and Yscl = 1.

29.
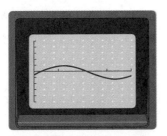

30.
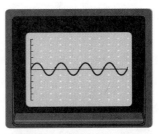

31.
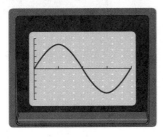

32.
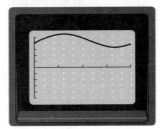

a. $y = 4 \sin x$
c. $y = \sin x + 4$

b. $y = \sin 4x$
d. $\sin (x - 30)$

Match each graph with its equation. The viewing windows have $Xscl = \dfrac{\pi}{2}$ and $Yscl = 1$.

33.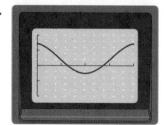

34.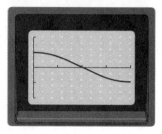

35.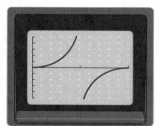

36.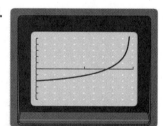

a. $y = -0.5 \cos x$

b. $y = \cos 0.5x$

c. $y = \cos\left(x + \dfrac{\pi}{4}\right)$

d. $y = \cos x + 0.5$

Match each graph with its equation. The viewing windows have $Xscl = 45$ and $Yscl = 1$.

37.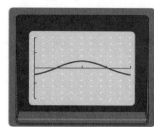

38.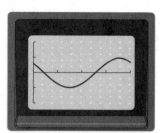

39.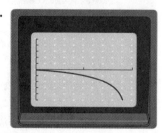

40.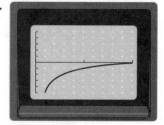

a. $y = -\tan x$

b. $y = \tan x - 2$

c. $y = \tan 2x$

d. $y = \tan(x + 90)$

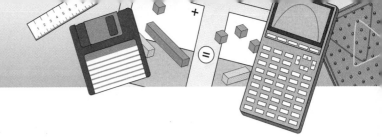

Summarize

41. **WRITING MATHEMATICS** Describe how you could create the graph of the function $y = 5 \cos (4x + 30°) + 2$ from the graph of $y = \cos x$.

42. **THINKING CRITICALLY** Write three different equations that will generate the graph. The viewing window has $Xscl = \dfrac{\pi}{2}$ and $Yscl = 1$. Two of the equations should involve the sine function. The third equation should involve the cosine function.

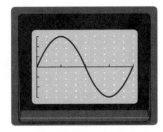

43. **GOING FURTHER** One way to graph trigonometric functions is by plotting points.

a. Copy and complete the table to determine points on the graph of $y = \sin x + \cos x$. Write values to the nearest hundredth.

x	0°	45°	90°	135°	180°	225°	270°	315°	360°
sin x									
cos x									
sin x + cos x									

b. Use graph paper to graph $y = \sin x + \cos x$ in the interval $[0°, 360°]$. Check your results with a graphing utility.

c. Make a table of values for the function $y = 2 \sin x + \cos 2x$ on the interval $0° \le x \le 360°$. Then graph the function on graph paper. Check your results with a graphing utility.

44. **GOING FURTHER** Examine the graphs of sin x, cos x, and tan x and decide if they are odd or even.

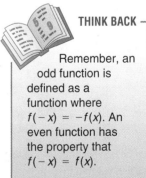

THINK BACK

Remember, an odd function is defined as a function where $f(-x) = -f(x)$. An even function has the property that $f(-x) = f(x)$.

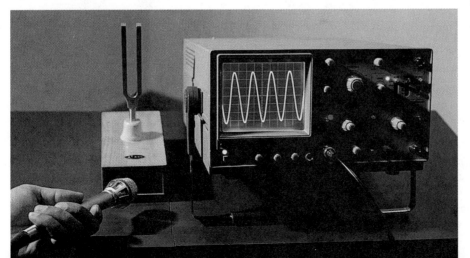

769

Explore

The equation $y = 20 \sin 45t$ can be used to model the displacement of the pendulum on a certain clock, where y represents the displacement of the tip of the pendulum in centimeters and t represents time in seconds.

1. Make a table of values for the function, using integral values of t from 0 to 24 seconds. Round displacements to the nearest tenth of a centimeter. What simplifies your calculations after $t = 8$?

2. Use the table to graph the function. Check your work using a graphing utility.

3. Describe the graph.

Build Understanding

A wide variety of natural and artificial phenomena produce smoothly increasing and decreasing data that repeat continuously and predictably. Rhythms that can be modeled and analyzed using the sine and cosine functions include the pattern of average daily temperatures and lengths of daylight throughout the year, the slow swing of a pendulum, the rapid beating of a car's pistons, the ebb and flow of tides, rodent populations, and house buyers.

Recall the graphs of $y = \sin x$ and $y = \cos x$ repeat themselves every 2π radians and have maximum and minimum values of $+1$ and -1.

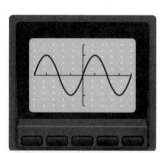

Any function whose graph exhibits a continuously repeating pattern is said to be **periodic**. The shortest repeating section of the graph is a **cycle**. The graph shows two cycles of the sine curve. The horizontal length of each cycle is called the **period** of the function.

Changes in the equations $y = \sin x$ and $y = \cos x$ produce predictable changes in the graphs of the equations.

Shown are the graphs of $y = \sin x$, $y = 2 \sin x$, and $y = 3 \sin x$. The coefficients 2 and 3 change the maximum and minimum values of the graphs to ± 2 and ± 3. The **amplitude** of a sine or cosine graph is related to the amount the graph varies above or below the x-axis. This amount can be expressed mathematically as half the difference between the maximum and minimum values of the function.

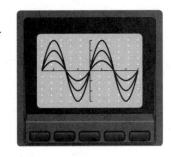

amplitude of $y = 3 \sin x$: $\frac{1}{2}(3 - [-3]) = \frac{1}{2}(6) = 3$

> **AMPLITUDE OF A SINE OR COSINE FUNCTION**
>
> **The amplitude of the function of $y = a \sin x$ or $y = a \cos x$ is $|a|$.**

The graphs of $y = \cos x$ and $y = \cos \frac{1}{2}x$, shown from 0 to 4π, illustrate that changing the coefficient of x in the equations $y = \sin x$ and $y = \cos x$ changes the period of a function. In a horizontal interval of 2π, $y = \cos \frac{1}{2}x$ completes only half of a cycle. The period of the curve is $\dfrac{2\pi}{\frac{1}{2}} = 4\pi$.

> **PERIOD AND CYCLE OF A SINE OR COSINE FUNCTION**
>
> **The period of the function $y = \sin bx$ or $y = \cos bx$ is $\dfrac{2\pi}{|b|}$. The graph completes $|b|$ cycles in 2π.**

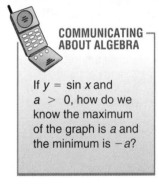

COMMUNICATING ABOUT ALGEBRA

If $y = \sin x$ and $a > 0$, how do we know the maximum of the graph is a and the minimum is $-a$?

EXAMPLE 1

Determine the amplitude and period of each function. Describe each cycle.

a. $y = 0.7 \cos(-6x)$

b. $y = 4 \sin 2x$ Graph on the interval $0 \le x \le 2\pi$.

Solution

a. $a = 0.7$, $b = -6$

amplitude: $|0.7| = 0.7$ $|a|$

period: $\dfrac{2\pi}{|-6|} = \dfrac{\pi}{3}$ $\dfrac{2\pi}{|b|}$

The function completes 6 cycles in 2π radians.

b. amplitude $= 4$,   period $= \pi$

Use $y = \sin x$ as your pattern. Increase the maximum and minimum values to ± 4. Compress two complete cycles into 2π radians. ◀

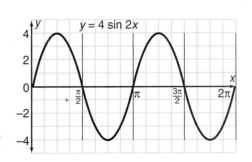

Like other graphs, the graphs of sine and cosine functions can be translated horizontally and vertically. The graph below on the left shows $y = \sin x$ and $y = \sin x + 3$. The equation of the graph of $y = \sin x$ translated down 3 units is $y = \sin x - 3$.

CHECK — UNDERSTANDING

What is the difference in evaluating $y = \sin x + 3$ and $y = \sin (x + 3)$?

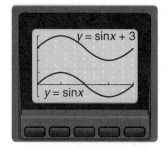

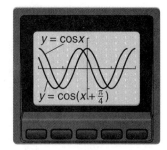

The graph above on the right shows $y = \cos x$ and $y = \cos\left(x + \dfrac{\pi}{4}\right)$.
The equation of the graph of $y = \cos x$ translated to the right $\dfrac{\pi}{4}$ radians is $y = \cos\left(x - \dfrac{\pi}{4}\right)$.

THINK BACK

Recall that $x^2 + y^2 = 25$ is the equation of a circle with radius 5 and center at (0, 0). Then $(x - 2)^2 + (y + 3)^2 = 25$ is the equation of a circle with radius 5 and center translated 2 units right and 3 units down to $(2, -3)$.

TRANSLATIONS OF A SINE OR COSINE GRAPH

To graph $y = a \sin (bx + c) + d$ **or**
$y = a \cos (bx + c) + d$, **translate the graph of**
$y = a \sin bx$ **or** $y = a \cos bx$

Vertical
  • d **units up if** $d > 0$ • $|d|$ **units down if** $d < 0$

Hortizontal
  • $\dfrac{c}{b}$ **units to the left if** $\dfrac{c}{b} > 0$ • $\left|\dfrac{c}{b}\right|$ **units to the right if** $\dfrac{c}{b} < 0$

CHECK — UNDERSTANDING

How can you use the graph of $y = a \sin bx$ to obtain the graph of $y = a \sin (bx + c)$?

EXAMPLE 2

Use the graphs of $y = \sin x$ and $y = \cos x$ to transform each graph.

a. Describe how the graph of $y = \cos x$ can be translated to produce the graph of $y = \cos\left(x - \dfrac{\pi}{6}\right) - 3.5$

b. Graph $y = 2 \sin\left(0.5x + \dfrac{\pi}{4}\right) - 1$ on the interval $-2\pi \le x \le 2\pi$

c. Change the graph in Part b to produce the graph of
$y = -\left(2 \sin\left(0.5x + \dfrac{\pi}{4}\right) - 1\right)$ on the interval $-2\pi \le x \le 2\pi$

Solution
a. $b = 1, c = -\dfrac{\pi}{6}, d = -3.5$

Since $d < 0$, translate $y = \cos x$ down 3.5 units.

Since $\dfrac{c}{b} = -\dfrac{\pi}{6} < 0$, translate $y = \cos x$ to the right $\dfrac{\pi}{6}$.

b. The graph is obtained by tranforming the graph of $y = \sin x$.

amplitude: $a = 2$

period: $\dfrac{2\pi}{0.5} = 4\pi$

horizontal translation: left $\dfrac{\pi}{6}$

vertical translation: down 1

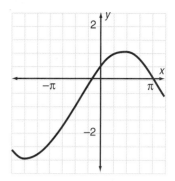

c. For any function $y = f(x)$, the graph of $y = -f(x)$ is the reflection of the graph of $y = f(x)$. Compare this graph with the previous one.

amplitude: $a = 2$

period: 4π

horizontal translation: left $\dfrac{\pi}{6}$

vertical translation: up 1   ◀

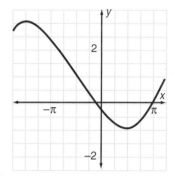

Graphs of the tangent, cotangent, cosecant, and secant functions are shown below in the viewing window with Xscl $= \dfrac{\pi}{2}$ and Yscl $= 1$.

$y = \tan x$

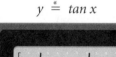

$y = \cot x$

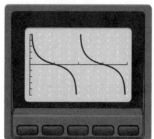

$y = \csc x$

$y = \sec x$

COMMUNICATING ABOUT ALGEBRA

Explain why the tangent, cotangent, cosecant, and secant functions are undefined at various points. Why should that prevent them from having amplitudes?

As you can see, the functions are undefined at various points and therefore do not have amplitudes. However, the methods you have learned for determining the periods and translations of sine and cosine functions can be applied to the tangent, cotangent, cosecant, and secant as well.

EXAMPLE 3

Determine the period and translation of the function $y = \sec 2x - 5$ from $y = \sec 2x$.

Solution

Since $b = 2$, the period is $\frac{2\pi}{2} = \pi$. There is no horizontal translation. Since $d = -5$, the graph of $y = \sec 2x$ must be translated 5 units down for the graph of $y = \sec 2x - 5$. ◀

Simple harmonic motion takes place when an object is shifted from its rest position and the force attempting to return it to rest is directly proportional to the shift. You can model simple harmonic motion using an equation of the form $y = a \sin b(t + c) + d$ or $y = a \cos b(t + c) + d$.

EXAMPLE 4

MEDICAL RESEARCH Human heart rates vary periodically over a 24-hour span. The function $y = 12 \sin (15t - 135) + 62$ models the rate y (beats/min) of a subject in a study of heart disease, where t is the number of hours after midnight.

a. Determine the patient's heart rate at 7 P.M.
b. Determine the patient's maximum heart rate and when it occurs.

Solution

a. 7 P.M. = 19 hours after midnight

$$y = 12 \sin (15 [19] - 135) + 62$$
$$= 12 \sin 150 + 62 = 68$$

The subject's heart rate is 68 beats/min at 7 P.M.

b. The maximum value of $\sin x$ is 1 when $x = 90°$. So, $15t - 135 = 90$, which gives $t = 15$. At that time, $y = 12(1) + 62$, or 74.

The maximum rate of 74 beats/min is at $t = 15$, or 3 P.M. ◀

TRY THESE

Determine the amplitude and period of the function.

1. $y = \sin 3x$

2. $y = 3 \cos x$

3. $y = -2 \cos \pi x$

4. $y = \frac{1}{4} \sin \frac{1}{2} x$

Describe how the graph of $y = \sin x$ or $y = \cos x$ can be translated to produce the graph of each function.

5. $y = \sin x - 3$

6. $y = 5 + \cos x$

7. $y = \sin(x - 200°)$

8. $y = (\cos (2x + \pi)) - 6$

Graph the interval $0 \le x \le 2\pi$.

9. $y = 2 \cos x$ **10.** $y = \sin \frac{1}{2}x$ **11.** $y = \cos x - 2$ **12.** $y = -3 \sin 2x$

13. PHYSICS The up-and-down motion of a weighted spring is modeled by $y = 4 \sin \left(\pi t - \frac{\pi}{2} \right)$, where y is the weight's distance above or below its equilibrium point of zero, in inches, and t is the time in seconds.

   a. What is the weight's position when the experiment begins?

   b. What is the highest point reached by the spring?

   c. When does the spring first reach its equilibrium point?

   d. What is the period of the function?

   e. Draw a graph showing one cycle of the function.

14. WRITING MATHEMATICS A sine curve with an amplitude of 5 and a period of 180° is translated 90° left and 3 units down from the graph of $y = \sin x$. Find the equation of the function and explain the steps you took to find it.

PRACTICE

Determine the amplitude and period of the function.

1. $y = \sin x$ **2.** $y = -2 \sin x$ **3.** $y = \cos 2x + 2$ **4.** $y = 1.5 \cos 4x$

5. $y = -3 \sin 2\pi x$ **6.** $y = 0.4 \cos 0.5x$ **7.** $y = -9 \sin \frac{1}{5}x$ **8.** $y = 4 \cos (-45x)$

Describe how the graph of $y = \sin x$ or $y = \cos x$ can be translated to produce the graph of each function.

9. $y = \cos x - 2$ **10.** $y = 6 + \sin x$ **11.** $y = \cos (x + \pi)$ **12.** $y = \sin (x - 15°)$

13. $y = \sin (2x + 150°)$ **14.** $y = \cos \left(x + \frac{\pi}{2} \right) + 3$ **15.** $y = \sin (3x - 90°) - 4$

Graph each on the interval $0 \le x < 2\pi$.

16. $y = \cos x$ **17.** $y = 2 \sin x$ **18.** $y = \cos \frac{1}{2}x$ **19.** $y = \sin 2x$

20. $y = -\sin x$ **21.** $y = 3 \cos 4x$ **22.** $y = -(2 \cos x - 1)$ **23.** $y = \sin \left(x - \frac{\pi}{2} \right) + 2$

24. WILDLIFE BIOLOGY Wildlife populations often grow and decline periodically. The function $n = 8000 + 5000 \sin \pi t$ gives n, the number of lemmings on an Arctic island, where t is the number of years after the beginning of a study on January 1, 1996.

   a. Determine the lemming population on January 1 of 1996, 1997, and 1998.

   b. What is the maximum lemming population? When was it first reached?

   c. What is the minimum population? When was it first reached?

   d. Determine the period of the population function.

   e. Draw a graph showing population as a function of time for one cycle.

25. MUSIC A pianist struck the lowest note on a piano, producing a sound wave with the equation $y = 4 \sin 55\pi t$, where t is measured in seconds. A second pianist struck the same note, producing the sound wave $y = 5 \sin 55\pi t$.

a. The loudness of a sound varies directly with the amplitude of the sound wave. Which of the two notes was loudest?

b. The *frequency* of a musical note is the number of times the sound wave vibrates in one second. Frequency is the reciprocal of period. Determine the period and frequency of the lowest note on a piano.

26. WRITING MATHEMATICS Explain how you can determine the amplitude, period, and translation of a sine curve from the graph of the curve. Give an example.

EXTEND

Use the function $y = -4 + 3 \sin 2x$ in the interval $0 \leq x \leq 2\pi$.

27. At what value(s) of x, , will the function reach its maximum? What is the maximum?

28. At what value(s) of x will the function reach its minimum? What is the minimum?

Solve each on the interval $0 \leq x \leq 2\pi$. Use a graphing utility if necessary.

29. $2 \cos x + \sqrt{3} = 0$

30. $-2 \sin x = 1$

31. $2 \tan x - 7 = 0$

32. $(\sec x)^2 - 2 = 0$

THINK CRITICALLY

Use the information in the table to write the equation of each sine function in the form $y = a(\sin bx + c) + d$.

	Amplitude	Period	Translation Up or Down	Translation Left or Right
33.	3	π	0	0
34.	1	$\dfrac{\pi}{2}$	6 down	0
35.	2	$\dfrac{3\pi}{2}$	4 up	0
36.	4	$\dfrac{\pi}{4}$	3 down	$\dfrac{\pi}{6}$ left
37.	1	$\dfrac{5\pi}{2}$	1 up	$\dfrac{5\pi}{6}$ right

38. METEOROLOGY The average daily temperature T (° F) in Atlanta, Georgia can be modeled by $T = 57.3 + 24.8 \sin\left(\frac{\pi}{6}m + 4\right)$, where m is the date in months since the beginning of the year. (January $1 = 0.0$, February $1 = 1.0$)

   a. Determine the average daily temperature for September 15 predicted by the model.

   b. Determine the maximum average daily temperature and the date it is achieved.

   c. Determine the minimum average daily temperature and the date it is achieved.

MIXED REVIEW

Solve the formula for the given variable.

39. $A = \frac{1}{2}h(b_1 + b_2)$, for b_2 **40.** $A = P + Prt$, for P **41.** $F = \frac{mv^2}{r}$, for m

Factor.

42. $x^2 - 81$ **43.** $x^2 + 6x + 9$

44. $6x^2 + 7x - 3$ **45.** $3x^2 + 9x - 30$

46. Determine the center and radius of the circle $x^2 + 4x + y^2 - 6y = 3$.

47. STANDARDIZED TESTS Choose the letter that gives the period of the function $y = \cos(3x + 6) + 8$.

   A. 120°   **B.** 60°   **C.** 45°   **D.** 30°

48. GEOMETRY The point $L(x, y)$ is equidistant from the points $M(-2, 1)$ and $N(4, 3)$. Find L.

PROJECT *Connection*   A *sextant* is a device that measures the angle of separation between two objects. You can make a sextant by gluing a photocopy of a protractor onto a cardboard rectangle. Then glue the cardboard onto the flat end of a dowel 1 m in length so it resembles a one-legged table. You will also need three straight pins. Place pin X at the 0° mark and pin V at the center of the protractor.

  1. To measure the angle of separation, plant the sextant securely into the ground. Twist the dowel so that one of the two objects aligns with pins X and V. Without moving the dowel, along the curved edge of the protractor, align the third pin, pin Y, with the second object and pin V. The angle of separation is angle XVY.

  2. Have two friends, A and B, each stand about 20–30 ft from the base of a reference point and from each other. At A's location, use the quadrant and the height of the reference point from the Project Connection on page 752 to determine the angle of separation between the reference point and B. Similarly, determine B's distance and the angle of separation between A and the reference point.

  3. From these data use the law of cosines to determine the distance between A and B. Verify this distance using a tape measure.

Explore

1. Use a graphing utility to graph $y = \sin x$ for $0° \leq x \leq 360°$. Then use the TRACE feature to find all values of x in the interval you have graphed for which $\sin x = 0.5$.

2. For each angle in your answer to Question 1, list four angles that are coterminal with it.

3. Explain why you cannot find a unique solution to the equation $\sin x = c$, for a given constant c where $-1 \leq c \leq 1$.

4. Write a set of four ordered pairs (x, y) that satisfies the equation $x = \sin y$ for $x = 0.5$. Does the set describe a function? Explain.

Build Understanding

Recall a relation is a *function* if each member of the domain is paired with exactly one member of the range. The relation $\{(1, 3), (4, 5), (4, 3)\}$ is not a function because 4 is paired with both 5 and 3.

By interchanging domain and range, you obtain the *inverse* of a function. The inverse may or may not be a function. The inverse of the function $\{(6, 9), (8, 2), (7, 9)\}$, for example, is not a function: $\{(9, 6), (2, 8), (9, 7)\}$.

Inverse trigonometric relations are obtained when the domain and range of a trigonometric function are interchanged. Look at angles coterminal with 0 and their cosines to determine if the inverse trigonometric relation is a function.

cosine function: $\{(0, 1), (2\pi, 1), (4\pi, 1), (6\pi, 1), \ldots\}$

inverse relation: $\{(1, 0), (1, 2\pi), (1, 4\pi), (1, 6\pi), \ldots\}$

The inverse of the cosine function is clearly not a function. For similar reasons, none of the inverse trigonometric relations are functions.

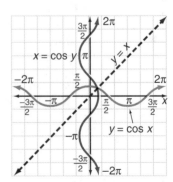

Recall that by reflecting the graph of a function over the line $y = x$, you obtain the graph of the inverse of the function. You can see that the graph of $x = \cos y$ which is the inverse of $y = \cos x$, is not a function. Notice the graph fails the vertical line test.

Notice what happens if the domain of $y = \cos x$ is restricted to values of x from 0 to π. Now the graph of the inverse passes the vertical line test, because coterminal angles of rotation have been eliminated from the domain of the cosine function. Restricting the domain of the trigonometric functions is the key to creating inverse trigonometric relations that are functions.

Let **Cosine x** (note the capital letter) be the cosine function with the restricted domain $0 \le x \le \pi$. Let **Sine x** be the sine function with the restricted domain $-\frac{\pi}{2} \le x \le \frac{\pi}{2}$, and **Tangent x** be the tangent function with restricted domain $-\frac{\pi}{2} < x < \frac{\pi}{2}$.

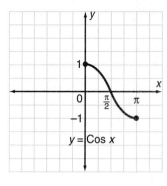

$y = \text{Cos } x$

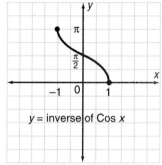

$y = $ inverse of Cos x

INVERSE TRIGONOMETRIC FUNCTIONS

- The inverse Sine function is defined by
  $y = \text{Arcsin } x$, where $x = \text{Sin } y$ and $-1 \le x \le 1$.

- The inverse Cosine function is defined by
  $y = \text{Arccos } x$, where $x = \text{Cos } y$ and $-1 \le x \le 1$.

- The inverse Tangent function is defined by
  $y = \text{Arctan } x$ where $x = \text{Tan } y$.

The inverse functions are also denoted $\text{Cos}^{-1}x$, $\text{Sin}^{-1}x$, and $\text{Tan}^{-1}x$. In the same manner as $f^{-1}(x)$, the -1 is inverse notation and is not meant to signify an exponent or reciprocal.

THINK BACK

Use the SIN^{-1}, COS^{-1}, and TAN^{-1} keys on a calculator to determine an angle with a trigonometric function equal to a given value.

EXAMPLE 1

Evaluate each expression.

  a. $\text{Sin}^{-1} 1$ **b.** $\tan\left(\text{Arccos } \frac{3}{4}\right)$

Solution

  a. $\text{Sin}^{-1} 1$ is read "the angle whose sine is 1." For the sine function, there are an infinite number of angles with sines of 1. The capital letter here indicates that the solution must fall in the interval $-\frac{\pi}{2} \le x \le \frac{\pi}{2}$. The graph of the sine curve shows that the only angle in that interval with a sine of 1 is $\frac{\pi}{2}$, or 90°.

  b. You can solve Part a above using your knowledge that the angle with a sine of 1 measures $\frac{\pi}{2}$. It is unlikely that you know "the angle whose cosine is $\frac{3}{4}$." You can sketch the angle, however, making sure that you locate it in the restricted region for the Cosine function, $0 \le x \le \pi$. By the Pythagorean theorem, the side opposite x measures $\sqrt{7}$. So, $\tan\left(\text{Arccos } \frac{3}{4}\right) = \frac{\sqrt{7}}{3}$ ◄

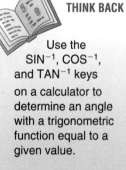

Real world problems involving inverse trigonometric functions usually require that you determine the measure of an angle rather than the value of a function.

EXAMPLE 2

A 76-ft Ponderosa pine casts a 98-ft shadow. Determine the angle of elevation of the sun.

Solution

$$\angle A = \text{Arctan}\,\frac{76}{98}$$

Use your calculator in degree mode.

$$\text{Arctan}\,\frac{76}{98} \approx 37.793943$$

The angle measures approximately 37.8°.

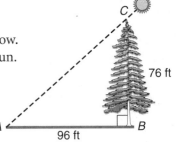

96 ft

76 ft

TRY THESE

1. Write the equation $a = \tan b$ in the form of an inverse function.

Determine $\angle A$ in radians.

2.
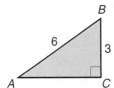

3.
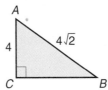

4.

Evaluate each expression.

5. $\text{Arcsin}\,\frac{\sqrt{2}}{2}$

6. $\text{Tan}^{-1}1$

7. $\text{Arccos}\,\frac{1}{2}$

8. $\text{Sin}^{-1}(-1)$

9. $\tan\left(\text{Arcsin}\,\frac{5}{13}\right)$

10. $\cos\,(\text{Cos}^{-1}\,0.3)$

11. $\text{Sin}^{-1}(\sin 42°)$

12. $\cos\left(\text{Arcsin}\,\frac{3}{5}\right)$

13. NAVIGATION A plane flying at an altitude of 6 mi is directly over point A. The navigator spots an airport 16 mi from and at the same elevation as A. Determine the angle of depression of the airport as seen from the plane. Round to the nearest tenth of a degree.

14. WRITING MATHEMATICS Explain the meaning of Arcsin x.

PRACTICE

Evaluate each expression.

1. $\text{Tan}^{-1}\sqrt{3}$

2. $\text{Arcsin}\,1$

3. $\text{Cos}^{-1}\left(-\frac{\sqrt{3}}{2}\right)$

4. $\text{Arccos}\,0$

5. $\text{Arccos}\left(-\frac{\sqrt{2}}{2}\right)$

6. $\tan\left(\text{Arctan}\,\frac{5}{6}\right)$

7. $\text{Cos}^{-1}(\cos 23.6°)$

8. $\sin\left(\text{Arctan}\,\frac{\sqrt{3}}{3}\right)$

9. $\text{Sin}^{-1}\left(\cos\frac{\pi}{2}\right)$

10. $\text{Arcsin}\left(\cos\frac{\pi}{6}\right)$

11. $\sin\left(2\,\text{Sin}^{-1}\,\frac{1}{2}\right)$

12. $\text{Arccos}\left(\tan\frac{3\pi}{4}\right)$

Determine the measures of the angles in a right triangle with the given side lengths.

13. 3, 4, 5   **14.** 5, 12, 13   **15.** 8, 15, 17   **16.** 2.8, 4.5, 5.3

17. CONSTRUCTION The top of an 18-ft ladder rests against the side of a building 13 ft 5 in. above the ground. Determine the angle the ladder makes with the ground.

18. TRANSPORTATION At its maximum angle, Baldwin Street in Dunedin, New Zealand, the world's steepest street, rises 1 m for every 1.266 m of horizontal distance. Determine, to the nearest tenth of a degree, the maximum angle of the street.

19. WRITING MATHEMATICS Explain why the interval chosen for restricting the domain of the Cosine function differs from that chosen for the Sine and the Tangent.

EXTEND

Evaluate each expression.

20. $\sin\left(\text{Arcsin } 1 - \text{Arccos } \frac{1}{2}\right)$

21. $\tan\left(\text{Cos}^{-1}0 + \text{Tan}^{-1}1\right)$

22. $\cos\left(\text{Sin}^{-1}0 - \text{Arccos } 1\right)$

23. $\cos\left(\text{Cos}^{-1}\left[-\frac{\sqrt{2}}{2}\right] - \frac{\pi}{2}\right)$

THINK CRITICALLY

24. SPORTS The maximum height h, in ft, reached by an object projected at a velocity v, in ft/sec, at an angle θ is given by $h = \dfrac{(v \sin \theta)^2}{64}$ Find the angle at which you must shoot a basketball at a velocity of 40 ft/sec for it to reach a maximum height of 13.5 ft.

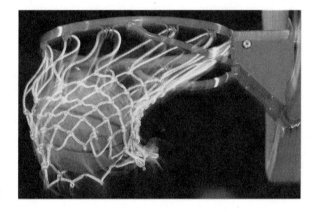

Expressions such as $(\tan x)^2$ or $(\sin q)^3$ are written $\tan^2 x$ and $\sin^3 q$. Use inverse trigonometric functions to determine the solution(s) of each equation.

25. $5 \sin^2 \theta + 3 \sin \theta - 1 = 0, \quad -\frac{\pi}{2} \le \theta \le \frac{\pi}{2}$ **26.** $2 \tan^2 x + 9 \tan x + 3 = 0, \quad -\frac{\pi}{2} < x < \frac{\pi}{2}$

MIXED REVIEW

Name the field property illustrated by each equation.

27. $4(5 - 3) = 4 \cdot 5 - 4 \cdot 3$ **28.** $k + 0 = k$ **29.** $m + n = n + m$

Solve for x. Check your answer.

30. $\sqrt{x} - 5 = 2$ **31.** $\sqrt{3x - 5} = 4$ **32.** $1 + \sqrt{2x + 7} = \sqrt{x + 15}$

33. STANDARDIZED TESTS Find $\cos\left(\text{Arcsin}\left[\frac{5}{13}\right]\right)$. Choose the letter of the correct answer.

   A. $\frac{5}{12}$ **B.** $\frac{12}{13}$ **C.** $30°$ **D.** $\text{Tan}^{-1}\frac{5}{13}$

• • • • CHAPTER REVIEW • • • •

Match the letter of the word in the right column with the description at the left.

1. ratio of the lengths of the legs of a right triangle   **a.** oblique

2. triangle used to determine trigonometric functions of angles of rotation   **b.** period

3. a triangle with no right angles   **c.** radian

4. angle measure determined by the quotient of the arc length and radius   **d.** reference

5. the length of one cycle of a graph   **e.** tangent

Lessons 15.1 and 15.2   ANGLE MEASURE AND ANGLES OF ROTATION   pages 737–742, 743–745

- An **angle of rotation** is in **standard position** when its **vertex** is at the origin, its **initial side** is on the positive x-axis, and its **terminal side** is in the plane. The **measure** of the angle is determined by the amount of rotation of the terminal side.

- In a circle of radius r, the **radian measure** of a central angle θ that intercepts an arc of length s is $\frac{s}{r}$. Use 1 radian $= \left(\frac{180°}{\pi}\right)$ and $1° = \left(\frac{\pi}{180}\right)$ radians to convert between radians and degrees.

Convert to radian measure or degree measure.

6. $90°$ 7. $225°$ 8. $-150°$ 9. $900°$

10. $\frac{3\pi}{4}$ 11. 0.7π 12. $\frac{2\pi}{3}$ 13. $-\frac{8\pi}{5}$

For each angle, find a coterminal angle θ where $0° \le \theta < 360°$.

14. $500°$ 15. $-13°$ 16. $1035°$ 17. $-1199°$

Lesson 15.3   THE TRIGONOMETRIC FUNCTIONS   pages 746–753

- If θ is an angle in standard position, $P(x, y)$ is a point (not the origin) on the terminal side of the angle, and r is the length of the hypotenuse of the **reference triangle** formed by drawing a perpendicular from P to the x-axis, then the six **trigonometric functions** of θ are:

     sine θ (written **sin θ**) $= \frac{y}{r}$ cosine θ (written **cos θ**) $= \frac{x}{r}$ tangent θ (written **tan θ**) $= \frac{y}{x}$

     cosecant θ (written **csc θ**) $= \frac{r}{y}$ secant θ (written **sec θ**) $= \frac{r}{x}$ cotangent θ (written **cot θ**) $= \frac{x}{y}$

Evaluate without using a calculator. Give answers in simplest form.

18. $\sin 30°$ 19. $\cos 180°$ 20. $\tan \frac{2\pi}{3}$ 21. $\sec(-45°)$

Evaluate using a calculator. Round answers to the nearest ten-thousandth.

22. $\cot 44°$ 23. $\sin 28°$ 24. $\csc \frac{-6\pi}{7}$ 25. $\cos \frac{2\pi}{5}$

- To solve a real-world problem in trigonometry, draw a sketch and label it. Note the angles and sides that you know and use them to write an equation.

26. Find the length of a guy wire to the top of a 460-ft antenna that makes a 57° angle with the ground to the nearest tenth of a foot.

- An **oblique** triangle has no right angles. The following relationships hold in any oblique triangle with sides a, b, and c opposite vertices A, B, and C, respectively.

Law of sines: $\dfrac{\sin A}{a} = \dfrac{\sin B}{b} = \dfrac{\sin C}{c}$

Law of cosines:
$$a^2 = b^2 + c^2 - 2bc \cos A \qquad b^2 = a^2 + c^2 - 2ac \cos B \qquad c^2 = a^2 + b^2 - 2ab \cos C$$

Solve △ABC. Round to the nearest tenth.

27. $b = 10, A = 85.9°, B = 30°$ **28.** $a = 14, b = 6.5, c = 11.1$

- The **period** of the graph of $y = a \sin bx$ or $y = a \cos bx$ is $\dfrac{2\pi}{|b|}$. The **amplitude** is $|a|$.

- To graph $y = a \sin (bx + c) + d$ or $y = a \cos (bx + c) + d$, translate the graph of $y = a \sin bx$ or $y = a \cos bx$ vertically d units and horizontally $\dfrac{c}{b}$ units.

Determine the amplitude and period of the graph of the function.

29. $y = 3 \sin x$ **30.** $y = -2 \cos x$ **31.** $y = \cos 4x$ **32.** $y = 5 \sin \dfrac{1}{3} x$

Graph on the interval $0 \le x \le 2\pi$.

33. $y = \sin x$ **34.** $y = 3 \sin 0.5 x$ **35.** $y = -2 \cos 2x$ **36.** $y = 0.5 \cos \dfrac{1}{4} x$

- The **inverse Cosine function** ($\text{Cos}^{-1}x$) is defined by Arccos $x = y$, where Cos $y = x$ and $-1 \le x \le 1$.

- The **inverse Sine function** ($\text{Sin}^{-1}x$) is defined by Arcsin $x = y$, where Sin $y = x$ and $-1 \le x \le 1$.

- The **inverse Tangent function** ($\text{Tan}^{-1}x$) is defined by Arctan $x = y$, where Tan $y = x$.

Evaluate each expression.

37. $\text{Sin}^{-1}(0.5)$ **38.** $\text{Cos}^{-1}\left(-\dfrac{\sqrt{2}}{2}\right)$ **39.** $\tan\left(\text{Arcsin}\dfrac{2}{3}\right)$

CHAPTER ASSESSMENT

CHAPTER TEST

Convert to radian measure.

1. 270°

2. −60°

Convert to degree measure.

3. $\dfrac{\pi}{4}$

4. $\dfrac{11\pi}{6}$

For each angle, find a coterminal angle θ where $0° \leq \theta \leq 360°$.

5. 779°

6. −644°

Evaluate without using a calculator.

7. cos 45°

8. sin 90°

9. $\tan \dfrac{5\pi}{6}$

10. $\csc \left(-\dfrac{2\pi}{3}\right)$

11. With the sun at an angle of 30° to the ground, a tree casts an 80-ft shadow. How tall is the tree?

12. The roof of a hiker's hut slants up from the ground to a height of 4.2 m at an angle of 56°. Find the length of the roof.

13. The angle of incidence i of a light ray entering a piece of glass is 45° and the angle of refraction r is 30°. Find the index of refraction $n = \dfrac{\sin i}{\sin r}$ of the glass.

Solve $\triangle ABC$.

14. $a = 4, b = 3, c = 4.5$

15. $a = 32, c = 45, C = 98°$

Determine the area of $\triangle ABC$.

16. $a = 7.8, c = 9.6, \angle B = 77.1°$

17. $b = 11.6, c = 15.1, \angle A = 58.2°$

Determine the amplitude and period of each function.

18. $y = \cos x$

19. $y = 3 \sin 2x$

20. $y = 0.7 \cos \dfrac{1}{2}x$

21. $y = -5 \sin 3x$

Describe how the graph of $y = \sin x$ can be translated to produce the graph of each function.

22. $y = 3 + \sin x$

23. $y = \sin \left(x - \dfrac{\pi}{4}\right)$

24. $y = \sin (3x + 90°) - 2$

Graph in the interval $0 \leq x \leq 2\pi$.

25. $y = 2 \cos x$

26. $y = \sin 2x$

27. $y = -\cos \dfrac{1}{2}x$

28. $y = 3 \cos 0.25x$

Evaluate each expression.

29. $\text{Tan}^{-1} 1$

30. Arccos 0.5

31. $\sin^{-1} -\dfrac{\sqrt{3}}{2}$

32. $\cos \left(\text{Arcsin} \dfrac{4}{5}\right)$

33. Arctan (tan 0.39)

34. WRITING MATHEMATICS Is the inverse of the sine function a function? Explain.

35. STANDARDIZED TESTS What time is it when the minute hand of a clock has moved through $\dfrac{2\pi}{5}$ radians past 3:00?

    A. 3:12　　　　　**B.** 3:24
    C. 4:12　　　　　**D.** 5:30

36. When a 14-ft ladder leans against a wall, its base is 6 feet from the wall. Determine the angle that the ladder makes with the ground.

PERFORMANCE ASSESSMENT

FUNCTIONAL TRIANGLES Find a right triangle in your classroom and sketch it. Measure the sides of the triangle and label them on your sketch. Then find each of the following.

a. The measures of all the angles of the triangle in degrees and in radians
b. The six trigonometric functions of each of the angles of the triangle

PHOTOGRAPHIC TRIANGLES In a newspaper, book, or magazine, find a photograph of an oblique triangle that exists somewhere in the world. Be sure the planes of the photo and the triangle are parallel so that the triangle does not appear distorted.

a. Measure the angles of the triangle.
b. Estimate the length of one side of the triangle. Explain how you made your estimate.
c. Find the lengths of the other sides of the triangle, assuming your estimate of the first side is accurate. Explain how you found the lengths.

CHOOSE YOUR METHOD Sketch an oblique triangle on graph paper. Measure the sides and angles of the figure and the length of the altitude to one side. Then determine the area of the triangle using each of the following methods.

a. Estimate the area by counting squares.
b. Calculate the area using the formula for the area of a triangle.
c. Calculate the area using the lengths of two sides and the sine of the included angles.
d. Calculate the area using Heron's formula.

Compare your results. Explain any discrepancies you observe. What is the best value for the area of the triangle? Why?

CURVY EQUATIONS The amplitude of the graph is 1 and the period is 2π. Express the equation of the graph in as many ways as you can in terms of both the sine function and the cosine function.

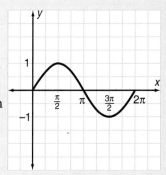

PROJECT ASSESSMENT

PROJECT *Connection* For reference, prepare a brief summary of the Project Connections organizing your observations and data. Include any comments or tips you have for using the quadrant and sextant effectively. You may also wish to research additional information about early navigational instruments such as the astrolabe or the Jacob's staff.

1. Decide on a locale, such as a large backyard, a section of small park, or school grounds that you wish to map.

2. Select the landmarks in this locale that you will map.

3. Map the landmarks to scale, using your quadrant, sextant, and only one distance measurement.

4. Display your finished map to the class and have your classmates determine what locale you mapped.

1. In the diagram below, AP represents the distance between house P on an island in a small lake and house A on the shore of the lake. To find the length of $\overline{AP}$, a surveyor took $\overline{ABC}$ as a base line and made the following measurements: $\angle A = 50°$, $\angle PBC = 70°$, $\angle C = 90°$, and $BC = 150$ yds. Determine the distance between the two houses to the nearest yard.

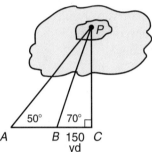

Evaluate each expression.

2. Arc cos $\frac{1}{2}$

3. $\tan\left(\text{Arc sin } \frac{8}{17}\right)$

4. Solve graphically: $\begin{cases} y + 2x > 10 \\ y \geq x + 1 \end{cases}$

5. Write and solve an equation for the problem. A park department wants to place a distance marker at K, the beginning of two straight paths, KT and KS, through a rectangular park, $PQRS$, as shown at the right. The length of the park, RS, is 12 km and the total length of the two paths, $KT + KS$, is 18 km. If K is equidistant from T and R, determine the distances KT and KS to be listed on the marker.

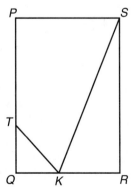

6. The colorful marine life in a stream in Hawaii can be viewed from a slow-moving glass-bottomed boat. Cynthia and Sera were in a boat that traveled 9 mi upstream in the same time it took to go 15 mi downstream. If the rate of the boat in still water was 8 mi per hour, what was the rate of the stream?

7. Determine the equations of the asymptotes for the curve $4x^2 - 9y^2 + 4y = 51$.

8. **WRITING MATHEMATICS** If you are given the measure of acute $\angle A$, side a, and side b in $\triangle ABC$, describe the circumstances under which $\triangle ABC$ will not be uniquely determined.

9. **STANDARDIZED TESTS** For the right triangle shown, 2 log c equals.

   A. $\log 2a + \log 2b$
   B. $2 \log a + 2 \log b$
   C. $2 \log (a + b)$
   D. $\log (a^2 + b^2)$

10. Express in degrees the measure of an angle of $\frac{5\pi}{12}$ radians.

11. Write an equation of the axis of symmetry of the curve $y = 3x^2 - 5x + 8$.

12. A *tsunami* (commonly called a *tidal wave*) is a fast-moving ocean wave caused by an underwater earthquake. The depth (d) of the water can be modeled by the equation $d = 9 - 19 \sin \frac{2\pi}{15} t$ where d is in meters and t is in minutes. Determine the depth of the water 2 min after the tsunami strikes.

13. Determine the coordinates of the center of the conic section whose equation is $x^2 + y^2 - 6x + 4y = 51$

14. Determine the solution of $x = 1 + \sqrt{x + 5}$.

15. Evaluate $\frac{r + s}{1 - rs}$ for $r = \frac{1}{3}$ and $s = \frac{1}{7}$.

STANDARDIZED TEST

STANDARD FIVE-CHOICE Select the best choice for each question.

1. In circle O, the length of radius $\overline{OB}$ is 5 cm and the length of $\overset{\frown}{AB}$ is 5 cm. The measure of $\angle AOB$ is

   A. 60° **B.** 1 radian **C.** 90°
   D. > 60° **E.** π radians

2. $\dfrac{\log_4 64}{\log_4 8}$ is equal to

   A. $\log_4 8$ **B.** $\dfrac{1}{2}$ **C.** 2
   D. $\log_4 2$ **E.** $\log_4 56$

3. The solution of
   $$\dfrac{x}{x-2} - \dfrac{2}{x+4} = \dfrac{12}{x^2+2x-8}$$ is

   A. $\{-4, 2\}$ **B.** $\{-4\}$ **C.** $\{2\}$
   D. $\{4, 2\}$ **E.** no solution

4. In right $\triangle PQR$ with the right angle at R, $\angle P > \angle Q$. Which of the following statements are true?

   I. $\sin P > \sin Q$ **II.** $\cos P > \cos Q$
   III. $\sin P > \cos Q$

   A. I only **B.** II only **C.** III only
   D. I and II **E.** I and III

5. The graph below shows three points in the complex plane. These points represent the zeros of which polynomial equation

   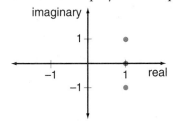
   imaginary

   A. $x^3 - 1 = 0$ **B.** $x^3 + 1 = 0$
   C. $x^3 - x^2 - 2 = 0$
   D. $x^3 - 3x^2 + 4x - 2 = 0$
   E. $x^3 - 3x^2 + 20 = 0$

6. To play Minnesota's Daily 3 game, a player chooses three digits each from 0 through 9 and must match a 3-digit number. The probability of matching the three winning digits is

   A. $\dfrac{1}{720}$ **B.** $\dfrac{3}{720}$ **C.** $\dfrac{1}{1000}$
   D. $\dfrac{3}{1000}$ **E.** $\dfrac{9}{1000}$

7. The equation of the graph below is

   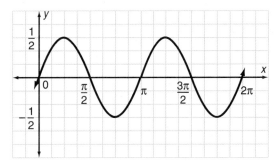

   A. $y = 2\sin\dfrac{1}{2}x$ **B.** $y = \dfrac{1}{2}\sin 2x$
   C. $y = \dfrac{1}{2}\cos 2x$ **D.** $y = \dfrac{1}{2}\sin\dfrac{1}{2}x$
   E. $y = \dfrac{1}{2}\cos\dfrac{1}{2}x$

8. The one real zero of $3x^3 - 7x^2 - 7x - 10 = 0$ lies between

   A. -1 and 0 **B.** 0 and 1
   C. 1 and 2 **D.** 2 and 3
   E. 3 and 4

9. In the expansion of $(1 + a)^n$, the coefficient of the fourth terms equals the coefficient of the sixth term. The value of n is

   A. 7 **B.** 8 **C.** 9 **D.** 10 **E.** 12

10. The number of points of intersection of the graphs in the following system is
    $$\begin{cases} 25x^2 + 9y^2 = 225 \\ xy + 12 = 0 \end{cases}$$

    A. 8 **B.** 4 **C.** 2 **D.** 1 **E.** 0

Take a Look
AHEAD

Make notes about things that look new.

- Locate and copy a trigonometric expression that is equal to 1. Does this expression involve any new notation?

- What are some different identities that are covered in this chapter?

- In what type of coordinate system do you usually work? What new coordinate system is introduced in Lesson 16.5?

Make notes about things that look familiar.

- Use the word identity in some sentences. What are some common meanings of this word? Which of these meanings do you think is closest to what the word identity indicates in mathematics?

- What are the six trigonometric functions of an angle θ in standard position? What is the definition of each function in terms of x, y, and r?

- What is an extraneous solution to an equation?

DATA*Activity*

Moving at the Speed of Sound

The rate at which sound travels depends on the composition and temperature of the medium it passes through. Although air is the most common medium, sound can travel through any type of matter. Liquids and solids are better transmitters of sound than air. As the temperature of the medium rises, so does the speed of sound. In a vacuum, sound cannot travel at all.

SKILL FOCUS

- Add, subtract, multiply, and divide real numbers.

- Determine median and range for a data set.

- Convert metric units.

- Identify a functional relationship.

- Determine the rate of change and apply the result.

SOUND

In this chapter you will see how:

- **PIANO TUNERS** make use of the properties of sound waves to model specific musical notes.
  (Lesson 16.3, page 811)

- **AERODYNAMIC ENGINEERS** determine the Mach number of an aircraft and locate the area where a sonic boom is heard.
  (Lesson 16.4, page 820)

SPEED OF SOUND	
Substance	**Speed at 0°C (m/s)**
air	332
hydrogen	1284
alcohol	1207
water	1498
aluminum	5000
brick	3650
copper	3810
cork	500
glass	4540
iron (steel)	5200
maple (wood)	4110

Temperature (°C)	Speed in Air (m/s)
0	332
20	344
50	362

Use the table to answer the following questions.

1. What is the median speed for the substances shown? What is the range of speeds for this group of substances?

2. In 0.25 s, how much farther would a sound travel in maple wood than in brick?

3. About how long would it take sound to be transmitted through 10 km of copper wire?

4. Based on the data shown, what type of relationship exists between speed of sound and temperature?

5. Determine the increase in the speed of sound for every 1°C rise in temperature. What do you predict the speed to be at 65°C?

6. **WORKING TOGETHER** Make two graphs showing the speed of sound in aluminum, in glass, and in steel. Both graphs must be accurate, but should give different visual impressions of how the speeds compare. One graph should exaggerate the differences while the other makes the differences seem small.

PROJECT

Listen Here!

If you stop and listen, you can hear that your everyday world is filled with sounds—music, noise, talk, and laughter. Do the sounds you hear have something in common besides your hearing them?

PROJECT GOAL

To build a device to demonstrate properties of sound and sound waves.

Getting Started

Work in groups of four or five students.

1. Brainstorm a list of features about sounds. For example, what are some sources of sounds? How do you describe sounds?

2. You will build a sonameter, which is a simple device you will use to investigate sound in the Project Connections. For one sonameter, you will need the following materials: a wood board (about 10 cm × 50 cm), 2 eye screws, 2 long round pencils, 100 cm of nylon string or fishing line.

3. To construct a sonameter, screw the eye screws into opposite ends of the board. Securely tie one end of the piece of nylon string to one eye screw. Stretch the string until it is tight and tie it to the other screw. At each end of the board, slide a pencil beneath the string.

4. Pluck the string of the sonameter and listen to the sound. Describe the sound using the terms you brainstormed in Question 1.

PROJECT Connections

Lesson 16.2, page 803
Use trigonometric functions to represent sound waves.

Lesson 16.3, page 810
Relate the pitch and frequency of sound to the length of a vibrating string.

Lesson 16.4, page 819
Investigate factors affecting intensity of sound.

Chapter Assessment, page 835
Design and conduct an investigation of how string characteristics affect pitch.

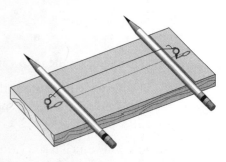

Internet Connection

www.swpco.com/
swpco/algebra2.html

790

Prove Trigonometric Identities

Explore

• Use the definitions of the six trigonometric functions defined in terms of x, y, and r. Simplify each product or quotient.

1. $\sin \theta \csc \theta$

2. $\cos \theta \sec \theta$

3. $\tan \theta \cot \theta$

4. $\dfrac{\sin \theta}{\cos \theta}$

5. $\dfrac{\cos \theta}{\sin \theta}$

6. $\dfrac{\sec \theta}{\csc \theta}$

7. Use your results from Questions 1–3 to define $\csc \theta$ in terms of $\sin \theta$, $\sec \theta$ in terms of $\cos \theta$, and $\cot \theta$ in terms of $\tan \theta$.

8. Which trigonometric function can be represented by your solution to Question 4?

9. Which trigonometric function can be represented by your solution to Exercise 5?

10. Use your calculator. Copy and complete the following table. Round entries to four decimal places.

θ	$\sin \theta$	$\cos \theta$	$\tan \theta$
36°			
−36°			
164°			
−164°			
285°			
−285°			

11. Use your results from Question 9 to express $\sin(-\theta)$ in terms of $\sin \theta$, $\cos(-\theta)$ in terms of $\cos \theta$, and $\tan(-\theta)$ in terms of $\tan \theta$.

12. If $\sin^2 \theta$ is defined to be $(\sin \theta)^2$ and $\cos^2 \theta$ is defined to be $(\cos \theta)^2$, determine the value of $\sin^2 \theta + \cos^2 \theta$ for each angle in the table in Question 9. Make a prediction about the value of $\sin^2 \theta + \cos^2 \theta$.

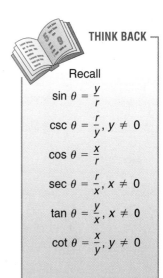

THINK BACK

Recall

$$\sin \theta = \frac{y}{r}$$

$$\csc \theta = \frac{r}{y}, y \neq 0$$

$$\cos \theta = \frac{x}{r}$$

$$\sec \theta = \frac{r}{x}, x \neq 0$$

$$\tan \theta = \frac{y}{x}, x \neq 0$$

$$\cot \theta = \frac{x}{y}, y \neq 0$$

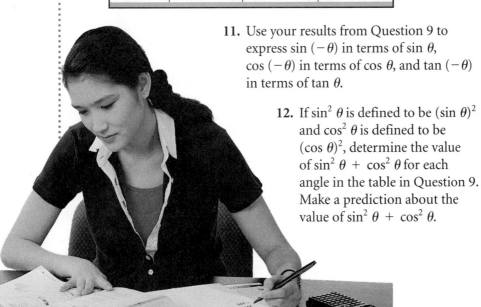

PROBLEM — SOLVING TIP

Along with θ, the Greek letters α (alpha) and β (beta) are commonly used as angle variables.

Build Understanding

- An **identity** is an equation that is true for all values of the variable, except those for which either side of the equation is undefined. Trigonometry has many identities. You have already developed some of the *fundamental identities* in Explore.

THE RECIPROCAL IDENTITIES

$$\sin \theta = \frac{1}{\csc \theta} \qquad \csc \theta = \frac{1}{\sin \theta} \qquad \sin \theta \csc \theta = 1$$

$$\cos \theta = \frac{1}{\sec \theta} \qquad \sec \theta = \frac{1}{\cos \theta} \qquad \cos \theta \sec \theta = 1$$

$$\tan \theta = \frac{1}{\cot \theta} \qquad \cot \theta = \frac{1}{\tan \theta} \qquad \tan \theta \cot \theta = 1$$

THE RATIO IDENTITIES

$$\tan \theta = \frac{\sin \theta}{\cos \theta} \qquad \cot \theta = \frac{\cos \theta}{\sin \theta}$$

THE COFUNCTION IDENTITIES

$$\sin \theta = \cos (90° - \theta) \qquad \cos \theta = \sin (90° - \theta)$$

$$\tan \theta = \cot (90° - \theta) \qquad \cot \theta = \tan (90° - \theta)$$

$$\csc \theta = \sec (90° - \theta) \qquad \sec \theta = \csc (90° - \theta)$$

THE PYTHAGOREAN IDENTITIES

$$\sin^2 \theta + \cos^2 \theta = 1$$

$$1 + \cot^2 \theta = \csc^2 \theta$$

$$1 + \tan^2 \theta = \sec^2 \theta$$

THE NEGATIVE ANGLE IDENTITIES

$$\sin (-\theta) = -\sin \theta \qquad \csc (-\theta) = -\csc \theta$$

$$\cos (-\theta) = \cos \theta \qquad \sec (-\theta) = \sec \theta$$

$$\tan (-\theta) = -\tan \theta \qquad \cot (-\theta) = -\cot \theta$$

You can use these identities, and substitution to simplify trigonometric expressions. You can also use the fundamental identities to simplify trigonometric expressions.

EXAMPLE 1

Simplify using the fundamental identities.

  a. $\cos (90° - \beta) \cot \beta$

  b. $(1 + \sin \theta)(1 - \sin \theta)$

  c. $\tan (-\alpha) \cos (-\alpha)$

Solution

  a. $\cos (90° - \beta) \cot \beta = \sin \beta \cot \beta$ Cofunction identity
  $$= \sin \beta \left(\frac{\cos \beta}{\sin \beta}\right) \quad \text{Ratio identity}$$
  $$= \cos \beta$$

  b. $(1 + \sin \theta)(1 - \sin \theta) = 1 - \sin^2 \theta$ Multiply.
  $$= \cos^2 \theta \quad \text{Pythagorean identity}$$

  c. $\tan (-\alpha) \cos (-\alpha) = (-\tan \alpha)(\cos \alpha)$ Negative angle identity
  $$= \left(-\frac{\sin \alpha}{\cos \alpha}\right)(\cos \alpha) \quad \text{Ratio identity}$$
  $$= -\sin \alpha \qquad \blacktriangleleft$$

To prove that an equation is an identity, write equivalent expressions for one side of the equation until you obtain an expression that is identical to the other side of the equation.

EXAMPLE 2

Prove each identity.

  a. $\sin^2 \theta + \sin \theta \cos \theta \cot \theta = 1$

  b. $\sec \alpha - \sin \alpha \tan \alpha = \cos \alpha$

Solution

  a.

$\sin^2 \theta + \sin \theta \cos \theta \cot \theta$	1	
$\sin^2 \theta + \sin \theta \cos \theta \dfrac{\cos \theta}{\sin \theta}$		Ratio identity
$\sin^2 \theta + \cos^2 \theta$		
	1	Pythagorean identity

  b.

$\sec \alpha - \sin \alpha \tan \alpha$	$\cos \alpha$	
$\dfrac{1}{\cos \alpha} - \sin \alpha \cdot \dfrac{\sin \alpha}{\cos \alpha}$		Reciprocal and ratio identities
$\dfrac{1}{\cos \alpha} - \dfrac{\sin^2 \alpha}{\cos \alpha}$		Multiply.
$\dfrac{1 - \sin^2 \alpha}{\cos \alpha}$		Add.
$\dfrac{\cos^2 \alpha}{\cos \alpha}$		Pythagorean identity
$\cos \alpha$		Simplify.

$\blacktriangleleft$

PROBLEM SOLVING TIP

When you are proving an identity by writing equivalent expressions for one side of the equation, try to transform the more complicated side into the simpler side.

PROBLEM SOLVING TIP

When tangent, cotangent, secant or cosecant occur, try to express them in terms of sines and cosines.

CHECK UNDERSTANDING

Show how the Pythagorean identity, $\sin^2 \theta + \cos^2 \theta = 1$ can be derived from the Pythagorean theorem.

Another way to prove that a trigonometric equation is an identity is to write equivalent expressions for *both* sides of the equation.

The vertical line that separates the left side from the right side of a trigonometric equation indicates that you must work on each side separately. You may not perform operations such as multiplication, addition, or squaring on both sides of the equation.

EXAMPLE 3

Prove: $\csc \beta - \sin \beta = \dfrac{\cos \beta}{\tan \beta}$

Solution

$$
\begin{array}{c|c}
\csc \beta - \sin \beta & \dfrac{\cos \beta}{\tan \beta} \\
\hline
\dfrac{1}{\sin \beta} - \sin \beta & \dfrac{\cos \beta}{\dfrac{\sin \beta}{\cos \beta}} \\
\end{array}
$$
$\tan \beta = \dfrac{\sin \beta}{\cos \beta}$

Common denominator $\dfrac{1 - \sin^2 \beta}{\sin \beta}$ $\Big|$ $\dfrac{\cos^2 \beta}{\sin \beta}$ Simplify.

Pythagorean identity $\dfrac{\cos^2 \beta}{\sin \beta}$ $\Big|$ $\dfrac{\cos^2 \beta}{\sin \beta}$ ◄

Identities are used in real world situations.

EXAMPLE 4

OPTICS For any substance, there is an angle at which the reflected light is polarized. This angle, known as the *polarizing angle* can be found using Brewster's law. Brewster's law states the relationship between the two indices of refraction, n_1 and n_2 and the polarizing angle θ_p as $\tan \theta_p = \dfrac{n_2}{n_1}$. Use the law of refraction $n_1 \sin \theta_p = n_2 \sin \theta_r$ to prove Brewster's law.

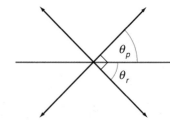

Solution

As shown, $\theta_r + \theta_p = 90°$. The reflected and refracted beams of light are perpendicular. So, $\theta_r = 90° - \theta_p$.

$$
\begin{aligned}
n_1 \sin \theta_p &= n_2 \sin \theta_r && \text{Law of refraction} \\
&= n_2 \sin (90° - \theta_p) && \text{Substitution} \\
&= n_2 \cos \theta_p && \text{Cofunction identity} \\
\dfrac{\sin \theta_p}{\cos \theta_p} &= \dfrac{n_2}{n_1} && \text{Multiply both sides by } \dfrac{1}{n_1 \cos \theta_p}. \\
\tan \theta_p &= \dfrac{n_2}{n_1} && \text{Ratio identity} \quad ◄
\end{aligned}
$$

Use the numbers 1 and −1 to make each equation into an identity.

1. $\csc^2 \theta - \cot^2 \theta = \blacksquare$

2. $\tan \theta \cot \theta = \blacksquare$

3. $\dfrac{\sin \theta}{\sin (-\theta)} = \blacksquare$, $\sin (-\theta) \neq 0$

4. $\dfrac{\cos (-\theta)}{\cos \theta} = \blacksquare$, $\cos (\theta) \neq 0$

5. WRITING MATHEMATICS The negative angle identities are often called odd-even identities. Why do you think this is so? Which trigonometric functions are even? Which trigonometric functions are odd?

Write each expression in terms of sine or in terms of cosine.

6. $\cot \theta \sin \theta$

7. $\dfrac{\cos \theta}{\cot \theta}$

8. $\dfrac{\tan \theta}{\cos \theta}$

9. $(\sec^2 \theta - 1) (\cos^2 \theta)$

Simplify each expression.

10. $(1 - \cos \alpha) (1 + \cos \alpha)$

11. $\sin (-\beta) \cot (-\beta)$

12. $\cot (90° - \theta) \cos (-\theta)$

13. $(1 - \sec \beta) (1 + \sec \beta)$

14. ENGINEERING A 15 ft walkway between the Taylors' house and their garage is at a higher elevation near the garage. The sine of the angle of elevation from the house to the garage is 0.2222.

  a. Use a Pythagorean identity to determine the cosine of the angle of elevation.

  b. Determine the distance between the house and the garage.

  c. Determine the angle of elevation to the nearest tenth of a degree.

Prove each identity.

15. $\dfrac{\cos \alpha \tan \alpha}{\sin \alpha} = 1$

16. $\csc \beta (1 - \cos^2 \beta) = \sin \beta$

17. $\cos \alpha \csc \alpha = \dfrac{1}{\tan \alpha}$

18. $\dfrac{\cos \beta - \sin \beta - \cos^3 \beta}{\cos \beta} = \sin^2 \beta - \tan \beta$

Use the numbers 1 and −1 to make each equation into an identity.

1. $-\cos^2 \theta - \sin^2 \theta = \blacksquare$

2. $\sec (-\theta) \cos (-\theta) = \blacksquare$

3. $\dfrac{\cot (-\theta)}{\cot \theta} = \blacksquare$ $\cot \theta \neq 0$

4. $\dfrac{\cos \theta}{\sin \theta \cot \theta} = \blacksquare$, $\sin \theta \cot \theta \neq 0$

5. WRITING MATHEMATICS Compile a list of helpful strategies for proving identities. Compare your list with those of other students, and add to it as needed.

Write each expression in terms of sine or in terms of cosine.

6. $\sec \theta \tan \theta$

7. $\dfrac{\cos \theta}{\sec \theta}$

8. $\dfrac{\sin \theta}{\cot \theta}$

9. $(\csc^2 \theta - 1) (\sin^2 \theta)$

Simplify each expression.

10. $\dfrac{\cot (-\alpha)}{\csc (\alpha)}$

11. $\cos (-\beta) \tan (-\beta)$

12. $\tan (90° - \theta) \sin (-\theta)$

13. $(1 - \csc \beta) (1 + \csc \beta)$

14. **ENGINEERING** An escalator has a length of 415 ft and a vertical rise of 89.6 ft.

   a. Determine the cosecant of the angle of elevation.

   b. Use a Pythagorean identity to determine the cotangent of the angle of elevation.

   c. Find the angle of elevation to the nearest tenth of a degree.

Prove each identity.

15. $\sec^2 \alpha - 3 = \tan^2 \alpha - 2$

16. $\csc^2 \beta + 6 = \cot^2 \beta + 7$

17. $\dfrac{1}{\sin^2 \theta} + \dfrac{1}{\cos^2 \theta} = \dfrac{1}{\sin^2 \theta \cos^2 \theta}$

18. $\cos^2 \alpha = 1 - \dfrac{1}{\csc^2 \alpha}$

19. $\dfrac{\sin^2 \theta - 4 \sin \theta + 4}{\sin^2 \theta - 4} = \dfrac{\sin \theta - 2}{\sin \theta + 2}$

20. $\dfrac{\cos^2 \beta - 10 \cos \beta + 25}{\cos^2 \beta - 25} = \dfrac{\cos \beta - 5}{\cos \beta + 5}$

21. $\dfrac{1}{\cot^2 \alpha} - \dfrac{1}{\csc^2 \alpha} = \dfrac{\sin^4 \alpha}{\cos^2 \alpha}$

22. $\cot \beta + \tan \beta = \csc \beta \cdot \sec \beta$

23. $(\tan \theta + 1)^2 = \sec^2 \theta + 2 \tan \theta$

24. $\dfrac{1 - \cos^2 \alpha}{\cos \alpha - 1} = \dfrac{-\sec \alpha - 1}{\sec \alpha}$

25. $\dfrac{\sec \beta \sin \beta}{\csc \beta \cos \beta} = \sec^2 \beta - 1$

26. $\dfrac{1 - \cos \theta}{1 + \cos \theta} = (\cot \theta - \csc \theta)^2$

27. $\dfrac{\sin \alpha}{\csc \alpha + \cot \alpha} = 1 - \cos \alpha$

28. $\dfrac{\cos \theta}{\sec \theta + \tan \theta} = 1 - \sin \theta$

29. **GEOMETRICAL OPTICS** When a ray of light passes from one medium into a second, an angle of incidence, known as the polarizing angle, can be found experimentally such that the reflected light is completely polarized. Recall $\tan \theta_p = \dfrac{n_2}{n_1}$, where θ_p represents the polarizing angle, n_2 is the index of refraction of medium two and n_1 is the index of refraction of medium one.

   a. Express this equation in terms of $\cot \theta_p$.

   b. The number of refraction of air is 1.0 and the index of refraction of a diamond is 2.42. Find the polarizing angle if air is medium 1.

EXTEND

You can use a graphing utility to help you determine if a trigonometric equation is not an identity. Graph the left side of the equation and the right side of the equation on the same set of axes. If the graphs do not coincide, the equation is not an identity. If the graphs of the equations do coincide, the equation may be an identity. You must then verify the identity algebraically, because the two graphs can appear to be the same even though they are not.

Use a graphing utility to determine if each equation may be an identity. Write *yes* or *no*.

30. $\sin \theta \cos \theta \tan \theta = \cos^2 \theta$

31. $\cos \theta \csc \theta \tan \theta = 1$

32. $\sec \beta \csc \beta = \tan \beta + \cot \beta$

33. $\dfrac{\sin \alpha}{\cos \alpha} = \dfrac{\csc \alpha}{\sec \alpha}$

You can also prove that an equation is *not* an identity by finding a *counterexample*. For example, you can show that $\sqrt{1 - \cos^2 \alpha} = \sin \alpha$ is not an identity by finding a value of α for which the statement is false. If you substitute $\alpha = 270°$, the left side of the equation is 1, and the right side is -1. Since $1 \neq -1$, $\sqrt{1 - \cos^2 \alpha} = \sin \alpha$ is not an identity.

Find a counterexample to prove that each equation is not an identity.

34. $\sqrt{\cos^2 \theta} = \cos \theta$

35. $\sqrt{\sin^2 \theta} = \sin \theta$

36. $\sqrt{1 + \tan^2 \theta} = \sec \theta$

37. $\sqrt{1 + \cot^2 \theta} = \csc \theta$

THINK CRITICALLY

38. Find three values of k for which $k \sin \beta = \sin k \beta$ is an identity.

39. Write a trigonometric equation that is not an identity, but appears to be when graphed.

40. A statement such as $\cos \alpha = 2$ that is false for all values of α is called a *contradiction*. Write three contradictions using other trigonometric functions.

Prove each identity.

41. $\sec^2 \theta - \sin^2 \theta = \cos^2 \theta + \tan^2 \theta$

42. $4 \sec^2 \beta - 5 = 4 \tan^2 \beta - 1$

MIXED REVIEW

Solve each logarithmic equation.

43. $\log_5 x = 4$

44. $\log_x 81 = 4$

45. STANDARDIZED TESTS The geometric mean between -36 and -100 is:

A. -60 **B.** -60 **C.** -100 **D.** -60 or 60

46. How many different 8-card hands can you select from a deck of 52 playing cards?

Prove each identity.

47. $2 \csc^2 \theta = \dfrac{1}{1 - \cos \theta} + \dfrac{1}{1 + \cos \theta}$

48. $\cos \beta = \dfrac{\sin \beta}{\tan \beta}$

Solve Trigonometric Equations

Explore

You have learned to use a graphing utility to solve an equation of the form $3x^2 = \sqrt{3} + x^2$ by graphing $y_1 = 3x^2$ and $y_2 = \sqrt{3} + x^2$ on the same set of axes and then finding their points of intersection.

1. Consider the equation $3 \sin x = \sqrt{3} + \sin x$. What expressions would you enter for y_1 and y_2 to solve this equation graphically?

2. Use the procedure you developed in Question 1 to graph the equation $3 \sin x = \sqrt{3} + \sin x$ on the interval $-360° \le x < 360°$.

3. Use your graphing utility to find the x-values of the points at which the graphs of y_1 and y_2 intersect on the interval $-360° \le x \le 360°$.

4. Continue to trace to the right of 360°. Name the x-values of the next four points at which the graphs of y_1 and y_2 intersect.

5. How are the values –300°, 60°, 420°, and 780° related?

6. How are the values –240°, 120°, 480°, and 840° related?

7. Use your results from Questions 5 and 6 to predict the x-values of the first two points to the left of –360° at which the graphs of y_1 and y_2 intersect.

Build Understanding

Recall trigonometric identities are true for all values of the variable, except those for which either side of the equation is undefined. Most trigonometric equations are true for only some values of the variable. For example, $\sin \alpha = \cos (90° - \alpha)$ is true for all values of α, but $\cos \beta = -1$ is true for only some values of β.

EXAMPLE 1

Solve $2 \cos x + 1 = 0$ on the interval $0 \le x < 2\pi$.

Solution

$$2 \cos x + 1 = 0$$
$$2 \cos x = -1$$
$$\cos x = -\frac{1}{2}$$

The graphs of $y_1 = \cos x$ and $y_2 = -\frac{1}{2}$ are displayed on the same set of axes at the top of the next page.

Any point at which they intersect is a solution to $\cos x = -\frac{1}{2}$.

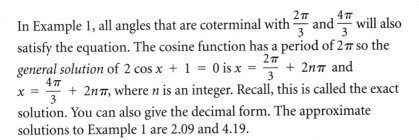

The solutions are $\frac{2\pi}{3}$ and $\frac{4\pi}{3}$ on the interval $0 \le x < 2\pi$. ◀

In Example 1, all angles that are coterminal with $\frac{2\pi}{3}$ and $\frac{4\pi}{3}$ will also satisfy the equation. The cosine function has a period of 2π so the *general solution* of $2 \cos x + 1 = 0$ is $x = \frac{2\pi}{3} + 2n\pi$ and $x = \frac{4\pi}{3} + 2n\pi$, where n is an integer. Recall, this is called the exact solution. You can also give the decimal form. The approximate solutions to Example 1 are 2.09 and 4.19.

You can combine algebra with trigonometric identities to solve many trigonometric equations.

EXAMPLE 2

Solve $\sec^2 x - 4 = 0$ on the interval $0° \le x < 360°$.

Solution

$$\sec^2 x - 4 = 0$$
$$\sec^2 x = 4$$
$$\sec x = \pm 2 \qquad \text{Take the square root of each side.}$$
$$\sec x = 2 \text{ and } \sec x = -2$$

The solutions are 60°, 120°, 240°, 300°. ◀

You can use factoring and the zero product property to solve some trigonometric equations.

EXAMPLE 3

Solve $2 \sin^2 x + 3 \sin x + 1 = 0$ on the interval $0° \le x < 360°$.

Solution

$$2 \sin^2 x + 3 \sin x + 1 = 0$$
$$(2 \sin x + 1)(\sin x + 1) = 0 \qquad \text{Factor.}$$
$$2 \sin x + 1 = 0 \text{ or } \sin x + 1 = 0 \qquad \text{Zero product property.}$$
$$2 \sin x = -1 \text{ or } \sin x = -1$$
$$\sin x = -\frac{1}{2} \text{ or } \sin x = -1$$

Use a reference triangle to find the solutions 210°, 270°, and 330°. ◀

COMMUNICATING ABOUT ALGEBRA

The solution to the equation $\tan \theta = -1$, $0 \le \theta < \pi$, is $\frac{3\pi}{4}$. What is the general solution? Why?

PROBLEM SOLVING TIP

The mode of your calculator should always match the parameters of the problem. In Example 3, use degree mode.

THINK BACK

Using a reference triangle, you can find $x = 60°, 300°$ and $x = 120°, 240°$.

THINK BACK

Recall the zero product property, which states that for all real numbers a and b, $ab = 0$ if and only if $a = 0$ or $b = 0$.

CHECK ────
UNDERSTANDING

Why can you solve trigonometric equations by taking the square root of both sides but not prove identities this way?

COMMUNICATING ─┐
ABOUT ALGEBRA

In Example 4, the equation was written in terms of a single function, sine. Would you get the same solutions if the equation was written in terms of cosine? Explain.

When a trigonometric equation contains more than one function, use the trigonometric identities to write it in terms of a single function.

EXAMPLE 4

Solve $\csc^2 x \sin x = 2 - \sin x$ on the interval $0 \leq x < 2\pi$.

$$\csc^2 x \sin x = 2 - \sin x$$

$$\frac{1}{\sin^2 x} \cdot \sin x = 2 - \sin x$$

$$\frac{1}{\sin x} = 2 - \sin x \qquad \sin^2 x \neq 0 \text{ or } \sin x \neq 0$$

$$1 = 2 \sin x - \sin^2 x \qquad \text{Multiply both sides by } \sin x.$$

$$\sin^2 x - 2 \sin x + 1 = 0$$

$$(\sin x - 1)(\sin x - 1) = 0 \qquad \text{Factor.}$$

$$\sin x - 1 = 0 \qquad \text{Zero product property}$$

$$\sin x = 1$$

$$x = \frac{\pi}{2}$$

The solution is $\frac{\pi}{2}$. Check your answer by substitution and make sure that it does not violate the restriction that $\sin x \neq 0$. ◀

The equations in Examples 1–4 have exact solutions. Since most trigonometric equations do not have exact solutions, you will have to approximate them.

EXAMPLE 5

ROBOTICS You can model the angle x formed by the "knee" of a robot leg using the formula $2.3 \cos^2 x + 1.3 \cos x - 1.2 = 0$. Use a graphing utility to determine x, where $0 \leq x < \frac{\pi}{2}$. Round your answer to the nearest hundredth of a radian.

Solution
The graph crosses the x-axis at about 1.06, so $x \approx 1.06$ radians.

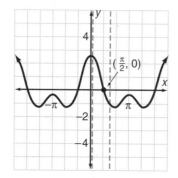

If $0° \leq x < 360°$, how many solutions does each equation have?

1. $5 \cos x = 4$ **2.** $\tan x = 5$ **3.** $\tan^2 x = 6$ **4.** $6 \cos x + 4 = -2$

5. WRITING MATHEMATICS Describe the difference between a trigonometric equation and a trigonometric identity.

Solve for x, where $0° \leq x < 360°$.

6. $2 \sin x + \sqrt{3} = 0$ **7.** $\sqrt{3} \csc x - 2 = 0$

8. $2 \sin^2 x = 3 \cos x$ **9.** $3 \sec^2 x - 7 \sec x + 2 = 0$

Solve for x, where $0 \leq x < 2\pi$. Give the exact solution if possible, otherwise, round to the nearest hundredth of a radian.

10. $\sqrt{2} \sin x \cos x - \sin x = 0$ **11.** $\tan^2 x - 4 \sec x + 5 = 0$

12. $6 \cos^2 x + 7 \sin x = 3$ **13.** $16 \sec^2 x - 42 \tan x = 11$

14. MUSIC You can represent a specific sound wave generated by a tuning fork using the equation $\sin x + \cos x = \dfrac{\sqrt{3}}{3}$. Use a graphing utility to determine the solution to this equation on the interval $0 \leq x < 2\pi$. Round your answer to the nearest hundredth of a radian.

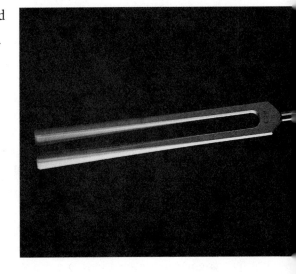

Use a graphing utility to solve each equation for all values of x if x is measured in radians.

15. $\tan x = \sqrt{3}$

16. $\cos x = \cot x$

17. $20 \sin^2 x - 13 \sin x + 2 = 0$

18. $10 \cos^2 x - 23 \cos x - 5 = 0$

PRACTICE

If $0° \leq x < 360°$, how many solutions does each equation have?

1. $\sin x - \csc x = 0$ **2.** $\cot^2 x = 7$ **3.** $\tan x = -1$ **4.** $\cos x - 1 = -2$

5. WRITING MATHEMATICS Compile a list of helpful strategies for solving trigonometric equations. Compare your list with those of other students, and add to it as needed.

Solve for x, where $0° \leq x < 360°$. Check your answer.

6. $2 \cos x - \sqrt{3} = 0$ **7.** $\sqrt{2} \csc x = 2$

8. $-2 \cos^2 x = 3 \sin x$ **9.** $4 \csc^2 x + 5 \csc x = 6$

Solve for x, where $0 \leq x < 2\pi$. Give the exact solution, otherwise, round to the nearest hundredth of a radian.

10. $3 \tan^2 x - 1 = 0$

11. $4 \sin^2 x - 1 = 0$

12. $15 \sin^2 x - 17 \cos x - 11 = 0$

13. $18 \cos^2 x + 27 \sin x - 22 = 0$

14. $3 \cot x \cos x + \cos x = 0$

15. $5 \tan x \sin x + \sin x = 0$

16. GEOMETRY The area of an isosceles triangle can be modeled by the equation $A = \frac{1}{2}s^2 \sin \alpha$ where s is the length of each of the two congruent sides and α is the angle formed by these sides. If the length of a side is 12 cm and the area is 55 cm^2, determine the angle formed by the congruent sides to the nearest hundredth of a radian.

Solve each equation for all values of x if x is measured in radians. Check by substituting into the original equation.

17. $9 \sin^2 x - 6 \sin x + 1 = 0$

18. $4 \sin x - 3 = 0$

19. $4 \cos^2 x - 4 \cos x + 1 = 0$

20. $5 \sec x + 9 = 0$

Use a graphing utility to solve for x, where $0 \leq x < 2\pi$.

21. $5 \sin^2 x - 4 \sin x - 2 = 0$

22. $6 \tan^2 x + \tan x - 3 = 0$

23. $3 \cos^2 x + 6 \cos x + 2 = 0$

24. $4 \cot^2 x + \cot x - 1 = 0$

25. PHYSICS The work W done by a girl pulling an 8 lb sled along a horizontal surface at a constant speed can be calculated using the equation $W = Pd \cos \theta$, where P is the pulling force, d is the distance, and θ is the angle the rope makes with the horizontal. If the work is 27 foot-pounds, the pulling force is 1.9 lb, and the distance is 20 ft, determine θ to the nearest degree.

EXTEND

Using certain algebraic techniques, such as squaring both sides of the equation or dividing both sides of the equation by a trigonometric function, to solve trigonometric equations may cause you to introduce an extraneous solution or omit a solution. Therefore, you must use substitution or a graphing utility to check your solutions.

Solve for x algebraically, where $0° \leq x < 360°$. Then check using substitution or a graphing utility.

26. $1 - \tan x = \sec x$ **27.** $1 - \cot x = \csc x$ **28.** $\sin x \tan x = \sin x$ **29.** $\cos x \cot x = \cos x$

SIMPLE HARMONIC MOTION The displacement y (in inches) of a weight that is oscillating up and down from its rest position can be modeled by the equation $y = 8 \cos \frac{\pi}{2}(t - 1)$ where t is measured in seconds. Round answers to 30–31 to nearest hundredth of a second.

30. Find the first time at which the displacement will be 2 in.

31. Find the first time at which the displacement will be 4 in.

32. WRITING MATHEMATICS Explain why you are asked to solve trigonometric equations on the interval $0° \leq x < 360°$ instead of $0° \leq x \leq 360°$.

33. For what value of k is $(\cos x + \sin x)^2 + (\cos x - \sin x)^2 = k$ an identity?

Solve for x, where $0° \leq x < 360°$.

34. $2 \sin x \leq 1$ **35.** $2 \cos x \leq \sqrt{3}$ **36.** $|\cos x| \leq \dfrac{\sqrt{3}}{2}$ **37.** $|\sin x| \geq \dfrac{1}{2}$

MIXED REVIEW

Determine each product.

38. $\begin{bmatrix} 2 & -1 & 4 \\ 0 & -2 & -3 \\ 1 & 1 & 1 \end{bmatrix} \begin{bmatrix} 5 & 6 \\ -4 & 3 \\ 2 & 0 \end{bmatrix}$

39. $\begin{bmatrix} 8 \\ 6 \\ 3 \end{bmatrix} [2 \ 5]$

Simplify each expression.

40. $e^8 \cdot e^{-2}$

41. $(27e^{18})^{\frac{1}{3}}$

42. STANDARDIZED TESTS The number of ways a cup with two different flavors of frozen yogurt can be made from a selection of 12 flavors is

    A. 479,001,600   **B.** 132   **C.** 66   **D.** 6

Use a graphing utility to solve for x, where $0 \leq x < 2\pi$.

43. $\cos^2 x - 4 \sin x - \dfrac{1}{2} = 0$

44. $2 \sin^2 x + 5 \cos x = 0$

Simplify.

45. $(3 - 4i) + (5 - i)$ **46.** $i^{23} - i$ **47.** $\dfrac{8}{2i}$

PROJECT *Connection*   Pluck the string of the sonameter. Your ear detects sounds as small changes of air pressure on the ear drum. The pressure changes are *sound waves* and can be described by the function $y = a \sin bt$ where t represents the time measured in seconds.

The *period* of the sound wave is the time for one cycle of the wave to occur. The *frequency* of the sound wave, measured in hertz (Hz), describes the number of cycles that occur each second. A sound wave described by the function $y = a \sin bt$ has a period of $\dfrac{2\pi}{|b|}$ s and a frequency of $\dfrac{|b|}{2\pi}$ Hz.

  1. If the sound wave of a musical tone is described by the function $y = 4 \sin 880\pi t$, determine the period and frequency of the sound wave.

  2. Humans can hear sound waves with frequencies between 16 and 20,000 Hz. What is the range of periods for these sound waves?

16.3 The Sum and Difference Identities

Explore

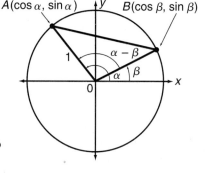

Angles α and β are shown in standard position on the unit circle.

1. What is the measure of $\angle$BOA?

2. Use the law of cosines to find (AB)

3. Now use the distance formula to find AB.

4. Set your results from Question 3 and 4 equal to one another and simplify the resulting equation.

Build Understanding

The identities you studied in previous lessons involved only *one* angle. In Explore, you derived the identity for the cosine of the difference of *two* angles: $\cos(\alpha - \beta) = \cos\alpha\cos\beta + \sin\alpha\sin\beta$. The following identities can be used to evaluate trigonometric functions of the sum or difference of two angles.

> **THE SUM AND DIFFERENCE IDENTITIES**
>
> $\cos(\alpha - \beta) = \cos\alpha\cos\beta + \sin\alpha\sin\beta$
>
> $\cos(\alpha + \beta) = \cos\alpha\cos\beta - \sin\alpha\sin\beta$
>
> $\sin(\alpha - \beta) = \sin\alpha\cos\beta - \cos\alpha\sin\beta$
>
> $\sin(\alpha + \beta) = \sin\alpha\cos\beta + \cos\alpha\sin\beta$
>
> $\tan(\alpha - \beta) = \dfrac{\tan\alpha - \tan\beta}{1 + \tan\alpha\tan\beta}$
>
> $\tan(\alpha + \beta) = \dfrac{\tan\alpha + \tan\beta}{1 - \tan\alpha\tan\beta}$

You can use the sum and difference identities to prove other identities.

EXAMPLE 1

Prove $\quad 2\sin\alpha\cos\beta = \sin(\alpha - \beta) + \sin(\alpha + \beta)$

Solution

$2\sin\alpha\cos\beta$	$\sin(\alpha - \beta) + \sin(\alpha + \beta)$
	$\sin\alpha\cos\beta - \cos\alpha\sin\beta + \sin\alpha\cos\beta + \cos\alpha\sin\beta$
	$2\sin\alpha\cos\beta$ ◄

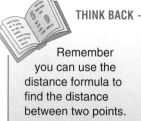

THINK BACK

Remember you can use the distance formula to find the distance between two points.

The sum and difference identities enable you to use angles with known trigonometric ratios to determine values for other angles.

EXAMPLE 2

Determine the exact value of each expression using the sum and difference identities.

   a. $\cos 75°$ **b.** $\sin \dfrac{\pi}{12}$ **c.** $\tan \dfrac{5\pi}{12}$

Solution

a. $\cos 75° = \cos(45° + 30°)$ *Use two angles with known cosines.*

$$= \cos 45° \cos 30° - \sin 45° \sin 30°$$

$$= \frac{\sqrt{2}}{2}\left(\frac{\sqrt{3}}{2}\right) - \frac{\sqrt{2}}{2}\left(\frac{1}{2}\right)$$

$$= \frac{\sqrt{6} - \sqrt{2}}{4}$$

b. $\sin \dfrac{\pi}{12} = \sin\left(\dfrac{\pi}{4} - \dfrac{\pi}{6}\right) = \sin \dfrac{\pi}{4} \cos \dfrac{\pi}{6} - \cos \dfrac{\pi}{4} \sin \dfrac{\pi}{6}$

$$= \frac{\sqrt{2}}{2}\left(\frac{\sqrt{3}}{2}\right) - \frac{\sqrt{2}}{2}\left(\frac{1}{2}\right)$$

$$= \frac{\sqrt{6} - \sqrt{2}}{4}$$

c. $\tan \dfrac{5\pi}{12} = \tan\left(\dfrac{\pi}{4} + \dfrac{\pi}{6}\right) = \dfrac{\tan \dfrac{\pi}{4} + \tan \dfrac{\pi}{6}}{1 - \tan \dfrac{\pi}{4} \tan \dfrac{\pi}{6}}$

$$= \frac{1 + \dfrac{\sqrt{3}}{3}}{1 - 1\left(\dfrac{\sqrt{3}}{3}\right)}$$

$$= \frac{3 + \sqrt{3}}{3 - \sqrt{3}}$$

$$= \frac{3 + \sqrt{3}}{3 - \sqrt{3}} \cdot \frac{3 + \sqrt{3}}{3 + \sqrt{3}}$$

$$= \frac{12 + 6\sqrt{3}}{6}$$

$$= 2 + \sqrt{3}$$

If you are given the exact values of one trigonometric function of α and one trigonometric function of β, then you can determine the exact value of a trigonometric function of the sum or difference of these angles. Note that you do not have to determine the measures of α and β.

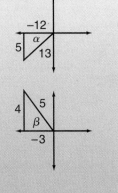

EXAMPLE 3

Determine $\cos(\alpha + \beta)$ given $\sin \alpha = -\dfrac{12}{13}$, $\cos \beta = -\dfrac{3}{5}$, $180° < \alpha < 270°$, and $90° < \beta < 180°$.

Solution

First determine $\cos \alpha$ and $\sin \beta$ using the identity $\sin^2 \theta + \cos^2 \theta = 1$.

$$\cos^2 \alpha = 1 - \left(-\frac{12}{13}\right)^2 \qquad \sin^2 \beta = 1 - \left(-\frac{3}{5}\right)^2$$

$$\cos^2 \alpha = \frac{25}{169} \qquad\qquad \sin^2 \beta = \frac{16}{25}$$

$$\cos \alpha = \pm\frac{5}{13} \qquad\qquad \sin \beta = \pm\frac{4}{5}$$

Since α is in Quadrant III, $\cos \alpha = -\dfrac{5}{13}$.

Since β is in Quadrant II, $\sin \beta = \dfrac{4}{5}$.

$$\cos(\alpha + \beta) = \cos \alpha \cos \beta - \sin \alpha \sin \beta$$

$$= -\frac{5}{13}\left(-\frac{3}{5}\right) - \left(-\frac{12}{13}\right)\left(\frac{4}{5}\right)$$

$$= \frac{63}{65} \qquad \blacktriangleleft$$

You can use the sum and difference identities to solve trigonometric equations.

EXAMPLE 4

Solve $\cos\left(x + \dfrac{\pi}{6}\right) = 1 + \cos\left(x - \dfrac{\pi}{6}\right)$ on the interval $0 \le x < 2\pi$.

Solution

$$\cos\left(x + \frac{\pi}{6}\right) = 1 + \cos\left(x - \frac{\pi}{6}\right)$$

Sum identity $\quad \cos x \cos \dfrac{\pi}{6} - \sin x \sin \dfrac{\pi}{6} = 1 + \cos x \cos \dfrac{\pi}{6} + \sin x \sin \dfrac{\pi}{6} \quad$ Difference identity

$$-\sin x \sin \frac{\pi}{6} = 1 + \sin x \sin \frac{\pi}{6}$$

$$-2 \sin x \sin \frac{\pi}{6} = 1$$

$$-2 \sin x \left(\frac{1}{2}\right) = 1$$

$$\sin x = -1$$

$$x = \frac{3\pi}{2} \qquad \blacktriangleleft$$

You can also use the sum and difference identities to solve real world problems.

EXAMPLE 5

AVIATION A jet is flying at an altitude of 8 mi. Its radar can detect objects on the ground up to 17 mi ahead and up to 10 mi behind. Determine the exact value of the cosine of θ and the value of θ to the nearest hundredth radian.

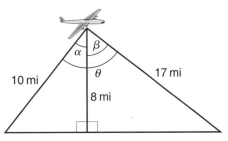

Solution

$$\cos \alpha = \frac{8}{10} = \frac{4}{5} \qquad \cos \beta = \frac{8}{17}$$

$$\sin \alpha = \frac{6}{10} = \frac{3}{5} \qquad \sin \beta = \frac{15}{17}$$

$$\theta = \alpha + \beta$$

$$\cos \theta = \cos \alpha \cos \beta - \sin \alpha \sin \beta$$

$$= \frac{4}{5}\left(\frac{8}{17}\right) - \frac{3}{5}\left(\frac{15}{17}\right)$$

$$= -\frac{13}{85}$$

$$\theta = \text{Arccosine}\left(-\frac{13}{85}\right) \approx 1.72 \text{ radians}$$

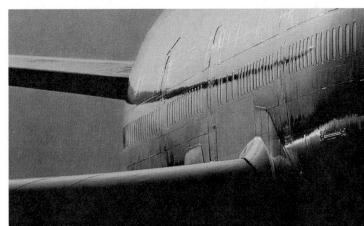

TRY THESE

Express in terms of the sine, cosine, or tangent of one angle.

1. $\sin 65° \cos 55° - \cos 65° \sin 55°$

2. $\cos 99° \cos 22° + \sin 99° \sin 22°$

3. $\dfrac{\tan \frac{5\pi}{7} + \tan \frac{\pi}{7}}{1 - \tan \frac{5\pi}{7} \tan \frac{\pi}{7}}$

4. $\dfrac{\tan \frac{4\pi}{3} - \tan \frac{\pi}{3}}{1 + \tan \frac{4\pi}{3} \tan \frac{\pi}{3}}$

Determine the exact value of each expression.

5. $\sin 75°$ **6.** $\cos 105°$ **7.** $\tan 195°$ **8.** $\sin \frac{11\pi}{12}$ **9.** $\cos \frac{19\pi}{12}$ **10.** $\tan \frac{9\pi}{4}$

11. WRITING MATHEMATICS Is $\cos (\alpha + \beta)$ equal to $\cos \alpha + \cos \beta$? Explain. Provide a numerical example to illustrate your answer.

Determine the exact value for each trigonometric expression, given $\cos \alpha = \frac{4}{5}$, $\sin \beta = -\frac{5}{13}$, $270° < \alpha < 360°$, and $180° < \beta < 270°$.

12. $\cos (\alpha - \beta)$ **13.** $\sin (\alpha - \beta)$ **14.** $\tan (\alpha + \beta)$

15. **SWIMMING** Damon is a swimmer who is hoping to compete in the Olympics. He practices in a 75-ft pool at the nearby college. So that Damon can monitor his performance, his coach mounted a video camera 10 ft above the ground on the back wall, which is 15 ft from the end of the pool. Through what angle θ must the video camera rotate, to the nearest hundredth of a radian?

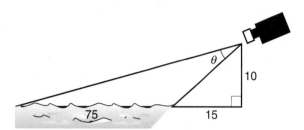

Solve for x, where $0 \le x < 2\pi$. Give exact solutions if possible, however, round to the nearest hundredth of a radian.

16. $\sin\left(x + \dfrac{7\pi}{3}\right) + \sin\left(x - \dfrac{7\pi}{3}\right) = 1$

17. $\cos\left(x + \dfrac{\pi}{6}\right) = \cos\left(x - \dfrac{\pi}{6}\right) + 1$

Use a sum or difference identity to prove each identity.

18. $\cos(180° + \alpha) = -\cos\alpha$

19. $-\sin\beta = \cos(270° - \beta)$

PRACTICE

Express in terms of the sine, cosine, or tangent of one angle.

1. $\sin\dfrac{7\pi}{9}\cos\dfrac{5\pi}{9} - \cos\dfrac{7\pi}{9}\sin\dfrac{5\pi}{9}$

2. $\cos\dfrac{6\pi}{5}\cos\dfrac{2\pi}{5} - \sin\dfrac{6\pi}{5}\sin\dfrac{2\pi}{5}$

3. $\dfrac{\tan 62° + \tan 75°}{1 - \tan 62° \tan 75°}$

4. $\dfrac{\tan 111° - \tan 87°}{1 - \tan 111° \tan 87°}$

Determine the exact value of each expression.

5. $\sin 345°$ 6. $\tan 255°$ 7. $\sin(-75°)$ 8. $\cos(-285°)$ 9. $\cos\dfrac{7\pi}{12}$ 10. $\tan\dfrac{\pi}{12}$

11. **WRITING MATHEMATICS** Josie claims that since $\sin 2\alpha = \sin(\alpha + \alpha)$, $\sin 2\alpha = 2\sin\alpha$. Do you agree or disagree? Explain.

Determine the exact value for each trigonometric expression, given $\cos\alpha = -\dfrac{\sqrt{2}}{2}$, $\sin\beta = \dfrac{15}{17}$, $180° < \alpha < 270°$, and $90° < \beta < 180°$.

12. $\cos(\alpha - \beta)$ 13. $\cos(\alpha + \beta)$ 14. $\sin(\alpha - \beta)$

15. **SECURITY** A video security monitor is mounted 8 ft above the floor on the front wall of a music store. It scans through an angle β in a vertical plane along the aisle. The aisle begins 7 ft from the front wall and is 35 ft long. Determine the measure of angle β to the nearest hundredth of a radian that will allow the entire aisle to be scanned.

Solve for x, where $0 \le x < 2\pi$. Give exact solutions if possible, otherwise, round to the nearest hundredth of a radian.

16. $\sin\left(x + \dfrac{\pi}{4}\right) = \sin\left(x - \dfrac{\pi}{4}\right) - 1$

17. $\cos\left(x + \dfrac{\pi}{4}\right) = 1 - \cos\left(x - \dfrac{\pi}{4}\right)$

18. $\cos\left(x + \dfrac{\pi}{8}\right) = \cos\left(x - \dfrac{\pi}{8}\right)$

19. $\sin\left(x + \dfrac{\pi}{7}\right) = \sin\left(x - \dfrac{\pi}{7}\right) - \dfrac{1}{2}$

20. $\tan\left(x + \dfrac{3\pi}{4}\right) = \cos\left(x - \dfrac{3\pi}{4}\right)$

21. $\tan\left(x - \dfrac{\pi}{4}\right) = \sin\left(x + \dfrac{\pi}{4}\right)$

Use a sum or difference identity to prove each identity.

22. $\sin(\pi + \theta) = -\sin\theta$

23. $\cos\left(\dfrac{3\pi}{2} - \alpha\right) = -\sin\alpha$

24. $\cos\left(\dfrac{3\pi}{4} + \beta\right) = -\dfrac{\sqrt{2}}{2}(\cos\beta + \sin\beta)$

25. $\sin\left(\dfrac{\pi}{4} + \theta\right) = \dfrac{\sqrt{2}}{2}(\cos\theta + \sin\theta)$

26. $\tan\left(\dfrac{\pi}{4} - \alpha\right) = \dfrac{1 - \tan\alpha}{1 + \tan\alpha}$

27. $\tan\left(\dfrac{\pi}{4} + \beta\right) = \dfrac{1 + \tan\beta}{1 - \tan\beta}$

28. Use a ratio identity to derive the sum identity for tangent.

29. Use your result from Exercise 28 to derive the difference identity for tangent.

30. LIGHTING DESIGN Suzanne placed a spotlight on the floor 10 ft from a wall that has a lithograph hanging on it. The bottom edge of the lithograph is 5 ft from the floor. Determine the measure of angle θ, to the nearest hundredth of a radian, which illuminates the entire lithograph if the lithograph is 4 ft high.

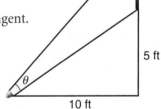

4 ft

5 ft

θ

10 ft

EXTEND

You can derive the **product/sum identities** by adding or subtracting the sum and difference identities.

$$\begin{array}{l} \cos\alpha\cos\beta + \sin\alpha\sin\beta = \cos(\alpha - \beta) \\ \underline{\cos\alpha\cos\beta - \sin\alpha\sin\beta = \cos(\alpha + \beta)} \\ 2\cos\alpha\cos\beta \qquad\qquad = \cos(\alpha - \beta) + \cos(\alpha + \beta) \end{array}$$

Add $\cos(\alpha + \beta)$ to $\cos(\alpha - \beta)$.

$$\begin{array}{l} \cos\alpha\cos\beta + \sin\alpha\sin\beta = \cos(\alpha - \beta) \\ \underline{\cos\alpha\cos\beta - \sin\alpha\sin\beta = \cos(\alpha + \beta)} \\ \qquad\quad 2\sin\alpha\sin\beta = \cos(\alpha - \beta) - \cos(\alpha + \beta) \end{array}$$

Subtract $\cos(\alpha + \beta)$ from $\cos(\alpha - \beta)$.

31. Derive a product/sum identity by adding $\sin(\alpha - \beta)$ to $\sin(\alpha + \beta)$.

32. Derive a product/sum identity by subtracting $\sin(\alpha - \beta)$ from $\sin(\alpha + \beta)$.

Express each product as a sum or difference.

33. $2\sin 60° \cos 12°$

34. $2\sin 42° \sin 14°$

35. $2\cos 72° \cos 15°$

36. $2\cos 100° \sin 16°$

Prove each identity.

37. $\tan\theta = \dfrac{\sin 5\theta - \sin 3\theta}{\cos 5\theta + \cos 3\theta}$

38. $\cot\theta = \dfrac{\cos 3\theta + \cos\theta}{\sin 3\theta - \sin\theta}$

THINK CRITICALLY

Use a sum or difference identity to prove each identity.

39. $\cos (\alpha + \beta) = \dfrac{1 - \tan \alpha \tan \beta}{\sec \alpha \sec \beta}$

40. $\sin (\alpha + \beta) = \dfrac{\cot \alpha + \cot \beta}{\csc \alpha \csc \beta}$

41. $\cot (\alpha - \beta) = \dfrac{-1 - \cot \alpha \cot \beta}{\cot \alpha - \cot \beta}$

42. $\tan (\alpha + \beta) = \dfrac{\cot \alpha + \cot \beta}{\cot \alpha \cot \beta - 1}$

MIXED REVIEW

43. STANDARDIZED TESTS Lines m and n are perpendicular. If the slope of line n is -4, then the slope of line m is

    A. 4   **B.** -4 **C.** $\dfrac{1}{4}$ **D.** $-\dfrac{1}{4}$

Solve each equation by completing the square.

44. $x^2 + 8x - 5 = 0$ **45.** $2x^2 - 7x + 1 = 0$

46. Use the binomial theorem to expand $(2x + 3)^6$.

Express in terms of the tangent of one angle.

47. $\dfrac{\tan 24° + \tan 65°}{1 - \tan 24° \tan 65°}$ **48.** $\dfrac{\tan 122° - \tan 22°}{1 + \tan 122° \tan 22°}$

PROJECT *Connection* One way of contrasting two sounds is by *pitch*. For example, the sound made by a piccolo has a high pitch and the sound made by a bass is a low pitch. The pitch of a sound is a function of the frequency of the sound wave. A high-pitched sound has a greater frequency than a low-pitched sound.

1. Using your sonameter, vary the distance between the pencils and determine how the length of the string between the two pencils affects the pitch of the sound made when the string is plucked.

2. Use your observations from Question 1 to describe how the frequency of the sound waves produced by the sonameter is affected by the length of string vibrating between the pencils.

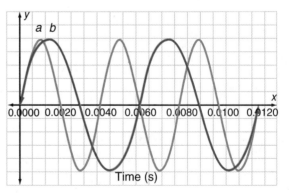

3. Which of the two sound waves shown will produce a sound with a higher pitch?

Music is dependent on *pitch*, which is the quality of sound. Pitch is determined by the frequency of the vibrations that yield a sound wave. A piano tuner adjusts the strings of a piano to produce a desired pitch in relation to a given pitch. He or she uses a tuning fork, which is a U-shaped, two-pronged instrument that produces a sound of a fixed pitch when struck.

A sound wave can be represented by a cosine or sine function. The sound waves produced by specific musical notes can be modeled by the equations $y = \cos 2a\pi t$ or $y = \sin 2a\pi t$, where a represents the frequency in hertz and t represents time in seconds.

Decision Making

Write an equation to represent each sound wave.

1. middle C (256 Hz)   **2.** A above middle C (440 Hz)

3. Determine the periods of the sound waves in Questions 1 and 2.

An alternate form of the product/sum identity for cosine is
$$\cos \alpha + \cos \beta = 2 \cos \left(\frac{\alpha + \beta}{2}\right) \cos \left(\frac{\alpha - \beta}{2}\right).$$

4. If middle C and G below middle C (196 Hz) are played together, an equation for the resulting sound wave is $y = \cos 512\pi t + \cos 392\pi t$. Use a product/sum identity to rewrite the right side of this equation as a product.

5. If C above middle C (512 Hz) and C below middle C (128 Hz) are played together, an equation for the resulting sound wave is $y = \cos 1024\pi t + \cos 256\pi t$. Use a product/sum identity to rewrite the right side of this equation as a product.

6. Derive the alternate form of the product/sum identity for cosine used above. Let $x = \alpha + \beta$ and $y = \alpha - \beta$. Add these equations to solve for α and subtract these equations to solve for β. Then substitute the values you obtain into the identity for $\cos (\alpha - \beta) + \cos (\alpha + \beta)$.

16.4 Double- and Half-Angle Identities

Explore

SPOTLIGHT ON LEARNING

WHAT? In this lesson you will learn
- to use the double- and half-angle identities to find the exact values of trigonometric expressions.
- to use the double- and half-angle identities to solve trigonometric equations.

WHY? Knowing how to use the double- and half-angle identities can help you solve problems in aviation, geometry, and aerodynamics.

1. What angle must you substitute for β in the identity $\sin(\alpha + \beta) = \sin \alpha \cos \beta + \cos \alpha \sin \beta$ to find an identity for $\sin 2\alpha$?

2. Use your result from Question 1 to derive an identity for $\sin 2\alpha$.

3. Use the identity $\cos(\alpha + \beta) = \cos \alpha \cos \beta - \sin \alpha \sin \beta$ to derive an identity for $\cos 2\alpha$.

4. Use your result from Question 3 and the Pythagorean identity $\sin^2 \alpha + \cos^2 \alpha = 1$ to derive two alternate forms for $\cos 2\alpha$.

5. Use the identity $\tan(\alpha + \beta) = \dfrac{\tan \alpha + \tan \beta}{1 - \tan \alpha \tan \beta}$ to derive an identity for $\tan 2\alpha$.

Build Understanding

The double-angle identities which you developed in Explore are used to find the values of trigonometric functions.

> **DOUBLE-ANGLE IDENTITIES**
>
> $\sin 2\alpha = 2 \sin \alpha \cos \alpha$
>
> $\cos 2\alpha = \cos^2 \alpha - \sin^2 \alpha$
>
> $\cos 2\alpha = 2 \cos^2 \alpha - 1$
>
> $\cos 2\alpha = 1 - 2 \sin^2 \alpha$
>
> $\tan 2\alpha = \dfrac{2 \tan \alpha}{1 - \tan^2 \alpha}, \tan \alpha \neq \pm 1$

CHECK UNDERSTANDING

In Explore, what other method can you use to find tan 2α? Does it produce the same result?

EXAMPLE 1

If $\sin \alpha = -\dfrac{12}{13}$ and $270° < \alpha < 360°$, determine the exact values of

a. $\sin 2\alpha$ **b.** $\cos 2\alpha$ **c.** $\tan 2\alpha$

Solution

Before solving for the double-angle, calculate $\cos \alpha$ and $\tan \alpha$. To determine $\cos \alpha$, use the identity $\sin^2 \alpha + \cos^2 \alpha = 1$.

$$\cos^2 \alpha = 1 - \left(-\frac{12}{13}\right)^2 = \frac{25}{169} = \pm \frac{5}{13}$$

Since α is in Quadrant IV, $\cos \alpha = \dfrac{5}{13}$.

$$\tan \theta = \frac{\sin \theta}{\cos \theta} = \frac{-\frac{12}{13}}{\frac{5}{13}} = -\frac{12}{5}$$

a. $\sin 2\alpha = 2 \sin \alpha \cos \alpha = 2\left(-\frac{12}{13}\right)\left(\frac{5}{13}\right) = -\frac{120}{169}$

b. $\cos 2\alpha = 1 - 2 \sin^2 \alpha = 1 - 2\left(-\frac{12}{13}\right)^2 = -\frac{119}{169}$

c. $\tan 2\alpha = \frac{2 \tan \alpha}{1 - \tan^2 \alpha} = \frac{2\left(-\frac{12}{5}\right)}{1 - \left(-\frac{12}{5}\right)^2} = \frac{120}{119}$

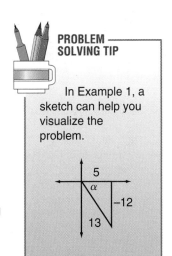
You can derive half-angle identities by substituting $\frac{\theta}{2}$ for α in two of the formulas for $\cos 2\alpha$.

$$\cos 2\alpha = 1 - 2 \sin^2 \alpha \qquad\qquad \cos 2\alpha = 2 \cos^2 \alpha - 1$$

$$\cos 2\left(\frac{\theta}{2}\right) = 1 - 2 \sin^2 \frac{\theta}{2} \qquad\qquad \cos 2\left(\frac{\theta}{2}\right) = 2\cos^2 \frac{\theta}{2} - 1$$

$$\cos \theta = 1 - 2 \sin^2 \frac{\theta}{2} \qquad\qquad \cos \theta = 2 \cos^2 \frac{\theta}{2} - 1$$

$$2 \sin^2 \frac{\theta}{2} = 1 - \cos \theta \qquad\qquad 1 + \cos \theta = 2 \cos^2 \frac{\theta}{2}$$

$$\sin^2 \frac{\theta}{2} = \frac{1 - \cos \theta}{2} \qquad\qquad \cos^2 \frac{\theta}{2} = \frac{1 + \cos \theta}{2}$$

$$\sin \frac{\theta}{2} = \pm \sqrt{\frac{1 - \cos \theta}{2}} \qquad\qquad \cos \frac{\theta}{2} = \pm \sqrt{\frac{1 + \cos \theta}{2}}$$

Neither $\sin \frac{\theta}{2}$ or $\cos \frac{\theta}{2}$ has two values. You must select $+$ or $-$ by determining the quadrant in which $\frac{\theta}{2}$ lies. You can derive the half-angle identity for tangent from the half-angle identities for sine and cosine. There are three equivalent forms of the tangent identity.

HALF-ANGLE IDENTITIES

$$\sin \frac{\theta}{2} = \pm \sqrt{\frac{1 - \cos \theta}{2}} \qquad \tan \frac{\theta}{2} = \pm \sqrt{\frac{1 - \cos \theta}{1 + \cos \theta}}, \cos \theta \neq -1$$

$$\cos \frac{\theta}{2} = \pm \sqrt{\frac{1 + \cos \theta}{2}} \qquad \tan \frac{\theta}{2} = \frac{1 - \cos \theta}{\sin \theta}, \sin \theta \neq 0$$

$$\tan \frac{\theta}{2} = \frac{\sin \theta}{1 + \cos \theta}, \cos \theta \neq -1$$

You can use the half-angle identities to find the value of trigonometric functions.

EXAMPLE 2

Use the half-angle identities to determine the exact value of

a. $\cos 105°$ **b.** $\sin \dfrac{\pi}{8}$

Solution

a. Since 105° is in Quadrant II, cos 105° is negative.

$$\cos 105° = -\sqrt{\frac{1 + \cos 210°}{2}} \qquad 105° = \frac{1}{2}(210°)$$

$$= -\sqrt{\frac{1 + \left(-\dfrac{\sqrt{3}}{2}\right)}{2}} \qquad \cos 210° = -\frac{\sqrt{3}}{2}$$

$$= -\sqrt{\frac{2 - \sqrt{3}}{4}}$$

$$= -\frac{\sqrt{2 - \sqrt{3}}}{2}$$

Check your answer by evaluating $-\dfrac{\sqrt{2 - \sqrt{3}}}{2}$ and cos 105°.

b. Since $\dfrac{\pi}{8}$ is in Quadrant I, $\sin \dfrac{\pi}{8}$ is positive.

$$\sin \frac{\pi}{8} = \sqrt{\frac{1 - \cos \dfrac{\pi}{4}}{2}} \qquad \frac{\pi}{8} = \frac{1}{2}\left(\frac{\pi}{4}\right)$$

$$= \sqrt{\frac{1 - \dfrac{\sqrt{2}}{2}}{2}} \qquad \cos \frac{\pi}{4} = \frac{\sqrt{2}}{2}$$

$$= \sqrt{\frac{2 - \sqrt{2}}{4}}$$

$$= \sqrt{\frac{2 - \sqrt{2}}{2}}$$

Check your answer by evaluating $\dfrac{\sqrt{2 - \sqrt{2}}}{2}$ and $\sin \dfrac{\pi}{8}$. ◄

You can use the half-angle identities to determine the values of $\sin \dfrac{\theta}{2}$, and $\cos \dfrac{\theta}{2}$ if you are given the value of one trigonometric function of θ and the quadrant in which it lies.

EXAMPLE 3

If $\sin \theta = -\dfrac{4}{5}$ and $270° < \theta < 360°$, determine the exact values of

a. $\sin \dfrac{\theta}{2}$ **b.** $\cos \dfrac{\theta}{2}$

Solution

To find $\cos \theta$, use the identity $\sin^2 \theta + \cos^2 \theta = 1$.

Since $\sin \theta = -\frac{4}{5}$, $\cos^2 \theta = 1 - \left(-\frac{4}{5}\right)^2 = \frac{9}{25}$.

Thus, $\cos \theta = \pm \frac{3}{5}$.

Since θ is in Quadrant IV, $\cos \theta = \frac{3}{5}$.

Since $270° < \theta < 360°$, $135° < \frac{\theta}{2} < 180°$.

a. $\sin \frac{\theta}{2} = \sqrt{\dfrac{1 - \cos \theta}{2}}$ $\frac{\theta}{2}$ is in Quadrant II, so sine is positive.

$= \sqrt{\dfrac{1 - \dfrac{3}{5}}{2}}$

$= \sqrt{\dfrac{1}{5}}$

$= \dfrac{\sqrt{5}}{5}$ Rationalize denominator.

b. $\cos \frac{\theta}{2} = -\sqrt{\dfrac{1 + \cos \theta}{2}}$ Cosine is negative in Quadrant II.

$= -\sqrt{\dfrac{1 + \dfrac{3}{5}}{2}}$

$= -\sqrt{\dfrac{4}{5}}$

$= -\dfrac{2\sqrt{5}}{5}$ Rationalize denominator. ◄

COMMUNICATING ABOUT ALGEBRA

Would your answers to Examples 3a and 3b change if $630° < \theta < 720°$? Explain.

You can use the double- and half-angle identities to solve trigonometric equations.

EXAMPLE 4

Solve $\sin 2x + \sin x = 0$, $0 \leq x < 2\pi$

Solution

$$\sin 2x + \sin x = 0$$
$$2 \sin x \cos x + \sin x = 0 \qquad \sin 2x = 2 \sin x \cos x$$
$$\sin x \,(2\cos x + 1) = 0 \qquad \text{Factor.}$$
$$\sin x = 0 \qquad \text{or} \qquad 2\cos x + 1 = 0$$
$$\sin x = 0 \qquad \text{or} \qquad \cos x = -\frac{1}{2}$$
$$x = 0, \pi \qquad \text{or} \qquad x = \frac{2\pi}{3}, \frac{4\pi}{3}$$

The solutions are $0, \pi, \frac{2\pi}{3}$ and $\frac{4\pi}{3}$. ◄

You can model the horizontal distance x traveled by a projectile with the equation $x = \dfrac{v^2 \sin 2\theta}{g}$, where $g = 32$ ft/s^2, v is the initial velocity in ft/s, and θ is the measure of the launch angle.

EXAMPLE 5

SPORTS A football player kicks the ball with an initial velocity of 79 ft/s at an angle θ whose sine is $\dfrac{8}{17}$ and whose cosine is $\dfrac{15}{17}$. How far is the ball from the football player when it hits the ground? Round your answer to the nearest foot.

Solution

$$x = \frac{v^2 \sin 2\theta}{g} = \frac{2v^2 \sin \theta \cos \theta}{g} \qquad \sin 2\theta = 2 \sin \theta \cos \theta$$

$$= \frac{2 \cdot 79^2 \cdot \dfrac{8}{17} \cdot \dfrac{15}{17}}{32}$$

$$\approx 162 \text{ ft}$$

The ball is 162 ft from the football player when it hits the ground. ◄

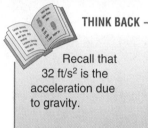

THINK BACK

Recall that 32 ft/s^2 is the acceleration due to gravity.

TRY THESE

Use a half-angle identity to determine the exact value of each expression.

1. $\sin 75°$ **2.** $\cos 67.5°$ **3.** $\tan 22.5°$

4. $\sin \dfrac{7\pi}{12}$ **5.** $\cos \dfrac{7\pi}{8}$ **6.** $\tan \dfrac{11\pi}{12}$

7. WRITING MATHEMATICS Explain how to determine $\sin \dfrac{\theta}{2}$ if you know $\sin \theta$ and the quadrant in which θ lies.

Let $\sin \alpha = \dfrac{7}{25}$ and $90° < \alpha < 180°$. Determine the exact values of each expression.

8. $\sin 2\alpha$ **9.** $\cos 2\alpha$ **10.** $\tan 2\alpha$

11. $\sin \dfrac{\alpha}{2}$ **12.** $\cos \dfrac{\alpha}{2}$ **13.** $\tan \dfrac{\alpha}{2}$

14. INDIRECT MEASUREMENT When a pole casts a shadow of 30 ft, the angle of elevation of the Sun is θ. When it casts a shadow of 14 ft, the angle of elevation is 2θ. Determine the height of the pole to the nearest foot.

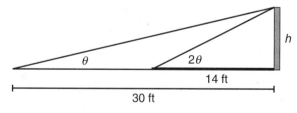

Solve for x, where $0 \le x < 2\pi$.

15. $\cos 2x + 3 \cos x + 2 = 0$ **16.** $\cos 2x - 3 \sin x - 2 = 0$

17. $2 \sin x + \sin 2x = 0$ **18.** $2 \cos x + \sin 2x = 0$

Use a double- or half-angle identity to prove each identity.

19. $2 \sin^2 \alpha = 1 - \cos 2\alpha$

20. $(\sin \beta + \cos \beta)^2 - 1 = \sin 2\beta$

21. SPORTS You can model the maximum height h reached by a projectile with the equation $h = \dfrac{v^2 \sin^2 \theta}{2g}$ where $g = 32$ ft/s^2, v is the velocity in ft/s, and θ is the measure of the launch angle. Determine the maximum height reached by a soccer ball kicked with an initial velocity of 60 ft/s at a launch angle of 40°.

PRACTICE

Use a half-angle identity to determine the exact value of each expression.

1. $\sin 105°$

2. $\cos 112.5°$

3. $\tan 255°$

4. $\sin \dfrac{5\pi}{8}$

5. $\cos \dfrac{11\pi}{8}$

6. $\tan \dfrac{7\pi}{12}$

7. WRITING MATHEMATICS Why do two soccer balls kicked from ground level with the same initial speed and launch angles of 30° and 60°, respectively, travel the same distance before hitting the ground?

Let $\sin \alpha = -\dfrac{9}{41}$ and $270° < \alpha < 360°$. Determine the exact values of each expression.

8. $\sin 2\alpha$

9. $\cos 2\alpha$

10. $\tan 2\alpha$

11. $\sin \dfrac{\alpha}{2}$

12. $\cos \dfrac{\alpha}{2}$

13. $\tan \dfrac{\alpha}{2}$

Let $\sin \alpha = -\dfrac{12}{13}$ and $180° < \alpha < 270°$. Determine the exact values of each expression.

14. $\sin 2\alpha$

15. $\cos 2\alpha$

16. $\tan 2\alpha$

17. $\sin \dfrac{\alpha}{2}$

18. $\cos \dfrac{\alpha}{2}$

19. $\tan \dfrac{\alpha}{2}$

20. SPORTS You can model the total time t, in seconds, of the flight of a projectile with the equation $t = \dfrac{2v \sin \theta}{g}$, where $g = 32$ ft/s^2, v is the velocity in ft/s, and θ is the measure of the launch angle. Determine the time that a football is aloft, to the nearest tenth of a second, if it is kicked with an initial velocity of 65 ft/s at a launch angle of 41°.

Solve for x, where $0 \le x < 2\pi$.

21. $\cos 2x + 5 \cos x + 3 = 0$

22. $\cos 2x + 9 \sin x + 4 = 0$

23. $2 \tan x - \tan 2x = 0$

24. $\cos x - \cos 2x = 0$

Use a double- or half-angle identity to prove each identity.

25. $\cot \dfrac{\beta}{2} = \dfrac{1 + \cos \beta}{\sin \beta}$

26. $\cot \dfrac{\alpha}{2} = \dfrac{\sin \alpha}{1 - \cos \alpha}$

27. $\tan^2 \dfrac{\theta}{2} = \dfrac{1 - \cos \theta}{1 + \cos \theta}$

28. $\left(\cos \dfrac{\beta}{2} + \sin \dfrac{\beta}{2}\right)^2 = \sin \beta + 1$

29. $\cot \alpha = \dfrac{1 + \cos 2\alpha}{\sin 2\alpha}$

30. $\cos 2\theta = \dfrac{\cot \theta - \tan \theta}{\cot \alpha + \tan \theta}$

31. SPORTS A soccer player kicks the ball with an initial velocity of 60 ft/s at an angle of 39°. How far is the ball from the soccer player when it hits the ground? Round your answer to the nearest foot.

EXTEND

You can use the double- and half-angle identities, as well as the sum and difference identities, to find identities for triple-angles.

32. Derive an identity for $\sin 3\theta$. Express your answer in terms of sine only.

33. Derive an identity for $\cos 3\theta$. Express your answer in terms of cosine only.

34. Use your results from Exercises 32–33 to determine the exact values of $\sin 3\theta$ and $\cos 3\theta$ if $\sin \theta = \dfrac{4}{5}$ and $90° < \theta < 180°$.

35. GEOMETRY You can express the area A of an isosceles triangle with the formula $A = \left(s \sin \dfrac{\alpha}{2}\right)\left(s \cos \dfrac{\alpha}{2}\right)$, where s is the length of each of the congruent sides and α is the angle formed by these sides.

　a. Determine the area of an isosceles triangle to the nearest square inch if $s = 12$ in. and $\alpha = 38°$.

　b. Determine the area of an isosceles triangle to the nearest square meter if $s = 4$ m and $\alpha = 67°$.

　c. Use a graphing utility to graph the formula for the area of an isosceles triangle with $s = 4$ in. What is the maximum area and the angle α at which the maximum area occurs?

36. NATURE A fish jumps out of the water for 1 s with an initial velocity of 18 ft/s. At what launch angle did the fish jump? Use the formula introduced in Exercise 20.

37. SPORTS Use a graphing utility to graph the equation $h = \dfrac{(60)^2 \sin^2 \theta}{2(32)}$.
What is the angle θ which results in the greatest height?

THINK CRITICALLY

38. Determine two different radical expressions for cos 75°. Then show that they are equal. (Hint: Write one expression using a half-angle identity and another expression using a difference identity.)

39. At what launch angle will a projectile with a given initial velocity travel the maximum distance?

40. Derive an identity for cot 2α and express it in terms of cotangent only. Note any restrictions on the variable.

41. Prove $\frac{1}{2}(\cot\theta - \tan\theta) = \frac{\cos 2\theta}{\sin 2\theta}$.

MIXED REVIEW

Determine $(f \circ g)(-1)$ and $(g \circ f)(-1)$ for each pair of functions.

42. $f(x) = -x^2, g(x) = 2x + 4$

43. $f(x) = 2x^2, g(x) = x - 7$

Use a graphing utility to solve each equation to the nearest thousandth.

44. $13 = 5^x$

45. $2 = 3^x$

46. STANDARDIZED TESTS The product of $(2 + 5i)$ and $(6 - 2i)$ is

   A. $26 + 22i$ **B.** $22 - 26i$ **C.** $-22 + 26i$ **D.** $22 + 26i$

47. Determine S_7 if $a_1 = 80$ and $r = 2$.

48. If $\sin\alpha = \frac{3}{5}$ and $0° < \alpha < 90°$, determine $\sin 2\alpha$.

PROJECT *Connection*

One way of contrasting two sounds is by *sound level*.

The amplitude of the sound wave determines the sound level. For example, sound waves that produce a louder sound have greater amplitudes than those that produce softer sound. Instruments used to measure sound levels are calibrated in decibels (dB). The table lists the sound levels in decibels of several familiar sounds.

Sound	Sound Level (dB)
whisper	15
conversation	65
busy street traffic	75
rock concert	100
threshold of pain	120

1. Estimate the sound level of some other familiar sounds. Where would you place them in the table?

2. Using your sonameter, gently pluck the string. Observe the amplitude of the vibrating string and listen to its sound. Now forcefully pluck the string and repeat your observations. How does the force with which you pluck the string affect its amplitude and the sound level of the sound it produces?

3. Rank the sound wave functions by increasing levels of the sounds they produce.

   a. $y = 2\sin 300t$ **b.** $y = 300\sin 200t$ **c.** $y = 20\sin 3000t$ **d.** $y = 0.3\sin 300t$

Career
Aerodynamic Engineer

Aerodynamic engineers study the dynamics of gases. They analyze the atmospheric interactions of gases with moving objects, such as airplanes. Sound is a pressure wave traveling through the air at "the speed of sound." When an airplane travels slower than the speed of sound, the pressure waves move ahead of the airplane and the air can move out of the way. When an airplane travels faster than the speed of sound, the air ahead gets no advance warning of the airplane's approach because the airplane is moving faster than its own pressure waves. The sound waves form a cone behind the airplane with vertex angle θ. As the air moves abruptly aside to let the airplane pass, a "shock wave" results. A listener in the area will hear a sonic boom.

The Mach number, M, of an airplane is the ratio of its speed to the speed of sound. It is related to the vertex angle of the cone by

$$\sin \frac{\theta}{2} = \frac{1}{M}.$$

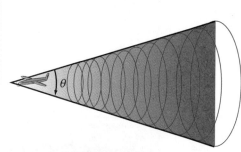

Decision Making

1. If a plane is traveling at 1200 mi/h, determine its Mach number to the nearest tenth. (The speed of sound at 32°F is approximately 743 mi/h.)

2. The hypersonic region corresponds to a Mach number of 5 or more. At what speeds would an object have to travel to achieve this Mach number?

3. If $\theta = 50°$, determine the Mach number to the nearest tenth.

4. If $\theta = 25°$, determine the Mach number to the nearest tenth.

5. If the Mach number is 4, determine the measure of angle θ to the nearest degree.

6. If the Mach number is 2, determine the measure of angle θ to the nearest degree.

7. You may have learned a way to estimate the distance of lightning by counting the seconds from the flash until the thunder is heard. Using the speed of sound from Question 1, find the speed of sound in miles per second and explain the lightning-distance rule.

Explore

● You have graphed ordered pairs, relations, and functions using the rectangular coordinate system. Another method can be used to describe the location of a point in a plane called the polar coordinate system.

The **polar coordinate system** consists of a point O, called the **pole**, and a horizontal ray with the pole as its left endpoint, called the **polar axis**. The **polar coordinates** of point P are ordered pair (r, θ), where $|r|$ is the distance from the pole to point P and θ is the measure of the angle formed by the polar axis and $\overrightarrow{OP}$.

1. On polar coordinate graph paper, graph the point $P(3, 30°)$ by first locating the terminal side $\overrightarrow{OP}$ so the angle θ between the polar axis and $\overrightarrow{OP}$ is 30°. Then plot the point 3 units from the pole on $\overrightarrow{OP}$. Name two other points that represent the same point. (*Hint*: use coterminal angles.)

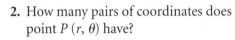

2. How many pairs of coordinates does point $P(r, \theta)$ have?

If r is positive, then point $P(r, \theta)$ is located on $\overrightarrow{OP}$, the terminal side of θ. If r is negative, then point $P(r, \theta)$ is located on the ray opposite $\overrightarrow{OP}$. The points $Q(2, 120°)$, $R(-3, 60°)$, and $S(-4, -30°)$ are shown.

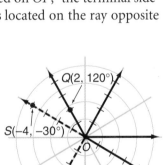

3. Graph the point $P(3, 300°)$.

4. Is there an ordered pair with a negative r-value that represents the same point? If so, name the point. If not, explain.

COMMUNICATING ABOUT ALGEBRA

What are the polar coordinates of the pole, 0?

THINK BACK

Recall that an angle is positive if it is measured in a counterclockwise direction and negative if it is measured in a clockwise direction.

5. Is there an ordered pair with a negative θ-value that represents the same point? If so, name the point. If not, explain why not.

6. Is there an ordered pair with a negative r-value and a negative θ-value that represents the same point? If so, name the point. If not, explain why not.

Build Understanding

There is a relationship between the polar and rectangular coordinate systems that allows you to express the coordinates of a point in one system with corresponding coordinates in the other system. Let the polar axis coincide with the nonnegative x-axis, as shown in the diagram.

By definition of the sine and cosine functions, $x = r\cos\theta$, $y = r\sin\theta$, where $r = \sqrt{x^2 + y^2}$, and $\tan\theta = \frac{y}{x}$.

You can use the following formulas to convert between rectangular and polar coordinates.

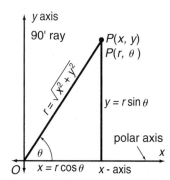

POLAR-RECTANGULAR CONVERSION FORMULAS

From rectangular to polar coordinates:

$r = \pm\sqrt{x^2 + y^2}$

$\theta = \text{Arctan}\,\frac{y}{x}$ if $x > 0$

$\theta = \text{Arctan}\,\frac{y}{x} + 180°$ if $x < 0$

From polar to rectangular coordinates:

$x = r\cos\theta$

$y = r\sin\theta$

EXAMPLE 1

Convert

a. $P(-5, 12)$ to polar coordinates.

b. $Q(3, 150°)$ to rectangular coordinates.

Solution

a. $r = \sqrt{(-5)^2 + (12)^2} = 13$ $r = \sqrt{x^2 + y^2}$

$\theta = \text{Arctan}\left(\frac{12}{-5}\right) + 180°$ $x < 0$

$\theta \approx -67° + 180° \approx 113°$ Round to the nearest degree.

The polar coordinates of P are $(13, 113°)$

b. $x = 3 \cos 150° = 3\left(-\dfrac{\sqrt{3}}{2}\right) = -\dfrac{3\sqrt{3}}{2}$ $x = r \cos \theta$

$y = 3 \sin 150° = 3\left(\dfrac{1}{2}\right) = \dfrac{3}{2}$ $y = r \sin \theta$

The rectangular coordinates of Q are $\left(-\dfrac{3\sqrt{3}}{2}, \dfrac{3}{2}\right)$.

You can also use the conversion formulas to change equations from polar form to rectangular form or vice versa. Sometimes one form may be more convenient than the other to graph or solve.

PROBLEM SOLVING TIP

Use the rectangular to polar conversion feature of your graphing utility to verify your answer to Question 1a. Use the polar to rectangular conversion feature of your graphing utility to verify your answer to Question 1b.

EXAMPLE 2

PHYSICS The radiation pattern of a certain antenna can be represented by the equation $x^2 + y^2 = 40\sqrt{x^2 + y^2} + 40x$. An engineer has been asked to plot this equation. To do this, she will first convert to polar form.

Solution

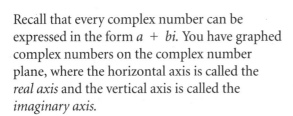

$x^2 + y^2 = 40\sqrt{x^2 + y^2} + 40x$

$r^2 = 40\sqrt{x^2 + y^2} + 40x$ $r^2 = x^2 + y^2$

$r^2 = 40r + 40r \cos \theta$ $r = \sqrt{x^2 + y^2}, x = r \cos \theta$

$r = 40 + 40 \cos \theta$

$r = 40(1 + \cos \theta)$

To plot the equation, make sure your graphing utility is in polar mode and radian mode.

Recall that every complex number can be expressed in the form $a + bi$. You have graphed complex numbers on the complex number plane, where the horizontal axis is called the *real axis* and the vertical axis is called the *imaginary axis*.

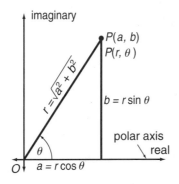

By using polar coordinates on the complex plane, you can express the real and imaginary parts of the complex number $a + bi$ in terms of its distance and direction from the origin. The point P representing $a + bi$ has rectangular coordinates (a, b) and polar coordinates (r, θ). Let the polar axis coincide with the nonnegative real axis, as shown in the diagram.

Let r be positive. Then $a = r \cos \theta$, $b = r \sin \theta$, $r = \sqrt{a^2 + b^2}$, $\tan \theta = \dfrac{b}{a}$, and θ is the angle with (a, b) on its terminal side. Thus the complex number $a + bi$ can be written

$$r \cos \theta + i \,(r \sin \theta) = r(\cos \theta + i \sin \theta).$$

In polar form, r is the absolute value or **modulus** of $a + bi$ and represents the distance from P to the pole. It is assumed to be nonnegative. Angle θ is called the **argument** of $a + bi$. Complex numbers are often represented as $z = a + bi$, so

$$r = |z| = \sqrt{a^2 + b^2}.$$

EXAMPLE 3

Express in rectangular form.

   a. $4 \, (\cos 45° + i \sin 45°)$ **b.** $2\left(\cos \dfrac{11\pi}{6} + i \sin \dfrac{11\pi}{6}\right)$

Solution

   a. $4 \, (\cos 45° + i \sin 45°) = 4\left(\dfrac{\sqrt{2}}{2} + \dfrac{i\sqrt{2}}{2}\right) = 2\sqrt{2} + 2i\sqrt{2}$

      The rectangular form is $2\sqrt{2} + 2i\sqrt{2}$.

   b. $2\left(\cos \dfrac{11\pi}{6} + i \sin \dfrac{11\pi}{6}\right) = 2\left(\dfrac{\sqrt{3}}{2} - \dfrac{i}{2}\right) = \sqrt{3} - i$

      The rectangular form is $\sqrt{3} - i$.   ◄

When converting a complex number from rectangular to polar form, it is customary to use a value of θ in the interval $0° \leq \theta < 360°$.

EXAMPLE 4

Express in polar form with $0° \leq \theta < 360°$

   a. $3 - 4i$ **b.** $-8 + 15i$

Solution

   a. $r = \sqrt{3^2 + (-4)^2} = 5$ $r = \sqrt{x^2 + y^2}$

      $\theta = \text{Arctan}\left(-\dfrac{4}{3}\right)$

      $\theta \approx -53°$ Round to the nearest degree.

      Since $0° \leq \theta < 360°$, let $\theta = -53° + 360° = 307°$. The polar form of $3 - 4i$ is $5(\cos 307° + i \sin 307°)$

   b. $r = \sqrt{(-8)^2 + (15)^2} = 17$ $r = \sqrt{x^2 + y^2}$

      $\theta = \text{Arctan}\left(-\dfrac{15}{8}\right) + 180°$ $x < 0$

      $\theta \approx -62° + 180° = 118°$ Round to the nearest degree.

      The polar form of $-8 + 15i$ is $17(\cos 118° + i \sin 118°)$ ◄

Graph each of the following points in the same polar coordinate system.

1. $P(1, 30°)$

2. $Q(2, -45°)$

3. $R(-3, 300°)$

4. $S(-2, -210°)$

Express each point with polar coordinates. Round θ to the nearest degree.

5. $P(2, -2)$

6. $Q(-4, 4\sqrt{3})$

7. $R(9, 40)$

8. $S(-7, 24)$

9. WRITING MATHEMATICS If you obtain the equation Arctan $\theta = 0$ when representing a point P given in rectangular coordinates with polar coordinates, what is the value of θ if $x > 0$? if $x < 0$? Explain. Then determine the polar coordinates of $U(3, 0)$ and $V(-7, 0)$.

Express each point with rectangular coordinates.

10. $P(5, 45°)$

11. $Q(-6, -60°)$

12. $R\left(3, -\dfrac{2\pi}{3}\right)$

13. $S\left(-2, -\dfrac{4\pi}{3}\right)$

Express in polar form with $0° \leq \theta < 360°$.

14. $-4 + 3i$

15. $5 - 12i$

16. -8

17. 10

18. ASTRONOMY The path of a certain comet can be modeled by the equation $r = \dfrac{3}{2 - \sin \theta}$. Write the equation in rectangular form.

Express in rectangular form.

19. $\cos 60° + i \sin 60°$

20. $5(\cos 135° + i \sin 135°)$

21. $6\left(\cos \dfrac{5\pi}{6} + i \sin \dfrac{5\pi}{6}\right)$

22. $4\left(\cos \dfrac{7\pi}{6} + i \sin \dfrac{7\pi}{6}\right)$

23. DESIGN Use a graphing utility to graph the equation $r = 4 - 4 \sin \theta$ in polar mode. What shape is the curve?

PRACTICE

Graph each of the following points in the same polar coordinate system.

1. $P(2, 75°)$

2. $Q(3, -60°)$

3. $R(-4, 195°)$

4. $S(-1, -30°)$

Express each point with polar coordinates. Round θ to the nearest degree.

5. $P(5, 5)$

6. $Q(9, -9)$

7. $R(8, 0)$

8. $S(-7, 0)$

9. $T(-2, 2\sqrt{3})$

10. $U(3, -3\sqrt{3})$

11. $V(-9, -40)$

12. $W(-12, -5)$

13. WRITING MATHEMATICS Describe the polar form of a real number a.

Express each point with rectangular coordinates.

14. $P(2, 135°)$

15. $Q(-2, -120°)$

16. $R\left(4, -\dfrac{3\pi}{4}\right)$

17. $S\left(-5, \dfrac{7\pi}{4}\right)$

Express in polar form with $0° \le \theta < 360°$.

18. $-8 + 15i$

19. $9 - 40i$

20. $-7 + 24i$

21. $6 + 8i$

22. 12

23. -15

24. $3i$

25. $-4i$

26. ACOUSTICS The pick-up pattern of a unidirectional microphone can be modeled by the equation $r = 2 + 2\cos\theta$. Write the equation in rectangular form.

Express in rectangular form.

27. $8\,(\cos 225° + i \sin 225°)$

28. $5(\cos 30° + i \sin 30°)$

29. $2\left(\cos\dfrac{5\pi}{3} + i \sin\dfrac{5\pi}{3}\right)$

30. $3\left(\cos\dfrac{7\pi}{4} + i \sin\dfrac{7\pi}{4}\right)$

31. $6\,(\cos 0 + i \sin 0)$

32. $7\,(\cos \pi + i \sin \pi)$

33. SPACE EXPLORATION The orbit of a certain satellite can be modeled by the equation $r = \dfrac{225{,}000}{40 - 2\cos\theta}$. Write the equation in rectangular form.

EXTEND

You can calculate the product of two complex numbers expressed in polar form using the formula

$$r_1(\cos \alpha + \sin \alpha) \cdot r_2(\cos \beta + i \sin \beta) = r_1 r_2[\cos(\alpha + \beta) + i \sin(\alpha + \beta)]$$

Determine each product.

34. $5\,(\cos 80° + i \sin 80°) \cdot 6(\cos 20° + i \sin 20°)$

35. $8\,(\cos 122° + i \sin 122°) \cdot 9(\cos 18° + i \sin 18°)$

36. $9\left(\cos\dfrac{\pi}{3} + i \sin\dfrac{\pi}{3}\right) \cdot 10\left(\cos\dfrac{\pi}{4} + i \sin\dfrac{\pi}{4}\right)$

37. $7\left(\cos\dfrac{5\pi}{3} + i \sin\dfrac{5\pi}{3}\right) \cdot 7\left(\cos\dfrac{3\pi}{4} + i \sin\dfrac{3\pi}{4}\right)$

You can calculate the quotient of two complex numbers expressed in polar form using the formula $r_1(\cos \alpha + i \sin \alpha) \div r_2(\cos \beta + i \sin \beta) = \dfrac{r_1}{r_2}(\cos(\alpha - \beta) + i \sin (\alpha - \beta))$

38. What restriction must be placed on this formula?

Determine each quotient.

39. $12(\cos 65° + i \sin 65°) \div 3(\cos 25° + i \sin 25°)$

40. $15(\cos 54° + i \sin 54°) \div 5(\cos 38° + i \sin 38°)$

41. $20\left(\cos \dfrac{2\pi}{3} + i \sin \dfrac{2\pi}{3}\right) \div 4\left(\cos \dfrac{\pi}{4} + i \sin \dfrac{\pi}{4}\right)$

42. $18\left(\cos \dfrac{4\pi}{5} + i \sin \dfrac{4\pi}{5}\right) \div 9\left(\cos \dfrac{\pi}{7} + i \sin \dfrac{\pi}{7}\right)$

DESIGN Use a graphing utility to graph each equation in polar mode. Determine the number of "petals" that each "rose" has.

43. $r = \cos 2\theta$ **44.** $r = \cos 3\theta$ **45.** $r = \cos 4\theta$

THINK CRITICALLY

46. In Exercises 43–45, each equation is of the form $r = \cos n\theta$. Determine the relationship between n and the number of petals. (*Hint*: Consider n is even and n is odd.)

47. About what line are the graphs of cosine roses symmetric?

48. Derive: $\quad r_1(\cos \alpha + i \sin \alpha) \cdot r_2(\cos \beta + i \sin \beta) = r_1 r_2[\cos (\alpha + \beta) + i \sin (\alpha + \beta)]$

49. Graph a few polar equations of the form $r = \sin n\theta$. About what line are the sine roses symmetric?

50. If $r(\cos \alpha + i \sin \alpha)$ and $r(\cos \beta + i \sin \beta)$ are representations of the same number. Determine the relationship between α and β.

MIXED REVIEW

51. Eight books written by authors with different last names are placed at random on a bookshelf. What is the probability that they will be in alphabetical order?

Express in expanded form.

52. $\log_4 6f^2$

53. $\log_5 \dfrac{u}{v}$

54. **STANDARDIZED TESTS** The solutions to the equation $x^2 + 8x + 25 = 0$ are

     A. -1 and 7 **B.** $-4 \pm 3i$ **C.** $4 \pm 3i$ **D.** 1 and -7

Determine the axis of symmetry of each graph.

55. $y = \dfrac{1}{2}x^2 - 8x - 5$

56. $y = 2x^2 + 100x - 10$

Express in rectangular form.

57. $12\left(\cos \dfrac{5\pi}{6} + i \sin \dfrac{5\pi}{6}\right)$

58. $15\left(\cos \dfrac{3\pi}{4} + i \sin \dfrac{3\pi}{4}\right)$

Use Vectors

A **vector quantity** is a quantity, such as velocity or force, that has both magnitude and direction. A **vector** which is a directed line segment is used to represent a vector quantity. The **norm** or length of a vector represents the magnitude and the direction of the vector indicates the direction.

Vectors **u**, **v**, and **w** are displayed at the right. Note that vectors **u** and **w** are **equivalent vectors** because they have the same length and direction. Vectors **u** and **v** are **opposite vectors** because they have the same length but opposite directions. You can determine the length of a vector by using the distance formula. The length of **v** is written $\|\mathbf{v}\|$

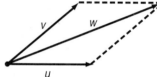

To determine the sum vector **w** of two vectors **u** and **v** that share the same initial point, use **u** and **v** to form a parallelogram. The diagonal is the desired sum, or **resultant**.

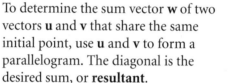

To add vectors **a** and **b** that do not share the same initial point, move the vectors parallel to their original directions and use head-to-tail addition.

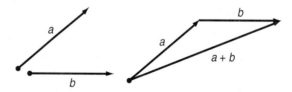

To subtract vector **d** from vector **c**, add vector −**d** to vector **c**. To multiply a vector **v** by a real number or **scalar** r, multiply the length of **v** by $|r|$. Reverse the direction if r is negative.

You can **resolve** vector **v** into two perpendicular component vectors. Place the tail of **v** at the origin and the head of **v** at $P(x, y)$. Let θ represent the direction of **v** with respect to the x-axis and r represent the norm of the vector. The x-component of the vector is $r \cos \theta$, and the y-component is $r \sin \theta$.

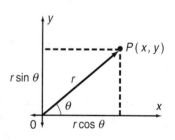

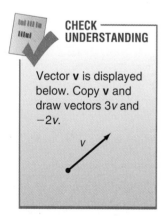

Problem

Ricardo and Hakeem are riding in a blimp from Middlebury to the far side of a waterfall. Because there is no wind, they can maintain a straight-line course 50° north of east at a constant speed of 16 mi/h. Gregory loves to look down at trains from the air, and he knows that the DWB Railroad runs in a straight line from south to north 32 miles east of Middlebury. After how many hours will they reach the railroad?

Explore The Problem

● **Round to the nearest tenth.**

1. How long would it take them to reach the railroad tracks if they were traveling due east?

2. Draw a diagram that shows the velocity vector **v**, the straight line of the railroad tracks, and the east and north components of the velocity, which should be labeled $\mathbf{v}_e$ and $\mathbf{v}_n$ respectively.

3. Express $\mathbf{v}_e$ in terms of **v**.   4. Express $\mathbf{v}_n$ in terms of **v**.

5. Write an equation that expresses time in terms of distance and speed. Then use the equation and your result from Question 3 to determine how many hours it will take them to reach the railroad tracks.

6. How far will they have traveled on their straight-line course when they reach the railroad tracks?

Investigate Further

● The **bearing** of a vector is the angle measured clockwise from the north to the vector. The *heading* of an airplane is the bearing of its path in still air. The *course* of an airplane is its actual path under the influence of wind.

An airplane with an airspeed of 300 mi/h is flying on a heading of 61°. The wind is blowing from the north at 28 mi/h. What are the groundspeed and course of the plane?

7. Draw a diagram that shows the airspeed vector. Then connect the tail of the wind vector to the head of the airspeed vector. Draw the vector that represents the groundspeed vector.

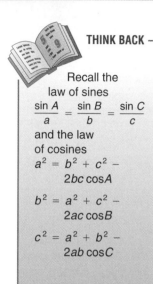

THINK BACK

Recall the law of sines

$$\frac{\sin A}{a} = \frac{\sin B}{b} = \frac{\sin C}{c}$$

and the law of cosines

$a^2 = b^2 + c^2 - 2bc \cos A$

$b^2 = a^2 + c^2 - 2ac \cos B$

$c^2 = a^2 + b^2 - 2ab \cos C$

8. Use the law of cosines to find the groundspeed of the airplane.

9. Use your result from Question 8 and the law of sines to find the measure of angle θ, the angle between the airspeed vector and the groundspeed vector.

10. Use your result from Question 9 to determine the course of the airplane.

11. **WRITING MATHEMATICS** Explain how and when you can use the law of cosines and the law of sines to solve problems involving vectors.

Apply The Strategy

12. **SPORTS** In a baseball game, the baseball comes at the player in center field at 40 ft/s at an angle of 45°. Determine the x- and y-components of the velocity vector.

13. **PHYSICS** Forces of 15 lb and 20 lb are applied to a crate at right angles. Determine the magnitude of the resultant force to the nearest pound.

14. **PHYSICS** Forces of 16 lb and 22 lb make an angle of 42° with each other and are applied to a bookcase at the same point. Determine the magnitude of the resultant force to the nearest pound.

15. **PHYSICS** Determine the measure of the angle between forces of 65 lb and 75 lb applied at the same point, if the magnitude of the resultant force is 92 lb.

16. If the magnitude of the resultant force in Question 15 was 140 lb, what is the measure of the angle between the forces? Explain.

17. What angle between the force vectors in Question 15 would give minimum resultant force? What would that force be? Explain.

18. **GEOMETRY** The **inner product** or dot product of vectors $\mathbf{u} \cdot \mathbf{v}$ is a scalar defined as follows: $\mathbf{u} \cdot \mathbf{v} = \|\mathbf{u}\| \|\mathbf{v}\| \cos \theta$ where θ is the angle between the vectors. If $\mathbf{u} \cdot \mathbf{v} = 0$, what conclusion can you draw about the vectors? Explain.

REVIEW PROBLEM SOLVING STRATEGIES

Cap Colors

1. Three baseball fans are sitting one behind the other in the first three rows of the bleachers. They are shown five baseball caps, two blue and three white. One of the caps is placed on the head of each of the three fans. None of the fans can see their own cap. They are told that, if they can tell the color of their own cap, they will win a season pass. The fan in the third row looks at the two fans in front of him and says he cannot tell the color of his own cap. The fan in the second row hears this, looks at the fan in front of him, turns and looks at the fan in back of him, and then says that he can't tell the color of his cap either. The fan in the first row declares that she doesn't even need to turn around to tell what color her cap is. What color is her cap?

Walk My Way

2. Ryan was returning home from the beach. His friend took him half the distance by car, and Ryan noticed they traveled at a rate that was 15 times as fast as he could walk. Ryan took the bus for the second half of the trip and, because of heavy traffic, the rate of travel was half as fast as he could walk. Would Ryan have saved time if he had simply walked home from the beach? If so, how much time (in relation to the time for the whole trip)?

THE THREE SISTERS

3. Three sisters shared a gift of $48 from their aunt, each getting a sum equal to her age 6 years before. Alice, the youngest, being mathematically clever, suggested the following switch. "I'll keep half the money I received and divide the rest equally between the two of you. But then Betty, our middle sister, must do the same, keeping half her new total and dividing the rest equally between our oldest sister, Cindy, and me. Finally, Cindy must do the same." The sister agreed, and they each ended up with $16. How old were the sisters?

· · · CHAPTER REVIEW · · ·

VOCABULARY

Choose the word from the list that completes each statement.

1. A(n) __?__ is an equation that is true for all values of the variable, except those for which either side of the equation is undefined.

2. In polar form, r represents the distance from P to the pole and is called the __?__.

3. The __?__ is a horizontal ray with the pole as its left endpoint.

4. A(n) __?__ has both magnitude and direction.

a. vector quantity

b. polar axis

c. identity

d. modulus

Lesson 16.1   PROVE TRIGONOMETRIC IDENTITIES   pages 791–797

- To prove an identity, you can transform one side of the equation into the other side or transform *both* sides of the equation until you obtain the same expression on both sides.

Prove each identity.

5. $\csc \theta = \cos \theta \cot \theta + \sin \theta$

6. $\dfrac{\csc \beta + 1}{-\csc \beta} = \dfrac{1 - \sin^2 \beta}{\sin \beta - 1}$

Lesson 16.2   SOLVE TRIGONOMETRIC EQUATIONS   pages 798–803

- You can combine algebra with trigonometric identities to find the exact solutions to many trigonometric equations. If a trigonometric equation does not have an exact solution, you can find an approximate solution in decimal form.

Solve for x, where $0 \le x < 2\pi$. Give the exact solution if possible, otherwise, round to the nearest hundredth of a radian.

7. $\csc^2 x - 4 = 0$

8. $4 \sec^2 x + 5 \sec x = 6$

9. $12 \sin^2 x - 9 \cos x - 7 = 0$

10. $9 \cos^2 x - 14 \sin x - 11 = 0$

Lesson 16.3   THE SUM AND DIFFERENCE IDENTITIES   pages 804–811

- You can use the sum and difference identities to evaluate trigonometric expressions and prove identities.

$$\sin (\alpha \pm \beta) = \sin \alpha \cos \beta \pm \cos \alpha \sin \beta \qquad \cos (\alpha \pm \beta) = \cos \alpha \cos \beta \pm \sin \alpha \sin \beta$$

Determine the exact value for each trigonometric expression, given
$\cos \alpha = \dfrac{8}{17}, \quad \sin \beta = \dfrac{3}{5}, \quad 270° < \alpha < 360°, \quad$ and $\quad 90° < \beta < 180°.$

11. $\cos (\alpha + \beta)$

12. $\sin (\alpha - \beta)$

13. $\tan (\alpha - \beta)$

Determine the exact value of each expression.

14. $\cos 345°$ **15.** $\sin(-15°)$ **16.** $\tan 195°$

Use a sum or difference identity to prove each identity.

17. $\sin\left(\dfrac{3\pi}{2} + \alpha\right) = -\cos\alpha$ **18.** $\tan\left(\dfrac{3\pi}{4} - \beta\right) = \dfrac{\tan\beta + 1}{\tan\beta - 1}$

Lesson 16.4   DOUBLE- AND HALF-ANGLE IDENTITIES   pages 812–820

- You can use the double- and half-angle identities to evaluate trigonometric expressions and prove identities.

Determine the exact value of each expression if $\cos\alpha = -\dfrac{40}{41}$ and $180° < \alpha < 270$.

19. $\sin 2\alpha$ **20.** $\tan 2\alpha$ **21.** $\cos\dfrac{\alpha}{2}$

Determine the exact value of each expression.

22. $\cos 157.5°$ **23.** $\sin 285°$ **24.** $\tan 112.5°$

Lesson 16.5   POLAR COORDINATES AND COMPLEX NUMBERS   pages 821–827

- To write the coordinates of a point $P(x, y)$ in polar coordinates, use $r = \sqrt{x^2 + y^2}$, $\cos\theta = \dfrac{x}{r}$, $\sin\theta = \dfrac{y}{r}$. To change from polar to rectangular coordinates, use $x = r\cos\theta$, $y = r\sin\theta$. The polar form of a complex number $a + bi$ is $r(\cos\theta + i\sin\theta)$ where $r = \sqrt{a^2 + b^2}$.

Express with polar coordinates.   **Express with rectangular coordinates.**

25. $P(-9, 9\sqrt{3})$ **26.** $Q(12, -12)$ **27.** $R(5, 30°)$ **28.** $S(8, 135°)$

Express in rectangular form.

29. $8(\cos 120° + i\sin 120°)$ **30.** $4(\cos 150° + i\sin 150°)$

Express in polar form with $0° < \theta < 360°$.

31. $-15 + 8i$ **32.** $9 + 40i$ **33.** -5 **34.** 12

Lesson 16.6   USE VECTORS   pages 828–831

- You can resolve a vector into two perpendicular component vectors. The x-component of the vector is $r\cos\theta$, and the y-component is $r\sin\theta$.

35. Lauren is traveling in a balloon on a straight-line course 42° north of east at a constant speed of 18 mi/h. Express $\mathbf{v}_e$ and $\mathbf{v}_n$ in terms of $\mathbf{v}$.

36. An airplane with an airspeed of 325 mi/h is flying on a heading 59°. The wind is blowing from the north at 32 mi/h. What are the groundspeed and course of the plane?

CHAPTER ASSESSMENT

CHAPTER TEST

1. **STANDARDIZED TESTS** Which expression(s) is(are) equivalent to $\cos \theta$?

   I. $\cos(-\theta)$ **II.** $\dfrac{1}{\csc \theta}$ **III.** $\sin(90° - \theta)$

   A. I only   **B.** II and III
   C. I and III   **D.** I, II, and III

2. **WRITING MATHEMATICS** Describe different strategies you can use to prove a trigonometric identity. Include specific techniques.

Prove each identity.

3. $\sec \alpha \csc \alpha = \sin \alpha \sec \alpha + \cot \alpha$

4. $\sec^2 \beta - \tan^2 \beta = \dfrac{\sin \beta}{\csc \beta} + \dfrac{\cos \beta}{\sec \beta}$

5. $\sin(270° + \theta) = -\cos \theta$

Solve for x, where $0 \le x < 2\pi$. Give the exact solution if possible, otherwise, round to the nearest hundredth of a radian.

6. $\sin 2x = 2 \cos x$

7. $\cos 2x + \sin x = 1$

8. $\cos \left(x + \dfrac{\pi}{4} \right) = \sin \left(x - \dfrac{\pi}{4} \right)$

9. $8 \sin^2 x - 2 \cos x - 5 = 0$

10. **STANDARDIZED TESTS** Which of the following does not represent the exact value of $\sin 75°$?

   A. $\dfrac{\sqrt{2} + \sqrt{3}}{2}$ **B.** $\dfrac{\sqrt{6} + \sqrt{2}}{4}$

   C. $\dfrac{\sqrt{2}}{2}\left(\dfrac{\sqrt{3}}{2} \right) + \dfrac{\sqrt{2}}{2}\left(\dfrac{1}{2} \right)$ **D.** 0.9659

Determine the exact value of each expression first by using the sum and difference identities and then by using the half-angle identities.

11. $\sin 105°$ 12. $\cos 285°$ 13. $\cos(-15°)$

14. Find $\cos(\alpha - \beta)$ if $\sin \alpha = \dfrac{\sqrt{3}}{2}$, $\tan \beta = -1$, $90° < \alpha < 180°$, and $90° < \beta < 180°$.

15. Find $\tan(\alpha + \beta)$ if $\sin \alpha = \dfrac{40}{41}$, $\cos \beta = -\dfrac{5}{13}$, $0° < \alpha < 90°$, and $180° < \beta < 270°$.

If $\cos \alpha = -\dfrac{8}{17}$ and $90° < \alpha < 180°$, determine the exact value of each expression.

16. $\sin 2\alpha$ 17. $\cos \dfrac{\alpha}{2}$ 18. $\tan 2\alpha$

19. A football player kicks the ball with an initial velocity of 68 ft/s at a launch angle of 27°. How far does the ball travel before it hits the ground?

Express each point with polar coordinates.

20. $\left(-3, 3\sqrt{3} \right)$ 21. $(-5, 0)$

Express each point with rectangular coordinates.

22. $(6, -135°)$ 23. $(-8, 315°)$

Express in polar form with $0° \le \theta < 360°$.

24. $-5 + 12i$ 25. $4 + 3i$

Express in rectangular form.

26. $4\left(\cos \dfrac{4\pi}{3} + i \sin \dfrac{4\pi}{3} \right)$ 27. $5\left(\cos \dfrac{5\pi}{4} + i \sin \dfrac{5\pi}{4} \right)$

28. The orbit of a certain satellite can be modeled by the equation $r = \dfrac{210{,}000}{38 - 2 \cos \theta}$. Write the equation in rectangular form.

29. An airplane with an airspeed of 260 mi/h is flying at a heading of 55°. The wind is blowing from the north at 25 mi/h. What are the ground-speed and course of the plane?

30. Forces of 58 lb and 67 lb act on a body at right angles. Determine the magnitude of the resultant force to the nearest pound.

PERFORMANCE ASSESSMENT

LIVING AT THE POLE Work with a partner. Find a map of your city, town, or village. Locate your street on the map. Superimpose a polar coordinate system on the map with the pole at your home. Determine the approximate locations of three of your friend's homes or places you like to visit, and express these points with polar coordinates. Then have your partner convert these points to rectangular coordinates. Repeat this exercise with the pole located at your partner's home.

EQUATION CREATION Work with a partner. Write three trigonometric equations that have $x = 57°$ as one solution. Have your partner graph these equations and determine the other solution(s) for x on the interval $0° \leq x < 360°$. Then switch roles with your partner and repeat for $x = 36°$.

EQUATION ART Use a graphing utility in polar and radian modes to graph several functions of the forms $r = a + b \cos \theta$ and $r = a + b \sin \theta$, whose graphs are called limaçons. Discuss their symmetries. How do the graphs change as a and b change? A special case of the limaçon occurs when $a = b$. Graph several functions of the form $r = a + a \cos \theta$ and $r = a + a \sin \theta$. Why do you think the graphs of these equations are called cardioids?

TABLE YOUR POOL The angle at which the ball reflects off the surface is equal to the angle at which it strikes the edge. A ball is struck at the indicated angle. Copy the diagram and illustrate the next five reflections on the pool table.

What factors might influence the reflections?

PROJECT ASSESSMENT

PROJECT *Connection*   Prepare a brief summary of the Project Connections. Organize your observations and data into tables and graphs.

1. Decide on a procedure to investigate how the frequency of a sound wave is affected by some characteristic of the sonameter string, such as its tension, diameter, or composition. Remember that you must keep all other conditions exactly the same while you change the one variable under investigation.

2. Make a prediction of the outcome of your experiment. Conduct your investigation with the sonameter and summarize your conclusion. If the result was not what you expected, decide if you need to repeat the investigation or change your original ideas.

3. Choose a member of your team to demonstrate your conclusions to the class.

4. There are many other interesting investigations about sound. For example, try wrapping a playing radio in different materials to determine which one reduces sound the most. Or, experiment to determine how well you can detect the direction from which a sound is coming. Try it with your eyes open and then with your eyes closed.

1. If $\angle x$ is in Quadrant II and $\cos x = -\frac{4}{5}$, find the value of $\sin 2x$.

2. **STANDARDIZED TESTS** The equation of the reflection of the graph of $y = 4^x$ in the line $y = x$ is.

   A. $y = \log x$ **B.** $y = 4^{-x}$
   C. $y = -4^x$ **D.** $y = 4^x + 1$

3. If the sum of the solutions of the quadratic equation $6x^2 + kx - 3 = 0$ is equal to the product of the solutions, what is the value of k?

4. **WRITING MATHEMATICS** Explain why you cannot use the Distributive Property to claim that $\sin (\alpha + \beta) = \sin \alpha + \sin \beta$. Refer to the case where $\alpha = 30°$ and $\beta = 45°$ as an example. Give another nontrigonometric example in which the Distributive Property does not apply to the situation.

5. **STANDARDIZED TESTS** A number cube is rolled once and a coin is tossed once. Which of the following are true about these events?

   I.   The events are mutually exclusive
   II. The events are independent
   III. The events are complimentary

   A. I only   **B.** II only   **C.** III only
   D. I and II   **E.** I, II, and III

6. Evaluate: $\dfrac{(4.6 \times 10^3) \times (3 \times 10^{-2})}{2 \times 10^4}$

Solve each trigonometric equation for all values of θ for $0° \le \theta < 360°$.

7. $|\sin \theta - 3| = 2$

8. $2 \sec^2 \theta - 3 \tan \theta - 5 = 0$

9. Determine the equation of the circle passing through the points $(7, 1)$, $(6, 8)$, and $(-1, 7)$.

10. Nahib drops a ball from a height of 6 ft. Each time the ball bounces, it rises to a height of 50% of the distance it fell. How far does it travel by the end of its fourth bounce?

11. A school computer system has a scanner that receives and processes grades in 45 min. The school received a new scanner that can do the same job in 30 min. If a programmer in the data processing office makes it possible for the system to work with both scanners simultaneously, how long would it take to do the job?

12. The waterwheel shown has a radius of 7 ft and rotates at the rate of 6 revolutions per min. The path of point P on the rim of the wheel is sinusoidal and can be modeled by the 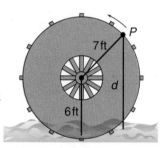 function $d = 6 + 7 \cos \frac{\pi}{5}(t - 2)$ where d is the distance of point P from the surface of the water and t is the number of seconds that the wheel rotates after you begin tracking point P. Determine how far P is from the water after 5 s.

13. Divide $(x^4 + y^4)$ by $(x + y)$.

14. **STANDARDIZED TESTS** The graph shown below is a reflection over the y-axis of the graph of which of the following equations?

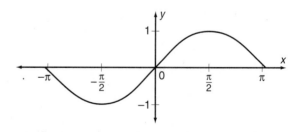

   A. $y = \sin x$ **B.** $y = \cos x$
   C. $y = \sin 2x$ **D.** $y = \cos 2x$

··· STANDARDIZED TEST ···

QUANTITATIVE COMPARISON In each question, compare the quantity in Column 1 with the quantity in Column 2. Select the letter of the correct answer from these choices:

A. The quantity in Column 1 is greater.
B. The quantity in Column 2 is greater.
C. The two quantities are equal.
D. The relationship cannot be determined by the information given.

Notes: In some questions, information which refers to one or both columns is centered over both columns. A symbol used in both columns has the same meaning in each column. All variables represent real numbers. Most figures are not drawn to scale.

	Column I	**Column II**
1.	$0° < \theta < 360°$	
	The value of θ if $\tan \theta \cos \theta - \tan \theta = 0$	The value of θ if $|2 \cos \theta - 3| = 5$
2.	The common ratio of the geometric sequence $\log 81, \log 9, \log 3, \ldots$	The common difference of the arithmetic sequence $4.5, 4, 3.5, \ldots$
3.	$\begin{cases} 5x - 5y = 50 \\ z = y^2 \end{cases}$	
	x	z
4.	$(x + 1)(x + 2) = 0$	
	$x(x + 1)$	$x(x + 2)$
5.	A complex number and its conjugate are each expressed in polar form.	
	The r-value of the number	The r-value of the conjugate

	Column I	**Column II**
6.	$A = \begin{bmatrix} 2 & 0 & 0 \\ 0 & 4 & 0 \\ 0 & 0 & 5 \end{bmatrix}$	$A^{-1} = \begin{bmatrix} a_{11} & a_{12} & a_{13} \\ a_{21} & a_{22} & a_{23} \\ a_{31} & a_{32} & a_{33} \end{bmatrix}$
	a_{22}	1
7.	The compliment of $\frac{3\pi}{8}$	The supplement of $\frac{7\pi}{8}$
8.	$N = 50r\left(\frac{1}{p} + \frac{1}{q}\right)$	
	N when $r = 5$, $p = 10$, and $q = 25$	N when $r = 3$, $p = 5$, and $q = 10$
9.	The period of $y = \sin x$	The period of $y = \tan \frac{1}{2}x$
10.	Two hockey players simultaneously hit the puck, each applying a force of 18 pounds.	
	The magnitude of the resultant if the direction of the forces form an angle of 120°.	The magnitude of the resultant if the direction of the forces form an angle of 60°.
11.	$\log_8 8^5$	$8^{\log_8 5}$
12.	The number of vertical asymptotes of the function $xy = 3$	The number of vertical asymptotes of the function $y - \frac{x - 2}{x^3 - x - 6}$
13.	$x^2 + x + k$	
	The value of k if $(x + 5)$ is a factor of the polynomial	The value of k if $(x + 4)$ is a factor of the polynomial

Geometry
Quick Notes

● **Geometry Basics**

All geometric figures are made up of at least one point.

point   line   ray   line segment (or segment)   angle

● **About Lines**

Lines in a plane can be either parallel to each other or they can intersect each other.

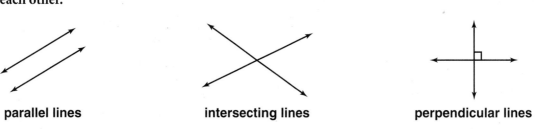

parallel lines   intersecting lines   perpendicular lines

● **About Angles**

Angles are measured in degrees.

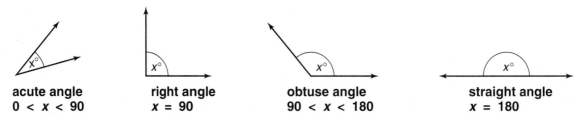

acute angle   right angle   obtuse angle   straight angle
$0 < x < 90$ $x = 90$ $90 < x < 180$ $x = 180$

● **Complementary and Supplementary Angles**

Two angles are complementary if the sum of their measures is exactly 90°.

Two angles are supplementary if the sum of their measures is exactly 180°.

About Triangles

Triangles are three-sided plane figures. They can be classified according to the measures of their sides or their angles.

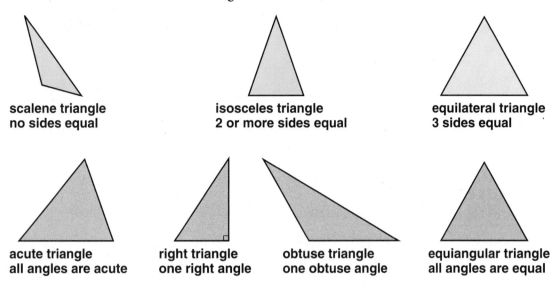

scalene triangle
no sides equal

isosceles triangle
2 or more sides equal

equilateral triangle
3 sides equal

acute triangle
all angles are acute

right triangle
one right angle

obtuse triangle
one obtuse angle

equiangular triangle
all angles are equal

About Quadrilaterals

Quadrilaterals are four-sided plane figures. Each figure in the diagram has all the properties of the figures preceding it, including the properties listed with that figure.

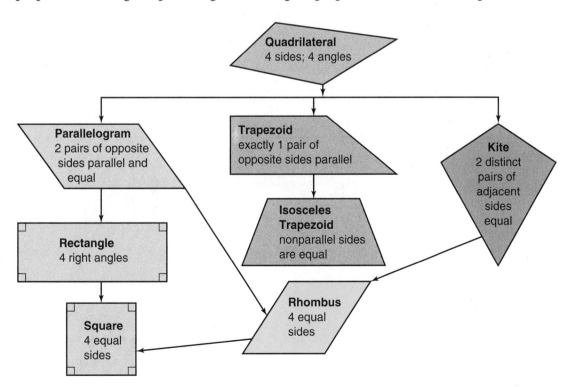

Quadrilateral
4 sides; 4 angles

Parallelogram
2 pairs of opposite sides parallel and equal

Trapezoid
exactly 1 pair of opposite sides parallel

Kite
2 distinct pairs of adjacent sides equal

Rectangle
4 right angles

Isosceles Trapezoid
nonparallel sides are equal

Square
4 equal sides

Rhombus
4 equal sides

About Other Polygons

Polygons are plane figures made up of segments and angles. Triangles and four-sided figures are also polygons.

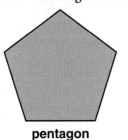
pentagon

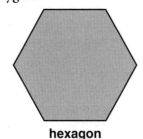
hexagon

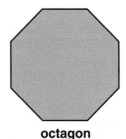
octagon

Perimeter Formulas

In the following formulas, l = length, w = width, s = side, and P = perimeter.

Perimeter of a rectangle	$P = 2l + 2w$
Perimeter of a square	$P = 4s$

Area Formulas

In the following formulas, b = base, B = long base, h = height, l = length, w = width, s = side, and A = area.

Area of a parallelogram	$A = bh$
Area of a rectangle	$A = lw$
Area of a square	$A = s^2$
Area of a trapezoid	$A = \frac{1}{2}(B + b)h$
Area of a triangle	$A = \frac{1}{2}bh$

About Circles and Spheres

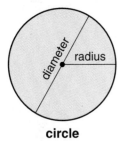
circle

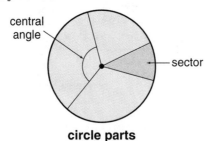
circle parts

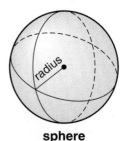
sphere

Circle Formulas

Circumference of a circle	$C = 2\pi r$ or $C = \pi d$
Area of a circle	$A = \pi r^2$

Sphere Formulas

Area of a sphere	$A = 4\pi r^2$
Volume of a sphere	$V = \frac{4}{3}\pi r^3$

About Geometric Solid Figures

Geometric solid figures are made up of plane polygons. Below are some geometric solid right figures.

cone

pyramid

cylinder

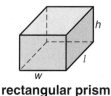

rectangular prism

cube

Base

The cone and the pyramid have one base. The cylinder and the prism have two bases and they are parallel. The cone and cylinder have circular bases. The base of a pyramid or prism can be any polygonal shape.

Lateral Surface

The lateral surface is the side or sides of the solid figure other than a base. The cone and cylinder have one lateral surface. The lateral surface of a pyramid is made up of triangles. The lateral surface of a right prism is made up of rectangles.

Slant Height

The slant height of a cone is measured from the vertex of the cone to the edge of its base. The slant height of a pyramid is measured from the vertex to the center of one side of the base.

Formulas

Total surface area of a right circular cone	$T = \pi r(l + r)$
Volume of a cone	$V = \dfrac{1}{3}\pi r^2 h$
Total surface area of a right cylinder	$T = 2\pi r(r + h)$
Volume of a cylinder	$V = \pi r^2 h$
Total surface area of a rectangular prism	$T = 2(lw + lh + wh)$
Volume of a rectangular prism	$V = lwh$
Total surface area of a cube	$T = 6s^2$
Volume of a cube	$V = s^3$

Technology
Quick Notes

Using the Texas Instruments TI-82 Graphing Calculator

- **For additional features and instructions, consult your user's manual.**

  THE KEYBOARD The feature accessed when a key is pressed is shown in white on the key. To access the features in blue above each key, first press the 2ND key. To access what is in white above the keys, first press ALPHA.

  CALCULATIONS Calculations are performed in the Home Screen. This screen may be returned to at any time by pressing 2ND QUIT. The calculator evaluates according to the order of operations. Press ENTER to calculate. For $3 + 4 \times 5$ ENTER, the result is 23. You can replay the previous line by pressing 2ND ENTER. Use the arrow keys to edit.

Displaying Graphs

- **GRAPH FEATURE** To enter an equation, press Y=. Enter an equation such as $y = 2x + 3$ using the X, T, θ key for x. Then press GRAPH to display the graph in the viewing window.

  VIEWING WINDOW To set range values for the viewing window, press WINDOW. Press ▶ to access FORMAT where you can choose features such as Grid Off or Grid On.

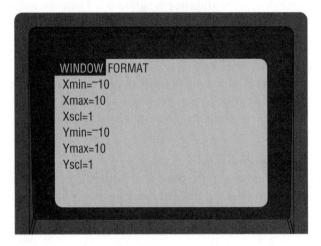

```
WINDOW FORMAT
Xmin=-10
Xmax=10
Xscl=1
Ymin=-10
Ymax=10
Yscl=1
```

  ZOOM FEATURE Press ZOOM and then 6 (Standard) to set a standard viewing window. Press ZOOM 1 (Box) to highlight a particular area and zoom in on that part of the graph. Press ZOOM 8 (Integer) to set values for a friendly window.

  TRACE FEATURE Pressing TRACE places the cursor directly on the graph and shows the x- and y-coordinates of the point where the cursor is located. You can move the cursor along the graph using the right and left arrow keys.

  TABLE FEATURE Press 2ND TBLSET to set up the table. Then press 2ND TABLE to see a table of values for each equation.

INTERSECTION FEATURE To determine the coordinates of the point of intersection of two graphs, press 2ND CALC, then 5 (Intersect). The calculator will then prompt you to identify the first graph. Use the right and left arrow keys to move the cursor to the first graph, close to the point of intersection. Repeat to identify the second graph and get the coordinates of the point of intersection.

Statistics

ENTERING DATA Enter data into lists by pressing STAT 1 (Edit).

CALCULATING STATISTICS Return to the Home Screen by pressing 2ND QUIT. To calculate the mean of List 1, press 2ND LIST ▶ (MATH) 3 (Mean) 2ND L1 ENTER. To calculate the median, choose 4 instead of 3. To calculate statistics, press STAT ▶ (CALC) 1 (1 - Var Stats) ENTER. You can also choose 2 to calculate two variable statistics and 5 to calculate linear regression. To see the lower quartile of a boxplot, press VARS 5 (Statistics) ▶ ▶ ▶ (BOX) 1 (Q1). Use 2 for median and 3 for the upper quartile.

GRAPHING DATA To graph your data, press 2ND STATPLOT and choose a scatter plot, a line graph, a boxplot, or a histogram. Then choose the data list. Then press GRAPH to draw the graph.

Using the Casio CFX-9800G Graphing Calculator

For additional features and instructions, consult your user's manual.

THE KEYBOARD The feature accessed when a key is pressed is shown in white on the key. To access the features in gold above each key, first press the SHIFT key. To access what is in red above the keys, first press ALPHA.

THE MAIN MENU This is the screen you see when you first turn the calculator on. Highlight and press EXE or press the number to choose the menu item. You can access the Main Menu at any time by pressing the MENU key.

PERFORMING CALCULATIONS Calculations are performed by pressing 1 (for COMPutations) in the Main Menu. The calculator evaluates according to the order of operations. Press EXE (for EXEcute) to calculate. For 3 + 4 × 5 EXE, the result is 23. You can replay the previous line by pressing ◀. Use the arrow keys to edit.

Displaying Graphs

COMP MODE Press 1 in the Main Menu. To enter an equation, press GRAPH. Enter an equation such as $y = 2x + 3$ using the X, θ, T key for x. Then press EXE to display the graph in the viewing window.

GRAPH MODE Press 6 (GRAPH) in the Main Menu. Then press AC. Use the up or down arrows to choose a location to store the equation. Enter the equation as above. Then press F6 (DRW) to display the graph in the viewing window.

VIEWING WINDOW To set range values for the viewing window, press RANGE. You can use F1 (INIT) to set standard values for a viewing window.

ZOOM FEATURE Press SHIFT F2 (ZOOM) to access this feature. Press F1 (BOX) to highlight a particular area and zoom in on that part of the graph. Press F5 (AUT) to set range values for a friendly window.

TRACE FEATURE Pressing SHIFT F1 (TRACE) places the cursor directly on the graph and shows the *x*- and *y*-coordinates of the point where the cursor is located. You can move the cursor along the graph using the right arrow keys.

TABLE FEATURE Press 8 (TABLE) in the Main Menu. Press AC to clear the screen. Then press F1 (RANGE FUNC). Select the function. Then press F5 (RNG) to set up the table. Press F6 (TBL) to see the table of values. You can press SHIFT QUIT to return to a previous screen.

INTERSECTION FEATURE To determine the coordinates of the point of intersection of two graphs, press SHIFT, then 9 (G-SOLV). Then press F5 (ISCT). The calculator will then prompt you to identify the graphs.

Statistics

ENTERING DATA From the Main Menu, press 3 (SD). Press SHIFT SET UP and select STOre for S-data. Then press EXIT AC to clear the screen. Enter data, pressing F1 after each entry.

CALCULATING STATISTICS To calculate the mean ($\overline{x}$), press F4 (DEV) F1 ($\overline{x}$) EXE. To calculate the median, press F4 (DEV) F4 ▼ and F2 (Med) EXE.

GRAPHING DATA To graph your data, press SHIFT SET UP and select DRAW for S-graph. Press EXIT and then GRAPH EXE.

Using the Hewlett-Packard 38G Graphing Calculator

- **For additional features and instructions, consult your user's manual.**

THE KEYBOARD The feature accessed when a key is pressed is shown in yellow on the key. To access the features in green above each key, first press the green key. To access what is in red below the keys, first press A...Z. The blank keys at the top of the keyboard are used for the menu items at the bottom of the screen. Pressing the Menu key at the far right will return menu items to the screen.

CALCULATIONS Calculations are performed in the Home Screen. This screen may be returned to at any time by pressing HOME. The calculator evaluates according to the order of operations. Press ENTER to calculate. For 3 + 4 × 5 ENTER, the result is 23. You can edit a previous line by using the arrow keys to choose a line you want to edit. Press the Copy key. Use the arrow keys to edit.

Displaying Graphs

- **GRAPH FEATURE** To enter an equation, press LIBrary. Select Function. Press ENTER. Press the Edit key to enter an equation such as $y = 2x + 3$, using the X, T, θ key for x. Then press PLOT to display the graph in the viewing window.

VIEWING WINDOW Press the green key and VIEWS. Select Auto Scale to get a friendly window, or you can press the green key and PLOT to change the range in the viewing window. Next, press PLOT to show the graph.

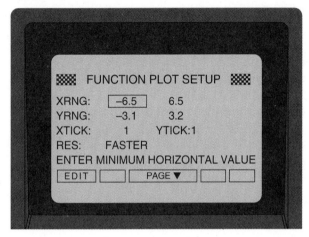

ZOOM FEATURE Press the Menu key. Press the Zoom key. Then select Box... to highlight a particular area and zoom in on that part of the graph.

TRACE FEATURE Pressing the Trace key puts the cursor directly on the graph and shows the x- and y-coordinates of the point of its location. You can move the cursor along the graph using the right and left arrow keys.

TABLE FEATURE To display a table of values for the graphed equation, press NUMber.

INTERSECTION FEATURE To determine the coordinates of the point of intersection of two graphs, move the cursor close to the point of intersection. Press the Function key and select Intersection. This verifies one of your equations. Press ENTER. Then press the Function key and select Intersection to verify the other equation. Press ENTER.

Statistical Features

- **ENTERING DATA** Data may be entered into a list by pressing LIB and selecting Statistics.

CALCULATING STATISTICS To calculate mean, median, upper, and lower quartiles of the data entered in C1, press the Stats key and use the arrow keys to access all the information.

GRAPHING DATA To graph your data, press the green key and then PLOT. Then press the Choos key. Select BoxWhisker or Histogram and ENTER. Then press the green key and VIEWS. Select Auto Scale.

GLOSSARY

• • A • •

absolute value (p. 82)   For any real number a, the *absolute value*, written $|a|$, is defined to be $|a| = a$ if $a \geq 0$ and $|a| = -a$ if $a < 0$.

amortization period (p. 7)   The length of time in which a borrower must repay a loan.

amplitude (p. 771)   Half the difference between the maximum and minimum values of a periodic function. For functions of the form $y = a \sin x$ or $y = a \cos x$, the amplitude is $|a|$.

angle of depression (p. 756)   The angle formed between a horizontal line and an observer's line of sight to an object below the horizon.

angle of elevation (p. 756)   The angle formed between a horizontal line and an observer's line of sight to an object above the horizon.

angle of rotation (p. 743)   The angle formed by the rotation of the terminal side about the vertex.

antilogarithm (of log x) (p. 437)   If $\log x$ is known, then x is the antilogarithm of $\log x$.

arc (p. 738)   Part of a circle.

argument (p. 824)   The *argument* of point $P(r, \theta)$ is the angle that $\overrightarrow{OP}$ makes with the polar axis.

arithmetic means (p. 634) The terms between any two nonconsecutive terms of an arithmetic sequence. A single arithmetic mean between any two given numbers in a sequence is called the *arithmetic mean* of the numbers.

arithmetic sequence (p. 633)   A sequence in which each term after the first is obtained by adding a constant to the previous term.

arithmetic series (p. 640)   The sum of the terms of an arithmetic sequence. The sum of the first n terms of an arithmetic series is given by $S_n = \frac{n}{2}(a_1 + a_n)$.

asymptote (p. 514, 587)   A line that a graph approaches more and more closely but does not intersect.

augmented matrix (p. 183)   A matrix formed by writing the constants of the system of equations in a column and then attaching this column to the matrix of coefficients.

axis of symmetry (p. 107, 287, 592)   A line which divides a figure such that if the figure is folded over that line, the two parts coincide; the vertical line that passes through the vertex of a parabola.

• • B • •

bar graph (p. 90)   A series of bars used to represent values.

bearing (of a vector) (p. 829)   The angle measured clockwise from the north to a vector.

biased results (p. 724)   Results that are not representative of a general population.

binomial (p. 230)   A polynomial with two terms.

binomial experiment (p. 710)   An experiment where there are a fixed number of independent trials, exactly two possible outcomes for any trial, and the same probability of success on each trial.

binomial theorem (p. 654)   If n is a positive integer, then $(a + b)^n + a^n + \frac{n}{1}a^{n-1}b^1 + \frac{n(n-1)}{1 \cdot 2}a^{n-2}b^2 + \frac{n(n-1)(n-2)}{1 \cdot 2 \cdot 3}a^{n-3}b^3 + \cdots + b^n$

bond (p. 358)   A loan to a government or corporation. The bondholder receives interest during the term of the bond and the amount is repaid at maturity.

boxplot (p. 34)   A visual display that uses the three quartiles and the least and greatest values of a set of data.

branches (p. 514, 587)   Distinct parts of a discontinuous, non-linear graph such as a hyperbola or rational function.

• • C • •

cell (p. 13)   An area of a spreadsheet designated by a column and row.

center (of a circle) (p. 575)   The point from which all points on the circle are equidistant.

center (of an ellipse) (p. 581)   The point at which the major and minor axes of an ellipse intersect.

center (of a hyperbola) (p. 587) The point at which the transverse and the conjugate axes of a hyperbola intersect.

change-of-base formula (p. 438) $\log_b x = \dfrac{\log_a x}{\log_a b}$

circle (p. 575) The set of all points that are equidistant from a given point, called the center of the circle.

circle graph (pie chart) (p. 91) A graph that relates two or more values by representing them as portions or sectors of a circle. Each portion represents part of the whole.

closed formula (p. 627) See *explicit formula*.

closed half-plane (p. 322) A half-plane that includes the boundary line.

closed set (p. 339) A region that includes the boundary.

clustered sample (p. 724) A sample selected from a particular segment of the population.

coefficient of correlation (p. 42) A statistical measure of how closely data fits a line.

coefficients matrix (p. 203) Matrix of coefficients of a system.

combination (p. 679) A selection of elements from a set without regard to order.

common difference (p. 633) The constant added to each term of an arithmetic sequence to obtain the next term. To determine the common difference, d, of an arithmetic sequence, subtract any two consecutive terms: $d = a_{n+1} - a_n$.

common logarithm (p. 437) A base 10 logarithm, written $\log x$.

common ratio (p. 635) The constant by which each term of a geometric sequence is multiplied to obtain the next term. To determine the common ratio, r, of a geometric sequence, divide any term by the previous term: $r = \dfrac{a_{n+1}}{a_n}$.

complementary events (p. 686) The events E and *not E* are complementary events. Complementary events have probabilities whose sum is 1.

completing the square (p. 275) A method that can be used to solve any quadratic equation in which a constant is added to both sides of the equation to make one side of the equation a perfect square trinomial.

complex fractions (p. 543) Fractions that have one or more fractions in either the numerator or the denominator.

complex number (p. 402) A number of the form $a + bi$ where a and b are real numbers and $i = \sqrt{-1}$.

complex plane (p. 405) A coordinate system for graphing complex numbers with a horizontal real axis and a vertical imaginary axis.

complex rational expression (p. 543) A rational expression that contains one or more rational expressions in its numerator or denominator.

composition of functions (p. 144) A way to combine functions by successive application of the functions in a specific order. Given two functions f and g, the composite function $f \circ g$ can be described by $(f \circ g)(x) = f(g(x))$.

compound inequality (p. 77) Two inequalities joined by *and* or by *or*.

compound interest (p. 460) Interest paid on the principal and on previously earned interest if all interest is left in the account.

compound probability (p. 688) Probability that involves two or more events.

conditional probability (p. 697) The probability that event B will occur given that event A has already occured, symbolized $P(B|A)$.

conjugate axis (p. 587) The line segment perpendicular to the transverse axis at the center of a hyperbola.

conjugate zero theorem (p. 493) If $F(x)$ is a polynomial with real coefficients and if a and b are real numbers if $a + bi$ is a zero, then $a - bi$ is a zero; also if $a + \sqrt{b}$ is a zero, then $a - \sqrt{b}$ is a zero.

conjugates (p. 386, 406) 1. Two binomials with the same first terms and opposite last terms. 2. Two complex numbers with the same real part and opposite imaginary parts.

conjunction (p. 77) A compound inequality joined by *and*.

consistent system (p. 159) A system of equations with at least one solution.

constants matrix (p. 203) Contains the constants of a system.

constant of variation (p. 529) The constant $k \neq 0$ in the equation of a direct variation $y = kx$, an inverse variation $xy = k$, or a joint variation $y = kwx$.

constant term (p. 473) The term of a polynomial that is not multiplied by a variable.

constraints (p. 350)   Conditions that limit available resources. In linear programming, constraints are represented by inequalities.

continuous function (p. 517)   A function whose graph has no holes or breaks in it.

converge (p. 647)   An infinite series converges if the sequence of partial sums approaches some number as the number of terms gets very large.

correlation (p. 41)   The relationship between two sets of data represented by a number from -1 (a perfect negative correlation) to 1 (a perfect positive correlation).

corresponding elements (p. 20)   Elements in the same position in each of two matrices with the same dimensions.

cosecant (p. 746)   If θ is an angle in standard position and (x, y) is a point distinct from the origin on the terminal side of θ, then $\csc \theta = \frac{r}{y}$, where $r = \sqrt{x^2 + y^2}$.

cosine (p. 746)   If θ is an angle in standard position and (x, y) is a point distinct from the origin on the terminal side of θ, then $\cos \theta = \frac{x}{r}$, where $r = \sqrt{x^2 + y^2}$.

cotangent (p. 746)   If θ is an angle in standard position and (x, y) is a point distinct from the origin on the terminal side of θ, then $\cot \theta = \frac{x}{y}$, where $r = \sqrt{x^2 + y^2}$.

coterminal angles (p. 744)   Angles of rotation in standard position with coinciding terminal sides.

Cramer's rule (p. 212)   A method for solving systems of equations using determinants.

cycle (p. 770)   The shortest repeating section of the graph of a periodic function.

$$\bullet \ \bullet \ \mathbf{D} \ \bullet \ \bullet$$

deferred payment price (p. 6)   The amount of money, including interest, that a borrower must repay for an item bought on installment.

degenerate conic sections (p. 600)   The point, the line, and two intersecting lines.

degree (of a polynomial) (p. 473)   The value of the largest exponent that appears in any term of the polynomial.

dense (p. 64)   The set of real numbers is *dense* because between any two real numbers, another real number can be found.

dependent events (p. 696)   If the outcome of one event is affected by the outcome of another event, the events are dependent. If A and B are dependent events, then
$P(A \text{ and } B) = P(A) \cdot P(B \text{ after } A)$.

dependent system (p. 159)   A consistent system with an infinite number of solutions.

depression polynomial (p. 493)   A polynomial with a degree less than the degree of the original polynomial which is found by dividing the polynomial by a binomial of the form $x - a$.

Descarte's rule of signs (p. 494)   The number of positive real zeros of a polynomial $F(x)$ is equal to the number of sign changes of the coefficients of $F(x)$, or is less than this number by an even number. The number of negative real zeros is equal to the number of sign changes of the coefficients of $F(-x)$, or is less than this number by an even number.

determinant (p. 210)   A numerical value associated with a square matrix.

diagonals method (p. 211)   A method for evaluating a determinant.

difference of two squares (p. 243)
$$a^2 - b^2 = (a + b)(a - b)$$

difference or sum of two cubes (p. 249)
$$a^3 - b^3 = (a - b)(a^2 + ab + b^2)$$
$$a^3 + b^3 = (a + b)(a^2 - ab + b^2)$$

dilation (p. 128)   A transformation that produces a graph with a different slope than that of the original graph.

dimensions (of a matrix) (p. 13)   The number of horizontal rows and the number of vertical columns in a matrix.

directrix   See parabola. (12.6)

direct variation function (p. 152)   A function of the form $y = kx$, where k is the constant of variation.

directed graph (p. 216)   A graph representing connections between directed points.

discontinuous function (p. 598)   A function whose graph has holes or breaks in it.

discriminant (p. 297)   The expression $b^2 - 4ac$ found under the radical symbol in the quadratic formula.

disjoint (sets) (p. 689)   Sets that have no elements in common. The intersection of disjoint sets is empty.

distance formula (p. 567)   The distance between two points with coordinates (x_1, y_1) and (x_2, y_2) is given by $d = \sqrt{(x_2 - x_1)^2 + (y_2 - y_1)^2}$.

diverge (p. 647)   A series diverges if the sequence of partial sums increases (or decreases) without limit as the number of terms gets very large.

domain (of a rational function) (p. 513)   The set of all values of x for which the demoninator $q(x)$ is not equal to zero.

domain (of a relation) (p. 112)   The set of first coordinates of the ordered pairs.

down payment (p. 12)   A portion of the price that a buyer pays at the time of purchase before financing the remaining portion.

double solution (p. 257)   When the solution is the same number in either case, when there are two solutions.

• • E • •

element (of a matrix) (p. 13)   Each number or entry in a matrix.

elimination method (p. 157)   An algebraic method for solving a system of linear equations in which the equations in the system, or multiples thereof, are added to or subtracted from each other to eliminate all but one of the variables.

ellipse (p 581)   The set of all points in the plane the sum of whose distances from two fixed points F_1 and F_2, the foci, is constant.

ellipsis (p. 627)   Three dots used to show that a sequence continues indefinitely. For example, the sequence of even numbers is written as $2, 4, 6, 8, \ldots$.

ellipsoid (p. 586)   A three-dimensional figure obtained by rotating an ellipse around its major axis.

equal matrices (p. 20)   Matrices in which all elements in corresponding positions are equal.

equilalent expressions (p. 532)   Expressions that have the same simplified form.

equivalent vectors (p. 828)   Vectors that have the same length and direction.

even function (p. 114)   A function for which $f(-x) = f(x)$ for all values of x in its domain.

event (p. 25)   An outcome or combination of outcomes.

expansion by minors (p. 211)   A method for evaluating any third or higher order determinant.

experimental probability (p. 27)   A measure of what actually happens when an experiment is performed, a survey is conducted, or historical records are tabulated.
$$P(E) = \frac{\text{Number of times event occurs}}{\text{Number of trials}}$$

explicit formula (p. 627)   A rule that defines the nth term of a sequence using the position n of the term within the sequence. Also called a *closed formula*.

exponential decay curve (p. 431)   The graph of an exponential equation of the form $y = ab^x$, where $0 < b < 1$ and $a > 0$.

exponential equation (p. 392)   An equation containing a variable raised to a rational power.

exponential function (p. 423, 428)   A function that can be expressed in the form $f(x) = b^x$, where $b > 0$ and $b \neq 1$.

exponential growth curve (p. 430)   The graph of an exponential equation of the form $y = ab^x$, where $b > 1$ and $a > 0$.

extraneous solution (p. 82, 393)   An invalid solution that is sometimes introduced when both sides of an equation are raised to a power.

• • F • •

factor by grouping (p. 250)   When some terms have one common factor and other terms have different common factors you group them according to this factor.

factored form (of a polynomial) (p. 242)   A polynomial is in *factored form* when it is expressed as the product of its prime factors.

factorial notation (p. 654)   The notation n!, read n factorial, represents the product of all the positive integers from n to 1:
$$n! = n\,(n-1)\,(n-2) \cdots (2)\,(1)$$

factor theorem (p. 482)   A polynomial $F(x)$ has a factor $(x - r)$ if and only if $F(r) = 0$.

feasible region (p. 350) The intersection of the graphs of a system of linear inequalities or constraints that includes all possible solutions to the system.

field (p. 70)   When two operations and a set of numbers display the six field properties, they form a *field*.

field properties (p. 70)   Closure property, commutative property, associative property, identity property, inverse property, and distributive property of multiplication over addition.

finite sequence (p. 627)   A sequence that stops at a particular term.

finite series (p. 639)   The sum of the terms of a finite sequence.

first quartile (p. 34)   The median of the lower half of a set of data.

formula (p. 5)   A type of literal equation.

finance charge (p. 7)   The amount over the loan amount that a borrower must pay to a lender for a loan.

focus   See ellipse (12.4), hyperbola (12.5), and parabola (12.6).

fractal geometry (p. 407)   Study of nonlinear dimensions.

function (p. 112)   A relation in which each element in the domain is paired with one and only one element in the range.

function notation (p. 114)   The notation for representing a rule that associates an input value (independent variable) with an output value (dependent variable). The most commonly used function notation is the "f of x" notation, written "$f(x)$."

fundamental counting principle (p. 671)   If one event can occur in m different ways and for each of these ways a second event can occur n different ways, then together the events can occur in $m \cdot n$ different ways.

fundamental theorem of algebra (p. 492)   Every polynomial equation of degree 1 or more has at least one solution in the set of complex numbers.

• • G • •

general form (of the graph of a second degree equation) (p. 599)   An equation of the form $Ax^2 + Bxy + Cy^2 + Dx + Ey + F = 0$.

general term (p. 624) The nth term, symbolized a_n, of a sequence.

geometrical probability (p. 26)   A method of calculating or visualizing theoretical probability.

geometric means (p. 635)   The terms between any two nonconsecutive terms of a geometric sequence. A single geometric mean between any two given numbers in a sequence is called the *geometric mean* of the numbers.

geometric sequence (p. 635)   A sequence in which each term after the first is a constant multiple of the previous term.

geometric series (p. 640)   The sum of the terms of a geometric sequence. The sum of the first n terms of a geometric sequence is given by
$$S_n = \frac{a_1(1 - r^n)}{1 - r}, r \neq 1.$$

graph of inequality (p. 322)   The graph of all the solutions of the inequality.

Greatest Common Factor (GCF) (p. 242)   One type of factoring is to factor this from each term of the polynomial.

greatest integer function (p. 130)   A function that assigns to each real number the greatest integer that is less than or equal to that number.

• • H • •

half-planes (p. 322)   The two regions into which the graph of a linear equation divides the coordinate plane.

Heron's formula (p. 761)   The area A of a triangle with sides measuring a, b, and c is given by $A = \sqrt{s(s - a)(s - b)(s - c)}$, where $s = \frac{1}{2}(a + b + c)$.

histogram (p. 703)   A vertical bar graph showing intervals of equal width and without spaces between the bars.

horizontal asymptote (p. 424)   The distance between the x-axis and a graph where $x > 1$ gets closer to 0 as x gets smaller.

horizontal line test (p. 138)   A test used on the graph of a function to determine whether the graph represents a one-to-one function. If no horizontal line can be drawn that intersects the graph in more than one point, then the graph represents a one-to-one function and the function has an inverse that is also a function.

horizontal translation (p. 127)   A shift left or right that produces the same graph in a new position.

horizontal shift (p. 127) Moves the graph to the right or left.

hyperbola (p. 587) The set of all points in the plane such that the absolute value of the difference of whose distances from two fixed points F_1 and F_2, the foci, is a constant.

hyperboloid (p. 597) Three-dimensional figure obtained by rotating a hyperbola about its conjugate axis.

• • ▌ • •

identity (p. 792) An equation that is true for all values of the variable, except those for which either side of the equation is undefined.

imaginary axis (p. 405) The vertical axis of a complex plane.

imaginary number (p. 399) A pure imaginary is a number of the form bi where b is a real number and $b \neq 0$ and $i = \sqrt{-1}$.

inconsistent system (p. 154) A system of equations with no solutions.

independent events (p. 695) If the outcome of one event is not affected by the outcome of another event, the events are independent. Events A and B are independent if and only if $P(A \text{ and } B) = P(A) \cdot P(B)$.

independent system (p. 159) A system of equations that has exactly one solution.

index (p. 379) The number which tells the root of the expression. In the expression $\sqrt[n]{b}$, n is the index.

index of summation (p. 641) The variable used below Σ (sigma). In the notation $\sum\limits_{n=1}^{7} 2^n$, n is the index of summation, 1 is the *lower limit*, and 7 is the *upper limit*.

infinite sequence (p. 627) A sequence that does not have a last term.

infinite series (p. 639) The sum of the terms of an infinite sequence.

inner product (p. 830) The *inner product* of vectors **u** and **v** is a scalar defined as:
u · **v** = ‖**u**‖ ‖**v**‖ cos **θ**.

initial side (p. 737) The ray, in its original position, that is rotated about its vertex to form an angle.

initial value (p. 407) The first value sustituted into the function defining an iterative process.

intersection (of sets) (p. 688) The intersection of sets A and B, symbolized $A \cap B$, contains the elements that belong to both sets.

interquartile range (p. 36) The difference between the third and fourth quartile.

inverse function (p. 139) An inverse of a function that is also a function denoted by the symbol f^{-1}.

inverse of a function (p. 137) The relation that results from interchanging the first and second coordinates of each ordered pair in a given function represented by a set of ordered pairs.

inverse of a matrix (p. 196) Two matrices are *multiplicative inverses* of each other if their product is the identity matrix.

inverse of a relation (p. 137) The relation that results from interchanging the first and second coordinates of each ordered pair in a given relation represented by a set of ordered pairs.

inverse square variation (p. 525) One variable varies inversely as the *square* of the other variable. If y varies inversely as the square of x, the inverse square relationship can be written $y = \dfrac{k}{x^2}$, where $x \neq 0$ and $k \neq 0$.

inverse trigonometric function (p. 779) A function obtained when the restricted domain and range of a trigonometric function are interchanged.
$y = \text{Arcsin } x$, where $x = \text{Sin } y$ and $-1 \leq x \leq 1$.
$y = \text{Arccos } x$, where $x = \text{Cos } y$ and $-1 \leq x \leq 1$.
$y = \text{Arctan } x$, where $x = \text{Tan } y$.

inverse trigonometric relation (p. 779) A relation obtained when the domain and range of a trigonometric function are interchanged.

inverse variation (p. 524) A relationship between x and y that can be written in the form $y = \dfrac{k}{x}$, or $xy = k$, where the constant of variation $k \neq 0$. In an inverse square variation, y varies inversely as the square of x, so $\dfrac{k}{x^2}$.

irrational number (p. 63) A number that cannot be expressed as the ratio of integers $\dfrac{a}{b}$ and whose decimal representation neither repeats nor terminates.

iterate (p. 407) Each successive output in a continually recycling process.

iteration (p. 407)   A repetitive process in which the output of the previous step is used as the input of the next step.

• • J • •

joint variation (p. 526)   A relationship in which one variable varies directly as the product of two or more other variables. If y varies jointly as x and z, this is expressed by $y = kxz$, where $k \neq 0$ is the constant of variation.

• • L • •

law of cosines (p. 760)   For any triangle ABC with sides measuring a, b, and c opposite the vertices A, B, and C, respectively, $a^2 = b^2 + c^2 - 2bc \cos A$, $b^2 = a^2 + c^2 - 2ac \cos B$, and $c^2 = a^2 + b^2 - 2ab \cos C$.

law of sines (p. 759)   For any triangle ABC with sides measuring a, b, and c opposite the vertices A, B, and C, respectively, $\dfrac{\sin A}{a} = \dfrac{\sin B}{b} = \dfrac{\sin C}{c}$.

leading coefficient (p. 473)   The coefficient of the term in the polynomial with the greatest exponent.

like radicals (p. 387)   Radicals with the same index and radicand.

line graph (p. 90)   A graph consisting of a line or lines that is effective for showing trends and fluctuations in data.

linear equation (p. 38)   An equation in which no variable has an exponent greater than one.

linear function (p. 114)   A function that can be written in the form $f(x) = mx + b$, where m and b are constants.

linear programming (p. 350)   A method used by business and government to allocate resources in a way that maximizes or minimizes quantities such as profit or cost.

line of best fit (p. 40)   The line that approximates a trend for the data on a scatter plot.

line of reflection (p. 126)   The line over which the graph is reflected.

line of symmetry (p. 107)   A line which divides a figure so that when the figure is folded over that line, the two parts coincide.

literal equation (p. 5)   An equation with more than one letter or variable.

location theorem (p. 494)   If $F(x)$ is a polynomial function with terms in descending degrees of the variable, such that $F(a)$ and $F(b)$ have opposite signs, then there is at least one real zero between a and b.

logarithm (p. 436)   An exponent.

logarithmic function (p. 436)   A function that can be expressed in the form $y = \log_b x$, where $b > 0$ and $b \neq 1$, that is the inverse of the exponential function $y = b^x$.

• • M • •

major axis (p. 581)   The longer axis of an ellipse.

mapping (p. 112)   Demonstrates how each element of the domain of a relation is paired with a member of the range.

matrix (p. 14)   A rectangular arrangement of numbers enclosed by brackets.

matrix equation (p. 198)   An equation of the form $AX = B$, where A, X, and B are matrices.

matrix multiplication (p. 189)   The process by which you multiply matrices. The operation is not commutative.

maximum point (p. 287)   The highest point on the graph of a quadratic equation.

mean (p. 32)   The arithmetic average of a set of data that is calculated by dividing the sum of the data by the number of items.

measure of an angle (p. 737)   The amount of rotation from the initial side to the terminal side.

measures of central tendency (p. 32)   Values used to give information about the overall "center" of a set of data, including the mean, the median, and the mode.

measures of dispersion (p. 34)   Values used to describe the way in which data is clumped together or spread out, including the range and quartiles.

median (p. 32)   The middle value when data are arranged in numerical order.

method of finite differences (p. 500)   A method that can be applied to determine whether the relationship between paired data can be modeled by a polynomial function.

midpoint (p. 568)   The point on a line segment that is equidistant from the endpoints of the segment.

midpoint formula (p. 568)   The midpoint of the line segment with endpoints (x_1, y_1) and (x_2, y_2) is the point $\left(\dfrac{x_1 + x_2}{2}, \dfrac{y_1 + y_2}{2}\right)$.

minor axis (p. 581)   The shorter axis of an ellipse.

minimum point (p. 287)   The lowest point on the graph of a quadratic equation.

minor (p. 211)   The determinant formed when the row and column containing the element are eliminated.

mode (p. 32)   The number that occurs most often in a set of data.

modulus (p. 824)   The *modulus* of the point $P(r, \theta)$ representing $a + bi$ is the distance from $a + bi$ to the pole, which is assumed to be nonnegative.

monomial (p. 229)   A real number, a variable, or a product of a real number and one or more variables.

multiplicative inverses (p. 196) When the product of two matrices is the identitiy matrix.

mutually exclusive events (p. 689)   Events that cannot occur simultaneously.

• • N • •

natural logarithm (p. 449)   A base e logarithm, written $\ln x$.

norm (p. 828) The length of a vector.

normal distribution (p. 703)   A distribution whose graph is a normal, or bell-shaped, curve.

• • O • •

objective function (p. 350) An equation that represents a quantity such as profit or cost which is to be maximized or minimized using the linear programming method.

oblique triangle (p. 758)   A triangle that does not contain a right angle.

odd function (p. 114)   A function for which $f(-x) = -f(x)$ for all values of x in its domain.

odds (p. 25)   The ratio of the number of ways an event can occur to the number of ways in which it can fail to occur.

one-to-one function (p. 117)   A function f such that, for any two elements x_1 and x_2 in the domain, $f(x_1) = f(x_2)$ if and only if $x_1 = x_2$.

open half-plane (p. 322)   A half-plane that does not include the boundary line.

open set (p. 340)   A region that does not include the boundary.

opposite vectors (p. 828)   Vectors that have the same length but opposite directions.

order of operations (p. 68)   The order in which mathematical operations are performed.

ordered triple (p 164)   The solution of an equation in three variables.

outliers (p. 34)   A value far to the left of the first quartile or to the the right of the third quartile in a boxplot for a data set.

• • P • •

paraboloid (p. 597)   A three dimensional surface formed by rotating a parabola about its axis of symmetry.

parabola (p. 287, 592)   The U-shaped graph of a quadratic function. The set of all points in a plane that are equidistant from a given line, called the directrix, and a given point not on the line, called the focus.

parallel lines (p. 121)   Two lines in the same coordinate plane whose slopes are equal or are both undefined.

partial sum (p. 639)   The sum of the first n terms of a sequence. Partial sums are symbolized by S_n.

Pascal's triangle (p. 653)   A number pattern for determining the coefficients of the terms of a binomial expansion.

perfect square trinomial (p. 244)
$(a + b)^2 = a^2 + 2ab + b^2$
$(a - b)^2 = a^2 - 2ab + b^2$

period (p. 770, 771)   The horizontal length of each cycle of the graph of a periodic function. The least positive number h such that $f(x + h) = f(x)$. For functions of the form $y = \sin bx$ or $y = \cos bx$, the period is $\dfrac{2\pi}{|b|}$.

periodic function (p. 770)   Any function whose graph exhibits a continuously repeating pattern such that $f(x + h) = f(x)$ for every x in the domain of x.

permutation (p. 672)   An arrangement of some or all of the elements of a set in a specific order. If the arrangement is in a line, the permutation is linear; if the arrangement is in a circle, the permutation is circular.

perpendicular lines (p. 121)   Two lines in the same coordinate plane whose slopes are negative reciprocals of each other. A vertical line and a horizontal line are also perpendicular.

piecewise function (p. 130)   A function defined differently over various parts of its domain.

point discontinuity (p. 518)   A single point at which a function is discontinuous.

point-slope form of a linear equation (p. 120)   A linear equation in the form $(y - y_1) = m(x - x_1)$, where m is the slope of the line and (x_1, y_1) are the coordinates of a given point on the line.

polar coordinates (p. 821)   The polar coordinates of a point P are an ordered pair (r, θ), where $|r|$ is the distance from the pole to point P and θ is the measure of the angle formed by the polar axis and $\overrightarrow{OP}$

polar coordinate system (p. 821)   Consists of a point O, called the pole, and a horizont ay with the pole as its left endpoint, called the polar axis.

poll (p. 724)   A survey for gathering data.

polygon (p. 346)   A closed plane figure formed by at least three line segments.

polygonal region (p. 346)   A region in the coordinate plane bounded by a polygon.

polynomial (p. 229, 473)   A monomial or a sum of monomials. An expression of the form $a_n x^n + a_{n-1} x^{n-1} + \cdots + a_1 x + a_0$

polynomial function (p. 473)   A monomial or a sum of monomials. A polynomial expression written in the form $F(x) = a_n x^n + a_{n-1} x^{n-1} + \ldots + a_1 x + a_0$ or $y = a_n x^n + a_{n-1} x^{n-1} + \ldots + a_1 x + a_0$

principal (p. 460)   An amount of money that is invested or borrowed.

principal *n*th root (p. 374)   If $b^n = a$, then $b = \sqrt[n]{a}$ is the principal nth root of a. If a has more than one nth root, the nonnegative root is the principal root.

principal root (p. 374)   The nonnegative root of a number.

principal square root (p. 63, 374)   The nonnegative square root of a number.

probability (p. 25)   A number from 0 to 1 that expresses the likelihood that an event E will occur.

properties of logarithms (p. 443)   If m, n, and b are positive real numbers, $b \neq 1$, and p is any real number, then

$$\log_b mn = \log_b m + \log_b n \quad \text{Product Property}$$
$$\log_b \frac{m}{n} = \log_b m - \log_b n \quad \text{Quotient Property}$$
$$\log_b m^p = p \log_b m \quad \text{Power Property}$$

Pythagorean theorem (p. 59) For a right triangle with legs a and b and hypotenuse c, $a^2 + b^2 = c^2$.

• • Q • •

quadrantal angle (15.3)   An angle whose terminal side falls on an axis.

quadratic equation (p. 255)   A polynomial equation of degree 2 of the general form $ax^2 + bx + c = 0$.

quadratic form (p. 303)   If an equation $f(x) = 0$ can be rewritten in the form $a(f(x))^2 + b(f(x)) + c = 0, a \neq 0$, it is said to be in *quadratic form*.

quadratic formula (p. 296)   For a quadratic equation of the form $ax^2 + bx + c = 0$, where a, b, and c are real numbers and $a \neq 0$, $x = \frac{-b \pm \sqrt{b^2 - 4ac}}{2a}$.

quadratic function (p. 287)   A function that can be written in the form $f(x) = ax^2 + bx + c$, where a, b, and c are real numbers and $a \neq 0$.

quadratic inequality (p. 332) An inequality that results when the equal symbol in a quadratic equation is replaced by $<, >, \leq, \geq,$ or $\neq$.

quadratic system (p. 602)   A set of equations that involves at least one quadratic equation.

• • R • •

radian (p. 738)   A unit of angle measure; the measure of a central angle θ whose sides intercept an arc of length r in a circle of radius r.

radical equation (p. 392)   An equation containing a variable in a radicand.

radicand (p. 63)   In the expression $\sqrt[n]{b}$, b is the radicand.

radius (p. 575)   The distance of any point on the circle and the center of the circle.

random numbers (p. 719)   A large set of numbers that have no detectable pattern associated with them.

range (of a relation) (p. 112)   The set of second coordinates of the ordered pairs.

range (of values) (p. 34)   The difference between the greatest and least values in a set of data.

rational equation (p. 531)   An equation that contains one or more rational expressions.

rational exponents (p. 376)   If m and n are positive integers, a is a real number, then the properties of integral exponents are also true for rational exponents.

rational expression (p. 531)   An expression that can be written in the form $\frac{p(x)}{q(x)}$ where $p(x)$ and $q(x)$ are polynomials and $q(x) \neq 0$.

rational function (513)   A function of the form $y = f(x) = \frac{p(x)}{q(x)}$, where $p(x)$ and $q(x)$ are polynomials in x and $q(x) \neq 0$.

rational number (p. 63)   A number that can be written in the form $\frac{a}{b}$, where a and b are integers and b is not equal to zero.

rationalizing the denominator (p. 386)   The process of removing radicals from the denominator of a radical expression.

rational zero theorem (p. 492)   If a polynomial function with integral coefficients has a rational zero $\frac{p}{q}$ (in simplest form), then p is a factor of the constant term and q is a factor of the leading coefficient.

ratio test (p. 659)   A method for determining whether a non-geometric infinite series converges. If the algebraic expression representing the ratio of successive terms approaches 1 as n gets very large, the series converges.

real axis (p. 405) In a complex axis, the horizontal axis.

real numbers (p. 63)   The union of the set of rational and irrational numbers.

real zeros (p. 486)   The x-intercepts of the graph of a function.

reciprocal functions (p. 747)   The sine and cosecant functions, the cosine and secant functions, and the tangent and cotangent functions are pairs of *reciprocal functions*.

rectangular hyperbola (p. 589)   A hyperbola with its center at the origin, the coordinate axes as asymptotes, and an equation of the form $xy = k$, $k \neq 0$.

recursive formula (p. 627)   A rule that defines the nth term of a sequence by using the $(n-1)$ term.

reference angle (p. 749)   The smallest positive acute angle determined by the x-axis and the terminal side of an angle.

reference triangle (p. 746)   The right triangle formed when a perpendicular is drawn from any point on the terminal side of an angle to the x-axis.

reflection (p. 126)   A transformation in which a figure is flipped over a line that is called the *line of reflection*.

regression line (p. 42)   The line of best fit.

relation (p. 112)   A set of ordered pairs.

remainder theorem (p. 481)   If a polynomial $F(x)$ is divided by $(x - r)$, then the remainder is equal to $F(r)$.

resistance (p. 56)   Opposition to the flow of electricity (measured in ohms).

resolve (p. 828)   To determine the x- and y-components of a vector.

resultant (p. 828)   The sum of two vectors.

Richter scale (p. 442)   A logarithmic scale used to measure the magnitude of earthquakes.

roots (p. 247)   The solutions to a quadratic equation.

row operations (p. 184)   Produces new matrices that leads to systems having the same solution as the original system.

• • S • •

sample (p. 724)   A representative part of the general population.

sample space (p. 25)   The set of all possible outcomes.

scalar (p. 828)   A real number.

scalar multiplication (p. 20)   Multiplication of a matrix by a real number.

scatter plot (p. 40)   Shows the relationship between two real world quantities on a coordinate plane.

scientific notation (p. 47)   A number expressed in the form $a \times 10^n$, where $1 < a < 10$ and n is any integer.

secant (p. 746)   If θ is an angle in standard position and (x, y) is a point distinct from the origin on the terminal side of θ, then $\sec \theta = \frac{r}{x}$, where $r = \sqrt{x^2 + y^2}$.

second quartile (p. 34)   The median of the set of data.

semi-major axis (of an ellipse) (p. 610)   One-half the major axis of an ellipse.

semi-minor axis (of an ellipse) (p. 610)   One-half the minor axis of an ellipse.

sequence (p. 624)   An ordered set of numbers that are related mathematically.

series (p. 639)   The sum of the terms of a sequence.

sigma notation (p. 641)   A shorthand method of writing a sum that uses the uppercase Greek letter sigma, Σ.

simple interest (p. 460)   Interest paid once per year at the end of the year on the total balance in a savings account.

simplest form (p. 376)   A radical expression is in simplest form when it is written using the least possible index, the radicand contains no fractions, and no denominator contains a radical.

simplest form (of a rational expression) (p. 531)   A rational expression is in simplest form when its numerator and denominator contain no common factors. Expressions which have the same simplified form are equivalent expressions.

sign graph (p. 333)   An organized method of analyzing the signs of the factors of the trinomial related to a quadratic inequality.

simulation (p. 717)   A representation of a real-world event that possesses the same mathematical characteristics as the real situation and from which probabilities can be estimated.

sine (p. 746)   If θ is an angle in standard position and (x, y) is a point distinct from the origin on the terminal side of θ, then $\sin \theta = \frac{y}{r}$, where $r = \sqrt{x^2 + y^2}$.

slope (of a line) (p. 39)   The steepness of the line, which is measured by the ratio of the change in the vertical distance to the change in the horizontal distance.

slope-intercept form (of a linear equation) (p. 39, 119)   A linear equation in the form $y = mx + b$, where m is the slope and b is the y-intercept of the graph.

solutions (of a polynomial equation) (p. 474, 486)   The values of x that make the polynomial equation true.

square matrix (p. 20)   A matrix with the same number of rows and columns.

standard deviation (p. 704)   A measure of dispersion for a set of data. The standard deviation, symbolized σ, is the positive square root of the variance.

standard form (of a linear equation) (p. 38)   A linear equation in the form $Ax + By = C$, where A, B, and C are integers and A and B are not both zero.

standard position (p. 737)   The position of an angle with vertex at the origin, initial side on the positive x-axis, and terminal side in the plane.

stratified sample (p. 724)   A sample that reflects the makeup of the population with regard to an influential factor.

substitution method (p. 156)   An algebraic method for solving a system of linear equations in which an equation is solved for one variable in terms of the other(s).

sum and product of the solutions (p. 298)   If s_1 and s_2 are the solutions of a quadratic equation of the form $ax^2 + bx + c = 0, a \neq 0$, then $s_1 + s_2 = -\frac{b}{a}$ and $s_1 s_2 = \frac{c}{a}$.

summation notation (p. 641)   See *sigma notation*.

synthetic division (p. 480)   An algebraic method for dividing polynomials.

system of linear equations (p. 156)   Two or more linear equations that are considered together. Also called a *linear system*.

system of linear inequalities (p. 324)   Two or more inequalities that are considered together.

system of linear equations in three variables (p. 165)   consists of three equations. The system has exactly one solution when the three planes intersect in a single point.

• • **T** • •

tangent (p. 746)   If θ is an angle in standard position and (x, y) is a point distinct from the origin on the terminal side of θ, then $\tan \theta = \frac{y}{x}$ where $r = \sqrt{x^2 + y^2}$.

term (p. 624)   Each value in a sequence. See also *polynomial*.

terminal side (p. 737)   The ray that is formed by rotating the initial side of the angle about its vertex.

theoretical probability (p. 26)
$$P(E) = \frac{\text{Number of outcomes in an event}}{\text{Number of outcomes in sample space}}$$

third quartile (p. 34)   The median of the upper half of a set of data.

transition matrix (14.7)   A matrix used to compute probabilities for problems involving linked chains of events (Markov chains).

translation (of a graph) (p. 127)   A shift that produces the same graph in a new position.

transverse axis (p. 587)   The line segment joining the two vertices of a hyperbola.

tree diagram (p. 26)   A diagram that shows all the possible outcomes in a sample space.

trichotomy property (p. 63)   For real numbers a and b, exactly one of the following is true: $a < b$, $a = b$, or $a > b$.

trigonometric functions (p. 746)   The sine, cosine, tangent, cosecant, secant, and cotangent functions.

trigonometry (p. 737)   The study of the relations between the sides and angles of triangles.

trinomial (p. 230)   A polynomial with three terms.

triple solution (257)   When a solution has the multiplicity of three.

• • U • •

unit circle (p. 750)   A circle centered at the origin whose radius is 1 unit.

upper and lower bound theorem (p. 495)   For a positive number p, if $F(x)$ is divided by $(x - p)$ and the remainder and the coefficients of the quotient are all positive or all negative, then the zeros of $F(x)$ cannot be greater than p. The value of p is an *upper bound* for $F(x)$. If q is an upper bound for $F(-x)$, then $-q$ is a *lower bound* for $F(x)$.

union (of sets) (p. 688)   The union of sets A and B, symbolized $A \cup B$, contains the elements that belong to either or both sets.

• • V • •

variable matrix (p. 203)   Contains the variables of a system.

variance (p. 709)   A measure of dispersion for a data set. The variance is the average of the squares of the mean deviations.

vector (p. 753, 828)   A directed line segment used to represent a vector quantity.

vector quantity (p. 828)   A quantity that has both magnitude and direction.

Venn diagram (p. 688)   A diagram that displays the relationship between sets.

vertex (of an angle) (p. 743)   The common endpoint of the two rays that form an angle.

vertex (of a parabola) (p. 287, 592)   The minimum or maximum point on the graph of a quadratic equation. The point midway between the focus and the directrix of the parabola.

vertical line test (p. 112)   A test used on the graph of a relation to determine whether the relation is also a function. If any vertical line drawn through the graph of a relation intersects the graph in more than one point, the graph does not represent a function.

vertical shift (p. 127) Moves the graph up or down.

vertical translation (p. 121)   An upward or downward shift that produces the same graph in a new position.

vertices (of an ellipse) (p. 581)   The points at which the major or minor axes intersect the graph of an ellipse.

vertices (of a hyperbola) (p. 587)   The point on each branch of the hyperbola nearest the center.

• • W • •

whiskers (p. 34)   Horizontal lines in a boxplot showing the range of the data and the first and fourth quartile.

• • X • •

x-intercept (p. 38)   The x-coordinate of the point where a graph crosses the x-axis.

• • Y • •

y-intercept (p. 38)   The y-coordinate of the point where a graph crosses the y-axis.

• • Z • •

zero matrix (p. 20)   A matrix in which all elements are zero.

zero product property (p. 254)   For any real numbers a and b, if $ab = 0$, then $a = 0$, or $b = 0$, or both $a = 0$ and $b = 0$.

zeros of a polynomial function (p. 474, 486)   The values of x for which $F(x)$, or y, is 0. The zeros of the polynomial function are the *solutions* of the corresponding polynomial equation when $F(x) = 0$.

• • • SELECTED ANSWERS • • •

Chapter 1 Modeling and Predicting

Lesson 1.1, pages 5–11

TRY THESE **1.** division **3.** subtraction, multiplication **5.** 11 **7.** -4 **9.** 9 **11.** 4

13. $3x - 4 = 5$. Add 4 unit blocks to each side: $3x - 4 + 4 = 5 + 4$. Remove zero pairs: $3x = 9$. Group x-blocks with unit blocks: $x = 3$.

15. $178 **17a.** $371.96 **17b.** $44,635.20
17c. $14,635.20

PRACTICE **1.** 23 **3.** -2.5 **5.** 42 **7.** 3 **9.** 5
11. 21 **13.** $R = P + C$ **15** $b = \dfrac{2A}{h}$
17. $w = \dfrac{P}{2} - l$ **19a.** $408 = 342 + m$
19b. The markup is $66. **21a.** $161.45
21b. $9687.00 **21c.** $1687.00

EXTEND **23.** a, c, d **25.** -10 **27.** 10 **29.** 0.5
31. 13 **33.** 18% **35.** $4200

THINK CRITICALLY

37. The division property of equality excludes 0 as a divisor. In this case x can have any value, including 0. Thus a false or nonequivalent equation results.

MIXED REVIEW **39.** 7 **40.** -7 **41.** 1 **42.** 3
43. 40.5 **44.** 240% **45.** 116 **46.** 1 **47.** -7
48. 5 **49.** 14 **50.** -6

Lesson 1.2, pages 12–18

TRY THESE **1.** $3 * n - 5$ **3.** $(a + h) / 2$
5. 10; 13 **7.** 0.75; 8.19
9. $= A2 * B2$; 54; $2 * A2 + 2 * B2$; 30
11. $= A4 * B4$; 28.08; $2 * A4 + 2 * B4$; 22.2
13. $= A4 + A4 * (B4 / 100)$; 6.095

15.

	A	B	C
1	Sticker	Dealer's	Markup
2	Price	Cost	
3	276	240	=B3–A3; 36
4	276	225	=B4–A4; 51
5	276	208	=B5–A5; 68

PRACTICE **1.** $p / (x - y)$ **3.** $4 * a - 3 / c$
5. 22; 72 **7.** 13; 58.32 **9.** $= 12 * A3 * B3$; 4551.36
11. $= 12 * A5 * B5$; 5527.68
13. $= 4A * (4B / 100)$; 1851; $4A - 4C$; 10489

15. $= 0.01 * B3 / 12$; 0.005; 131.95 D3: $= A3 * C3 * (1 + C3) ^\wedge 60 / ((1 + C3) ^\wedge 60 - 1)$
17. D5: $= A5 * C5 * (1 + C5) ^\wedge 60 / ((1 + C5) ^\wedge 60 - 1)$

19.

	A	B	C	D
1	Base 1	Base 2	Height	Area
2	4	6	10	=0.5*C2*(A2+B2); 50
3	4	6	15	=0.5*C3*(A3+B3); 75
4	4	6	20	=0.5*C4*(A4+B4); 100

EXTEND **21.** The greater down payment and the shorter repayment period; 30% down, 3 years to pay

23.

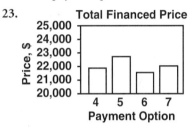

MIXED REVIEW **25.** $3x - 15y$ **26.** $24a^2$
27. $-2ne + 6n^2$ **28.** $a^2 + 2ab + b^2$ **29.** 2; 8
30. 8; 1 **31.** B

Lesson 1.3, pages 19–24

TRY THESE **1.** 3×2

3. $\begin{bmatrix} -1 & -15 \\ 8 & 15.5 \\ 31 & -3 \end{bmatrix}$ **5.** $\begin{bmatrix} 73.2 & 90 \\ 0 & 12 \\ 6 & -93.6 \end{bmatrix}$

7. $\begin{bmatrix} 3 & 6 \\ 1 & 8 \end{bmatrix} + \begin{bmatrix} 2 & 9 \\ 4 & 7 \end{bmatrix} = \begin{bmatrix} 5 & 15 \\ 5 & 15 \end{bmatrix} = \begin{bmatrix} 2 & 9 \\ 4 & 7 \end{bmatrix} + \begin{bmatrix} 3 & 6 \\ 1 & 8 \end{bmatrix}$

PRACTICE **1.** $0, -5, -6, 20, 5, 2$

3. $\begin{bmatrix} 7 & 2 & -2 \\ 33 & 3 & -6 \end{bmatrix}$ **5.** $\begin{bmatrix} 8 & -7 & -18 \\ -6 & 14 & 5 \end{bmatrix}$

7. $\begin{bmatrix} 15 & 5 & -8 \\ -13 & 7 & -5 \end{bmatrix}$ **9.** $\begin{bmatrix} 7 & 7 & 4 \\ 13 & -2 & -8 \end{bmatrix}$

13. $\begin{bmatrix} 71 & 71 & 53 \\ 76 & 25 & 30 \end{bmatrix}$ **15.** $\begin{bmatrix} 3120 & 3744 & 2028 \\ 5148 & 3276 & 2964 \end{bmatrix}$

EXTEND **17.** no

19. $\begin{bmatrix} 2.8 & -1.6 & 4 \\ 11 & -5.6 & -12 \\ 0.4 & 0 & 20 \end{bmatrix}$ **21.** $\begin{bmatrix} 0 & 4 & -1 \\ 3 & -1 & -7 \end{bmatrix}$

23. No, because subtraction of real numbers, the operation on which matrix subtraction is based, is not commutative. $\begin{bmatrix} 5 & 5 \\ 5 & 5 \end{bmatrix} - \begin{bmatrix} 2 & 2 \\ 2 & 2 \end{bmatrix} = \begin{bmatrix} 3 & 3 \\ 3 & 3 \end{bmatrix}$, but

$\begin{bmatrix} 2 & 2 \\ 2 & 2 \end{bmatrix} - \begin{bmatrix} 5 & 5 \\ 5 & 5 \end{bmatrix} = \begin{bmatrix} -3 & -3 \\ -3 & -3 \end{bmatrix}$

25. b_{75}

29. Answers will vary. Possible answer:

$\begin{bmatrix} 5 & 2 \\ 3 & -2 \\ 9 & 4 \end{bmatrix} + \begin{bmatrix} 4 & 3 \\ -3 & 3 \\ 6 & 1 \end{bmatrix} = \begin{bmatrix} 9 & 5 \\ 0 & 1 \\ 15 & 5 \end{bmatrix}$

$\begin{bmatrix} 4 & 3 \\ -3 & 3 \\ 6 & 1 \end{bmatrix} + \begin{bmatrix} 5 & 2 \\ 3 & -2 \\ 9 & 4 \end{bmatrix} = \begin{bmatrix} 9 & 5 \\ 0 & 1 \\ 15 & 5 \end{bmatrix}$

THINK CRITICALLY

33. It is similar to the original triangle but has sides twice as long.

MIXED REVIEW

35. 3 **36.** −6 **37.** $\frac{3}{4}$ **38.** 3

39. [15 22 211 8 2 3]
40. [233 12 18 0 215 26]
41. [7 26 21 28 8 1]
42. [41 28 228 16 9 8]
43. C

Lesson 1.4, pages 25–31

TRY THESE **1.** 5, 6 **3.** $\frac{1}{4}$ **5.** $\frac{1}{8}$ **7.** 1 **9.** 1 : 3

11.

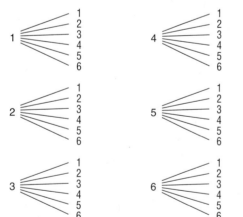

First roll Second roll First roll Second roll

13. $\frac{1}{27}$

PRACTICE

1–3. Examples will vary. 0: There will be 13 inches in a foot tomorrow morning; 0.5: a randomly chosen person will be female; 1: A tossed coin will come up heads or tails.

5. $\frac{7}{25}$ or 0.28 **7.** numbers 1–20 **9.** $\frac{1}{4}$ or 0.25

11. 0 **13.** 1 **15.** 0.16 or $\frac{4}{25}$ **17.** $\frac{23}{276}$ or $\frac{1}{12}$

EXTEND

21. $\frac{6}{11}$

THINK CRITICALLY

23. Twelve. The tail could come up on the first toss, second, third, and so on to the twelfth toss, a total of twelve outcomes.

MIXED REVIEW

25. 8 **26.** −0.5 **27.** −3 **28.** 19 **29.** $\frac{1}{6}$ **30.** $\frac{1}{2}$

31. $\frac{1}{3}$ **32.** 0 **33.** D

Lesson 1.5, pages 32–37

TRY THESE **1.** 41, 46, none, 25 **3.** 492, 492, none, 177
5a. 3.35, 3.25, 2.6, 3.7 **7.** 70 **9.** 60 **11.** 25%
13.

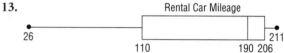

Rental Car Mileage
26   110   190 206   211

PRACTICE

1. 19, 20, none, 25 **3.** 332.875, 328.5, 314 and 361, 47
5. 6.76, 6.79, none, 0.18 **7.** 57.625, 56, 51 and 53, 33
9.

Pulse Rates
46   51   56   61.5   79

11. Class 1, Class 1, 98, 42 **13.** Class 2

EXTEND

17. 46, 55 **21.** 13

THINK CRITICALLY

23. 160 or 161 **25.** Answers will vary. For example, {3, 3, 3, 4, 5, 5, 6, 7, 7, 7}

MIXED REVIEW

28. $\begin{bmatrix} -7 & -3 & 9 \\ 3 & 1 & -4 \end{bmatrix}$ **29.** $\begin{bmatrix} -5 & 1 & 8 \\ -2 & 4 & 3 \end{bmatrix}$

30. $\begin{bmatrix} -6 & -12 & 3 \\ 15 & -9 & -21 \end{bmatrix}$ **31.** $\begin{bmatrix} 12 & 2 & -17 \\ -1 & -5 & 1 \end{bmatrix}$

32. 8.8, 8.5, 6, 6, 13 **33.** 29, 30.5, 31, 25.5, 31
34. E

Lesson 1.6, pages 38–45

TRY THESE

1. $5, 2, -\dfrac{2}{5}$ **3.** $50, -2.5, 0.05$

5.

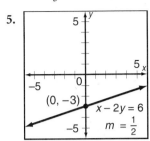

7. Where $y = 90$ on the graph, $x = 20$, so 20 adult tickets were sold.

9. Equations will vary. $y = 0.4x + 11$, where $x = $ Olympic number $(1904 = 1)$ and $y = $ winning pole vault seconds, in feet

11. 20.2 ft; 19.4 ft; answers will vary

13. Answers will vary. about 19.7 ft

15. negative

PRACTICE

1. $4, -2, \dfrac{1}{2}$ **3.** $-\dfrac{3}{2}, \dfrac{3}{4}, \dfrac{1}{2}$

5. $x + y = 9$

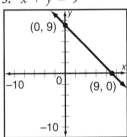

7. $y = 2x - 3$ $m = 2$

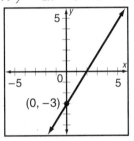

9. $2x - 5y + 15 = 0$ $m = \dfrac{2}{5}$

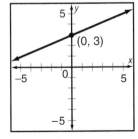

11. Where $x = 750$ on the graph, $y = 1500$, so 1500 paperbacks were sold.

13. Equations will vary. $y = -0.3x + 49.5$, where $x = $ Olympic number $(1904 = 1)$ and y = winning men's 400-meter run time, in seconds

15. 42.6 seconds; 43.2 seconds; answers will vary

17. Answers will vary, about 43.0s.

19. positive **21.** negative

EXTEND

25. $3x + 4y = -20$ **27a.** 24.98 mi/gal, 18.83 mi/gal

27b. 0.58 mi/gal, 6.53 mi/gal, 2.4%, 53.1%

THINK CRITICALLY

29. slope $= \dfrac{2}{0}$, undefined expression

31. $y = 600{,}000 - 2500x$ **33.** \$450,000

MIXED REVIEW **35.** A **36.** 53, 56, 57, 47.5, 57.5

37. 3, 3, 1 and 3 and 4, 1, 4 **38.** $6, -5, \dfrac{5}{6}$

39. $2, -10, 5$ **40.** $-4, -4, -1$

Lesson 1.7, pages 46–49

APPLY THE STRATEGY

15. $\approx 5.87 \times 10^{12}$ miles **17.** $\approx 2{,}300{,}000$ years

19. ≈ 109 persons per square mile

Chapter Review, pages 50–51

1. c **2.** b **3.** d **4.** e **5.** a **6.** -2 **7.** -20 **8.** $5\dfrac{1}{2}$ **9.** 10 **10.** \$251.23 **11.** \$12,059.11

12. Cell C3: A3*B3; 37.03; Cell D3: C3*1.05; 38.88; Cell C4: A4*B4; 47.68; Cell D4: C4*1.05; 50.06

13. $\begin{bmatrix} -4 & -15 & 6 \\ 13 & 36 & -2 \end{bmatrix}$ **14.** $\begin{bmatrix} 8 & 1 & 4 \\ 13 & 6 & -20 \end{bmatrix}$

15. $\begin{bmatrix} -30 & -40 & 5 \\ 0 & 75 & 45 \end{bmatrix}$ **16.** $\begin{bmatrix} -10 & 6 & -9 \\ -26 & -27 & 31 \end{bmatrix}$

17. 1, 2, 3, 4, 5, 6 **18.** $\dfrac{1}{3}$ **19.** $\dfrac{1}{2}$

20. 6, 4, 2, 11 **21.** 173.75, 166.5, none, 60

22. 3.55, 3.6, 3.2 and 3.7, 0.8 **23.** $\dfrac{5}{7}, -5, 7$

24. $3, 6, -2$ **25.** $-\dfrac{2}{3}, 0.2, 0.3$

26. positive **27.** negative **28.** 4.56×10^{7}

29. 3.07×10^{-4} **30.** 9.1×10^{0}, or 9.1

31. 7.777×10^{-1}

Chapter 2 Real Numbers, Equations and Inequalities

Lesson 2.2, pages 62–67

Try These 1. real, rational, integer 3. real, rational 5. real, rational 7. real, irrational 9. Irrational; 11 is not a perfect square; the number cannot be written as the ratio of two integers. 11. Irrational; 24 is not a perfect square; the number cannot be written as the ratio of two integers. 13. $<$ 15. $<$ 23. $3.14084507 < 3.141592654 < 3.142857143$

Practice 1. real, rational 3. real, irrational 5. real, rational 7. real, rational, integer 9. Rational; it can be written as the ratio of integers, $-\frac{15}{1}$ 11. Rational; it can be written as the ratio of integers, $\frac{3}{4}$. 13. Rational; it can be written as the ratio of integers, $\frac{126}{25}$.

15. irrational 17. $=$ 19. $>$

25. 3.1459 is less than either $\frac{256}{81}$, or 3.16050 or the square root of 10, 3.16228.

Extend
27. $\sqrt{10},\ -2\frac{2}{3},\ -2.6,\ -\sqrt{5},\ -2.2162158214,\ -\frac{215}{98},$ $2.\overline{23},\ \sqrt{5},\ 2.24$

29. $-5.678,\ -0.5,\ -0.055,\ 0.5,\ \sqrt{\frac{49}{100}},\ \frac{3}{4},\ \sqrt{11},\ 5.\overline{5}$

33. wrought iron $<$ steel $<$ cast iron

Think Critically

35. No. Select any two consecutive whole numbers. No other whole numbers exist between them. Therefore, the set of whole numbers is not dense.

37. Because $b = a$, $b - a = 0$, so dividing both sides of the equation by $b - a$ is division by 0. Division by 0 is undefined.

Mixed Review

38. 10:35 39. $\frac{2}{45}$ 40. C 41. B 42. $\ell = \frac{V}{wh}$ 43. 6.5m
44. Rational; it can be written as $-\frac{2}{5}$.
45. Irrational; 89 is not a square number, it cannot be written as the ratio of two integers.
46. Rational; it is a repeating decimal.
47. Irrational; it cannot be written as the ratio of two integers.

Lesson 2.3, pages 68–73

Try These 1. 12 3. 6.75 5. 9 7. identify property of multiplication 9. associative property of multiplication 11. inverse property of multiplication 15. Remove the right parentheses after 16.

Practice 1. 28.75 3. $196\frac{15}{16}$ 5. -6
7. identity property of addition 9. inverse property of multiplication 11. associative property of multiplication
13. no; no; $(a - b) - c \neq a - (b - c)$ and $(a \div b) \div c \neq a \div (b \div c)$ 15. about 1017.9 ft^2

Extend 17. $8x^2 - 19x$ 19. $2r + 33$ 21. $2a + 2$
23. sample answer:
$$40 \times 36 - 2\left[\frac{1}{2}(10 \times 10)\right] - 2\left[\frac{1}{2}(26 \times 10)\right]$$
25. 7.5 ft^2

Think Critically
27. $(1 + 2)^4 + 8 - 6\sqrt{2.2 + 1.8} \div (3 + 9)$

31. $(9 \cdot 3 \cdot 3) + (4 \cdot 4 \cdot 3) + \left(\frac{1}{2} \cdot 3 \cdot 3\right) +$ $\left(\frac{1}{2} \cdot 4 \cdot 3\right) - \left(\frac{2}{3} \cdot 3 \cdot 3\right) - \left(\frac{1}{3} \cdot 3 \cdot 3\right) -$ $\left(\frac{2}{3} \cdot 3 \cdot 3\right) - \left(\frac{1}{6} \cdot 3 \cdot 3\right)$
33. $4 \cdot 123$

Mixed Review
35. $\begin{bmatrix} -5 & -3 \\ 0 & -2 \end{bmatrix}$ 36. $\begin{bmatrix} -12 & 14 \\ 8 & -5 \end{bmatrix}$

37. D 38. real, rational, integer 39. real, rational 40. real, irrational 41. real, rational, integer, whole number, natural number

Lesson 2.4, pages 74–81
Try These
3. 2 5. 20 7. $-\frac{45}{17}$ 9. $x \geq 3$
11. $C \leq -\frac{9}{5}$ 13. $n \geq 1$ or $n < \frac{3}{5}$
15. yes 17. $t \geq 90°$
19. $40 \leq p \leq 60$ 21. $b \leq \$3.61$
Practice
1. subtraction property of inequality
3. commutative property of addition
5. -5 7. 97 9. 60 11. $x < 3$
13. $x < -1$ or $x > 1$
15. $x > 28$ 19. a 21. d 23. $f \leq 40$
25. $2(6.25) + x \leq 18$; $x \leq \$5.50$
Extend
29. $\frac{6}{7} < x < 2\frac{3}{7}$
31. $55 \leq BNN \leq 56\frac{1}{2}$ 33. \$4000

35. true **37.** true
39. Examples will vary. All numbers satisfy one inequality or the other.

MIXED REVIEW
40. 0 **41.** 1 **42.** $\frac{1}{a}$ **43.** $-a$ **44.** D **45.** -2

46. 11 **47.** $\frac{25}{11}$ **48.** Range \$35 to \$150, Median \$90 $\frac{1}{4}$
of the rooms are \$50 or less per night and $\frac{1}{4}$ are \$105
per night or more.

Lesson 2.5, pages 82–88

TRY THESE
1. $-9 < x + 4$ and $x + 4 < 9$
3. $2.5x - 1 \leq -5.2$ or $2.5x - 1 \geq 5.2$
5. 40, 24 **7.** 11, -7.5 **9.** no solution
11. $-1, 7$ **15.** $-3 < x < 9$ **17.** $-5 < x < \frac{5}{3}$
19. $x \leq 4$ or $x \geq 12$ **21.** $-3 \leq x \leq 2$
23. C **25.** $|w - 420| \leq 2$ and $|l - 594| \leq 2$

PRACTICE
1. $-1\frac{1}{4} \leq z + \frac{1}{2}$ and $z + \frac{1}{2} \leq 1\frac{1}{4}$
3. $4p - 7 \leq -22$ or $4p - 7 \geq 22$
5. 5, -1 **7.** $\frac{22}{3}, -\frac{12}{5}$ **9.** -4 **11.** 0
13. $-1 < x < 5$ **15.** all nonzero real numbers
17. $x \geq 4$ or $x \leq \frac{2}{3}$ **19.** all reals except $p = -\frac{5}{3}$
21. no solution **23.** $-12 \leq n \leq 7$ **25.** b **27.** a
29. $|p - 19.4| \leq 1.25; 18.15 \leq p \leq 20.65$
EXTEND **31.** $|x - 4| \geq 2$ **33.** $|5x - 6| > 5$
35. $|3x - 2| < 5$ **37.** $x \geq -3$ **39.** $\frac{8}{5} < x < 16$
41. $x > 2$ **43.** $-1 \leq x \leq 9$; midpoint is 4
45. $-18 \leq x \leq 42$; midpoint is 12
47. The solution is $(b - c) \leq x \leq (b + c)$; the midpoint is b.
49. No. Examples will vary. If $|x - a| = b$, then $x = a + b$ or $-b + a$. If $|x| - a = b$, then $x = a + b$ or $-(a + b)$, $-(a + b) \neq -b + a$.
51. $|t - 11.8| \leq 0.2$
53a. $64.9 \leq h \leq 70.1$ **53b.** $h > 70.1$ or $h < 64.9$

THINK CRITICALLY
55.–57. Answers will vary; sample answers are given.
55.

```
<-+--+--+--+--+--+--+--●--+--+--+--+--+--+--+--●--+->
  -8 -7 -6 -5 -4 -3 -2 -1  0  1  2  3  4  5  6  7
```
$|x - 1| \leq 5$
57.

```
<--+--+--+--+--+--○--+--+--+--+--+--○--+--+--+->
  -6 -5 -4 -3 -2 -1  0  1  2  3  4  5  6  7  8  9
```
$|x - 2| > 3$
59. abs $(x - 3) > 5$

MIXED REVIEW
60. 0 **61.** -62 **62.** 1 **63.** $\frac{5}{17}$ **64.** $g = \frac{2s}{t^2}$

Lesson 2.6, pages 89–93

TRY THESE **1.** graphs will vary. one possibility: double line graph, one line showing production, the other showing consumption

EXTEND **7.** Total Production + Total Imports > Total Consumption **9.** possible graph: total energy consumption as the sum of each type of consumption

THINK CRITICALLY **11.** Answers will vary; possible answer: Sources Vary as Energy Production Increases
13. The graph shows energy consumption.
15. answers will vary; possible answer: 32 quadrillion BTUs **17.** Students may choose to use a different scale or a different graph type.

MIXED REVIEW **18.** B **19.** D **20.** $h = \dfrac{V}{\pi r^2}$
21. $h = 9$ **22.** disjunction **23.** conjunction

Lesson 2.7, pages 94–97

APPLY THE STRATEGY **13.** \$10,380 **15.** 30
17. \$2600 **19.** \$65

Chapter Review, pages 98–99

1. c **2.** e **3.** b **4.** a **5.** d **6.** $\sqrt{3}$ **7.** Mark off twice AB. **8.** rational, $\frac{1}{8}$ **9.** rational $\frac{17}{99}$
10. rational, $\frac{25}{1}$ **11.** irrational **12.** 66 **13.** 97
14. associative property of mult. **15.** commutative property of add. **16.** 0.2 **17.** $\frac{3}{2}$ **18.** -2
19. $x \leq -3$

```
<-----●--+--+--+--+--+--+--+--+--+->
   -4    -2    0     2     4
```
20. $-4 \leq y < -2$

```
<--+--●--○--+--+--+--+--+--+--+->
   -4    -2    0     2     4
```
21. $t \leq -3$ or $t > 3$

```
<-----●--+--+--+--+--+--○--------->
   -4    -2    0     2     4
```
22. $-1, \frac{7}{3}$ **23.** $-3 \leq y \leq 4$ **24.** 13; reject 3
25.

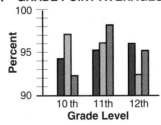

GRADE POINT AVERAGES

26. GRADE POINT AVERAGES

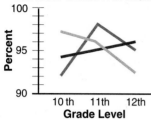

27. Although all 3 students have the same grade point average at the end of the 3-year period, the line graph appears to be more effective in displaying the trend over the 3-year period. Blue shows a steady rise, while both of the other students' averages have declined in the 12th grade.

28. 50 days

Chapter 3 **Functions and Graphs**

Lesson 3.2, pages 112–118

TRY THESE

1.

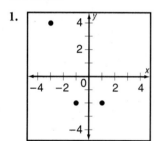

D: $\{-3, -1, 1\}$;
R: $\{-2, 4\}$;
II, III, IV

3. not a function.

7. $Q = \{(x, y): y = x - 3, x = -2, 0, 5\}$;
$D: \{-2, 0, 5\}; R: \{-5, -2, 2\}$

9. $S = \{(x, y): y = 2x, x = -2, 0, 4\}$;
$D: \{-2, 0, 4\}; R: \{-4, 0, 8\}$

11. function **13.** 7

15.

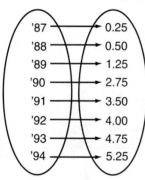

17.

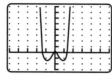

even

19.

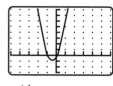

neither

PRACTICE

1. function **3.** not a function

5. $P = \{(x, y): y = x + 3, x = -2, 1, 2\}$;
$D: \{-2, 1, 2\}; R: \{1, 4, 5\}$

7. $R = \left\{(x, y): y = \frac{1}{2}x, x = -2, -\frac{1}{2}, 1\right\}$;
$D: \left\{-2, -\frac{1}{2}, 1\right\}; R: \left\{-1, -\frac{1}{4}, \frac{1}{2}\right\}$

9. $T = \{(x, y): y = x^2 - 1, x = -1, 3, 5\}$;
$D: \{-1, 3, 5\}; R: \{0, 8, 24\}$

13. function **15.** not a function

17. not a function, -10 **19.** 1215 **21.** -10.125

23.

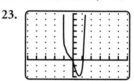

neither

25.

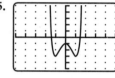

even

27. Function; each element of domain (Division) corresponds to one range (Pct) element; not a function, one element of the domain (69.5) corresponds to two elements of the range. (Atlantic and Central)

29. $(2, -1)$

31.

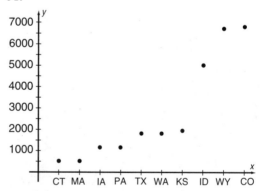

EXTEND

33. function, one-to-one

35a. $100 - x$ **35.** $x(100 - x)$; A is a function of x.
35c. 475; 2176;
35d. $0 < x < 100$ **35e.** 2500 ft^2 **35f.** no

THINK CRITICALLY

41. 3183 ft^2

42. C

43. $P = \{(x, y) : y = -3x, \; x = -5, -1, 3\};$
$D : \{-5, -1, 3\}; \; R = \{-9, 3, 15\}$

44. $Q = \{(x, y) : y = \frac{1}{4}x, \; x = 0, 12, 20\};$
$D : \{0, 12\}; \; R : \{0, 3, 5\}$

Lesson 3.3, pages 119–125

TRY THESE

1. $y = -x + 2$ **3.** $y = \frac{1}{2}x + 4$

5. $y - 6 = 5(x - 3)$
$y = 5x - 9$
$-5x + y = -9$

7. $y = -\frac{5}{8}x$ **9.** $y = -3x - 8$ **13.** $y = 3x - 2$

15. yes, down $1\frac{1}{2}$ units **19.** 43.9; 35.3; 26.8 years

PRACTICE

1. 1 **3.** $\frac{1}{5}$

5. $y - 3 = -3(x + 2)$
$y = -3x - 3$
$3x + y = -3$

7. $y + 1 = -\frac{1}{2}(x - 6)$
$y = -\frac{1}{2}x + 2$
$x + 2y = 4$

9. $2x + 5y = 0$ **11.** $y = -10x + 700; 500$

13. $y = -4x - 7$ **15.** no **17.** yes, down 1 unit

19. $y = 9.81x - 10.33$; yes, the correlation coefficient is 0.87.

EXTEND

21. Since $m_{AB} = -1, m_{AC} = \frac{4}{5}$, and $m_{AB} = 0$ no product is −1 and there is no right angle. Thus, triangle ABC is not a right triangle.

23. $m_{AB} = 1, m_{BC} = -2, m_{CD} = 1$, and $m_{AD} = -2$. Since the opposite sides are parallel, *ABCD* is a parallelogram.

THINK CRITICALLY

25a. $-\frac{A}{B}$ **25b.** $\frac{C}{B}$ **25c.** $\frac{C}{A}$

27. $Bx - Ay = 0$

29. $244.13, $11,718.20 **30.** $284.61, $10,245.87

31. C

Lesson 3.6, pages 137–144

TRY THESE

1. $\{(2, 1), (4, 3), (6, 5), (8, 7)\}$

3. $\{(5, 2), (7, 3), (5, -4), (9, 2)\}$

5. $y = x + 5$

7. $y = -2x + 4$

11. $f^{-1}(x) - \frac{1}{3}x - \frac{7}{3}$ **13.** $f^{-1}(x) = -\frac{1}{3}x + 1$

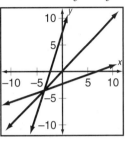
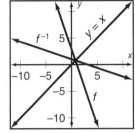

15. function **17.** not a function

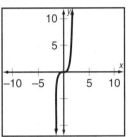
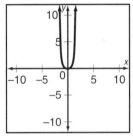

19. c **21.** a

PRACTICE

1. $\{(1, 4), (0, 5), (6, 6), (3, 7)\}$

3. $\{(4, -1), (2, -3), (1, -1), (2, 2)\}$

5. $y = x - 2$ **7.** $y = -x + 2$

9. $y = \frac{1}{8}x + \frac{1}{2}$ **11.** $y = \frac{1}{7}x - \frac{15}{7}$

15. $f^{-1}(x) = \frac{1}{2}x + 6$ **17.** $f^{-1}(x) = -\frac{1}{3}x + 2$

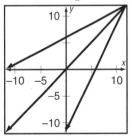
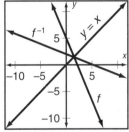

19. not a function **21.** function

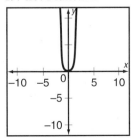
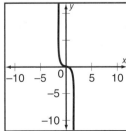

23. b   **25.** d

27. 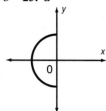   **29.**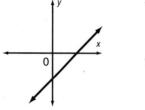

31. $m = \dfrac{E}{c^2}$, inverse is a function

33. $r = \sqrt[3]{\dfrac{3V}{4\pi}}$ yes

EXTEND

35. $\{x : x \geq 0\}$ or $\{x : x \leq 0\}$;
$\{x : x \geq 0\}$

37. $\{x : x \geq -3\}$ $\{x : x \geq 0\}$

THINK CRITICALLY

39. They are perpendicular to $y = x$.

41. $f^{-1}(x) = \dfrac{2}{x - 1} (x \neq 1)$

MIXED REVIEW

45. $y = 5x - 3$ **46.** $y = -\dfrac{1}{5}x + 4$

47. 9   **48.** 0   **49.** B

Lesson 3.7, pages 143–151

TRY THESE

1. $(f + g)(x) = 5x + 6$;
$(f - g)(x) = 3x - 6$;
$(f \circ g)(x) = 4x^2 + 24x$

3. $(f + g)(x) = 3x^2 - 2x - 1$;
$(f - g)(x) = -3x^2 + 2x - 1$;
$(f \circ g)(x) = 6x^3 - 3x^2$

5. $\dfrac{\sqrt{x}}{x + 3}; x \geq 0$ and $x \neq -3$ **7.** $3x - 3$ **9.** 81; 27

11. $4x^2; -4x^2$ **13.** $-x^2 + 1; -x^2 - 2x - 1$

17. yes   **19.** yes

PRACTICE   **1.** $(f + g)(x) = 4x - 4$;
$(f - g)(x) = 2x + 4$;
$(f \circ g)(x) = 3x^2 - 12x$

3. $(f + g)(x) = 4x^2 + 5x - 2$;
$(f - g)(x) = -4x^2 + 5x - 2$;
$(f \circ g)(x) = 20x^3 - 8x^2$

5. $\left(\dfrac{f}{g}\right)(x) = \dfrac{3x + 1}{2x + 1}; \left\{x : x \neq -\dfrac{1}{2}\right\}$;
$\left(\dfrac{g}{f}\right)(x) = \dfrac{2x + 1}{3x + 1}; \left\{x : x \neq -\dfrac{1}{3}\right\}$

7. $\left(\dfrac{f}{g}\right)(x) = \dfrac{\sqrt{x - 3}}{x^2}; \{x : x \geq 3\}$
$\left(\dfrac{g}{f}\right)(x) = \dfrac{x^2}{\sqrt{x - 3}}; \{x : x \geq 3\}$

9. $4x - 3$ **11.** $x^2 + x - 1$

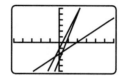   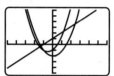

13. $5x + 2$

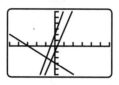

15. $-578; -127$

17. $(f \circ g)(x) = 6x - 12$ {all reals}
$(g \circ f)(x) = 6x - 4$ {all reals}

19. $(f \circ g)(x) = \dfrac{3}{2}x^6$ {all reals}
$(g \circ f)(x) = 108x^6$ {all reals}

21. $5x^2 + 30x + 45$; {all reals}; $5x^2 + 3$; {all reals}

23. yes   **25.** yes   **27.** no   **29a.** $v(x) = 0.05x$

29b. $b(x) = 0.08x$ **31a.** $r(x) = x - 200$

31b. $C(x) = 0.85x$ **31c.** $r(C(x)) = 0.85x - 200$

31d. $C(r(x)) = 0.85x - 170$

31e. $4050; $4080; take discount first.

EXTEND

35. $f(x) = x - 6; g(x) = \sqrt{x}$

37. $f(x) = x^2 + x + 1; g(x) = x^2$

39. $p(f) = 8.0407f$, where f represents the number of French francs.   **41.** $3.00

43. $S(10) = 282{,}743.3$ mm^2

THINK CRITICALLY

45. Let $f(x) = h$ and $g(x) = k$, where h and k are constants, then $f(g(x)) = f(k) = h$.

47. If $f(x) = 0$ and/or $g(x) = 0$ **49.** $3x^4 + 5$

51. $9x + 20$

52. $-32 < x < 44$ **53.** $x < -58$ or $x > 42$
54. \$627,870.09   **55.** 0.115   **56.** A
57. a: positive b: positive c: negative n: odd
58. a: positive b: negative c: negative n: odd
59. a: negative b: negative c: positive n: even

Lesson 3.8, pages 152–155

APPLY THE STRATEGY

17. 80   **19.** 50   **21a.** $k = 2.54$ or $k = 0.3937$
21b. 304.8 cm   **23.** 5013 bracelets   **25a.** Yes

Chapter Review, pages 156–157

1. d   **2.** c   **3.** b   **4.** a   **5.** $(2, -3)$ **6.** $(-2, -3)$
7. $(-2, 3)$ **8.** not a function   **9.** function
10. 4   **11.** 0   **12.** 18   **13.** -92
14. $y = -\frac{3}{2}x$; $3x + 2y = 0$
15. $y = -x$; $x + y = 0$
16. $y = -\frac{1}{4}x + \frac{13}{4}$; $x + 4y = 13$ **17.** $y = \frac{1}{6}x - 7$
18. $g(x) = |x - 7| - 4$ **19.** $g(x) = -|x|$
20. right 2 units and up 9 units
21. left 5 units and up 7 units
22. right 3 units and down 6 units
23. $y = x + 5$ **24.** $y = \frac{1}{2}x + 2$
25. $y = -\frac{1}{4}x + 2$ **26.** $y = -x + 4$
27. $\{(3, 9), (5, 1), (6, 2), (-3, -3)\}$; yes
28. $f^{-1}(x) = -\frac{x}{3} + \frac{5}{3}$ **29.** -3

30. -18 **31.** 108   **32.** -2 **33.** \$1170.00   **34.** 122.5

Chapter 4 Systems of Linear Equations

Lesson 4.2, pages 168–175

TRY THESE   **1.** $(2, 1)$ **3.** $(-12, -60)$ **5.** $(3.5, 4.5)$
7a. $1{,}600x + 2{,}400y = 67{,}200$, $2{,}000x + 2{,}400y$, $= 73{,}200$ **7b.** lawn seats $= \$15.00$, reserved seats $= \$18.00$ **9.** infinite solutions, dependent

PRACTICE   **1.** $(8, 2)$ **3.** $(8, 4)$ **5.** $(10, 2)$
7. $\left(\frac{8}{5}, -\frac{1}{5}\right)$ **13.** no solutions inconsistent
15. $(-2.5, 4.5)$, independent

EXTEND   **17.** 53° and 37°   **19.** plane speed $= 550$ miles per hour, wind speed $= 50$ miles per hour
21. $\left(\frac{40}{7}, \frac{20}{7}\right)$

THINK CRITICALLY   **25.** $(a + b, a - b)$ **27.** 1   **29.** 6
MIXED REVIEW   **31.** $m = 2, b = 3$ **32.** $m = -3$, $b = 5$ **33.** $m = \frac{3}{4}, b = -3$ **34.** $m = 0$, $b = -\frac{10}{3}$ **35.** C

Lesson 4.3, pages 176–182

TRY THESE   **1.** yes   **3.** no   **5.** $\left(\frac{1}{2}, \frac{1}{3}, \frac{1}{6}\right)$

7. a. $a + b + c = 5700$, $a + b = 3400$, $b + c = 4200$.   **7b.** mill A = 1500 board-feet, mill B = 1900 board-feet, mill C = 2300 board-feet
9. no solution, inconsistent   **11.** infinite solutions, dependent

PRACTICE   **3.** $(3, -1, 2)$ **5.** $(4, -5, -3)$
7. $\left(\frac{1}{2}, \frac{3}{2}, -\frac{1}{2}\right)$ **9.** $(0, -2, 5)$ **11.** 60 grams of mix A, 50 grams of mix B, and 40 grams of mix C.
13. no solutions inconsistent

EXTEND   **15.** Thursday $= 20$ quarts, Friday $= 35$ quarts, Saturday $= 32$ quarts   **17.** $a = 4, b = 3$

THINK CRITICALLY   **19.** 50 grams of mix A, 40 grams of mix B, 30 grams of mic C   **21.** A $= 2$, B $= 4$, C $= 8$
23. Answers will vary. Possible answers: 3 planes intersecting in a line and 3 planes coinciding, 2 coinciding planes so all 3 planes intersect in a line.
25a.

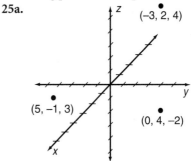

25b. $(6, 0, 0)$ $(0, 3, 0)$ $(0, 0, 2)$

25c.

MIXED REVIEW   **26.** $y = x - 3$ **27.** $y = 8 - x$
28. $y = 7x + 27$ **29.** $y = -1$ **30.** 1×3
31. 3×1 **32.** 2×2 **33.** 2×3
34. inconsistent   **35.** dependent   **36.** B
37. $8x^2 + 3$ **38.** $16x^2 + 48x + 36$ **39.** $\frac{x - 3}{2}$
40. $4x^2 + 2x + 3$ **41.** 2 units right, 6 units up
42. same vertex at $(0, 0)$, narrow   **43.** 8 units left, 1 unit down

Lesson 4.4, pages 183–188

TRY THESE

1. $\begin{bmatrix} 1 & 2 & | & 7 \\ 1 & -1 & | & -2 \end{bmatrix}$ **3.** $\begin{bmatrix} 2 & -3 & 0 & | & 12 \\ 0 & 3 & 1 & | & -12 \\ 5 & 0 & -3 & | & 3 \end{bmatrix}$

5. $\begin{cases} x = .65 \\ y = 1.2; \ (.65, 1.2, .43) \\ z = .43 \end{cases}$

7. $\begin{bmatrix} 2 & 4 & | & 22 \\ 6 & -3 & | & -9 \end{bmatrix}$ **9.** $\begin{bmatrix} 6 & -3 & | & -9 \\ 0 & 5 & | & 25 \end{bmatrix}$

11. $(-2, 3, 1)$; independent **13.** no solution; inconsistent

15a. $.07x + .08y + .09z = 212$, $x + y + z = 2500$, $z - y = 1100$

15b. $\begin{bmatrix} 0.07 & 0.08 & 0.09 & | & 212 \\ 1 & 1 & 1 & | & 2500 \\ 0 & -1 & 1 & | & 1100 \end{bmatrix}$

15c. amount at 7% = $400, amount at 8% = $500, amount at 9% = $1600

PRACTICE

1. $\begin{bmatrix} 3 & 4 & | & 7 \\ -5 & 2 & | & 10 \end{bmatrix}$ **3.** $\begin{bmatrix} 1 & -3 & 0 & | & 6 \\ 0 & 1 & 2 & | & 2 \\ 7 & -3 & -5 & | & 14 \end{bmatrix}$

5. $\begin{cases} x + 2y + 3z = 4 \\ \quad\ y + 2z = 4; \ (-2, 0, 2) \\ \qquad\quad z = 2 \end{cases}$

9. no solution; inconsistent **11.** infinite solution; dependent **13.** $(1, 2, -1)$; independent
15. $x - 2y = 0$, $-y + z = 1$, $.70x + 1.50y + 0.8z = 8.20$; pounds of apples = 4, pounds of cheese = 2, and pounds of tomatoes = 3
17. $(7, -3, -4)$

EXTEND **21.** $(16, 18, 22)$

MIXED REVIEW **25.** 12.5% **26.** D

Lesson 4.5, pages 189–195

TRY THESE **1.** $AB_{2 \times 2}$ **3.** not defined

5. $AB = [15]$, $BA = \begin{bmatrix} -3 & 7 & 2 \\ -12 & 28 & 8 \\ 15 & -35 & -10 \end{bmatrix}$

7. $AB = \begin{bmatrix} 4 & 8 \\ -18 & 11 \end{bmatrix}$, $BA = \begin{bmatrix} 3 & -4 & 4 \\ -5 & 2 & 2 \\ -51 & -6 & 10 \end{bmatrix}$

11. not defined

13. $\begin{bmatrix} 17 & 2 & 8 \\ 5 & 11 & -6 \\ 16 & 1 & 10 \end{bmatrix}$ **15.** not defined **17.** $\begin{bmatrix} 4 & 9 & 1 \\ 14 & 22 & -16 \end{bmatrix}$

PRACTICE **1.** not defined **3.** $AB_{3 \times 5}$

7. $[0]$ **9.** $\begin{bmatrix} 4 \\ 12 \\ -1 \end{bmatrix}$ **11.** $\begin{bmatrix} 3 & -20 & -11 \\ 2 & 10 & -4 \\ 15 & -13 & 1 \end{bmatrix}$ **13.** yes

EXTEND
17. no

19a. $\begin{bmatrix} -1 & 3 & -2 \\ -1 & 0 & 4 \end{bmatrix}$ **19b.** $(-1, -1), (3, 0), (-2, 4)$

19c.

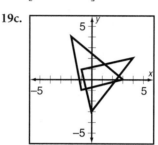

Original triangle is $(-1, 1) \ (0, -3) \ (4, 2)$. Rotated triangle is $(-1, -1) \ (3, 0) \ (-2, 4)$.

THINK CRITICALLY
21. $x = 38, y = 23$ **23.** $x = 1, y = 4$

27a. $\begin{bmatrix} 1 & 2 \\ 0 & 0 \end{bmatrix}$; $A^2 = A$; they are the same.

27b. $A^5 = A \cdot A \cdot A \cdot A \cdot A$; working from the right, the product of each pair of matrices is A. Similarly, $A^n = A$.

27c. Answers will vary. $\begin{bmatrix} \frac{1}{2} & \frac{1}{2} \\ \frac{1}{2} & \frac{1}{2} \end{bmatrix}$

MIXED REVIEW **28.** additive identity
29. multiplicative inverse **30.** additive inverse
31. multiplicative identity

32. ◄─┼─┼─┼─┼─┼─┼─●─┼─┼─┼─●─┼─┼─►
$-6\ -5\ -4\ -3\ -2\ -1\ \ 0\ \ 1\ \ 2\ \ 3\ \ 4\ \ 5\ \ 6$

33. ◄─┼─┼─┼─○━━━━━━━○─┼─┼─►
$-6\ -5\ -4\ -3\ -2\ -1\ \ 0\ \ 1\ \ 2\ \ 3\ \ 4\ \ 5\ \ 6$

34. ◄━━━━●─┼─┼─┼─┼─┼─●━━━━━►
$-6\ -5\ -4\ -3\ -2\ -1\ \ 0\ \ 1\ \ 2\ \ 3\ \ 4\ \ 5\ \ 6$

35. ◄─┼─○━━━━━━━○─┼─┼─┼─►
$-6\ -5\ -4\ -3\ -2\ -1\ \ 0\ \ 1\ \ 2\ \ 3\ \ 4\ \ 5\ \ 6$

36a. 0.86 ft **36b.** 1.14 ft **36c.** 3.64 ft **37.** B

Lesson 4.6, pages 203–208

TRY THESE **1.** true **3.** false

5. $A^{-1} = \begin{bmatrix} -3 & 5 \\ 5 & -8 \end{bmatrix}$ **7.** $A^{-1} = \begin{bmatrix} 0.3 & 0.4 \\ -0.1 & 0.2 \end{bmatrix}$

9. $A = \begin{bmatrix} \frac{2}{3} & 0 & \frac{1}{6} \\ \frac{1}{3} & 0 & \frac{1}{3} \\ -2 & 1 & \frac{1}{2} \end{bmatrix}$ **11.** $X = \begin{bmatrix} -\frac{3}{22} \\ \frac{5}{22} \end{bmatrix}$

PRACTICE **1.** The product is equal to A. **3.** yes
5. no **7.** does not exist

9. $\begin{bmatrix} \frac{1}{2} & -1 \\ -\frac{3}{4} & \frac{7}{4} \end{bmatrix}$ **11.** does not exist

13. Only square matrices can have inverses.

15. $X = \begin{bmatrix} \frac{42}{11} \\ -\frac{25}{11} \end{bmatrix}$

EXTEND

17. $\begin{bmatrix} 2 & -7 \\ -1 & 4 \end{bmatrix}$ **19.** $\begin{bmatrix} 11 & -9 \\ -6 & 5 \end{bmatrix}$

THINK CRITICALLY

23. $A^{-1} = \begin{bmatrix} \frac{1}{a} & 0 \\ 0 & \frac{1}{b} \end{bmatrix}$

25. Multiplication property; Associative property; Definition of multiplication inverse; Definition of multiplicative identity

MIXED REVIEW **27.** rational **28.** irrational
29. rational **30.** irrational **31.** B
32. 9 subcompacts, 5 compacts, 6 full-size

Lesson 4.7, pages 203–209

TRY THESE

1. yes **3.** yes
5. $\begin{bmatrix} 3 & -7 \\ 7 & 3 \end{bmatrix}\begin{bmatrix} x \\ y \end{bmatrix} = \begin{bmatrix} 7 \\ 3 \end{bmatrix}$ **7.** $\begin{bmatrix} -1 & 0 & 1 \\ 0 & 4 & 3 \\ 1 & -1 & 0 \end{bmatrix}\begin{bmatrix} x \\ y \\ z \end{bmatrix} = \begin{bmatrix} 6 \\ -1 \\ 0 \end{bmatrix}$

9. $(-44, 18)$ **11.** $(1, 2)$ **13.** $\left(\frac{1}{3}, -\frac{1}{3}, \frac{13}{3}\right)$
15a. acres of gladiolas $= 100$, acres of irises $= 50$, acres of tulips $= 50$
15b. acres of gladiolas $= 50$, acres of irises $= 25$, acres of tulips $= 175$

PRACTICE

3. $\begin{cases} x + 5z = 12 \\ -8x + 4y = 0 \\ 2z = 4 \end{cases}$

5. $(28, -46)$ **7.** $(2, 1, -3)$ **9.** $(-3, 1)$
11. $(-2, 12, -20)$
EXTEND **13.** $(11, 3)$

15. $\begin{cases} 2w + 4x - 5y + 12z = 2 \\ 4w - x + 12y - z = 5 \\ -w + 4x + 2z = 13 \\ 2w + 10x + y = 5 \end{cases}$ **17.** $\left(\frac{1}{5}, \frac{6}{5}\right)$

THINK CRITICALLY **23.** $\left(\frac{6}{b}, \frac{-4}{a}\right)$ **25.** $(1, 2, 3, 4)$

MIXED REVIEW
27. $y = 8 - x$ **28.** $y = 7 + \frac{7}{5}x$ **29.** $y = -1$
30. $(-4, -2)$ **31.** $(-3, 0)$ **32.** $(10, 8)$

Lesson 4.8, pages 210–215

TRY THESE **1.** -14 **3.** 0 **5.** -3 **9.** 5
11. Using the definition of the determinant,
$a_1 b_2 - b_1 a_2$ gives $y(1) - m(x) = b$;
$y - mx = b$; $y = mx + b$
13. $(0, 0)$ **15.** $(-1, 2, 0)$ **17.** $(20, -13, -12)$

PRACTICE **1.** 24 **3.** 1 **5.** -228 **9.** $(3, -4)$
11. not possible **13.** not possible

EXTEND
17. -34 **21.** area $= \frac{23}{2}$

THINK CRITICALLY
23. $[0]$ **25.** $4x - 3y = 18$ **27.** $\left(-\frac{1}{a - b}, \frac{1}{a - b}\right)$

MIXED REVIEW
29. A **30.** 38, 42, 46, 50; yes, $f^{-1}(x) = x - 32$

Lesson 4.9, pages 216–219

APPLY THE STRATEGY

11. $\begin{array}{c} \\ A \\ B \\ C \\ D \end{array}\begin{array}{c} A \quad B \quad C \quad D \\ \begin{bmatrix} 2 & 1 & 0 & 2 \\ 0 & 2 & 1 & 1 \\ 0 & 2 & 1 & 1 \\ 1 & 0 & 0 & 1 \end{bmatrix} \end{array}, \begin{array}{c} \\ L \\ M \\ N \\ O \end{array}\begin{array}{c} L \quad M \quad N \quad O \\ \begin{bmatrix} 1 & 1 & 1 & 1 \\ 0 & 3 & 2 & 0 \\ 1 & 0 & 1 & 1 \\ 1 & 1 & 1 & 1 \end{bmatrix} \end{array};$

Each matrix is the square of the corresponding original matrix. System 2 has more one-stop connections than System 1: sixteen as compared to fifteen.

$$
\mathbf{13.}\quad
\begin{array}{c}
\\T\\F\\R\\B\\C
\end{array}
\begin{array}{c}
T\;F\;R\;B\;C\\
\begin{bmatrix}
0 & 1 & 0 & 1 & 1\\
1 & 0 & 1 & 0 & 1\\
0 & 1 & 0 & 1 & 1\\
0 & 0 & 1 & 0 & 0\\
0 & 0 & 1 & 0 & 0
\end{bmatrix}
\end{array}
$$

$$
\mathbf{15.}\quad
\begin{array}{c}
\\T\\F\\R\\B\\C
\end{array}
\begin{array}{c}
T\;F\;R\;B\;C\\
\begin{bmatrix}
1 & 0 & 3 & 0 & 1\\
0 & 2 & 1 & 2 & 2\\
1 & 0 & 3 & 0 & 1\\
0 & 1 & 1 & 1 & 1\\
0 & 1 & 1 & 1 & 1
\end{bmatrix}
\end{array}\; ; 23
$$

17.

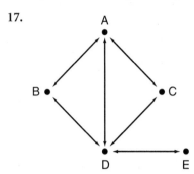

Chapter Review, pages 220–221

1. d   **2.** a   **3.** e
4. $(1, -1)$ **5.** $(-2, 5)$

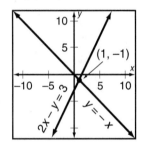   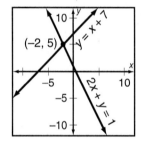

6. no solution

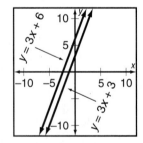

7. $(3, 8)$ **8.** $(6, 5)$ **9.** $\left(2, -1, \dfrac{1}{2}\right)$ **10.** $(2, -2)$

11. $(0, -1)$ **12.** $(4, -3, 2)$ **13.** $[32]$

14. $\begin{bmatrix} 4 & 8 & 12\\ 5 & 10 & 15\\ 6 & 12 & 18 \end{bmatrix}$

15. $[14 \quad 32]$ **16.** not defined   **17.** $\begin{bmatrix} 0.1 & 0.2\\ -0.15 & 0.2 \end{bmatrix}$

18. No A **19.** $\begin{bmatrix} -0.4 & 0.6 & 0.2\\ 0.6 & 0.6 & 0.2\\ -0.4 & 1.6 & 0.2 \end{bmatrix}$

20. $(-9, -4)$ **21.** $\left(\dfrac{13}{3}, \dfrac{19}{9}\right)$ **22.** $(1, -6, -2)$

23. -22 **24.** 249 **25.** $(1, 3)$ **26.** $\left(\dfrac{1}{3}, -\dfrac{1}{3}\right)$

27.
$$
\begin{array}{c}
\\A\\B\\C\\D\\E\\F\\G\\H
\end{array}
\begin{array}{c}
A\;\;B\;\;C\;\;D\;\;E\;\;F\;\;G\;\;H\\
\begin{bmatrix}
0 & 1 & 1 & 1 & 0 & 0 & 0 & 0\\
1 & 0 & 1 & 1 & 0 & 0 & 0 & 0\\
1 & 1 & 0 & 1 & 0 & 0 & 0 & 0\\
1 & 1 & 1 & 0 & 1 & 1 & 1 & 1\\
0 & 0 & 0 & 1 & 0 & 1 & 1 & 1\\
0 & 0 & 0 & 1 & 1 & 0 & 1 & 0\\
0 & 0 & 0 & 1 & 1 & 1 & 0 & 1\\
0 & 0 & 0 & 1 & 1 & 0 & 1 & 0
\end{bmatrix}
\end{array}
$$

Chapter 5 Polynomials and Factoring

Lesson 5.2, pages 234–241

TRY THESE   **1.** $27x^8$ **3.** $3x^4$ **5.** $\dfrac{1}{y^4}$ **7.** $3x^3$ **9.** 2

11. x **13.** $2x^2 - 5x + 2$

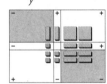

15. $3x^2 + 3x + 2$ **17.** $4y^2 - 2y + 13$
19. $6t^2 - 5t - 6$ **23.** $2x^2 + 13x$

PRACTICE   **1.** -81 **3.** $16m^4n^6$ **5.** $\dfrac{1}{2^3}$ **7.** $-4x^4$
9. -3 **11.** x **13.** $2s^2 - 8s$

15. $3x^2 - 11x + 6$

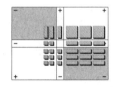

17. $4z^3 - 2z^2 + 4$ **19.** $-6m^2 - 3m + 8$
21. $6a^3b + 12a^2b^2 - 3ab^3$ **23.** $8m^2 + 2m - 3$
25. $2 - 2r - 4r^2$ **27.** $9z^2 - 16$
29. $4m^2 + 12m + 9$ **31.** $r^3 - r^2t - 3rt^2 + 2t^3$

EXTEND   **33.** $11x^2 + 4x - 10$ **35.** $12x^2 - 7x - 12$
37. x^7, y^5 **39.** x^2, y^{-13}

MIXED REVIEW   **43.** 3 **44.** 4 **45.** -2

46. $f^{-1}(x) - \dfrac{3}{2}x$ **47.** $f^{-1}(x) = 2x - 1$

48. $y = \dfrac{x - 2}{3}$ **49.** 2 **50.** 0 **51.** 1 **52.** B

Lesson 5.3, pages 242–248

TRY THESE

1. $3y(y - 2)$ 3. $(r - t)(3c + 2d)$
5. 2200   7. $0.01(4m + 5)(4m - 5)$
11. $(x + 4)^2$ 13. $(z - 2)^2$
15. $(x + 1)(x + 3)$ 17. $(m + 4)(m - 3)$
19. $x^2 - 36 = (x + 6)(x - 6)$
21. $(2z + 1)(z + 8)$

PRACTICE

1. $p(1 + rt)$ 3. $(y + z)(4r + 7s)$
5. $3a(3b^2 - 2b + 1)$ 7. $\frac{1}{2}(153 + 47) = 100$
9. $(p + 10)(p - 10)$
11. $(u + v + 13)(u + v - 13)$
13. $0.36(2r + s^2)(2r - s^2)$
17. $(t + 1)^2$ 19. $(w - 5)^2$ 21. $(4 + m)^2$
23. $(2x + 1)^2$ 25. $(1 + 3z)^2$
27. $(z + 4)(z + 3)$ 29. $(n + 1)(n - 6)$
31. $(5 - n)(3 - n)$
33. $5 + 8x - 4x^2 = (5 - 2x)(1 + 2x)$

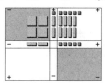

35. $(2a + 3)(a + 2)$ 37. $(3 + 2x)(3 + x)$
39. $(4 + a)(3 - 4a)$ 41. $2\pi r^2 + 2\pi rh$
$= 2\pi r(r + h)$

EXTEND

43. $(2mn - 5)^2$ 45. $(7x^2 + y)(x^2 - y)$
47. $(3cd + 2f)(5cd + f)$ 49. 30,000

THINK CRITICALLY

51. 25   53. $(3k^a - 1)(k^a - 3)$
55. $(2x^2 + x - 2)(x - 2)$

MIXED REVIEW

56. $x = \dfrac{d - b}{a + c}$

57. $c = \dfrac{S}{\pi a} - b$ or $\dfrac{S - \pi ab}{\pi a}$

58. D   59. C

Lesson 5.4, pages 249–253

TRY THESE

1. $3(x + 3)(x - 3)$ 3. $3(2z + 1)(z - 2)$
5. $(b^2 + 25)(b + 5)(b - 5)$
7. 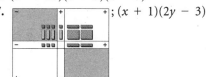 ; $(x + 1)(2y - 3)$

9. $(x - 5)(x - 3y)$ 11. $(z + 4 + x)(z + 4 - x)$
13. $(r - s)(r^2 + rs + s^2)$
15. $(4x - 1)(16x^2 + 4x + 1)$

PRACTICE

1. $5(y + 4)(y - 4)$ 3. $3(x - 2)(x - 3)$
5. $4(x + 2)^2$ 9. $(t^2 + 9)(t + 3)(t - 3)$
11. $(a^4 + 4)(a^2 + 2)(a^2 - 2)$
13.

$(x + 1)(3y - 2)$

15. $(x + 3)(2a + 1)$ 17. $(2 - x)(5y - 3x)$
19. $(t - 5 + u)(t - 5 - u)$ 21. $(7 + r)(3 - r)$
23. $(2a - b + c)$

EXTEND   27. $(3x^2 + 2xy + y^2)(3x^2 - 2xy + y^2)$

THINK CRITICALLY

29. $(x + 2)(x^2 - 8x + 12)$ 31. $3(x - 2)(3x + 1)$
33. $(x - y - 3)(x^2 - 2xy + y^2 + 3x - 3y + 9)$

MIXED REVIEW   34. 1612   35. 2007   36. 1758   37. D
38. $y = \dfrac{5}{2}x + \dfrac{5}{2}$ 39. $y = -\dfrac{2}{5}x + \dfrac{11}{5}$ 40. B

Lesson 5.5, pages 255–261

TRY THESE

1. $0, 7$ 3. $2, -2$ 5. $-4, -3$ 7. $0, 8$ 9. $-3, 8$
11. $x =$ width of side strip; $(2x + 3)(4x + 5) = 45$;
1 in.   13. $1, 5, -5$ 15. $0, -2, 3$
17. $x^2 - 5x + 6 = 0$ 19. $x^3 - 6x^2 + 9x = 0$
21. $x^3 + 5x^2 + 2x - 8 = 0$

PRACTICE   1. $0, 8$ 3. $0, -3$ 5. $3, -3$ 7. $3, 4$
9. $-2, 8$ 11. $\dfrac{3}{2}, -1$ 13. $-\dfrac{3}{2}, \dfrac{2}{3}$ 15. $2, -5$
17. $-4, 7$ 21. 2 in.   23. $-1, 6, -6$ 25. $0, -6, -6$
27. $0, -3, 4$ 29. 16 in., 36 in.
31. $x^2 - 3x = 0$ 33. $x^2 - 5x - 14 = 0$
35. $x^2 - 8x + 16 = 0$ 37. $x^3 - 12x^2 + 36x = 0$
39. $x^3 - 12x^2 + 45x - 50 = 0$
41. $x^3 - 15x^2 + 66x - 80 = 0$

EXTEND   43. $x = -\dfrac{1}{3}, -2$ 45. $2, -2, 3, -3$
47. 3 in., 4 in., 5 in.   49. 27 and 29

THINK CRITICALLY

51. $(-2, 7)(1, 4)$
53. The graph touches the x-axis at 5.

MIXED REVIEW

54. $\begin{bmatrix} 4 & -4 \\ -1 & 8 \\ 7 & 1 \end{bmatrix}$ 55. $\begin{bmatrix} -18 & 40 \\ -1 & 5 \end{bmatrix}$

56. C   57. no, $a \not> a$ 58. yes   59. B

Lesson 5.6, pages 262–265

APPLY THE STRATEGY

17. 16 ft; 10 ft^2 **19.** 24 ft; 20 ft^2

21. $p = z + r, z = 8, r = 4$

23. $z + 2r + 2z + r = 3z + 3r$
$\quad = 3(z + r); 3(8 + 4) = 36$ ft

25. area was multiplied by 4 or 2^2; area will be multiplied by 9 or 3^2; area will be multiplied by n^2.

27. $a^2 + (1 - a) = (1 - a)^2 + a$; If you multiply each side of the equation, you get $a^2 - a + 1$.

Chapter Review, pages 266–271

1. c   **2.** d   **3.** a   **4.** e   **5.** b   **6.** $x^3 - 2x^2 + 2x - 5$

7. $x^3 + 2x^2 - 4x + 4$ **8.** $2x^3 - 2x - 1$

9. $(x^3 - 2x^2 + 2x - 5) - (-x^3 - 2x^2 + 4x - 4)$

10. Subtraction is the same as addition of an opposite.

11. $(3x - 2)(2x - 1)$ **12.** $6x^2 - 7x + 2$

13. $(6x^2 - 7x + 2) \div (3x - 2) = 2x - 1$
$(6x^2 - 7x + 2) \div (2x - 1) = 3x - 1$

14. $6x^2 - 7x + 2$ **15.** $-6x^7$ **16.** $16y^6$ **17.** $4m^3$

18. 3   **19.** $\frac{1}{4}$ **20.** $4x^2 - 6x - 9$

21. $-2x^3 - 3x^2 + x - 4$ **22.** $8x^2 + 10x - 3$

23. $6k^4 - 15k^3 + 6k^2$ **24.** $1 - 4x^2$

25. $8xy(3x^2 - y^3)$ **26.** $(2x + 5)(2x - 5)$

27. $(y - 5)^2$ **28.** $(x - 1)(2x + 3)$

29. $3a(x - 3)(2x + 1)$ **30.** $(z^2 + 4)(z + 2)(z - 2)$

31. $(a + 1)(b - 2)$ **32.** $(x - 4)(x^2 + 4x + 16)$

33. $0, 5$ **34.** ± 4 **35.** $3, 3$ **36.** $1, 2$ **37.** $\frac{4}{3}, 2$

38. $-2, 0, 5$ **39.** $-2, -1, 2$ **40.** $\pm 2, 0, 0, 0$

41. The sum of any seven consecutive whole numbers always has a factor of 7.

$m + (m + 1) + (m + 2) +$
$(m + 3) + (m + 4) + (m + 5) +$
$(m + 6) = 7m + 21 = 7(m + 3)$

Chapter 6 Quadratic Functions and Equations

Lesson 6.1, pages 275–281

TRY THESE   **1.** $x^2 + 10x + 25$ **3.** $x^2 - 13x + \frac{169}{4}$

5. ± 12 **7.** ± 14 **9.** $5, 3$ **11.** $-4 \pm 3\sqrt{2}$

13. $9 \pm \sqrt{91}$ **15.** $-10, 2$ **17.** $4 \pm \sqrt{26}$ **19.** 3 yd by 5 yd

PRACTICE   **1.** $x^2 + 40x + 400$ **3.** $x^2 - 11x + \frac{121}{4}$

5. $x^2 - \frac{10}{3x} + \frac{25}{9}$ **7.** 24   **9.** ± 24

13. $-2 \pm 2\sqrt{3}$ **15.** $12 \pm \sqrt{147}$ **17.** $-3 \pm \sqrt{3}$

19. $-5 \pm \sqrt{15}$ **21.** length: 12 in., width: 9 in.

23. Forgot to add 1 to both sides of the equation; correct solution is $1 \pm \sqrt{6}$

25. $-9a, 2a$ **27.** $-4a, \frac{1}{4}a$

29. $(x + 2)^2 + (y + 3)^2 = 18$

THINK CRITICALLY   **35.** $\dfrac{-a + a\sqrt{13}}{6}$

MIXED REVIEW   **36.** C   **37.** $4(3x + 5)(3x + 5)$

38. $2x(x - 44)$ **39.** $f^{-1}(x) = 2x + 16$

40. $f^{-1}(x) = 4x - 8$

Lesson 6.3, pages 287–295

TRY THESE   **1.** $(6, 34); x = 6$; downward

3. $(0, 10); x = 0$; upward   **5.** $(0.5, 0.5); x = 0.5$; downward   **9.** $(5, 11), (4, 6), (1, 3)$ **11.** $(7, -12), (6, 2), (2, 18)$

13.

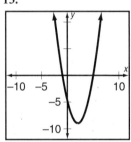

15.

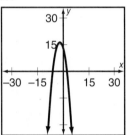

17. $y = -2x^2 + 4x + 2$ **19.** $y = x^2 - 9$

PRACTICE   **1.** $(2, 5); x = 2$; downward   **3.** $(0, 18); x = 0$; upward   **5.** $\left(\frac{1}{3}, -\frac{2}{3}\right); x = \frac{1}{3}$; downward

9. $(13, 14), (12, 1), (8, -31)$ **11.** $(5, -28), (4, 2), (1, 20)$

13.

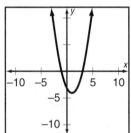

15.

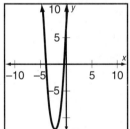

17.

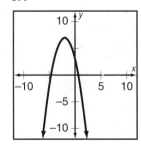

19. $y = x^2 + 4x - 1$
21. $y = -2x^2 + 8x - 3$
23. 4h; 94

EXTEND **25.** $y = x^2 + 5x + 10$
27. $y = 3x^2 + 4x - 7$
THINK CRITICALLY **31.** $x = 0$ **35.** $a = 2, b = 16$
MIXED REVIEW **37.** 4 **38.** 0 **39.** C **40.** 11.7898
41. 23.5584 **42.** $(4, -46); x = 4$ **43.** $(-4, 22);$
$x = -4$ **44.** $3, -3$ **45.** $11, 1$ **46.** no solution
47. $x > -z$ or $x < -4$ **48.** $\frac{1}{4}x$ **49.** $\frac{1}{3}x - 2$
50. $\sqrt{x}$ if $x > 0$ **51.** $\frac{1}{x} - 1$

Lesson 6.4, pages 296-302

TRY THESE **1.** $-1 \pm \frac{\sqrt{6}}{2}$ **3.** $\frac{5}{3}$ **5.** no real solutions
7. two real, irrational solutions **9.** 0 **11.** 4 **13.** 16
15. 12; 4 **17.** $\frac{4}{3}; -\frac{7}{3}$ **19.** $x^2 + 13x + 42 = 0$
21. -9 or 15
PRACTICE **1.** $\frac{3}{2} \pm \frac{\sqrt{69}}{6}$ **3.** $-5, -\frac{1}{3}$ **5.** $-\frac{8}{3}, 1$
7. $\sqrt{3}$ s **9.** two real, irrational solutions **11.** one
real, rational solution **13.** two real, rational solutions
15. 2 **17.** 0 **19.** 2 **21.** ± 30 **25.** $-8; -12$
27. $\frac{5}{4}; -\frac{9}{4}$ **29.** 1; 6 **31.** $11x^2 + 101x - 90 = 0$ or
$x^2 + \frac{101}{11}x - \frac{90}{11} = 0$ **33.** $27x^2 + 30x + 8 = 0$ or
$x^2 + \frac{30}{27}x + \frac{8}{27} = 0$ **35.** $x^2 + 3x - 40 = 0; -8, 5$
37. $x^2 - 35x - 750 = 0; 50; -15$
EXTEND **39.** $-\frac{2}{3}q, \frac{1}{2}q$ **41.** $x^2 - 8fx + 12f^2 = 0$
43. $x^2 + 10h + 21h^2 = 0$
THINK CRITICALLY **45.** $k < -16$
MIXED REVIEW **46.** $(-1, -2, 4)$ **47.** $(2, -3, 5)$
48. $-5, 4$ **49.** $-2.5, 6$ **50.** B

Lesson 6.5, pages 303-307

TRY THESE **1.** no **3.** $(x^2)^2 + 4(x^2) - 5 = 0$
5. $(\sqrt{x})^2 - 7(\sqrt{x}) + 10 = 0$ **9.** $x^2 + 5$
11. $x^2 + 8x - 9$ **13.** no **15.** $\pm\sqrt{5}, \pm 3$
17. 25 **19.** 5, 13
PRACTICE
1. $(x^2)^2 - 5(x^2) + 15 = 0$ **3.** no
5. $(\sqrt{x})^2 - 3(\sqrt{x}) + 1 = 0$ **9.** $2x^2 + 1$
11. $x^2 + 6x - 3$ **13.** no **15.** $\pm 2\sqrt{2}$
17. $\pm\sqrt{\frac{2}{3}}, \pm\sqrt{\frac{4}{3}}$ **19.** 16 **21.** 16, 64 **23.** 13, 16
25. $-6, 6$ **27.** length: 48 in., width: 36 in., height: 42 in.
EXTEND **29.** 7, 9 **31.** 29 m
THINK CRITICALLY **33.** $-9, -3, 1, 7$ **35.** $-\frac{5}{3}, 4$
MIXED REVIEW **37.** $-5, 15$ **38.** $-18, 2$ **39.** C
40. $(x - 5)(x^2 + 5x + 25)$
41. $(x - 10)(x^2 + 10x + 100)$ **42.** $\pm\sqrt{6}, \pm 3$
43. 64

Lesson 6.6, pages 308-311

APPLY THE STRATEGY
13. uniform; 7 ft/sec **15.** nonuniform; 24 ft/sec

Chapter Review, pages 312-313

1. e **2.** d **3.** a **4.** c **5.** b **6.** 32 **7.** 40 **8.** $\frac{5}{3}$
9. $-3, -6$ **10.** $5 \pm \sqrt{27}$ **11.** $\frac{3}{4} \pm \sqrt{\frac{29}{16}}$
12. right 3 units and down 5 units **13.** left 6 units
14. left 4 units and down 8 units **15.** c, a, b, d
16. $(-1, -4); x = 1$ **17.** $(2, 0); x = 2$

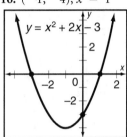
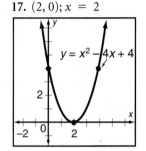

18. $(-2, 9); x = -2$ **19.** 324 ft

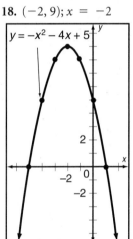
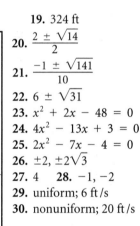

20. $\frac{2 \pm \sqrt{14}}{2}$
21. $\frac{-1 \pm \sqrt{141}}{10}$
22. $6 \pm \sqrt{31}$
23. $x^2 + 2x - 48 = 0$
24. $4x^2 - 13x + 3 = 0$
25. $2x^2 - 7x - 4 = 0$
26. $\pm 2, \pm 2\sqrt{3}$
27. 4 **28.** $-1, -2$
29. uniform; 6 ft/s
30. nonuniform; 20 ft/s

Chapter 7 Inequalities and Linear Programming

Lesson 7.2, pages 324-331

TRY THESE **1a.** yes **1b.** no **1c.** yes **1d.** no **1e.** no
1f. no
3.
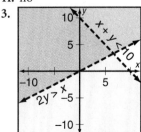
5.
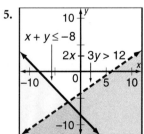

7.

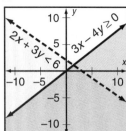

9.

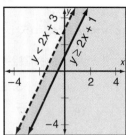

11. $x + y < 18$ and $6x + 4y \geq 100$

PRACTICE   **1a.** both   **1b.** one; $y < 3x - 5$
1c. one; $x + 2y \geq 7$ **1d.** neither

3.

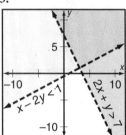

5.

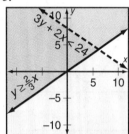

7.

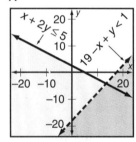

9.

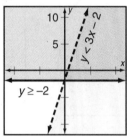

11a. $y > 0.72(220 - x)$; $y < 0.87(220 - x)$.
11b.

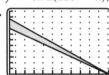

11c. $130 < y < 157$

13a. $x + y \leq 30$
$0.2x + 0.05y \geq 2$
$x \geq 0 \qquad y \geq 0$

13b.

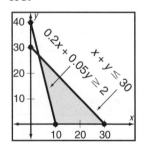

13e. 4.5 million, 6 million

EXTEND   **15.** $y > x - 4$ $y < 3x + 2$

17.

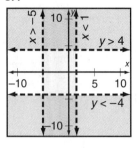

19.

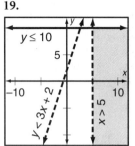

21.

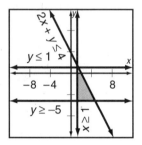

THINK CRITICALLY

23. Since absolute value is never negative, $|3x - 4y| \neq -4$ for any x and y. The entire coordinate plane is the region of solutions.

MIXED REVIEW

27. $17, 1, 1.\overline{32}, 1.3, \sqrt{64}, \frac{3}{8}, 0$ **28.** B

29. $x > 6$

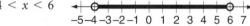

30. $\frac{9}{4}$ **31.** 3 and 5

32. False; the graph lies in Quadrants III and IV.

33. False; $(2x + 5)^2 = 4x^2 + 20x + 25$

Lesson 7.3, pages 332–338

TRY THESE

1. no   **3.** no   **5.** both ≥ 0 or both ≤ 0
7. one < 0 and one > 0
9. $-4 < x < 6$

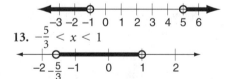

11.

$x < -1$ or $x > 5$

13. $-\frac{5}{3} < x < 1$

15. $0 \leq t \leq 4$ sec

PRACTICE

1. yes   **3.** yes

5. $1 < x < 3$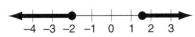

7. $x < 1$ or $x > 4$

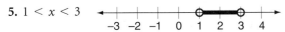

9. $x \leq -2$ or $x \geq \frac{1}{2}$

11. $x = -4$

13. $-3 < x < \frac{1}{2}$

15. width ≥ 30 in.   **17.** $x < -2.75$ **19.** $x \leq -2.75$
21. $0.5 < x < 4$ **23.** $0.5 \leq x \leq 4$ **27.** $x > -0.2$
29. $10 \leq x \leq 50$ bracelets per week

EXTEND   **31.** $1 \leq x \leq 4$ **33.** $x \leq -1$ and $x \geq 1$
35. $-2 \leq x \leq 6$

THINK CRITICALLY   **37.** $x^2 - 4 \geq 0$
39. $x^2 + 3x > 0$ **41.** $3 \leq x \leq 4$ cm

MIXED REVIEW   **42.** 42   **43.** $-5t^3 - 8t^2 + 5t - 8$
44. 180 km   **45.** $-5 < y < 5$ **46.** $-\frac{1}{3} \leq k \leq 3$
47. $4w^5 - 8w^3$ **48.** C

Lesson 7.4, pages 339–345

TRY THESE   **1.** yes   **3.** yes   **5.** yes   **7.** no
9.

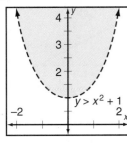

11.

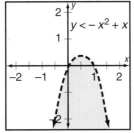

13.

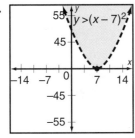

PRACTICE   **1.** yes   **3.** no   **5.** no
7.

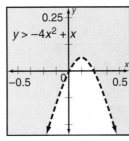

9.

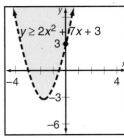

11.

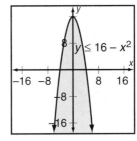

13.

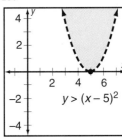

15a. $60 - 2x$ **15b.** $50 - 2x$ **15c.** 3000 ft^2
15d. $(60 - 2x)(50 - 2x)$
15e. $y < 3000 - (60 - 2x)(50 - 2x)$
17. $w < 91$ pounds

EXTEND
19a.

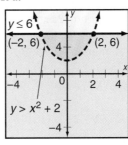

19b. $(2, 6)$ and $(-2, 6)$
19c. It is not; the second inequality is not a true statement when $x = 1$ and $y = 2$.

21.

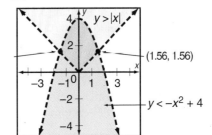

THINK CRITICALLY

25a. $y \leq x^2$ and $y \geq x^2 + 10$
25b. $y \geq x^2$ and $y \leq x^2 + 10$

MIXED REVIEW   29. B **30.** 1, 4 **31.** $x - 7$
32. $\{x \mid x \geq 4\}$ **33.**

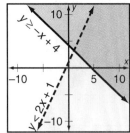

34. 78.5 ft^2 **35.** (5, 13) **36.** $\frac{71}{100}$ **37.** $\frac{71}{99}$

38.   **39.**

40. $0 < x < 1$

Lesson 7.6, pages 350 – 357

TRY THESE

1. no **3.** yes **5.** yes
7. maximum 240 at (8, 10) **9.** minimum 22 at (2, 8)
11. maximum 79 at (6, 11); minimum 33 at (2, 5)
13. maximum 3540 at (13, 11); minimum 1800 at (6, 6)
15. $x \geq 15$; $y \geq 10$; $x + y \leq 50$
17. They should sell 15 short- and 10 long-sleeved shirts.

PRACTICE

1. (0, 0), (0, 7), (9, 0) **3.** (0, 6), $\left(2\frac{2}{3}, 7\frac{1}{3}\right)$, (10, 0)
5. maximum 36 at (12, 0); minimum -32 at (0, 16)
7. maximum 264 at (12, 12); minimum 148 at (4, 9)
9. minimum 18 at (2, 4) **11.** maximum 30 at (0, 5)
13. maximum 5 at (5, 5) **15.** minimum 0 at (0, 0)
17. minimum 26 at (6, 2)
19. Plant 1100 acres of Iceberg and 2500 acres of Romaine to yield a maximum profit of $845,000.

21. 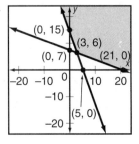   3 statisticians and 6 interviewers for a cost of $6000

EXTEND

25. $0.05x + 0.10y \geq 15$
27. 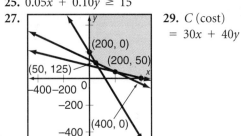   **29.** C (cost) $= 30x + 40y$

MIXED REVIEW   31. B **32.** -3 **33.** $y \geq -7$
34. $x = 3$ or $x = -6$ **35.** (2, 5) **36.** $(-1, 3)$
37. $y = -\frac{1}{2}x + 10$

Lesson 7.7, pages 358 – 361

APPLY THE STRATEGY

21a. $m + b \leq 31{,}500$
21b. $10{,}000 \leq m \leq 20{,}000$; $b \leq 0.75m$;
$b \geq 0, m \geq 0$
21c. $R = 0.075m + 0.105b$
21d. Rhea should invest $18,000 in money markets and $13,500 in bonds for a maximum return of $2767.50.
23a. She should invest $45,000 in Bond A and $15,000 in Bond B for a return of $3750.
23b. An investment of $22,500 in Bond A and $7500 in Bond B would minimize the fees.

Chapter Review, pages 362 – 363

1. e **2.** f **3.** b **4.** c **5.** a **6.** d **7.** no **8.** yes
9. no

10. 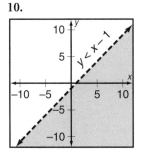 **11.**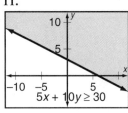

12. 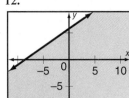 **13.**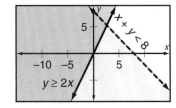

14.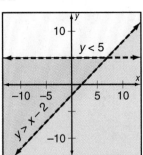

15.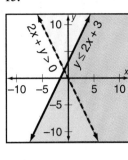

16. $(6, 1), (7, 1), (8, 1), (7, 2)$ **17.** yes   **18.** no
19. no   **20.** yes   **21.** yes   **22.** no   **23.** yes   **24.** no
25. vertex $= (2, -3)$;   **26.** vertex $= (0, 0)$; $(0, 0)$;
    $(0, 1), (2 - \sqrt{3}, 0)$;

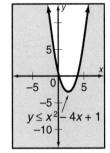

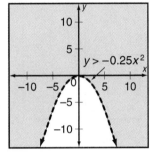

27. vertex $= (0, 9)$; $(0, 9), (-0.75, 0), (0.75, 0)$;

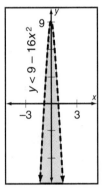

28. $(0, 0), (0, 5), (7, 3), (4, 0)$
29. $20; (4, 8)$
30. \$45,000 in bond A, \$45,000 in Bond B

Chapter 8 Exponents and Radicals

Lesson 8.2, pages 374–380

TRY THESE   **1.** -9 **3.** no real solution   **5.** $3\sqrt{5}$
7. $-2|x|\sqrt{6}$ **9.** 4   **11.** ≈ 11.85558 **13.** $\sqrt{2}$
15. $\sqrt[6]{5^5}$ **17.** $2^5\sqrt{6}$ or $32\sqrt{6}$ ft/s

PRACTICE
1. $\sqrt{3}$ **3.** $\sqrt[10]{(x + y)^3}$ **5.** $15^{\frac{1}{4}}$ **7.** $n^{-\frac{1}{5}}$ **9.** -11
11. 0.3 **13.** 0   **15.** 36   **17.** $7\sqrt{2}$ **19.** $|h|\sqrt{2}$
21. $0.8|x^3|$ **23.** $4b^2|a|$ **25.** $2x^5$ **27.** $x + 4$ **29.** 2
31. b^{4n-5} **33.** 2   **35.** 8   **37.** 343   **39.** -625 **41.** 27
43. ≈ 6126.808 **45.** $\sqrt{7}$ **47.** $\sqrt[4]{3}$ **49.** $\sqrt[12]{2^{13}}$

51. $\sqrt[15]{m^9 n^{10}}$ **53a.** $t = \left(\dfrac{d}{16}\right)^{\frac{1}{2}}$
53b. $\sqrt{\dfrac{864}{16}} = \sqrt{54} = 3\sqrt{6}$ s
55a. $P = (2.7)^{\frac{t}{16}}$ **55b.** $\approx \$13,649$
EXTEND   **57.** $x \geq 0$ **59.** $x \geq -3$ **63.** 125   **65.** $0, -1$
THINK CRITICALLY   **67.** $\sqrt[n]{a}$ **69.** $\sqrt[z]{a^{x+y}}$ **71.** $0, 1$
MIXED REVIEW   **72.** $(y + 9)(y - 9)$
73. $(3a + 2)(3a - 2)$ **74.** $2d(c^2 + 4)(c + 2)(c - 2)$
75. $(x - 1)(x^2 + x + 1)$ **76.** B

Lesson 8.4, pages 385–391

TRY THESE
1. $2 - \sqrt{6}$ **3.** $-\sqrt{3.5} - 5$
5. $\dfrac{\sqrt{6}}{\sqrt{6}}$ **7.** $\dfrac{\sqrt[3]{100n}}{\sqrt[3]{100n}}$ **9.** $\sqrt{15}$ **11.** $\dfrac{1}{4}$ **13.** $\dfrac{2\sqrt[3]{3n^2}}{3n}$
15. $-12\sqrt{10}$ **17.** $-3 - 2\sqrt{2}$

PRACTICE
1. $\sqrt{77}$ **3.** 2   **5.** $6|x|y\sqrt{z}$ **7.** 7   **9.** $\dfrac{1}{8}$ **11.** $2^{\frac{9}{4}}$
13. $\dfrac{\sqrt{15}}{5}$ **15.** $x^3\sqrt[5]{-2}$ **17.** $\dfrac{5x\sqrt{x}}{|y|}$ **19.** $\dfrac{\sqrt{5a}}{3a^2 b}$
21. $v = \dfrac{\sqrt{2Km}}{m}$ **23.** $38\sqrt{5}$ **25.** $2(m - 2)\sqrt[4]{3m}$
27. $16 - 4\sqrt{15}$ **29.** $4x + 36y - 24\sqrt{xy}$
31. $\dfrac{5\sqrt{2} - 4}{2}$ **33.** $\dfrac{3\sqrt{30} + 4\sqrt{15} - 9\sqrt{3} - 6\sqrt{6}}{3}$
35a. ≈ 1376 kg   **35b.** $m_v = \dfrac{cm(c^2 - v^2)^{-\frac{1}{2}}}{c^2 - v^2}$
EXTEND
37. $\dfrac{5\sqrt{6}}{6}$ **39.** $-15\sqrt{3}$ **41.** $\dfrac{x^2}{x + 1}$
THINK CRITICALLY
47. $\sqrt{6} + \sqrt{5}$ **49.** $\sqrt{7} - \sqrt{6}$
MIXED REVIEW
51. $(2, 2, 1)$ **52.** $(0, -2, 1)$ **53.** $5, -3$
54. $\dfrac{-5 \pm \sqrt{17}}{4}$ **55.** $\pm \dfrac{\sqrt{10}}{2}$ **56.** D

Lesson 8.5, pages 392–398

TRY THESE   **1.** 53   **3.** -64 **5.** -2187 **7.** 8   **9.** 4
PRACTICE   **1.** 121   **3.** 89   **5.** 324
7. -27 **9.** 13   **11.** 3125
13. 64   **15.** 100   **17.** 2.329775
19a. $F_1 = \sqrt{R^2 - F_2^2}$ **19b.** 600 lb.   **21.** 8
23. -1 **25.** -7 **27.** 15   **29.** $-4, -\dfrac{20}{9}$
31. $\sqrt{\dfrac{4Af^2}{\pi}}$ **33.** $\sqrt{a^3}$ **35.** ≈ 200 mi

EXTEND **37.** ≈ 24.6 m **39.** 9, 16 **41.** 4
THINK CRITICALLY **43.** 3
MIXED REVIEW **47.** D **48.** $w = \dfrac{P - 2l}{2}$ or $w = \dfrac{P}{2} - l$

49. $F = \dfrac{9}{5}C + 32$ or $F = \dfrac{9C + 160}{5}$

50. $a = \dfrac{2A}{h} - b$ or $a = \dfrac{2A - bh}{h}$

51. $-6x + 4$ **52.** $x^2 + 2x + 1$ **53.** $\dfrac{\sqrt{x}}{x}$ **54.** $\dfrac{1}{8}$
55. $x = 6$

Lesson 8.6, pages 399–404

TRY THESE **1.** $5i$ **3.** $\dfrac{i\sqrt{2}}{2}$ **5.** i **7.** $8i$ **9.** $-4i$

11. $-\sqrt{15}$ **13.** $\pm 9i$ **15.** $\dfrac{3 \pm i\sqrt{3}}{3}$

17. imaginary **19.** $2.16i$ ohms

PRACTICE **1.** $7i\sqrt{2}$ **3.** $-6ni$ **5.** $-24i$ **7.** -11
9. $\dfrac{2}{3}i$ **11.** $-\dfrac{2i\sqrt{5}}{3}$ **13.** i **15.** $-i$ **17.** -3
19. $9i$ **21.** $1475i$ **23.** $-7i$ **25.** 72 **27.** 12
31. $-i\sqrt{2}$ **33.** $31i\sqrt{2}$ **35.** 24 **37.** $-10i\sqrt{10}$
39. $-\dfrac{5i}{3}$ **41.** $\dfrac{\sqrt{2}}{2}$ **43.** $\pm i\sqrt{11}$ **45.** $\dfrac{-1 \pm i\sqrt{3}}{2}$

47. $\dfrac{3 \pm i\sqrt{23}}{4}$ **49.** real, unequal **51.** real, equal

53. real, unequal

EXTEND **57.** -1 **59.** 1
61. The pattern consists of the continuously repeating sequence $-i, -1, i, 1.$
63. $7, -6$

MIXED REVIEW **69.** $\begin{bmatrix} 5 & 10 \\ 20 & 15 \end{bmatrix}$ **70.** $\begin{bmatrix} -3 & 3 \\ 3 & -3 \end{bmatrix}$

71. $\begin{bmatrix} 1 & 3 \\ -6 & 17 \end{bmatrix}$ **72.** $\begin{bmatrix} -8 & 8 \\ 3 & -3 \end{bmatrix}$ **73.** $-10 - 7\sqrt{2}$

74. $6 - 4\sqrt{2}$ **75.** 2 **76.** $1 - 2\sqrt{2}$ **77.** B

Lesson 8.7, pages 405–409

TRY THESE **1.** $-1 - 3i$ **3.** $12 - 8i$ **5.** 5
6.–9. **11.**

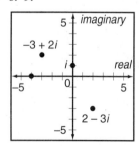

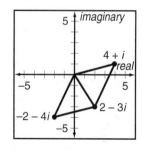

13. voltage: $-16 + 2i; \ |Z| = \sqrt{13}$ ohms
PRACTICE **1.** $8 + 11i$ **3.** $-1 + 8i$ **5.** $-19 + 3i$
7. $-24 + 28i$ **9.** $153 + 107i$ **11.** $\dfrac{24}{25} - \dfrac{7}{25}i$

14.–17. **19.** $\sqrt{58}$ **21.** 29

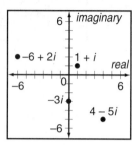

23. **25.**

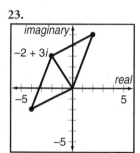
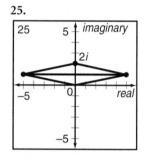

27. $-2i, -4, 16, 256$
EXTEND **31.** $x^2 - 2x + 5 = 0$
33. $2x^2 + 5x + 4 = 0$ **35.** $(2 - 3i)(2 - 3i) = 4 - 6i - 6i + 9i^2 = -5 - 12i$
THINK CRITICALLY **37.** $\sqrt{a^2 + b^2}$
39. $(a + bi)(a + bi)$

MIXED REVIEW **41.** no **42.** yes **43.** no
44. -2 **45.** -1 **46.** -4 **47.** $\approx 16.0\%$ **48.** A

Lesson 8.8, pages 410–413

APPLY THE STRATEGY **9a.** \$45,000 **9b.** \$169,924.81
11. \$27,155.42 **13.** \$886.41

Chapter Review, pages 414–415

1. c **2.** d **3.** a **4.** b **5.** $k^{\frac{1}{3}}$ **6.** $9^{\frac{5}{6}}$ **7.** $\sqrt{b}$
8. $\sqrt[9]{(m + 4)^2}$ **9.** x **10.** $2x^2\sqrt[5]{2}$ **11.** $\dfrac{1}{5}$ **12.** $11^{\frac{1}{2}}$
13. **14.**

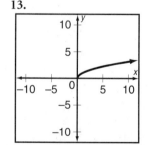
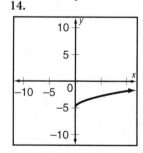

15.

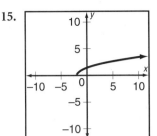

16. $13\sqrt{6}$ **17.** 10
18. $\frac{\sqrt{3}}{3}$ **19.** $\frac{1}{2}$
20. $2 - \sqrt{2}$ **21.** 58
22. -27 **23.** 16
24. 10 **25.** $7i$
26. $2i\sqrt{5}$ **27.** i
28. $6i\sqrt{3}$ **29.** -10
30. $2i$

31. $-\sqrt{6}$ **32.** $-3i$ **33.** $15 - 2i$ **34.** $1 + 3i$
35. $-3 + 11i$ **36.** $\frac{7}{25} - \frac{24}{25}i$

37.

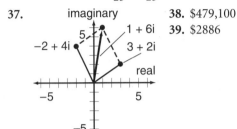

38. $479,100
39. $2886

Chapter 9 Exponential and Logarithmic Functions

Lesson 9.2, pages 428–435

TRY THESE **1.** 278.378 **3.** 0.012 **5.** 1 **7.** $\frac{1}{3}$
9. $2,103.83

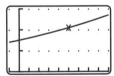

11. $\frac{1}{4}$ **13.** -4 **15.** -3
17. ± 3 **19.** 2.096
21. 0.774

PRACTICE **1.** 39.874 **3.** 0.001 **5.** 16 **7.** $\frac{1}{2}$
11. $\frac{7}{2}$ **13.** $-\frac{1}{2}$ **15.** $-\frac{4}{3}$ **17.** $\pm\sqrt{3}$ **19.** $-\frac{1}{3}$
21. $-\frac{2}{3}$ **23.** 1.292 **25.** 0.356
27a. $A = 18,000(1 + (-0.15))^t = 18,000(0.85)^t$
27b.   **27c.** $9396.11

x scl = 3 y scl = 5000

EXTEND **29a.** $y = 33.396 \cdot 1.076^x$ **29c.** $301 billion
33. Answers will vary. An example is $5^6 = 25^{x^2}$.
35. The graph gets steeper. **37.** -8

MIXED REVIEW **39.** $x = 2$ **40.** $x = -8$ **41.** C
42. A **43.** $y = \frac{1}{4}x + \frac{3}{2}$ **44.** $y = -x - 3$
45. 1.338 **46.** 1.456
47. $\begin{bmatrix} 5 & 13 & -1 & 3 \\ 6 & 13 & 8 & 17 \end{bmatrix}$ **48.** $\begin{bmatrix} 23 & 39 \\ 35 & 119 \end{bmatrix}$

Lesson 9.3, pages 436–442

TRY THESE **1.** $7^2 = 49$ **3.** $3^{-3} = \frac{1}{27}$
5. $\log_3 243 = 5$ **7.** $\log_{10} 0.001 = -3$ **11.** -4
13. 0 **15.** 1.5911 **17.** 0.3562 **19.** 0.0377 **21.** 7.5
23. 256 **25.** 2 **27.** -9

PRACTICE **1.** $9^3 = 729$ **3.** $\frac{1}{256}$ **5.** $\pi^1 = \pi$
7. 11 hours **9.** $\log_{625} 5 = \frac{1}{4}$ **11.** $\log_{32} \frac{1}{2} = -\frac{1}{5}$
13. 5 **15.** 0 **17.** 2.0755 **19.** $\frac{3}{2}$ **21.** 1.1073
23. 0.8271 **25.** 215.4269 **27.** 7.6243×10^{11}
29. 3125 **31.** 4 **33.** 36 **35.** -8 **37.** 64

EXTEND **39a.** 2 units left **39a.** 3 units right **41.** 7.8

THINK CRITICALLY **43.** -0.9420
47. $f(g(x)) = f(b^x) = \log_b b^x = x$

MIXED REVIEW **49.** $4x - 4$ **50.** $3x^2 - 12x$
51. $-1, 9$ **52.** $-3, 5$ **53.** C **54.** A

Lesson 9.4, pages 443–448

TRY THESE **1.** $\log_7 3 + \log_7 x$ **3.** $7\log_8 x + 3\log_8 y$
7. $\log_8 81x^5$ **9.** $\log_b \frac{49d^6}{e^4}$ **11.** 4 **13.** ± 4 **15.** 16
19. 10

PRACTICE **1.** $\log_4 5 + \log_4 y$ **3.** $\log_8 x - \log_8 6$
5. $6\log_2 x + 7\log_2 y$ **7.** $2\log_7 x + \log_7 y$
9a. $D(I) = 10\log I - 10\log I_0$ **9b.** 70 dB
11. $\log_3 u^3 v^8$ **13.** $\log_d \frac{q^2 r^6}{t^4}$ **15.** $\log_5 \sqrt{w^7}$ **17.** 125
19. ± 7 **21.** 8

EXTEND **25.** 1.6 **27.** 7.5
29. $\log x = \frac{1}{2}\log 2 + \frac{1}{2}\log c + \frac{1}{2}\log d - \frac{1}{2}\log e$
31. $\log p = \log 2 + \log \pi + \log m - \frac{1}{2}\log m - \frac{1}{2}\log k$

THINK CRITICALLY
33. By the definition of logarithms, $y = \log_b x$ if and only if $b^y = x$. Substitute $\log_b x$ for y in $b^y = x$ to obtain $b^{\log_b x} = x$.
35. No. The domain of y_1 is all real numbers except $x = 3$. The domain of y_2 is $x : x > 3$.

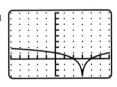

MIXED REVIEW **37.** C **38.** $[4 \; 27]$
39. $\begin{bmatrix} 44 & -10 & 24 \\ -33 & 9 & -38 \\ 26 & -10 & 12 \end{bmatrix}$
40. $26 + 2i$ **41.** $-19 - 71i$
42. $\log_8 4 + \log_8 t$ **43.** $\log_9 q - \log_9 2$
44. $\log_3 4 + 2\log_3 x + 3\log_3 y$

Lesson 9.5, pages 449–453

TRY THESE **1.** 148.4132 **3.** 2.5494 **5.** e^3 **7.** $0.2e^{-4}$
11. $\ln 3 + 4\ln u + 7\ln v$ **13.** $\ln 4x$ **15.** 28
17. 1.5000 **19.** 0.7782

PRACTICE **1.** 1.2840 **3.** 0.6065 **5.** 8.2269
9. e^9 **11.** $2e^5$ **13.** $\ln 8 + \ln t$ **15.** $6\ln m - 5\ln n$
17. $\ln 10jk$ **19.** $\ln \dfrac{7}{z}$ **21.** 1.1073 **23.** 0.8155

EXTEND **25a.** $y = -5.3945 + 19.1266 \ln x$
25c. 86% **27a.** exponential curve
27b. $y = 3{,}732{,}113(0.9931)^x$, where $x = 0$ represents
1950, $x = 10$ represents 1960, and so on
27c. very closely because the correlation coefficient is
-0.9708

THINK CRITICALLY
29. Since $\log_b b = 1, \log_e e = 1$, and $\ln e = 1$.
31. $\ln e^x = x\ln e = x(1) = x$
MIXED REVIEW **34.** B **35.** $-\dfrac{3}{2} \pm \dfrac{\sqrt{19}}{2}$ **36.** $\dfrac{4}{3} \pm \dfrac{\sqrt{10}}{3}$
37. 2.0244 **38.** 0.4421

Lesson 9.6, pages 454–459

TRY THESE **1.** 2.0704 **3.** 2.4849 **7.** 2.3979
9. 1.8261 **11.** -0.7133 **13.** 21.5443
15. 3,333,333.3333 **17.** 2.7183

PRACTICE **1.** 1.8271 **3.** -0.6167 **5.** -3.4340
7. 16.0293 **9.** b **11.** 2.3979 **13.** 1.8808
15. -0.4909 **17.** 1000 **19.** 112.4683 **21.** 2.0427
23. 1998

EXTEND **25.** 6.7549 **27.** 4 **29.** 1500 years
31. 18,300 years

THINK CRITICALLY **35.** $x = 0.3958a$

MIXED REVIEW **37.** $-3 \pm 4i$ **38.** $-1 \pm 2i$
39. 0.115 **40.** $F = \{(x, y): y = x + 3, x = 2, -2, 5\}$;
Domain: $\{2, -2, 5\}$; Range: $\{5, 1, 8\}$
41. $E = \{(x, y): y = 6x, x = 0, 1, -3\}$; Domain: $\{0, 1, -3\}$; Range: $\{0, 6, -18\}$

42. D **43.** B

Lesson 9.7, pages 460–463

APPLY THE STRATEGY **11.** $9096.21 **13.** 7 years
15. 5.54%

Chapter Review, pages 464–465

1. e **2.** c **3.** d **4.** a **5.** b **6.** move f left 1 unit
7. move f down 2 units **8.** reflect f over y-axis
9. reflect f over x-axis **10.** $\dfrac{1}{2}$ **11.** -3 **12.** $\dfrac{5}{2}$
13. ± 4 **14.** -2 **15.** $\pm\sqrt{3}$ **16.** 1.3181 **17.** 0.4719

18. 1.5629 **19.** 1.3023 **20.** 0.7374 **21.** 0.8982
22. 9 **23.** 625 **24.** 1024 **25.** $\dfrac{7}{5}$ **26.** $\log_6 j - \log_6 3$
27. $2\log_9 x - 4\log_9 y$ **28.** $\log_4 \dfrac{x^3}{5}$ **29.** $\log a^{\frac{1}{2}} b$
30. ± 15 **31.** 10 **32.** $\ln 3 + \ln h + \ln k$
33. $2\ln x + 3\ln y$ **34.** $\ln \dfrac{9x^4}{y^2}$ **35.** 2.6856
36. 1.0238 **37.** 1.9243 **38.** 100 **39.** 20.0855
40. 4.3468 **41.** $6686.76 **42.** $6880.28

Chapter 10 Polynomial Functions

Lesson 10.1, pages 473–478

TRY THESE
1. quotient $= 2x^2 + x + 1$, remainder $= 9$,
dividend $= 10x^3 + x^2 + 3x + 7$, divisor $= 5x - 2$
3. $x - 7 + \dfrac{-7}{x - 2}$ **5.** $a^2 - 4a + 2 + \dfrac{-4}{2a + 3}$
7. $x - 2$ **11.** $(x + 4)$ cm
PRACTICE **3.** $x + 2 + \dfrac{-10}{3x + 3}$
5. $2m^2 - 2m + 4 + \dfrac{-7}{2m + 1}$
7. $x^2 - 2x + 4 + \dfrac{8}{x + 2}$
9. $y^3 + y^2 + y + 1$ **11.** $2x^2 - 2x + 1$
13. $(y + 3)(y + 2)(y - 1)$
15. $(x + 1)(x - 5)(x + 4)$
17. $(a - 2)(-a + 1)(a - 3)$ **19.** $(x - 2)$ cm
21. $(x + 4)(x - 1)(2x + 1)(x - 2)$
EXTEND **23.** $\dfrac{1}{2}x - 1$ **25.** $5x - 4y + \dfrac{13y^2}{2x + y}$
27. $2a^2 - 5b^2 + \dfrac{2b^4}{2a^2 + 3b^2}$

THINK CRITICALLY
29. yes **31.** no **33.** no **35.** yes **37.** true
39. true **41.** false
MIXED REVIEW

43. $a = \dfrac{2A - hb}{h}$ **44.** $m = \dfrac{Ft}{v - v_0}$

45. $x = 3, y = 5$ **46.** $x = -1, y = 6$

Lesson 10.2, pages 479–485

TRY THESE

3. $2y^2 - 3y + 4 + \dfrac{-7}{y + 1}$

5. $-2a^3 - a^2 - a + 1 + \dfrac{2}{a - 1}$

7. $6x^3 - 12x^2 + 3 + \dfrac{2}{x + \frac{1}{3}}$

9. $c^3 + 3c^2 + 2c - 2 + \dfrac{1}{3c - 2}$

11. $(3x^2 - 10x - 8)$ h   **13.** $F(-3) = -20$

15. $F\left(-\dfrac{5}{2}\right) = 0$ **17.** yes   **19.** yes   **21.** yes

PRACTICE

1. $x^2 + 6x + 10 + \dfrac{18}{x - 3}$ **3.** $y^2 + 3y + 6 + \dfrac{5}{y - 1}$

5. $3z^2 - 6z + 10 + \dfrac{-18}{z + 2}$

7. $w^2 + 2w + 2 + \dfrac{3}{3w - 1}$ **9.** $2t^2 + t + 2 + \dfrac{-5}{2t + 1}$

11. $x + 2$ and $3x + 1$ **13.** $F(-3) = 25$

15. $F(2) = 264$ **17.** $F\left(\dfrac{2}{3}\right) = 0$ **21.** yes

23. no   **25.** yes   **27.** yes

EXTEND   **29a.** $x^2 + 3x - 4$
29b. $(2x - 1)(3x + 2)(x + 4)(x - 1)$

31. $\dfrac{4x^3 + 2x^2 - 5}{x - \frac{1}{2}} = 4x^2 + 4x + 2 + \dfrac{-4}{x - \frac{1}{2}}$

33. $k = 17$

THINK CRITICALLY   **35.** $t = 4$ **37.** $3a + b = 27$
39. $R = 18$

MIXED REVIEW   **40.** 384   **41.** 4096   **42.** 3   **43.** 4
44. B   **45.** $x = 3$ **46.** $x = 1$ **47.** $x = -3, -1$
48. $2 \log x - \log y$ **49.** $\dfrac{1}{2}(x + y)$

Lesson 10.4, pages 491–499

TRY THESE   **1.** $x = 1, 3, -3$ **3.** $x = -\dfrac{1}{3}, -1, 2$

5. $x = 1, \dfrac{1 + i\sqrt{7}}{2}$ **7.** between -2 and -1, between 0
and 1   **9.** 2 or 0 positive 1 negative   **11.** 4 cm

PRACTICE   **1.** $x = -1, 1, -5$ **3.** $x = -1, -1, -2, 4$
7. $x = -2, -1 \pm \sqrt{3}$
9. between -3 and -2, between 0 and 1
11. 2 or 0 positive   1 negative
13. upper bound $= 2$ lower bound $= -1$
15. yes; if each dimension is increased by 1 in. to 3 in.
by 4 in. by 5 in., the volume

EXTEND   **17.** 5.61 cm

THINK CRITICALLY   **19.** $(x - r_1)(x - r_2) = 0$
21a. $x^2 - x - 6 = 0$ **21b.** $x^2 - 2x - 4 = 0$
21c. $x^2 - 2x + 5 = 0$

MIXED REVIEW   **25.** $(x + 1)^2$ **26.** 16   **27.** $x^2 + 1$
28. 10   **29.** C

Lesson 10.5, pages 500–503

APPLY THE STRATEGY   **13.** $y = 11x + 8$
17. $y = -16x^2 + 180x + 5$
19. between 11 and 12 seconds (11.28 s)
21. $y = -3x^2 + 58x - 47$

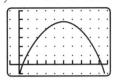

23. \$23,333

Chapter Review, pages 504–505

1. d   **2.** a   **3.** e   **4.** c   **5.** b

6. $2x^2 - x + 3 + \dfrac{4}{3x + 2}$ **7.** $x - 6 + \dfrac{72}{x + 6}$

8. $3x^2 + 7x + 21 + \dfrac{64}{x - 4}$

9. $x^2 - 2x + 4 + \dfrac{-16}{x + 2}$

10. $R = -19$; $(x + 2)$ is not a factor.

11. $R = 0$; $(x - 3)$ is a factor.

12. 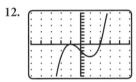   **13.**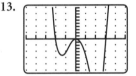

14. valley.   **15.** hill.

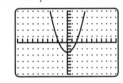   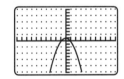

16. hill, then valley.   **17.** valley, then hill, then valley.

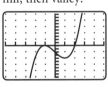   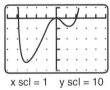

x scl = 1   y scl = 10

18. $-3, -1, 5$ **19.** $-2, -\dfrac{1}{2}, 1$ **20.** $-1, 1 \pm i$
21. $-2; 1 \pm \sqrt{5}$ **22.** Pos: 2 or 0, Neg: 3 or 1
23. Pos: 3 or 1, Neg: 0   **24.** UB $= 2$; LB $= -1$
25. UB $= 5$; LB $= -3$
26. yes; x-values constant difference; degree 2
27. No; x-values not constant difference
28. $y = x^2 + 1$ **29.** $y = -x^2 + 2x - 1$

Chapter 11 Rational Expressions and Equations

Lesson 11.2, pages 517–523

TRY THESE **1.** $x \neq -1$ **3.** $x \neq 4, x \neq -4$
5. $x \neq 0$ **7.** none **9.** $(-5, -6), (3, 2)$
11. $f(x) = 0, x = 3$ **13.** $f(x) = 3, x = -2$
15.

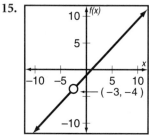

17a. \$40,000 **17b.** \$8 **17c.** $C(n) = \dfrac{30,000 + 2n}{n}$

PRACTICE **1.** $x \neq 0$ **3.** $x \neq 5$ **5.** $x \neq 10, -10$
7. $x \neq 7$ **9.** $x \neq 2, -2, 3, -3$ **11.** none
13. $-1, -3, -19, -199, -1999, 2001, 201, 21, 5, 3$
15. $f(x) = 0, x = 5, x = -5$ **17.** $f(x) = 6, x = 3$
19. $f(x) = 0, x = 4, x = -2, x = 2$
21.   **23.**

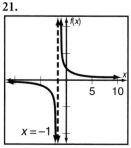

25.   **27.**

29.

EXTEND **31.** Answers will vary; $f(x) = \dfrac{2x^2}{(x-5)(x+2)}$
33a. $9, 9\frac{1}{8}$ **33b.** $p = 9$

33c.

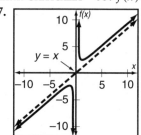

33d. \$9.50, one hour after opening
THINK CRITICALLY **35.** $f(x) \approx x$
37.

MIXED REVIEW
39. $\begin{bmatrix} -1 & 6 & 23 \\ 0 & -6 & 0 \end{bmatrix}$ **40.** $\begin{bmatrix} -9 & 20 & -1 \\ -16 & -12 & 22 \end{bmatrix}$
41. $\begin{bmatrix} -15 & 39 & 33 \\ -24 & -27 & 33 \end{bmatrix}$ **42.** $\begin{bmatrix} 18 & -40 & 2 \\ 32 & 24 & -44 \end{bmatrix}$
43. no **44.** yes **45.** yes **46.** C

Lesson 11.3, pages 524–530

TRY THESE **1.** inversely **3.** inversely **5.** directly
7. jointly, 0.5 **9.** directly, 12 **11.** 55.44
13. 7.5 feet **15.** $81\dfrac{W}{m^2}$
PRACTICE **1.** directly **3.** inversely **5.** directly
7. directly; k = 7.99 **9.** inversely; k = 20
11. 16 **13.** 2.4 hours
EXTEND **17.** $\approx$ 33.5 mph **19.** 10 tons
21.

THINK CRITICALLY **23.** 0.75

Lesson 11.4, pages 531–535

TRY THESE **1.** $\dfrac{3x^3}{2}, x \neq 0$ **3.** $\dfrac{x+1}{4}, x \neq 4$ **5.** $\dfrac{3x}{2}$

7. $\dfrac{6(x + 1)}{x + 10}$ **9.** $-\dfrac{(x + 1)(x - 3)}{x}$ **11.** $\dfrac{a^2}{a + 1}$ cents

PRACTICE **1.** $\dfrac{b}{3ac}, a \neq 0, c \neq 0$ **3.** $x, x \neq -4$

5. $5x - 8, x \neq 0$ **7.** $\dfrac{a + b}{a}, a \neq 0, a \neq b$

9. $\dfrac{4x^3}{5y}$ **11.** $\dfrac{3(x + 6)}{x}$ **13.** $\dfrac{x + 4}{2}$ **15.** $y(y + 1)$

17. 1 **19.** $\dfrac{3(x + 4)}{2(y - 3)}$ **21.** $\dfrac{(x - 3)^2}{x + 3}$ **23.** $\dfrac{x - 7}{x - 6}$

27a. $\dfrac{m}{300}, \dfrac{m}{400}$ **27b.** $\dfrac{7m}{1200}$ **27c.** $\dfrac{2400}{7}$ mph

EXTEND **29.** $x^{5m}y^{4n}$ **31.** $\dfrac{x^{2m} - 7}{x^{2m} - 6}$ **33.** $2x + h$

THINK CRITICALLY **35.** $x^2 - 4x - 12$

MIXED REVIEW **37.** $(3, 6)$ **38.** $(-1, 4)$ **39.** $(2, 5)$
40. $(-2, 4)$ **41.** $2, -2, 3, -3$ **42.** 64 **43.** $8, 125$ **44.** B

Lesson 11.5, pages 536–542

TRY THESE **1.** x^2y^3 **3.** $4(x - y)(x + y)$

5. $\dfrac{-1 - k}{m}$ **7.** 2 **9.** $\dfrac{4n + 6}{(n - 3)(n + 2)(n + 3)}$

PRACTICE **1.** $24x^3y$ **3.** $m(m + n)$

5. $3(x + 4)(x + 5)$ **9.** $\dfrac{-2}{p + 1}$ **11.** $\dfrac{x(6 - x)}{2(2 - x)}$

13. $\dfrac{-x - 1}{(x - 1)(x - 2)}$ **15.** $\dfrac{m + 2}{m}$ **17.** $\dfrac{2y - 14}{15y + 75}$

19. $\dfrac{x^2 - x - 3}{x(x - 4)}$ **21.** $\dfrac{9x^2 - 42x + 81}{(x - 3)^2(x + 3)}$ **23.** $\dfrac{2x - 1}{x^2 - y^2}$

25a. $\dfrac{h - 8}{16}, \dfrac{h - 8}{12}, \dfrac{h - 8}{h}$ **25b.** $\dfrac{7h^2 - 8h - 384}{48h}$ **25c.** $\dfrac{59}{120}$

EXTEND **27.** $\dfrac{m^2 - 3m - 8}{2m^2 + 7m + 6}$ **29a.** 8 **29b.** 3

29c. 7.5 **29d.** $\dfrac{a + b}{2ab}$

THINK CRITICALLY **31.** 1

MIXED REVIEW **33.** 9 **34.** 23 **35.** 4 **36.** 40
37. $(1, 2, 3)$ **38.** $(2, 4, 1)$ **39.** $(0, -6, 1)$
40. B **41.** 1 **42.** $\dfrac{2x^2 - 2x + 2}{x^2 - 1}$ **43.** $-\dfrac{x - 1}{x - 3}$

Lesson 11.6, pages 543–547

TRY THESE **1.** $\dfrac{5}{4}$ **3.** $\dfrac{3y + 1}{2y - 2}$ **5.** $\dfrac{1}{a - 1}$ **7.** $\dfrac{ab}{a + b}$

9. $\dfrac{x + 1}{3}$

PRACTICE **1.** 4 **3.** $\dfrac{2}{3}$ **5.** $\dfrac{10x + 10}{4x + 3}$ **7.** $\dfrac{2x + 2y - 1}{3x + 3y - 2}$

9. x **11.** $\dfrac{5x - 19}{3x - 10}$ **13.** $\dfrac{2}{2x - 5}$ **15.** $\dfrac{n - 3}{n + 5}$

17. $\dfrac{5x}{5x - 16}$ **19a.** $\dfrac{100}{n^2 - 4}, \dfrac{40}{n + 2}, \dfrac{50}{n + 2}, \dfrac{25}{n - 2}$

19b. $\dfrac{\dfrac{100}{n^2 - 4} + \dfrac{40}{n + 2}}{\dfrac{50}{n + 2} + \dfrac{25}{n - 2}}; \dfrac{8n + 4}{15n - 10}$

EXTEND **21.** $\dfrac{2 - c}{c}$ **23.** $\dfrac{1}{2y}$

THINK CRITICALLY **25.** $1, 1.5, 1.4, 1.4166667,$
$1.4137931, 1.4142857$ **27.** It equals $\sqrt{2}$.

MIXED REVIEW **28.** $\begin{bmatrix} -3 & 5 \\ 5 & -8 \end{bmatrix}$ **29.** does not exist

30. $\begin{bmatrix} \dfrac{3}{11} & -\dfrac{5}{11} \\ \dfrac{1}{11} & \dfrac{2}{11} \end{bmatrix}$ **31.** $\begin{bmatrix} 0 & 1 \\ -1 & 0 \end{bmatrix}$ **32.** $x \leq -6$ or $x \geq 4$

33. $x \geq -1$ or $x \leq -3$ **34.** $x < -5$ or $x > 5$
35. all real numbers **36.** C

Lesson 11.7, pages 548–553

TRY THESE **1.** 24 **3.** 38 **5.** $-5, -1$ **7.** $-\dfrac{3}{5}$

9. no solution **11a.** $\dfrac{2x^3 - x^2 + 5}{x^2 + 2} = 5$

11b. Sales surpassed 5000/day about 3.2 months after the skates were introduced.

PRACTICE **1.** 16 **3.** 3 **5.** 7 **7.** -8 **9.** 7 **11.** 3
13. no solution **15.** no solution **17.** 75 mi/h and
186 mi/h **19.** $2 + \sqrt{3}$ and $2 - \sqrt{3}$

EXTEND **23.** $r = \dfrac{E - IR_2}{I}$ **25.** $AB = 24$ $EF = 5$
THINK CRITICALLY **27.** $x = n$

MIXED REVIEW **29.** $-2 < x < 2$
30. $x \leq -2$ or $x \geq 1$ **31.** $x < -\dfrac{3}{2}$ or $x > \dfrac{17}{2}$
32. all real numbers **33.** -1 **34.** $-i$ **35.** $4i$
36. $-2i$ **37.** A **38.** 7 **39.** $1, 3$ **40.** 8 **41.** $1, -\dfrac{1}{2}$

Lesson 11.8, pages 554–557

APPLY THE STRATEGY **15a.** 6.25
15b. $4.25 < AA < 6.25$

Chapter Review, pages 558–559

1. d **2.** a **3.** e **4.** c **5.** b
6. $x \neq 4$ **7.** $x \neq 5; -5$ **8.** all real numbers
9. $(8, 16)$ **10.** $x = 1, y = 4$ **11.** 3 **12.** 392 **13.** $\dfrac{x}{y}$
14. $\dfrac{5(a - b)}{2a}$ **15.** $\dfrac{15f}{2e}$ **16.** $2n + 1$ **17.** $\dfrac{12x}{x^2 - 9}$

18. $\dfrac{23}{y^2 - 7y}$ **19.** $\dfrac{17}{14}$ **20.** $\dfrac{5xy}{y - x}$ **21.** 3   **22.** 12

23. 8   **24.** no solution   **25.** $2.75 < AA < 3.75$

Chapter 12 Conic Sections

Lesson 12.1, pages 567–571

Try These   **1.** 13; (8, 9.5)   **3.** $\sqrt{40}$; (10, 0)
5. $(-4, 7)$ **7.** 15.6 mi

Practice   **1.** 10; (6, 1)   **3.** $\sqrt{50}$; (2.5, 2.5)
5. 27; (1.7, 7)   **7.** 11.6; (2.3, 0.7)
9. $2\sqrt{26}$; $\left(3\sqrt{6}, 5\sqrt{5}\right)$ **11.** $(-1, -1)$ **13.** 1 or -23
15. The sides measure $\sqrt{53}$, $\sqrt{53}$, and $\sqrt{106}$.   **17.** 14°

Extend   **19.** 92.9 miles   **21.** 13

Mixed Review   **25.** 32   **26.** 5   **27.** 343   **28.** D

Lesson 12.3, pages 575–580

Try These   **1.** $x^2 + y^2 = 25$
3. $(x - 3.7)^2 + (y + 6.5)^2 = \pi^2$
5. (0, 0); 10   **7.** $(-12, 5)$, $2\sqrt{5}$

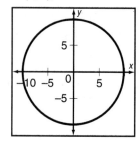 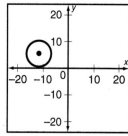

9. $x^2 + y^2 = (26{,}260)^2$

11. $y = -1 \pm \sqrt{\dfrac{25}{4} - \left(x - \dfrac{3}{2}\right)^2}$

Practice   **1.** $x^2 + y^2 = 196$
3. $(x + 9)^2 + (y + 3)^2 = 49$
5. $(x - 4.3)^2 + (y + 7.7)^2 = 9\pi^2$
7. (0, 0); 3   **9.** $(-1, 0)$; 5

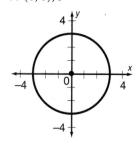 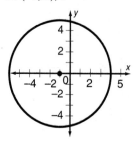

11. $(-3, 2)$; $2\sqrt{7}$ **13.** $(0, -7)$; 1

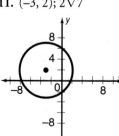

15. $x^2 + y^2 = 144$ **17.** 0.1π hours
19. $x^2 + y^2 = 6.76$, $x^2 + y^2 = 60.84$, $x^2 + y^2 = 169$
$x^2 + y^2 = 331.24$, $x^2 + y^2 = 547.56$

Extend
21. $x^2 + y^2 = 169$

23. $\left(x - \dfrac{1}{21}\right)^2 + \left(y + \dfrac{3}{2}\right)^2 = 25$
25a. $(x + 2)^2 + (y - 5)^2 = 40$ **25b.** $(0, 11)$,
$(0, -1) + (-2, \sqrt{15}, 0) + (-2, -\sqrt{15}, 0)$
27. $(x - 1)^2 + (y - 2)^2 = 34$

Mixed Review   **31.** 3; -5 **32.** 2; $\dfrac{5}{2}$ **33.** $\dfrac{1}{3}, \dfrac{1}{6}$ **34.** $\dfrac{x}{5}$
35. $x - 4$ **36.** $\dfrac{1 - x}{x}$ **37.** B

Lesson 12.4, pages 581–586

Try These   **1.** 8   **3.** $(-5, 1)$, $(3, 1)$, $(-1, 2)$, $(-1, 0)$
5. $(-1 - \sqrt{15}, 1)$, $(-1 + \sqrt{15}, 1)$
7.

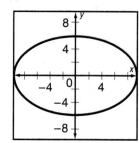

foci: $(-8, 0)$, $(8, 0)$;
vertices: $(-10, 0)$, $(10, 0)$,
$(0, -6)$, $(0, 6)$; center:
$(0, 0)$; length of major
axis: 20; length of minor
axis: 12

9.

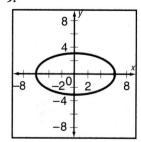

foci: $(-3\sqrt{3}, 0)$, $(3\sqrt{3}, 0)$;
vertices: $(-6, 0)$,
$(6, 0)$, $(0, -3)$, $(0, 3)$;
center: $(0, 0)$; length of
major axis: 12; length
of minor axis: 6

11. $\dfrac{(x + 3)^2}{16} + \dfrac{(y - 1)^2}{9} = 1$; center: $(-3, 1)$; vertices:
$(-7, 1)$, $(1, 1)$, $(-3, -2)$, $(-3, 4)$;
foci: $(-3 - \sqrt{7}, 1)$, $(-3 + \sqrt{7})$

PRACTICE 1. $\dfrac{x^2}{25} + \dfrac{y^2}{4} = 1$

3. $\dfrac{(x-2)^2}{9} + \dfrac{(y+3)^2}{1} = 1$

5. 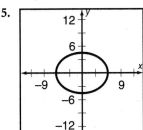 foci: $(-4, 0)$, $(4, 0)$; vertices: $(-6, 0)$, $(6, 0)$, $(0, -2\sqrt{5})$, $(0, 2\sqrt{5})$; center: $(0, 0)$; length of major axis: 12; length of minor axis: $4\sqrt{5}$

7. 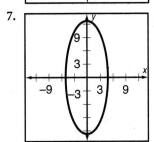 foci: $(0, -12)$, $(0, 12)$; vertices: $(-5, 0)$, $(5, 0)$, $(0, -13)$, $(0, 13)$; center: $(0, 0)$; length of major axis: 26; length of minor axis: 10

9. 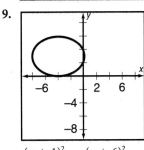 foci: $(-8, 3)$, $(0, 3)$; vertices: $(-9, 3)$, $(1, 3)$, $(-4, 0)$, $(-4, 6)$; center: $(-4, 3)$; length of major axis: 10; length of minor axis: 6

13. $\dfrac{(x+1)^2}{25} + \dfrac{(y+6)^2}{4} = 1$

EXTEND 15. 0.38 17. 0 19. $e \approx 0.25$

THINK CRITICALLY 21. $\dfrac{x^2}{25} + \dfrac{y^2}{25} = 1$

MIXED REVIEW 22. $x + 3$ 23. $x^2 - 9x + 5$
24. $3x^3 + 9x^2 + 2x + 6$ 25. $x^2 + x + 1$ 26. C

Lesson 12.5, pages 587–591

TRY THESE 1. $\dfrac{x^2}{16} - \dfrac{y^2}{9} = 1$ 3. $xy = -9$

5.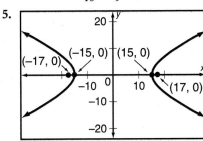

7.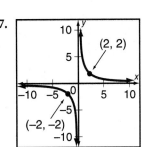

9. $y_1 = 3 + \sqrt{\dfrac{8x^2}{15} - 8}$,

$y_2 = 3 - \sqrt{\dfrac{8x^2}{15} - 8}$

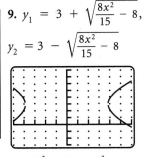

PRACTICE 1. $xy = 36$ 3. $\dfrac{(x-3)^2}{9} - \dfrac{(y-4)^2}{1} = 1$

5. $xy = -64$ 7. $\dfrac{y^2}{324} - \dfrac{x^2}{100} = 1$

9.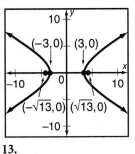

11.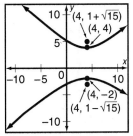

13.

15. center: $(2, -4)$; vertices: $(5, -4)$, $(-1, -4)$; foci: $(2 + \sqrt{34}, -4)$, $(2 - \sqrt{34}, -4)$; slopes: $\pm\dfrac{5}{3}$

17. $y = 3 \pm \sqrt{16 + \dfrac{4}{5}(x + 5)^2}$,

EXTEND 21. hyperbola
23. Place the ranger stations at $(-4400, 0)$ and $(4400, 0)$. The strike took place on one branch of the hyperbola $\dfrac{x^2}{2750^2} - \dfrac{y^2}{3435^2} = 1$. 25. $a = b$

THINK CRITICALLY 27. $\dfrac{(x-3)^2}{16} - \dfrac{(y+2)^2}{9} = 1$

MIXED REVIEW 29. 0.125 30. 0.5 31. 0.25
32. 0.375 33. B

Lesson 12.6, pages 592–597

TRY THESE 1. $(x - 0)^2 = 4(-5)(y - 0)$
3. $(x - 6)^2 = 4(2)(y - 3)$

5. $(0, 0)$, $(0, 3)$, $y = -3$

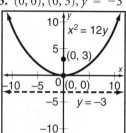

7. $(0, 0)$, $(2, 0)$, $x = -2$

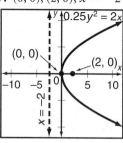

9. $x^2 = -20y$ **11.** $x^2 = 36y$ **13.** vertex: $(-2, 2)$;
focus: $(-2, 4)$; directrix: $y = 0$

PRACTICE

1. $(0, 0)$, $(6, 0)$, $x = -6$

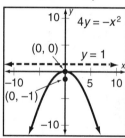

3. $(0, 0)$, $(0, 5)$, $y = -5$

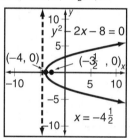

5. $(0, 0)$, $(0, -1)$, $y = 1$

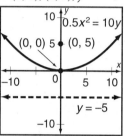

7. $(-4, 0)$, $(-3\frac{1}{2}, 0)$, $x = -4\frac{1}{2}$

9. $(-1, 2)$, $(2, 2)$, $x = -4$

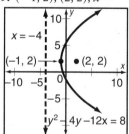

11. $y^2 = 12x$
13. $y^2 = -24x$
15. $(x - 1)^2 = 32(y + 1)$
17. $-8(y - 6) = (x - 8)^2$
21. $(2, 3)$, $(3, 3)$, $x = 1$
23. $(-2, -3)$, $\left(-2, \frac{11}{4}\right)$,
$\qquad y = -\frac{13}{4}$

EXTEND **25.** $x = h$ **27.** $x = 1$ **29.** $x = 4$
31a. $x^2 = -\frac{80}{3}y$ **31b.** 12.6 feet
THINK CRITICALLY **33.** $\frac{17}{5}$
MIXED REVIEW
36. $\begin{bmatrix} 3 & -1 \\ 6 & 9 \end{bmatrix}$ **37.** $\begin{bmatrix} 11 & -2 \\ 11 & 1 \end{bmatrix}$ **38.** $\begin{bmatrix} 2 & 3 \\ -1 & 4 \end{bmatrix}$ **39.** $\begin{bmatrix} 2 & 3 \\ -1 & 4 \end{bmatrix}$
40. $-9, 20$ **41.** $5, -13$ **42.** $-3, \frac{2}{3}$ **43.** $\frac{1}{5}, -\frac{3}{5}$ **44.** 49
45. no solution **46.** 11 **47.** 9 **48.** A

Lesson 12.7, pages 598–602

TRY THESE

1. $5x^2 + 6xy + 4y^2 - 3x - 2y + 1 = 0$;
$5, 6, 4, -3, -2, 1$
3. $4x^2 + 9y^2 - 8x + 36y + 4 = 0$; $4, 0, 9, -8, 36, 4$
5. ellipse **7.** hyperbola **9.** ellipse **11.** hyperbola
13. The graph is the line $y = x + 3$, a degenerate parabola.

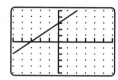

PRACTICE **1.** parabola **3.** hyperbola **5.** parabola
7. parabola **9.** hyperbola **11.** hyperbola
13. circle **15.** hyperbola **17.** parabola
19.

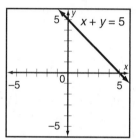

21.

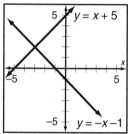

a degenerate parabola   a degenerate hyperbola

EXTEND **27.** ±12

THINK CRITICALLY

31. $y = \dfrac{-E \pm \sqrt{E^2 - 4CF}}{2C}$

MIXED REVIEW **33.** no **34.** yes **35.** no **36.** $5^{\frac{1}{2}}$
37. $x^{\frac{1}{2}}$ **38.** $2^{\frac{1}{3}}n^{\frac{5}{3}}$ **39.** $h^{\frac{3}{5}}k^3$ **40.** 4 **41.** $\frac{3}{2}$ **42.** 0.01
43. 8 **44.** D

Lesson 12.8, pages 603–609

TRY THESE **1.** line, ellipse; $(2, 3)$, $(-1.5, -4)$
3. ellipse, hyperbola; $(0, 2)$, $(0, -2)$
5. hyperbola, line; $(4, 5)$, $(0, -3)$ **7.** 108 yd^2
PRACTICE **1.** line, circle; $(-8, -6)$, $(6, 8)$
3. hyperbola, line; $(6, 9)$, $(-6, -9)$
5. parabola, line; $\left(\frac{1}{2}, -3\right)$, $\left(0, -\frac{10}{3}\right)$
7. ellipse, line; $(1, 7)$, $(7, 1)$
9. line, ellipse; $(0, 2)$, $\left(\frac{20}{13}, -\frac{24}{13}\right)$
11. hyperbola, hyperbola; $(4, -5)$, $\left(-27, \frac{1}{6}\right)$
13. 9 cm, 40 cm **15.** $(-32°, -10°)$ or $(32°W, 10°S)$
EXTEND **17.** $(2, 4)$ **21.** 7, 17 and $-7, -17$
THINK CRITICALLY **25.** $l = \frac{1}{4}(P + \sqrt{P^2 - 16A})$,
$w = \frac{1}{4}(P - \sqrt{P^2 - 16A})$

MIXED REVIEW **26.** 16; 14; 23 **27.** 462.25; 465.5; none **28.** -3.08; -5.6; none **29.** $\frac{7}{12}$; $\frac{7}{12}$; none
30. $y = 3x - 2$ **31.** $y = 2x + 10$
32. $y = 0.5x - 1$ **33.** $y = 3x - 2$ **34.** A
35. $(-1, -1)$ **36.** $(1, 1), (2, 4)$ **37.** $(0, 5), (0, -5)$

Lesson 12.9, pages 610–613

APPLY THE STRATEGY **15.** Sun; 1.88 **17.** Sun. 851.0
19. Sun; 4.59 **21.** 4523 mi; 4109 mi
23. 39,100 min; 27.1 d

Chapter Review, pages 614–615

1. c **2.** d **3.** a **4.** b **5.** 5, $(7, 0.5)$
6. 17, $(-8, -0.5)$ **7.** $2\sqrt{13}$, $(0, 1)$
8. $(x - 3)^2 + (y + 7)^2 = 5$ **9.** $(1, -3)$, 7
10. foci: $\left(5 - \sqrt{7}, 3\right)$,
$\left(5 + \sqrt{7}, 3\right)$; vertices:
$(1, 3), (9, 3), (5, 6), (5, 0)$;
center: $(5, 3)$;

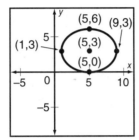

11. $(-2, 5)$; $(-2, 13)$ and $(-2, -3)$; $(-2, 15)$ and $(-2, -5)$; $\pm\frac{4}{3}$ **12.** $\frac{x^2}{25} - \frac{y^2}{4} = 1$ **13.** $(0, 0)$; $(12, 0)$; $x = -12$ **14.** $(-5, 3)$; $(-5, 6)$; $y = 0$
15. hyperbola **16.** parabola **17.** $(-3, 4), (4, -3)$
18. $(2, 3), (-1.5, -4)$ **19.** $(3, 3), (-3, 3)$
20. 1.9 years

Chapter 13 Sequences and Series

Lesson 13.2, pages 626–632

TRY THESE **1.** explicit **3.** recursive **5.** explicit
7. 0, 1, 2, 3, 4 **9.** $-2, 4, -8, 16, -32$ **11.** 2, 3, 1, 5, -3
13a. \$5.65, \$5.90, \$6.15, \$6.40, \$6.65, \$6.90
13b. $a_1 = 5.65$ and $a_n = a_{n-1} + 0.25$ **15.** 6655
19. $a_n = \frac{n+1}{n+2}$
PRACTICE **1.** 9, 13, 17, 21, 25 **3.** $-1, 1, -1, 1, -1$
5. 2, 4, 16, 256, 65536 **7.** 74 **9.** 1125
11. $a_n = 2n - 1$ **13.** $a_n = \frac{1}{3^n}$ **15.** $a_1 = 1$, $a_n = a_{n-1} + n$
EXTEND **21.** $a_n = \frac{x^n}{2^{n-1}}$ **23.** 0.4 AU, 0.7 AU, 1 AU, 1.6 AU; 37,200,000 mi; 65,100,000 mi; 93,000,000 mi; 148,800,000 mi

THINK CRITICALLY **27.** an exponent function
MIXED REVIEW **29.** $-2, 1, 3$ **30.** $5n - 3$; $-n - 7$; $6n^2 - 11n - 10$ **31.** $a - 6b + 2c + d$; $a - 6b - 2c - d$; $2ac - 12bc + ad - 6bd$ **32.** C

Lesson 13.3, pages 633–638

TRY THESE **1.** arithmetic; $d = 5$ **3.** geometric; $r = 5$
5. arithmetic; $d = \frac{1}{2}$ **7.** 68 seats
9. $a_n = -14 + (n - 1)(4)$; 50
11. $a_n = 5(-3)^{n-1}$; 3645
13. arithmetic: $169\frac{1}{3}$, $334\frac{2}{3}$, geometric: 20,100
15. $5\frac{4}{5}, 7\frac{3}{5}, 9\frac{2}{5}, 11\frac{1}{5}$
PRACTICE **1.** 3, 14 **3.** 2.15; 12.02 **5.** $-\frac{1}{3}, \frac{1}{9}$ **9.** -16
11. 65 **13.** -2.5 **15.** -0.75 **17.** -1536
19. $a_n = 300(0.1)^{n-1}$ **21.** $a_n = 16 + (n + 1)$; 32; 304 ft **23.** 8, 32, 128 or $-8, 32, -128$
EXTEND **25.** c **27.** 3; 9
29. $\frac{1}{2048}$% or approximately 0.049%
THINK CRITICALLY **33.** 14.9%
MIXED REVIEW **34.** y-axis **35.** both axes, origin
36. none **37.** origin **38.** $19^{\frac{1}{2}}$ **39.** $c^{\frac{3}{4}}$ **40.** $11\frac{5}{6}$
41. $(n - 1)^{\frac{7}{3}}$ **42.** A

Lesson 13.4, pages 639–645

TRY THESE **1.** 380 **3.** 793 **5.** 990 **9.** $\frac{341}{256}$
11. 1,048,575 **13.** $-\frac{11}{8}$
15. $S_n = \frac{n}{2}[a_1 + a_n]$, where $n = 15$, $a_1 = 24$, and $a_{15} = 10$; 255 logs
17. $\sum_{n=1}^{7}(2n - 1)$; 49 **19.** $\sum_{n=1}^{6} 2^n$; 126
PRACTICE **1.** 169 **3.** -68 **5.** 217.5 **7.** 351 **9.** 525
11. $\frac{511}{256}$ **13.** $62\sqrt{2} + 62$ **15.** 216.8 **17.** 765
19. $\frac{279}{4}$ or $\frac{99}{4}$ **21.** \$217,661.21 **23.** $\sum_{n=1}^{5}(4n - 3)$; 45
25. $\sum_{n=1}^{21}(3n + 2)$; 735 **27.** $\sum_{n=1}^{9}[5(2^n)]$; 5110 **29.** 9
EXTEND **33.** n^2 **35.** $-\frac{24863}{29}$
THINK CRITICALLY **37.** 2, 8, 18
39. 12, 14, 16 or 22, 14, 6
MIXED REVIEW **40.** $x - 8$ **41.** $x^2 - 2x - 1 + \frac{-8}{x - 2}$
42. $x^2 + 6$ **43.** $x^3 + x + 1$ **44.** C

Lesson 13.5, pages 646–652

TRY THESE 1. converges 3. diverges 5. $5\frac{1}{3}$
7. none 9. $\frac{100}{11}$ 11. $\frac{29}{9}$ 13. $\frac{428}{333}$
PRACTICE 1. converges 3. converges 5. converges
7. 8 9. none 11. $\frac{8}{3}$ 13. none 15. $\frac{432}{7}$ 17. $\frac{81}{4}$
19. 2 21. $\frac{500}{3}$ 23. $\frac{31}{33}$ 25. $\frac{25}{33}$ 27. $\frac{73}{111}$ 29. $\frac{13}{45}$
31. $\approx 3{,}191{,}489$ 33. 240 ft
EXTEND 35. $\frac{1}{5}$ 37. $\frac{3}{2}$
THINK CRITICALLY 39. $\frac{41}{99}$ 41. $\frac{89}{99}$
43. Write the two digits over 99.
45. $\frac{a_1}{1-r}$ for $|r| < 1$
MIXED REVIEW 47. 16 48. 81 49. $\frac{1}{16}$ 50. 4
51. $\frac{2\sqrt{3}}{3}$ 52. $\frac{\sqrt{2}}{2}$ 53. $-2 + \sqrt{7}$ 54. $\frac{2\sqrt[3]{6}}{3}$
55. B 56. $x = 3, 4$ 57. $x = -\frac{1}{3}, \frac{1}{2}$
58. $x = -\frac{5}{6}, \frac{2}{3}$ 59. $x = \frac{5 + \sqrt{61}}{6}, \frac{5 - \sqrt{61}}{6}$

Lesson 13.6, pages 653–657

TRY THESE 1. 18 3. 6 5. 84
7. $d^6 - 12d^5 + 60d^4 - 160d^3 + 240d^2 - 192d + 64$
9. $10x^2y^3$ 11. $4032c^5d^4$
13. $(x^3 + 12x^2 + 48x + 64)\,\text{in.}^3$
PRACTICE
1. $a^6 + 6a^5b + 15a^4b^2 + 20a^3b^3 + 15a^2b^4 + 6ab^5 + b^6$
3. $c^4 - 20c^3d + 150c^2d^2 - 500cd^3 + 625d^4$
5. $16 - 96b + 216b^2 - 216b^3 + 81b^4$
7. $b^6 - 3ab^4c + 3a^2b^2c^2 - a^3c^3$
9. $30^3 + 3(30)^2(1) + 3(30)(1)^2 + 1^3 (= 29{,}791)$
11. $a^4 - \frac{12a^3}{b} + \frac{54a^2}{b^2} - \frac{108a}{b^3} + \frac{81}{b^4}$ 13. $35a^4b^3$
15. $2268x^6$ 17. $84a^2b^5$ 19. $-745{,}472n^3$
21. $1215a^4b^2$ 23. $4860m^4n^8$
25. $1.21896; (1.02)^{10} = 1.21899442$; the sum of the remaining terms is 0.00003442.
EXTEND 29. 85, 140 31. 36! 33. $(p + 6)!$
35. $x - y$ 37. $4032c^4d^5$ 39. $-4 + 4i$
THINK CRITICALLY 41. $1 + \frac{1}{2}b - \frac{1}{8}b^2 + \frac{1}{16}b^3$
MIXED REVIEW 43. $x^2 + 3x + 1$
44. $2x^3 + 3x^2 + x$ 45. $2x^2 + 2x + 1$
46. $4x^2 + 6x + 2$ 47. 4 48. 144 49. $\frac{2}{3}$
50. 864 51. $(-3, 1)$ 52. B 53. $(1, 2, 3)$
54. $(2, 4, 1)$ 55. $(0, -6, 1)$

Lesson 13.7, pages 658–661

APPLY THE STRATEGY
25a.

n	1	2	3	4	5	6	7	8
S_n	1	2	2.5	2.667	2.708	2.717	2.718	2.718

25b.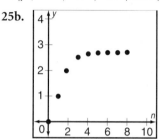

25c. ≈ 2.718
25d. e

Chapter Review, pages 662–663

1. b 2. e 3. c 4. a 5. d
6. explicit 7. recursive 8. 0, 3, 8, 15, 24
9. 3, 2, 1, 0, -1 10. $a_n = 2^{n-1}$
11. $a_n = -(2n - 1)$ 12. $a_1 = 4, a_n = a_{n-1} + 4$
13. $a_1 = 1, a_{n-1} + a_{n-2}$ 14. 59 15. 13 16. 3
17. 2 18. 81 19. 11 20. arithmetic 2205
21. geometric 1462 22. arithmetic 20,100
23. geometric 21.3125 24. geometric 1093
25. geometric 1706.25
26. $\sum_{i=1}^{7}(3i + 1)$ 27. $\sum_{i=1}^{6}(2^i - 1)$
28. c 29. d 30. c 31. $85\frac{1}{3}$ 32. 8 33. ≈ 2.14
34. $\frac{5}{11}$ 35. $a^4 + 4a^3b + 6a^2b^2 + 4ab^3 + b^4$
36. $21m^5n^2$ 37. $-56a^3b^5$ 38. $10x^2y^{12}$

Chapter 14 Probability and Statistics

Lesson 14.1, pages 671–678

TRY THESE 1. $4 \cdot 3 \cdot 2 = 24$ 3. 60,480 5. n
7. 720 9. 720 11. $8! = 40{,}320$ 13. $_4P_2 = 12$
15a. $5! = 120$ 15b. $5^5 = 3125$
17. $\frac{12!}{2! \, 2! \, 3! \, 2!} = 9{,}979{,}200$
PRACTICE 1. $4 \cdot 3 = 12$ 3. $5 \cdot 5 = 25$
5. $5 \cdot 4 \cdot 3 = 60$ 7. 30,240 9. 1
13. 8 15. $\frac{(n + 1)!}{2}$ 17. $(5 - 1)! = 24$
19. $_6P_4 = 360$ 21. $6! = 720$ 23. $\frac{10!}{3! \, 2! \, 4!} = 12{,}600$
25. $_3P_2 = 6$ 27. $_5P_3 = 60$ 29. $\frac{14!}{5! \, 3! \, 6!} = 168{,}168$

EXTEND   31. $1 \cdot {}_5P_5 = 120$
33. $1 \cdot 5 \cdot 4 \cdot 3 \cdot 2 \cdot 1 \cdot 1 = 120$
35. $1 \cdot 4 \cdot 3 \cdot 2 \cdot 1 \cdot 1 = 48$
37. $2 \cdot 5 \cdot 4$ or $2 \cdot {}_5P_2$ or 40
39. $3 \cdot 3 \cdot 2$ or ${}_3P_1 \cdot {}_3P_1 \cdot {}_2P_1$ or 18
41. $1 \cdot 6! = 720$
43. $4 \cdot 5 \cdot 4 \cdot 3 \cdot 2 \cdot 1 \cdot 3$ or ${}_4P_2 \cdot {}_5P_5$ or 1440
45. $2(3 \cdot 3 \cdot 2 \cdot 2 \cdot 1 \cdot 1) = 72$
MIXED REVIEW   52. $\frac{3}{6}$ or $\frac{1}{2}$ or 0.5
53. $\frac{2}{6}$ or $\frac{1}{3}$ **54.** C

Lesson 14.2, pages 679–685

TRY THESE   1. 35   **3.** 1   **5.** 1   **7.** ${}_7C_2$ or 21
9. ${}_9C_4$ or 126   **11.** ${}_8C_4$ or 70   **13.** ${}_4C_2 \cdot {}_{48}C_{11}$
15. ${}_{39}C_{13}$ **17.** ${}_9C_3 + {}_9C_4 + {}_9C_5 + {}_9C_6 + {}_9C_7 + {}_9C_8 + {}_9C_9$ or 466

PRACTICE   1. 28   **3.** 495   **5.** 1000   **7.** ${}_{22}C_4$ or 7315
9. ${}_{20}C_6$ or 38,760   **11.** ${}_{12}C_4$ or 495   **13.** ${}_{10}C_4$ or 210
15. ${}_{12}C_6$ or 924   **17.** ${}_{11}C_6$ or 462
19. ${}_5C_3 \cdot {}_7C_3$ or 350   **21.** ${}_4C_3 \cdot {}_6C_2 = 60$
23. ${}_4C_2 \cdot {}_{48}C_{11}$ **25.** ${}_{13}C_4 \cdot {}_{13}C_3 \cdot {}_{26}C_6$
27. ${}_{10}C_0 + {}_{10}C_1 + {}_{10}C_2 + {}_{10}C_3 + {}_{10}C_4$ or 386
29. ${}_{10}C_0 + {}_{10}C_1 + {}_{10}C_2 + {}_{10}C_3 + {}_{10}C_4 + {}_{10}C_5$ or 638

EXTEND
31. ${}_4C_2({}_{13}C_5 \cdot {}_{13}C_5) \cdot {}_2C_1({}_{13}C_3)$ or 5,684,658,408
35. 30   **37.** 1 8 28 56 70 56 28 8 1

39a. $\binom{8}{3} = 56$ $\binom{7}{3} = 35$ $\binom{7}{2} = 21$

39b. $\binom{8}{3} = \binom{7}{3} + \binom{7}{2}$ **39d.** $\binom{n}{r} = \binom{n-1}{r} + \binom{n-1}{r-1}$

THINK CRITICALLY   41. n **43.** $n + 3$

MIXED REVIEW   48. $\frac{3}{6}$ or $\frac{1}{2}$ **49.** $\frac{26}{52}$ or $\frac{1}{2}$ **50.** A
51. D   **52.** yes
53. No. Suppose George sits between Ethel and Karyn. Then Ethel sits next to George and George sits next to Karyn, but Ethel does not sit next to Karyn.

54. yes   **55.** $\begin{bmatrix} 8 & -14 \\ -6 & -10 \\ 0 & -2 \end{bmatrix}$ **56.** $\begin{bmatrix} 19 & 3 \\ -3 & -2 \end{bmatrix}$

Lesson 14.3, pages 686–694

TRY THESE   3. 1   **5.** 4:8 or 1:2   **7.** $\frac{1}{4}$ **9.** 0   **11.** $\frac{5}{8}$
13. 31%

PRACTICE   3. $\frac{2}{13}$ **5.** $\frac{1}{26}$ **7.** 1:1   **9.** 1:25   **11.** 6:7
13. $\frac{2}{3}$ **15.** $\frac{1}{2}$ **17.** $\frac{1}{5}$ **19.** $\frac{1}{10}$ **21.** $\frac{1}{208}$ **23.** $\frac{3}{104}$
25. $\frac{17}{26}$ **27.** $\frac{2}{5}$ **29.** $\frac{19}{30}$

EXTEND   31. $\frac{14}{25}$ **33.** $\frac{3}{25}$ **35.** $\frac{4}{5}$ **37.** $\frac{1}{5}$ **39.** $\frac{2}{3}$
THINK CRITICALLY   41. 3   **43.** $\frac{2}{25}$ **45.** $\frac{7}{9}$ **47.** $\frac{1}{3}$
MIXED REVIEW   48. $m = -9;\ n = -4$
49. $a = 1;\ b = -6;\ c = -2$ **50.** C

Lesson 14.4, pages 695–702

TRY THESE   3. $\frac{1}{4}; \frac{13}{51}$ **5.** $\frac{9}{25}; \frac{3}{10}$ **7.** 0; 0   **9.** $\frac{2}{3}$
11. $\frac{1}{2}$ **13.** $P(B|A) = \frac{1}{2}$ **15.** $\frac{1}{4}$ **17.** $\frac{1}{4}$

PRACTICE   3. $\frac{2}{7}$ **5.** $\frac{23}{595}$ **7.** $\frac{24}{595}$ **9.** $\frac{5}{506}$ **11.** $\frac{45}{1012}$
13. $\frac{14}{39}$ **15.** $\frac{1}{10}$ **17.** $\frac{3}{5}$ **19.** 0   **21.** 28 tiles   **23.** $\frac{13}{28}$
25. $\frac{3}{14}$

EXTEND   27. $\frac{8}{15}$ **29.** $\frac{11}{12}$ **31.** 0.612

THINK CRITICALLY   33. $\frac{1}{3}$ **35.** $P(B|A) = \frac{4}{11}$

MIXED REVIEW   37. $-1, 7$ **38.** $-3 \pm \sqrt{13}$ **39.** D
40. $x < -8$ or $x > -1$ **41.** $-4 < x < 0$ **42.** 11
43. $\frac{7}{8}$ **44.** $2(x + 5)(x - 5)$ **45.** $a(x + 4)(x + 3)$
46. $(g - 4)(g - 3h)$ **47.** B

Lesson 14.5, pages 703–709

TRY THESE   3. 15–20   **5.** $\frac{17}{35}$
7.

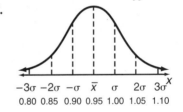

9. 8 cartons
11.

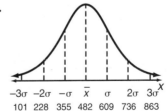

13. 84th   **15.** 0.3413, about 34%
17. 0.1587, about 15.9%
PRACTICE   3. 4.0–4.4   **5.** $\frac{7}{20}$ **7.** 47.5%   **9a.** 13.5%
9b. 15.9%   **9c.** 90.65%   **11.** A   **13.** A   **17.** 68%
19. Yes. The 90th percentile is equivalent to a score under 710.5; Cesar's score was higher.
21. 0.4641, about 46%   **23.** 0.1587, about 15.9%

25.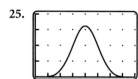

Xmin = 67	Ymin = 0
Xmax = 77	Ymax = 0.2
Xscl = 2.5	Yscl = 0.05

THINK CRITICALLY **27.** 37 **29.** 239.5

MIXED REVIEW **31.** C

Lesson 14.6, pages 710–716

TRY THESE

7.

Number of Odd Outcomes	0	1	2	3
Probability	$\frac{8}{125}$	$\frac{36}{125}$	$\frac{54}{125}$	$\frac{27}{125}$

11. 14.5% **13.** 2.1% **15.** 0.4%

PRACTICE **7.** $\frac{25}{216}$ **9.**

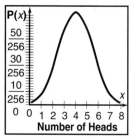

11. $\frac{1}{2^7}, \frac{7}{2^7}, \frac{21}{2^7}, \frac{35}{2^7}, \frac{35}{2^7}, \frac{21}{2^7}, \frac{7}{2^7}, \frac{1}{2^7}$ **13.** $\frac{26}{27} = 0.963$

15. $\frac{80}{243} \approx 0.329$ **17.** 0.420 **19.** 0.704 **21.** 0.183

23. 20% **25.** 0.889 **27.** 0.112 **29.** $\frac{29}{36}$

THINK CRITICALLY

31. P (3 green) = 0.216, P (1 red) = 0.432, P (2 red) = 0.288, P (3 red) = 0.064

35. $\frac{648}{3125}$ or 0.207 or 21%; 6 games

MIXED REVIEW **36.** $12\sqrt{3}$ **37.** $6a^3b^4 \cdot \sqrt{2b}$

38. $10\sqrt[3]{2}$ **39.** C **40.** $x = 25$ **41.** $x = 27$
42. $x = 3$

Lesson 14.7, pages 717–723

TRY THESE **1.** One method is to let the outcomes 1, 3, 5 represent "the drug works" and the outcomes 2, 4 represent "the drug does not work." Disregard the outcome 6. One trial is to roll one cube twice.
2. Toss 6 coins at once. For each coin, let heads represent black and tails represent white.
5. One method is to represent the number of days in

December by choosing 31 digits at random from a table of random numbers, accepting only the digits 1–7 as choices. Let the occurrence of a 1 or a 2 represent catching a cold and the occurance of 3, 4, 5, 6, or 7 represent not catching a cold.

PRACTICE

3. $P = {}_5C_4\left(\frac{4}{5}\right)^4\left(\frac{1}{5}\right)^1$ or 0.41 **5.** ${}_{20}C_{10}\left(\frac{2}{5}\right)^{10}\left(\frac{3}{5}\right)^{10}$ or 0.12

7. ${}_{30}C_{20}\left(\frac{3}{5}\right)^{20}\left(\frac{2}{5}\right)^{10}$ or 0.12

13. $\frac{{}_{10}C_7}{2^{10}}$ or ${}_{10}C_7\left(\frac{1}{2}\right)^7\left(\frac{1}{2}\right)^3 \approx 0.117$

15. ${}_{96}C_0\left(\frac{5}{6}\right)^{96} + {}_{96}C_1\left(\frac{1}{6}\right)\left(\frac{5}{6}\right)^{95} + \ldots + {}_{96}C_{10}\left(\frac{1}{6}\right)^{10}\left(\frac{5}{6}\right)^{86}$

THINK CRITICALLY **19.** for i: 1 . . . 100
21. randint (coin, 0, 1)
MIXED REVIEW **23.** $4\sqrt{3}$ **24.** $3.5\sqrt[3]{3}$ **25.** B

Lesson 14.8, pages 724–727

APPLY THE STRATEGY **7.** Poll the whole population. The data can be found by a show of hands in each senior homeroom. **9.** too small a sample; people walking by building may have business in building and may not be representative. **11.** not random sampling; if knowing that every 20th appliance is to be inspected, workers take more care with these particular appliances.

Chapter Review, pages 728–729

1. d **2.** a **3.** b **4.** e **5.** c **6.** 720 **7.** 15

8. $\frac{7}{10}$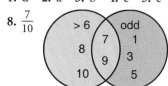

9. $\frac{1}{4}$ **10.** $\frac{25}{102}$ **11.** $\frac{2}{5}$ **12.** 68% **13.** 13.5%

14. $\frac{25}{216}$ or 11.6% **15.** $\frac{171}{1296}$ or 13.2% **16.** 0.001

17. One model is to use a standard deck of cards. Let getting a spade represent "does not cure" and getting a card or any other deck represent "does cure". One trial is to pick one card from the full deck.
18. 0.144 **19.** 30, 15, 5
20.

Sample of Size 2	Mean
{7, 10}	8.5
{7, 13}	10.0
{7, 18}	12.5
{10, 13}	11.5
{10, 18}	14.0
{13, 18}	15.5

For the set of sample means, the mean is 12.0 and the standard deviation is 2.345.

Chapter 15 Trigonometric Functions and Graphs

Lesson 15.1, pages 737–742

TRY THESE 1. $180° < \theta < 270°$ 3. I 5. $\frac{\pi}{6}$ 7. $\frac{10\pi}{9}$
9. $225°$ 11. 17.0 in.

PRACTICE 1. $270° < \theta < 360°$ 3. III 5. $\frac{\pi}{2}$
7. $\frac{11\pi}{6}$ 9. $\frac{5\pi}{4}$ 11. $\frac{17\pi}{9}$ 13. $\frac{3\pi}{10}$ 15. $\frac{9\pi}{20}$ 17. $360°$
19. $270°$ 21. $45°$ 23. $247.5°$ 25. $320°$ 27. $171°$
29. $42°$ 31. $\frac{\pi}{18}$ 33. $\frac{7\pi}{12}$ 35. $\frac{5\pi}{6}$, 5π in.
37a. 0.017 37b. $\approx 1{,}600{,}000$ mi

EXTEND
41. $80.4°$ 43. 48.77; $\approx 0.271\pi$
45. 672.27 revolutions

THINK CRITICALLY 47. Equations may vary but
should be in the form $\frac{r}{\pi} = \frac{d}{180}$.

MIXED REVIEW
49. $27.8, 28, 19$ 50. $0.24, 0.2$, none
51. $4\frac{1}{4}, 4\frac{1}{8}$, none
52. $f^{-1}(x) = \frac{x}{5}$ 53. $f^{-1}(x) = x + 6$
54. $f^{-1}(x) = -3x$ 55. $f^{-1}(x) = \frac{4x - 2}{3}$
56. 13 units 57. $(3, 1.5)$ 58. B

Lesson 15.3, pages 746–753

TRY THESE
1. $\sin -\frac{5}{13}$; $\cos \frac{12}{13}$; $\tan -\frac{5}{12}$; $\csc -\frac{13}{5}$; $\sec -\frac{13}{12}$; $\cot -\frac{12}{5}$
3. $-\sqrt{2}$ 5. 0 7. 0.9998 9. 1.5557 11. positive
13. negative

PRACTICE
1. $\sin -\frac{8}{17}$, $\cos \frac{15}{17}$, $\tan -\frac{8}{15}$, $\csc -\frac{17}{8}$, $\sec \frac{17}{15}$, $\cot -\frac{15}{8}$
3. $\sin \frac{2\sqrt{5}}{5}$, $\cos \frac{\sqrt{5}}{5}$, $\tan 2$, $\csc \frac{\sqrt{5}}{2}$, $\sec \sqrt{5}$, $\cot \frac{1}{2}$
5. $-\frac{\sqrt{2}}{2}$ 7. $\sqrt{3}$ 9. $-\sqrt{3}$ 11. $-\frac{2\sqrt{3}}{3}$ 13. $-\sqrt{3}$
15. 1 17. 0.9877 19. -0.4960 21. ≈ 1.49

EXTEND 25. $\frac{13}{5}, -\frac{13}{12}$ 27. $135°, 225°$

THINK CRITICALLY 31a. positive 31b. positive if θ is
in Quad. I, negative if θ is in Quad. III
MIXED REVIEW 33. $\frac{\pi}{40}$ 34. -4 35. undefined
36. 0 37. $\frac{1}{2}$ 38. $\frac{5}{2}$ 39. -1 40. A

Lesson 15.4, pages 754–757

APPLY THE STRATEGY 15. 3.2 mi 17. 183.7 ft

Lesson 15.5, pages 758–764

TRY THESE 1. $b = 15.2$, $c = 20.3$, $\angle C = 96°$
3. 5.4 cm^2 5. 52.9 yd

PRACTICE
1. $a = 4.1$, $c = 4.9$, $\angle A = 24.8°$ 3. $\angle A = 43.9°$,
$\angle B = 38.3°$, $\angle C = 97.8°$ 5. $b = 10.3$, $\angle A = 48.4°$,
$\angle C = 57.3°$ 7. (ambiguous case) $m\angle C = 41.8°$,
$m\angle B = 108.2°$, $b = 11.4$, OR $m\angle C = 138.2°$,
$m\angle B = 11.8$, $b = 2.5$ 9. 60.2 11. 160.5 mi
13. $\$45.82$ 15. 24 mi or 106 mi

EXTEND 17. $24, 24$ 19. $60, 60$ 21. $36, 36$

THINK CRITICALLY 25. none

MIXED REVIEW 26. $\$1275.36$ 27. $5, 8$ 28. $\frac{9}{5}, -\frac{3}{5}$
29. $\frac{5}{4}, \frac{3}{2}$ 30. $y = (x - 3)^2 - 1$; vertex $(3, -1)$,
y-intercept $(0, 8)$, x-intercepts $(2, 0)$ and $(4, 0)$ 31. C

Lesson 15.7, pages 770–777

TRY THESE 1. $1, \frac{2\pi}{3}$ 3. $2, \pi$ 5. down 3 units
7. right $\frac{10\pi}{9}$

9.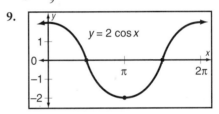

11.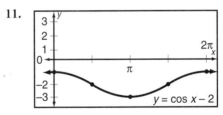

13a. -4 in 13b. 4 in 13c. 0.5 sec 13d. 2 radians
13e.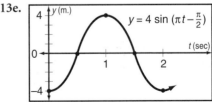

PRACTICE
1. $1, 2\pi$ 3. $1, \pi$ 5. $3, \frac{\pi}{2}$ 7. $9, 10\pi$
9. down 2 units 11. left π units 13. left $75°$
15. right $30°$ and down 4 units

17.

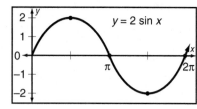

19.

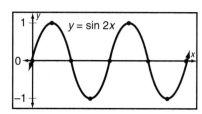

21.

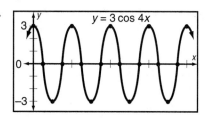

23.

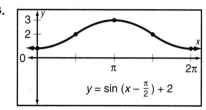

25a. the second   **25b.** $\frac{2}{55}$ sec; 27.5 vibrations/sec

EXTEND

27. $\frac{\pi}{4}$ and $\frac{5\pi}{4}$; -1 **29.** $\frac{5\pi}{6}, \frac{7\pi}{6}$ **31.** $\approx 1.29, 4.34$

THINK CRITICALLY

33. $y = 3 \sin 2x$ **35.** $y = 2 \sin \frac{4}{3}x + 4$

37. $y = \sin\left(\frac{4}{5}x - \pi\right) + 1$

MIXED REVIEW

39. $\frac{2A}{h} - b_1$ **40.** $\frac{A}{1 + rt}$ **41.** $\frac{Fr}{v^2}$

42. $(x + 9)(x - 9)$ **43.** $(x + 3)^2$

44. $(2x + 3)(3x - 1)$ **45.** $3(x + 5)(x - 2)$

46. $(-2, 3)4$ **47.** A   **48.** $(1, 2)$

Lesson 15.8, pages 778–781

TRY THESE   **1.** $a = \text{Tan}^{-1}b$ **3.** $\frac{\pi}{4}$ **5.** $\frac{\pi}{4}$ **7.** $\frac{\pi}{3}$

9. $\frac{5}{12}$ **11.** 42°   **13.** $\approx 20.6°$

PRACTICE   **1.** $\frac{\pi}{3}$ **3.** $\frac{5\pi}{6}$ **5.** $\frac{3\pi}{4}$ **7.** 23.6°   **9.** 0

11. $\frac{\sqrt{3}}{2}$ **13.** 36.9°, 53.1°   **15.** 61.9°, 28.1°   **17.** 48.2°

EXTEND   **21.** -1 **23.** $\frac{\sqrt{2}}{2}$

THINK CRITICALLY   **25.** 0.2408, -0.9946

MIXED REVIEW   **27.** distributive   **28.** additive identity
29. commutative of addition   **30.** 49   **31.** 7   **32.** 1
33. B

Chapter Review, pages 782–783

1. e   **2.** d   **3.** a   **4.** c   **5.** b   **6.** $\frac{\pi}{2}$ **7.** $\frac{5\pi}{4}$

8. $\frac{-5\pi}{6}$ **9.** 5π **10.** 135°   **11.** 126°   **12.** 120°

13. $-288°$ **14.** 140°   **15.** 347°   **16.** 315°
17. 241°   **18.** 0.5   **19.** -1 **20.** $-\sqrt{3}$ **21.** $\sqrt{2}$
22. 1.0355   **23.** 0.4695   **24.** -2.3048 **25.** 0.3090
26. ≈ 548.5 ft   **27.** $a = 19.9, c = 18.0, C = 64.1°$
28. $A = 102.2°, B = 27.0°, C = 50.8°$

29. $3, 2\pi$ **30.** $2, 2\pi$ **31.** $1, \frac{\pi}{2}$ **32.** $5, 6\pi$

33.

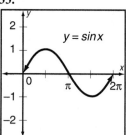

34.

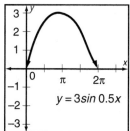

35.

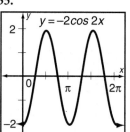

36.

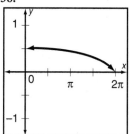

37. $\frac{\pi}{6}$ **38.** $\frac{3\pi}{4}$ **39.** 0.8944

Chapter 16 Trigonometric Identities and Equations

Lesson 16.1, pages 791–797

TRY THESE   **1.** 1   **3.** -1 **7.** $\sin \theta$ **9.** $\sin^2 \theta$
11. $\cos \beta$ **13.** $-\tan^2 \beta$

PRACTICE **1.** -1 **3.** -1 **7.** $\cos^2 \theta$ **9.** $\cos^2 \theta$

11. $-\sin \beta$ **13.** $-\cot^2 \beta$

29a. $-\cot \theta_p = \dfrac{n_1}{n_2}$ **29b.** $67.5°$

EXTEND **31.** Yes **33.** no **35.** $\theta = 300°$

37. $\theta = 330°$

MIXED REVIEW **43.** 625 **44.** 3 **45.** D

46. 752,538,150

47. $\dfrac{1}{1 - \cos \theta} + \dfrac{1}{1 + \cos \theta} =$

$\dfrac{1 + \cos \theta}{(1 - \cos \theta)(1 + \cos \theta)} = \dfrac{1 - \cos \theta}{(1 - \cos \theta)(1 + \cos \theta)} =$

$\dfrac{2}{1 - \cos^2 \theta} = \dfrac{2}{\sin^2 \theta} = 2 \csc^2 \theta$

48. $\dfrac{\sin \beta}{\tan \beta} = \dfrac{\sin \beta}{\dfrac{\sin \beta}{\cos \beta}} = \dfrac{\cos \beta \sin \beta}{\sin \beta} = \cos \beta$

Lesson 16.2, pages 798–803

TRY THESE **1.** 2 **3.** 4 **7.** $60°, 120°$ **9.** $60°, 300°$

11. $\dfrac{\pi}{3}, \dfrac{5\pi}{3}$ **13.** $0.12, 1.19, 3.27, 4.33$ **15.** $\dfrac{\pi}{3} + n\pi$

17. $0.41 + 2n\pi, 2.89 + 2n\pi$

PRACTICE **1.** 2 **3.** 2 **7.** $45°, 135°$ **9.** $210°, 330°$

11. $\dfrac{\pi}{6}, \dfrac{5\pi}{6}, \dfrac{7\pi}{6}, \dfrac{11\pi}{6}$ **13.** $0.17, 2.97$

15. $0, 2.94, \pi, 6.09$ **17.** $0.34 + 2n\pi, 2.80 + 2 n\pi$

19. $\dfrac{\pi}{3} + 2n\pi, \dfrac{5\pi}{3} + 2n\pi$ **21.** $3.50, 5.93$

23. $2.01, 4.28$ **25.** $45°$

EXTEND **27.** $90°$ **29.** $45°, 90°, 225°, 270°$ **31.** 0.33 s

THINK CRITICALLY **33.** 2 **35.** $30° \leq x \leq 330°$

37. $30° \leq x \leq 150°$ and $210° \leq x < 330°$

MIXED REVIEW **38.** $\begin{bmatrix} 22 & 9 \\ 2 & -6 \\ 3 & 9 \end{bmatrix}$ **39.** $\begin{bmatrix} 16 & 40 \\ 12 & 30 \\ 6 & 15 \end{bmatrix}$ **40.** e^6

41. $3e^6$ **42.** C **43.** $0.12, 3.02$ **44.** $1.93, 4.35$

45. $8 - 5i$ **46.** $-i$ **47.** $-4i$

Lesson 16.3, pages 804–811

TRY THESE **1.** $\sin 10°$ **3.** $\tan \dfrac{6\pi}{7}$ **5.** $\dfrac{-\sqrt{6} + \sqrt{2}}{4}$

7. $2 - \sqrt{3}$ **9.** $\dfrac{\sqrt{6} - \sqrt{2}}{4}$ **13.** $\dfrac{56}{65}$ **15.** 0.47

17. $\dfrac{3\pi}{2}$

PRACTICE **1.** $\sin \dfrac{2\pi}{9}$ **3.** $\tan 137°$ **5.** $\dfrac{-\sqrt{6} + \sqrt{2}}{4}$

7. $\dfrac{-\sqrt{6} - \sqrt{2}}{4}$ **9.** $\dfrac{\sqrt{2} - \sqrt{6}}{4}$ **13.** $\dfrac{23\sqrt{2}}{34}$

15. 0.66 **17.** $\dfrac{\pi}{4}, \dfrac{7\pi}{4}$ **19.** $2.18, 4.10$ **21.** $1.45, 3.26$

EXTEND

33. $\sin 72° + \sin 48°$ **35.** $\cos 57° + \cos 87°$

MIXED REVIEW **43.** C **44.** $4 \pm \sqrt{21}$

45. $\dfrac{7 \pm \sqrt{41}}{4}$

46. $64x^6 + 576x^5 + 2160x^4 + 4320x^3$
$+ 4860x^2 + 2916x + 729$

47. $\tan 89°$ **48.** $\tan 100°$

Lesson 16.4, pages 812–820

TRY THESE **1.** $\dfrac{\sqrt{2 + \sqrt{3}}}{2}$ **3.** $\sqrt{2} - 1$ **5.** $-\dfrac{\sqrt{2 + \sqrt{2}}}{2}$

9. $\dfrac{527}{625}$ **11.** $\dfrac{7\sqrt{2}}{10}$ **13.** 7 **15.** $\dfrac{2\pi}{3}, \pi, \dfrac{4\pi}{3}$ **17.** $0, \pi$

21. 23 ft

PRACTICE **1.** $\dfrac{\sqrt{2 + \sqrt{3}}}{2}$ **3.** $2 + \sqrt{3}$ **5.** $-\dfrac{\sqrt{2 - \sqrt{2}}}{2}$

9. $\dfrac{1519}{1681}$ **11.** $\dfrac{\sqrt{82}}{82}$ **13.** $-\dfrac{1}{9}$ **15.** $-\dfrac{119}{169}$ **17.** $\dfrac{3\sqrt{13}}{13}$

19. $-\dfrac{3}{2}$ **21.** $\dfrac{2\pi}{3}, \dfrac{4\pi}{3}$ **23.** $0, \pi$ **31.** 110 ft

EXTEND **33.** $4\cos^3 \theta - 3\cos \theta = \cos 3\theta$ **35a.** 44 in.^{2}

35b. 7 m^2 **35c.** 8 in.^{2}; $\dfrac{\pi}{2}$ or $90°$ **37.** $90°$ or $\dfrac{\pi}{2}$

THINK CRITICALLY **39.** $45°$

MIXED REVIEW **42.** $-4; 2$ **43.** $128; -5$ **44.** 1.594

45. 0.631 **46.** D **47.** $10,160$ **48.** $\dfrac{24}{25}$

Lesson 16.5, pages 821–827

TRY THESE

1–4.

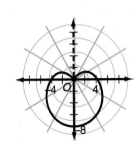

5. $P(2\sqrt{2}, 315°)$

7. $R(41, 77°)$

11. $Q(-3, 3\sqrt{3})$

13. $S(1, \sqrt{3})$

15. $13(\cos 293° + i \sin 293°)$

17. $10(\cos 0° + i \sin 0°)$ **19.** $\dfrac{1}{2} + \dfrac{i\sqrt{3}}{2}$

21. $-3\sqrt{3} + 3i$

23. heart-shaped

1–4.

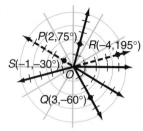

5. $P(5\sqrt{2}, 45°)$ **7.** $R(8, 0°)$ **9.** $T(4, 120°)$
11. $V(41, 257°)$ **15.** $Q(1, \sqrt{3})$ **17.** $S\left(-\dfrac{5\sqrt{2}}{2}, \dfrac{5\sqrt{2}}{2}\right)$
19. $41(\cos 283° + i\sin 283°)$
21. $10(\cos 53° + i\sin 53°)$
23. $15(\cos 180° + i\sin 180°)$
25. $4(\cos 270° + i\sin 270°)$ **27.** $-4\sqrt{2} - 4i\sqrt{2}$
29. $1 - i\sqrt{3}$ **31.** 6
33. $40\sqrt{x^2 + y^2} - 2x = 225{,}000$

EXTEND   **35.** $72(\cos 140° + i\sin 140°)$
37. $49\left(\cos \dfrac{19\pi}{12} + i\sin \dfrac{19\pi}{12}\right)$
39. $4(\cos 40° + i\sin 40°)$
41. $5\left(\cos \dfrac{5\pi}{12} + i\sin \dfrac{5\pi}{12}\right)$

43. 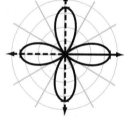   **45.**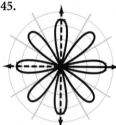

THINK CRITICALLY
47. The polar axis.   **49.** The line $\theta = \dfrac{\pi}{2}$
MIXED REVIEW   **51.** $\dfrac{1}{40}$, 320 **52.** $\log_4 6 + 2\log_4 f$
53. $\log_5 u - \log_5 v$ **54.** B   **55.** $x = 8$
56. $x = -25$ **57.** $-6\sqrt{3} + 6i$
58. $-\dfrac{15\sqrt{2}}{2} + \dfrac{15i\sqrt{2}}{2}$

Lesson 16.6, pages 828–831
APPLY THE STRATEGY   **13.** 25 lb   **15.** $82°$

Chapter Review, pages 832–833
1. c   **2.** d   **3.** b   **4.** a
5. $\cos\theta \cot\theta + \sin\theta = \cos\theta \left(\dfrac{\cos\theta}{\sin\theta}\right) + \sin\theta =$
$\dfrac{\cos^2\theta}{\sin\theta} + \dfrac{\sin^2\theta}{\sin\theta} = \dfrac{1}{\sin\theta} = \csc\theta$

6.

$\dfrac{\csc\beta + 1}{-\csc\beta}$	$\dfrac{1 - \sin^2\beta}{\sin\beta - 1}$
$\dfrac{\dfrac{1}{\sin\beta} + 1}{-\dfrac{1}{\sin\beta}}$	$\dfrac{(1 - \sin\beta)(1 + \sin\beta)}{\sin\beta - 1}$
$\dfrac{1 + \sin\beta}{\sin\beta}(-\sin\beta)$	$-1(1 + \sin\beta)$
$-1 - \sin\beta$	$-1 - \sin\beta$

7. $\dfrac{\pi}{6}, \dfrac{5\pi}{6}, \dfrac{7\pi}{6}, \dfrac{11\pi}{6}$ **8.** $\dfrac{2\pi}{3}, \dfrac{4\pi}{3}$
9. $1.19, 5.09$ **10.** $3.30, 6.12$
11. $\dfrac{13}{85}$ **12.** $\dfrac{36}{85}$ **13.** $-\dfrac{36}{77}$ **14.** $\dfrac{\sqrt{2} + \sqrt{6}}{4}$
15. $\dfrac{\sqrt{2} - \sqrt{6}}{4}$ **16.** $\dfrac{3 - \sqrt{3}}{3 + \sqrt{3}}$
17. $\sin\left(\dfrac{3\pi}{2} + \alpha\right) = \sin\dfrac{3\pi}{2} + \cos\alpha +$
$\cos\dfrac{3\pi}{2}\sin\alpha = (-1)\cos\alpha + (0)\sin\alpha = -\cos\alpha$
18. $\tan\left(\dfrac{3\pi}{4} - \beta\right) = \dfrac{\tan\dfrac{3\pi}{4} - \tan\beta}{1 + \tan\dfrac{3\pi}{4}\tan\beta} =$
$\dfrac{-1 - \tan\beta}{1 + (-1)\tan\beta} = \dfrac{-1 - \tan\beta}{1 - \tan\beta} = \dfrac{\tan\beta + 1}{\tan\beta - 1}$
19. $\dfrac{720}{1681}$ **20.** $\dfrac{720}{1519}$ **21.** $-\dfrac{\sqrt{82}}{82}$
22. $-\dfrac{\sqrt{2 + \sqrt{2}}}{2}$ **23.** $-\dfrac{\sqrt{2 + \sqrt{3}}}{2}$ **24.** $-1 - \sqrt{2}$

25. $P(18, 120°)$ **26.** $Q(12\sqrt{2}, 315°)$ **27.** $R\left(\dfrac{5\sqrt{3}}{2}, \dfrac{5}{2}\right)$
28. $S(-4\sqrt{2}, 4\sqrt{2})$ **29.** $-4 + 4i\sqrt{3}$
30. $-2\sqrt{3} + 2i$ **31.** $17(\cos 152° + i\sin 152°)$
32. $41(\cos 77° + i\sin 77°)$
33. $5(\cos 180° + i\sin 180°)$
34. $12(\cos 0° + i\sin 0°)$
35. $v_e = 13.4$ mi/h, $v_n = 12.0$ mi/h
36. 309.7 mi/h, $64°$

INDEX

food, 191, 456, 693

fractal geometry, 407–408

genetics, 248, 714

geography, 117, 570, 675, 749, 752

geology, 534

geometry, 16, 24, 67, 115, 125,
141–142, 173–174, 188, 193, 195,
207, 215, 238–239, 243, 246–248,
251, 258–259, 279, 299, 305–306,
338, 378, 388, 397, 476–477, 483,
484, 496–497, 527, 534, 535, 553,
568–571, 591, 607–608, 655,
682–683, 741, 763, 777, 796, 802,
818, 830

graphing, 489

health, 30, 328, 528, 700, 720

history, 64–65

housing, 122, 452

international standards, 84–85

linguistics, 693

manufacturing, 21, 91, 170, 208, 250,
278, 292, 351–352, 395, 451, 502,
534, 544, 551, 637, 693, 762

measurement, 154, 756, 762, 816

mechanics, 497, 651

medicine, 79, 692, 701, 774

metallurgy, 66, 356

meteorology, 35, 528, 692, 720, 777

mixture problems, 49, 172–173, 178,
180–181, 187, 204–206, 357

modeling, 61, 66, 70, 109, 136,
181–182, 218, 246–247, 323, 348,
354, 373, 516, 574, *see also*
Algeblocks Models

money and finance, 80, 148–149,
154, 186, 188, 207, 326, 360,
411–412, 431, 433, 462, 477, 630,
638, 685

mortality rates, 433

motion, 476, 483, 502, 629, 635, 637,
802

music, 452, 671, 683, 776, 801

navigation, 389, 577, 579, 606–608,
739, 741–742, 750, 754, 756,
762–763, 780

news reporting, 92, 683

number theory, 258–259, 264,
277–278, 300–301, 306, 328, 361,
434, 481, 483–484, 552, 608, 656,
660, 675

oceanography, 525, 591

office supplies, 78

optics, 751, 794, 796

ore production, 194

Pascal's triangle, 684

personal finance, 6–11, 14, 17, 23, 75,
79, 125, 173, 523

pharmacology, 343, 451, 720

photography, 522, 526, 539, 545, 699

physics, 48, 141, 154, 291, 293, 307,
336, 378, 389, 397, 439, 446–447,
452, 457, 529, 650–651, 740, 751,
775, 802, 823, 830

politics, 329, 682, 691, 715

population, 86, 167, 430, 433, 452,
457

pricing, 5, 9, 15, 34, 40, 43, 95–96,
147–149, 187

probability, 686–701, 704, 706–715,
717–722, 727–733

quality control, 690, 712, 721

safety, 683, 720

sales, 22–23, 116, 172, 201

seismology, 445, 607

semaphore, 676, 700

space science, 71, 450, 453, 826

speed, 27, 37, 43, 173, 538

sports and recreation, 16, 27, 29, 33,
35–36, 43–44, 72, 79, 117, 192,
301, 342, 388, 527, 539, 552, 561,
579, 636, 675–676, 680–684,
700–701, 708, 713–715, 720–721,
740, 781, 808, 816, 817–818, 830

statistics, 27, 29, 36

surveying, 570, 698, 756, 761

technology, 64, 67, 71, 117, 199, 260,
600, 721

three-dimensional coordinate
system, 182

transportation, 79, 342, 522, 535,
538, 571, 589, 715, 781

travel, 78, 113, 279, 584, 721, 741

urban planning, 291

velocity problems, 48, 308–310, 313,
335–336, 377, 389, 391, 447, 499,
528, 540, 552, 562, 611, 751, 781,
816–819, 829–830, 834

wages, 201, 214, 327, 628, 642–643

weather, 78, 140, 214

wildlife, 756, 818

See also Careers

Approach as a limit, 647

Arc, 738, 739, 846

Argument, 824, 846

Arithmetic
mean, 634, 638, 846
sequences, 633–638, 662, 846
series, 639–645, 846

Assessed value, 111

Assessment
Chapter Assessment, 52–53,
100–101, 158–159, 222–223,
268–269, 314–315, 364–365,
416–417, 466–467, 506–507,

560–561, 616–617, 664–665,
730–731, 784–785, 834–835

Chapter Test, 52, 100, 158, 222, 268,
314, 364, 416, 466, 506, 560, 616,
664, 731, 784, 834

Performance Assessment, 53, 101,
159, 223, 269, 315, 365, 417, 467,
507, 561, 617, 665, 731, 785, 835

Project Assessment, 53, 101, 159,
223, 269, 315, 365, 417, 467, 507,
561, 617, 665, 731, 785, 835

Standardized (Cumulative) Test, *see*
Standardized Tests

Associative property, 69, 98, 190

Asymptote, 514, 518, 587, 767, 846
horizontal, 424, 519
slant, 522
vertical, 519

Augmented matricies, 183–188, 846

Average, 51

Axis, 405, 415, 581, 587, 592, 823, 846

• • **B** • •.

Bar graph, 90, 846

Base, 234

Bearing, 829, 846

Bernoulli experiment, 711

Binomial(s), 230, 846
distribution, 710–716
experiment, 710, 846
multiplying FOIL, 237
probability, formula for, 712
theorem, 653–657, 663, 846

Bond, 363, 846

Boundary of two half-planes, 362

Box-and-whiskers plot, 34

Boxplot, 34, 846

Boyle's law, 525, 591

Branches, 514, 587, 846

Brewster's law, 794

British thermal unit, 89

• • **C** • •

Calculator, 6–11, 13–24, 27–31, 35–37,
40–45, 47–48, 60, 63–64, 67, 112,
119, 204, 276, 287, 307, 371–372,
374, 376–378, 410, 429, 431–432,
434, 437, 441, 443, 449, 451, 487,
491–492, 495, 547, 557, 568, 626,
630, 656, 673, 684, 719, 725–726,
749, 751, 760, 762, 780, 782, 791,
792, 799, 805, 824
factorial key, 449
second function key, 449
See also Graphing calculator

• • E • •

• • F • •

INDEX

trigonometric, 748
Rational equations, 510–561, 855
Rational exponents, 374–380, 855
Rational expressions, 510–561, 855
  adding, 536–541
  complex, 543
  dividing, 531–535
  multiplying, 531–535
  simplest form, 531
  simplifying, 531–535, 543–547
  subtracting, 536–541
Rational functions, 513, 558, 855
  graphing, 517–523
Rationalizing the denominator, 386, 414, 855
Rational number(s), 62, 98, 855
Rational zero theorem, 492, 855
Real
  axis, 415, 823
  part, 415
  zeros, 486, 504, 855
Real number(s), 56–102, 855
Reciprocal, 69, 121, 747
  functions, 747, 855
  identities, 792
Rectangular hyperbola, 589, 855
Recursive formula, 627, 662, 855
Reference
  angle, 749, 855
  triangle, 746, 782, 855
Reflection, 126, 855
Reflexive property, 74
Regression line, 42, 855
Relations, 112–118, 156, 778, 855
  inverse, 137, 157, 778
Remainder theorem, 481, 855
Resolve, 828, 855
Restricted, 517
Resultant, 828, 855
Review
  Chapter Review, 50–51, 98–99, 156–157, 220–221, 266–267, 312–313, 362–363, 414–415, 464–465, 504–505, 558–559, 614–615, 662–663, 728–729, 782–783, 832–833
  Cumulative Review, 54, 102, 160, 224, 270, 316, 366, 418, 468, 508, 562, 618, 666, 732, 786, 836
  Mixed Review, 10, 17, 24, 30, 37, 45, 67, 72, 80, 87, 93, 118, 125, 142, 150, 174, 182, 188, 195, 201, 208, 215, 240, 248, 253, 260, 280, 294, 302, 307, 320, 338, 345, 356, 379, 390, 398, 404, 409, 434, 441, 448, 453, 458, 478, 485, 498, 523, 535,

541, 547, 553, 571, 580, 585, 591, 596, 602, 608, 631, 638, 644, 651, 657, 677, 685, 693, 702, 709, 715, 723, 742, 752, 763, 777, 781, 797, 803, 810, 819, 827
  Review Problem Solving Strategies, 49, 97, 155, 219, 265, 311, 361, 413, 463, 503, 557, 613, 661, 727, 757, 831
Richter scale, 442, 855
Rotation, angle of, 737, 782
Row operations, 184, 855

• • S • •

Sample, 25, 721, 724, 729, 855
Sampling, 724–727, 855
Scalar, 20, 828, 855
Scalar multiplication, 20, 855
Scatter plots, 40, 855
Scientific notation, 46–48, 855
Secant, 746, 782, 855
Second degree equation, general form of, 599
Self-similar, 644
Semicircle, 580
Semi-major axis, 610, 855
Semi-minor axis, 610, 856
Semiperimeter, 391, 761
Sequences, 620–665, 856
Series, 620–665, 856
Sigma, 641
  notation, 654, 856
Signs, Descartes' rule of, 494
Simple harmonic motion, 774
Simple interest, 460, 856
Simplest form, 376, 385, 531, 559, 856
Simulations, 717–723, 729, 856
Sine, 746, 758, 759, 771, 772, 779, 783, 782, 783, 856
Skill Focus, 2, 56, 104, 162, 226, 272, 318, 368, 420, 470, 510, 564, 620, 668, 734, 788
Slant asymptote, 522
Slope, 39, 51, 856
Slope-intercept form, 39, 119, 156, 347, 856
Solution, 257, 486, 856
Spreadsheet(s), 12–18, 89–93, 856
Square matrix, 19, 856
Square root, 63, 374
  perfect square, 63
  principal, 63
Square window, 577
Standard deviation, 704, 709, 725, 729, 856

Standard form, 120, 156, 856
  of the equation of a circle, 576, 614
  of the equation of a hyperbola, 587
  of the equation of an ellipse, 582–583
  of the equation of a parabola, 592–593
Standardized Tests, 17, 30, 45, 52, 67, 72, 80, 93, 100–102, 118, 125, 142, 150, 158, 160, 174, 182, 188, 195, 201, 208, 215, 222, 224–225, 240, 248, 253, 259–260, 268, 270–271, 280, 294, 302, 307, 314, 317, 330, 338, 345, 356, 364, 366–367, 379, 390, 398, 404, 409, 416, 418–419, 434, 441, 448, 453, 458, 466, 468–469, 485, 498, 506, 507–509, 523, 535, 541, 547, 553, 562, 563, 571, 580, 585, 591, 596, 602, 608, 616, 618–619, 631, 638, 644, 651, 657, 664, 666–667, 677, 685, 693, 702, 708–709, 715, 723, 730, 732–733, 742, 752, 763, 777, 781, 784, 786–787, 797, 803, 810, 819, 827, 834, 836–837
  Quantitative Comparison, 225, 419, 619, 619, 837
  Standard Five-Choice, 55, 161, 271, 367, 469, 563, 667, 787, 837
  Student Produced Answers, 103, 317, 509, 733
Standard normal
  curve, 705
  distribution, 705, 729
Standard position, 737, 782, 856
Statistics, 668–731
Stratified sample, 724, 729, 856
Substitution method, 168, 220, 856
Substitution property, 74
Subtraction property
  of equality, 5, 74
  of inequality, 75
Sum and difference identities, 804–811
Summation notation, 641, 856
Sum of solutions of a quadratic equation, 298
Supplementary angles, 739
Symmetric, 74, 107, 156
Symmetry
  axis of, 107, 287, 592
  lines of, 107
Synthetic division, 479–485, 856
System(s)
  consistent, 171, 179
  dependent, 171, 179
  inconsistent, 171, 179
  independent, 171, 179

INDEX

Photo Credits

p. 186: ©Charles Thatcher/Tony Stone Images; p. 187: ©Bob Daemmrich/Stock, Boston; p. 189: ©Rob Crandall/Stock, Boston; p. 191: FMC Corporation; p. 194: College of Mount St. Joseph, Cincinnati, Ohio; p. 199: ©Greg Grosse/Courtesy of Service Merchandise, Inc.; p. 202: ©Kaluzny/Thatcher/Tony Stone Images; p. 204: ©Marvy!/The Stock Market; p. 208: Alan Brown/Photonics; p. 209: ©Walter Hodges/Tony Stone Images (top); p. 214: Sygma Photo News; p. 218: ©Steven Hunt/The Image Bank.

CHAPTER 5

p. 226: ©Steven Peters/Tony Stone Images (top); ©Alan Brown/Photonics (bottom); p. 227: ©Telegraph Colour Library/FPG International; p. 228: ©G & V Chapman/The Image Bank; p.230: ©Greg Grosse: p. 233: ©Greg Grosse; p. 237: ©Randy Duchaine/The Stock Market; p. 238: ©Phyllis Picardi/Stock, Boston; p. 241: ©David Woods/The Stock Market; p. 246: ©Greg Grosse; p. 248: ©Telegraph Colour Library/FPG International; p. 253: ©James Marshall/The Stock Market; p. 258: ©Greg Grosse (top); ©Cathlyn Melloan/Tony Stone Images (bottom); p. 261: Courtesy Hewlett-Packard Company (top); Courtesy Hewlett-Packard Company (bottom); p. 263: ©Paul Barton/The Stock Market.

CHAPTER 6

p. 272: The Image Bank (top); ©Poulides/Thatcher/Tony Stone Images (bottom); p. 273: ©David Sailors/The Stock Market; p. 274: The Image Bank; p. 275: ©Greg Grosse; p. 277: ©J. Buehner/H. Armstrong Roberts; p. 278: Courtesy of BATUS Group of Companies—Marshall Field & Co.; p. 279: Photograph by Globus-Gateway; p. 281: ©Dag Sunberg/The Image Bank; p. 283: ©Greg Grosse; p. 285: ©Greg Grosse; p. 286: ©Bob Daemmrich/Stock,Boston; p. 289: ©Karl Hentz/The Image Bank; p. 291: © Marc Romanelli/The Image Bank; p. 293: ©James D'Addio/The Stock Market; p. 295: ©Jon Riley/Tony Stone Images (right); p. 297: Courtesy of The Oriental Institute Museum of The University of Chicago; p. 299: The Phoenix, Cincinnati, Ohio; p. 301: ©Uniphoto, Inc.; p. 303: ©Greg Grosse; p. 305: Steelcase, Inc.; p. 307: ©Robert C. Hayes.

CHAPTER 7

p. 318: ©George Chan/Tony Stone Images (top); ©Bob Daemmrich/Stock, Boston (bottom); 319: ©Bob Daemmrich/Stock, Boston; p. 320: ©Greg Grosse; p. 322: ©Greg Grosse; p. 328: ©Jeff Kaufman/FPG International; p. 331: ©Bruce Ayres/Tony Stone Images; p. 333: ©J. Pinderhughes/The Stock Market; p. 336: ©Andy Sacks/Tony Stone Images; p. 337: ©Juan M. Silva/The Image Bank; p. 343: ©D. Young-Wolff/PhotoEdit; p. 347: ©Greg Grosse; p. 349: ©Reuters/Bettmann; p. 356: ©Paul Chesley/Tony Stone Images; p. 357: ©Gabe Palmer/The Stock Market; p. 358: ©Alan Brown/Photonics; p. 360: ©Frank Siteman/Stock, Boston.

CHAPTER 8

p. 368: ©Index Stock Photography, Inc. (top); ©Tom Carroll/FPG International (bottom); p. 369: ©Brownie Harris/The Stock Market; p. 370: NASA; p. 371: ©Greg Grosse; p. 376: ©Mike Vines/Tony Stone Images; p. 378: ©Jon Feingersh/The Stock Market; p. 380: ©Lawrence Migdale/Tony Stone Images (top); p. 381: ©Greg Grosse; p. 388: ©Ariel Skelley/The Stock Market; p. 391: UPI/Bettmann (top); NASA (bottom); p. 395: NASA; p.396:©Kent Knudson/FPG International; p. 397: ©Albert Normandin/The Image Bank; p. 398: NASA; p. 402: Hewlett-Packard Company; p.403: ©Stuart McCall/Tony Stone Images; p. 407: ©Vladimir Pcholkin/FPG International (top); ©Joe Devenney/The Image Bank (bottom); p. 410: ©The Image Bank; p. 412: ©Jim Cummins/FPG International.

CHAPTER 9

p. 420: ©Ed Honowitz/Tony Stone Images (top); ©Ed Honowitz/Tony Stone Images; p. 421: ©David Weintraub/Stock, Boston; ©Sally Grotta/The Stock Market (bottom); p. 422: ©Penny Gentieu/Tony Stone Images; p. 424: ©Greg Grosse; p. 425: ©Greg Grosse; p. 428: ©Superstock, Inc. p. 431: ©Alan Brown/Photonics; p. 432: ©The Telegraph Colour Library/FPG International; p. 435: ©The Telegraph Colour Library/FPG International; (top); ©David Joel/Tony Stone Images (bottom); p. 438: ©Ken Straiton/The Stock Market; p. 440: ©Charles Thatcher/Tony Stone Images; p. 442: ©1991 Michael Salas/The Image Bank (top); ©FPG International (bottom); p. 446: NASA; ©Alan Brown/Photonics; p. 451: ©Bruce Ayres/Tony Stone Images; p. 452: ©John Lamb/Tony Stone Images; ©Richard Pasley/Stock, Boston; p. 459:

©Nawrocki Stock Photo, Inc. (top); ©Murray & Associates/Tony Stone Images (bottom); p. 461: ©Greg Grosse; p. 462: ©Murray & Associates/Tony Stone Images.

CHAPTER 10

p. 470: ©Derek Berwin/The Image Bank; p. 471: ©William Taufic/The Stock Market; p.472: ©Doris De Witt/Tony Stone Images; p. 473: ©Greg Grosse; p. 478: ©Greg Grosse; p. 483: ©Gabe Palmer/The Stock Market; p. 486: ©Greg Grosse; p. 487: ©Anne Rippy/The Image Bank; p. 489: ©FMC Corporation; p. 499: ©Ulf Sjostedt/FPG International; p.500: ©Jose Carrillo/Stock, Boston; p. 501: Grant Heilman Photography, Inc.; p. 502: ©Lawrence Migdale/Stock, Boston.

CHAPTER 11

p. 510: ©Jose Luis Pelaez/The Stock Market (top); p. 511: ©G&M. David de Lossy/The Image Bank; p. 512: ©Chris Sorensen/The Stock Market; p. 513: ©Greg Grosse; p. 516: ©Jon Davison/The Image Bank; p. 517: ©Bob Daemmrich/Stock, Boston; p. 519: ©The New York Stock Exchange; p. 521: ©Heinley and Savage/Tony Stone Images; p. 522: ©Desmond Burden/Tony Stone Images; p. 523: ©Roy Morsch/The Stock Market (top); ©Greg Grosse (bottom); p. 527: ©Roy Morsch/The Stock Market; p. 530: ©A.M. Rosario/The Image Bank; p. 531: ©Greg Grosse; p. 533: Courtesy Silicon Graphics; p. 534: ©Martin Land/Science Photo Library/Photo Researchers, Inc.; p. 536: ©Charles Krebs/The Stock Market; p. 537: ©David Stoecklein/The Stock Market; p. 539: ©Michael Newman/PhotoEdit; p. 540: ©Joe Towers/The Stock Market; p. 542: ©David Burnett/Contact/The Stock Market; p. 543: ©Erik Von Fischer/Photonics; p. 545: ©Jeff Isaac Greenberg; p. 546: ©Bob Daemmrich/Stock, Boston; p. 548: ©Greg Grosse; p. 550: ©W.K. Almond/Stock, Boston; p. 551: ©David Young-Wolff/PhotoEdit.

CHAPTER 12

p. 564: ©Lightscapes/The Stock Market (top); ©Joseph Nettis/Stock, Boston (bottom); p. 565: ©The Image Bank; p. 567: ©Peter Timmermans/Tony Stone Images; p. 571: ©Sally Mayman/Tony Stone Images; p. 572: ©Greg Grosse; p. 576: ©Martin Roberts/Tony Stone Images; p. 579: ©Steven Marks/The Image Bank; p. 581: ©Stephen Marks/The Image Bank; p. 582: ©Masahiro Sano/The Stock Market; p. 586: ©FPG International; p. 589: ©Ford Motor Company; p. 590: ©The Bettmann Archive; p. 594: ©Patti McConville/The Image Bank; p. 597: ©Jay Freis/The Image Bank; p. 599: ©Glen Allison/Tony Stone Images; p. 602: ©Greg Grosse; p. 606: ©Tom Tracy/Tony Stone Images; p. 609: ©Lonny Kalfus/Tony Stone Images; p. 611: ©TRW, Inc.; p. 612: ©UPI/Bettmann.

CHAPTER 13

p. 620: ©Kimball Hall/The Image Bank (top); The Image Bank (bottom); p. 621: ©Andy Sacks/Tony Stone Images; p. 622: ©Vladimir Pcholkin/FPG International; p. 623: ©Charles C. Place/The Image Bank (top); ©Robert Fried/Stock, Boston (bottom); p. 625: ©Greg Grosse; p. 626: Bob Daemmrich/Stock, Boston; p. 627: ©Renee Lynn/Tony Stone Images; p. 629: ©Jose L. Pelaez/Stock, Boston; p. 630: ©Tim Barnwell/Stock, Boston; p. 632: ©Spencer Jones/FPG International; p. 633: ©Mimi Ostendorf-Smith/Photonics; p. 635: ©Greg Grosse; p. 636: ©Walter Schmid/Tony Stone Images; p. 638: Arizona Tourist Bureau (bottom); p. 639: ©Greg Grosse; p. 642: ©Peter Vandermark/Stock, Boston; p. 643: ©Peter Menzel/Stock,Boston; p. 645: ©Paul Merideth/Tony Stone Images; p. 646: ©Greg Grosse; p.649: National Geographic Society; p. 650: ©Lawrence Migdale; p. 652: ©Glen Allison/Tony Stone Images; p. 655: ©Randy O'Rourke/The Stock Market; p. 657: ©Vladimir Pcholkin/FPG International; p. 658: ©Phil Degginger/Tony Stone Images; p. 660: ©Hal H. Harrison/Grant Heilman.

CHAPTER 14

p. 668: ©Ed Bock/The Stock Market (top); p. 669: ©Steven Peters/Tony Stone Images; p. 670: ©AP Photo/Gary Tramontina; p. 671: ©Alan Schein/The Stock Market; p. 675: D. Rawclife/FPG International; p. 678: ©G&VChapman/The Image Bank; p. 679: ©Tony Freeman/PhotoEdit; p. 680: ©Richard Hutchings/PhotoEdit; p. 683: ©Robert Frerck/Tony Stone Images; p. 686: ©Greg Grosse; p. 692: ©Rob Boudreau/Tony Stone Images; p. 694: William Taufic/The Stock Market; p. 696: ©D&I MacDonald/Photo Network; p. 703: ©Greg Grosse; p. 707: ©David Young-Wolff/PhotoEdit; p. 710: ©Steve Bronstein/The Image Bank; p. 713: ©C/B Productions/The Stock Market; p. 716: ©David Young-Wolff/PhotoEdit; p. 718: ©David Young-Wolff/PhotoEdit; p. 720:

ON THE INTERNET

WWW: http://www.thomson.com
EMAIL: findit@kiosk.thomson.com

Access South-Western Educational Publishing's complete catalog online through thomson.com. Internet users can search catalogs, examine subject-specific resource centers, and subscribe to electronic discussions lists. In addition, you'll find new product information and information on upcoming events in Mathematics.

For information on our products and services, point your web browser to
http://www.swpco.com/swpco.html

For technical support, you may email: hotline_education@kdc.com

South-Western Algebra 1: An Integrated Approach Internet Connection is a web site which accompanies the chapter projects. You can access the Internet Connection at
www.swpco.com/swpco/algebra1.html

A service of I⒯P®